The Boundary Element Method

The Boundary Element Method

Volume 2

Applications in Solids and Structures

M. H. Aliabadi
Queen Mary, University of London, UK

JOHN WILEY & SONS, LTD

Other Wiley Editorial Offices

John Wiley & Sons, Inc., 605 Third Avenue, New York, NY 10158-0012, USA

Jossey-Bass, 989 Market Street, San Francisco, CA 94103-1741, USA

Wiley-VCH Verlag GmbH, Pappelallee 3, D-69469 Weinheim, Germany

John Wiley & Sons Australia, Ltd., 33 Park Road, Milton, Queensland 4064, Australia

John Wiley & Sons (Asia) Pte Ltd., 2 Clementi Loop #02-01, Jin Xing Distripark, Singapore 129809

John Wiley & Sons Canada, Ltd., 22 Worcester Road, Etobicoke, Ontario, Canada M9W 1L1

British Library Cataloguing in Publication Data

A catalogue record for this book is available from the British Library

ISBN 0-470-84298-9

Produced from LaTeX PostScript files supplied by the author.
Printed and bound in Great Britain by Antony Rowe, Chippenham, Wiltshire.
This book is printed on acid-free paper responsibly manufactured from sustainable forestry, in which at least two
trees are planted for each one used for paper production.

Contents

Preface xiii
Preface to Volume 2 xv
Acknowledgements xvii

1 Introduction   **1**
  1.1 Applications 4
  1.2 Mathematical Preliminaries 5
      1.2.1 Cartesian Tensor Notation 5
      1.2.2 Divergence Theorem 8
      1.2.3 Dirac Delta Function 8
  1.3 Overview of the Book 8
      1.3.1 Volume 1 9
      1.3.2 Volume 2 11

2 Elastostatics   **15**
  2.1 Introduction 15
  2.2 Basic Elasticity 16
      2.2.1 Stress 16
      2.2.2 Equilibrium 16
      2.2.3 Plane Stress 19
      2.2.4 Principal Stress 21
      2.2.5 Strain 22
      2.2.6 Compatibility Equations 23
      2.2.7 Stress-Strain Relationship 24
      2.2.8 Governing Equations of Elasticity 26
  2.3 Boundary Element Formulations 26
      2.3.1 Reciprocal Theorem 26
      2.3.2 Betti's Reciprocal Theorem 28
      2.3.3 Boundary Integral Equations 29
      2.3.4 Fundamental Solutions 30
      2.3.5 Boundary Displacement Equation 34
      2.3.6 Axisymmetric Problems 37
      2.3.7 Infinite Regions 38
      2.3.8 Semi-infinite Regions 39
      2.3.9 Boundary Stress Equation 40
      2.3.10 Numerical Discretization 42
  2.4 Stress and Displacements at Interior Points 56
  2.5 Boundary Stresses 56

	2.5.1	Indirect Approach	57
	2.5.2	Direct Approach	61
2.6	Body Forces		61
	2.6.1	Domain Discretization	62
	2.6.2	Galerkin Vector	63
	2.6.3	Particular Integrals	65
	2.6.4	Multiple Reciprocity Method	67
	2.6.5	Dual Reciprocity Method	67
2.7	Multi-region Formulation		70
2.8	Corner Problems		71
2.9	Anisotropic Elasticity		74
	2.9.1	Basic Equations	74
	2.9.2	Fundamental Solutions	75
2.10	Galerkin Formulation		76
	2.10.1	Regularization	79
	2.10.2	The Uniqueness Problem	80
2.11	Coupling Finite Elements and Boundary Elements		81
	2.11.1	Stiffness Matrix Approach	81
	2.11.2	Multi-region Approach	82
2.12	Examples		83
	2.12.1	Pressurized Cylinder	83
	2.12.2	Stress Concentration	87
	2.12.3	Plate with an Elliptical Hole	87
	2.12.4	A Strip Under Self-Weight and Centrifugal Loads	88
	2.12.5	Orthotropic Circular Ring	90
	2.12.6	Hole in a Strip	92
	2.12.7	Analysis of Human Tibia	92
2.13	Summary		95

3 Plates and Shells   **103**
3.1	Introduction		103
3.2	Shear Deformable Plates		105
	3.2.1	Basic Theory	105
	3.2.2	Boundary Integral Formulations	109
	3.2.3	Multi–region Formulation	113
	3.2.4	Foundation Plates	114
3.3	Shear Deformable Shallow Shells		115
	3.3.1	Basic Theory	115
	3.3.2	Integral Equation Formulations	123
	3.3.3	Boundary Integral Equations	124
	3.3.4	Numerical Discretization	126
	3.3.5	Stress Resultants Integral Equations	130
	3.3.6	Hypersingular Integral Equations	131
	3.3.7	Boundary Stress Resultants	133
3.4	Examples		134
	3.4.1	Simply Supported Thin Square Plate	134
	3.4.2	L-shaped Plate Structure	136
	3.4.3	Cantilever Beam	136
	3.4.4	X-core Structure	138

      3.4.5   Shallow Spherical Shell 138

      3.4.6   Simply Supported Spherical Shell 145

      3.4.7   Cylindrical Shell 147

  3.5 Kirchhoff Plate Theory 148

      3.5.1   Basic Concepts 148

      3.5.2   Boundary Integral Formulation 150

      3.5.3   Example: Rectangular Plate with an Opening 151

  3.6 Summary 154

4 Thermoelasticity   161

  4.1 Introduction 161

  4.2 Basic Concepts 162

      4.2.1   Heat Conduction Equation 162

      4.2.2   Thermoelasticity Equations 162

  4.3 Boundary Integral Formulations 163

      4.3.1   Steady State Thermoelasticity 163

      4.3.2   Transient Thermoelasticity 174

  4.4 Examples 186

      4.4.1   Thick Hollow Cylinder 186

      4.4.2   Transient Thermal Problems 187

  4.5 Summary 191

5 Elastodynamics   195

  5.1 Introduction 195

  5.2 Transient Elastodynamics 196

      5.2.1   Basic Concepts 196

      5.2.2   Time Domain Integral Equations 196

      5.2.3   Numerical Discretization 198

  5.3 Laplace Transform Method 201

      5.3.1   Basic Concepts 201

      5.3.2   Integral Equation Formulations 201

      5.3.3   Numerical Discretization 202

      5.3.4   Laplace Inversion Method 203

  5.4 Dual Reciprocity Method 204

      5.4.1   Numerical Discretization 206

      5.4.2   Dual Reciprocity in the Laplace Domain 207

      5.4.3   Time Integration 208

  5.5 Examples 209

      5.5.1   A Strip Subjected to Pure Tension and Shear Loads 209

      5.5.2   Plate with a Hole 211

      5.5.3   Rectangular Bar Subjected to Tensile Load 211

      5.5.4   Comparison of DRM, TDM and LTM 213

  5.6 Acoustic Scattering in Fluid-Solid Problems 220

      5.6.1   Example : Cylinder Immersed in Fluid 221

  5.7 Summary 222

6   Elastoplasticity   **229**
    6.1   Introduction 229
    6.2   Basic Concepts 230
        6.2.1   Elastoplastic Relationships 232
        6.2.2   Multiaxial Loading 233
        6.2.3   Plastic Stress-Strain Relationships 235
    6.3   The Governing Elastoplastic Equations 237
    6.4   Initial Strain Boundary Integral Formulation 239
        6.4.1   Internal Stresses 241
        6.4.2   Stresses at Boundary Points 243
        6.4.3   Alternative Approaches 244
    6.5   Numerical Discretization 245
        6.5.1   Treatment of the Integrals 246
        6.5.2   System Matrices Assembly 250
    6.6   Non-linear Solution Algorithm 252
        6.6.1   Implicit Procedures 252
        6.6.2   Explicit Procedure 253
    6.7   Examples 254
        6.7.1   Two-dimensional Benchmarks 254
        6.7.2   Three-dimensional Benchmarks 260
    6.8   Summary 261

7   Contact Mechanics   **269**
    7.1   Introduction 269
    7.2   Basic Contact Mechanics 270
        7.2.1   Friction 270
        7.2.2   Classification of Contact 270
        7.2.3   Modes of Contact 271
        7.2.4   Local Coordinate System 272
        7.2.5   Definition of Normal Gap 272
        7.2.6   Numerical Modelling Concept 273
    7.3   Elastostatic Contact 274
        7.3.1   Numerical Modelling 277
        7.3.2   Fully Incremental Loading Technique 279
        7.3.3   Examples 284
    7.4   Elastoplastic Contact 290
        7.4.1   Incremental Formulation 291
        7.4.2   Examples 292
    7.5   Non-conforming Discretization Methods 301
        7.5.1   Non-Conforming Discretization Formulations 301
        7.5.2   Examples 306
    7.6   Three-dimensional Contact problems 308
        7.6.1   Contact Conditions 311
        7.6.2   Example 311
    7.7   Summary 313

8 Fracture Mechanics **319**
  8.1 Introduction 319
  8.2 Basic Fracture Mechanics 320
      8.2.1 Stress Intensity Factor 320
      8.2.2 Modes of Fracture 320
      8.2.3 Energy Balance 321
      8.2.4 Stress and Displacements Fields 322
      8.2.5 Residual Strength 324
      8.2.6 Fatigue Crack Growth 325
      8.2.7 Criteria for Crack Growth Direction 326
  8.3 Modelling Co-planar Surfaces 328
      8.3.1 Displacement Integral Equation 328
      8.3.2 Traction Integral Equation 329
  8.4 Dual Boundary Element Method 330
      8.4.1 Crack Modelling Strategy 331
      8.4.2 DBEM with Continuous Elements 341
  8.5 Multi-region Method 343
  8.6 Crack Green's Function Method 343
  8.7 Displacement Discontinuity Method 345
  8.8 Methods for Evaluating Stress Intensity Factors 346
      8.8.1 Displacement Extrapolation Method 346
      8.8.2 Subtraction of Singularity Method 357
      8.8.3 Energy Release Rate Method 358
      8.8.4 Path Independent Integrals 358
      8.8.5 Energy Domain Integral 367
      8.8.6 Weight Function Method 370
  8.9 Examples 372
      8.9.1 Edge Crack 372
      8.9.2 Centre Crack 373
      8.9.3 Centre Crack Plate Loaded by Tension and Bending 373
      8.9.4 Double Edge Crack in a Composite Laminate 375
      8.9.5 Penny-Shaped Crack 376
      8.9.6 Inclined Penny-Shaped Crack 378
      8.9.7 Centre Crack Panel Elastoplastic Analysis 378
      8.9.8 Elliptical Crack Subjected to Impact Load 380
      8.9.9 Interface Cracks 382
  8.10 Crack Growth 383
      8.10.1 Two-dimensional Modelling 384
      8.10.2 Three-dimensional Modelling 392
  8.11 Summary 404

9 Sensitivity Analysis and Shape Optimization **425**
  9.1 Introduction 425
  9.2 Shape Optimization Methods 426
  9.3 Design Sensitivity Analysis 427
      9.3.1 Derivative Potential Formulation 427
      9.3.2 Derivative Displacement Formulation 429
  9.4 Numerical Examples 431
      9.4.1 Sensitivities of a Square Plate with an Elliptical Hole 431

9.4.2   Sensitivities of Elastic Ring 432
    9.4.3   A Plate Subjected to Uniaxial Tension 432
9.5   Structural Optimization 435
    9.5.1   Optimum Shape of a Tank with Internal Pressure 435
9.6   Shape Identification 435
    9.6.1   Identification of an Elliptical Cavity 437
    9.6.2   Acoustic Scattering in Fluid-Solid 439
9.7   Flaw Identification 443
    9.7.1   Potential Formulation 445
    9.7.2   Elasticity Formulation 446
9.8   Examples 448
    9.8.1   Curved Crack Identification 448
    9.8.2   Identification of an Elliptical Crack 448
9.9   Bone Remodelling 449
    9.9.1   Medullary Pin 452
9.10 Summary 453

10 Assembled Structures   461
10.1 Introduction 461
10.2 Structures Reinforced with Beams 462
    10.2.1 Two-dimensional Formulations 462
    10.2.2 Plate Bending Formulations 466
10.3 Patch Attachments 470
    10.3.1 Two-dimensional Riveted Joints Model 471
    10.3.2 Plate Bending Formulations 472
    10.3.3 Adhesively Bonded Patches 475
10.4 Examples 478
    10.4.1 Three-point Bending Concrete Beam 478
    10.4.2 Single or Double Circular Patch Bonded to a Sheet 478
    10.4.3 Simply Supported Plate with a Stiffener 481
    10.4.4 Reinforced Cylindrical Shell 483
    10.4.5 Fastened Single Patch Repair 483
    10.4.6 Cylindrical Shell Panels with a Patch. 485
10.5 Summary 487

11 Numerical Integration   491
11.1 Integration of Regular Integrals 491
11.2 Near Singular Integration 493
    11.2.1 Transformation of Variable Technique 493
    11.2.2 Singularity Subtraction Technique 497
11.3 Weakly Singular Integration 497
    11.3.1 Two-dimensional Problems 497
    11.3.2 Three-dimensional Problems 502
11.4 Strongly Singular 508
    11.4.1 Basic Definitions 508
    11.4.2 Two-dimensional Problems 510
    11.4.3 Three-dimensional Problems 513
11.5 Hypersingular Integrals 517
    11.5.1 Basic Definitions 517

11.5.2 Two-dimensional Problems 518
11.5.3 Three-dimensional Problems 519
11.6 Summary 520

A Complex Stress Functions   **525**
A.1 Airy Stress Functions 525
A.2 Muskhelishvili Stress Functions 525
A.3 Complex Stress Function Method 527

B Limiting Process for Boundary Stress Equation   **531**
B.1 Enriched Element Approach 535

C Limit Integrals For Shallow Shells   **539**
C.1 The Displacement Integral Equations 539
C.2 The Bending Stress Resultant Integral Equations 541
C.3 The Shear Stress Resultant Integral Equation 543

D Particular Solutions For Shear Deformable Plates   **547**
D.1 Plate Bending 547
D.2 Two-dimensional Plane Stress 549

E Fundamental Solutions for Elastodynamics   **553**
E.1 Fundamental Solutions for Time Domain 553
E.2 Convoluted Fundamental Solutions 556
     E.2.1 Two-dimensional Problems 556
     E.2.2 Three-dimensional Problems 558
E.3 Fundamental Solutions for Laplace Transform Domain 562
     E.3.1 Behaviour of Transformed Fundamental Solutions 564

F Shape Functions of Brick Cell Elements   **569**
F.1 Continuous cells 569
F.2 Face-discontinuous cells 571

G Fast Solver For Contact Problems   **573**
G.1 LU decomposition 573
     G.1.1 High Speed Solver for Contact Problems 573

Index   **577**

Preface

Engineering problems are commonly described by physical laws which can be mathematically represented in terms of partial differential equations. In many cases, an alternative (and equivalent) mathematical representation of the problem can be found in terms of integral equations. With advances in numerical modelling and ever increasing computer power, modelling techniques based on integral equations can now be used in the actual simulation of many practical engineering problems. The most general and effective numerical technique for solving integral equations is the Boundary Element Method (BEM).

This two-volume book presents a comprehensive coverage of the BEM and its engineering applications. Our initial intention was to produce a jointly authored single volume of around 600 pages, with a modern and up-to-date approach to the technique, emphasizing some traditional and recent applications. However, as the project began to unfold, we realized that the wealth of material available on the development, implementation and application of the method, particularly over the last decade, was such that it required two separate but complementary volumes, which were then written in parallel. An effort was made to use a consistent format and notation throughout both volumes.

Each volume starts with a derivation of the basic differential and integral equations of potential theory (Volume 1) and elastostatics (Volume 2). These self-contained chapters were written with the purpose of introducing the reader to some important concepts such as fundamental solutions, different orders of singularity, the role of interpolation functions and numerical integration, solvers, etc, and can be used for self-study or as a teaching aid to final-year undergraduate or postgraduate courses in different branches of engineering.

The above initial chapters are followed by detailed developments of numerical formulations of the BEM and their applications to a number of engineering problems. This includes a substantial amount of material not previously covered by other text books on the subject. Volume 1 deals with applications to heat transfer, acoustics, electrochemistry and fluid mechanics problems, while Volume 2 concentrates on solids and structures, describing applications to elasticity, plasticity, elastodynamics, fracture mechanics and contact analysis. One important feature of both volumes is the inclusion of modern topics such as fast solvers, hypersingular formulations, the dual BEM, genetic algorithms for optimization and inverse analysis, and many others. We also tried, whenever possible, to present real-life engineering applications as well as simple problems which are useful for benchmarking new techniques and computer codes.

We are very grateful for the support provided by our institutions, Brunel University and Queen Mary, University of London, and by our publisher, John Wiley & Sons.

Luiz Wrobel   Ferri Aliabadi
Brunel University   Queen Mary
Uxbridge, UK   University of London, UK

Preface to Volume 2

This volume describes formulations and applications of the Boundary Element Method (BEM) in solids and structures. The text concentrates to a large extent, on the developments in the last decade, although fairly comprehensive historical reviews are also presented. The work is an integrated presentation of the modern Boundary Element Method, and contains a compilation of the work of many researchers as well as accounting for some of the author's most recent work on the subject.

The book consists of eleven chapters. Each chapter starts with a brief introduction to relevant basic theories followed by a detailed derivation and implementation of the BEM. The text as a whole is aimed at graduate level students with the first two chapters describing the essential material for the first course on boundary elements in stress analysis. Chapter 1 presents an overview of the BEM, its merits, limitations and future prospects. Chapter 2 describes boundary element formulations in elasticity. The chapter is written as a self-contained text that may be used as a teaching text for a final year graduate or postgraduate course. The text includes sufficient details to enable the readers to understand different formulations and their computer implementation as applied to engineering problems. To reinforce the theoretical parts many examples are presented to demonstrate the accuracy and efficiency of the methods. Examples presented include simple benchmark problems as well as practical engineering problems.

Chapter 3 contains a comprehensive description of the BEM formulations for plates and shells. The chapter highlights recent advances in BEM that allows the method to be used effectively for thin structures. Chapter 4 presents the application of the BEM to static and transient thermoelastic problems. Numerical implementations are developed for standard and hypersingular boundary element formulations.

Chapter 5 presents three methods of solving elastodynamic problems. They are the time domain methods, Laplace transform method and dual reciprocity method. As in other chapters, both standard and hypersingular formulations are presented. Applications of the BEM to elastoplastic problems are presented in Chapter 6. Here numerical implementations are presented for both two- and three-dimensional problems. Numerical examples include prediction of residual stress fields produced by prestressing and cold-expansion of open holes.

The next four chapters deal with some important applications of the BEM. Chapter 7 presents BEM formulations for elastic and elastoplastic frictional contact problems. In this chapter a detailed description of the numerical algorithms necessary for modelling different types of contact problems is given. Chapter 8 deals with crack problems in fracture mechanics. Initially different methods for modelling cracks are described. Next the Dual Boundary Element Method (DBEM) is presented in detail. Also presented are effective modelling strategies and algorithms necessary for automatic simulation of crack growth processes.

Structural optimization is dealt with in Chapter 9. Here, formulations are presented based on sensitivity analysis for modelling structural optimization and shape identification problems. Also presented are DBEM formulation for flaw identifications. In Chapter 10, application of the BEM to assembled structures is presented.

Here formulations are described for panels reinforced with beam or repair patches. The reinforcements are considered to be either mechanically attached with rivets or adhesively bonded. This type of problem occurs frequently in engineering, for example in aerospace structures.

Finally, Chapter 11 presents a comprehensive review of the different methods for the effective and accurate evaluation of regular, near singular, weakly singular, strongly singular and hypersingular integrals.

The content of the book reflects my experience over 20 years of research on the Boundary Element Method. During this time, I had pleasure of supervising 30 PhD students, numerous post-doctoral fellows as well as collaborating with other researchers in the field. The content of the book relies heavily on their efforts and contributions.

I would like to thank W.S. Hall and D.P. Rooke for introducing me to BEM and fracture mechanics respectively. Their friendship, support and encouragement over the years is much appreciated.

Thanks are also due to A. Cisilino, T. Dirgantara, P. Fedelinski, V.M.A. Leitão, J.J. Perez-Gavilan, P.H. Wen and A. Young for their comments and suggestions on parts of the manuscript. A very special thanks goes to T. Dirgantara for his tremendous help with the reproduction of many of the figures.

It has been a pleasure to collaborate with my friend Luiz Wrobel on this project. I have enjoyed immensely our many stimulating conversations over the last 18 months.

Finally, I would like to thank my wife Gail, and my sons Darius and Kouros for their support and patience during the writing of the book.

I dedicate the book to the memory of my father.

Ferri Aliabadi
Queen Mary, University of London
London, UK

Acknowledgements

The author would like to thank the publishers below, who allowed permission for the reproduction of their copyrighted figures. The numbers given in square brackets relate to the reference lists appearing at the end of each chapter.

Chapter 1: Figure 1.3 [40] is copyright Tech Science Press.

Chapter 2: Figures 2.42-2.46 [87] are reproduced by permission of John Wiley & Sons, Ltd. Figures 2.47-2.48 [50] are copyright Tech Science Press.

Chapter 3: Figures 3.18 and 3.27-3.31 [10] are reproduced by permission of John Wiley & Sons, Ltd.

Chapter 4: Figures 4.3-4.7 [23] are copyright WIT Press.

Chapter 5: Figures 5.2 and 5.3 [44] are reproduced by permission of John Wiley & Sons, Ltd. Figures 5.16-5.18 [36] are reproduced by permission of Academic Press Ltd.

Chapter 6: Figures 6.5, 6.16 and 6.21 [9] are reproduced by permission of Emerald (formerly MCB University Press Ltd).

Chapter 7: Figures 7.6, 7.7, 7.9, 7.11, 7.13 and 7.14 [19] and Figures 7.15-7.27 [24] are reproduced with permission from Elsevier Science. Figures 7.30 and 7.35 [5] are reproduced by permission of John Wiley & Sons, Ltd. Figures 7.39-7.44 [9] are reproduced by permission of WIT Press.

Chapter 8: Figure 8.31 [58] is copyright Kluwer. Figures 8.50 and 8.51 [180] and Figures 8.64-8.66 and 8.68-8.71 [54] are reproduced with permission from Elsevier Science. Figure 8.72 [200] is reproduced by permission of Blackwell Science Ltd.

Chapter 9: Figures 9.10, 9.12, 9.17 and 9.20 [28] and Figure 9.13 [23] are reproduced by permission of John Wiley & Sons, Ltd. Figures 9.23, 9.25 and 9.26 [24] are reproduced with permission from Elsevier Science. Figures 9.27, 9.28 and 9.30 [26] and Figure 9.29 [25] are reproduced by permission of WIT Press.

Chapter 10: Figures 10.1-10.3 [7] are reproduced with permission from Elsevier Science. Figures 10.11-10.13 [6] are copyright American Society of Civil Engineers. Figures 10.14-10.16 [8] are reproduced by permission of John Wiley & Sons, Ltd.

Chapter 1

Introduction

Engineering design processes are becoming more complex and multidisciplinary. In order to maintain a reasonable cost for large scale structures such as airframes, engines, vehicles, pressure vessels, etc., it is generally accepted that computer modelling and simulation must partially replace full scale and laboratory testings. Therefore, the ability to achieve a fully digital model of the product, process and testing has become an essential part of the design process.

Computational methods for structural analysis such as the Finite Element Method (FEM) and the Boundary Element Method (BEM) have attained a level of development that has made them essential tools for modern design engineers. The FEM is routinely used as a general analysis tool in industry. The BEM's applicability at present is not as wide ranging as FEM, however the method has become established as an effective alternative to FEM in several important areas of engineering analysis. $\leftarrow$ intro Although the BEM also known as the Boundary Integral Equation (BIE) method, is a relatively new technique for engineering analysis, the fundamentals can be traced back to classical mathematical formulations by Fredholm [26] and Mikhilin [39] in potential theory and Betti [15], Somigliana [43] and Kupradze [32] in elasticity. The development of the formulations in the context of boundary integral equations is due to Jaswon [36], Hess and Smith [35], Massonnet [38], Rizzo [42] and Cruse [21]. The work of Lachat [33] and Lachat and Watson [34] is perhaps the most significant early contribution towards BEM becoming an effective numerical technique. They developed an isoparametric formulation similar to those used in the finite element methods, and demonstrated that the BEM can be used as an efficient tool for solving problems with complex configurations. Around the same time, the first international symposium on the method [22] helped to bring the BEM to the attention of the engineering community and encourage further research on the topic.

The attraction of the BEM can largely be attributed to the reduction in the dimensionality of the problem; for two-dimensional problems, only the line-boundary of the domain needs to be discretized into elements (see Figure 1.1), and for three-dimensional problems only the surface of the problem needs to be discretized. This means that, compared to the FEM and other domain type analysis techniques, a boundary analysis results in a substantial reduction in modelling effort. Furthermore, as the unknown parameters such as displacements and stresses are approximated on the discretized surfaces, a much smaller system of equations is obtained compared to the FEM. In addition, the BEM allows a high level of integration between CAD and

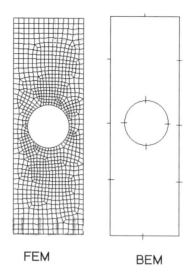

FEM   BEM

Figure 1.1: FEM and BEM Meshes for a plate with a hole.

analysis software due to the surface only modelling requirements of the two systems.

The coefficient matrix in the BEM is fully populated and non-symmetric. As the solution time for this type of matrix using direct solvers is proportional to the cubic power of the total degrees of freedom, the required computational time can become large for complex structural models. To overcome this problem, a number of techniques has been developed (see, for example, [9, 41, 20]). This simpler description of the body means that regions of high stress concentration can be modelled more efficiently as the necessary high concentration of grid points is confined to one less dimension. For example, as shown in Figure 1.1, it is possible to increase the mesh density around the hole without it affecting the discretization elsewhere.

Another important feature of the boundary element formulation is that it provides

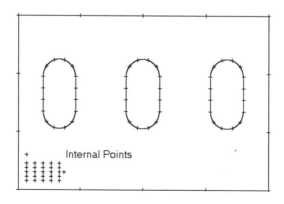

Figure 1.2: Internal points for boundary elements.

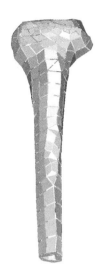

Figure 1.3: Boundary element model of a human tibia [40].

a continuous modelling of the interior, since no discretization of the interior is required; this leads to a high resolution of interior stresses and displacements. The internal point solutions are calculated after the boundary unknowns are calculated as a post-processing. The density, distribution and location of the internal points have no bearing on the boundary mesh or boundary unknowns (see Figure 1.2)

Other features of the BEM include automatic satisfaction of boundary conditions for infinite and semi-infinite domains, thus avoiding the need for the numerical discretization of remote boundaries.

As will be seen later in the book, the boundary element method requires the existence of so-called 'fundamental solutions'. A fundamental solution is the solution of the governing equations due to unit force. As such, the derivation of these solutions can be very difficult to achieve. For example, application of the BEM to problems with variable material properties has been limited due to the lack of suitable fundamental solutions. However, in years to come research on the BEM may result in ways of overcoming these difficulties.

Generally, solution times for the BEM are higher than the FEM for the same number of degrees of freedom. A common mistake is to design a BEM mesh similar to the FEM with the internal degrees of freedom removed. However, BEM models require much coarser surface discretization to achieve the same level of accuracy as FEM models.

Recent advances in computer technology have had a tremendous impact on the FEM and BEM solution times. Nowadays, FEM and BEM analyses for most 2D problems are achieved in several seconds, and 3D problems take minutes rather than hours. This increase in computing power has helped the BEM, as the difference in computing requirements between the two methods are now insignificant, and the important deciding factor is the time and effort taken by designers in establishing models. It must, however, be stated that unlike the BEM, finite element users enjoy having a lot of robust commercial software available to them. These codes are being

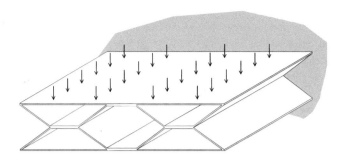

Figure 1.4: Assembled plate structure under uniform pressure.

used extensively in academia and industry. It is hoped that with an increasing number of engineers being informed of and trained in the use of the BEM, integrated FEM-BEM software will be developed in the near future by the IT sector. Applications of the boundary element method to problems in solids and structures can be found in [1–5, 10–19, 24–31].

This two volume book presents a comprehensive, up-to-date treatment of the boundary element method and its application to several engineering problems. Volume 1 covers applications to heat transfer, acoustics, electrochemistry and fluid mechanics problems, while Volume 2 concentrates on solids and structures, describing applications to elasticity, plasticity, elastodynamics, fracture mechanics and contact analysis. Both volumes reflect the experience of the authors over a period of more than 20 years of boundary element research.

1.1 Applications

Applications of the BEM to linear stress analysis of complex structures has been popular due to the simple modelling requirements of the method. Figure 1.3 shows a boundary element model of a tibial bone [40]. For this problem the CAD surfaces were meshed directly by a boundary element preprocessor. The mesh consists of quadrilateral and triangular elements similar to those used in two-dimensional finite elements.

Thin structures have historically posed a problem for the BEM. However, recent developments [23, 44] have resulted in the development of new boundary element methods which are capable of solving complex thin and thick plate and shell structures. Figure 1.4 shows an X-core wing section of an aircraft. The boundary element and the finite element models are shown in Figure 1.5. Notice that the BEM modelling of the structure is achieved with one-dimensional line elements. Another important application is modelling assembled structures such as reinforced panels and lap joints. Figure 1.6 shows FEM and BEM meshes for a reinforced curved fuselage panel. Notice how simple it is with the BEM model to change the location of the reinforcements or alter the number of attachments.

Perhaps the most appealing application of the boundary element method is crack growth simulation [7]. BEM is now being used in preference to the FEM for crack

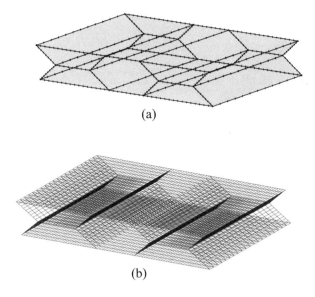

(a)

(b)

Figure 1.5: BEM mesh and FEM mesh for X-core structure.

growth simulation. The ability of the BEM to automatically follow the path of a growing crack with limited remeshing and no user intervention, together with its inherent accuracy, make the method a truly robust numerical tool for fracture mechanics. Figure 1.7 presents three steps in an incremental crack growth process. The BEM remeshing requirement as the crack grows is simply to add extra elements to the previous crack tip with no changes to the remaining mesh.

Nonlinear material behaviour such as plasticity are not generally regarded as desirable applications for the BEM. However, recent studies (see Chapter 6) have demonstrated that for a wide range of practical applications where the plastic zone is small compared to the overall volume of the component, the BEM can offer an efficient alternative to the FEM. For nonlinear problems the potential plastic zone is discretized into cells as shown in Figure 7.19.

1.2 Mathematical Preliminaries

In this section some basic mathematical concepts that will be utilized throughout the book are briefly reviewed. For a full detailed description of these concepts readers should consult textbooks on calculus and differential equations.

1.2.1 Cartesian Tensor Notation

The main notation used throughout this book is the Cartesian tensor notation. This is used to simplify and save time when writing long expressions, and is also useful for differentiation. However, vector notations are also used in some cases.

Tensor notation makes use of subscript indices (1, 2, 3) to represent the Cartesian directions (x, y, z), and renders summation symbols unnecessary when the same letter

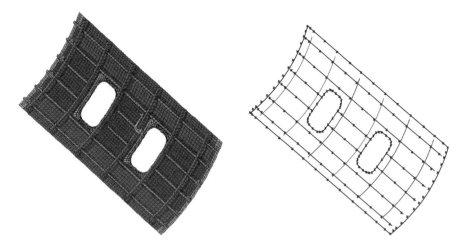

Figure 1.6: Finite element and boundary element model of a reinforced curved fuselage panel.

subscript occurs twice in a term. Hence, in three dimensions,

$$a_i a_i = a_1^2 + a_2^2 + a_3^2 \tag{1.1}$$

and

$$a_{ii} = a_{11} + a_{22} + a_{33} \tag{1.2}$$

Therefore, unless otherwise stated, subscripts will be assumed to have a range of 3 for three-dimensional problems and 2 for two-dimensional problems.

The distance r between two points $\mathbf{x}$ and $\mathbf{x}'$ can be written in the form

$$r = \left(r_i r_i\right)^{1/2} \tag{1.3}$$

with

$$r_i = x_i - x_i' \tag{1.4}$$

Its derivatives are of the form

$$\frac{\partial r}{\partial x_i} = r_{,i} = \frac{r_i}{r} \tag{1.5}$$

and

$$\frac{\partial r}{\partial x_i'} = -r_{,i} = -\frac{r_i}{r} \tag{1.6}$$

The normal derivative of the distance r with respect to the normal direction $\mathbf{n}$ will be represented by a product of the form

$$\frac{\partial r}{\partial n} = r_{,i}\, n_i \tag{1.7}$$

Other symbols which will be used throughout the text are the permutation tensor e_{ijk} and the Kronecker delta δ_{ij}, defined as follows:

$$e_{ijk} = \begin{cases} 1 & \text{for} \quad e_{123},\ e_{231},\ e_{312} \ \text{(for cyclic suffices order)} \\ -1 & \text{for} \quad e_{132},\ e_{213},\ e_{321} \ \text{(for anti-cyclic suffices order)} \\ 0 & \text{if any two suffices are equal} \end{cases} \tag{1.8}$$

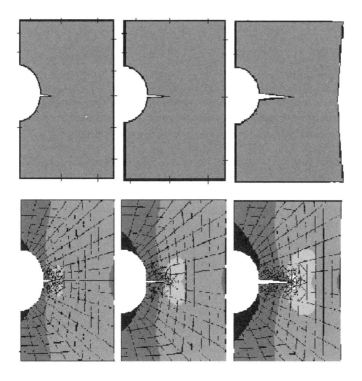

Figure 1.7: Crack growth simulation with BEM and FEM.

and

$$
\delta_{ij} = \begin{cases} 1 & \text{if} \quad i = j \\ 0 & \text{if} \quad i \neq j \end{cases}
\tag{1.9}
$$

Often we use the Kronecker delta function to manipulate expressions, for example

$$
\begin{aligned}
u_i \delta_{ij} &= u_j \\
u_i w_j \delta_{il} &= u_l w_j
\end{aligned}
\tag{1.10}
\tag{1.11}
$$

The matrix product of an N by N system of algebraic equations can be written as

$$
\sum_{j=1}^{N} A_{ij} x_j = y_i \qquad i = 1, ... N
\tag{1.12}
$$

The repeated indices in the above expression imply summation. The above equation can also be written as

$$
A_{ij} x_j = y_i \qquad i, j = 1, ... N
\tag{1.13}
$$

It is worth noting that $A_{ij} \neq A_{ji}$ unless A is symmetrical.

1.2.2 Divergence Theorem

The divergence theorem is a fundamental identity that relates a volume integral to a surface integral, that is

$$\int_V f_{i,i}dV = \int_S f_i n_i dS \tag{1.14}$$

where $f_{i,i} = \partial f_i/\partial x_i$, n_i denotes the components of the unit outward normal to the surface S.

1.2.3 Dirac Delta Function

The existence of fundamental solutions plays an important role in the formulation of the boundary element method. The fundamental solution is the solution of the governing differential equation due to a concentrated excitation such as a point force in solid and fluid mechanics, a point charge in electrostatics or an impulse force in acoustics.

The mathematical representation of the concentrated force can be achieved with the aid of the Dirac delta function $\Delta(\mathbf{X}', \mathbf{X})$. The main feature of the Dirac delta function is that it is zero at all $\mathbf{X}$ points, except at $\mathbf{X}' = \mathbf{X}$, where it becomes infinity. Thus, it represents a point singularity at the source point $\mathbf{X}'$, i.e.

$$\Delta(\mathbf{X}', \mathbf{X}) = \begin{cases} \infty & \text{for } \mathbf{X} = \mathbf{X}' \\ 0 & \text{otherwise} \end{cases} \tag{1.15}$$

The property of the Dirac delta function is

$$\int_a^b \Delta(\mathbf{X}', \mathbf{X}) f(\mathbf{X}) d\mathbf{X} = f(\mathbf{X}') \tag{1.16}$$

where $-\infty \leq a, b \leq \infty$.and $a < \mathbf{X}' < b$.

The Heaviside unit step function is related to the Dirac delta function via

$$\frac{dH(\mathbf{X} - \mathbf{X}')}{d\mathbf{X}} = \Delta(\mathbf{X}', \mathbf{X}) \tag{1.17}$$

where

$$H(\mathbf{X} - \mathbf{X}') = \begin{cases} 0 & \text{for } \mathbf{X} \leq \mathbf{X}' \\ 1 & \text{for } \mathbf{X} > \mathbf{X}' \end{cases} \tag{1.18}$$

1.3 Overview of the Book

This two volume book is designed to provide readers with a comprehensive and up-to-date account of the boundary element method and its application to the solution of engineering problems. Each volume is a self-contained book, with the first volume covering applications in thermo-fluids and acoustics and the second in applications in solids and structures. Chapter 2 in each volume, dealing with potential and elasto-statics problems, respectively, have been designed as introductory chapters that may be used as teaching material for final year undergraduate or postgraduate courses.

This two volume book covers a substantial amount of material previously not covered by other textbooks on the subject.

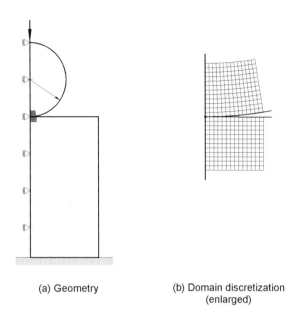

(a) Geometry

(b) Domain discretization
(enlarged)

Figure 1.8: Boundary element model for an elastoplastic contact problem.

1.3.1 Volume 1

This volume covers applications in heat transfer, acoustics, electrochemistry and fluid mechanics. Chapter 2 presents an introduction to the boundary integral equations of potential theory. The formulation is developed from very basic concepts, starting from Green's identities and the definitions of single- and double-layer potentials. The order of the singularities appearing in the integrals of these potentials is discussed in detail, together with their behaviour when an internal source point is taken to the boundary. The steps and approximations involved in the numerical solution of the boundary integral equations are presented for different types of elements, as well as the modifications required to extend the basic formulation to deal with non-isotropic and/or non-homogeneous media. Other features discussed in this chapter are different approximation procedures to treat body forces, Galerkin and hypersingular formulations, and novel fast solvers. Chapter 2 is a self-contained text that can be used as teaching material for a final year undergraduate or postgraduate course.

Heat transfer problems are presented in Chapters 3 and 4. Chapter 3 deals with steady problems, and includes BEM formulations for conduction, convection and radiation heat transfer. Nonlinear heat transfer problems are also discussed, as applied to nonlinear materials and heat sources, phase change and dendritic growth. Applications of the BEM to transient heat transfer problems are presented in Chapter 4. The chapter starts by presenting different formulations for transient heat conduction, namely the Laplace transform technique, dual reciprocity method and time-dependent fundamental solutions. The numerical implementation of the latter is discussed in great detail, including the consideration of non-zero initial conditions and internal heat generation. BEM formulations are also presented for complex transient heat transfer problems, such as convection-diffusion, bioheat transfer and hyperbolic heat conduction. Nonlinear heat transfer and coupled heat and mass transfer problems are also discussed in Chapter 4.

Chapter 5 presents a comprehensive treatment of acoustic problems, including formulations and applications for internal and external problems, both in the frequency and time domains. For internal problems, the presentation starts with the standard BEM formulation, and then proceeds to discuss hypersingular formulations, problems with a mean flow and eigenvalue analysis. The well known problem of irregular frequencies, characteristic of external problems, is presented next, followed by a formulation for the scattering of sound waves by thin bodies. The core of the chapter is a comprehensive treatment of standard, hypersingular and dual formulations for the propagation and attenuation of outdoor noise. Chapter 5 closes with a brief discussion of further BEM applications in acoustics, which includes problems such as tyre noise, underwater acoustics, bioacoustics, and the analysis of musical instruments.

Electrochemical problems are treated in Chapter 6. The two main classes of problems considered in this chapter are corrosion modelling and electrodeposition. Basic definitions are provided, together with the mathematical and BEM modelling. A number of practical applications are included, mainly in the area of cathodic protection of ship hulls, pipelines and offshore structures.

The next three chapters deal with different classes of fluid mechanics problems. Chapter 7 concentrates on the flow of ideal fluids, described by the Laplace or Poisson equation. The chapter starts with a brief review of the theory of irrotational flows, and proceeds to present BEM formulations and applications to a variety of problems such as cavity flow, salt water intrusion into coastal aquifers, flow in heterogeneous porous media, viscous fingering, bubble growth, capillary waves and highly nonlinear waves.

Problems of slow viscous flow, or creeping flow, are presented in Chapter 8. Following a brief introduction to Stokes flows, the chapter presents a standard and a hypersingular formulation for slow viscous flows. The completed double-layer BEM is presented next, applied to several problems involving different types of boundary conditions and deforming boundaries. Chapter 8 also includes formulations for Oseen flows, fast solution methods, non-Newtonian flow problems and transient flows.

Chapter 9 presents BEM formulations for general viscous flows, described by the Navier-Stokes equations. This includes velocity-pressure, penalty function, dual reciprocity and velocity-vorticity formulations, applied to steady and transient flows. An extension of the velocity-vorticity formulation for non-Newtonian flow problems is also discussed.

Chapter 10 deals with inverse analysis applied to all the fields covered in the previous chapters, i.e. heat transfer, acoustics, electrochemistry and fluid mechanics. Problems discussed include the identification of boundary conditions, material properties, and missing geometric definitions. Techniques applied range from standard conjugate gradient methods to more advanced evolutionary-type techniques.

Finally, Chapter 11 concentrates on numerical integration techniques, presenting different methods for the efficient and accurate evaluation of regular, near singular, weakly singular, strongly singular and hypersingular integrals.

1.3.2 Volume 2

This volume covers applications to linear, nonlinear and transient problems in solids and structures. Chapter 2 presents a comprehensive description of the BEM formulation for elastostatic problems. It covers basic concepts as well as more advanced topics of hypersingular formulation for the stress integral equations. As in Volume 1, the basic steps and approximations involved in the numerical solution of the boundary integral equation are presented in detail. Particular attention has been paid to advanced implementation of the method as necessary for solving practical engineering problems. Formulations cover both isotropic and anisotropic problems. Other features presented include a treatment of body force terms, techniques for coupling the BEM with the FEM, and Symmetric Galerkin boundary element formulation.

Chapter 3 presents the BEM formulation for plates and shells. The formulation for shear deformable plates is initially presented. Next, extension of the method to shear deformable shells is presented. Also presented are BEM formulations for the Kirchhoff plate theory. However, the main emphasis is placed on the shear deformable theory due to its applicability to a wide range of practical applications.

Chapter 4 deals with steady state and transient thermoelastic problems. In addition to the temperature and displacement boundary integrals, the stress and flux hypersingular integral equations are also presented. Chapter 5 presents boundary element formulations for elastodynamic problems. In the chapter, formulations are presented for the time domain, frequency domain and dual reciprocity methods. As in previous chapters, stress hypersingular integral equations are also presented.

Chapter 6 presents BEM for elastoplastic problems. Initially some basic concepts in plasticity are reviewed. Implementation of the method for two- and three-dimensional problems is described in detail. Application of the method to the evaluation of residual stress fields due to mechanical cold working is also described.

Chapter 7 presents the BEM formulation for analysis of contact problems. Initially some basic concepts in nonlinear contact mechanics are reviewed. The formulation presented deals with elastic and elastoplastic frictional contact problems. More recent formulations based on non-conforming discretization are also presented. Fracture mechanics is the subject of the next chapter. Here, emphasis is placed on the Dual Boundary Element Method (DBEM), although a fair account of other methods such as the sub-region, displacement discontinuity and crack Green's function is also given. Several established methods for the evaluation of stress intensity factors, including quarter-point elements and the J-integral method, are presented. Application of the DBEM to crack growth analysis is also described. Several examples of the method are presented for elastic, elastoplastic and dynamic crack growth problems.

Chapter 9 presents boundary element formulations for sensitivity and shape optimization. The direct differentiation method is adopted as an effective method for the evaluation of design sensitivities for two- and three-dimensional problems. Applications of the method to shape optimization and shape identification are also described, along with an application to remodelling analysis in bone mechanics. Also presented are shape identification for structures immersed in fluids.

Chapter 10 presents BEM formulations for assembled structures. Here, formulations are presented for panels reinforced by beam attachments. Also presented are formulations for mechanically fastened and adhesively bonded patched plates. The final chapter deals with numerical integration. Here, different methods for the accurate evaluation of boundary integrals are presented for regular, near-singular, weakly singular, strongly singular and hypersingular integrals.

References

[1] Aliabadi, M.H. and Rooke, D.P., *Numerical Fracture*, Kluwer Academic Press, Dordrecht and Computational Mechanics Publications, Southampton, 1991.

[2] Aliabadi, M.H. and Brebbia, C.A., *Advances in Boundary Elements for Fracture Mechanics*. Elsevier Science, Oxford, 1993.

[3] Aliabadi, M.H. and Brebbia, C.A., *Advanced Formulations in Boundary Elements*. Elsevier Science, Oxford, 1993.

[4] Aliabadi, M.H. and Brebbia, C.A., *Adaptive Finite and Boundary Element Methods*. Elsevier Science, Oxford, 1993.

[5] Aliabadi, M.H and Brebbia, C.A., *Computational Methods in Contact Mechanics*. Elsevier Applied Science, Oxford, 1993.

[6] Aliabadi, M.H., *Plate Bending Analysis with Boundary Elements*. Computational Mechanics Publications, 1998.

[7] Aliabadi, M.H., A new generation of boundary elements for fracture mechanics, *International Journal of Fracture*, **86**, 91-125, 1997.

[8] Antes, H. and Panagiotopoulos, P.D., *The Boundary Integral Approach to Static and Dynamic Contact Problems*, Birkhäuser, Basel, 1992,

[9] Araujo, F.C. and Martines, J.C., A study of efficient multi-zone BE/BE coupling algorithms based on iterative solvers- applications to 3D time-harmonic problems, *Advances in Boundary Element Techniques*, Hoggar, Geneva, 21-30, 2001.

[10] Balas, J., Saldek, J. and Sladek, V., *Stress Analysis by Boundary Element Methods*, Elsevier, Amsterdam,1989.

[11] Banerjee, P.K., *The Boundary Element Methods in Engineering*, McGraw-Hill, New York, 1992.

[12] Becker, A., *The Boundary Element Method in Engineering*, McGraw-Hill, London, 1992.

[13] Beer, G. and Watson, J.O., *Introduction to Finite and Boundary Element Methods for Engineers*, Wiley, Chichester, 1992.

[14] Beskos, D.E., *Boundary Element Analysis of Plates and Shells*, Springer-Verlag, Berlin, 1991.

[15] Betti, E., Teori dell elasticita, *II Nuovo Cimento*, 7-10, 1872.

[16] Bonnet, M., *Boundary Integral Equation Methods for Solids and Fluids*, Wiley, Chichester, 1999.

[17] Brebbia, C.A., Telles, J.C.F. and Wrobel, L.C., *Boundary Element Techniques*, Springer-Verlag, Berlin, 1984.

[18] Brebbia, C.A. and Aliabadi, M.H., *Industrial Applications of the Boundary Element Method.* Elsevier Science, London 1993.

[19] Bush, M.B., *Discontinuous Materials and Structures,* WIT Press, Southampton, 1999.

[20] Bucher, H.F. and Wrobel, L.C., A novel approach to applying fast wavelet transformations in the boundary element method, *Advances in Boundary Element Techniques,* Hoggar, Geneva, 3-12, 2001.

[21] Cruse, T.A., Numerical solutions in three-dimensional elastostatics, *International Journal of Solids and Structures,* **5,** 1259-1274, 1969

[22] Cruse, T.A. and Rizzo, F.J., *Boundary Integral Equation Method: Computational Applications in Applied Mechanics,* ASME, AMD-Vol No. 11, 1975.

[23] Dirgantara, T. and Aliabadi, M.H., A new boundary element formulation for shear deformable shell analysis, *International Journal for Numerical Methods in Engineering,* **45,** 1257-1275, 1999.

[24] Dominguez, J., *Boundary Elements in Dynamics,* Computational Mechanics Publications, Southampton, 1993.

[25] El-Zafraney, A., *Techniques of the Boundary Element Method,* Ellis Horwood Series in Mathematics and its Applications, Ellis Horwood, New York, 1993.

[26] Fredholm, I., Sur une classe d'equations fonctionelles, *Acta Mathematica,* **27,** 365-390, 1903.

[27] Hall, W.S., *The Boundary Element Method,* Kluwer Academic Publishers, Dordrecht, 1994.

[28] Ingham, D.B. and Wrobel, L.C., *Boundary Integral Formulations for Inverse Analysis,* Computational Mechanics, Southampton, 1997.

[29] Kane, J.H., *Boundary Element Analysis in Engineering Continuum Mechanics,* Prentice-Hall, New York, 1994.

[30] Kassab, A. and Aliabadi, M.H., *Coupled Fields Problems*, WIT Press, Southampton 2001.

[31] Kitahara, M., *Boundary Integral Equation Methods to Eigenvalue Problems of Elastodynamics and Thin Plates,* Elsevier, 1985.

[32] Kupradze, V.D., *Potential Methods in the Theory of Elasticity*, Israel Programme for Scientific Translations, Jerusalem, 1965.

[33] Lachat, J.C., A further development of the boundary integral technique for elastostatics., PhD thesis, University of Southampton, 1975.

[34] Lachat, J.C. and Watson, J.O., Effective numerical treatment of boundary integral equations, *International Journal for Numerical Methods in Engineering,* **10,** 991-1005, 1976.

[35] Hess, J.L. and Smith, A.M.O., Calculation of potential flows about arbitrary bodies, *Progress in Aeronautical Sciences*, **8**, Pergamon Press, London, 1967.

[36] Jaswon, M.A., Integral equation method in potential theory, I, *Proceedings of the Royal Society of London, Series A*, **275**, 23-32, 1963.

[37] Jaswon, M. and Symm, G.T., *Integral Equation Methods in Potential Theory and Elastostatics*, Academic Press, London, 1977.

[38] Massonnet, C.E., Numerical use of integral procedures, In *Stress Analysis*, Chapter 10, 198-235, Wiley, London, 1965.

[39] Mikhlin, S.G., *Integral Equations,* Pergamon Press, London, 1957.

[40] Muller-Karger, C.M., Gonzalez, C., Aliabadi, M.H., Cerrolaza, M., Three dimensional BEM and FEM stress analysis of the human tibia under pathological conditions, *Computer Methods in Engineering and Sciences*, **2**, 1-13, 2001.

[41] Rigby, R and Aliabadi, M.H., Out of core solver for large multi-zone boundary element matrices, *International Journal for Numerical Methods in Engineering*, **38**, 1507-1533 (1995).

[42] Rizzo, F.J., An integral equation approach to boundary-value problems of classical elastostatics, *Quarterly Journal of Applied mathematics*, **25**, 83-95, 1967.

[43] Somigliana, C., Sopra l'equilibrio di un corpo elastico isotrope, *Il Nuovo Ciemento, serie III*, **20**, 181-185, 1886.

[44] Wen,P.H., Aliabadi,M.H. and Young,A., Boundary element analysis of assembled multilayered structures, *Advances in Boundary Element Techniques II*, Hoggar Press, Geneva, 555-562, 2001.

Chapter 2

Elastostatics

2.1 Introduction

The boundary integral equation fundamentals in elastostatics can be traced back to classical mathematical formulations by Betti [9], Somigliana [81], Muskhelishvili [52] and Kupradze [31]. The development of these formulations in the context of numerical methods was subsequently developed in the mid-sixties by Massonnet [40], Rizzo [70] and Cruse [18] who introduced the first formulation for three-dimensional elasticity. Later developments in the seventies introduced parametric representation to the formulation similar to that used in the finite element method [19, 34, 66]. Ricardella [66] and Cruse [19] improved the numerical implementation of the BEM to a certain degree by the introduction of linear variation over elements. Perhaps the most significant contribution to the numerical development of BEM resulted from the work of Lachat [34] and Lachat and Watson [33], who developed an isoparametric representation of the BEM. They also developed an accurate method for the integration of singular integrals in the BEM, and proposed a block based solver for the solution of large system of equations in the BEM.

The boundary element formulations may be divided into two different but closely related categories. The first and perhaps the most popular is the so-called *direct* formulation, in which the unknown functions appearing in the formulation are actual physical variables of the problem. In elasticity these unknown functions are the displacement and traction fields. The other approach is called the *indirect* formulation, in which unknown functions are represented by fictitious source densities. Once these source densities are found, the values of physical parameters can be obtained by simple integrations. In this book the *direct* boundary element formulation will mostly be described. The *indirect* formulation will be briefly discussed in Chapter 8 for applications to fracture problems.

This chapter presents an introduction to the boundary element method as applied to elastostatics. Initially the basic equations of elasticity that are necessary for the derivation of the BEM are presented. Betti's reciprocal theorem is used to derive the displacement and stress integral equations. The boundary displacement and stress integral equations are derived in detail by considering the limiting forms of Somigliana's identities. Two-dimensional, axisymmetric and three-dimensional problems are described. Methods for the derivation of the fundamental solutions are presented using complex stress functions and the Galerkin vector. Particular attention is paid to accurate methods for numerical implementation of the BEM. Isoparametric formulations are introduced for constant and higher order elements, and different types of elements commonly used in the application of the BEM to solids and structures are described. Different methods for dealing with body force terms, including the Galerkin vectors, particular integrals and the dual reciprocity method, are presented. Also presented are formulations for non-homogenous, orthotropic and anisotropic problems.

Symmetric Galerkin formulations are briefly presented, and the difficulty with the formulation for multi-connected bodies is discussed. Finally, different methods for coupling the BEM to the FEM are discussed.

2.2 Basic Elasticity

In this section, the basic ideas and relationships for the theory of elasticity that are necessary for development of the boundary element method and its applications to linear stress analysis are presented.

2.2.1 Stress

Consider an arbitrary shaped body, as illustrated in Figure 2.1. The body is in equilibrium under the action of externally applied forces $P_1, P_2, \ldots$. It is assumed the body is continuously deformable so that the forces are transmitted throughout its volume. At an internal point, say O, there is a resultant force δP, which may be considered as being uniformly distributed over a small area δA. The internal *force per unit area* is called the *stress* σ, and is given by

$$\sigma = \frac{\delta P}{\delta A} \tag{2.1}$$

In the limit,

$$\sigma = \lim_{\delta A \to 0} \left(\frac{\delta P}{\delta A} \right) = \frac{dP}{dA} \tag{2.2}$$

Generally, the force can be resolved into normal δP_n and tangential δP_s components, as shown in Figure 2.2. The normal or *direct stress* associated with these components is defined as

$$\sigma = \lim_{\delta A \to 0} \left(\frac{\delta P_n}{\delta A} \right) = \frac{dP_n}{dA} \tag{2.3}$$

and a *shear stress* is defined by

$$\tau = \lim_{\delta A \to 0} \left(\frac{\delta P_s}{\delta A} \right) = \frac{dP_s}{dA} \tag{2.4}$$

The resultant stress is given as $\sqrt{\sigma^2 + \tau^2}$.

2.2.2 Equilibrium

The state of stress at a point can be described by stress components formed on an element sides $\delta x, \delta y, \delta z$ formed at O by cutting planes, shown in Figure 2.3.

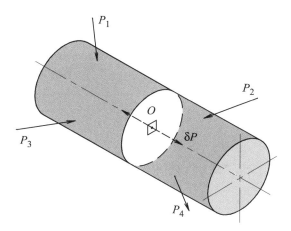

Figure 2.1: Internal force at a point in a body.

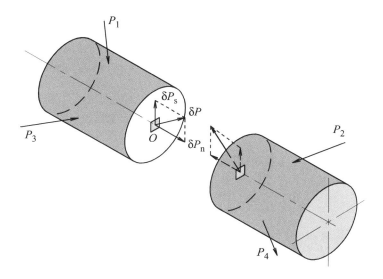

Figure 2.2: Internal force components at a point.

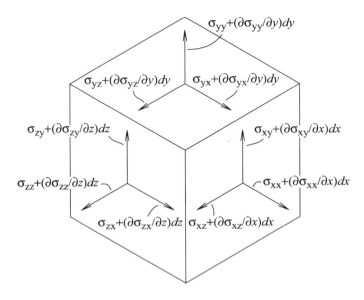

Figure 2.3: Stresses on the faces of an element.

The state of stress in an elemental volume of a loaded three-dimensional body, shown in Figure 2.3, can be defined in terms of six components of stress (σ_{xx}, σ_{yy}, σ_{zz}, σ_{xy}, σ_{xz}, σ_{yz}); they have a symmetry such that $\sigma_{xy} = \sigma_{yx}$, $\sigma_{xz} = \sigma_{zx}$ and $\sigma_{yz} = \sigma_{zy}$. The first suffix indicates the direction of the normal to the plane on which the stress acts, and the second suffix indicates the direction in which it acts. In tensor notation, if $1, 2$ and 3 directions coincide with the Cartesian directions, x, y and z, the stress components are normal components σ_{11}, σ_{22} and σ_{33} and shear components σ_{12}, σ_{13} and σ_{23}.

The equations of equilibrium for the three-dimensional system subjected to external forces and body forces b_x, b_y and b_z are given by

$$\frac{\partial \sigma_{xx}}{\partial x} + \frac{\partial \sigma_{xy}}{\partial y} + \frac{\partial \sigma_{xz}}{\partial z} + b_x = 0$$

$$\frac{\partial \sigma_{yx}}{\partial x} + \frac{\partial \sigma_{yy}}{\partial y} + \frac{\partial \sigma_{yz}}{\partial z} + b_y = 0$$

$$\frac{\partial \sigma_{zx}}{\partial x} + \frac{\partial \sigma_{zy}}{\partial y} + \frac{\partial \sigma_{zz}}{\partial z} + b_z = 0 \tag{2.5}$$

which can also be written in a tensor notation as

$$\frac{\partial \sigma_{ii}}{\partial x_i} + \frac{\partial \sigma_{ij}}{\partial x_j} + \frac{\partial \sigma_{ik}}{\partial x_k} + b_i = 0 \tag{2.6}$$

for $i \neq j \neq k$, $i, j, k = 1, 2, 3$, or in a more compact form as

$$\sigma_{ij,j} + b_i = 0 \tag{2.7}$$

where $i, j = 1, 2, 3$ and subscript $_{,j}$ denotes differentiation with respect to x_j.

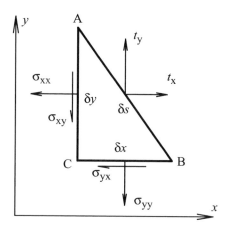

Figure 2.4: Stresses on the face of an element.

2.2.3 Plane Stress

The equations of equilibrium can be simplified by the two-dimensional specialization of plane stress conditions. In this condition, which is applicable to thin plates, it is assumed that stresses across the thickness of the sheet are negligible. Assuming, say, that the *z-axis* is in the direction of the thickness, then the stress components σ_{zz}, σ_{xz} and σ_{yz} are all zero. Then non-zero components σ_{xx}, σ_{yy} and σ_{xy} are averaged over the thickness and assumed to be independent of z. This state of stress is commonly referred to as the *generalized plane stress*. The equilibrium equations then simplify to

$$\frac{\partial \sigma_{xx}}{\partial x} + \frac{\partial \sigma_{xy}}{\partial y} + b_x = 0$$

$$\frac{\partial \sigma_{yx}}{\partial x} + \frac{\partial \sigma_{yy}}{\partial y} + b_y = 0 \qquad (2.8)$$

The above equations satisfy the requirements of equilibrium at all internal points of the body. Equilibrium must also be satisfied at all locations on the boundary of the body where the components of tractions (surface forces per unit area) are t_x and t_y. For the triangular element of Figure 2.4 at the boundary of a two-dimensional body of unit thickness, the tractions on element AB of the boundary must be in equilibrium with internal forces on internal faces AC and CB.

Summing up the forces in the x−direction, we have

$$t_x \delta s - \sigma_{xx} \delta y - \sigma_{yx} \delta x = 0$$

Taking the limit as δx approaches zero gives

$$t_x = \sigma_{xx} \frac{dy}{ds} + \sigma_{yx} \frac{dx}{ds} \qquad (2.9)$$

The derivatives dy/ds and dx/ds are the directional cosines l and m of the angles which a normal to AB makes with the x and y axes, respectively. Hence

$$t_x = \sigma_{xx} l + \sigma_{yx} m \qquad (2.10)$$

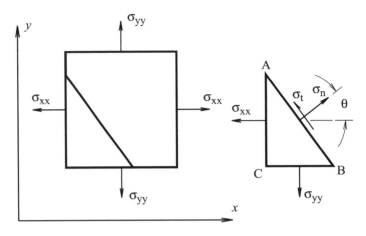

Figure 2.5: Stress on an inclined plane at a point.

and, by similarly considering the summation of forces in the y-direction it can be, shown that

$$t_y = \sigma_{yy}m + \sigma_{xy}l \tag{2.11}$$

The extension to three-dimensional bodies is relatively simple, and can be shown to give

$$
\begin{aligned}
t_x &= \sigma_{xx}l + \sigma_{yx}m + \sigma_{zx}n \\
t_y &= \sigma_{xy}l + \sigma_{yy}m + \sigma_{zy}n \\
t_z &= \sigma_{xz}l + \sigma_{yz}m + \sigma_{zz}n
\end{aligned} \tag{2.12}
$$

where l, m and n are the directional cosines of the angles (l, m, n are components of the outward unit normal) that a normal to the surface makes with the x, y and z axes, respectively. The expression for stresses on inclined planes can be derived by considering a corner cut off element by the plane AB inclined at angle θ to the y-axis (see Figure 2.5).

For equilibrium of ABC, the forces on AB, BC and CA must also be in equilibrium. As the element is a constant unit thickness, the area of the faces is proportional to the lengths of the sides of the triangle. Resolving forces normal to the plane AB,

$$\sigma_n AB - \sigma_{xx}AC\cos\theta - \sigma_{yy}BC\sin\theta = 0 \tag{2.13}$$

Dividing through by AB,

$$\sigma_n - \sigma_{xx}\frac{AC}{AB}\cos\theta - \sigma_{yy}\frac{BC}{AB}\sin\theta = 0 \tag{2.14}$$

Therefore,

$$
\begin{aligned}
\sigma_n &= \sigma_{xx}\cos^2\theta + \sigma_{yy}\sin^2\theta \\
&= \frac{1}{2}(\sigma_{xx} + \sigma_{yy}) + \frac{1}{2}(\sigma_{xx} - \sigma_{yy})\cos 2\theta
\end{aligned} \tag{2.15}
$$

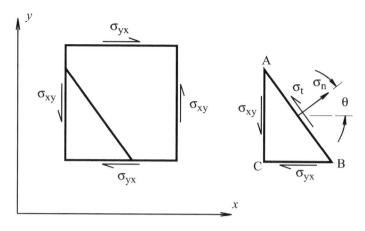

Figure 2.6: Shear stresses on an inclined plane.

Similarly, the shear stress component can be obtained by resolving forces parallel to AB and dividing by AB to give

$$\sigma_t = -\sigma_{xx}\cos\theta\sin\theta + \sigma_{yy}\sin\theta\cos\theta$$

$$= -\frac{1}{2}(\sigma_{xx} - \sigma_{yy})\sin 2\theta \qquad (2.16)$$

The normal component of stress for an element subjected to shear stress (see Figure 2.6) can be obtained by resolving forces normal to the plane AB, and dividing AB as before to give

$$\sigma_n = \sigma_{xy}\sin 2\theta \qquad (2.17)$$

Similarly, for shear stress components, by resolving forces parallel to the plane AB, and dividing by AB, we have

$$\sigma_t = \sigma_{xy}\cos 2\theta \qquad (2.18)$$

Now, for a general two-dimensional stress system, the normal and shear stresses can be obtained by combining (2.15) and (2.17) to give

$$\sigma_n = \frac{1}{2}(\sigma_{xx} + \sigma_{yy}) + \frac{1}{2}(\sigma_{xx} - \sigma_{yy})\cos 2\theta + \sigma_{xy}\sin 2\theta \qquad (2.19)$$

and, from (2.16) and (2.18),

$$\sigma_t = -\frac{1}{2}(\sigma_{xx} - \sigma_{yy})\sin 2\theta + \sigma_{xy}\cos 2\theta \qquad (2.20)$$

2.2.4   Principal Stress

Given that σ_n varies with the angle θ and will attain a maximum or minimum value when $d\sigma_n/d\theta = 0$, from (2.19)

$$\frac{d\sigma_n}{d\theta} = -(\sigma_{xx} - \sigma_{yy})\sin 2\theta + 2\sigma_{xy}\cos 2\theta = 0 \qquad (2.21)$$

or

$$\tan 2\theta = \frac{2\sigma_{xy}}{\sigma_{xx} - \sigma_{yy}} \tag{2.22}$$

The two planes θ and $\theta + \pi/2$ on which the direct stress is either a maximum or minimum are called 'principal planes' and the normal stresses acting on them are termed *principal stresses*. It can be seen by comparing (2.21) and (2.20) that shear stress σ_t is equal to zero on these planes. From (2.21),

$$\sin 2\theta = \frac{2\sigma_{xy}}{\sqrt{(\sigma_{xx} - \sigma_{yy})^2 + 4\sigma_{xy}^2}}$$

$$\cos 2\theta = \frac{\sigma_{xx} - \sigma_{yy}}{\sqrt{(\sigma_{xx} - \sigma_{yy})^2 + 4\sigma_{xy}^2}} \tag{2.23}$$

and

$$\sin 2(\theta + \frac{\pi}{2}) = \frac{-2\sigma_{xy}}{\sqrt{(\sigma_{xx} - \sigma_{yy})^2 + 4\sigma_{xy}^2}}$$

$$\cos 2(\theta + \frac{\pi}{2}) = \frac{-(\sigma_{xx} - \sigma_{yy})}{\sqrt{(\sigma_{xx} - \sigma_{yy})^2 + 4\sigma_{xy}^2}} \tag{2.24}$$

Substituting the above relationships into equation (2.19) gives

$$\sigma_{\max} = \frac{1}{2}\left[(\sigma_{xx} + \sigma_{yy}) + \sqrt{(\sigma_{xx} - \sigma_{yy})^2 + 4\sigma_{xy}^2}\right] \tag{2.25}$$

and

$$\sigma_{\min} = \frac{1}{2}\left[(\sigma_{xx} + \sigma_{yy}) - \sqrt{(\sigma_{xx} - \sigma_{yy})^2 + 4\sigma_{xy}^2}\right] \tag{2.26}$$

The maximum and minimum shear stresses at a point can be obtained in a similar manner from consideration of $d\sigma_t/d\theta = 0$, to give

$$\tau_{\max,\min} = \pm\sqrt{(\sigma_{xx} - \sigma_{yy})^2 + 4\sigma_{xy}^2} \tag{2.27}$$

Comparing (2.27) with (2.25) and (2.26) shows that

$$\tau_{\max} = \frac{1}{2}(\sigma_{\max} - \sigma_{\min})$$

2.2.5 Strain

The external and internal forces described earlier result in linear and angular displacements in a deformable body. These displacements are generally defined in terms of *strains*. *Direct longitudinal strains* are associated with direct stresses and related to changes in length. If a line element of length L at a point in the body undergoes an extension in length ΔL, then the longitudinal strains at that point are defined as

$$\varepsilon = \lim_{L \to 0} \frac{\Delta L}{L} \tag{2.28}$$

The six strain components corresponding to the components of the stress can be written in tensor notation as $\varepsilon_{ij} = \varepsilon_{ji}$, $i,j = 1,2,3$; in Cartesian directions $\varepsilon_{xy}, \varepsilon_{yy}$ and ε_{zz} are direct strains and $\varepsilon_{xy}, \varepsilon_{yz}$ and ε_{zx} are shear strains. Denoting the displacements u_x, u_y and u_z (u_i in tensor notation), the strains are related to displacements via

$$\varepsilon_{xx} = \frac{\partial u_x}{\partial x} \qquad \varepsilon_{yy} = \frac{\partial u_y}{\partial y} \qquad \varepsilon_{zz} = \frac{\partial u_z}{\partial z} \qquad (2.29)$$

for direct strains, and

$$\begin{aligned}
\varepsilon_{xy} &= \frac{1}{2}\left(\frac{\partial u_y}{\partial x} + \frac{\partial u_x}{\partial y}\right) \\
\varepsilon_{xz} &= \frac{1}{2}\left(\frac{\partial u_z}{\partial x} + \frac{\partial u_x}{\partial z}\right) \\
\varepsilon_{yz} &= \frac{1}{2}\left(\frac{\partial u_z}{\partial y} + \frac{\partial u_y}{\partial z}\right)
\end{aligned} \qquad (2.30)$$

for shear strains. Alternatively, in tensor notation they can be written as

$$\varepsilon_{ij} = \frac{1}{2}(u_{i,j} + u_{j,i}) \qquad (2.31)$$

2.2.6 Compatibility Equations

In this section, the relationship between the six strains are defined in terms of three displacements u_x, u_y and u_z. Differentiating (2.30) with respect to x and y gives

$$2\frac{\partial^2 \varepsilon_{xy}}{\partial x \partial y} = \frac{\partial^2}{\partial x \partial y}\left(\frac{\partial u_y}{\partial x}\right) + \frac{\partial^2}{\partial x \partial y}\left(\frac{\partial u_x}{\partial y}\right)$$

which may be rearranged to give

$$-2\frac{\partial^2 \varepsilon_{xy}}{\partial x \partial y} + \frac{\partial^2 \varepsilon_{yy}}{\partial x^2} + \frac{\partial^2 \varepsilon_{xx}}{\partial y^2} = 0 \qquad (2.32)$$

Similarly, the other compatibility equations can be obtained as

$$-2\frac{\partial^2 \varepsilon_{yz}}{\partial y \partial z} + \frac{\partial^2 \varepsilon_{yy}}{\partial z^2} + \frac{\partial^2 \varepsilon_{zz}}{\partial y^2} = 0 \qquad (2.33)$$

$$-2\frac{\partial^2 \varepsilon_{xz}}{\partial x \partial z} + \frac{\partial^2 \varepsilon_{zz}}{\partial x^2} + \frac{\partial^2 \varepsilon_{xx}}{\partial z^2} = 0 \qquad (2.34)$$

Now differentiating ε_{xy} with respect to x and z and adding the results to ε_{zx}, differentiation with respect to y and x gives

$$\frac{\partial^2 \varepsilon_x}{\partial y \partial z} = \frac{\partial}{\partial x}\left(-\frac{\partial \varepsilon_{yz}}{\partial x} + \frac{\partial \varepsilon_{xz}}{\partial y} + \frac{\partial \varepsilon_{xy}}{\partial z}\right) \qquad (2.35)$$

$$\frac{\partial^2 \varepsilon_y}{\partial x \partial z} = \frac{\partial}{\partial y}\left(-\frac{\partial \varepsilon_{yz}}{\partial x} - \frac{\partial \varepsilon_{xz}}{\partial y} + \frac{\partial \varepsilon_{xy}}{\partial z}\right) \qquad (2.36)$$

$$\frac{\partial^2 \varepsilon_z}{\partial x \partial y} = \frac{\partial}{\partial x}\left(\frac{\partial \varepsilon_{yz}}{\partial x} + \frac{\partial \varepsilon_{xz}}{\partial y} - \frac{\partial \varepsilon_{xy}}{\partial z}\right) \qquad (2.37)$$

Equations (2.32) to (2.37) are six equations of strain compatibility. Alternatively, they can be written in tensor notation as

$$-2\frac{\partial^2 \varepsilon_{ij}}{\partial x_i \partial x_j} + \frac{\partial^2 \varepsilon_{ii}}{\partial x_j^2} + \frac{\partial^2 \varepsilon_{jj}}{\partial x_i^2} = 0 \tag{2.38}$$

$$\frac{\partial^2 \varepsilon_{ij}}{\partial x_j \partial x_k} - \frac{\partial}{\partial x_i}\left(-\frac{\partial \varepsilon_{jk}}{\partial x_i} + \frac{\partial \varepsilon_{ik}}{\partial x_j} + \frac{\partial \varepsilon_{ij}}{\partial x_k}\right) = 0 \tag{2.39}$$

where $i, j, k = 1, 2, 3$ and $i \neq j \neq k$.

2.2.7 Stress-Strain Relationship

In the preceding sections, the derivation of the equilibrium, strain-displacement and compatibility equations were derived without any assumption as to the stress-strain behaviour of the body. Consider a material possessing the same properties at all points (i.e. homogeneous), for which stress is directly proportional to strain (i.e. linear elastic), and whose properties at all points are the same in all directions (isotropic). Experimental studies of material behaviour made by Robert Hooke's work in 1678 showed that, up to a certain limit, the extension of a bar subjected to an axial tensile loading was often directly proportional. In other words, the application of a uniform direct stress, say σ_{xx}, will only produce strains given by the equation

$$\varepsilon_{xx} = \frac{\sigma_{xx}}{E} \qquad \varepsilon_{yy} = -\nu\frac{\sigma_{xx}}{E} \qquad \varepsilon_{zz} = -\nu\frac{\sigma_{xx}}{E} \tag{2.40}$$

where E is a constant known as the modulus of elasticity or Young's modulus; ν is a constant termed Poisson's ratio. Equation (2.40) is an expression of Hooke's law. For an element subjected to general loading, the strains are given as

$$\varepsilon_{xx} = \frac{1}{E}\left[\sigma_{xx} - \nu(\sigma_{yy} + \sigma_{zz})\right] \qquad \varepsilon_{xy} = \frac{1+\nu}{E}\sigma_{xy}$$

$$\varepsilon_{yy} = \frac{1}{E}\left[\sigma_{yy} - \nu(\sigma_{xx} + \sigma_{zz})\right] \qquad \varepsilon_{yz} = \frac{1+\nu}{E}\sigma_{yz}$$

$$\varepsilon_{zz} = \frac{1}{E}\left[\sigma_{zz} - \nu(\sigma_{xx} + \sigma_{yy})\right] \qquad \varepsilon_{zx} = \frac{1+\nu}{E}\sigma_{zx} \tag{2.41}$$

Alternatively, Hooke's law can be written as

$$\sigma_{xx} = \lambda e + 2\mu\varepsilon_{xx} \qquad \sigma_{xy} = 2\mu\varepsilon_{xy}$$

$$\sigma_{yy} = \lambda e + 2\mu\varepsilon_{yy} \qquad \sigma_{yz} = 2\mu\varepsilon_{yz}$$

$$\sigma_{zz} = \lambda e + 2\mu\varepsilon_{zz} \qquad \sigma_{xz} = 2\mu\varepsilon_{xz} \tag{2.42}$$

where $\lambda = 2\nu\mu/(1 - 2\nu)$ is the *Lamé constant*, $\mu = E/2(1 + \nu)$ is the shear modulus of elasticity and $e = (\varepsilon_{xx} + \varepsilon_{yy} + \varepsilon_{zz})$ is the volumetric strain. The stress-strain or constitutive equations of elasticity can be written in tensor notation as

$$\sigma_{ij} = \lambda\delta_{ij}\varepsilon_{kk} + 2\mu\varepsilon_{ij} \tag{2.43}$$

where δ_{ij} is the Kronecker delta whose properties are

$$\delta_{ij} = \begin{cases} 0 & i \neq j \\ 1 & i = j \end{cases}$$

Plane Stress Condition

For thin plates, if no loadings are applied perpendicular to the surface of the plate, the stresses σ_{xx}, σ_{yy} and σ_{zz} are averaged over their thickness and assumed to be independent of Z. This state of stress is commonly referred to as *generalized plane stress*. The strain components ε_{yz} and ε_{zx} also vanish on the surface, and the component ε_{zz} is given by

$$\varepsilon_{zz} = \frac{-\nu}{1-\nu}(\varepsilon_{xx} + \varepsilon_{yy}) \tag{2.44}$$

The stress-strain relationship simplifies considerably to

$$\begin{aligned}
\varepsilon_{xx} &= \frac{1}{E}(\sigma_{xx} - \nu\sigma_{yy}) \\
\varepsilon_{yy} &= \frac{1}{E}(\sigma_{yy} - \nu\sigma_{xx}) \\
\varepsilon_{xy} &= \frac{1+\nu}{E}\sigma_{xy}
\end{aligned} \tag{2.45}$$

Plane Strain Condition

Plane strain conditions are assumed applicable to thick plates, that is those bodies in which geometry and loading do not vary significantly in, say, the z direction. In these problems, the dependent variables are assumed to be functions of the (x, y) coordinates only. The displacement component in the $z-$direction u_z is zero at every cross-section, and the strain components ε_{zz}, ε_{yz} and ε_{zx} will therefore vanish. Thus the only non-zero strains are

$$\varepsilon_{xx} = \frac{\partial u_x}{\partial x} \qquad \varepsilon_{yy} = \frac{\partial u_y}{\partial y} \qquad \varepsilon_{xy} = \frac{1}{2}\left(\frac{\partial u_y}{\partial x} + \frac{\partial u_x}{\partial y}\right) \tag{2.46}$$

Furthermore, as $\varepsilon_{zz} = \varepsilon_{xz} = \varepsilon_{yz} = 0$ and $\varepsilon_{xx}, \varepsilon_{yy}$ and ε_{xy} are functions of x and y only, the only compatibility equation for two-dimensional plane strain case is

$$-2\frac{\partial^2 \varepsilon_{xy}}{\partial x \partial y} + \frac{\partial^2 \varepsilon_{yy}}{\partial x^2} + \frac{\partial^2 \varepsilon_{xx}}{\partial y^2} = 0 \tag{2.47}$$

Hooke's law in (2.41) reduces to

$$\begin{aligned}
\varepsilon_{xx} &= \frac{1-\nu^2}{E}\left(\sigma_{xx} - \frac{\nu}{1-\nu}\sigma_{yy}\right) \\
\varepsilon_{yy} &= \frac{1-\nu^2}{E}\left(\sigma_{yy} - \frac{\nu}{1-\nu}\sigma_{xx}\right) \\
\varepsilon_{xy} &= \frac{1+\nu}{E}\sigma_{xy}
\end{aligned} \tag{2.48}$$

Bear in mind that for plane strain, since $\varepsilon_{zz} = 0$, the stress σ_{zz} is now given as

$$\sigma_{zz} = \nu(\sigma_{xx} + \sigma_{yy}) \tag{2.49}$$

2.2.8   Governing Equations of Elasticity

The governing equations of elasticity, known as Navier's equations, are the conditions of equilibrium expressed in terms of displacements. They can be obtained by substituting the stress-strain relationship (2.42) into equations of equilibrium (2.5) using the strain-displacement relationships (2.29) and (2.30), to give

$$\mu\nabla^2 u_x + (\lambda + \mu)\frac{\partial e}{\partial x} + b_x = 0$$

$$\mu\nabla^2 u_y + (\lambda + \mu)\frac{\partial e}{\partial y} + b_y = 0$$

$$\mu\nabla^2 u_z + (\lambda + \mu)\frac{\partial e}{\partial z} + b_z = 0 \qquad (2.50)$$

where $\nabla^2 = (\frac{\partial^2}{\partial x^2} \quad \frac{\partial^2}{\partial y^2} \quad \frac{\partial^2}{\partial z^2})$ is the Laplacian operator. The Navier equation can be written in tensor notation as

$$\mu u_{i,jj} + (\mu + \lambda)u_{j,ji} + b_i = 0 \qquad (2.51)$$

for $i, j = 1, 2, 3$.

2.3   Boundary Element Formulations

There are several methods of deriving the BEM formulations: the reciprocal theorem, the weighted residual concept and the variational approach. Here, BEM formulations will be derived using the reciprocal theorem.

2.3.1   Reciprocal Theorem

The reciprocal theorem, accredited to Maxwell, Betti and Rayleigh, is a powerful method of analysis for linear elastic systems. In the next section, the reciprocal theorem is derived in its general form, from the consideration of equilibrium, and used to derive the boundary integral equation. Here, we shall consider a useful property of linear elastic systems resulting from the principle of superposition. The principle allows us to express the deflection at any point in a structure in terms of a constant coefficient and the applied load. For example, a load, say P_1, applied at a point 1 in a linear elastic body will produce a deflection Δ_1 at the point given by

$$\Delta_1 = f_{11} P_1$$

where the *flexibility* coefficient f_{11} is defined as the deflection at point 1 in the direction of P_1, produced by a unit load at point 1. Consider a body subjected to a system of loads as shown in Figure 2.7. Here, each of the loads will contribute to the deflection at point 1. Therefore, the total deflection Δ_1 in the direction of the load P_1 produced by all the loads is given as

$$\Delta_1 = f_{11} P_1 + f_{12} P_2 + \text{........} + f_{1n} P_n$$

where f_{12} is the deflection at point 1 in the direction of P_1, produced by unit load at point 2 in the direction of the load P_2, and so on. The corresponding deflections

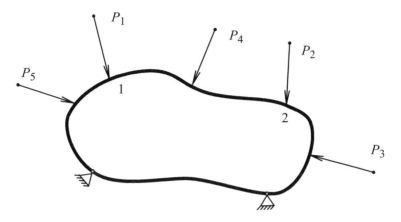

Figure 2.7: Linear elastic body subjected to loads $P_1, P_2, ... P_n$.

at the points of application of the complete system of loads can be written in matrix form as

$$
\left\{
\begin{array}{c}
\Delta_1 \\
\Delta_2 \\
\cdot \\
\Delta_n
\end{array}
\right\}
=
\left[
\begin{array}{cccc}
f_{11} & f_{12} & \cdot & f_{1n} \\
f_{21} & \cdot & \cdot & f_{2n} \\
\cdot & \cdot & \cdot & \cdot \\
f_{n1} & \cdot & \cdot & f_{nn}
\end{array}
\right]
\left\{
\begin{array}{c}
P_1 \\
P_2 \\
\cdot \\
P_n
\end{array}
\right\}
$$

Assuming that the body is subjected to P_1 at a point 1, and then, while P_1 is kept in position, a force P_2 is applied at point 2, the total strain energy U_1 of the body is given by

$$
U_1 = \frac{P_1}{2}(f_{11}P_1) + \frac{P_2}{2}(f_{22}P_2) + P_1(f_{12}P_2) \tag{2.52}
$$

Notice that the third term on the right-hand side of (2.52) is due to additional work done by P_1 as it is displaced through further by $f_{12}P_2$ by the action of P_2. Similarly, we apply P_2 followed by P_1, we have

$$
U_2 = \frac{P_1}{2}(f_{11}P_1) + \frac{P_2}{2}(f_{22}P_2) + P_2(f_{21}P_1) \tag{2.53}
$$

By the principle of superposition, the strain energy stored in the body is independent of the order in which the loads are applied. Hence

$$
U_1 = U_2
$$

and it follows that

$$
f_{12} = f_{21}
$$

Therefore the reciprocal theorem states: *The deflection at a point 1 in a given direction due to a unit load at point 2 in a second direction is equal to the deflection at point 2 in the second direction due to a unit load at point 1 in the first direction.*

2.3.2   Betti's Reciprocal Theorem

The direct boundary element formulation for elastostatics problems can be derived from Betti's reciprocal work theorem for two self-equilibrated states (u_i, t_i, b_i) and (u_i^*, t_i^*, b_i^*); u_i and u_i^* are displacements; t_i and t_i^* are tractions (i.e. $t_i = \sigma_{ij} n_j$, where n_j are the components of the outward normal to the boundary); and b_i and b_i^* are body forces.

From the equilibrium equations in (2.7), it is possible to write the following relationship:

$$\int_V (\sigma_{ij,j} + b_i) u_i^* dV = 0 \qquad (2.54)$$

where repeated suffix summation is assumed and V denotes the domain with boundary S of the problem (see Figure 2.8). The stresses, body forces and displacements are a function of $\mathbf{X} \epsilon V$ ($\mathbf{X} \equiv x, y, z$) for a three-dimensional body. The integral involving the $\sigma_{ij,j} u_i^*$ term can be rewritten as

$$\int_V \sigma_{ij,j} u_i^* dV = \int_V (\sigma_{ij} u_i^*)_{,j} dV - \int_V \sigma_{ij} \varepsilon_{ij}^* dV \qquad (2.55)$$

since

$$\begin{aligned} \sigma_{ij} \varepsilon_{ij}^* &= \frac{1}{2}(\sigma_{ij} u_{i,j}^* + \sigma_{ij} u_{j,i}^*) \\ &= \frac{1}{2}(\sigma_{ij} u_{i,j}^* + \sigma_{ji} u_{j,i}^*) \\ &= \sigma_{ij} u_{i,j}^* \end{aligned} \qquad (2.56)$$

From the divergence theorem (i.e. $\int_V f_{i,i} dV = \int_S f_i n_i dS$) we have

$$\int_V (\sigma_{ij} u_i^*)_{,j} dV = \int_S \sigma_{ij} n_j u_i^* dS = \int_S t_i u_i^* dS \qquad (2.57)$$

therefore

$$\int_V \sigma_{ij,j} u_i^* dV = \int_S t_i u_i^* dS - \int_V \sigma_{ij} \varepsilon_{ij}^* dV \qquad (2.58)$$

Using (2.58) in (2.54) gives

$$\int_S t_i u_i^* dS + \int_V b_i u_i^* dV = \int_V \sigma_{ij} \varepsilon_{ij}^* dV \qquad (2.59)$$

The integral on the right-hand side of (2.59) can be written (see equation (2.43)) as

$$\int_V \sigma_{ij} \varepsilon_{ij}^* dV = \int_V \left[\lambda \delta_{ij} \varepsilon_{ij}^* \varepsilon_{kk} + 2\mu \varepsilon_{ij} \varepsilon_{ij}^* \right] dV \qquad (2.60)$$

Since $\delta_{ij} \varepsilon_{ij}^* = \varepsilon_{mm}^*$ and $\varepsilon_{kk} = \delta_{ij} \varepsilon_{ij}$, it follows that

$$\int_V \sigma_{ij} \varepsilon_{ij}^* dV = \int_V \left[\lambda \delta_{ij} \varepsilon_{mm}^* + 2\mu \varepsilon_{ij}^* \right] \varepsilon_{ij} dV = \int_V \sigma_{ij}^* \varepsilon_{ij} dV \qquad (2.61)$$

From (2.59) and (2.61), it can be seen that the following relationship holds:

$$\int_S t_i u_i^* dS + \int_V b_i u_i^* dV = \int_S t_i^* u_i dS + \int_V b_i^* u_i dV \qquad (2.62)$$

which is known as Betti's reciprocal work theorem.

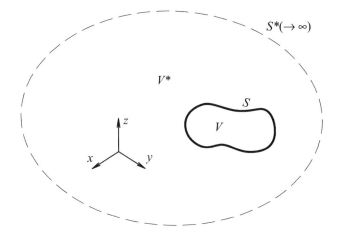

Figure 2.8: An infinite region V^* containing the actual body V.

2.3.3 Boundary Integral Equations

The boundary integral equation for elastostatic problems can be derived from Betti's reciprocal work theorem (2.62) by taking the body force b_i^* to correspond to a point force in an infinite sheet, represented by the Dirac delta function $\Delta(X - X')$ as

$$b_i^* = \Delta(\mathbf{X} - \mathbf{X}')e_i$$

where the unit vector component e_i corresponds to a unit positive force in the i direction applied at $\mathbf{X}'$ and $\mathbf{X}, \mathbf{X}' \in V$. In two-dimensional problems, e_i is a force per unit thickness, and in three-dimensional problems it is a pure concentrated force.

The Dirac delta function has the property

$$\int_V g(\mathbf{X})\Delta(\mathbf{X} - \mathbf{X}')dV = g(\mathbf{X}')$$

Using this property, the last integral in equation (2.62) can be written as

$$\int_V b_i^* u_i dV = \int_V \Delta(\mathbf{X} - \mathbf{X}')e_i u_i dV = u_i(\mathbf{X}')e_i \tag{2.63}$$

The displacement and traction fields corresponding to the point force solution can be written as

$$u_i^* = U_{ij}(\mathbf{X}', \mathbf{X})e_j \tag{2.64}$$

and

$$t_i^* = T_{ij}(\mathbf{X}', \mathbf{X})e_j \tag{2.65}$$

From the above solutions and equation (2.63) it, can be seen that equation (2.62) can be written as

$$u_i(\mathbf{X}') = \int_S U_{ij}(\mathbf{X}', \mathbf{x})t_j(\mathbf{x})dS - \int_S T_{ij}(\mathbf{X}', \mathbf{x})u_j(\mathbf{x})dS + \int_V U_{ij}(\mathbf{X}', \mathbf{X})b_j(\mathbf{X})dV \tag{2.66}$$

where $\mathbf{x} \in S$. The above equation is known as the Somigliana's identity for displacements. It relates the value of displacements at an internal point $\mathbf{X}'$ to boundary values of the displacements and tractions. Equation (2.66) can be written in matrix form as

$$\left\{ \begin{array}{c} u_1 \\ u_2 \end{array} \right\} = \int_S \left[\begin{array}{cc} U_{11} & U_{12} \\ U_{21} & U_{22} \end{array} \right] \left\{ \begin{array}{c} t_1 \\ t_2 \end{array} \right\} dS$$

$$- \int_S \left[\begin{array}{cc} T_{11} & T_{12} \\ T_{21} & T_{22} \end{array} \right] \left\{ \begin{array}{c} u_1 \\ u_2 \end{array} \right\} dS + \int_V \left[\begin{array}{cc} U_{11} & U_{12} \\ U_{21} & U_{22} \end{array} \right] \left\{ \begin{array}{c} b_1 \\ b_2 \end{array} \right\} dV$$

for two-dimensional problems, and

$$\left\{ \begin{array}{c} u_1 \\ u_2 \\ u_3 \end{array} \right\} = \int_S \left[\begin{array}{ccc} U_{11} & U_{12} & U_{13} \\ U_{21} & U_{22} & U_{23} \\ U_{31} & U_{32} & U_{33} \end{array} \right] \left\{ \begin{array}{c} t_1 \\ t_2 \\ t_3 \end{array} \right\} dS$$

$$- \int_S \left[\begin{array}{ccc} T_{11} & T_{12} & T_{13} \\ T_{21} & T_{22} & T_{23} \\ T_{31} & T_{32} & T_{33} \end{array} \right] \left\{ \begin{array}{c} u_1 \\ u_2 \\ u_3 \end{array} \right\} dS + \int_V \left[\begin{array}{ccc} U_{11} & U_{12} & U_{13} \\ U_{21} & U_{22} & U_{23} \\ U_{31} & U_{32} & U_{33} \end{array} \right] \left\{ \begin{array}{c} b_1 \\ b_2 \\ b_3 \end{array} \right\} dV$$

for three-dimensional problems. The strains at any interior point can be obtained by differentiating the displacements in equation (2.66) with respect to the source point $\mathbf{X}'$ to give

$$\begin{aligned} u_{i,k}(\mathbf{X}') &= \int_S U_{ij,k}(\mathbf{X}', \mathbf{x}) t_j(\mathbf{x}) dS - \int_S T_{ij,k}(\mathbf{X}', \mathbf{x}) u_j(\mathbf{x}) dS \\ &\quad + \int_V U_{ij,k}(\mathbf{X}', \mathbf{X}) b_j(\mathbf{X}) dV \end{aligned} \tag{2.67}$$

where $U_{ij,k}$ and $T_{ij,k}$ are derivatives of the fundamental solutions. Finally, Somigliana's identity for stresses can be obtained by substituting equation (2.67) into Hooke's law (2.43), to give

$$\sigma_{ij}(\mathbf{X}') = \int_S D_{kij}(\mathbf{X}', \mathbf{x}) t_k(\mathbf{x}) dS - \int_S S_{kij}(\mathbf{X}', \mathbf{x}) u_k(\mathbf{x}) dS + \int_V D_{kij}(\mathbf{X}', \mathbf{X}) b_k(\mathbf{X}) dV \tag{2.68}$$

where D_{kij} and S_{kij} are obtained from $U_{ij,k}$ and $T_{ij,k}$, and the application of Hooke's law. In the next section, derivation of the above point force solutions is described.

2.3.4 Fundamental Solutions

As shown in previous sections, the existence of the point force solution plays an important role in the formulation of the boundary element method. Navier's equation can now be written for a unit point force applied to the body at a point $\mathbf{X}'$, as

$$\mu u^*_{i,jj} + \frac{\mu}{1 - 2\nu} u^*_{j,ji} + \Delta(\mathbf{X} - \mathbf{X}') e_i = 0 \tag{2.69}$$

The solutions of the governing equations due to a point force are commonly referred to as 'fundamental solutions'. There are several ways of obtaining a solution to the above governing equation. Here, methods based on the Galerkin vector are presented. An alternative method based on complex stress functions is presented in Appendix A, together with stress functions for half-plane, circular cavity and an isolated crack.

Galerkin Vector

The most popular technique for deriving the fundamental solutions is through the use of the Galerkin vector, G_i. In this section, the Galerkin vector approach will be used to evaluate the fundamental solutions due to a unit point force in an infinite medium. The point force solution in an infinite medium was originally derived by Lord Kelvin, and is known as Kelvin's fundamental solution. The displacements are expressed in terms of the Galerkin vector as

$$u_i^* = G_{i,kk} - \frac{1}{2(1-\nu)} G_{k,ik} \tag{2.70}$$

Substituting (2.70) into (2.69) gives

$$\mu G_{i,kkjj} - \frac{\mu}{2(1-\nu)} G_{k,ikjj} + \frac{\mu}{(1-\nu)} \left(G_{j,kkij} - \frac{1}{2(1-\nu)} G_{k,jkij} \right) + \Delta(\mathbf{X} - \mathbf{X}')e_i = 0 \tag{2.71}$$

which can be simplified to

$$\mu G_{i,kkjj} + \Delta(\mathbf{X} - \mathbf{X}')e_i = 0 \tag{2.72}$$

since $G_{k,ikjj} = G_{k,jjki}$, $G_{j,kkjj} = G_{k,jjki}$ and $G_{k,jkij} = G_{k,jjki}$. Equation (2.72) can also be written as

$$\mu \nabla^2 (\nabla^2 G_i) + \Delta(\mathbf{X} - \mathbf{X}')e_i = 0 \tag{2.73}$$

Now, let $F_i = \nabla^2 G_i$, then (2.74) can be written as

$$\nabla^2 F_i + \frac{1}{\mu} \Delta(\mathbf{X} - \mathbf{X}')e_i = 0 \tag{2.74}$$

Three-dimensional Problems The solution of (2.74) is well known from the potential theory and is given by

$$F_i = \frac{1}{4\pi\mu r} e_i \tag{2.75}$$

for three-dimensional problems. The Galerkin vector is given by

$$G_i = \frac{1}{8\pi\mu} r e_i \tag{2.76}$$

Substituting the derivatives of (2.76) into (2.70) gives

$$u_i^* = \frac{1}{8\pi\mu} \left[r_{,kk} e_i - \frac{1}{2(1-\nu)} r_{,ik} e_k \right] \tag{2.77}$$

Now noting that $r_{,ik} = (\delta_{ik} - r_{,i} r_{,k})/r$ and $r_{,kk} = 2/r$, equation (2.77) can be rewritten as

$$u_i^* = \frac{1}{16\pi\mu(1-\nu)r} [(3 - 4\nu)\delta_{ij} + r_{,i} r_{,j}] e_j \tag{2.78}$$

From (2.64), we have

$$U_{ij}(\mathbf{x}', \mathbf{x}) = \frac{1}{16\pi\mu(1-\nu)r} [(3 - 4\nu)\delta_{ij} + r_{,i} r_{,j}] \tag{2.79}$$

where $U_{ij}(\mathbf{x'}, \mathbf{x})$ represents the displacement in the j direction at point $\mathbf{x}$ due to a unit point force acting in the i direction at $\mathbf{x'}$. The traction fundamental solution is obtained from (2.78), through the usual displacement-strain and strain-stress relationships, and by noting that $t_i^* = \sigma_{ij}^* n_j$, to give

$$t_i^* = \frac{-1}{8\pi(1-\nu)r^2} \left\{ \frac{\partial r}{\partial n}[(1-2\nu)\delta_{ij} + 3r_{,i}r_{,j} - (1-2\nu)(n_j r_{,i} - n_i r_{,j}) \right\} e_j \quad (2.80)$$

where n_j denotes the components of the outward normal at the field point $\mathbf{x}$. Again from (2.65), we have

$$T_{ij}(\mathbf{x'}, \mathbf{x}) = \frac{-1}{8\pi(1-\nu)r^2} \left\{ \frac{\partial r}{\partial n}[(1-2\nu)\delta_{ij} + 3r_{,i}r_{,j} - (1-2\nu)(n_j r_{,i} - n_i r_{,j}) \right\}$$
$$(2.81)$$

where $T_{ij}(\mathbf{x'}, \mathbf{x})$ represents the traction in the j direction at point $\mathbf{x}$ due to a unit point force acting in the i direction at $\mathbf{x'}$. Differentiating (2.79) and (2.81) with respect to $\mathbf{x'}$ gives

$$U_{ij,k}(\mathbf{x'}, \mathbf{x}) = \frac{1}{16\pi\mu(1-\nu)r^2} \left\{ (3-4\nu)\delta_{ij}r_{,k} + 3r_{,i}r_{,j}r_{,k} - \delta_{jk}r_{,i} - \delta_{ki}r_{,j} \right\} \quad (2.82)$$

$$T_{ij,k}(\mathbf{x'}, \mathbf{x}) = \frac{-1}{8\pi(1-\nu)r^3} \left\{ [(1-2\nu)\delta_{ij}r_{,k} + 3r_{,i}r_{,j} n_k \right. \quad (2.83)$$
$$+ (1-2\nu)[\delta_{jk} - 3r_{,j}r_{,k} n_i + (1-2\nu)[\delta_{ki} - 3r_{,k}r_{,i} n_j$$
$$\left. + [3\delta_{jk}r_{,i} + 3\delta_{ki}r_{,j} - 3(1-2\nu)\delta_{ij}r_{,k} - 15r_{,i}r_{,j}r_{,k} r_{,l}n_{,l,k} \right\}$$

Using the stress-strain relationships, gives

$$D_{kij} = \frac{1}{8\pi(1-\nu)r^2} [(1-2\nu)(\delta_{ik}r_{,j} + \delta_{jk}r_{,i} - \delta_{ij}r_{,k}) + 3r_{,i}r_{,j}r_{,k} \quad (2.84)$$

$$S_{kij} = \frac{\mu}{4\pi(1-\nu)r^3} \left\{ 3\frac{\partial r}{\partial n}[(1-2\nu)\delta_{ij}r_{,k} + \nu(\delta_{ik}r_{,j} + \delta_{jk}r_{,i}) - 5r_{,i}r_{,j}r_{,k} \right.$$
$$+ 3\nu(n_i r_{,j}r_{,k} + n_j r_{,i}r_{,k}) + (1-2\nu)(3n_k r_{,i}r_{,j} + n_j\delta_{ik} + n_i\delta_{jk})$$
$$\left. - (1-4\nu)n_k\delta_{ij} \right\} \quad (2.85)$$

Two-dimensional Problems   A similar procedure can be followed for two-dimensional problems. The solution of (2.74) for two-dimensional problems is given by

$$F_i = -\frac{1}{2\pi\mu} \ln(r)e_i \quad (2.86)$$

The corresponding Galerkin vector is

$$G_i = -\frac{1}{8\pi\mu} r^2 \ln(r)e_i \quad (2.87)$$

The displacement and traction fundamental solutions for the two-dimensional plane strain condition can be obtained by following the same procedure as for 3D, and are given by

$$U_{ij}(\mathbf{X}', \mathbf{x}) = \frac{1}{8\pi\mu(1-\nu)} \left\{ (3-4\nu)\ln(\frac{1}{r})\delta_{ij} + r_{,i}r_{,j} \right\} \tag{2.88}$$

$$T_{ij}(\mathbf{X}', \mathbf{x}) = \frac{-1}{4\pi(1-\nu)r} \left\{ \frac{\partial r}{\partial n} [(1-2\nu)\delta_{ij} + 2r_{,i}r_{,j}] - (1-2\nu)(r_{,i}n_j - r_{,j}n_i) \right\} \tag{2.89}$$

The displacement fundamental solution for two-dimensional problems listed above contains logarithmic functions. The existence of these logarithmic terms may lead to non-uniqueness in the solution of the boundary elements. Generally, this difficulty can be circumvented by normalizing the distance r within the logarithmic term, with respect to some dimension, say the maximum dimension of the structure $l_{\max}$; therefore, the term $\ln(1/r)$ becomes $\ln(l_{\max}/r)$. This normalization procedure only adds a rigid body displacement, of magnitude $\ln(l_{\max})$, into the formulation, and hence does not affect the solution.

The fundamental solutions for the stress integral equation can be obtained from differentiation of (2.88) and (2.89) with respect to the source point $\mathbf{x}'$, to give

$$U_{ij,k}(\mathbf{x}', \mathbf{x}) = -\frac{1+\nu}{4\pi(1-\nu)E} \frac{1}{r} \left\{ (3-4\nu)\delta_{ij}r_{,k} - \delta_{jk}r_{,i} - \delta_{ik}r_{,j} + 2r_{,i}r_{,j}r_{,k} \right\} \tag{2.90}$$

$$
\begin{aligned}
T_{ij,k}(\mathbf{x}', \mathbf{x}) = \ &-\frac{1}{4\pi(1-\nu)} \frac{1}{r^2} \left\{ 2\frac{\partial r}{\partial n} [\delta_{ik}r_{,j} + \delta_{jk}r_{,i} - r_{,k}((1-2\nu)\delta_{ij} + 4r_{,i}r_{,j})] \right. \\
&+ n_k [(1-2\nu)\delta_{ij} + 2r_{,i}r_{,j}] - n_j(1-2\nu)[\delta_{ik} - 2r_{,i}r_{,k}] \\
&\left. + n_i(1-2\nu)[\delta_{jk} - 2r_{,j}r_{,k}] \right\}
\end{aligned}
\tag{2.91}
$$

and use of the stress-strain relationships to give

$$D_{kij}(\mathbf{X}', \mathbf{x}) = \frac{1}{4\pi(1-\nu)r} \left\{ (1-2\nu)(-r_{,k}\delta_{ij} + r_{,j}\delta_{ki} + r_{,i}\delta_{jk}) + 2r_{,i}r_{,j}r_{,k} \right\} \tag{2.92}$$

$$
\begin{aligned}
S_{kij}(\mathbf{X}', \mathbf{x}) = \ &\frac{\mu}{2\pi(1-\nu)} \frac{1}{r^2} \left\{ 2\frac{\partial r}{\partial n} [(1-2\nu)\delta_{ij}r_{,k} + \nu(r_{,j}\delta_{ik} + r_{,i}\delta_{jk}) - 4r_{,i}r_{,j}r_{,k}] \right. \\
&+ 2\nu(n_i r_{,i}r_{,k} + n_j r_{,i}r_{,k}) + (1-2\nu)(2n_k r_{,i}r_{,j} + n_j\delta_{ik} + n_i\delta_{jk}) \\
&\left. - (1-4\nu)n_k\delta_{ij} \right\}
\end{aligned}
\tag{2.93}
$$

The fundamental solutions for plane stress can be obtained by introducing the modified Poisson's coefficient ν' and Young's modulus E', defined as

$$\nu' = \frac{\nu}{1+\nu} \quad \text{and} \quad E' = E\left[1 - \left(\frac{\nu}{1+\nu}\right)^2\right] \tag{2.94}$$

Axisymmetric Problems

There are two approaches to obtain axisymmetric fundamental solutions [3, 20, 29]. The first is to derive fundamental solutions based on ring loads and the second is to integrate the three-dimensional solutions with respect to the hoop direction. Consider an arbitrary axisymmetric domain shown in Figure 2.9. The displacement fundamental solutions are given as:

$$
\begin{aligned}
U_{rr}(\mathbf{x}',\mathbf{x}) \;=\; & \frac{1}{\Lambda R(\mathbf{x})R(\mathbf{x}')C}\left\{\left[(3-4\nu)(R^2(\mathbf{x})+R^2(\mathbf{x}')+4(1-\nu)(Z-z)^2\right]\right. \\
& \times K\left(m,\frac{\pi}{2}\right) \\
& \left.-\left[C^2(3-4\nu)+\frac{(Z-z)^2\left[R^2(\mathbf{x})+R^2(\mathbf{x}')+(Z-z)^2\right]}{(R(\mathbf{x}')-R(\mathbf{x}))^2+(Z-z)^2}\right]E\left(m,\frac{\pi}{2}\right)\right\}
\end{aligned}
$$

$$
U_{rz}(\mathbf{x}',\mathbf{x})=\frac{(Z-z)}{\Lambda R(\mathbf{x}')C}\left\{K\left(m,\frac{\pi}{2}\right)-\frac{R^2(\mathbf{x})-R^2(\mathbf{x}')+(Z-z)^2}{(R(\mathbf{x}')-R(\mathbf{x}))^2+(Z-z)^2}E\left(m,\frac{\pi}{2}\right)\right\}
$$

$$
U_{zr}(\mathbf{x}',\mathbf{x})=\frac{(Z-z)}{\Lambda R(\mathbf{x}')C}\left\{-K\left(m,\frac{\pi}{2}\right)+\frac{R^2(\mathbf{x})-R^2(\mathbf{x}')+(Z-z)^2}{(R(\mathbf{x}')-R(\mathbf{x}))^2+(Z-z)^2}E\left(m,\frac{\pi}{2}\right)\right\}
$$

$$
U_{zz}(\mathbf{x}',\mathbf{x})=\frac{2}{\Lambda C}\left\{(3-4\nu)K\left(m,\frac{\pi}{2}\right)+\frac{(Z-z)^2}{(R(\mathbf{x}')-R(\mathbf{x}))^2+(Z-z)^2}E\left(m,\frac{\pi}{2}\right)\right\}
$$

$$(2.95)$$

where $\Lambda = 16\pi^2\mu(1-\nu)$, $C = \sqrt{(R(\mathbf{x}')-R(\mathbf{x}))^2-(Z-z)^2}$, $m = 2\sqrt{R(\mathbf{x}')R(\mathbf{x})}/C$ and, $K\left(m,\frac{\pi}{2}\right)$ and $E\left(m,\frac{\pi}{2}\right)$ are elliptical integrals of first and second kind, respectively.

The traction fundamental solutions can be obtained from (2.95) by utilizing the displacement-strain and strain-stress relationships. The fundamental solutions for interior stresses are given in Balas *et al.* [5].

2.3.5 Boundary Displacement Equation

The boundary integral equation (2.66) is valid for any source point within the domain V. In order to obtain a solution for the points on the boundary, it is necessary to consider the limit as $\mathbf{X}'\to\mathbf{x}'\in S$.

Let us consider the problem domain to have been augmented around the boundary source point $\mathbf{x}'$ by a semi-circular region for two-dimensional problems and a hemispherical region for three-dimensional problems, with boundary S_ε^+, radius ε and centred at $\mathbf{x}'$, as illustrated in Figures 2.10 and 2.11. The augmented problem boundary S^A is now represented by

$$S^A = (S - S_\varepsilon) + S_\varepsilon^+ \tag{2.96}$$

where S_ε is the portion of the original boundary that has been removed. The behaviour of the singularities can be studied by taking the limit as $\varepsilon \to 0$, and therefore $S^A \to S$.

Consider the first boundary integral on the right-hand side of equation (2.66). In the limit, when $\varepsilon \to 0$, it can be written as

$$
\int_S U_{ij}(\mathbf{X}',\mathbf{x})t_j(\mathbf{x})dS
$$

$$
= \lim_{\varepsilon\to0}\int_{S-S_\varepsilon} U_{ij}(\mathbf{x}',\mathbf{x})t_j(\mathbf{x})dS + \lim_{\varepsilon\to0}\int_{S_\varepsilon^+} U_{ij}(\mathbf{x}',\mathbf{x})t_j(\mathbf{x})dS \tag{2.97}
$$

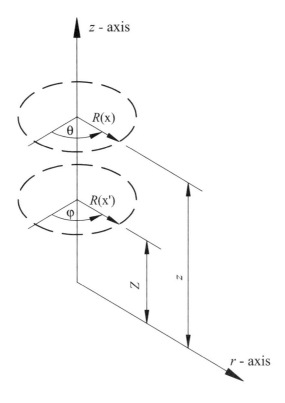

Figure 2.9: Cylindrical coordinate system for axisymmetric problems.

in which the second integral on the right hand side contains singular integrands of order $O(\ln r)$ in two-dimensions and $O(r^{-1})$ in three-dimensions; however, it can be shown that the integral tends to zero as $\varepsilon \to 0$. The first integral on the right-hand side of (2.97) is integrable as an improper integral, and care must be taken in its numerical evaluation.

Consider now the limiting form of the second boundary integral on the right-hand side of (2.66). It can be written as

$$
\int_S T_{ij}(\mathbf{X'}, \mathbf{x}) u_j(\mathbf{x}) dS
$$

$$
= \lim_{\varepsilon \to 0} \int_{S-S_\varepsilon} T_{ij}(\mathbf{x'}, \mathbf{x}) u_j(\mathbf{x}) dS + \lim_{\varepsilon \to 0} \int_{S_\varepsilon^+} T_{ij}(\mathbf{x'}, \mathbf{x}) u_j(\mathbf{x}) dS \qquad (2.98)
$$

in which both limiting expressions on the right-hand expressions on the right-hand side contain a strongly singular integrand of order $O(r^{-1})$ in two-dimensions and $O(r^{-2})$ in three-dimensions. The first integral is treated in a Cauchy principle value sense. The second integral is regularized by the first term of a Taylor series expansion of the displacements[1] about the source point, to give

$$
\lim_{\varepsilon \to 0} \int_{S_\varepsilon^+} T_{ij}(\mathbf{x'}, \mathbf{x}) u_j(\mathbf{x}) dS \qquad (2.99)
$$

---

[1] Displacements are assumed to be Hölder continuous. That is $\mathbf{u} \in \mathbf{C}^{0,\alpha}$ if there exist $C > 0$ and $0 < \alpha \le 1$ such that $\mid u(\mathbf{x'}) - u(\mathbf{x}) \mid \le Cr^\alpha$.

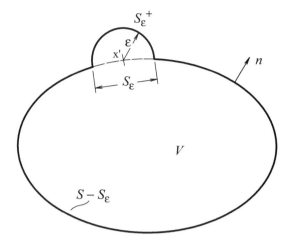

Figure 2.10: Source point $\mathbf{x}'$ located on the boundary surrounded by a semi-circular arc.

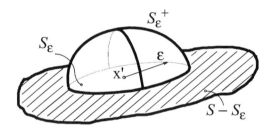

Figure 2.11: Source point $\mathbf{x}'$ located on a boundary surrounded by a hemisphere.

$$= \lim_{\varepsilon \to 0} \left\{ \int_{S_\varepsilon^+} T_{ij}(\mathbf{x}',\mathbf{x}) \left[u_j(\mathbf{x}) - u_j(\mathbf{x}') \right] dS \right\} + u_j(\mathbf{x}') \lim_{\varepsilon \to 0} \left\{ \int_{S_\varepsilon^+} T_{ij}(\mathbf{x}',\mathbf{x}) dS \right\}$$

where the first integral on the right-hand side of equation (2.99) vanishes to zero because of the requirement of displacement continuity. The second integral leads to a jump term in the displacements as follows:

$$u_j(\mathbf{x}') \lim_{\varepsilon \to 0} \int_{S_\varepsilon^+} T_{ij}(\mathbf{x}',\mathbf{x}) dS = \alpha_{ij}(\mathbf{x}') u_j(\mathbf{x}') \tag{2.100}$$

Equation (2.98) can now be written as

$$\int_S T_{ij}(\mathbf{x}',\mathbf{x}) u_j(\mathbf{x}) dS = \fint_S T_{ij}(\mathbf{x}',\mathbf{x}) u_j(\mathbf{x}) dS + \alpha_{ij}(\mathbf{x}') u_j(\mathbf{x}') \tag{2.101}$$

where $\fint$ stands for Cauchy principle value integral. Therefore, the displacement boundary integral equation can be written as

$$\circledast \quad C_{ij}(\mathbf{x}') u_j(\mathbf{x}') + \fint_S T_{ij}(\mathbf{x}',\mathbf{x}) u_j(\mathbf{x}) dS = \int_S U_{ij}(\mathbf{x}',\mathbf{x}) t_j(\mathbf{x}) dS$$

$$+ \int_V U_{ij}(\mathbf{x'}, \mathbf{X}) b_j(\mathbf{X}) dV \tag{2.102}$$

where the free term $C_{ij}(\mathbf{x'}) = \delta_{ij}(\mathbf{x'}) + \alpha_{ij}(\mathbf{x'})$. Consider a smooth boundary as shown in Figure 2.10, and denoting the r as the distance between the source point $\mathbf{x'}$ and the field point $\mathbf{x}$ on the arc S_ε^+. Using the polar coordinate system, it can be shown that

$$\mathbf{r} = \varepsilon \cos \theta \, \hat{\imath} + \varepsilon \sin \theta \, \hat{\jmath}$$

$$dS_\varepsilon^+ = \varepsilon d\theta$$

$$\frac{\partial r}{\partial n} = \frac{\mathbf{n.r}}{r} = \frac{\varepsilon}{\varepsilon} = 1 \ (\text{Note } \mathbf{r} = \varepsilon \mathbf{n} \text{ as } \mathbf{n} \text{ is in the same direction as } \mathbf{r})$$

$$\mathbf{r}_{,1} = \cos \theta \qquad \mathbf{r}_{,2} = \sin \theta$$

Using the above relationships, the constant α_{ij} can be evaluated as follows:

$$\alpha_{11} = \frac{-1}{4\pi(1-\nu)} \lim_{\varepsilon \to 0} \int_{S_\varepsilon^+} \frac{1}{\varepsilon} [(1 - 2\nu) + 2\mathbf{r}_{,1}\mathbf{r}_{,1}] \, \varepsilon d\theta$$

$$= \frac{-1}{4\pi(1-\nu)} \int_0^\pi [(1 - 2\nu) + 2\cos^2 \theta] \, d\theta = -\frac{1}{2}$$

and similarly

$$\alpha_{12} = \frac{-1}{4\pi(1-\nu)} \int_0^\pi 2 \sin \theta \cos \theta d\theta = 0$$

$$\alpha_{21} = \frac{-1}{4\pi(1-\nu)} \int_0^\pi 2 \sin \theta \cos \theta d\theta = 0$$

$$\alpha_{22} = \frac{-1}{4\pi(1-\nu)} \int_0^\pi [(1 - 2\nu) + 2\sin^2 \theta] \, d\theta = -\frac{1}{2}$$

For three-dimensional problems the procedure is similar; the integral is over a hemisphere S_ε^+ as shown in Figure 2.11. Using the spherical coordinate system, then

$$\mathbf{r} = \varepsilon \cos \phi \sin \theta \, \hat{\imath} + \varepsilon \sin \phi \sin \theta \, \hat{\jmath} + \varepsilon \cos \theta \, \hat{k}$$

$$dS_\varepsilon^* = \varepsilon^2 \sin \theta d\theta d\phi$$

$$\frac{\partial r}{\partial n} = \frac{\varepsilon}{\varepsilon} = 1$$

Using the above relationships,

$$\alpha_{11} = \frac{-1}{8\pi(1-\nu)} \lim_{\varepsilon \to 0} \int_0^{2\pi} \int_0^{\frac{\pi}{2}} \frac{1}{\varepsilon^2} [(1 - 2\nu) + 3\sin^2 \theta \cos^2 \theta] \, \varepsilon^2 \sin \theta d\theta d\phi = -\frac{1}{2}$$

Similarly, it can be shown in general that $\alpha_{ij} = -\delta_{ij}/2$, as in two-dimensions.

For non-smooth boundaries, the common practice is to evaluate the free terms from consideration of rigid body motion. This approach is described in Section 2.3.10.

2.3.6 Axisymmetric Problems

The axisymmetric fundamental solutions were presented in Section 2.3.3 The displacement boundary integral equation can be rewritten in the transformed cylindrical coordinate system as

$$\left\{ \begin{array}{c} C_r u_r \\ \\ C_z u_z \end{array} \right\} + 2\pi \int_\Gamma \left[\begin{array}{cc} T_{rr} & T_{rz} \\ \\ T_{zr} & T_{zz} \end{array} \right] \left\{ \begin{array}{c} u_r \\ \\ u_z \end{array} \right\} R(\mathbf{x})d\Gamma$$

$$= 2\pi \int_\Gamma \left[\begin{array}{cc} T_{rr} & T_{rz} \\ \\ T_{zr} & T_{zz} \end{array} \right] \left\{ \begin{array}{c} u_r \\ \\ u_z \end{array} \right\} R(\mathbf{x})d\Gamma \tag{2.103}$$

where Γ denotes the generator of the axisymmetric body. The treatment of body forces is described in [85].

2.3.7 Infinite Regions

One of the special features of the BEM is the automatic satisfaction of boundary conditions for infinite regions, thus avoiding the numerical approximation of remote boundaries. The Kelvin fundamental solution described in Section 2.3.3 for bounded domains can also be used for infinite regions provided certain regularity conditions are satisfied on the remote boundary. Consider the cavity with surface S in an infinite medium shown in Figure 2.12. Let r_o be the radius of a sphere (or a circle for two-dimensional problems) with surface S_∞ and centred at point $\mathbf{x}'$. The displacement boundary integral equation (2.102), in the absence of body forces, can be written as

$$C_{ij}(\mathbf{x}')u_j(\mathbf{x}') + \int_S T_{ij}(\mathbf{x}',\mathbf{x})u_j(\mathbf{x})dS + \int_{S_\infty} T_{ij}(\mathbf{x}',\mathbf{x})u_j(\mathbf{x})dS$$

$$= \int_S U_{ij}(\mathbf{x}',\mathbf{x})t_j(\mathbf{x})dS + \int_{S_\infty} U_{ij}(\mathbf{x}',\mathbf{x})t_j(\mathbf{x})dS \tag{2.104}$$

where $\mathbf{x}' \in S$. It can be shown that, as $r_o \to \infty$, the above equation can be expressed in terms of S only, i.e.

$$\lim_{r_o \to \infty} \int_{S_\infty} [T_{ij}(\mathbf{x}',\mathbf{x})u_j(\mathbf{x}) - U_{ij}(\mathbf{x}',\mathbf{x})t_j(\mathbf{x})] \, dS = 0 \tag{2.105}$$

For three-dimensional problems, the following properties hold:

$$\begin{array}{rcl} U_{ij}(\mathbf{x}',\mathbf{x}) & = & O(r_o^{-1}) \\ T_{ij}(\mathbf{x}',\mathbf{x}) & = & O(r_o^{-2}) \\ dS(\mathbf{x}) & = & |J|d\phi d\theta \end{array} \tag{2.106}$$

where $|J| = O(r_o)^2$ denotes the Jacobian of transformation. The regularity conditions in (2.105) are satisfied if the displacements u_j and tractions t_j have $1/r$ and $1/r^2$ behaviour, respectively. Hence, the two terms in (2.105) vanish, respectively.

For two-dimensional problems, the fundamental solutions behave as $\ln r$ and $1/r$ for displacements and tractions, respectively. The regularity conditions at infinity for this case imply that U_{ij} also behaves as $\ln r_o$ and T_{ij} as $1/r_o$. Hence, the two terms in the integral (2.105) do not approach zero separately, but cancel each other as $r_o \to \infty$.

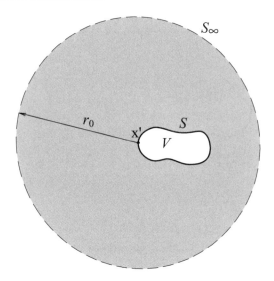

Figure 2.12: An infinite domain with a cavity.

2.3.8 Semi-infinite Regions

As stated before, it may sometimes be appropriate to utilize fundamental solutions other than Kelvin's solutions for infinite regions. In Appendix A, several fundamental solutions were presented, namely semi-infinite, circular and elliptical cut-outs and an isolated crack. The advantage of using these fundamental solutions instead of Kelvin lies in the fact that special geometrical features incorporated into these fundamental solutions will no longer be required to be approximated with the BEM. In this section, the BEM formulation for semi-infinite regions is further elaborated upon, due to its importance in several engineering applications such as soil mechanics.

Consider, the semi-infinite region shown in Figure 2.13, with the remote boundary denoted as S_∞. The integrals over the infinite boundary S_∞ vanish due to the regularity conditions as described in Section 2.3.5. The displacement integral equation (2.102), in the absence of body forces can be written as

$$C_{ij}^S(\mathbf{x}')u_j(\mathbf{x}') + \int_{S_o} T_{ij}^S(\mathbf{x}',\mathbf{x})u_j(\mathbf{x})dS = \int_{S_o} U_{ij}^S(\mathbf{x}',\mathbf{x})t_j(\mathbf{x})dS$$
$$+ \int_{S_H} U_{ij}^S(\mathbf{x}',\mathbf{x})t_j(\mathbf{x})dS \qquad (2.107)$$

where $U_{ij}^S(\mathbf{x}',\mathbf{x})$ and $T_{ij}^S(\mathbf{x}',\mathbf{x})$ are fundamental solutions for a semi-infinite region. It is also worth noting that the traction fundamental solution $T_{ij}^S(\mathbf{x}',\mathbf{x}) = 0$ on the half-plane boundary S_H and the fundamental solutions satisfy the regularity conditions (2.105).

If there are no loads applied on the half-plane S_H, the above equation reduces to

$$C_{ij}^S(\mathbf{x}')u_j(\mathbf{x}') + \int_{S_o} T_{ij}^S(\mathbf{x}',\mathbf{x})u_j(\mathbf{x})dS = \int_{S_o} U_{ij}^S(\mathbf{x}',\mathbf{x})t_j(\mathbf{x})dS \qquad (2.108)$$

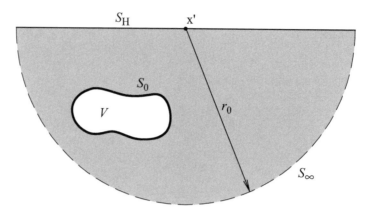

Figure 2.13: Semi-infinite region.

The fundamental solutions for two- and three-dimensional semi-infinite regions can be found in [47] and [82], respectively.

2.3.9   Boundary Stress Equation

In a similar way to displacements, the Somigliana's identity for stresses can be used to derive the boundary stress integral equations. This new integral equation can be utilized to evaluate the stresses on the boundary. However, a simpler numerical technique is often employed for this purpose (see Section 2.5.2). The stress boundary integral equation is nevertheless very important in modelling crack problems with the boundary element method, as described in Chapter 8.

As in the case of the displacement integral equation, the domain integrals containing the body force terms are not singular, and so the derivation can be confined to the boundary integrals. Consider the first integral on the right-hand side of equation (2.68). In the limit, as $\varepsilon \to 0$, it can be written as

$$\int_S D_{kij}(\mathbf{x}',\mathbf{x})t_k(\mathbf{x})dS$$

$$= \lim_{\varepsilon \to 0} \int_{S-S_\varepsilon} D_{kij}(\mathbf{x}',\mathbf{x})t_k(\mathbf{x})dS + \lim_{\varepsilon \to 0} \int_{S_\varepsilon^+} D_{kij}(\mathbf{x}',\mathbf{x})t_k(\mathbf{x})dS \quad (2.109)$$

in which the two limiting expressions on the right-hand side contain a strongly singular integrand of order $O(r^{-1})$ for two-dimensional problems and $O(r^{-2})$ for three-dimensional problems. The former results in an improper integral which is integrable in a Cauchy principle-value sense. The latter can be regularized by isolating the first term of Taylor series expansion of the tractions around the source point, resulting in

$$\lim_{\varepsilon \to 0} \int_{S_\varepsilon^+} D_{kij}(\mathbf{x}',\mathbf{x})t_k(\mathbf{x})dS \quad (2.110)$$

$$= \lim_{\varepsilon \to 0} \int_{S_\varepsilon^+} D_{kij}(\mathbf{x}',\mathbf{x})\left[t_k(\mathbf{x})-t_k(\mathbf{x}')\right]dS + t_k(\mathbf{x}')\lim_{\varepsilon \to 0} \int_{S_\varepsilon^+} D_{kij}(\mathbf{x}',\mathbf{x})dS$$

The first integral on the right-hand side of equation (2.110) is integrable and vanishes in the limiting process. The second integral leads to a jump term on the tractions, given by

$$t_k(\mathbf{x}') \lim_{\varepsilon \to 0} \int_{S_\varepsilon^+} D_{kij}(\mathbf{x}',\mathbf{x})dS = \sigma_{km}(\mathbf{x}') \lim_{\varepsilon \to 0} \int_{S_\varepsilon^+} D_{kij}(\mathbf{x}',\mathbf{x})n_m dS$$

$$= \gamma_{ij}(\mathbf{x}') \qquad (2.111)$$

where stresses are assumed to be Hölder continuous. If $\mathbf{x}'$ is assumed to be a smooth boundary point, $\gamma_{ij}(\mathbf{x}')$ can be given explicitly as

$$\gamma_{ij}(\mathbf{x}') = \frac{(5\nu - 7)\sigma_{ij} + (1 - 5\nu)(\sigma_{mm} - \sigma_{ij})}{-30(1-\nu)}\delta_{ij} + \frac{(8-10\nu)}{30(1-\nu)}\sigma_{ij}(1-\delta_{ij}) \qquad (2.112)$$

Therefore equation (2.109) can be rewritten as

$$\int_S D_{kij}(\mathbf{x}',\mathbf{x})t_k(\mathbf{x})dS = \int_S D_{kij}(\mathbf{x}',\mathbf{x})t_k(\mathbf{x})dS + \gamma_{ij}(\mathbf{x}') \qquad (2.113)$$

Consider now the second boundary integral on the right-hand side of equation (2.68). In the limit, as $\varepsilon \to 0$, it can be written as

$$\int_S S_{kij}(\mathbf{x}',\mathbf{x})u_k(\mathbf{x})dS$$

$$= \lim_{\varepsilon \to 0} \int_{S-S_\varepsilon} S_{kij}(\mathbf{x}',\mathbf{x})u_k(\mathbf{x})dS + \lim_{\varepsilon \to 0} \int_{S_\varepsilon^+} S_{kij}(\mathbf{x}',\mathbf{x})u_k(\mathbf{x})dS \qquad (2.114)$$

Both terms on the right-hand side of equation (2.114) contain hypersingular integrands or order $O(r^{-d})$, where $d = 2$ for two-dimensional problems and $d = 3$ for three-dimensional problems. The second limit expression can be regularized with the first two terms of a Taylor series expansion of the displacements around the source point. The displacements u_k are assumed to belong to $C^{1,\alpha}$ (i.e. the derivatives are Hölder continuous). Equation (2.115) can therefore be re-written by utilizing a Taylor series expansion u_k as

$$\int_S S_{kij}(\mathbf{x}',\mathbf{x})u_k(\mathbf{x})dS \qquad (2.115)$$

$$= \lim_{\varepsilon \to 0} \left\{ \int_{S_\varepsilon^+} S_{kij}(\mathbf{x}',\mathbf{x})[u_k(\mathbf{x}) - u_k(\mathbf{x}') - u_{k,m}(\mathbf{x}')(x_m - x'_m)]dS \right.$$

$$\left. + u_k(\mathbf{x}') \int_{S_\varepsilon^+} S_{kij}(\mathbf{x}',\mathbf{x})dS + u_{k,m}(\mathbf{x}') \int_{S_\varepsilon^+} S_{kij}(\mathbf{x}',\mathbf{x})(x_m - x'_m)dS \right\}$$

It is worth noting that, because of the assumption that $u_k \in C^{1,\alpha}$, we have $| u_k(\mathbf{x}) - u_k(\mathbf{x}') - u_{k,m}(\mathbf{x}')(x_m - x'_m) | = O(\varepsilon^\alpha)$, together with

$$S_{kij}(\mathbf{x}',\mathbf{x}) = O\left(\frac{1}{\varepsilon^d}\right) \quad \text{and} \quad \int_{S_\varepsilon^+} dS = O(\varepsilon^{d-1})$$

and hence the second term on the right-hand side of (2.115) tends to zero in the limit. The integrand of the last integral in (2.115) is of order $O(\varepsilon^{1-d})$, and the limit of the integral can be written as

$$\lim_{\varepsilon \to 0} u_{k,m}(\mathbf{x}') \int_{S_\varepsilon^+} S_{kij}(\mathbf{x}',\mathbf{x})(x_m - x'_m)dS = \beta_{ij}(\mathbf{x}')$$

For a smooth boundary point $\mathbf{x}'$, $\beta_{ij}(\mathbf{x}')$ is given as

$$\beta_{ij}(\mathbf{x}') = \frac{(8 - 10\nu)\sigma_{ij} + (1 - 5\nu)(\sigma_{mm} - \sigma_{ij})}{-30(1-\nu)}\delta_{ij} + \frac{(-7+5\nu)}{30(1-\nu)}(1-\delta_{ij}) \qquad (2.116)$$

for three-dimensional problems. The free terms $\beta_{ij}(\mathbf{x}')$ and $\gamma_{ij}(\mathbf{x}')$ yield a free term in the stress boundary condition, that is $\beta_{ij}(\mathbf{x}') - \gamma_{ij}(\mathbf{x}') = -\sigma_{ij}(\mathbf{x}')/2$.

The third integral on the right-hand side of (2.115) can be written as

$$u_k(\mathbf{x}')\int_{S_\varepsilon^+} S_{kij}(\mathbf{x}', \mathbf{x})dS = u_k(\mathbf{x}')\bar{b}_{kij}(\mathbf{x}')O(\frac{1}{\varepsilon^2})\int_{S_\varepsilon^+} dS$$

$$= u_k(\mathbf{x}')\frac{b_{kij}}{\varepsilon} \qquad (2.117)$$

which is unbounded as $\varepsilon \to 0$. The terms in (2.117) and the first integral on the right-hand side of (2.115) are collected together to give a Hadamard principal value integral, i.e.

$$ {=\!\!\!\!\!\!\int}_S S_{kij}(\mathbf{x}', \mathbf{x})u_k(\mathbf{x})dS = \lim_{\varepsilon \to 0}\left\{\int_{S-S_\varepsilon} S_{kij}(\mathbf{x}', \mathbf{x})u_k(\mathbf{x})dS + u_k(\mathbf{x}')\frac{b_{kij}(\mathbf{x}')}{\varepsilon}\right\} \qquad (2.118)$$

Finally, equation (2.115) can be written as

$$\sigma_{ij}(\mathbf{x}') - (\beta_{ij}(\mathbf{x}') - \gamma_{ij}(\mathbf{x}')) = {\int\!\!\!\!\!-}_S D_{kij}(\mathbf{x}', \mathbf{x})t_k(\mathbf{x})dS - {=\!\!\!\!\!\!\int}_S S_{kij}(\mathbf{x}', \mathbf{x})u_k(\mathbf{x})dS$$

$$+ \int_V D_{kij}(\mathbf{x}', \mathbf{X})b_k(\mathbf{X})dV \qquad (2.119)$$

where the free terms $\sigma_{ij}(\mathbf{x}') - (\beta_{ij}(\mathbf{x}') - \gamma_{ij}(\mathbf{x}')) = \frac{1}{2}\sigma_{ij}(\mathbf{x}')$, so that the final form of the boundary stress equation for a source point on a smooth part of the boundary is given by

$$\frac{1}{2}\sigma_{ij}(\mathbf{x}') = {\int\!\!\!\!\!-}_S D_{kij}(\mathbf{x}', \mathbf{x})t_k(\mathbf{x})dS - {=\!\!\!\!\!\!\int}_S S_{kij}(\mathbf{x}', \mathbf{x})u_k(\mathbf{x})dS$$

$$+ \int_V D_{kij}(\mathbf{x}', \mathbf{X})b_k(\mathbf{X})dV \qquad (2.120)$$

For non-smooth boundaries, the derivation is more complicated due to continuity requirements and it will described in Section 2.5.1.

2.3.10 Numerical Discretization

It is only possible to solve the boundary element formulation in equation (2.102) analytically for very simple problems. In this section, we concentrate on numerical methods of solution which can be used to solve general problems. The Boundary Element Method (BEM) is the numerical method of solution for boundary integral equations, based on a discretization procedure. The first step in the discretization is to divide the boundary S into N_e elements, as shown in Figure 2.14, so that in the absence of body forces equation (2.102) becomes

$$C_{ij}(\mathbf{x}')u_j(\mathbf{x}') + \sum_{n=1}^{N_e}\int_{S_n} T_{ij}(\mathbf{x}', \mathbf{x})u_j(\mathbf{x})dS = \sum_{n=1}^{N_e}\int_{S_n} U_{ij}(\mathbf{x}', \mathbf{x})t_j(\mathbf{x})dS \qquad (2.121)$$

where $S = \sum_{n=1}^{N} S_n$.

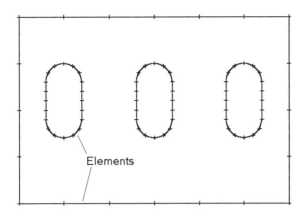

Figure 2.14: Discretization of the boundary into elements.

Constant Elements

In this case the geometry is divided into N_e straight elements, and the values of u_j and t_j are assumed to be constant on each element and equal to their values at its mid-side node. The BEM equation for constant elements can be written as

$$C_{ij}(\mathbf{x}^c)u_j(\mathbf{x}^c) + \sum_{n=1}^{N_e} u_j^c \int_{S_n} T_{ij}(\mathbf{x}^c, \mathbf{x})dS_n = \sum_{n=1}^{N_e} t_j^c \int_{S_n} U_{ij}(\mathbf{x}^c, \mathbf{x})dS_n \qquad (2.122)$$

where u_j^c and t_j^c are values of displacements and tractions at nodal points in the centre of the elements. The source point $\mathbf{x}'$ is also taken at the nodal points, that is $\mathbf{x}' = \mathbf{x}^c$, $c = 1, 2, ..., N_e$. In this way, N_e equations are generated for the N_e unknown displacements and tractions. It is worth noting that as $\mathbf{x}'$ is always located on a smooth boundary in this case, $C_{ij}(\mathbf{x}') = 0.5\delta_{ij}$.

Isoparametric Elements

One of the most significant improvements in the development of the boundary element method was the introduction of the parametric representation of both geometry and unknown functions [34] similar to the *isoparametric* formulation in the Finite Element Method. In this type of formulation, the boundary parameter $\mathbf{x}$ (components x_j), the unknown displacement fields $u_j(\mathbf{x})$ and traction fields $t_j(\mathbf{x})$ are approximated using interpolation functions, in the following manner:

$$x_j = \sum_{\alpha=1}^{m} N_\alpha(\eta)x_j^\alpha$$

$$u_j = \sum_{\alpha=1}^{m} N_\alpha(\eta)u_j^\alpha$$

$$t_j = \sum_{\alpha=1}^{m} N_\alpha(\eta)t_j^\alpha \qquad (2.123)$$

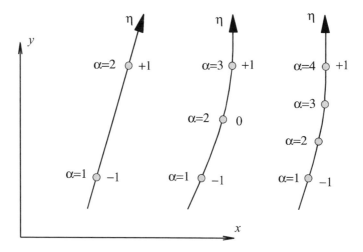

Figure 2.15: Two-dimensional boundary elements.

where N_α, which are called shape functions, are polynomials of degree $m - 1$, and have the property that they are equal to 1 at node α and zero at all other nodes. x_j^α, u_j^α and t_j^α are the values of the functions at node α. These shape functions are defined in terms of non-dimensional coordinates η $(-1 \leq \eta \leq 1)$, as shown in Figure 2.15.

For linear elements $(m = 2)$, we have

$$N_1 = \frac{1}{2}(1 - \eta) \qquad N_2 = \frac{1}{2}(1 + \eta) \tag{2.124}$$

For quadratic elements $(m = 3)$, we have

$$N_1 = \frac{1}{2}\eta(\eta - 1) \qquad N_2 = 1 - \eta^2 \qquad N_3 = \frac{1}{2}\eta(1 + \eta) \tag{2.125}$$

In general, shape functions can be derived from the Lagrangian polynomials which are defined, for degree $(m - 1)$, as

$$N_\alpha(\eta) = \prod_{i=0,\ i\neq\alpha}^{m} \frac{\eta - \eta_i}{\eta_\alpha - \eta_i} \tag{2.126}$$

It can be seen that $N_\alpha(\eta)$ is given by the product of m linear factors. The Lagrangian shape functions can be shown to have the following properties: at node β, $N_\alpha(\eta_\beta) = \delta_{\alpha\beta}$, $\eta_0 \leq \eta \leq \eta_m$, and also that the sum of the shape functions is equal to unity (i.e. $\sum_{\alpha=1}^{m} N_\alpha = 1$) and sum of its first derivatives is equal to zero.

A discretized boundary element formulation can be obtained by substituting the expressions in (2.123), into the integral equation (2.121) to obtain

$$C_{ij}(\mathbf{x}')u_j(\mathbf{x}') + \sum_{n=1}^{N_e}\sum_{\alpha=1}^{m} P_{ij}^{n\alpha}u_j^{n\alpha} = \sum_{n=1}^{N_e}\sum_{\alpha=1}^{m} Q_{ij}^{n\alpha}t_j^{n\alpha} \qquad i,j = 1,2 \tag{2.127}$$

The coefficients $P_{ij}^{n\alpha}$ and $Q_{ij}^{n\alpha}$ are defined in terms of integrals over S_n, where $dS_n(\mathbf{x})$ becomes $J^n(\eta)d\eta$; that is

$$P_{ij}^{n\alpha} = \int_{-1}^{1} N_\alpha(\eta) T_{ij}[\mathbf{x}', \mathbf{x}(\eta)] J^n(\eta) d\eta$$

$$Q_{ij}^{n\alpha} = \int_{-1}^{1} N_\alpha(\eta) U_{ij}[\mathbf{x}', \mathbf{x}(\eta)] J^n(\eta) d\eta \qquad (2.128)$$

In general $J(\eta)$, the Jacobian of transformation, is given by

$$J(\eta) = \sqrt{\left(\frac{dx_1}{d\eta}\right)^2 + \left(\frac{dx_2}{d\eta}\right)^2} \qquad (2.129)$$

therefore

$$J^n(\eta) = \sqrt{\left(\sum_{\alpha=1}^{m} \frac{dN_\alpha}{d\eta} x_1^{n\alpha}\right)^2 + \left(\sum_{\alpha=1}^{m} \frac{dN_\alpha}{d\eta} x_2^{n\alpha}\right)^2} \qquad (2.130)$$

Different choices of N_α lead to different formulations [1]. If the same shape functions are used for approximation of geometry and functions, the formulation is referred to as *isoparametric;* if the shape functions for the function is a higher order polynomial than that used for the geometry, the formulation is referred to as *superparamtric;* and conversely, if the function is represented by a lower order polynomial than the geometry the formulation is referred to as subparametric.

Three-dimensional Formulation

In three-dimensional problems the surface S is divided into triangular or quadrilateral elements, as shown in Figure 2.16. As in two-dimensional problems, the geometry parameter x_j and the displacement u_j and traction fields t_j are approximated using shape functions $N_\alpha(\eta_1, \eta_2)$, that is

$$x_j = \sum_{\alpha=1}^{m} N_\alpha(\eta_1, \eta_2) x_j^\alpha$$

$$u_j = \sum_{\alpha=1}^{m} N_\alpha(\eta_1, \eta_2) u_j^\alpha$$

$$t_j = \sum_{\alpha=1}^{m} N_\alpha(\eta_1, \eta_2) t_j^\alpha \qquad (2.131)$$

where $N_\alpha(\eta_1, \eta_2)$ are now functions of two local variables η_1 and η_2 $(-1 \leq \eta_1, \eta_2 \leq 1)$ for quadrilateral elements.

This transformation is equivalent to transforming the boundary elements into a local square or triangle in the (η_1, η_2) plane, as shown in Figure 2.17.

Interpolation functions for three-dimensional boundary elements are analogous to those of two-dimensional finite elements. The most common elements are triangular and quadrilateral elements. The simplest triangular element has three nodes, and its geometry is defined with oblique coordinates η_1, η_2 varying from 0 to 1 over two sides, as shown in Figure 2.18. The coordinates of each point are given in the form

$$x_i = \eta_1 x_i^1 + \eta_2 x_i^2 + \eta_3 x_i^3$$

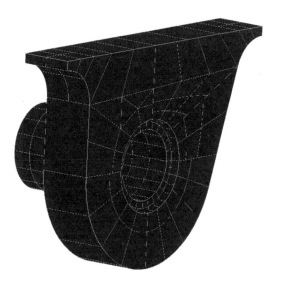

Figure 2.16: Discretization of the three-dimensional lug [55].

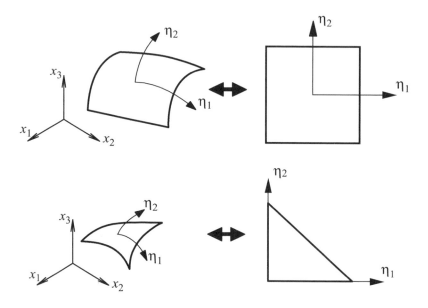

Figure 2.17: Transformation of boundary elements into a local system.

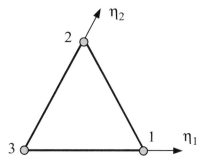

Figure 2.18: Three-noded triangle in a local coordinate system.

where x_i^1, x_i^2, x_i^3 are the coordinates of the nodal points, and $\eta_3 = 1 - \eta_1 - \eta_2$. The relationship between local coordinates and the Cartesian coordinates is given by

$$\eta_1 = \frac{1}{2A}(2A_1^0 + b_1 x_1 + a_1 x_2)$$

$$\eta_2 = \frac{1}{2A}(2A_2^0 + b_2 x_1 + a_2 x_2)$$

$$\eta_3 = \frac{1}{2A}(2A_3^0 + b_3 x_1 + a_3 x_2)$$

where $a_i = x_1^k - x_1^j$, $b_i = x_2^j - x_2^k$ and $2A_i^0 = x_1^j x_2^k - x_1^k x_2^j$ with $i = 1, 2, 3$ for $j = 2, 3, 1$ and $k = 3, 2, 1$. The area A is given by

$$A = \frac{1}{2}(b_1 a_2 - b_2 a_1)$$

which represents the area of the projection of the triangle over the plane x_1, x_2. The interpolation functions for a linear triangular element are given by

$$N_1 = \eta_1 \qquad N_2 = \eta_2 \qquad N_3 = \eta_3$$

For quadratic six-noded elements the shape functions are given by

$$
\begin{aligned}
N_1 &= \eta_1(2\eta_1 - 1) & N_4 &= 4\eta_1\eta_2 \\
N_2 &= \eta_2(2\eta_2 - 1) & N_5 &= 4\eta_2\eta_3 \\
N_3 &= \eta_3(2\eta_3 - 1) & N_6 &= 4\eta_3\eta_1
\end{aligned}
$$

and, for the cubic 10 noded element, we have

$$
\begin{aligned}
N_1 &= \frac{1}{2}\eta_1(3\eta_1 - 1)(3\eta_1 - 2) & N_6 &= \frac{9}{2}\eta_2\eta_3(3\eta_2 - 1) \\
N_2 &= \frac{1}{2}\eta_2(3\eta_2 - 1)(3\eta_2 - 2) & N_7 &= \frac{9}{2}\eta_2\eta_3(3\eta_3 - 1) \\
N_3 &= \frac{1}{2}\eta_3(3\eta_3 - 1)(3\eta_3 - 2) & N_8 &= \frac{9}{2}\eta_3\eta_1(3\eta_3 - 1) \\
N_4 &= \frac{9}{2}\eta_1\eta_2(3\eta_1 - 1) & N_9 &= \frac{9}{2}\eta_3\eta_1(3\eta_1 - 1) \\
N_5 &= \frac{9}{2}\eta_1\eta_2(3\eta_2 - 1) & N_{10} &= 27\eta_1\eta_2\eta_3
\end{aligned}
$$

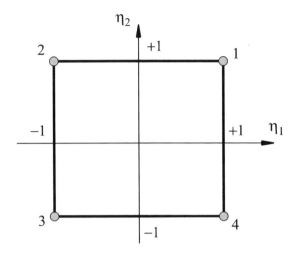

Figure 2.19: Linear 4-node quadrilateral element.

The shape functions for quadrilateral elements can be obtained from the Lagrangian polynomials. Although the polynomials in (2.126) are only one-dimensional, they may be generalized to two dimensions by simply forming products of the shape functions. For example, consider a quadrilateral element shown in Figure 2.19, defined by its four corners $(\pm 1, \pm 1)$; we may write the following function ϕ in the:

$$\phi(\eta_1, \eta_2) = N_1(\eta_1, \eta_2)\phi^1 + N_2(\eta_1, \eta_2)\phi^2 + N_3(\eta_1, \eta_2)\phi^3 + N_4(\eta_1, \eta_2)\phi^4 \qquad (2.132)$$

where

$$N_1(\eta_1, \eta_2) = N_1(\eta_1)N_1(\eta_2) = \frac{\eta - \eta_2}{\eta_1 - \eta_2} \times \frac{\eta - \eta_4}{\eta_1 - \eta_4} = \frac{1}{4}(\eta_1 + 1)(\eta_2 + 1)$$

$$N_2(\eta_1, \eta_2) = N_2(\eta_1)N_2(\eta_2) = \frac{\eta - \eta_1}{\eta_2 - \eta_1} \times \frac{\eta - \eta_3}{\eta_2 - \eta_3} = \frac{1}{4}(1 - \eta_1)(\eta_2 + 1)$$

$$N_3(\eta_1, \eta_2) = N_3(\eta_1)N_3(\eta_2) = \frac{\eta - \eta_4}{\eta_3 - \eta_4} \times \frac{\eta - \eta_2}{\eta_3 - \eta_2} = \frac{1}{4}(\eta_1 - 1)(\eta_2 - 1)$$

$$N_4(\eta_1, \eta_2) = N_4(\eta_1)N_4(\eta_2) = \frac{\eta - \eta_3}{\eta_4 - \eta_3} \times \frac{\eta - \eta_1}{\eta_4 - \eta_1} = \frac{1}{4}(1 + \eta_1)(1 - \eta_2)$$

Higher order shape functions for quadrilateral elements can be formulated in a similar fashion. In general, for Lagrangian elements we have

$$N_\alpha(\eta_1, \eta_2) = N_\alpha(\eta_1)N_\alpha(\eta_2) \qquad (2.133)$$

A commonly used set of quadrilateral elements is known as the *serendipity* family (see Figure 2.20). These elements do not contain any interior nodes and their interpolation functions were derived by inspection. In terms of the local coordinate system (η_1, η_2), the serendipity shape functions for linear elements (see Figure 2.19) are the same as those given in (2.132). For, quadratic 8-node elements shown in Figure 2.20 they are given as

$$N_\alpha(\eta_1, \eta_2) = \frac{1}{4}(1 + \eta_1\eta_1^\alpha)(1 + \eta_2\eta_2^\alpha)(\eta_1\eta_1^\alpha + \eta_2\eta_2^\alpha - 1), \text{ for nodes at } \eta_1^\alpha, \eta_2^\alpha = \pm 1$$

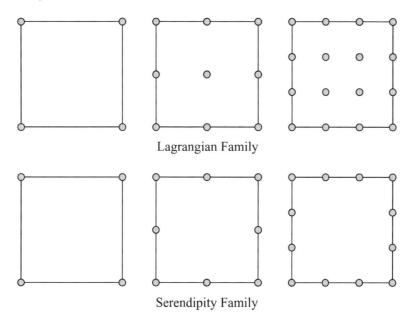

Figure 2.20: Lagrangian and *serendipity* elements.

$$N_\alpha(\eta_1, \eta_2) = \frac{1}{2}(1 - \eta_1^2)(1 + \eta_2\eta_2^\alpha), \text{ for nodes at } \eta_1^\alpha = 0, \eta_2^\alpha = \pm1$$

$$N_\alpha(\eta_1, \eta_2) = \frac{1}{2}(1 + \eta_1\eta_1^\alpha)(1 - \eta_2^2), \text{ for nodes at } \eta_1^\alpha = \pm1, \eta_2^\alpha = 0 \qquad (2.134)$$

For cubic elements, we have

$$N_\alpha(\eta_1, \eta_2) = \frac{1}{32}(1 + \eta_1\eta_1^\alpha)(1 + \eta_2\eta_2^\alpha)\left[9(\eta_1^2 + \eta_2^2) - 10\right], \text{ for nodes at } \eta_1^\alpha, \eta_2^\alpha = \pm1$$

$$N_\alpha(\eta_1, \eta_2) = \frac{9}{32}(1 + \eta_1\eta_1^\alpha)(1 - \eta_2^2)(1 + 9\eta_1\eta_1^\alpha), \text{ for nodes at } \eta_1^\alpha = \pm1, \eta_2^\alpha = \pm\frac{1}{3}$$

$$N_\alpha(\eta_1, \eta_2) = \frac{9}{32}(1 + \eta_1\eta_1^\alpha)(1 - \eta_1^2)(1 + 9\eta_1\eta_1^\alpha), \text{ for nodes at } \eta_1^\alpha = \pm\frac{1}{3}, \eta_2^\alpha = \pm1$$

$$(2.135)$$

For three-dimensional elements, the transformation from the global Cartesian system x_i to a local system (η_1, η_2) defined over the surface of the body is given by a Jacobian, similar to two-dimensional problems, which now takes the form $dS_n = \mid J^n(\eta_1, \eta_2) \mid d\eta_1 d\eta_2$, with the Jacobian of transformation given by

$$\mid J^n(\eta_1, \eta_2) \mid = \sqrt{J_1^2 + J_2^2 + J_3^2} \qquad (2.136)$$

where

$$J_1 = \frac{\partial x_2}{\partial \eta_1}\frac{\partial x_3}{\partial \eta_2} - \frac{\partial x_3}{\partial \eta_2}\frac{\partial x_2}{\partial \eta_1}$$

$$J_2 = \frac{\partial x_3}{\partial \eta_1}\frac{\partial x_1}{\partial \eta_2} - \frac{\partial x_1}{\partial \eta_2}\frac{\partial x_3}{\partial \eta_1}$$

$$J_3 = \frac{\partial x_1}{\partial \eta_1}\frac{\partial x_2}{\partial \eta_2} - \frac{\partial x_2}{\partial \eta_2}\frac{\partial x_1}{\partial \eta_1}$$

Notice that the components of the unit normal vector are given by $n_i = J_i / \mid J \mid$, ($i = 1, 2, 3$) and

$$\frac{\partial x_i}{\partial \eta_k} = \sum_{\alpha=1}^{m} \frac{\partial N_\alpha}{\partial \eta_k} x_i^\alpha$$

Substitution of (2.131) into the boundary element formulation (2.123) gives a similar equation to (2.127), i.e.

$$C_{ij}(\mathbf{x}')u_j(\mathbf{x}') + \sum_{n=1}^{N_e}\sum_{\alpha=1}^{m} P_{ij}^{n\alpha}u_j^{n\alpha} = \sum_{n=1}^{N_e}\sum_{\alpha=1}^{m} Q_{ij}^{n\alpha}t_j^{n\alpha} \qquad i,j = 1,2,3 \qquad (2.137)$$

where the terms $P_{ij}^{n\alpha}$ and $Q_{ij}^{n\alpha}$ are now given by double integrals, as follows:

$$P_{ij}^{n\alpha} = \int_{-1}^{1}\int_{-1}^{1} N_\alpha(\eta_1, \eta_2)T_{ij}[\mathbf{x}', \mathbf{x}(\eta_1, \eta_2)] J^n(\eta_1, \eta_2) d\eta_1 d\eta_2$$

$$Q_{ij}^{n\alpha} = \int_{-1}^{1}\int_{-1}^{1} N_\alpha(\eta_1, \eta_2)U_{ij}[\mathbf{x}', \mathbf{x}(\eta_1, \eta_2)] J^n(\eta_1, \eta_2) d\eta_1 d\eta_2 \qquad (2.138)$$

Assembly of System of Equations

The most straightforward method of solution for the integral equation (2.102) is point collocation; equations (2.127) or (2.137) are evaluated at nodal points $\mathbf{x}^c$, $c = 1, M$ (where M is the total number of nodes) to give

$$C_{ij}(\mathbf{x}^c)u_j(\mathbf{x}^c) + \sum_{n=1}^{N_e}\sum_{\alpha=1}^{m} P_{ij}^{n\alpha}(\mathbf{x}^c)u_j^{n\alpha} = \sum_{n=1}^{N_e}\sum_{\alpha=1}^{m} Q_{ij}^{n\alpha}(\mathbf{x}^c)t_j^{n\alpha} \qquad c = 1, M \quad (2.139)$$

The double sum in (2.139) must be evaluated bearing in mind that some nodes are shared between elements, and since the displacement values $u_j^{n\alpha}$ are uniquely defined at these nodes, they can be combined to give a sum over all nodes; so equation (2.139) can be written as

$$C_{ij}(\mathbf{x}^c)u_j(\mathbf{x}^c) + \sum_{\gamma=1}^{M} \bar{H}_{ij}^{c\gamma}u_j^\gamma = \sum_{n=1}^{N_e}\sum_{\alpha=1}^{m} G_{ij}^{cn\alpha}t_j^{n\alpha} \qquad c = 1, M \qquad (2.140)$$

where $\bar{H}_{ij}^{c\gamma}$ is made up from $P_{ij}^{n\alpha}(\mathbf{x}^c)$ and $G_{ij}^{cn\alpha}$ is equal to $Q_{ij}^{n\alpha}(\mathbf{x}^c)$. Collecting all the displacement unknowns in (2.140) gives

$$\sum_{\gamma=1}^{M} H_{ij}^{c\gamma}u_j^\gamma = \sum_{n=1}^{N_e}\sum_{\alpha=1}^{m} G_{ij}^{cn\alpha}t_j^{n\alpha} \qquad (2.141)$$

where $H_{ij}^{c\gamma} = C_{ij}(\mathbf{x}^c)\delta_{c\gamma} + \bar{H}_{ij}^{c\gamma}$ and $\delta_{c\gamma}$ is the Kronecker delta function. The discretized boundary element equation may now be written in matrix form as

$$\mathbf{Hu} = \mathbf{Gt} \qquad (2.142)$$

where $\mathbf{H}$ is a $2M \times 2M$ matrix and $\mathbf{G}$ is $2M \times 2N_e m$ matrix containing known integrals of the product of the shape functions, the Jacobians and the fundamental fields T_{ij} and U_{ij}, respectively. The vector $\mathbf{u}$ has $2M$ components and $\mathbf{t}$ has $2Nm$ components; both contain field unknowns and prescribed boundary conditions. From (2.140) it can be seen that the diagonal terms which contain C_{ij} can be evaluated directly from a consideration of rigid body motion; this leads to

$$C_{ij}(\mathbf{x}^c) = -\sum_{\gamma=1}^{M} \bar{H}_{ij}^{c\gamma} \tag{2.143}$$

Thus the diagonal terms in $H_{ij}^{c\gamma}$ can now be evaluated, since equation (2.143) can be rewritten as

$$C_{ij}(\mathbf{x}^c) + \bar{H}_{ij}^{cc} = -\sum_{\substack{\gamma=1 \ c\neq\gamma}}^{M} \bar{H}_{ij}^{c\gamma}$$

that is

$$H_{ij}^{cc} = -\sum_{\substack{\gamma=1 \ c\neq\gamma}}^{M} \bar{H}_{ij}^{c\gamma} \tag{2.144}$$

After substitution of the prescribed boundary conditions, the resulting system of algebraic equations may be written as

$$\mathbf{AX} = \mathbf{BY} = \mathbf{F} \tag{2.145}$$

The vector $\mathbf{X}$ contains all the unknown boundary displacements or tractions, $\mathbf{A}$ is a coefficient matrix which is a usually non-symmetric and densely populated, and $\mathbf{B}$ is a matrix which contains the coefficients corresponding to the prescribed boundary conditions $\mathbf{Y}$. The above system of equations is fully populated and non-symmetric. As the solution time for this type of matrix is proportional to the cubic power of the total degree of freedom, the required computational time can become large for complex structural models (particularly 3D). To overcome this problem, a number of techniques have been developed using block-base solvers [10, 69], lumping technique [28] and iterative solvers [38].

Expression (2.144) is valid for finite bodies. For infinite or semi-infinite bodies the regularity conditions at infinity are nolonger satisfied, since a rigid body movement assumes $u_j = O(1)$. In this situation, the following expression is considered:

$$\lim_{r_o \to \infty} \int_{S_\infty} (T_{ij}(\mathbf{x}',\mathbf{x})u_j(\mathbf{x}) - U_{ij}(\mathbf{x}',\mathbf{x})t_j(\mathbf{x})) \, dS_\infty = u_j(\mathbf{x}') \lim_{r_o \to \infty} \int_{S_\infty} T_{ij}(\mathbf{x}',\mathbf{x}) dS_\infty$$
$$= u_j(\mathbf{x}') \tag{2.146}$$

and the following relationship is obtained:

$$H_{ij}^{cc} = I - \sum_{\substack{\gamma=1 \ c\neq\gamma}}^{M} \bar{H}_{ij}^{c\gamma} \tag{2.147}$$

where I denotes the unity matrix.

In order to gain a better understanding of how the transformation from the system of equations (2.139) to (2.145) takes place, consider the configuration shown in Figure

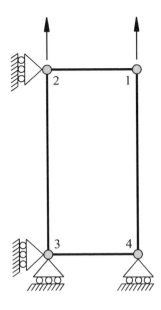

Figure 2.21: Test example.

2.21. The boundary conditions on the square plate, with corner nodes numbered 1 to 4 ($M = 4$), are prescribed as follows:

$$
\begin{aligned}
t_2 &= \sigma, & t_1 &= 0 & side\ 1-2; \\
u_1 &= 0, & t_2 &= 0 & side\ 2-3; \\
u_2 &= 0, & t_1 &= 0 & side\ 3-4; \\
t_1 &= 0, & t_2 &= 0 & side\ 4-1;
\end{aligned}
$$

Thus for this configuration of four elements with two nodes per element ($m = 2$), the system of equations in (2.139) can now be written as

$$
\begin{bmatrix}
C_{ij}^1 & & & \\
& C_{ij}^2 & & \\
& & C_{ij}^3 & \\
& & & C_{ij}^4
\end{bmatrix}
\begin{Bmatrix}
u_j^1 \\
u_j^2 \\
u_j^3 \\
u_j^4
\end{Bmatrix}
$$

$$
+
\begin{bmatrix}
P_{ij}^{111} & P_{ij}^{112} & P_{ij}^{121} & P_{ij}^{122} & P_{ij}^{131} & P_{ij}^{132} & P_{ij}^{141} & P_{ij}^{142} \\
P_{ij}^{211} & P_{ij}^{212} & P_{ij}^{221} & P_{ij}^{222} & P_{ij}^{231} & P_{ij}^{232} & P_{ij}^{241} & P_{ij}^{242} \\
P_{ij}^{311} & P_{ij}^{312} & P_{ij}^{321} & P_{ij}^{322} & P_{ij}^{331} & P_{ij}^{332} & P_{ij}^{341} & P_{ij}^{342} \\
P_{ij}^{411} & P_{ij}^{412} & P_{ij}^{421} & P_{ij}^{422} & P_{ij}^{431} & P_{ij}^{432} & P_{ij}^{441} & P_{ij}^{442}
\end{bmatrix}
\begin{Bmatrix}
u_j^{11} \\ u_j^{12} \\ u_j^{21} \\ u_j^{22} \\ u_j^{31} \\ u_j^{32} \\ u_j^{41} \\ u_j^{42}
\end{Bmatrix}
$$

$$
=
\begin{bmatrix}
Q_{ij}^{111} & Q_{ij}^{112} & Q_{ij}^{121} & Q_{ij}^{122} & Q_{ij}^{131} & Q_{ij}^{132} & Q_{ij}^{141} & Q_{ij}^{142} \\
Q_{ij}^{211} & Q_{ij}^{212} & Q_{ij}^{221} & Q_{ij}^{222} & Q_{ij}^{231} & Q_{ij}^{232} & Q_{ij}^{241} & Q_{ij}^{242} \\
Q_{ij}^{311} & Q_{ij}^{312} & Q_{ij}^{321} & Q_{ij}^{322} & Q_{ij}^{331} & Q_{ij}^{332} & Q_{ij}^{341} & Q_{ij}^{342} \\
Q_{ij}^{411} & Q_{ij}^{412} & Q_{ij}^{421} & Q_{ij}^{422} & Q_{ij}^{431} & Q_{ij}^{432} & Q_{ij}^{441} & Q_{ij}^{442}
\end{bmatrix}
\begin{Bmatrix}
t_j^{11} \\ t_j^{12} \\ t_j^{21} \\ t_j^{22} \\ t_j^{31} \\ t_j^{32} \\ t_j^{41} \\ t_j^{42}
\end{Bmatrix}
$$

where superscript c, n, α for $P^{cn\alpha}$ and $Q^{cn\alpha}$ denote the collocation node, element number and local node number (i.e. 1 or 2 for linear elements). For example, the entry P_{ij}^{221} corresponds to collocation at node 2, element 2 and node 1. In order to enforce the uniqueness of the displacements, the values of $u_j^{n\alpha}$ are combined to give a sum over all the nodes, that is

$$
\begin{bmatrix}
C_{ij}^{1} & & & \\
 & C_{ij}^{2} & & \\
 & & C_{ij}^{3} & \\
 & & & C_{ij}^{4}
\end{bmatrix}
\begin{Bmatrix}
u_j^{1} \\ u_j^{2} \\ u_j^{3} \\ u_j^{4}
\end{Bmatrix}
$$

$$
+ \begin{bmatrix}
P_{ij}^{111} + P_{ij}^{142} & P_{ij}^{112} + P_{ij}^{121} & P_{ij}^{122} + P_{ij}^{131} & P_{ij}^{132} + P_{ij}^{141} \\[6pt]
P_{ij}^{211} + P_{ij}^{242} & P_{ij}^{212} + P_{ij}^{221} & P_{ij}^{222} + P_{ij}^{231} & P_{ij}^{232} + P_{ij}^{241} \\[6pt]
P_{ij}^{311} + P_{ij}^{342} & P_{ij}^{312} + P_{ij}^{321} & P_{ij}^{322} + P_{ij}^{331} & P_{ij}^{332} + P_{ij}^{341} \\[6pt]
P_{ij}^{411} + P_{ij}^{442} & P_{ij}^{412} + P_{ij}^{421} & P_{ij}^{422} + P_{ij}^{431} & P_{ij}^{432} + P_{ij}^{441}
\end{bmatrix}
\begin{Bmatrix} u_j^1 \\[6pt] u_j^2 \\[6pt] u_j^3 \\[6pt] u_j^4 \end{Bmatrix}
$$

$$
= \begin{bmatrix}
Q_{ij}^{111} & Q_{ij}^{112} & Q_{ij}^{121} & Q_{ij}^{122} & Q_{ij}^{131} & Q_{ij}^{132} & Q_{ij}^{141} & Q_{ij}^{142} \\[6pt]
Q_{ij}^{211} & Q_{ij}^{212} & Q_{ij}^{221} & Q_{ij}^{222} & Q_{ij}^{231} & Q_{ij}^{232} & Q_{ij}^{241} & Q_{ij}^{242} \\[6pt]
Q_{ij}^{311} & Q_{ij}^{312} & Q_{ij}^{321} & Q_{ij}^{322} & Q_{ij}^{331} & Q_{ij}^{332} & Q_{ij}^{341} & Q_{ij}^{342} \\[6pt]
Q_{ij}^{411} & Q_{ij}^{412} & Q_{ij}^{421} & Q_{ij}^{422} & Q_{ij}^{431} & Q_{ij}^{432} & Q_{ij}^{441} & Q_{ij}^{442}
\end{bmatrix}
\begin{Bmatrix} t_j^{11} \\[4pt] t_j^{12} \\[4pt] t_j^{21} \\[4pt] t_j^{22} \\[4pt] t_j^{31} \\[4pt] t_j^{32} \\[4pt] t_j^{41} \\[4pt] t_j^{42} \end{Bmatrix}
$$

which can be rewritten using the notation employed in (2.140) as

$$
\begin{bmatrix}
C_{ij}^1 & & & \\
 & C_{ij}^2 & & \\
 & & C_{ij}^3 & \\
 & & & C_{ij}^4
\end{bmatrix}
\begin{Bmatrix} u_j^1 \\[4pt] u_j^2 \\[4pt] u_j^3 \\[4pt] u_j^4 \end{Bmatrix}
+ \begin{bmatrix}
\bar{H}_{ij}^{11} & \bar{H}_{ij}^{12} & \bar{H}_{ij}^{13} & \bar{H}_{ij}^{14} \\[6pt]
\bar{H}_{ij}^{21} & \bar{H}_{ij}^{22} & \bar{H}_{ij}^{23} & \bar{H}_{ij}^{24} \\[6pt]
\bar{H}_{ij}^{31} & \bar{H}_{ij}^{32} & \bar{H}_{ij}^{33} & \bar{H}_{ij}^{34} \\[6pt]
\bar{H}_{ij}^{41} & \bar{H}_{ij}^{42} & \bar{H}_{ij}^{43} & \bar{H}_{ij}^{44}
\end{bmatrix}
\begin{Bmatrix} u_j^1 \\[4pt] u_j^2 \\[4pt] u_j^3 \\[4pt] u_j^4 \end{Bmatrix}
$$

$$
= \begin{bmatrix}
G_{ij}^{111} & G_{ij}^{112} & G_{ij}^{121} & G_{ij}^{122} & G_{ij}^{131} & G_{ij}^{132} & G_{ij}^{141} & G_{ij}^{142} \\[6pt]
G_{ij}^{211} & G_{ij}^{212} & G_{ij}^{221} & G_{ij}^{222} & G_{ij}^{231} & G_{ij}^{232} & G_{ij}^{241} & G_{ij}^{242} \\[6pt]
G_{ij}^{311} & G_{ij}^{312} & G_{ij}^{321} & G_{ij}^{322} & G_{ij}^{331} & G_{ij}^{332} & G_{ij}^{341} & G_{ij}^{342} \\[6pt]
G_{ij}^{411} & G_{ij}^{412} & G_{ij}^{421} & G_{ij}^{422} & G_{ij}^{431} & G_{ij}^{432} & G_{ij}^{441} & G_{ij}^{442}
\end{bmatrix}
\begin{Bmatrix} t_j^{11} \\[4pt] t_j^{12} \\[4pt] t_j^{21} \\[4pt] t_j^{22} \\[4pt] t_j^{31} \\[4pt] t_j^{32} \\[4pt] t_j^{41} \\[4pt] t_j^{42} \end{Bmatrix}
$$

Since the collocation points are taken at the nodes, the coefficients of displacement can be collected together as described in (2.140) to give

$$
\begin{bmatrix}
\bar{H}_{ij}^{11} + C_{ij}^1 & \bar{H}_{ij}^{12} & \bar{H}_{ij}^{13} & \bar{H}_{ij}^{14} \\
\bar{H}_{ij}^{21} & \bar{H}_{ij}^{22} + C_{ij}^2 & \bar{H}_{ij}^{23} & \bar{H}_{ij}^{24} \\
\bar{H}_{ij}^{31} & \bar{H}_{ij}^{32} & \bar{H}_{ij}^{33} + C_{ij}^3 & \bar{H}_{ij}^{34} \\
\bar{H}_{ij}^{41} & \bar{H}_{ij}^{42} & \bar{H}_{ij}^{43} & \bar{H}_{ij}^{44} + C_{ij}^4
\end{bmatrix}
\begin{Bmatrix}
u_j^1 \\ u_j^2 \\ u_j^3 \\ u_j^4
\end{Bmatrix}
$$

$$
=
\begin{bmatrix}
G_{ij}^{111} & G_{ij}^{112} & G_{ij}^{121} & G_{ij}^{122} & G_{ij}^{131} & G_{ij}^{132} & G_{ij}^{141} & G_{ij}^{142} \\
G_{ij}^{211} & G_{ij}^{212} & G_{ij}^{221} & G_{ij}^{222} & G_{ij}^{231} & G_{ij}^{232} & G_{ij}^{241} & G_{ij}^{242} \\
G_{ij}^{311} & G_{ij}^{312} & G_{ij}^{321} & G_{ij}^{322} & G_{ij}^{331} & G_{ij}^{332} & G_{ij}^{341} & G_{ij}^{342} \\
G_{ij}^{411} & G_{ij}^{412} & G_{ij}^{421} & G_{ij}^{422} & G_{ij}^{431} & G_{ij}^{432} & G_{ij}^{441} & G_{ij}^{442}
\end{bmatrix}
\begin{Bmatrix}
t_j^{11} \\ t_j^{12} \\ t_j^{21} \\ t_j^{22} \\ t_j^{31} \\ t_j^{32} \\ t_j^{41} \\ t_j^{42}
\end{Bmatrix}
$$

The boundary conditions can be expressed in terms of the nodal variables as

$$
u_1^2 = u_1^3 = 0 \text{ and } u_2^3 = u_2^4 = 0
$$

for displacements, and

$$
\begin{aligned}
t_1^{11} &= t_1^{12} = 0 \\
t_2^{11} &= t_2^{12} = \sigma \\
t_2^{21} &= t_2^{22} = 0 \\
t_1^{31} &= t_1^{32} = 0 \\
t_j^{41} &= 0, \; j = 1, 2 \\
t_j^{42} &= 0, \; j = 1, 2
\end{aligned}
$$

for tractions. Therefore, the matrix $\mathbf{A}$ can be obtained by collecting all the coefficients from **H** and **G** which correspond to the unknown terms, to give for **AX** the following:

$$
\mathbf{AX} =
\begin{bmatrix}
H_{11}^{11} & H_{12}^{11} & H_{12}^{12} & -G_{11}^{121} & -G_{11}^{122} & -G_{12}^{131} & -G_{12}^{132} & H_{11}^{14} \\[6pt]
H_{21}^{11} & H_{22}^{11} & H_{22}^{12} & -G_{21}^{121} & -G_{21}^{122} & -G_{22}^{131} & -G_{22}^{132} & H_{21}^{14} \\[6pt]
H_{11}^{21} & H_{12}^{21} & H_{12}^{22} & -G_{11}^{221} & -G_{11}^{222} & -G_{12}^{231} & -G_{12}^{232} & H_{11}^{24} \\[6pt]
H_{21}^{21} & H_{22}^{21} & H_{22}^{22} & -G_{21}^{221} & -G_{21}^{222} & -G_{22}^{231} & -G_{22}^{232} & H_{21}^{24} \\[6pt]
H_{11}^{31} & H_{12}^{31} & H_{12}^{32} & -G_{11}^{321} & -G_{11}^{322} & -G_{12}^{331} & -G_{12}^{332} & H_{11}^{34} \\[6pt]
H_{21}^{31} & H_{22}^{31} & H_{22}^{32} & -G_{21}^{321} & -G_{21}^{322} & -G_{22}^{331} & -G_{22}^{332} & H_{21}^{34} \\[6pt]
H_{11}^{41} & H_{12}^{41} & H_{12}^{42} & -G_{11}^{421} & -G_{11}^{422} & -G_{22}^{431} & -G_{12}^{432} & H_{11}^{44} \\[6pt]
H_{21}^{41} & H_{22}^{41} & H_{22}^{42} & -G_{21}^{421} & -G_{21}^{422} & -G_{22}^{431} & -G_{22}^{432} & H_{21}^{44}
\end{bmatrix}
\begin{Bmatrix}
u_1^1 \\[6pt]
u_2^1 \\[6pt]
u_2^2 \\[6pt]
t_1^{21} \\[6pt]
t_1^{22} \\[6pt]
t_1^{31} \\[6pt]
t_2^{32} \\[6pt]
u_1^4
\end{Bmatrix}
$$

which is an 8×8 (i.e. $2M \times 2M$) matrix; the right-hand side vector **F** is given by

$$
\begin{Bmatrix}
\left(G_{i2}^{111} + G_{i2}^{112}\right)\sigma \\[10pt]
\left(G_{ij}^{211} + G_{i2}^{212}\right)\sigma \\[10pt]
\left(G_{i2}^{311} + G_{i2}^{312}\right)\sigma \\[10pt]
\left(G_{i2}^{411} + G_{i2}^{412}\right)\sigma
\end{Bmatrix}
$$

which is a $2M$ vector, since $i = 1, 2$.

2.4 Stress and Displacements at Interior Points

Evaluation of stresses and displacements at any point within the body can be obtained from the solution of the system of equations (2.145). Once the values of all displacements and tractions are known on all boundaries, then Somigliana's identities for the unknown interior displacements (2.66) and the stresses (2.68) can be discretized in a similar way to the boundary element formulation (2.140). For interior displacement, at $\mathbf{X}^i$,

$$
u_j(\mathbf{X}^i) + \sum_{\gamma=1}^{M} \bar{H}_{ij}^{c\gamma} u_j^\gamma = \sum_{n=1}^{N_e} \sum_{\alpha=1}^{m} G_{ij}^{cn\alpha} t_j^{n\alpha} \tag{2.148}
$$

and for interior stresses

$$
\sigma_{ij}(\mathbf{X}^i) + \sum_{\gamma=1}^{M} S_{ij}^{c\gamma} u_j^\gamma = \sum_{n=1}^{N_e} \sum_{\alpha=1}^{m} D_{ij}^{cn\alpha} t_j^{n\alpha} \tag{2.149}
$$

2.5 Boundary Stresses

The methods of obtaining boundary stresses from the boundary element solutions can be divided into two groups. The first relies on using recovered boundary tractions

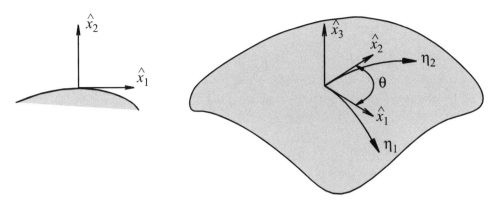

Figure 2.22: Coordinate system for two- and three-dimensional boundary stress calculations.

and displacements obtained from the BEM solution. Tangential strains are then calculated by differentiation of shape functions. These strains are converted by Hooke's law and Cauchy's formula to derive the stresses. The second group of methods is directly from the integral equation representation of the stresses. This approach is quite complicated, and is only recommended for special applications. The first approach is the most popular and economical, but, due to its local nature, it is prone to inaccuracies if coarse discretization is used.

2.5.1 Indirect Approach

In this section a relatively simple way of evaluating boundary stresses from tractions and tangential strains is described [2, 34, 76].

Two-dimensional Problems

Defining a local coordinate system such that $\hat{e}_{1j}(\equiv n_j, j = 1, 2)$ are the unit vectors in the normal directions and $\hat{e}_{2j}$ are the unit vector in the tangential directions to the boundary element (see Figure 2.22), so that

$$\hat{x}_i = \hat{x}_1 \hat{e}_{1i} + \hat{x}_2 \hat{e}_{2i} \tag{2.150}$$

Defining the displacements, strains, stresses and tractions as $\hat{u}_j$, $\hat{\varepsilon}_{ij}$, $\hat{\sigma}_{ij}$ and $\hat{t}_j$ respectively, in the local coordinates $\hat{x}_j$, the corresponding stress components $\hat{\sigma}_{ij}$ can be written as

$$\hat{\sigma}_{1i} = \hat{t}_i \tag{2.151}$$

and from Hooke's law for plane strain,

$$\hat{\sigma}_{22} = \frac{1}{1 - \nu} \left[\nu \hat{t}_1 + 2\mu \hat{\varepsilon}_{22} \right] \tag{2.152}$$

where $\hat{\varepsilon}_{22} = \partial \hat{u}_2 / \partial \hat{x}_2$, μ is the shear modulus and ν is Poisson's ratio.

The displacements and tractions (u_j and t_j) in the global coordinate system are related to those in the local system by the direction cosines $\hat{e}_{ij}$, as follows:

$$\hat{u}_i = \hat{e}_{ij} u_j \quad \text{and} \quad \hat{t}_j = \hat{e}_{ij} t_j \tag{2.153}$$

The rotation matrix $\hat{e}_{ij}$ can be written in terms of the known normal components as follows:

$$\begin{pmatrix} \hat{e}_{11} & \hat{e}_{12} \\ \hat{e}_{21} & \hat{e}_{22} \end{pmatrix} = \begin{pmatrix} n_1 & n_2 \\ -n_2 & n_1 \end{pmatrix} \tag{2.154}$$

In order to relate $\hat{\varepsilon}_{22}$, required in (2.152), to the global coordinates, it is necessary to recall that the boundary approximation using the shape functions $N_\alpha(\eta)$ is

$$x_i = \sum_{\alpha=1}^{m} N_\alpha(\eta) x_i^\alpha$$

The tangent vector evaluated at a boundary point (say $\eta = \eta'$) is given by

$$h_i(\eta') = \left(\frac{dx_i}{d\eta} \right)_{\eta=\eta'} = \sum_{\alpha=1}^{m} \frac{dN_\alpha(\eta')}{d\eta'} x_i^\alpha \tag{2.155}$$

Hence the unit tangent vector, at $\eta = \eta'$, is given by

$$\hat{e}_2 = \frac{h_i(\eta')}{h(\eta')} = \frac{h_i(\eta')}{J(\eta')} \tag{2.156}$$

where $h(\eta') = \sqrt{h_i h_i}$ and $J(\eta')$ is the Jacobian of transformation, defined in (2.129). The local coordinate $\hat{x}_2$ is related to the local intrinsic coordinate η, via

$$\hat{x}_2 = h(\eta')\eta' \tag{2.157}$$

so that

$$\frac{d\eta}{d\hat{x}_2} = \frac{1}{h(\eta')} = \frac{1}{J(\eta')} \tag{2.158}$$

The strain tensor in the local tangential direction is given by

$$\begin{aligned} \hat{\varepsilon}_{22} &= \frac{\partial \hat{u}_2}{\partial \hat{x}_2} = \frac{d\hat{u}_2}{d\eta} \frac{d\eta}{d\hat{x}_2} \\ &= \frac{1}{J(\eta')} \frac{d}{d\eta} (\hat{e}_{21} u_1 + \hat{e}_{22} u_2) \\ &= \sum_{\alpha=1}^{m} u_j^\alpha \sum_{\alpha=1}^{m} \left(\frac{dN_\alpha(\eta)}{d\eta} \right)_{\eta=\eta'} \frac{\hat{e}_{2i}}{J(\eta')} \end{aligned} \tag{2.159}$$

The local stress components can now be obtained from (2.151) and (2.152). The global stress components σ_{ij} are obtained from the local stress components $\hat{\sigma}_{ij}$ via transformation

$$\sigma_{ij} = \hat{e}_{ki} \hat{e}_{nj} \hat{\sigma}_{kn} \tag{2.160}$$

From the relationships between $\hat{e}_{ij}$ and n_j in (2.154), it follows that

$$\begin{aligned} \sigma_{11} &= n_1^2 \hat{\sigma}_{11} - 2n_1 n_2 \hat{\sigma}_{12} + n_2^2 \hat{\sigma}_{22} \\ \sigma_{12} &= n_1 n_2 \hat{\sigma}_{11} + (n_1^2 - n_2^2) \hat{\sigma}_{12} - n_1 n_2 \hat{\sigma}_{22} \\ \sigma_{22} &= n_2^2 \hat{\sigma}_{11} + 2n_1 n_2 \hat{\sigma}_{12} + n_1^2 \hat{\sigma}_{22} \end{aligned} \tag{2.161}$$

Thus, the global stress components at a boundary point can be expressed in terms of the displacements and tractions at that point and the components of the unit normal.

Three-dimensional Problems

For three-dimensional problems the procedure is similar to the two-dimensional problems described above. In general, the vector tangents $\partial x_i/\partial \eta_1$ and $\partial x_i/\partial \eta_2$ ($i = 1, 2, 3$) to the local coordinate lines of the quadrilateral surface elements are not orthogonal. Let e_{1j} and e_{2j} ($j = 1, 2, 3$) be the components of the unit vectors in the coordinate directions η_1 and η_2, respectively, and e_{3j} components of unit normal ($\equiv n_j$) to the element. Now let $\hat{e}_{ij}$ be the components (directional cosine) of the unit vector of the orthogonal system of axes defining a coordinate system $\hat{x}_i$ (i.e. $\hat{x}_i = x_1\hat{e}_{1i} + x_2\hat{e}_{2i} + x_3\hat{e}_{3i}$), as shown in Figure 2.22. This system is obtained by taking $\hat{e}_{1i} = e_{1i}$, $\hat{e}_{3i} = e_{3i}$ and e_{2i} as the vector product of e_1 and e_3. If $\hat{u}_j$, $\hat{\varepsilon}_{ij}$, $\hat{\sigma}_{ij}$ and $\hat{t}_j$ are the displacements, strain, stresses and tractions, respectively, in the local coordinates $\hat{x}_i$, then the stress components $\hat{\sigma}_{ij}$ can be written as

$$\hat{\sigma}_{3i} = \hat{t}_i \qquad i = 1, 2, 3 \tag{2.162}$$

and the remaining stress tensors can be expressed as follows:

$$
\begin{aligned}
\hat{\sigma}_{11} &= \frac{1}{1-\nu}\left(\nu\hat{t}_3 + 2\mu(\hat{\varepsilon}_{11} + \nu\hat{\varepsilon}_{22}\right) \\
\hat{\sigma}_{12} &= 2\mu\hat{\varepsilon}_{12} \\
\hat{\sigma}_{22} &= \frac{1}{1-\nu}\left(\nu\hat{t}_3 + 2\mu(\hat{\varepsilon}_{22} + \nu\hat{\varepsilon}_{11}\right)
\end{aligned}
\tag{2.163}
$$

The displacement and tractions in the global coordinate system are related to the local orthogonal system, via the transformation matrix $\hat{e}_{ij}$, as follows:

$$\hat{u}_i = \hat{e}_{ij}\hat{u}_j \qquad \hat{t}_i = \hat{e}_{ij}t_j \tag{2.164}$$

with $\hat{e}_{ij}$ given as

$$
\hat{e}_{ij} = \begin{pmatrix}
\hat{e}_{11} & \hat{e}_{12} & \hat{e}_{13} \\
\hat{e}_{21} & \hat{e}_{22} & \hat{e}_{23} \\
\hat{e}_{31} = n_1 & \hat{e}_{32} = n_2 & \hat{e}_{33} = n_3
\end{pmatrix}
$$

The two tangent vectors at a surface point (say $\eta_1 = \eta_1'$ and $\eta_2 = \eta_2'$) are given by

$$
\begin{aligned}
h_i(\eta_1', \eta_2') &= \left(\frac{\partial x_i}{\partial \eta_1}\right)_{\eta_1=\eta_1',\, \eta_2=\eta_2'} = \sum_{\alpha=1}^{m} \frac{\partial N_\alpha(\eta_1', \eta_2')}{\partial \eta_1'} x_i^\alpha \\
g_i(\eta_1', \eta_2') &= \left(\frac{\partial x_i}{\partial \eta_2}\right)_{\eta_1=\eta_1',\, \eta_2=\eta_2'} = \sum_{\alpha=1}^{m} \frac{\partial N_\alpha(\eta_1', \eta_2')}{\partial \eta_2'} x_i^\alpha
\end{aligned}
\tag{2.165}
$$

The outward normal vector is the vector product of these two vectors, that is

$$d_i(\eta_1', \eta_2') = h_i(\eta_1', \eta_2') \times g_i(\eta_1', \eta_2') \tag{2.166}$$

The local orthogonal vectors are defined by

$$\hat{e}_{i1} = \frac{h_i(\eta_1', \eta_2')}{h(\eta_1', \eta_2')}$$

$$\hat{e}_{2i} = \frac{g_i(\eta_1', \eta_2')}{g(\eta_1', \eta_2')}$$

$$\hat{e}_{3i} = \frac{1}{d(\eta_1', \eta_2')} \left[h(\eta_1', \eta_2')g_i(\eta_1', \eta_2') - h_i(\eta_1', \eta_2')g_i(\eta_1', \eta_2')\frac{h_i(\eta_1', \eta_2')}{h(\eta_1', \eta_2')} \right] \quad (2.167)$$

where $h(\eta_1', \eta_2') = \sqrt{h_i h_i}$, $g(\eta_1', \eta_2') = \sqrt{g_i g_i}$ and $d(\eta_1', \eta_2') = \sqrt{d_i d_i}$. From Figure 2.22, it is possible to relate the local coordinates in the surface tangential directions, $\hat{x}_1$ and $\hat{x}_2$, to the local intrinsic coordinates η_1' and η_2', via

$$\hat{x}_1 = g(\eta_1', \eta_2')\eta_2 \cos\theta + h(\eta_1', \eta_2')\eta_1$$
$$\hat{x}_2 = g(\eta_1', \eta_2')\eta_1 \sin\theta \quad (2.168)$$

where θ is the angle between e_1 and e_2. From (2.167) and (2.168),

$$\eta_1 = \frac{1}{h(\eta_1', \eta_2')}[\hat{x}_1 - \cot\theta\hat{x}_2]$$

$$\eta_2 = \frac{1}{g(\eta_1', \eta_2')\sin\theta}\hat{x}_2 \quad (2.169)$$

Differentiating η_1 and η_2 with respect to $\hat{x}_1$ and $\hat{x}_2$ gives

$$\frac{\partial\eta_1}{\partial\hat{x}_1} = \frac{1}{h(\eta_1', \eta_2')} \qquad \frac{\partial\eta_2}{\partial\hat{x}_1} = 0$$

$$\frac{\partial\eta_1}{\partial\hat{x}_2} = \frac{-1}{h(\eta_1', \eta_2')}\cot\theta \qquad \frac{\partial\eta_2}{\partial\hat{x}_2} = \frac{1}{g(\eta_1', \eta_2')\sin\theta} \quad (2.170)$$

Therefore, the strain tensors in the local orthogonal coordinate system can be evaluated as

$$\hat{\varepsilon}_{11} = \frac{\partial\hat{u}_1}{\partial\hat{x}_1} = \left[\frac{\partial\hat{u}_1}{\partial\eta_1}\frac{\partial\eta_1}{\partial\hat{x}_1} + \frac{\partial\hat{u}_1}{\partial\eta_2}\frac{\partial\eta_2}{\partial\hat{x}_1}\right] = \frac{1}{h(\eta_1', \eta_2')}\frac{\partial\hat{u}_1}{\partial\eta_1}$$

$$= \frac{1}{h(\eta_1', \eta_2')}\frac{\partial}{\partial\eta_1}(\hat{e}_{11}u_1 + \hat{e}_{12}u_2 + \hat{e}_{13}u_3) \quad (2.171)$$

$$= \sum_{\alpha=1}^{m} u_j^\alpha E_j^{11\alpha}(\eta_1', \eta_2') \quad (2.172)$$

where

$$E_j^{11\alpha}(\eta_1', \eta_2') = \left(\frac{\partial N_\alpha}{\partial\eta_1}\right)_{\eta_1=\eta_1', \eta_2=\eta_2'}\frac{\hat{e}_{1j}}{h(\eta_1', \eta_2')}$$

Similarly,

$$\hat{\varepsilon}_{22} = \frac{\partial\hat{u}_2}{\partial\hat{x}_2} = \sum_{\alpha=1}^{m} u_j^\alpha E_j^{22\alpha}(\eta_1', \eta_2') \quad (2.173)$$

$$\hat{\varepsilon}_{12} = \frac{1}{2}\left(\frac{\partial\hat{u}_1}{\partial\hat{x}_2} + \frac{\partial\hat{u}_2}{\partial\hat{x}_1}\right) = \sum_{\alpha=1}^{m} u_j^\alpha E_j^{12\alpha}(\eta_1', \eta_2') \quad (2.174)$$

where

$$E_j^{22\alpha}(\eta_1', \eta_2') = \hat{e}_{2j}\left[\frac{\partial N_\alpha}{\partial\eta_1'}\frac{-\cot\theta}{h(\eta_1', \eta_2')} + \frac{\partial N_\alpha}{\partial\eta_2'}\frac{1}{g(\eta_1', \eta_2')\sin\theta}\right]$$

$$E_j^{12\alpha}(\eta_1', \eta_2') \;=\; \hat{e}_{2j}\left[\frac{\partial N_\alpha}{\partial \eta_1'}\frac{1}{h(\eta_1', \eta_2')}\right]+$$

$$\hat{e}_{1j}\left[-\frac{\partial N_\alpha}{\partial \eta_1'}\frac{\cot\theta}{h(\eta_1', \eta_2')}+\frac{\partial N_\alpha}{\partial \eta_2'}\frac{1}{g(\eta_1', \eta_2')\sin\theta}\right]$$

The local stresses can be obtained by substituting (2.172)-(2.174) into (2.163). The global stress components σ_{ij} are obtained from the transformation

$$\sigma_{ij} = \hat{e}_{ki}\hat{e}_{nj}\hat{\sigma}_{kn} \tag{2.175}$$

2.5.2 Direct Approach

To derive the stress integral equation for a general boundary and not just the smooth one, it is convenient to consider the limiting case of the displacement derivative equation (2.67) as $\mathbf{X}' \to \mathbf{x}'$. The limiting process is similar to that used for the displacement integral equation, and is described in detail in Appendix B. The resulting integral equation can be written [89, 87], as

$$\hat{c}_{iljk}\hat{u}_{i,l} = I_{jk} \tag{2.176}$$

where

$$
\begin{aligned}
I_{jk} \;=\;& -d_{ijk}(\mathbf{x}')u_i' + \int_{S_R} U_{ij,k}(\mathbf{x}', \mathbf{x})\bar{t}_i(\mathbf{x})dS \\
&+ \int_{S_S} U_{ij,k}(\mathbf{x}', \mathbf{x})\sum_{\alpha=1} t_i^\alpha\left[N_\alpha(\eta_1, \eta_2) - L(\eta_1, \eta_2)\delta_{\alpha\alpha}\right]dS \\
&- \int_{S_R} T_{ij,k}(\mathbf{x}', \mathbf{x})\bar{u}_i dS \\
&+ \int_{S_S} T_{ij,k}(\mathbf{x}', \mathbf{x})\sum_{\alpha=1} u_i'^\alpha\left[\frac{\partial N_\alpha}{\partial \rho_l}\rho_l L(\eta_1, \eta_2)\right]dS
\end{aligned}
$$

and

$$\hat{c}_{iljk} = c_{iljk} + \delta_{ij}\delta_{lk} - \int_S U_{pj,k}(\mathbf{x}', \mathbf{x})E_{pqil}n_q'L(\eta_1, \eta_2)dS + \int_{S_S} T_{ij,k}(\mathbf{x}', \mathbf{x})\rho_l L(\eta_1, \eta_2)dS$$

Superscript $'$ denotes values at the source point, S_R and S_S denote regular and singular part of the boundary, respectively. The elastic constant E_{ijkl} is define as

$$\sigma_{ij}' = E_{ijkl}u_{k,l}' \tag{2.177}$$

2.6 Body Forces

In the previous, sections, body forces have been assumed to be zero; however, if this is not the case, the body force integral in equation (2.67) must be evaluated. There have been several methods developed for evaluation of this integral. They include:

- domain discretization

- Galerkin vector

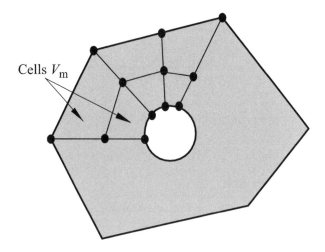

Figure 2.23: Discretization of the part of the domain into cells.

- particular integrals

- multiple reciprocity

- dual reciprocity.

Apart from the domain discretization approach, the other methods are used to transfer the domain body force integrals into equivalent boundary integrals. In this section, these methods will be described for gravitational and centrifugal body forces. Other more complicated body force terms such as thermal and inertia are described in, later chapters on thermoelasticity and elastodynamics, respectively.

2.6.1   Domain Discretization

In this approach, part of the domain where the body forces are applied is divided into M_e cells, as shown in Figure 2.23, so that

$$B_i = \int_V U_{ij}(\mathbf{x}', \mathbf{X})b_j(\mathbf{X})dV = \sum_{m=1}^{M_e} \int_{V_m} U_{ij}(\mathbf{x}', \mathbf{X})b_j(\mathbf{X})dV_m \qquad (2.178)$$

where

$$V = \sum_{m=1}^{M_e} V_m$$

The integral in (2.178) is generally well behaved, and numerical integration schemes such as the Gauss quadrature are used to carry out the integration. The matrix equation in (2.142) can be rewritten to include the body force contribution as

$$\mathbf{Hu} = \mathbf{Gt} + \mathbf{B} \qquad (2.179)$$

where $\mathbf{B}$ is the vector containing dM ($d = 2$ for 2D, $d = 3$ for 3D and $M =$ number of collocation points) contributions of the prescribed body force integral. After substitution of prescribed displacements and traction boundary conditions, the resulting system of equations can be rewritten as

$$\mathbf{AX} = \mathbf{RY} + \mathbf{B}$$
$$= \mathbf{F} \tag{2.180}$$

This approach, although the most general procedure available, has a drawback in that it is time consuming, especially for three-dimensional problems. However, it remains the only reliable method for the analysis of highly nonlinear problems.

2.6.2   Galerkin Vector

In this approach, the Galerkin vector is used to transform the domain body force integrals into related boundary integrals (see for example [22]). The displacement fundamental solution is related to the Galerkin vector as

$$U_{ij}(\mathbf{X}', \mathbf{X}) = G_{ij,kk}(\mathbf{X}', \mathbf{X}) - \frac{1}{2(1 - \nu)} G_{ik,kj}(\mathbf{X}', \mathbf{X}) \tag{2.181}$$

Substituting (2.181) into the body force integral gives

$$B_i = \int_V U_{ij}(\mathbf{x}', \mathbf{X}) b_j(\mathbf{X}) dV = \int_V \left[G_{ij,kk}(\mathbf{x}', \mathbf{X}) - \frac{1}{2(1 - \nu)} G_{ik,kj}(\mathbf{x}', \mathbf{X}) \right] b_j(\mathbf{X}) dV \tag{2.182}$$

To transfer the above integral into an equivalent boundary integral, we shall be using Gauss's theorem, that is

$$\int_V F_{ij,k} dV = \int_S F_{ij} n_k dS \tag{2.183}$$

where n_k denotes the components of the outward normal to the boundary S.

Gravitational Loads

A body force of constant mass density $\rho(\mathbf{X}) = Const.$ in a constant gravitational field $g_j(\mathbf{X}) = Const.$, and the body force b_j is given as

$$b_j(\mathbf{X}) = \rho(\mathbf{X}) g_j(\mathbf{X}) = Const. \tag{2.184}$$

As the body force term is constant, it can be taken outside the integral, and using (2.183) we have

$$B_i = b_j \int_S \left\{ G_{ij,k} - \frac{1}{2(1 - \nu)} G_{ik,j} \right\} n_k dS = \int_S P_i dS \tag{2.185}$$

The Galerkin vector corresponding to three- and two-dimensional problems are given in (2.76) and (2.87), respectively. Substituting (2.76) and (2.87) into (2.185) gives

$$P_i = \frac{1}{8\pi\mu} \left\{ b_i n_k r_{,k} - \frac{1}{2(1 - \nu)} b_k r_{,k} n_i \right\} \tag{2.186}$$

for three-dimensions, and

$$P_i = \frac{r}{8\pi\mu}\left\{\left[2\ln\frac{1}{r} - 1\right]\left(b_i n_k r_{,k} - \frac{1}{2(1-\nu)}b_k r_{,k} n_i\right)\right\} \qquad (2.187)$$

for two-dimensional plane strain conditions.

Similarly, the above expressions could be used to evaluate the body force integrals in the internal stress equation. That is

$$\int_V D_{kij}b_k dV = \int_S D_{ij} dS \qquad (2.188)$$

where

$$D_{ij} = \frac{1}{8\pi r}\left\{n_m r_{,m}(b_i r_{,j} + b_j r_{,i}) + \left[\frac{1}{1-\nu}\nu\delta_{ij}\left(n_m r_{,m} b_k r_{,k} - b_m n_m\right)\right.\right.$$
$$\left.\left. -\frac{1}{2}[b_m r_{,m}(n_i r_{,j} + n_j r_{,i}) + (1-2\nu)(b_i n_j + b_j n_i)]\right]\right\} \qquad (2.189)$$

for three-dimensions, and

$$D_{ij} = \frac{1}{8\pi}\left\{2n_m r_{,m}(b_i r_{,j} + b_j r_{,i})\right.$$
$$+\frac{1}{1-\nu}\left[\nu\delta_{ij}\left(2n_m r_{,m} b_k r_{,k} + \left[1 - 2\ln\frac{1}{r}\right]b_m n_m\right) - b_m r_{,m}(n_i r_j + n_j r_i)\right.$$
$$\left.\left. +\frac{1-2\nu}{2}\left(1 - 2\ln\frac{1}{r}\right)(b_i n_j + b_j n_i)\right]\right\} \qquad (2.190)$$

for two-dimensional plane strain.

Centrifugal Load

For a body rotating with angular velocity ω_i, the prescribed body force term can be written as

$$b_j = w_{ij}x_i \qquad (2.191)$$

where

$$w_{ij} = \rho\begin{bmatrix} \omega_2^2 + \omega_3^2 & -\omega_1\omega_2 & -\omega_3\omega_1 \\ -\omega_1\omega_2 & \omega_3^2 + \omega_1^2 & -\omega_2\omega_3 \\ -\omega_3\omega_1 & -\omega_2\omega_3 & \omega_1^2 + \omega_2^2 \end{bmatrix}$$

for three-dimensional problems, assuming that the axis of rotation passes through the origin of the coordinate system. For two-dimensional problems, the axis of rotation must either be (a) in the plane of the problem, which implies $\omega_3 = 0$, or (b) at a right-angle to the plane of the problem, which implies $-\omega_1 = \omega_2 = 0$.

The equivalent boundary integral for centrifugal load is obtained by substituting (2.191) into (2.182) and using (2.183) to give

$$B_i = w_{jk}\int_S\left\{x_j\left[G_{ik,m} - \frac{1}{2(1-\nu)}G_{im,k}\right]n_m - \frac{1-2\nu}{2(1-\nu)}G_{ik}n_j\right\}dS$$

$$= \int_S P_i dS \qquad (2.192)$$

where

$$P_i = \frac{1}{8\pi\mu}\left\{w_{ij}x_jr_{,k}n_k - \frac{1}{2(1-\nu)}\left[r_{,k}w_{jk}x_jn_i + (1-2\nu)rw_{ij}n_j\right]\right\} \qquad (2.193)$$

for three-dimensional problems, and

$$P_i = \frac{r}{8\pi\mu}\left\{\left(1-2\ln\frac{1}{r}\right)\left[x_jw_{ji}r_{,k}n_k - \frac{r_{,k}w_{kj}x_jn_i}{2(1-\nu)}\right] - \frac{(1-2\nu)}{2(1-\nu)}r\ln\frac{1}{r}w_{ki}n_k\right\} \qquad (2.194)$$

for two-dimension problems (i.e. both cases (a) and (b)). Similarly the expressions for the internal stresses are given by

$$D_{ij} = \frac{1}{8\pi}\left\{\left[n_kr_{,k}\frac{x_m}{r} + \frac{(1-2\nu)}{2(1-\nu)}n_m\right](w_{im}r_{,j} + w_{jm}r_{,i})\right. \qquad (2.195)$$

$$+ \frac{1}{(1-\nu)}\left[\nu\delta_{ij}\left(r_{,k}w_{km}\frac{x_m}{r}n_kr_{,k} - n_kw_{km}\frac{x_m}{r} + r_{,k}w_{km}n_m\right.\right.$$

$$\left.\left.\left. - \frac{x_m}{2r}\left(r_{,k}w_{km}\left[n_ir_{,j} + n_jr_{,i}\right] + (1-2\nu)\left[n_iw_{jm} + n_jw_{im}\right]\right)\right]\right\}\right.$$

for three-dimensional problems, and

$$D_{ij} = \frac{1}{8\pi}\left\{2n_mr_{,m}(x_kw_{ki}r_{,j} + x_kw_{kj}r_{,i}) + \frac{1}{(1-\nu)}\left[\nu\delta_{ij}\left(2r_{,k}w_{ks}r_sn_mr_{,m}\right.\right.\right.$$

$$\left.\left.+ \left(1-2\ln\frac{1}{r}\right)\left[x_mw_{mk}n_k - r_{,k}w_{mk}n_mr\right]\right) - r_mw_{mk}x_k(n_ir_{,j} + n_jr_{,i})\right.$$

$$\left.+ (x_mw_{mi}n_j + x_mw_{mj}n_i - \left[n_mw_{mi}r_{,j} + n_mw_{mj}r_{,i}\right]r)\right.$$

$$\left.\times\frac{(1-2\nu)}{2}\left(1-2\ln\frac{1}{r}\right)\right\} \qquad (2.196)$$

for two dimensional problems.

2.6.3   Particular Integrals

The particular integrals for elastostatic problems follow a similar procedure to the potential problems described in Volume 1. The governing equation (2.51) can be expressed as

$$\Im(u_i) + b_i = 0 \qquad (2.197)$$

where $\Im$ denotes Navier's operator. Assuming it is possible to construct a solution to the above equation, such that

$$u_i = \tilde{u}_i + \hat{u}_i$$
$$t_i = \tilde{t}_i + \hat{t}_i \qquad (2.198)$$

where $(\tilde{u}_i, \tilde{t}_i)$ satisfy the Navier's equation, $\Im(\tilde{u}_i) = 0$ and $(\hat{u}_i, \hat{t}_i)$ is a particular solution of

$$\Im(\hat{u}_i) + b_i = 0 \qquad (2.199)$$

Application to the Betti's reciprocal theorem of two distinct self-equilibrated states (u_i^*, t_i^*, b_i^*) and $(\hat{u}_i, \hat{t}_i, b_i)$ (see Section 2.3.1) gives

$$\int_S \hat{t}_iu_i^*dS + \int_V b_iu_i^*dV = \int_S t_i^*\hat{u}_idS + \int_V b_i^*\hat{u}_idV \qquad (2.200)$$

Selecting the (u_i^*, t_i^*, b_i^*) state to correspond to the point force solution as we did in the derivation of the boundary integral equation (2.102), we have

$$\int_V u_i^* b_i(\mathbf{x})dV = C_{ij}(\mathbf{x}')\hat{u}_j(\mathbf{x}') + \int_S T_{ij}(\mathbf{x}',\mathbf{x})\hat{u}_j(\mathbf{x})dS = \int_S U_{ij}(\mathbf{x}',\mathbf{x})\hat{t}_j(\mathbf{x})dS \quad (2.201)$$

Substituting (2.201) into (2.102) gives a totally boundary representation of the problem, i.e.

$$C_{ij}(\mathbf{x}')u_j(\mathbf{x}') + \int_S T_{ij}(\mathbf{x}',\mathbf{x})u_j(\mathbf{x})dS - \int_S U_{ij}(\mathbf{x}',\mathbf{x})t_j(\mathbf{x})dS$$
$$= C_{ij}(\mathbf{x}')\hat{u}_j(\mathbf{x}') + \int_S T_{ij}(\mathbf{x}',\mathbf{x})\hat{u}_j(\mathbf{x})dS - \int_S U_{ij}(\mathbf{x}',\mathbf{x})\hat{t}_j(\mathbf{x})dS$$

As before, the above integral equation can be discretized and then represented in a matrix form as

$$\mathbf{Hu} - \mathbf{Gt} = \mathbf{H\hat{u}} - \mathbf{G\hat{t}} \quad (2.202)$$

where $\hat{\mathbf{u}}$ and $\hat{\mathbf{t}}$ are vectors the containing values of $\tilde{u}_j$ and $\tilde{t}_j$, respectively. The above form is equivalent to (2.179), with the vector $\mathbf{B}$ given by

$$\mathbf{B} = \mathbf{H\hat{u}} - \mathbf{G\hat{t}} \quad (2.203)$$

The advantage of the above approach, as in the case of the Galerkin vector, is that if the particular solutions are known analytically, the method results in an exact transformation of the domain integrals into equivalent boundary integrals. An approximate method based on the particular integrals known as the Dual Reciprocity Method is described later in the chapter. Below, we shall present the particular solutions for the two cases of gravitational and centrifugal loads.

Gravitational Load

For three-dimensional problems, gravitational load $b_3 = \rho g$, and the particular displacement solutions are given by [79],

$$\hat{u}_1 = \frac{\rho g \lambda}{2\mu(3\lambda + 2\mu)} x_1 x_2$$

$$\hat{u}_2 = \frac{\rho g \lambda}{2\mu(3\lambda + 2\mu)} x_2 x_3$$

$$\hat{u}_3 = -\frac{(\lambda + \mu)\rho g}{2\mu(3\lambda + 2\mu)} x_3^2 - \frac{\rho g \lambda}{4\mu(3\lambda + 2\mu)}(x_1^2 + x_2^2) \quad (2.204)$$

For two-dimensional problems, the corresponding solutions (see, for example, [6]) are given by

$$\hat{u}_1 = \frac{\rho g \lambda}{4\mu(\lambda + \mu)} x_1 x_2$$

$$\hat{u}_2 = -\frac{\rho g}{8\mu(\lambda + \mu)}\left[(\lambda + 2\mu) x_2^2 + \lambda x_1^2\right] \quad (2.205)$$

The traction particular solutions can be obtained from the above displacement using the usual displacement-strain and stress-strain relationships.

Centrifugal Load

For a three-dimensional body rotating about the x_3 axis, the displacement fields are given by (see, for example, [6]):

$$\hat{u}_k = \frac{-\rho\omega^2}{8(\lambda+2\mu)}\left[\frac{5\lambda+4\mu}{4(\lambda+2\mu)}x_s x_s + \frac{\mu}{\lambda+\mu}x_3^2\right]x_k \qquad s,k=1,2$$

$$\hat{u}_3 = \frac{\rho\omega^3}{8(\lambda+2\mu)}x_k x_k x_3 \qquad k=1,2 \tag{2.206}$$

and, for two-dimensional body rotating about a fixed axis perpendicular to the plane of the body , we have

$$\hat{u}_i = \frac{-\rho\omega^3}{8(\lambda+2\mu)}x_k x_k x_i \qquad i=1,2$$

2.6.4 Multiple Reciprocity Method

The domain integral in (2.178) can be rewritten as [56]

$$B_i = \int_V U_{ij}^{(0)} b_j^{(0)} dV \tag{2.207}$$

From the original fundamental solutions presented in Section 2.3.3, so-called higher order fundamentals $U_{ij}^{(1)}$ can be derived through the use of the following equation:

$$\nabla^2 U_{ij}^{(1)} = U_{ij}^{(0)} \tag{2.208}$$

Hence the domain integral in (2.207) can now be written as

$$B_i = \int_V \nabla^2 U_{ij}^{(1)} b_j^{(0)} dV \tag{2.209}$$

Now applying the reciprocity theorem gives

$$B_i = \int_S \left(\frac{\partial U_{ij}}{\partial n}b_j^{(0)} - U_{ij}^{(1)}\frac{\partial b_j^{(0)}}{\partial n}\right) dS + \int_V U_{ij}^{(1)}\nabla^2 b_j^{(0)} dV \tag{2.210}$$

For gravitational and centrifugal loads, $\nabla^2 b_j^{(0)} = 0$, and the above equation reduces to

$$B_i = \int_S \left(\frac{\partial U_{ij}}{\partial n}b_j^{(0)} - U_{ij}^{(1)}\frac{\partial b_j^{(0)}}{\partial n}\right) dS \tag{2.211}$$

where for gravitational load $b_j^{(0)} = -\rho g_j$, $\partial b_j^{(0)}/\partial n = 0$ and for centrifugal load $b_j^{(0)} = w_{ij}x_i$, $\partial b_j^{(0)}/\partial n = w_{ij}n_i$.

2.6.5 Dual Reciprocity Method

In the Dual Reciprocity Method (DRM) body forces are approximated as [53]

$$b_i \simeq \sum_{k=1}^{N+L} \alpha_i^k f(\mathbf{x}^k, \mathbf{x}) \tag{2.212}$$

where α_i^k represents a set of initially unknown coefficients, f^k are approximating functions, N is the total number of boundary nodes in the discretization, and L is the total number of internal nodes. The function f^k can be compared to the interpolating functions N_α used to approximate the unknown fields and the geometry. However, the expansion (2.212) is a global expansion because it is valid over the entire domain.

After substitution of (2.212) the domain integral B_i can be written as

$$\int_V u_i^* b_i dV = \sum_{k=1}^{N+L} \alpha_i^k \int_V u_i f^k dV \tag{2.213}$$

Particular solutions $\hat{u}_{lj}$ and $\hat{t}_{lj}$ of Navier's equations can be found such that

$$\mu \hat{u}_{lj,ii} + \frac{\mu}{1-2\nu} \hat{u}_{lj,il} + \delta_{lj} f^k = 0 \tag{2.214}$$

Following the same procedure as in the particular integral approach described previously, we have

$$\int_V u_i^* b_i dV = \sum_{k=1}^{N+L} \alpha_l^k \left[C_{ij} \hat{u}_{lj}^k + \oint_S T_{ij}(\mathbf{x}', \mathbf{x}) \hat{u}_{lj}^k dS \right. \tag{2.215}$$

$$\left. = \int_S U_{ij}(\mathbf{x}', \mathbf{x}) \hat{t}_{lj}^k dS \right]$$

Substituting (2.215) into (2.102) gives a boundary representation of the problem, i.e.

$$C_{ij}(\mathbf{x}') u_j(\mathbf{x}') + \oint_S T_{ij}(\mathbf{x}', \mathbf{x}) u_j(\mathbf{x}) dS - \int_S U_{ij}(\mathbf{x}', \mathbf{x}) t_j(\mathbf{x}) dS \tag{2.216}$$

$$= \sum_{k=1}^{N+L} \alpha_l^k \left[C_{ij} \hat{u}_{lj}^k + \oint_S T_{ij}(\mathbf{x}', \mathbf{x}) \hat{u}_{lj}^k dS - \int_S U_{ij}(\mathbf{x}', \mathbf{x}) \hat{t}_{lj}^k dS \right]$$

The above integral equation can be represented in a discretized form by following the same steps as the displacement integral equation described in Section 2.3.10. It is worth noting that, since the particular solutions $\hat{u}_{lj}^k$ and $\hat{t}_{lj}^k$ are known functions once f^k is defined, there is no need to approximate their variation within each element with shape functions N_α. However, to do so will have the advantage that the same coefficient matrices as in (2.141) are obtained, as

$$\sum_{\gamma=1}^M H_{ij}^{c\gamma} u_j^\gamma - \sum_{n=1}^{N_e} \sum_{\alpha=1}^m G_{ij}^{cn\alpha} t_j^{n\alpha} = \sum_{k=1}^{N+L} \alpha_l^k \left[\sum_{\gamma=1}^M H_{ij}^{c\gamma} \hat{u}_{lj}^{\gamma k} - \sum_{n=1}^{N_e} \sum_{\alpha=1}^m G_{ij}^{cn\alpha} \hat{t}_{lj}^{n\alpha k} \right] \tag{2.217}$$

in which, as before, $H_{ij}^{c\gamma} = C_{ij} + \hat{H}_{ij}^{c\gamma}$ and $c = 1, M$. The above equation can be written in a matrix form as

$$\mathbf{Hu} - \mathbf{Gt} = \sum_{k=1}^{N+L} \alpha_l^k \left[\mathbf{H\hat{u}}^k - \mathbf{G\hat{t}}^k \right] \tag{2.218}$$

If each k term of the vectors $\hat{\mathbf{u}}^k$, $\hat{\mathbf{t}}^k$ is considered to correspond to one column of the matrices $\hat{\mathbf{U}}$ and $\hat{\mathbf{T}}$, respectively, then equation (2.218) can be written without the summation to give

$$\mathbf{Hu} - \mathbf{Gt} = \left(\mathbf{H\hat{U}} - \mathbf{G\hat{T}} \right) \alpha \tag{2.219}$$

Equation (2.219) is the basis for the application of the Dual Reciprocity Method (DRM), and only involves discretization of the boundary. The definition of interior nodes is not normally a necessary condition to obtain a solution; however, the solution will usually be more accurate if several internal points are used. The internal nodes may be defined in the desired number and location, depending on the distribution or complexity of the body force term b_i.

The vector $\boldsymbol{\alpha}$ in equation (2.219) is calculated as follows. By taking the values of b_i at $N + L$ different points, a set of equations similar to (2.212) is obtained, which may be expressed in a matrix form as

$$\mathbf{b} = \mathbf{F}\boldsymbol{\alpha} \tag{2.220}$$

where each column of $\mathbf{F}$ consists of a vector $\mathbf{f}_k$ containing the values of the function f_k at the $N + L$ DRM collocation points. The above equation may be inverted to obtain $\boldsymbol{\alpha}$, i.e.

$$\boldsymbol{\alpha} = \mathbf{F}^{-1}\mathbf{b} \tag{2.221}$$

Substituting into equation (2.219) gives

$$\mathbf{Hu} - \mathbf{Gt} = \mathbf{B} \tag{2.222}$$

where

$$\mathbf{B} = \left(\mathbf{H\hat{U}} - \mathbf{G\hat{T}}\right)\mathbf{F}^{-1}\mathbf{b}$$

The DRM is a general technique that can be applied to any type of body forces, known or unknown. The choice of the approximation function f^k can be problem dependent. However, the simple function

$$f^k = f(\mathbf{x}^k, \mathbf{x}) = c + r(\mathbf{x}^k, \mathbf{x}) \tag{2.223}$$

where c is a constant and $r(\mathbf{x}^k, \mathbf{x})$ is a distance between the points $\mathbf{x}^k$ and $\mathbf{x}$, has been used successfully for many elasticity problems (see, for example, [53, 73, 86]). Other functions such as thin plate splines have also been implemented successfully in elasticity [14].

For the approximating function (2.223), the particular displacements and tractions fields are given as

$$\hat{u}_{lj}^k = c\frac{(1-2\nu)}{(6-4\nu)\mu}r_{,l}r_{,j}r^2 + \frac{1}{48(1-\nu)\mu}\left[\left(\frac{11}{3} - 4\nu\right)\delta_{lj} - r_{,l}r_{,j}\right]r^3 \tag{2.224}$$

$$\hat{t}_{lj}^k = c\frac{(1-2\nu)}{(3-2\nu)}\left[\frac{1+2\nu}{1-2\nu}r_{,l}n_j + \frac{1}{2}r_{,j}n_l + \frac{1}{2}\delta_{lj}\frac{\partial r}{\partial n}\right]r + \frac{1}{24(1-\nu)} \tag{2.225}$$
$$\times \left[(5-6\nu)r_{,j}n_l - (1-6\nu)r_{,l}n_j + \frac{\partial r}{\partial n}\left((5-6\nu)\delta_{lj} - r_{,l}r_{,j}\right)\right]r^2$$

for three-dimensional problems, and

$$\hat{u}_{lj}^k = c\frac{(1-2\nu)}{(5-4\nu)\mu}r_{,l}r_{,j}r^2 + \frac{1}{30(1-\nu)\mu}\left[\left(3 - \frac{10\nu}{3}\right)\delta_{lj} - r_{,l}r_{,j}\right]r^3 \tag{2.226}$$

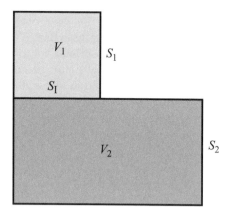

Figure 2.24: Multi-regions.

$$\hat{t}_{lj}^{k} = c\frac{2(1-2\nu)}{(5-4\nu)}\left[\frac{1+\nu}{1-2\nu}r_{,l}n_j + \frac{1}{2}r_{,j}n_l + \frac{1}{2}\delta_{lj}\frac{\partial r}{\partial n}\right]r + \frac{1}{15(1-\nu)} \quad (2.227)$$

$$\times \left[(4-5\nu)r_{,j}n_l - (1-5\nu)r_{,l}n_j + \frac{\partial r}{\partial n}((4-5\nu)\delta_{lj} - r_{,l}r_{,j})\right]r^2$$

for two-dimensional problems.

2.7 Multi-region Formulation

In some problems, as will be described later, it is necessary to consider more than one region, e.g. non-homogeneous problems. The multi-domain formulation is a straightforward extension of the BEM procedures described earlier in the chapter.

Consider a region V consisting of two subregions V_1 and V_2 which have boundaries S_1 and S_2 and an interface S_I, as shown in Figure 2.24. Over subregion V_1, the following variables are defined:

$\mathbf{U}^1, \mathbf{T}^1$ – nodal displacements and tractions at the external boundary S_1;

$\mathbf{U}_I^1, \mathbf{T}_I^1$ –nodal displacements and tractions at the interface S_I.

Similarly, for subregion V_2, we have:

$\mathbf{U}^2, \mathbf{T}^2$ – nodal displacements and tractions at the external boundary S_2;

$\mathbf{U}_I^2, \mathbf{T}_I^2$ –nodal displacements and tractions at the interface S_I.

The system of equations for subregion V_1 can be written as

$$\left[\begin{array}{cc} \mathbf{H}^1 & \mathbf{H}_I^1 \end{array}\right]\left\{\begin{array}{c} \mathbf{U}^1 \\ \mathbf{U}_I^1 \end{array}\right\} = \left[\begin{array}{cc} \mathbf{G}^1 & \mathbf{G}_I^1 \end{array}\right]\left\{\begin{array}{c} \mathbf{T}^1 \\ \mathbf{T}_I^1 \end{array}\right\} \quad (2.228)$$

For subregion V_2, we have

$$\left[\begin{array}{cc} \mathbf{H}^2 & \mathbf{H}_I^2 \end{array}\right]\left\{\begin{array}{c} \mathbf{U}^2 \\ \mathbf{U}_I^2 \end{array}\right\} = \left[\begin{array}{cc} \mathbf{G}^2 & \mathbf{G}_I^2 \end{array}\right]\left\{\begin{array}{c} \mathbf{T}^2 \\ \mathbf{T}_I^2 \end{array}\right\} \quad (2.229)$$

The compatibility and equilibrium conditions at the interfere S_I are

$$\begin{aligned}
\mathbf{U}_I^1 &= +\mathbf{U}_I^2 \equiv \mathbf{U}_I \\
\mathbf{T}_I^1 &= -\mathbf{T}_I^2 \equiv \mathbf{T}_I
\end{aligned} \tag{2.230}$$

Combining the coefficient matrices (2.228) and (2.229) for the two regions and applying the interface boundary conditions (2.230) gives

$$\begin{bmatrix}
\mathbf{H}^1 & \mathbf{H}_I^1 & 0 & 0 \\
0 & 0 & \mathbf{H}^2 & \mathbf{H}_I^2 \\
0 & 1 & 0 & -1 \\
0 & 0 & 0 & 0
\end{bmatrix}
\left\{\begin{array}{c}
\mathbf{U}^1 \\ \mathbf{U}_I^1 \\ \mathbf{U}^2 \\ \mathbf{U}_I^2
\end{array}\right\}
=
\begin{bmatrix}
\mathbf{G}^1 & \mathbf{G}_I^1 & 0 & 0 \\
0 & 0 & \mathbf{G}^2 & \mathbf{G}_I^2 \\
0 & 0 & 0 & 0 \\
0 & 1 & 0 & 1
\end{bmatrix}
\left\{\begin{array}{c}
\mathbf{T}^1 \\ \mathbf{T}_I^1 \\ \mathbf{T}^2 \\ \mathbf{T}_I^2
\end{array}\right\}$$

Rearranging the above system of equations in terms of $\mathbf{U}_I$ and $\mathbf{T}_I$ gives

$$\begin{bmatrix}
\mathbf{H}^1 & \mathbf{H}_I^1 & -\mathbf{G}_I^1 & 0 \\
0 & \mathbf{H}_I^2 & -\mathbf{G}_I^2 & \mathbf{H}^2
\end{bmatrix}
\left\{\begin{array}{c}
\mathbf{U}^1 \\ \mathbf{U}_I \\ \mathbf{T}_I \\ \mathbf{U}^2
\end{array}\right\}
=
\begin{bmatrix}
\mathbf{G}^1 & 0 \\
0 & \mathbf{G}^2
\end{bmatrix}
\left\{\begin{array}{c}
\mathbf{T}^1 \\ \mathbf{T}^2
\end{array}\right\} \tag{2.231}$$

Finally, after the substitution of the boundary conditions, the resulting system of equations can be written as

$$\begin{bmatrix}
\mathbf{A}^1 & \mathbf{H}_I^1 & -\mathbf{G}_I^1 & 0 \\
0 & -\mathbf{H}_I^2 & -\mathbf{G}_I^2 & \mathbf{A}^2
\end{bmatrix}
\left\{\begin{array}{c}
\mathbf{X}^1 \\ \mathbf{U}_I \\ \mathbf{T}_I \\ \mathbf{X}^2
\end{array}\right\}
=
\begin{bmatrix}
\mathbf{R}^1 & 0 \\
0 & \mathbf{R}^2
\end{bmatrix}
\left\{\begin{array}{c}
\mathbf{Y}^1 \\ \mathbf{Y}^2
\end{array}\right\} \tag{2.232}$$

It can be seen that the coefficient matrices are block-banded with one block for each region and overlaps between blocks in the $\mathbf{A}$ matrix at the common interface.

In construction of large, complex models it is often more economical to split the model into several smaller, simpler sub-models. These sub-models or regions, which may also have different materials properties, are modelled independently and then joined together along an interface as described above. At these interfaces the compatibility and equilibrium conditions are enforced for the interface elements. This strategy leads to an overall matrix system which has a blocked, sparse and unsymmetrical character. This character of multi-region formulation significantly extends the range of problems that can be solved, due to the large savings in storage and CPU required to solve the matrix [69]

2.8   Corner Problems

Difficulties arise in the boundary element formulation, if the prescribed boundary conditions on both sides of a corner are displacements in the same direction (either

u_1 or u_2). This difficulty is due to the non-uniqueness of the normal at the corner, as shown in Figure 2.25a. A number of procedures have been used to circumvent this difficulty. Generally, if the multi-valued tractions are assumed to be equal (i.e. $t_j^a = t_j^b$), then the errors are concentrated mainly at the corners and are not significant at other nodes. An approach which is simple to implement and is being used extensively is the use of discontinuous elements, in which the collocation points at the corners are taken inside the element, as shown in Figure 2.25b. In this approach, the geometry is still approximated using the continuous shape functions presented in Section 2.38, but the displacement and traction fields are now represented using semi-discontinuous elements. The shape function, for quadratic semi-discontinuous elements are given as

$$N_1 = \frac{1}{\lambda(1+\lambda)}\eta(-\lambda+\eta)$$

$$N_2 = 1 + \frac{1}{\lambda(1+\lambda)}\eta\left[(-1+\eta^2) - \eta(1+\lambda)\right]$$

$$N_3 = \frac{1}{\lambda(1+\lambda)}\eta(1+\eta)$$

for two-dimensional problems, and for the element quadrilateral element shown in Figure 2.26, where $\eta_2 = 1$ is taken to correspond to the discontinuous side, we have

$$N_1 = \frac{(1-\eta_1)(\lambda-\eta_2)(-\eta_1-\eta_2-1)}{2(\lambda+1)}$$

$$N_2 = \frac{(1-\eta_1^2)(\lambda-\eta_2)}{(\lambda+1)}$$

$$N_3 = \frac{(1+\eta_1)(\lambda-\eta_2)(\eta_1-\eta_2-1)}{2(\lambda+1)}$$

$$N_4 = \frac{(1-\eta_2)(\lambda-\eta_2)(1-\eta_1)}{2\lambda}$$

$$N_5 = \frac{(1+\eta_1)(1+\eta_2)(\lambda\eta_1+\eta_2-\lambda)}{2\lambda(1+\lambda)}$$

$$N_6 = \frac{(1-\eta_1^2)(1+\eta_2)}{(\lambda+1)}$$

$$N_7 = \frac{(1-\eta_1)(1+\eta_2)(-\lambda\eta_1+\eta_2-\lambda)}{2\lambda(\lambda+1)}$$

$$N_8 = \frac{(1+\eta_2)(\lambda-\eta_2)(1-\eta_1)}{2\lambda}$$

where λ is the parametric position of the collocation point $0 < \lambda \le 1$.

Another approach is based on introducing additional equations, so that every traction component is defined uniquely. At sharp corners the number of equations generated is not sufficient to cover the number of unknowns present. For example, if the tractions on all elements along an edge (in 3D) are unknown, then there will be more than three unknowns on that edge. As the BEM generates three equations per node then there are more unknowns than equations. Another example is if the displacements on all faces of a cube are prescribed, then all the tractions on the cube are unknown. As there are three faces at each corner, then there are nine traction unknowns at the corner but only three equations from BEM. This problem can also

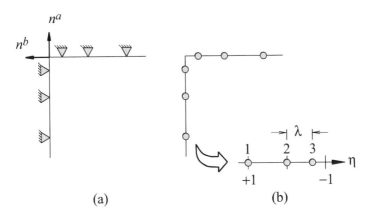

Figure 2.25: Discontinuous elements for corner nodes.

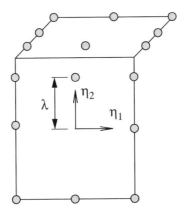

Figure 2.26: Discontinuous quadrilateral elements for corners.

occur at region interfaces- for example at the internal interface of three regions one has at least twelve unknowns (three displacements total and three tractions per node) but only nine equations (three per node). For sharp corners the philosophy is to eliminate traction unknowns until the number of unknowns equals the number of equations present [28, 69]. This elimination is carried out by finding extra equations which relate the traction unknowns on one element to tractions and/or displacements on another element.

Consider the case where the stresses on one element have been obtained using the indirect procedure presented in Section 2.5.3. Now consider another element adjoining that element, which has traction unknowns that must be eliminated at the same global position as (η_1', η_2'). Due to the continuity of the components of the stress tensor, the global stresses σ_{ij} are the same for both elements at this location, and are given by

$$\sigma_{ij} = \hat{e}_{pi}\hat{e}_{qj}\hat{\sigma}_{pq} \tag{2.233}$$

where $\hat{\sigma}_{pq}$ was derived in Section 2.5.2. Hence the tractions on the second element are determined by

$$t_i^{ei2} = \sigma_{ij} n_j^{el2} = \hat{e}_{pi} \hat{e}_{qj} \hat{\sigma}_{pq} n_j \tag{2.234}$$

The tractions on element 2 have been obtained in terms of the displacement and tractions of element 1.

2.9 Anisotropic Elasticity

Boundary element formulations for two-dimensional and three-dimensional anisotropic elasticity can be traced back to the early 1970s [19, 71, 83, 88]. However, most of the work since has been on the 2D applications (see, for example, [80]) with only a little dealing with three-dimensional problems (see, for example, [22]). The main difficulty with the 3D BEM for anisotropic material is with the complicated form of the fundamental solution and rather computationally intensive implementation. There have been other approaches developed based on the dual reciprocity methods[75], domain discretization [59], eigenfunction expansion [23] and series expansion [25] for the solution of anisotropic problems, however the direct approach still remains the most accurate and versatile method. Recently, by using the Radon transform explicit expressions for three-dimensional anisotropic Green's functions have been obtained [84]. A similar method has also been used to derive Green's functions for anisotropic piezoelectric solids [58].

2.9.1 Basic Equations

The equilibrium and compatibility equations are independent of the type of material. The stress-strain relationships, however, depend on the specific type of material behaviour. The compliance matrix for a completely anisotropic material subjected to a traixial stress system can be written as

$$
\begin{Bmatrix} \varepsilon_{xx} \\ \varepsilon_{yy} \\ \varepsilon_{zz} \\ \gamma_{yz} \\ \gamma_{zx} \\ \gamma_{xy} \end{Bmatrix} =
\begin{bmatrix}
S_{11} & S_{12} & S_{13} & S_{14} & S_{15} & S_{16} \\
S_{21} & S_{22} & S_{23} & S_{24} & S_{25} & S_{26} \\
S_{31} & S_{32} & S_{33} & S_{34} & S_{35} & S_{36} \\
S_{41} & S_{42} & S_{43} & S_{44} & S_{45} & S_{46} \\
S_{51} & S_{52} & S_{53} & S_{54} & S_{55} & S_{56} \\
S_{61} & S_{62} & S_{63} & S_{64} & S_{65} & S_{66}
\end{bmatrix}
\begin{Bmatrix} \sigma_{xx} \\ \sigma_{yy} \\ \sigma_{zz} \\ \tau_{yz} \\ \tau_{zx} \\ \tau_{xy} \end{Bmatrix} \tag{2.235}
$$

For two-dimensional problems, the above relationship can be written in a reduced form as

$$
\begin{Bmatrix} \varepsilon_{xx} \\ \varepsilon_{yy} \\ \gamma_{xy} \end{Bmatrix} =
\begin{bmatrix}
S_{11} & S_{12} & S_{16} \\
S_{21} & S_{22} & S_{26} \\
S_{61} & S_{62} & S_{66}
\end{bmatrix}
\begin{Bmatrix} \sigma_{xx} \\ \sigma_{yy} \\ \tau_{xy} \end{Bmatrix} \tag{2.236}
$$

where

$$
\begin{aligned}
S_{11} &= \frac{1}{E_1} & S_{12} &= -\frac{\nu_{12}}{E_1} = \frac{-\nu_{21}}{E_2} \\
S_{22} &= \frac{1}{E_2} & S_{16} &= \frac{\eta_{12,1}}{E_1} = \frac{\eta_{1,12}}{\mu_{12}} \\
S_{66} &= \frac{1}{\mu_{12}} & S_{26} &= \frac{\eta_{12,2}}{E_2} = \frac{\eta_{2,12}}{\mu_{12}}
\end{aligned}
$$

where E_k are the Young's moduli referring to axes x_k, μ_{12} is the shear modulus, ν_{ij} are Poisson's ratios and $\eta_{l,jk}$ are mutual coefficients of the first and second kind, respectively. For orthotropic materials, $S_{16} = S_{26} = 0$. The expression in (2.236) is valid for plane stress conditions. For plane strain $S_{ij} = S_{ij} - (S_{i3}S_{j3})/S_{33}$, $i,j = 1, 2, 6$.

2.9.2   Fundamental Solutions

In this section fundamental solutions for three- and two-dimensional problems are described. For more information on the implementation of these fundamental solutions in the BEM, readers should consult [75] and [80] for three-dimensional and two-dimensional problems, respectively.

Three-dimensional Problems

The solution of the governing equations of anisotropic elasticity for a point force in an infinite medium can be shown using Fourier transform to be [51]:

$$
U_{ij}(\hat{k}) = \frac{M_{ij}^{-1}(z)}{K^2}
$$

where $\hat{k} = \hat{K}|\,\mathbf{X}' - \mathbf{X}\,|$, $\hat{K}$ is the Fourier wave vector, $K = |\,\hat{K}\,|$, z is a unit vector in the direction of k, and the symmetric matrix $M_{ij}(z)$ has the following properties:

$$
\begin{aligned}
M_{ij}(z) &= C_{ijrz}z_r z_s \\
M_{ij}^{-1}(z)M_{jm}(z) &= \delta_{ij}
\end{aligned}
$$

Using the Fourier inversion, we have

$$
U_{ij}(\mathbf{X}', \mathbf{X}) = \frac{-1}{8\pi r} \int \int \int_{-\infty}^{\infty} \frac{M_{ij}^{-1}(z)}{k^2} \cos(\hat{k}z.\hat{T})d\hat{k} \tag{2.237}
$$

where $\hat{T}$ is a unit vector in the direction of $\mathbf{X}' - \mathbf{X}$, i.e. $\mathbf{X}' - \mathbf{X} = |\,\mathbf{X}' - \mathbf{X}\,|\hat{T}$. Consider the spherical coordinates system aligned with the direction of $\hat{T}$; the volume $d\hat{k}$ can be defined as

$$
\begin{aligned}
d\hat{k} &= k^2 \sin\alpha\, dk d\alpha d\phi \\
&= k^2 dk ds
\end{aligned}
$$

where ds denotes the surface of the unit sphere, $z.\hat{T} = \cos\alpha$ and ϕ is the polar angle in plane defined by $z.\hat{T} = 0$. Integrating with respect to k, it can be shown that (2.237) reduces to

$$
U_{ij}(\mathbf{X}', \mathbf{X}) = \int_0^{2\pi} M_{ij}^{-1}(z(\phi))d\phi \tag{2.238}
$$

where the above integral should be evaluated in the plane $\alpha = \pi/2$.

Two-dimensional Problems

The fundamental solutions for two-dimensional problems can be obtained using the complex stress approach described earlier in the chapter. The fundamental solutions for a source point

$$z'_k = x'_1 + \mu_k x'_2 \tag{2.239}$$

in a complex plane with $k = 1, 2$ and a field point defined by

$$z_k = x_1 + \mu_k x_2 \tag{2.240}$$

is given by

$$U_{ij}(z'_k, z_k) = 2\,\mathrm{Re}\left[p_{j1}A_{i1}\ln(z_1 - z'_1) + p_{j2}A_{i2}\ln(z_2 - z'_2)\right] \tag{2.241}$$

$$T_{ij}(z'_k, z_k) = 2\,\mathrm{Re}\left[\frac{1}{(z_1 - z'_1)}q_{j1}(\mu_1 n_1 - n_2)A_{i1} + \frac{1}{(z_2 - z'_2)}q_{j2}(\mu_2 n_1 - n_2)A_{i2}\right] \tag{2.242}$$

where

$$p_{ik} = \begin{bmatrix} S_{11}\mu_k^2 + S_{12} - S_{16}\mu_k \\[2mm] S_{12}\mu_k + S_{22}/\mu_k - S_{26} \end{bmatrix}$$

$$q_{jk} = \begin{bmatrix} \mu_1 & \mu_2 \\ -1 & -1 \end{bmatrix}$$

The complex coefficients A_{jk} are obtained from the requirements of unit load at z'_k and displacement continuity for the fundamental solution, by the linear system

$$\begin{bmatrix} 1 & -1 & 1 & -1 \\ \mu_1 & -\bar{\mu}_1 & \mu_2 & -\bar{\mu}_2 \\ p_{11} & -\bar{p}_{11} & p_{12} & -\bar{p}_{12} \\ p_{21} & -\bar{p}_{21} & p_{22} & -\bar{p}_{11} \end{bmatrix} \begin{Bmatrix} A_{j1} \\ \bar{A}_{j1} \\ A_{j2} \\ \bar{A}_{j2} \end{Bmatrix} = \begin{Bmatrix} \delta_{i2}/2\pi i \\ -\delta_{li}/2\pi i \\ 0 \\ 0 \end{Bmatrix}$$

where μ_k are the roots of the characteristic equation

$$S_{11}\mu^4 - 2S_{16}\mu^3 + (2S_{12} + S_{66})\mu^2 - 2S_{26}\mu + S_{22} = 0$$

The roots are always complex or pure imaginary and occur in conjugate pairs (μ_k and $\bar{\mu}_k$).

2.10   Galerkin Formulation

It was mentioned in Section 2.3.10 that the boundary element matrices generated by the point collocation method are fully populated and unsymmetric. An alternative method of solution for the integral equation is the use of the Galerkin method [4, 4, 61, 26, 27, 77, 78]. In the Galerkin approach, the displacement equation (2.66) is treated in a weighted residual sense (similar to the finite element method), using as

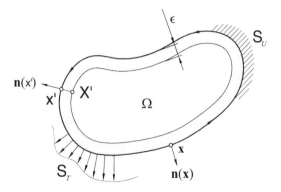

Figure 2.27: Two-dimensional model.

weights the interpolation functions employed in the discretization. The source point $\mathbf{x}'$ is no longer taken at nodal points, as the field point will become an integration variable.

Application of weighted residuals to the equilibrium equation (2.7) twice gives

$$\int_S \phi(\mathbf{x}') \int_V (\sigma_{ij,j}(\mathbf{X}) + b_i(\mathbf{X})) U_{ki}(\mathbf{X}', \mathbf{X}) dV \, dS' = 0 \qquad (2.243)$$

where the source point $\mathbf{X}'$ is defined by

$$\mathbf{X}' = \mathbf{x}' + \varepsilon \mathbf{n}(\mathbf{x}') \qquad (2.244)$$

with ε being the distance from $\mathbf{X}'$ to $\mathbf{x}'$ on the boundary S, $\mathbf{n}$ denoting the outward normal to the boundary at source point (see figure 2.27), and ϕ as the test function or weight in the weighted residual method. Integrating (2.243) by parts gives

$$\int_S \phi(\mathbf{x}') \left\{ \int_V (\sigma_{ij,j}(\mathbf{X}) U_{ki}(\mathbf{X}', \mathbf{X}))_{,j} dV - \int_V \sigma_{ij,j}(\mathbf{X}) U_{ki,j}(\mathbf{X}', \mathbf{X}) dV \right\} dS' = 0$$
$$(2.245)$$

in the absence of body forces, where $dS' = dS(\mathbf{x}')$. Using the divergence theorem and substituting the stress derivative of displacement relationship in (2.245) gives

$$\int_S \phi(\mathbf{x}') \left\{ \int_S \sigma_{ij}(\mathbf{x}) U_{ki}(\mathbf{X}', \mathbf{x}) n_j dS - \int_V E_{ijlm} u_{l,m}(\mathbf{X}) U_{ki,j}(\mathbf{X}', \mathbf{X}) dV \right\} dS' = 0$$
$$(2.246)$$

where elastic tensor E_{ijlm} is defined in (B.18). Substituting for tractions $t_i \, (= \sigma_{ij} n_j)$ into (2.246), and considering the symmetry of the elastic tensor, gives

$$\int_S \phi(\mathbf{x}') \left\{ \int_S U_{ki}(\mathbf{X}', \mathbf{x}) t_i(\mathbf{x}) dS - \int_V \Sigma_{klm}(\mathbf{X}', \mathbf{X}) u_{l,m} dV \right\} dS' = 0 \qquad (2.247)$$

Integrating the domain integral in (2.247) by parts yields

$$\int_S \phi(\mathbf{x}') \left\{ \int_S U_{ki}(\mathbf{X}', \mathbf{x}) t_i(\mathbf{x}) dS \right.$$
$$\left. - \int_V (\Sigma_{klm}(\mathbf{X}', \mathbf{X}) u_l)_{,m} dV + \int_V \Sigma_{klm,m}(\mathbf{X}', \mathbf{X}) u_l dV \right\} dS' = 0 \quad (2.248)$$

Further application of the divergence theorem, and selecting $\Sigma_{klm,m}$ to correspond to the stress gradient of the fundamental solution due to point force, reduces equation (2.248) to

$$\int_S \phi(\mathbf{x}')u_k(\mathbf{X}')dS' = \int_S \phi(\mathbf{x}') \int_S U_{ki}(\mathbf{X}',\mathbf{x})t_i(\mathbf{x})dS dS'$$
$$- \int_S \phi(\mathbf{x}') \int_S T_{ki}(\mathbf{X}',\mathbf{x})u_i(\mathbf{x})dS dS' \qquad (2.249)$$

where $T_{ki} = \Sigma_{kij}n_j$ denotes the Kelvin's traction fundamental solution. Taking the source point $\mathbf{X}'$ to the boundary gives

$$\int_S \phi(\mathbf{x}')u_k(\mathbf{x}')dS' = \int_S \phi(\mathbf{x}') \int_S U_{ki}(\mathbf{x}',\mathbf{x})t_i(\mathbf{x})dS dS'$$
$$- \lim_{\varepsilon \to 0} \int_S \phi(\mathbf{x}') \int_S T_{ki}(\mathbf{x}',\mathbf{x})u_i(\mathbf{x})dS dS' \qquad (2.250)$$

The Galerkin displacement integral equation in (2.250) leads to symmetric matrices only for pure displacement (Dirichelet) or pure traction (Neumann) boundary conditions. For mixed boundary conditions, the system of equations produced by (2.250) is unsymmetric. To overcome this difficulty [77, 78], the displacement integral equation is used for parts of the boundary with displacement boundary conditions, and the traction integral equation is used for other parts, that is parts where tractions are prescribed.

The boundary traction equation can be written as

$$\int_S \phi(\mathbf{x}')t_k(\mathbf{x}')dS' = \int_S \phi(\mathbf{x}') \int_S D_{ki}(\mathbf{x}',\mathbf{x})t_i(\mathbf{x})dS dS'$$
$$- \lim_{\varepsilon \to 0} \int_S \phi(\mathbf{x}') \int_S S_{ki}(\mathbf{x}',\mathbf{x})u_i(\mathbf{x})dS dS' \qquad (2.251)$$

where $D_{ki}(\mathbf{x}',\mathbf{x}) = D_{kij}(\mathbf{x}',\mathbf{x})n_i(\mathbf{x}')$ and $S_{ki}(\mathbf{x}',\mathbf{x}) = S_{kij}(\mathbf{x}',\mathbf{x})n_i(\mathbf{x}')$. It can be shown (see, for example, [65]) that $U_{ik}(\mathbf{x}',\mathbf{x}) = U_{ki}(\mathbf{x}',\mathbf{x})$, $D_{ki}(\mathbf{x}',\mathbf{x}) = T_{ki}(\mathbf{x}',\mathbf{x})$ and $S_{ik}(\mathbf{x}',\mathbf{x}) = S_{ki}(\mathbf{x}',\mathbf{x})$.

The test function $\phi(\mathbf{x})$ is taken to correspond to interpolation functions N_γ, described in Section 2.3.8, and approximating the displacements and tractions with the same interpolation function, i.e.

$$u_i(\mathbf{x}) = N_\gamma(\mathbf{x})u_i^\gamma \qquad (2.252)$$
$$t_i(\mathbf{x}) = N_\gamma(\mathbf{x})t_i^\gamma \qquad \gamma = 1,..M \qquad (2.253)$$

where M denotes the total number of nodes. Notice that the matrix N_γ includes the shape functions N_α for nodes belonging to the element under consideration, and has null entry for other nodes. Equations (2.250) and (2.251) can now be written as [62, 78]:

$$\int_S N_\gamma(\mathbf{x}')N_\gamma(\mathbf{x}')u_k^\gamma dS' = \int_S N_\gamma(\mathbf{x}') \int_S U_{ki}(\mathbf{x}',\mathbf{x})N_\gamma(\mathbf{x})t_i^\gamma dS dS' \qquad (2.254)$$
$$- \lim_{\varepsilon \to 0} \int_S N_\gamma(\mathbf{x}') \int_S T_{ki}(\mathbf{x}',\mathbf{x})N_\gamma(\mathbf{x})u_i^\gamma dS dS'$$

and

$$\int_S N_\gamma(\mathbf{x}')N_\gamma(\mathbf{x}')u_k^\gamma dS' = \int_S N_\gamma(\mathbf{x}') \int_S D_{ki}(\mathbf{x}',\mathbf{x})N_\gamma(\mathbf{x})t_i^\gamma dSdS' \qquad (2.255)$$

$$- \lim_{\varepsilon \to 0} \int_S N_\gamma(\mathbf{x}') \int_S S_{ki}(\mathbf{x}',\mathbf{x})N_\gamma(\mathbf{x})u_i^\gamma dSdS'$$

Rearranging the matrices according to the prescribed and unknown terms, we have

$$\lim_{\varepsilon \to 0} \begin{pmatrix} \mathbf{U}_{uu} & -\mathbf{T}_{ut} \\ -\mathbf{D}_{tu} & \mathbf{S}_{tt} \end{pmatrix} \begin{Bmatrix} \mathbf{t}_u \\ \mathbf{u}_t \end{Bmatrix} = \lim_{\varepsilon \to 0} \begin{pmatrix} -\mathbf{U}_{uu} & \mathbf{F}_u + \mathbf{T}_{ut} \\ -\mathbf{F}_t + \mathbf{D}_{tt} & -\mathbf{S}_{tu} \end{pmatrix} \begin{Bmatrix} \mathbf{t}_t \\ \mathbf{u}_u \end{Bmatrix}$$
$$(2.256)$$

The notation used for the sub-matrices coincides with the notation of the kernel that is being integrated (except $\mathbf{F}$, which denotes the integral of interpolation functions only), and the first and second sub-indices denote that part of the boundary on which the source and field points are located, respectively. The boundary values, of displacements and tractions are denoted in a similar way.

2.10.1   Regularization

The integrals can be evaluated by utilizing a Taylor series expansion about a singular point $\mathbf{x}'$ on the boundary [4, 62] to give

$$\int_S \phi(\mathbf{x}')u_k(\mathbf{X}')dS' = \int_S \phi(\mathbf{x}') \int_S U_{ki}(\mathbf{X}',\mathbf{x})t_i(\mathbf{x})dSdS'$$

$$- \int_S \phi(\mathbf{x}') \int_S T_{ki}(\mathbf{X}',\mathbf{x})[u_i(\mathbf{x})-u_i(\mathbf{x}')]dSdS'$$

$$+ \int_S \phi(\mathbf{x}')u_i(\mathbf{x}') \int_S T_{ki}(\mathbf{X}',\mathbf{x})dSdS' \qquad (2.257)$$

Using the property

$$\int_S T_{ki}(\mathbf{X}',\mathbf{x})dS = -\delta_{ki}$$

the last integral in (2.257) gives

$$\int_S \phi(\mathbf{x}') \int_S [u_i(\mathbf{X}')-u_i(\mathbf{x}')]dS' = \int_S \phi(\mathbf{x}') \int_S U_{ki}(\mathbf{X}',\mathbf{x})t_i(\mathbf{x})dSdS' \qquad (2.258)$$

$$- \int_S \phi(\mathbf{x}') \int_S T_{ki}(\mathbf{X}',\mathbf{x})[u_i(\mathbf{x})-u_i(\mathbf{x}')]dSdS'$$

Equation (2.257) now contains only weak singularities. Taking the source point to the boundary, that is $\mathbf{X}' \to \mathbf{x}'$ (i.e. $\varepsilon \to 0$), leads to

$$0 = \int_S \phi(\mathbf{x}') \int_S U_{ki}(\mathbf{x}',\mathbf{x})t_i(\mathbf{x})dSdS'$$

$$- \int_S \phi(\mathbf{x}') \int_S T_{ki}(\mathbf{x}',\mathbf{x})[u_i(\mathbf{x})-u_i(\mathbf{x}')]dSdS' \qquad (2.259)$$

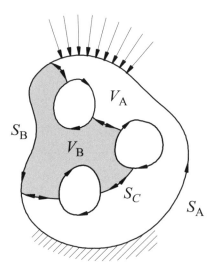

Figure 2.28: Multi-region modelling to overcome non-uniquness of symmetric Galerkin formulation.

For the singularities in the traction integral equation, it is necessary to consider the first two terms of the Taylor expansion of the displacement around the source point $\mathbf{x}'$, and one term of the traction for the kernel D_{ki}, so that

$$
\int_S \phi(\mathbf{x}')t_k(\mathbf{x}')dS' = \int_S \phi(\mathbf{x}') \int_S D_{ki}(\mathbf{x}',\mathbf{x})[t_i(\mathbf{x})-t_i(\mathbf{x}')]dSdS'
$$
$$
+ \int_S \phi(\mathbf{x}')t_i(\mathbf{x}') \int_S D_{ki}(\mathbf{x}',\mathbf{x})dSdS'
$$
$$
+ \int_S \phi(\mathbf{x}') \int_S S_{ki}(\mathbf{x}',\mathbf{x})[u_i(\mathbf{x})-u_i(\mathbf{x}') - u_{i,j}(\mathbf{x}')r_j]dSdS'
$$
$$
- \int_S \phi(\mathbf{x}')u_i(\mathbf{x}') \int_S S_{ki}(\mathbf{x}',\mathbf{x})dSdS'
$$
$$
\int_S \phi(\mathbf{x}')u_{i,j}(\mathbf{x}') \int_S S_{ki}(\mathbf{x}',\mathbf{x})r_jdSdS' \qquad (2.260)
$$

2.10.2   The Uniqueness Problem

Pérez-Gavilán and Aliabadi [63, 64] showed that the straightforward application of the symmetric Galerkin boundary elements when applied to multiple connected bodies leads to a non-unique solution. To demonstrate the non-uniqueness, they added a rigid body motion, rotation and translation, to the displacements of an unrestrained boundary, and showed that the system of equations remained unaffected, hence the solution to the system is not unique. The problem of non-uniqueness only arises when a closed boundary is unconstrained or only partially constrained.

To remedy the above difficulty, Pérez-Gavilán and Aliabadi [63, 64] proposed two different methods. The first method consisted of including sufficient displacement equations over the unrestrained boundary to constrain the rigid body motion. This approach will violate the symmetry of GSBEM.

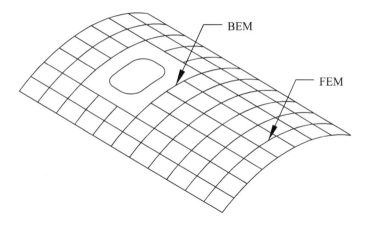

Figure 2.29: Coupling finite elements with boundary elements.

The second method is based on the multi-region formulation, as shown in Figure 2.28. Here, the idea is simply to avoid multiple connectivity altogether, by creating regions each of which are simply connected.

2.11 Coupling Finite Elements and Boundary Elements

For certain types of problem, it may be advantageous to have part of the problem modelled with the boundary elements and the reminder with the finite elements. It is generally accepted that the BEM is more efficient at modelling parts of the structure with special features such as stress concentrations, cracks, etc., and the FEM is more efficient at modelling the remaining parts, which may involve many degrees of freedom.

There are two main approaches for coupling BEM and FEM. The first approach, referred to here as the *stiffness matrix approach,* involves converting the BEM into an equivalent FEM stiffness matrix, and coupling it with the FEM part as if it was a super-element. The second approach is to join the two regions using a method similar to the multi-region formulation presented in Section 2.8.

2.11.1 Stiffness Matrix Approach

Consider a problem divided into two regions, as shown in Figure 2.29. Region A is finite elements and region B is boundary elements. For the finite element region A, the system of equations is given as

$$\mathbf{K}^A\mathbf{U}^A = \mathbf{F}^A \tag{2.261}$$

where $\mathbf{U}^A$ and $\mathbf{F}^A$ are vectors of nodal displacements and forces, respectively, and $\mathbf{K}^A$ is the stiffness matrix for region A. For the BEM region, we have

$$\mathbf{H}^B\mathbf{U}^B = \mathbf{G}^B\mathbf{T}^B \tag{2.262}$$

To couple the two regions, the BEM nodal tractions must be converted into equivalent nodal forces. The tractions over an element, say n, can be related to a nodal force F^β on that element, via

$$
\begin{aligned}
F_i^\beta &= \int_{S_n} N_\beta t_i dS_n \\
&= \int_{S_n} N_\beta \left(\sum_{\alpha=1} N_\alpha t_i^\alpha \right) dS_n
\end{aligned}
\tag{2.263}
$$

where the integral is carried out over the length of the element. The above equation can be written in a matrix form as

$$
\mathbf{F}^n = \mathbf{M}^n \mathbf{T}^n
\tag{2.264}
$$

The global system of equations for the BEM, using (2.264), can be written as

$$
\mathbf{M} \left(\mathbf{G}^B \right)^{-1} \mathbf{H}^B \mathbf{U}^B = \mathbf{F}^B
$$

or

$$
\mathbf{K}^B \mathbf{U}^B = \mathbf{F}^B
\tag{2.265}
$$

where $\mathbf{K}^B = \mathbf{M} \left(\mathbf{G}^B \right)^{-1} \mathbf{H}^B$. The BEM can now be treated as a super-element and incorporated into the global FEM system matrix. It is worth noting that if the collocation method is used, the stiffness matrix $\mathbf{K}^B$ will not be symmetric. To make the system matrix symmetric it has been suggested that it is multiplied by it's transpose [12], i.e. $\mathbf{K}^T \mathbf{K}$, which is equivalent to the least squares method of solution. The proposed matrix multiplication is proportional to cube of the degrees of freedom, and as such can be computationally expensive.

2.11.2 Multi-region Approach

In this approach the two regions are coupled using the multi-region approach presented in Section 2.8. The FEM nodal forces along the interface between the two regions I must now be converted to tractions, so that the system matrix can be written as

$$
\begin{bmatrix} \mathbf{K}^A & \mathbf{K}_I^A \end{bmatrix} \left\{ \begin{array}{c} \mathbf{U}^A \\ \mathbf{U}_I^A \end{array} \right\} = \left\{ \begin{array}{c} \mathbf{MT}^I \\ \mathbf{F}^A \end{array} \right\}
\tag{2.266}
$$

where superscript I denotes the interface. The two systems are coupled by applying the conditions of displacement compatibility and traction equilibrium (see (2.230)), to give

$$
\begin{bmatrix} \mathbf{K}^A & \mathbf{K}_I^A & -\mathbf{M} & 0 \\ 0 & \mathbf{H}_I^B & -\mathbf{G}_I^B & \mathbf{H}^2 \end{bmatrix} \left\{ \begin{array}{c} \mathbf{U}^1 \\ \mathbf{U}_I \\ \mathbf{T}_I \\ \mathbf{U}^2 \end{array} \right\} = \left\{ \begin{array}{c} \mathbf{F}^A \\ \mathbf{G}^B \mathbf{T}^B \end{array} \right\}
\tag{2.267}
$$

Applications of the above coupling techniques have been used to solve several interesting problems of mining excavations [8].

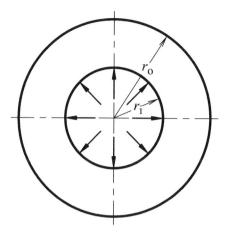

Figure 2.30: A pressurized cylinder.

Table 2.1: Radial displacements for the cylinder subjected to internal pressure.

$r(mm)$	BEM(24)	BEM(32)	BEM(64)	Analytical
50	0.01313	0.01314	0.01314	0.01314
60	0.01135	0.01135	0.01135	0.01135
70	0.01012	0.01013	0.01013	0.01013
80	0.00926	0.00927	0.00927	0.00927
90	0.00865	0.00865	0.00865	0.00865
100	0.00819	0.00820	0.00820	0.00820

2.12 Examples

2.12.1 Pressurized Cylinder

Consider the hollow cylinder shown in Figure 2.30. The inner radius is denoted as r_i and outer radius r_o. The cylinder is subjected to an internal pressure p. The analytical solutions for the problems are given as

$$\sigma_r = \frac{pr_i^2}{r_o^2 - r_i^2}\left(1 - \frac{r_o^2}{r^2}\right)$$

$$\sigma_\theta = \frac{pr_i^2}{r_o^2 - r_i^2}\left(1 + \frac{r_o^2}{r^2}\right)$$

for stresses, and

$$u_r = \frac{(1+\nu)pr}{E\left[\left(\frac{r_o}{r_i}\right)^2 - 1\right]}\left[\left(\frac{r_o}{r}\right)^2 + (1 - 2\nu)\right]$$

A BEM model was set up for $r_i = 50$ mm, $r_o = 100$ mm, $\nu = 0.32$ and $E = 730\,000$ MPa. Three BEM meshes were used with 24, 32 and 64 elements respectively.

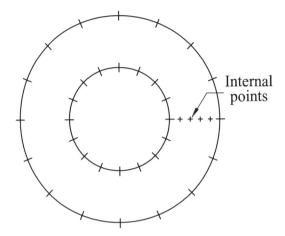

Figure 2.31: Boundary element mesh with 32 quadratic elements.

Table 2.2: σ_r/p for the cylinder subjected to internal pressure.

$r(mm)$	BEM(24)	BEM(32)	BEM(64)	Analytical
50	-1	-1	-1	-1
60	-0.5919	-0.5924	-0.5926	-0.5926
70	-0.3466	-0.3469	-0.3469	-0.3469
80	-0.1874	-0.1875	-0.1875	0.1875
90	-0.0782	-0.0782	-0.0782	-0.0782
100	0	0	0	0

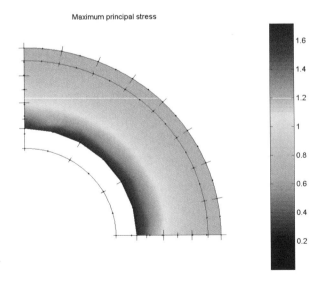

Figure 2.32: Maximum principal stress for the pressurized cylinder.

Table 2.3: σ_θ/r for the cylinder subjected to internal pressure.

$r(mm)$	BEM(24)	BEM(32)	BEM(64)	Analytical
50	1.6667	1.6667	1.6667	1.6667
60	1.2593	1.2593	1.2593	1.2593
70	1.0135	1.0136	1.0136	1.0136
80	0.8437	0.8542	0.8542	0.8542
90	0.7448	0.7445	0.7449	0.7449
100	0.6664	0.6667	0.6667	0.6667

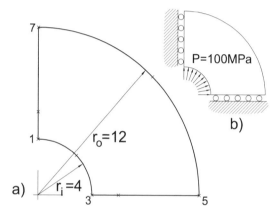

Figure 2.33: Pressurized cylinder (quarter model).

The boundary mesh for 24 quadratic elements is shown in Figure 2.31. The values of radial displacements and normalized stresses obtained are presented in Tables 2.1 to 2.3, together with the exact solutions. The maximum principal stresses are shown in Figure 2.32

In order to compare the collocation and Galerkin methods, the problem of an infinite cylinder with an internal pressure, as reported in [61], is presented. The analytical solutions for the cylinder have been presented in Section 2.11.1. To assess the accuracy of the boundary displacement solutions, the following norm was used:

$$R_N = \int_S |\, u_c - u_a \,|\, dS$$

where u_c and u_a denote the calculated and analytical solutions, respectively. Only a quarter of the problem was modelled, as shown in Figure 2.33. The accuracy of the stresses were checked against the total equilibrium of the system. Figures 2.34 and 2.35 present the displacement norm and the error in the equilibrium, respectively. As can be seen from the figures, both methods converge with a relatively small number of elements, but with the collocation method giving more accurate results for up to 16 elements. For more than 16 elements, errors were very small and differences were attributed to numerical rounding errors [61].

The time spent during integration and by the solution of a system of equations

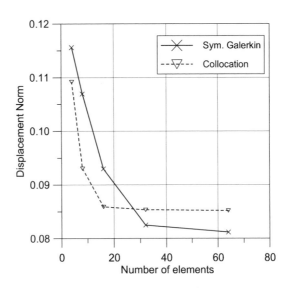

Figure 2.34: Displacement norms for quarter cylinder model.

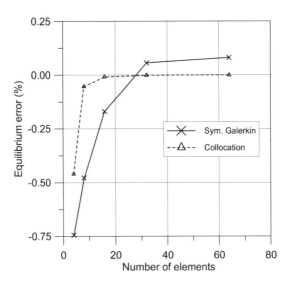

Figure 2.35: Equilibrium check for quarter cylinder model.

Table 2.4: Cylinder with internal pressure, execution times for collocation and Galerkin methods.

No Elem.	Collocation			Galerkin		
	Sys.	Solver	Total	Sys.	Solver	Total
24	0.16	0.11	0.27	3.74	0.05	3.79
48	0.55	0.77	1.32	7.85	0.39	8.24
96	1.98	5.99	7.97	22.41	3.35	25.76
192	6.70	51.57	58.27	72.22	28.45	100.7
384	24.17	320.4	344.6	265	244.8	509.84
768	90.94	2720	2812	1001	1993	2993.74

Table 2.5: Stress concentration factors ($K_t = \sigma_{max}/\sigma_o$) for a plate with a circular hole subjected to tensile stress σ_o at its ends.

R/W	$K_t(BEM)$	$K_t(FEM)$	$K_t[24]$
0.01	3.0	3.00	3.0
0.1	3.03	2.99	3.02
0.2	3.14	3.16	3.12
0.3	3.36	3.40	3.28

was also reported, and is presented in Table 2.4. As can be seen from the Table 2.4, for a small number of elements, the Galerkin execution time spent in the integration phase (setting up the system matrix) can be as much as 23 times greater than for the collocation method. On the other hand, for a very large number of elements, the time spent in the solution of the system of equations can be almost halved when exploiting the symmetry in the case of the Galerkin method.

The Galerkin boundary element method is still at its initial development stages as far as the BEM is concerned, and as such has been applied to a limited number of problems.

2.12.2 Stress Concentration

Consider a rectangular plate of width $2W$ and height $6W$ with a central circular hole of radius R subjected to uniform stress σ at its ends, as shown in Figure 2.36. This problem was solved using a BEM mesh of 48 quadratic elements. In Table 2.5 the BEM results for the stress concentration factors are presented for different ratios of R/W. Also presented in the table are the experimental results reported in [24] and finite element results obtained using 3626 nodes.

2.12.3 Plate with an Elliptical Hole

Consider a rectangular plate with an elliptical hole subjected to tensile stress σ at its ends, as shown in Figure 2.37. Analytical solutions for the case of an infinite plate

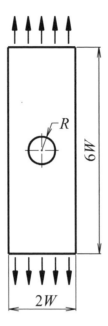

Figure 2.36: A rectangular plate with a central circular hole subjected to tension at its ends.

are given as

$$
u_1 = \frac{\sigma(a+b)}{16\mu} \left[-\frac{2+m-m^2}{1+m^2-2m\cos 2\alpha} \cos 3\alpha \right.
$$
$$
\left. + \left(-\kappa(m+1) + \frac{-(m^3+2m+3)+4(1+m)\cos 2\alpha}{1+m^2-2m\cos 2\alpha} \right) \cos \alpha \right]
$$

$$
u_2 = \frac{\sigma(a+b)}{16\mu} \left[-\frac{2+m-m^2}{1+m^2-2m\cos 2\alpha} \sin 3\alpha \right.
$$
$$
\left. + \left(\kappa(m+3) + \frac{m^3+4m^2+2m+5+4(1-m)\cos 2\alpha}{1+m^2-2m\cos 2\alpha} \right) \sin \alpha \right]
$$

$$
\sigma_\theta = \sigma \frac{1-m^2-2m+2\cos 2\alpha}{1+m^2-2m\cos 2\alpha}
$$

where $m = (a-b)/(a+b)$. For the BEM analysis the plate is modelled as a 800 mm×800 mm square with $a = 8$ mm and $b = 2$ mm. The BEM results for displacements and stresses at points A and B obtained using 80 and 120 quadratic elements [24] are shown in Tables 2.6 and 2.7, respectively, together with analytical results. The material constants are chosen as $\nu = 0.3$, $E = 3 \times 10^6$ MPa and the load as $\sigma = 100$ MPa.

2.12.4   A Strip Under Self-Weight and Centrifugal Loads

Consider a rectangular strip 2m wide and 4m long. The strip is under (a) its self-weight and (b) centrifugal load. Pérez-Gavilán and Aliabadi [60] studied this problem

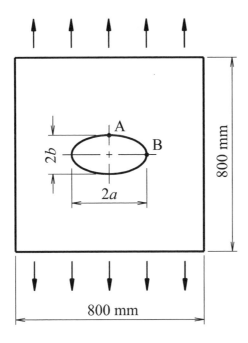

Figure 2.37: A square plate with an elliptical hole subjected to tensile stress.

Table 2.6: Displacements (mm) for points A and B of the elliptical hole.

Location	BEM(80)	BEM(120)	Analytical
A	0.54×10^{-4}	0.55×10^{-4}	0.55×10^{-4}
B	-0.236×10^{-4}	-0.236×10^{-4}	-0.236×10^{-4}

Table 2.7: Stresses (MPa) for points A and B of the elliptical hole.

Location	BEM(80)	BEM(120)	Analytical
A	102.5	101.5	100
B	865.3	899.9	900

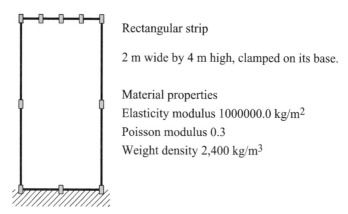

Rectangular strip

2 m wide by 4 m high, clamped on its base.

Material properties
Elasticity modulus 1000000.0 kg/m^2
Poisson modulus 0.3
Weight density 2,400 kg/m^3

Figure 2.38: A rectangular strip.

using the dual reciprocity technique with both collocation and symmetric Galerkin methods. The original mesh is shown in Figure 2.38. Additional analyses were carried out by subdividing the elements by half each time, and using 5, 9 and 12 internal points.

For the case of self-weight, the body force is given as $b_2 = \rho g$. The stresses for these problems are given by

$$\sigma_{11} = \sigma_{12} = 0 \text{ and } \sigma_{22} = \rho g x_2$$

The errors of the integrated tractions at the clamped edge relative to the total weights are presented in Figure 2.39. The curve denoted as *Body force at cntr pt* represents the value of the body force at the centre point as approximated by expression (2.212) with no internal points.

For the centrifugal load case, integrating the total vertical load at the clamped end gives $b_2 = 16\rho\omega^2$. The results for the centrifugal load are shown in Figure 2.40.

2.12.5   Orthotropic Circular Ring

Consider an orthotropic ring of inner radius r_i and outer radius r_o subjected to shear tractions along the inner body, as shown in Figure 2.41. Rizzo and Shippy [71] studied this problem for different rations of r_i/r_o. The compliance coefficient were selected to correspond to plywood, i.e.

$$
\begin{aligned}
S_{11} &= 8.33 \times 10^{-6} \text{ in}^2/\text{lb} \\
S_{22} &= 16.67 \times 10^{-6} \text{ in}^2/\text{lb} \\
S_{12} &= -0.6 \times 10^{-6} \text{ in}^2/\text{lb} \\
S_{66} &= 143 \times 10^{-6} \text{ in}^2/\text{lb}
\end{aligned}
$$

The hoop stress for θ (angle measured from the $x - axis$) obtained for $r_i/r_o = 0.0003$ are presented in Table 2.8, together with analytical solutions [37].

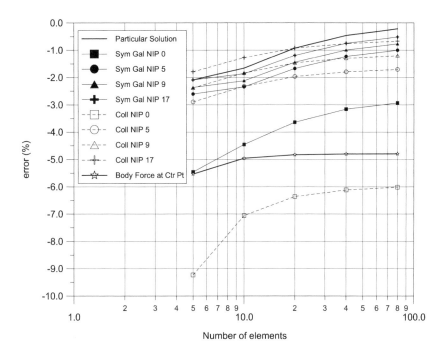

Figure 2.39: Percentage error for rectangular strip under self-weight.

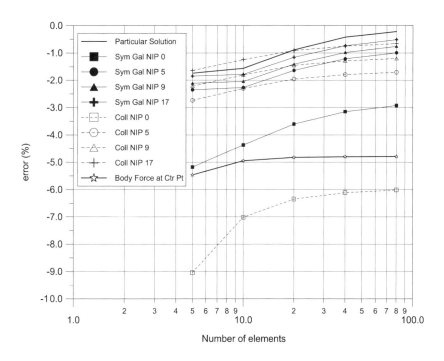

Figure 2.40: Percentage error for a rectagular strip under centrifugal load.

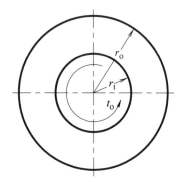

Figure 2.41: An orthoropic circular ring subjected to shear tractions t_o.

Table 2.8: Hoop stress distribution around the boundary of the hole.

θ(deg)	BEM [71]	Analytical [37]
0	0	0
15	-0.85	-0.84
30	-0.77	-0.78
45	-0.08	-0.08
60	0.69	0.69
90	0	0

2.12.6 Hole in a Strip

Consider a plate with a central hole subjected to uniform stress at the ends of the plate. Due to symmetry, only, a quarter of the problem can be modelled, as shown in Figure 2.42. The problem was solved by Wilde and Aliabadi [87] using the displacement integral equation and stress integral equation. The lateral surfaces were fixed to simulate plane strain conditions. The dimensions of the problems were set as $r = 2.5w$, $a = r$, $b = 9w$ and $T = 10$ units.

Boundary stress results were obtained using the Shape Function Derivative (SFD) approach described in Section 2.5.2 and the Singular Boundary Integral SBIE approach described in Section 2.5.1, Two localizing functions, M' (shape function associated with source point) and e^{-8r^2} ($r =| \eta - \eta' |$), were used. Figures 2.43 to 2.46 show the results for an off-centre cross-section ($z = 3w/4$). There is generally a good agreement between SFD and SBIE. However, as can be seen in Figure 2.46, the σ_{xx} calculated using the SFD technique deviate from the analytical solution near the hole.

2.12.7 Analysis of Human Tibia

In this example a three-dimensional boundary element method is used to analyze the proximal tibia of a human knee [50]. The method is used to evaluate stresses and displacements in the tibial plateau under static loadings. The geometry was

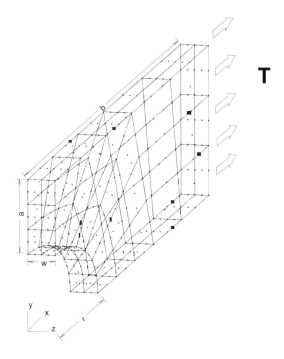

Figure 2.42: Quarter model of the hole in a strip problem.

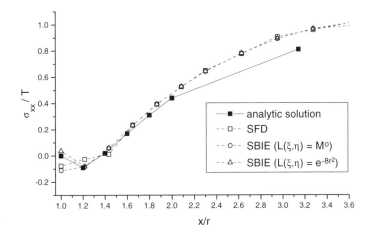

Figure 2.43: σ_{xx} along $z = 3w/4$, $y = 0$.

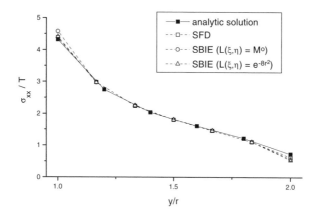

Figure 2.44: σ_{xx} along $z = 3w/4$, $x = 0$.

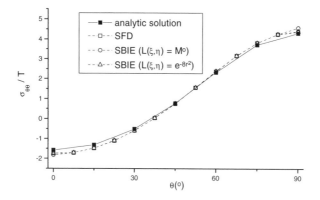

Figure 2.45: Angular stress along inner circumference ($z = 3w/4$).

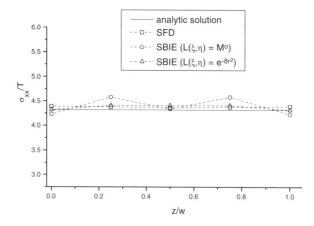

Figure 2.46: σ_{xx} along $x = 0$, $y = r$.

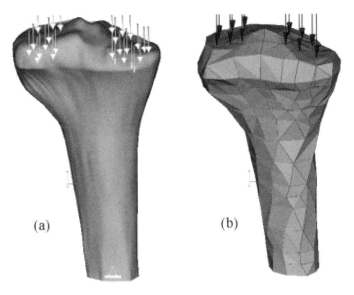

Figure 2.47: (a) Rendered model. (b) BEM discretization (1146 nodes).

generated via three-dimensional reconstruction of Computerized Tomographies (CT) and Magnetic Resonance Imaging (MRI). The rendered model and loading conditions of the tibia are shown in Figure 2.47. The BEM model for the tibia results in 413, 8-noded quadrilateral elements (see Figure 2.47). The BEM analysis yields a maximum Von Misses stress of 25.54 MPa in the distal part of the tibia and 11 MPa in the upper tibial, as shown in Figure 2.48. The maximum displacement occurs in the lateral upper part of the tibial with a maximum value of 0.27 mm, as shown in Figure 2.48. These results compare well with the FEM analysis using 1713 tetrahedral elements, giving a maximum value of Von Misses stress of 24.57 MPa and a maximum displacement of 0.246 mm.

2.13 Summary

In this chapter, boundary element formulations for linear elastostatic problems have been presented. Particular attention was given to the numerical implementation of the BEM using higher order elements. Generally, quadratic isoparametric elements are used in the application of the method to stress analysis due to their ability to accurately model curve boundaries and high variations of stresses. The successful implementation of the BEM relies heavily on accurate methods for the numerical integration of integrals. In Chapter 11, different methods for the evaluation of regular and singular integrals encountered in the BEM will be described in detail.

Many aspects of the formulations developed in this chapter will be utilized in later chapters. For example, the hypersingular stress integral equation is an essential part of the dual boundary elements formulation that will be presented in Chapter 8 for modelling crack problems.

Generally, BEM solutions are not highly sensitive to mesh distributions. Neverthe-

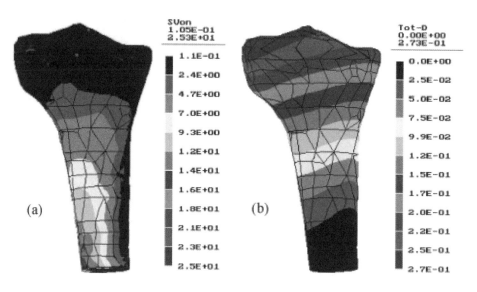

Figure 2.48: (a) Von Misses stresses (MPa). (b) Displacements (mm) calculated with BEM.

less, for complicated problems significant errors can occur if inadequate discretization is used. Several adaptive strategies have been developed to deal with the problem (see, for example, [15, 30, 35, 41, 49, 68]). Adaptive techniques, broadly speaking, require an error estimate and adaptive meshing strategy. Although much progress has been made, adaptive methods are still prohibitively computationally intensive for applications to three-dimensional problems. An alternative strategy is based on intelligent mesh design for the initial model. Conner [17], Portela et al. [67] and Salgado et al. [73] have developed knowledge-based approaches to the design of BEM meshes. The main idea is to use certain predefined rules to ensure that an optimal mesh is designed. An optimal mesh design is one that minimizes the total number of elements and the errors in the numerical solutions. Intelligent mesh design can serve as an ideal initial mesh for any adaptive strategy, and reduce the number of subsequent mesh refinements.

References

[1] Aliabadi, M.H. and Hall, W.S., Non-isoparametric formulations for three-dimensional boundary element method, *Engineering Analysis*, **5**, 198-204, 1988

[2] Aliabadi, M.H. and Rooke, D.P., *Numerical Fracture Mechanics*, Kluwer Academic Publishers, Dordrecht and Computational Mechanics, Southampton 1991.

[3] Bakr, A.A. and Fenner, R.T., Boundary integral equation analysis of axisymmetric thermoelastic problems, *Journal of Strain Analysis*, **18**, 29-251, 198.

[4] Balakrishna, C., Gray, L.J. and Kane, J.H., Efficient analytical integration of symmetric Galerkin boundary integrals over curved elements; elasticity formu-

lations, *Computer Methods Applied Mechanics and Engineering,* **117**, 157-179, 1994.

[5] Balas. J, Sladek, J. and V.Sladek., *Stress Analysis by Boundary Element Method,* Elsevier, Amsterdam, 1989.

[6] Banerjee, P.K., *The Boundary Element Methods in Engineering,* Mc-Graw Hill, Berkshire, 1994.

[7] Becker, A.A and Fenner, R.T., Use of the Hankel transform in boundary integral equation methods for axisymmetric problems, *International Journal for Numerical Methods in Engineering,* **19**, 1765-1769, 1983.

[8] Beer, G., and Watson, J.O., *Introduction to Finite and Boundary Element Methods for Engineers,* J.Wiley, Chichester 1992.

[9] Bettie, E., Teoria dell elasticita, *11 nuovo Ciemento,* 7-10, 1872.

[10] Bialecki, R A., Merkel, M., Mews, H. and Kuhn, G., In- and out-of core BEM equation solver with parallel and non-linear options, *International Journal for Numerical Methods in Engineering,* **39**, 4215-4242, 1996.

[11] Brebbia, C.A., Telles, J.C.F. and Wrobel, L.C., *Boundary Element Techniques,* Springer Verlag, Berlin 1984.

[12] Brebbia,C.A. and Georgiou,P., Combination of boundary and finite element methods, *Applied Matematical Modelling,* **3**, 212-220, 1979.

[13] Bonnet, M., Maier, G and Polizzotto, G., Symmetric Galerkin boundary element methods, *Applied Mechanics Review,* **51**, 669-704, 1998.

[14] Bridges, T and Wrobel, L.C., A dual reciprocity formulation for elasticity problems with body forces using augmented thin plate splines, *Communications in Numerical Methods in Engineering,* **12**, 209-220, 1996.

[15] Charafi, A., Neves, A.C. and Wrobel, L.C., *h-Hierarchical* adaptive boundary element method using local reanalysis. *International Journal for Numerical Methods in Engineering,* **38**, 2185-2207, 1995.

[16] Chen, G. and Zhou, J., *Boundary Element Methods,* Academic Press, London 1992.

[17] Conner, J., A knowledge based approach for boundary element mesh design, *Supercomputing in Engineering Structures,* Computational Mechanics Publications, Southampton and Springer-Verlag, Berlin, 1988.

[18] Cruse, T.A., Numerical solutions in three-dimensional elastostatics, *International Journal of Solids and Structures,* **5**, 1259-1274, 1969.

[19] Cruse, T.A., and Swedlow, J.L. Interactive program for analysis and design problems in advanced composite technology, Carnegie-Mellon University, Report AFML-TR-71-268, 1971.

[20] Cruse, T.A., Snow, D.W. and Wilson, R.B., Numerical solutions in axisymmetric elasticity, *Computers and Structures,* **7**, 445-451, 1977.

[21] Cruse, T.A. Mathematical formulations of the boundary integral equation method in solid mechanics, AFOSR-TR-77-1002, 1977.

[22] Danson, D.J., A boundary element formulation of problems in linear isotropic elasticity with body forces. In *Boundary Element Methods*,105-122, Springer Verlag, Berlin 1981.

[23] Debs, A., Henry, P.K. and Wilson, R.B., Alternative BEM formulation for 2- and 3-D anisotropic thermoelasticity, *International Journal of Solids and Structures*, **27**, 1721-1738, 1991.

[24] Flynn, P.D., Photoelastic comparison of stress concentrations due to semi-circular hole in tension bar, *Journal of Applied Mechanics, Trans. ASME*, 892-893, 1969.

[25] Gray, L.J., Ghosh, D. and Kaplan, T., Evaluation of the anisotropic Green's function in three-dimensional elasticity, *Computational Mechanics*, **17**, 255-261, 1996.

[26] Gray, L.J. and Griffith, B.E., A faster Galerkin boundary integral algorithm, *Communications in Numerical Methods in Engineering*, **14**, 1109-1117, 1998.

[27] Holzer, S.M., How to deal with hipersingular integrals in the symmetric BEM, *Communications in Numerical Methods for Engineering*, **9**, 219-232, 1993.

[28] Kane, J.H, Kashava Kumar, B.L. and Saigal, S., An arbitrary condensing, non-condensing solution strategy for large scale, multi-zone boundary element analysis, *Computer Methods in Applied Mechanics and Engineering*, **79**, 219-244, 1990.

[29] Kermanidis,T., A numerical solution for axially symmetrical elasticity problems, *International Journal of Solids & Structures*, **11**, 493-500, 1975.

[30] Kita, E. and Kamiya, N., Recent studies on adaptive boundary element methods, *Advances in Engineering Software*, **19**, 21-32, 1994.

[31] Kupradze, V.D., *Potential Methods in the Theory of Elasticity*, Israel Prog. Sci. Trans. 1965.

[32] Jeng, G. and Wexler, A., Isoparametric, finite element, variational solution of integral equations for three-dimensional fields, *International Journal for Numerical Methods in Engineering*, **11**, 1455-1471, 1977.

[33] Lachat, J.C., A further development of the boundary integral technique for elastostatics, PhD thesis, University of Southampton 1975.

[34] Lachat, J.C. and Watson, J.O., Effective numerical treatment of boundary integral equations: A formulation for three-dimensional elastostatics, *International Journal for.Numerical Methods in Engineering*, *10* 991-1005, 1976.

[35] Liapis, S., An adative boundary element method for the solution of potential flow problems, *Engineering Analysis*, **18**, 29-37, 1996.

[36] Le, M.F., Tonh, Y.F. and Wie, J.F., On a BEM solution of rotationally symmetrical body under nonsymmetrical surface load, *Proc. Int. Conf on BEM*, Beijing, 14-17, Pergamon Press, Oxford, 1986.

[37] Lekhnitskii, S.G., *Theory of Elasticity of an Anisotropic Elastic Body*, Holden-Day, San Francisco 1963.

[38] Mansur, W.J., Araujo, F.C. and Malaghini, J.E.B., Solution of BEM of equations via iterative techniques, *International Journal for Numerical Methods in Engineering*, **33**, 1823-1841, 1992.

[39] Martinez, M.J., A boundary element sensitivity formulation for bone remodelling, PhD Thesis, Wessex Institute of Technology, University of Wales, 1998.

[40] Massonnet, C.E., Numerical use of integral procedures, In *Stress Analysis*, Chapter 10, 98-235, Wiley, London, 1965.

[41] Mattheij, R.M.M. and Wang,K., A residual based adaptive BEM, *Advances in Boundary Element Techniques II*, Hoggar Press, 87-94, 2001.

[42] Mayr, M., Drexler, W. and Kuhn, G., A semi-analytical boundary integral approach for axisymmetric elastic bodies with arbitrary boundary conditions, *International Journal of Solids & Structures*, **16**, 863-871, 1980.

[43] Melan, E., Der spannungszustand der durch eine einzelkraft im Innern beanspruchten Halbscheibc; Z.Angew. Math. Mech., **12**, 343-346, 1932.

[44] Mi, Y., and Aliabadi, M.H., Dual boundary element method for three dimensional fracture mechanics analysis, *Engineering Analysis with Boundary Elements*, **10**, 161-171, 1993.

[45] Mi, Y., Three-dimensional dual boundary element analysis of crack growth, PhD Thesis, Wessex Institute of Technology, University of Portsmouth, 1995.

[46] Mi, Y., *Three-dimensional Analysis of Crack Growth*, Topics in Engineering Vol. 28, Computational Mechanics Publications, Southampton 1996.

[47] Mindilin, R.D., Force at a point in the interior of a semi-infinite solids, *Physics*, **7**, 195-202, 1936.

[48] Muci-Kuchler, K.H. and Miranda-Valenzuela, J.C., Error estimation in 3-D elasticity using Hermite-like boundary elements, *Advances in Boundary Element Techniques II*, Hoggar Press, Geneva, 95-106, 2001.

[49] Mukherjee, S., Error analysis and adaptivitiy in the boundary element and related methods, *Advances in Boundary Element Techniques II*, Hoggar Press, Geneva, 77-86, 2001.

[50] Muller-Karger, C.M., Gonzalez, C., Aliabadi, M.H. and Cerrolaza, M., Three dimensional BEM and FEM stress analysis of the human tibia under pathological conditions. *Computer Methods in Engineering and Sciences*, **2**, 1-13, 2001.

[51] Mura, T., *Micromechanics of Defects in Solids*, 2nd edition, Martinus Nijhoff Publishers 1978.

[52] Muskhelishvili, N.I., *Some Basic Problems of the Mathematical Theory of Elasticity*, Noordhoff, Leyden 1953.

[53] Nardini, N and Brebbia, C.A., A new approach to free vibration analysis using boundary elements, *Boundary Element Methods in Engineering,* Computational Mechanics Publications, Southampton, 312-326 1992.

[54] Nishimura, N. and Kobayashi, S. A boundary integral equation formulation for three-dimensional anisotropic elastostatics, *5th Int. Conf. Boundary Elements,* Springer Verlag, Berlin, 345-354, 1983.

[55] Noorozi,S., Vinney,J., Wait,D. and Sewell,P., The use of boundary element analysis as a design aid to analyse complex aspects of aircraft landing gear joint design, *Boundary Element Techniques, Proc. of the 1st International Conference,* Queen Mary and Westifield College, University of London, 425-436, 1999.

[56] Nowak, A. and Neves, A., *The Multiple Reciprocity Boundary Elements,* Computational Mechanics Publications, Southampton, 1994.

[57] Pan, E and Tonon, F., Three-dimensional Green's function in anisotropic piezoelectric solids, *International Journal for Numerical Methods in Engineering,* **37,** 943-958, 2000.

[58] Pan, Y.C. and Chou, T.W., Point force solution for an infinite transversely isotropic solid. Trans, ASME paper No WA/ADM-18, 608-612, 1976.

[59] Perez, M.W. and Wrobel, L.C., An integral-equation formulation for anisotropic elastostatics, *Journal of Applied Mechanics,* **63,** 891-902,1996.

[60] Pérez-Gavilán, J.J. and Aliabadi, M.H., A symmetric Galerkin formulation and dual reciprocity for 2D elastostatics, *Engineering Analysis with Boundary Elements,* **25,** 229-235, 2001.

[61] Pérez-Gavilán, J.J. and Aliabadi, M.H., Galerkin boundary element formulation with dual reciprocity for elastodynamics, *International Journal for Numerical Methods in Engineering,* **48,** 1331-1334, 2000.

[62] Pérez-Gavilán, J.J. and Aliabadi, M.H., Comparative study of boundary element Galerkin and collocation methods, *Proc. 2nd UK Conf. Boundary Element Method,* 269-279, 1999.

[63] Pérez-Gavilán, J.J. and Aliabadi, M.H, A symmetric Galerkin BEM for multiconnected bodies: a new approach, *Engineering Analysis with Boundary Elements,* **25,** 633-638, 2001

[64] Pérez-Gavilán, J.J. and Aliabadi, M.H., Symmetric Galerkin BEM for multiconnected bodies, *Communications in Numerical Methods for Engineering,* **17,** 761-770, 2001.

[65] Pérez-Gavilan, J.J., Galerkin Boundary Elements in Solid Mechanics, PhD thesis, Queen Mary, University of London, 2001.

[66] Riccardella, P.C., An implementation of the boundary-integral technique for planar problems of elasticity and elasto-plasticity, PhD thesis, Carnegie-Mellon University 1973.

[67] Portela, E., Adey, R.A. and Aliabadi, M.H., An object-oriented approach for boundary element mesh design, *Engineering with Computers*, 1994.

[68] Rencis, J.J., Mullen, R.L., A self-adaptive mesh refinment technique for boundary element solution of the Laplace equation, *Computational Mechanics*, **3**, 309-319, 1988.

[69] Rigby, R., and Aliabadi, M.H., Out-of-core solver for large, multi-zone boundary element matrices, *International Journal for Numerical Methods in Engineering*, **38**, 1507-1533, 1995.

[70] Rizzo, F.J., An integral equation approach to boundary value problems of classical elastostatics, *Quartely Journal of Applied Mathematics.*, **25**, 83-95, 1967.

[71] Rizzo, F.J. and Shippy, D.J., A method of stress determination in plane anisotropic bodies, *Journal of Composite Materials*, **4,**36-61, 1970.

[72] Rizzo, F.J. and Shippy, D.J., A boundary integral approach to potential and elasticity problems for axisymmetric bodies with arbitrary boundary conditions, *Mechanics Research Communications*, **6**, 99-103, 1979.

[73] Salgado, N.K, Aliabadi, M.H and Callen, R., Rule inferencing and object-orientation for boundary elements mesh design. An application to aircraft stiffened panels, *Artifical Intelligence in Engineering* **11**, 183-190, 1997.

[74] Salgado, N.K. and Aliabadi, M.H. ,A dual reciprocity method for the analysis of adhesively patched sheets, *Communications in Numerical Methods in Engineering* **13**, 397-405, 1997.

[75] Schclar, N.A. and Partridge, P., 3D anisotropic elasticity with BEM using the isotropic fundamental solution, *Engineering Analysis with Boundary Elements*, **11**, 137-144, 1993.

[76] Sladek, V. and Sladek, J., Improved computation of stresses using the boundary element method. *Applied Mathematical Modelling*, **10**, 249-255, 1986.

[77] Sirtori, S., General stress analysis method by means of integral equations and boundary elements, *Meccanica*, **14**, 210-218, 1979.

[78] Sitori, S., Maier, G., Novati, G. and Miccoli, S., A Galerkin symmetric boundary-element method in elasticity: Formulation and implementation, *International Journal for Numerical Methods in Engineering*, **35**, 255-282, 1992.

[79] Sokolnikoff, I., *Mathematical Theory of Elasticity*, McGraw-Hill, New York, 1956.

[80] Sollero, P. and Aliabadi, M.H., Fracture mechanics analysis of anisotropic plates by the boundary element method, *International Journal of Fracture*, **64**, 269-284, 1993.

[81] Somigliana, C., Sopra l'equilibrio di un corpo elastico isotropo, *Il Nuovo Cienmento*, 17-19, 1886.

[82] Telles, J.C.F., *The Boundary Element Method Applied to Inelastic Problems*, Springer-Verlag, Berlin 1983.

[83] Vogel, S.K. and Rizzo, F.J., An integral equation formulation of three-dimensional anisotropic elastostatic boundary value problems, *Journal of Elasticity*, **3**, 203-216, 1973.

[84] Wang, C.Y., Elastostatic fields produced by a point force in solids of general anisotropy, *Journal of Engineering Mathematics*, **32**, 41-52, 1997.

[85] Wang, H.C. and Banerjee, P.K., Multi-domain general axisymmetric stress analysis by BEM, *International Journal for Numerical Methods in Engineering*, **28**, 2065-2083, 1989.

[86] Wen, P.H., Aliabadi, M.H. and Young, A., Transformation of domain integrals to boundary integrals in BEM analysis of shear deformable plate bending problems, *Computational Mechanics*, **24**, 304-309, 1999.

[87] Wilde, A.J. and Aliabadi, M.H., Direct evalaution of boundary stresses in the 3D BEM of elastostatics, *Communications in Numerical Methods in Engineering*, **14**, 505-517, 1998.

[88] Wilson, R.B. and Cruse, T.A., Efficient implementation of anisotropic three-dimensional boundary integral equations stress analysis, *International Journal for Numerical Methods in Engineering*, **2**, 1383-1397, 1978

[89] Young, A., A single-domain boundary element method for 3D elastostatic crack analysis using continuous elements, *International Journal for Numerical Methods in Engineering*, **39**, 1265-1293, 1996.

Chapter 3

Plates and Shells

3.1 Introduction

Plate and shell type structures are widely used in engineering, for example aircraft wings and fuselage panels, storage tanks, pressure vessels, rocket motor casting, boiler drums, building slabs, building foundations, etc. There are two widely used theories for plates and shells. The first plate theory was developed by Kirchhoff [24] and is commonly referred to as the 'classical' theory; the other was developed by Reissner [41], and is known as the 'shear deformable' theory. The classical theory is adequate for analyzing some applications, however, for problems involving stress concentrations and cracks the theory has been shown not to be in agreement with experimental measurements (see references 5 and 6 in Reissner's paper). From the computational point of view, the classical plate theory presents the problem in terms of two independent degrees of freedom. Additional corner unknowns have to be included to account for the shear jump. Unlike classical theory, the Reissner theory takes into account the effect of the shear deformation, and presents the problem as a six-order boundary value problem. In Reissner plate theory, the problem is represented in terms of three degrees of freedom, involving generalized displacements (i.e. two rotations and deflection) and generalized tractions (i.e. normal and torsional moments, and transverse shear force). The extension of the Reissner theory to shells results in five degrees of freedom, as in-plane components of displacements and tractions would also have to be included in the model.

The application of the boundary element method to the classical plate theory was first proposed by Jaswon, Maiti and Symm [18]. The development of the direct BEM to plate theory based on the Kirchhoff theory was introduced by Forbes and Robinson [15], followed by the works of Bézine [7], Stern [46] and Tottenham [52]. Hartmann and Zormantel [16], further developed the method, and implemented it for plates with relatively complex geometries, loadings and supports. Following the above works, others (Abdel-Akher and Hartley [1], Karami *et al.* [21], Tanaka and Miyazaki [48], Venturini and Pavia [55] and Paiva and Aliabadi [34]) have further improved and applied the plate bending formulation. Application of the BEM to anisotropic materials is reported in [43].

The application of the direct BEM for analysis of the Reissner plate was presented by Vander Weeën [54]. Later, Karam and Telles [20] extended the formulation to account for infinite regions and also reported that Reissner's plate model is suitable for both thin and thick plates. Barcellos and Silva [5] presented a similar formulation to that of Vander Weeën [54] for the Mindlin plate model. Their formulation differs from the Reissner formulation in the shear factor constant. Westphal and Barcellos [66] investigated the importance of the neglected terms in the fundamental solutions derived by Vander Weeën [54]. They concluded that these terms have no effect on the results. In [12], a modified form of the fundamental solutions was derived by separating parts of the kernel representing the effect of transverse shear, to allow for the analysis of thin and thick plates.

One of the early applications of the BEM to thin plates resting on a Pasternak foundation is due to Balas *et al.* [4], who derived the fundamental solutions based on the work of Yu [68]. Fadhil and El-Zafrany [14] derived an alternative BEM formulation for thick plate resting on a Pasternak foundation. Rashed, Aliabadi and Brebbia [37, 39] presented BEM formulations for Reissner plates resting on Pasternak and Winkler foundations. Nonlinear BEM formulations can be found in Telles and Karam [49] for elastoplastic problems, and Xiao-Yan *et al.* [67] for geometrically nonlinear problems. Other contributions to the application of the BEM to large deformation and buckling can be found in [3, 19, 22, 23, 44, 47, 57]. More recently Wen, Aliabadi and Young [61] presented a boundary element formulation for Kirchhoff plate problems in the frequency domain. Recent advances in the boundary element method for plate bending analysis can be found in Aliabadi [2].

Newton and Tottenham [33] and Tottenham [52] developed a boundary integral equation for shallow shell problems, by decomposing of the fourth order governing equations into second order equations. Tosaka and Miyake [51] derived a direct BEM formulation for a shallow shell, starting from the reduced single complex-valued equation of the coupled $w - \phi$ formulation. More recently, Lu and Huang [27] developed a direct BEM formulation for shallow shells involving shear deformation.

An alternative approach has been developed to deal with shallow shell problems. The method is called the 'domain-boundary element method'. The work of Forbes and Robinson [15] was the first in which the direct domain-boundary element method for the static analysis of shallow shells was developed. Zhang and Atluri [69] have also derived formulations for the static and dynamic analysis of shallow shells based on a weighted residual method. Providakis and Beskos [35] and Beskos [6] extended the method to deal with static and vibration analysis of shallow shells. They derived the formulation with the aid of the reciprocal theorem. Jinmu and Shuyao [26] used this method for the analysis of geometrically nonlinear shallow shells. All the above formulations were derived by coupling the boundary element formulation of Kirchhoff plate bending and the two-dimensional plane stress elasticity.

Dirgantara and Aliabadi [8, 11] presented a new domain-boundary element formulation for the analysis of shear deformable shallow shells having a quadratic mid-surface. Wen, Aliabadi and Young [59, 60] developed two techniques, the direct integral method and the dual reciprocity method, to transfer the domain integrals in [8] into boundary integrals.

In this chapter, the derivation and implementation of boundary integral equations for the analysis of plates and shells involving shear deformation is initially presented. Initially, the formulation is derived for shear deformable plates. Next, the formulation is extended to shear deformable shallow shells [8, 59]. The shell formulation is shown to be formed by coupling boundary element formulations of shear deformable plate bending and two-dimensional plane stress elasticity. The domain integrals which

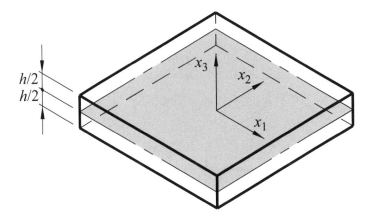

Figure 3.1: Plate geometry.

appear in this formulation are treated in two different ways: first, the integrals are evaluated numerically using constant cell discretization, and secondly, they are transformed into boundary integrals using the dual reciprocity technique. The extension of the formulation into assembled plate structures is described using a multi-region formulation. Several examples of various complexity are presented to assess the accuracy of the BEM, as well as demonstrating its efficiency and robustness compared to the finite element method. Applications of the method to Winkler and Pasternak models are described. Boundary integral equations for the Kirchhoff plate theory are also presented.

3.2 Shear Deformable Plates

3.2.1 Basic Theory

In this section, governing equations for a flat plate subjected to bending loads are presented. The governing equations for membrane loading, although already presented in Chapter 2, are also presented with slightly different notations. The two sets of equations are later used in a coupled form for the analysis of shells.

Consider an arbitrary plate of thickness h in the x_i space, as shown in Figure 10.4. The $x_1 - x_2$ plane is assumed to be located at the middle surface $x_3 = 0$, where $-h/2 \leqslant x_3 \leqslant +h/2$. The generalized displacements are denoted as w_i and u_α, where w_α denotes rotations (ϕ_{x_1} and ϕ_{x_2}), w_3 denotes the out-of-plane displacement w, and u_α denotes in-plane displacements (u_1 and u_2), as shown in Figure 3.2.

The generalized tractions are denoted as p_i and t_α, where p_α denotes tractions due to the stress couples (m_1 and m_2), p_3 denotes the traction due to shear stress resultant (t_3), and t_α denotes tractions due to membrane stress resultants (t_1 and t_2). Figure 3.2 shows the sign convention for the generalized tractions.

Throughout this chapter, Greek indices will vary from 1 to 2 and Roman indices from 1 to 3.

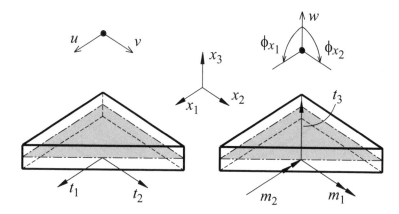

Figure 3.2: Sign convention for generalized displacements and tractions.

Stress Resultants and Stress Couples

The bending stress resultants $M_{\alpha\beta}$ are defined as

$$M_{\alpha\beta} = \int_{-h/2}^{+h/2} x_3 \sigma_{\alpha\beta} dx_3 \tag{3.1}$$

and the shearing stress resultants Q_α as

$$Q_\alpha = \int_{-h/2}^{+h/2} \sigma_{\alpha 3} dx_3 \tag{3.2}$$

where $\sigma_{\alpha\beta}$ are the three-dimensional components of the normal stresses through the plate thickness, and $\sigma_{\alpha 3}$ are the components of the transverse shear stresses.

The membrane stress resultants $N_{\alpha\beta}$ are defined as

$$N_{\alpha\beta} = \int_{-h/2}^{+h/2} \sigma_{\alpha\beta} dx_3 \tag{3.3}$$

Following the two-dimensional theory of elasticity, the normal stresses $\sigma_{\alpha\beta}$ due to membrane forces are assumed to be uniformly distributed over the thickness [41]. For shear deformable plate bending, the normal stresses due to bending and twisting moments $\sigma_{\alpha\beta}$ are assumed to vary linearly [42], that is

$$\sigma_{\alpha\beta} = \frac{1}{h} N_{\alpha\beta} + \frac{12 x_3}{h^3} M_{\alpha\beta} \tag{3.4}$$

and the transverse shear stresses $\sigma_{\alpha 3}$ vary parabolically over the thickness:

$$\sigma_{\alpha 3} = \frac{3}{2h} \left[1 - \left(\frac{2 x_3}{h} \right)^2 \right] Q_\alpha \tag{3.5}$$

The generalized bending and shear tractions at a boundary point are defined as

$$p_\alpha = M_{\alpha\beta} n_\beta \qquad \text{and} \qquad p_3 = Q_\alpha n_\alpha \tag{3.6}$$

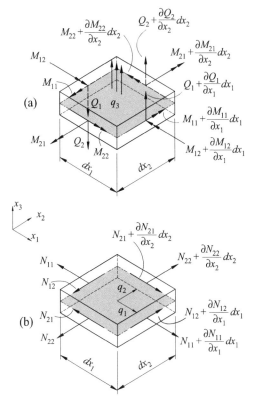

Figure 3.3: Stress resultant equilibrium of a plate element.

and membrane tractions as

$$t_\alpha = N_{\alpha\beta} n_\beta \tag{3.7}$$

where n_β are components of the outward normal vector to the plate boundary.

Strain-displacement Relationships

The transverse shear strains can be written as

$$\gamma_{\alpha 3} = \psi_\alpha = w_\alpha + w_{3,\alpha} \tag{3.8}$$

the curvature relationships or the flexural strain as

$$\kappa_{\alpha\beta} = 2\chi_{\alpha\beta} = w_{\alpha,\beta} + w_{\beta,\alpha} \tag{3.9}$$

and the in-plane strains of the element as

$$\varepsilon_{\alpha\beta} = \frac{1}{2} \left(u_{\alpha,\beta} + u_{\beta,\alpha} \right) \tag{3.10}$$

Equilibrium

The equilibrium equations for a typical plate element having dimensions of $dx_1 \times dx_2 \times h$ (see Figure 3.3) and under uniform load q_i (per unit area), can be written as follows:

$$M_{\alpha\beta,\beta} - Q_\alpha = 0 \tag{3.11}$$

$$Q_{\alpha,\alpha} + q_3 = 0 \tag{3.12}$$

$$N_{\alpha\beta,\beta} + q_\alpha = 0 \tag{3.13}$$

Stress Resultant-Strain Relationships

The stress resultant-strain relationships for plate bending are given by[1]

$$M_{\alpha\beta} = D\frac{1-\nu}{2}\left(2\chi_{\alpha\beta} + \frac{2\nu}{1-\nu}\chi_{\gamma\gamma}\delta_{\alpha\beta}\right) \tag{3.14}$$

$$Q_\alpha = C\psi_\alpha \tag{3.15}$$

where $D \ (= Eh^3/[12(1-\nu^2)])$ and $C \ (= D(1-\nu)\lambda^2/2)$ represent the bending stiffness and shear stiffness of the plate, respectively and $\lambda \ (= \sqrt{10}/h)$ is called the shear factor. For two-dimensional plane stress, membrane stress resultant-strain relationships (i.e. Hooke's law) can be written as

$$N_{\alpha\beta} = B\frac{1-\nu}{2}\left(2\varepsilon_{\alpha\beta} + \frac{2\nu}{1-\nu}\varepsilon_{\gamma\gamma}\delta_{\alpha\beta}\right) \tag{3.16}$$

where $B = Eh/(1-\nu^2)$. Equations (3.15)-(3.16) represent the generalized Hooke's law. Equations (3.15)-(3.16), together with equations (3.8 − 3.10) represent the stress resultant-displacement relationships as follows:

$$M_{\alpha\beta} = D\frac{1-\nu}{2}\left(w_{\alpha,\beta} + w_{\beta,\alpha} + \frac{2\nu}{1-\nu}w_{\gamma,\gamma}\delta_{\alpha\beta}\right) \tag{3.17}$$

$$Q_\alpha = C(w_\alpha + w_{3,\alpha}) \tag{3.18}$$

$$N_{\alpha\beta} = B\frac{1-\nu}{2}\left(u_{\alpha,\beta} + u_{\beta,\alpha} + \frac{2\nu}{1-\nu}u_{\gamma,\gamma}\delta_{\alpha\beta}\right) \tag{3.19}$$

Equilibrium Equations in Terms of Displacements

The equilibrium equations in terms of displacements are obtained by substituting (3.17)-(3.19) into (3.11)- (3.13), to give:

$$D\nabla^2 w_1 + \frac{D}{2}(1+\nu)\frac{\partial}{\partial x_2}\left(-\frac{\partial w_1}{\partial x_2} + \frac{\partial w_2}{\partial x_1}\right) - Cw_1 - C\frac{\partial w_3}{\partial x_1} = 0 \tag{3.20}$$

$$\frac{D}{2}(1+\nu)\frac{\partial}{\partial x_1}\left(\frac{\partial w_1}{\partial x_2} - \frac{\partial w_2}{\partial x_1}\right) + D\nabla^2 w_2 - Cw_2 - C\frac{\partial w_3}{\partial x_2} = 0 \tag{3.21}$$

---

[1] In Reissner [41], there is a term relating the effect of the transverse normal stresses on the bending stress resultants. According to Mindlin [28], this term has a negligible contribution to the results. For the sake of simplicity, and to be consistent with the stress resultant-strain relationships for shallow shells, this term will be ignored in this chapter.

$$CV^2 w_3 + C\frac{\partial w_1}{\partial x_1} + C\frac{\partial w_2}{\partial x_2} + q_3 = 0 \tag{3.22}$$

for plate bending, and

$$BV^2 u_1 + \frac{B}{2}(1+\nu)\frac{\partial}{\partial x_2}\left(-\frac{\partial u_1}{\partial x_2} + \frac{\partial u_2}{\partial x_1}\right) + q_1 = 0 \tag{3.23}$$

$$\frac{B}{2}(1+\nu)\frac{\partial}{\partial x_1}\left(\frac{\partial u_1}{\partial x_2} - \frac{\partial u_2}{\partial x_1}\right) + BV^2 u_2 + q_2 = 0 \tag{3.24}$$

for plane stress elasticity. Equations (3.20)-(3.22) can be written in a more compact form as

$$L^b_{ik} w_k + f^b_i = 0 \tag{3.25}$$

and similarly, equations (3.23) and (3.24) as

$$L^m_{\alpha\beta} u_\beta + f^m_\alpha = 0 \tag{3.26}$$

The Navier differential operator L^b_{ik} corresponds to shear deformable plate bending problems, and is given by

$$L^b_{\alpha\beta} = \frac{D}{2}\left[(1-\nu)\left(\nabla^2 - \lambda^2\right)\delta_{\alpha\beta} + (1+\nu)\frac{\partial}{\partial x_\alpha}\frac{\partial}{\partial x_\beta}\right] \tag{3.27}$$

$$L^b_{\alpha 3} = -\frac{(1-\nu)D}{2}\lambda^2\frac{\partial}{\partial x_\alpha} \tag{3.28}$$

$$L^b_{3\alpha} = -L^b_{\alpha 3} \tag{3.29}$$

$$L^b_{33} = \frac{(1-\nu)D}{2}\lambda^2\nabla^2 \tag{3.30}$$

with $f^b_\alpha = 0$ and $f^b_3 = q_3$.

The Navier differential operator $L^m_{\alpha\beta}$ corresponds to two-dimensional plane stress problems, and is given by

$$L^m_{\alpha\beta} = BV^2\delta_{\alpha\beta} + \frac{B(1+\nu)}{2}\frac{\partial}{\partial x_\alpha}\frac{\partial}{\partial x_\beta}(1 - 2\delta_{\alpha\beta}) \tag{3.31}$$

with $f^m_\alpha = q_\alpha$.

The boundary integral equation for the two-dimensional plane stress problem have already been described in Chapter 2. Next, we shall derive the boundary integral equations for the plate bending.

3.2.2   Boundary Integral Formulations

The boundary integral equations for rotations and out-of-plane displacements can be derived by utilizing Betti's reciprocal theorem in a way similar to that described in Chapter 2. Here, the state $(.)^*$ is defined for concentrated generalized loads: two bending moments $(i = \alpha = 1,2)$ and one concentrated shear force $(i = 3)$ at an arbitrary source point $\mathbf{X}' \in V$, as shown in Figure 3.4. Choosing the $(\cdot)^*$ state to represent the fundamental state, such as

$$M^*_{i\alpha\beta,\beta}(\mathbf{X}',\mathbf{X}) - Q^*_{i\alpha}(\mathbf{X}',\mathbf{X}) + \Delta(\mathbf{X}',\mathbf{X})\delta_{i\alpha} = 0$$

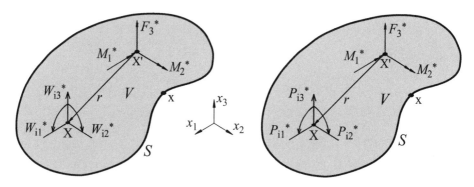

Figure 3.4: Fundamental state of displacements and tractions due to concentrated shear force and bending moments.

$$Q_{i\alpha,\alpha}^*(\mathbf{X'}, \mathbf{X}) + \Delta(\mathbf{X'}, \mathbf{X})\delta_{i3} = 0 \tag{3.32}$$

where $\Delta(\mathbf{X'}, \mathbf{X})$ is the Dirac delta, and making use of the following property:

$$\int_V \Delta(\mathbf{X'}, \mathbf{X})w_i(\mathbf{X})dV = w_i(\mathbf{X'}) \tag{3.33}$$

then utilizing the Betti's reciprocal theorem, the rotations and out-of-plane displacements for an internal source point $\mathbf{X'}$ can be written as [8]

$$w_i(\mathbf{X'}) + \int_S T_{ik}^b(\mathbf{X'}, \mathbf{x})w_k(\mathbf{x})\, dS = \int_S U_{ik}^b(\mathbf{X'}, \mathbf{x})p_k(\mathbf{x})\, dS$$

$$+ \int_V U_{ik}^b(\mathbf{X'}, \mathbf{X})f_k^b(\mathbf{X})dV \tag{3.34}$$

where $U_{ik}^b(\mathbf{X'}, \mathbf{x})$ and $T_{ik}^b(\mathbf{X'}, \mathbf{x})$ are the fundamental solutions for rotations and out-of-plane displacements and bending and shear traction, respectively. Substituting for f_k^b, we have

$$w_i(\mathbf{X'}) + \int_S T_{ij}^b(\mathbf{X'}, \mathbf{x})w_j(\mathbf{x})\, dS = \int_S U_{ij}^b(\mathbf{X'}, \mathbf{x})p_j(\mathbf{x})\, dS$$

$$+ \int_V U_{i3}^b(\mathbf{X'}, \mathbf{X})q_3(\mathbf{X})dV \tag{3.35}$$

As in the elasticity, if the point $\mathbf{X'}$ is taken to the boundary, $(\mathbf{X'} \to \mathbf{x'} \in S)$, and assuming that the displacements w_j and u_α are Hölder continuities, equation (3.35) for the source points on the boundary can be written as follows:

$$C_{ij}^b(\mathbf{x'})w_j(\mathbf{x'}) + \int_S T_{ij}^b(\mathbf{x'}, \mathbf{x})w_j(\mathbf{x})\, dS = \fint_S U_{ij}^b(\mathbf{x'}, \mathbf{x})p_j(\mathbf{x})\, dS$$

$$+ \int_V U_{i3}^b(\mathbf{x'}, \mathbf{X})q_3(\mathbf{X})dV \tag{3.36}$$

The limiting forms of the above integral and jump terms are described in detail in Appendix C.

The expressions for the kernels U_{ij}^b and T_{ij}^b are given by Vander Weeën [54] as follows:

$$U_{\alpha\beta}^b = \frac{1}{8\pi D(1-\nu)}\{[8B(z)-(1-\nu)(2\ln z - 1)]\delta_{\alpha\beta}$$
$$- [8A(z)+2(1-\nu)]r_{,\alpha}r_{,\beta}\}$$

$$U_{\alpha 3}^b = -U_{3\alpha}^b = \frac{1}{8\pi D}(2\ln z - 1)rr_{,\alpha}$$

$$U_{33}^b = \frac{1}{8\pi D(1-\nu)\lambda^2}[(1-\nu)z^2(\ln z - 1) - 8\ln z] \tag{3.37}$$

and

$$T_{\alpha\beta}^b = \frac{-1}{4\pi r}[(4A(z)+2zK_1(z)+1-\nu)(\delta_{\alpha\beta}r_{,n}+r_{,\beta}n_\alpha)$$
$$+ (4A(z)+1+\nu)r_{,\alpha}n_\beta - 2(8A(z)+2zK_1(z)+1-\nu)r_{,\alpha}r_{,\beta}r_{,n}]$$

$$T_{\alpha 3}^b = \frac{\lambda^2}{2\pi}[B(z)n_\alpha - A(z)r_{,\alpha}r_{,n}]$$

$$T_{3\alpha}^b = \frac{-(1-\nu)}{8\pi}\left[\left(2\frac{(1+\nu)}{(1-\nu)}\ln z - 1\right)n_\alpha + 2r_{,\alpha}r_{,n}\right]$$

$$T_{33}^b = \frac{-1}{2\pi r}r_{,n} \tag{3.38}$$

where $z = \lambda r$, λ is the shear factor, r is the absolute distance between the source and the field points, and

$$A(z) = K_0(z) + \frac{2}{z}\left[K_1(z) - \frac{1}{z}\right]$$

$$B(z) = K_0(z) + \frac{1}{z}\left[K_1(z) - \frac{1}{z}\right] \tag{3.39}$$

in which $K_0(z)$ and $K_1(z)$ are modified Bessel functions of the second kind. Expanding the modified Bessel functions for small arguments:

$$K_0(z) = \left[-a_o - \ln(\frac{z}{2})\right] + \left[-a_o + 1 - \ln(\frac{z}{2})\right]\frac{(z^2/4)}{(1!)^2}$$
$$+ \left[-a_o + 1 + \frac{1}{2} - \ln(\frac{z}{2})\right]\frac{(z^2/4)^2}{(2!)^2}$$
$$+ \left[-a_o + 1 + \frac{1}{2} + \frac{1}{3} - \ln(\frac{z}{2})\right]\frac{(z^2/4)^3}{(3!)^2} + \cdots \tag{3.40}$$

$$K_1(z) = \frac{1}{z} - \left[-a_o + \frac{1}{2} - \ln(\frac{z}{2})\right]\frac{(z^2/4)^{1/2}}{0!1!}$$
$$- \left[-a_o + 1 + \frac{1}{4} - \ln(\frac{z}{2})\right]\frac{(z^2/4)^{3/2}}{1!2!}$$
$$- \left[-a_o + 1 + \frac{1}{2} + \frac{1}{6} - \ln(\frac{z}{2})\right]\frac{(z^2/4)^{5/2}}{2!3!} + \cdots \tag{3.41}$$

where $a_o = 0.5772156649$ is the Euler constant. Substitute equations (3.40)-(3.41) into (3.39) and take the limit as $r \to 0$:

$$\lim_{r \to 0} A(z) = \frac{-1}{2}, \tag{3.42}$$

$$\lim_{r \to 0} B(z) = -\frac{1}{2}\left[\lim_{r \to 0}\ln(\frac{z}{2}) + a_o + \frac{1}{2}\right] \tag{3.43}$$

From (3.42) it can be seen that $A(z)$ is a smooth function, whereas $B(z)$ is a weakly singular $O(\ln r)$, as is apparent from (3.43). It follows that U_{ij}^b is weakly singular and T_{ij}^b has a strong (Cauchy principal value) singularity $O(r^{-1})$.

The last domain integral in (3.36) can be transferred to boundary integral [37] (by applying the divergence theorem), in the case of a uniform load ($q_3 = $ constant) to give

$$\int_V U_{i3}^b(\mathbf{x}',\mathbf{X})q_3(\mathbf{X})dV = q_3\int_S V_{i,\alpha}^b(\mathbf{x}',\mathbf{x})n_\alpha(\mathbf{x})dS \tag{3.44}$$

where V_i^b are the particular solutions of the equation $V_{i,\theta\theta}^b = U_{i3}^b$. The expressions for $V_{i,\beta}^b$ are

$$V_{\alpha,\beta}^b = \frac{r^2}{128\pi D}[(4\ln z - 5)\delta_{\alpha\beta} + 2(4\ln z - 3)r_{,\alpha}r_{,\beta}]$$

$$V_{3,\beta}^b = \frac{-rr_{,\beta}}{128\pi D(1-\nu)\lambda^2}[32(2\ln z - 1) - z^2(1-\nu)(4\ln z - 5)] \tag{3.45}$$

The transformation of the domain integral for other load distributions of q_3 can be carried out by using the dual reciprocity techniques, as described in Section 3.5.

The stress resultant components are obtained by differentiation of equation (3.35) with respect to the coordinate of the source point $\mathbf{X}'$, and then substituting them into the stress resultant-displacement relations in equations (3.18), to give

$$M_{\alpha\beta}(\mathbf{X}') = \int_S D_{\alpha\beta k}^b(\mathbf{X}',\mathbf{x})p_k(\mathbf{x})dS - \int_S S_{\alpha\beta k}^b(\mathbf{X}',\mathbf{x})w_k(\mathbf{x})dS$$

$$+q_3\int_S Q_{\alpha\beta}^b(\mathbf{X}',\mathbf{x})dS \tag{3.46}$$

$$Q_\beta(\mathbf{X}') = \int_S D_{3\beta k}^b(\mathbf{X}',\mathbf{x})p_k(\mathbf{x})dS - \int_S S_{3\beta k}^b(\mathbf{X}',\mathbf{x})w_k(\mathbf{x})dS$$

$$+ q_3\int_S Q_{3\beta}^b(\mathbf{X}',\mathbf{x})dS \tag{3.47}$$

The kernels $D_{i\beta k}^b$, $S_{i\beta k}^b$ and $Q_{i\beta}^b$, are given [54] as

$$S_{\alpha\beta\gamma}^b = \frac{1}{4\pi r}[(4A(z) + 2zK_1(z) + 1 - \nu)(\delta_{\beta\gamma}r_{,\alpha} + \delta_{\alpha\gamma}r_{,\beta})$$
$$- 2(8A(z) + 2zK_1(z) + 1 - \nu)r_{,\alpha}r_{,\beta}r_{,\gamma} + (4A(z) + 1 + \nu)\delta_{\alpha\beta}r_{,\gamma}]$$

$$S_{\alpha\beta3}^b = \frac{-(1-\nu)}{8\pi}\left[\left(2\frac{(1+\nu)}{(1-\nu)}\ln z - 1\right)\delta_{\alpha\beta} + 2r_{,\alpha}r_{,\beta}\right]$$

$$S_{3\beta\gamma}^b = \frac{\lambda^2}{2\pi}[B(z)\delta_{\gamma\beta} - A(z)r_{,\gamma}r_{,\beta}]$$

$$S_{3\beta3}^b = \frac{1}{2\pi r}r_{,\beta} \tag{3.48}$$

$$
\begin{aligned}
D^b_{\alpha\beta\gamma} &= \frac{D(1-\nu)}{4\pi r^2}\{(4A(z)+2zK_1(z)+1-\nu)(\delta_{\gamma\alpha}n_\beta+\delta_{\gamma\beta}n_\alpha) \\
&+ (4A(z)+1+3\nu)\delta_{\alpha\beta}n_\gamma - (16A(z)+6zK_1(z)+z^2K_0(z)+2-2\nu) \\
&\times [(n_\alpha r_{,\beta}+n_\beta r_{,\alpha})r_{,\gamma}+(\delta_{\gamma\alpha}r_{,\beta}+\delta_{\gamma\beta}r_{,\alpha})r_{,n}] \\
&- 2(8A(z)+2zK_1(z)+1+\nu)(\delta_{\alpha\beta}r_{,5}r_{,n}+n_\gamma r_{,\alpha}r_{,\beta}) \\
&+ 4(24A(z)+8zK_1(z)+z^2K_0(z)+2-2\nu)r_{,\alpha}r_{,\beta}r_{,\gamma}r_{,n}\} \\
D^b_{\alpha\beta 3} &= \frac{D(1-\nu)\lambda^2}{4\pi r}[(2A(z)+zK_1(z))(r_{,\beta}n_\alpha+r_{,\alpha}n_\beta) \\
&- 2(4A(z)+zK_1(z))r_{,\alpha}r_{,\beta}r_{,n}+2A(z)\delta_{\alpha\beta}r_{,n} \\
D^b_{3\beta\gamma} &= \frac{-D(1-\nu)\lambda^2}{4\pi r}[(2A(z)+zK_1(z))(\delta_{\gamma\beta}r_{,n}+r_{,\gamma}n_\beta) \\
&+ 2A(z)n_\gamma r_{,\beta}-2(4A(z)+zK_1(z))r_{,\gamma}r_{,\beta}r_{,n} \\
D^b_{3\beta 3} &= \frac{D(1-\nu)\lambda^2}{4\pi r^2}[(z^2B(z)+1)n_\beta-(z^2A(z)+2)r_{,\beta}r_{,n}]
\end{aligned}
\tag{3.49}
$$

and

$$
\begin{aligned}
Q^*_{\alpha\beta} &= \frac{-r}{64\pi}\{(4\ln z-3)[(1-\nu)(r_{,\beta}n_\alpha+r_{,\alpha}n_\beta)+(1+3\nu)\delta_{\alpha\beta}r_{,n}] \\
&+ 4[(1-\nu)r_{,\alpha}r_{,\beta}+\nu\delta_{\alpha\beta}r_{,n}]\} \\
Q^*_{3\beta} &= \frac{1}{8\pi}[(2\ln z-1)n_\beta+2r_{,\beta}r_{,n}]
\end{aligned}
\tag{3.50}
$$

3.2.3   Multi–region Formulation

The multi-region formulation for shear deformable plates was developed by Dirgantara and Aliabadi [10], and later extended to large scale assembled structures [63, 64]. Consider a part of the assembled plates joined along a straight line, as shown in Figure (3.5a), and suppose that the direction of the interface line is along the x_2-axis. The boundary values for each flat plate in the local coordinates (x^n_1, x^n_2, x^n_3) $(n = 1, 2, ...N$, with N being the total number of assembled plates along the interface), as shown in Figure (3.5a), are: $\{u^n_1, u^n_2\}$ the in-plane displacements, $\{w^n_1, w^n_2, w^n_3\}$, the rotations and out of plane deflection, $\{t^n_1, t^n_2\}$ and $\{p^n_1, p^n_2, p^n_3\}$ membrane tractions, moments and shear force. The compatibility equations for displacements along the interface are

$$
u^1_1\cos\theta_1 - w^1_3\sin\theta_1 = u^2_1\cos\theta_2 - w^2_3\sin\theta_2 = ... = u^N_1\cos\theta_N - w^N_3\sin\theta_N
$$

$$
u^1_1\sin\theta_1 + w^1_3\cos\theta_1 = u^2_1\sin\theta_2 + w^2_3\cos\theta_2 = ... = u^N_1\sin\theta_N + w^N_3\cos\theta_N
$$

$$
u^1_2 = u^2_2 = ... = u^N_2
\tag{3.51}
$$

$$
w^1_1 = w^2_1 = ... = w^N_1
$$

$$
w^1_2 = w^2_2 = ... = w^N_2 = 0
$$

and equilibrium equations for tractions are

$$
\sum_{n=1}^{N}(t^n_1\cos\theta_n - p^n_3\sin\theta_n) = 0
$$

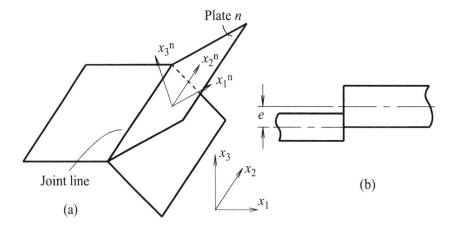

Figure 3.5: Assembled plate structures and local coordinates.

$$\sum_{n=1}^{N} (t_1^n \sin\theta_n + p_3^n \cos\theta_n) = 0$$

$$\sum_{n=1}^{N} t_2^n = 0 \qquad (3.52)$$

$$\sum_{n=1}^{N} p_\alpha^n = 0$$

where θ_n $(n = 1, 2, ..., N)$ represents the angle between axis x_1^n in the local coordinate system and x_1 in the global coordinate system, as shown in Figure (3.5a). In this case, the moments p_2^n $(n = 1, 2, ..., N)$ are unknown, as the rotations $w_2^n = 0$ on the joint line are fixed.

For the case of two offset plates joined as shown in Figure (3.5b), the displacement compatibility conditions are

$$u_\alpha^1 - ew_\alpha^1 = u_\alpha^2, \quad \alpha = 1, 2$$

$$w_k^1 = w_k^2 \quad k = 1, 2, 3$$

and the equilibrium equations for tractions are

$$t_\alpha^1 = t_\alpha^2, \quad p_3^1 = p_3^2$$

$$p_\alpha^1 - et_\alpha^1 = p_\alpha^2$$

where e denotes the gap between these two plates, as shown in Figure (3.5b).

3.2.4 Foundation Plates

Plates resting on elastic foundations have many applications in civil engineering. In this section, BEM formulations for Reissner plates resting on Winkler and Pasternak foundations are briefly described. Full details of the formulation can be found in Rashed *et al.* [37] and [39].

The displacement integral equation for the Pasternak model can be written as

$$C_{ij}^p(\mathbf{x}')w_j(\mathbf{x}') + \int_S T_{ij}^p(\mathbf{x}',\mathbf{x})w_j(\mathbf{x})dS$$

$$+G_f \int_S U_{i3,n}^p(\mathbf{x}',\mathbf{x})\mathbf{w}_3(\mathbf{x})dS = \int_S U_{ij}^p(\mathbf{x}',\mathbf{x})p_j dS$$

$$\int_S U_{i3}^p[p_f + G_f u_{3,n}(\mathbf{x})]dS + \int_V U_{i3}^p(\mathbf{x}',\mathbf{X})q(\mathbf{X})dV \qquad (3.53)$$

where G_f denotes the shear modulus of foundation and the shear traction $p_f = G_f(w_{f,n} - w_{3,n})$. For the Winkler model, $p_f = 0$ and $G_f = 0$.

For the Pasternak model, there is an additional unknown p_f in equation (3.53). For the case of free boundary (i.e. $p_f \neq 0$), an additional integral equation is required. The additional integral equation is generated from the governing equation for the foundation settlement (i.e. $(\nabla^2 - k_f/G_f)u_f(\mathbf{x}'') = 0$) as

$$C_f(\mathbf{x}'')w_f(\mathbf{x}'') + \int_{S_f} U_f(\mathbf{X}'',\mathbf{x})u_{f,n'}dS_f - \int_{S_f} U_{f,n'}(\mathbf{X}'',\mathbf{x})w_f(\mathbf{x})dS_f \qquad (3.54)$$

where S_f denotes the foundation, $\mathbf{x}'' \in S_f$, n' is the normal to S_f and U_f is a two-point fundamental solution of the modified Helmholtz equation, and is given by

$$U_f = \frac{1}{2\pi}K_0(\lambda_1 r)$$

in which $\lambda_1 = \sqrt{k_f/G_f}$, with k_f denoting the modulus of sub-grade reaction for the foundation. Along the boundary, $w_f = w_3$, we have

$$w_{f,n} = \frac{p_f}{G_f}\alpha_3 - u_\alpha n_\alpha \qquad (3.55)$$

where $\alpha_3 = (1 + G_f/C)$ with $C = D(1 - \nu)\lambda^2/2$. Substituting (3.55) into (3.54) and moving from S_f to S gives [37]

$$C_f(\mathbf{x}')w_f(\mathbf{x}') = \int_S U_{f,n}^p(\mathbf{x}',\mathbf{x})w_3 dS - \int_S U_f(\mathbf{x}',\mathbf{x})w_\alpha(\mathbf{x})n_\alpha dS$$

$$-\frac{\alpha_3}{G_f}\int_S U_f^p(\mathbf{x}',\mathbf{x})p_f(\mathbf{x})dS$$

with

$$U_{f,n}^p = -\frac{\lambda_1}{2\pi}K_1(\lambda_1 r)r_{,n}$$

3.3 Shear Deformable Shallow Shells

3.3.1 Basic Theory

Consider a sheet of material bounded by two curved surfaces whose thickness is small compared to other dimensions, but which is capable of resisting bending, in addition

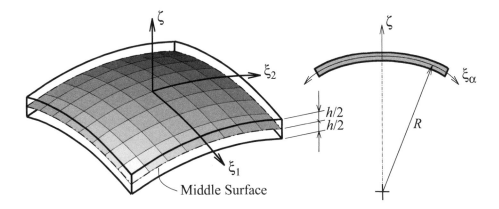

Figure 3.6: Curvilinear coordinate of the shell.

to membrane forces. Points on the shell's middle surface are defined using curvilinear coordinates ξ_1 and ξ_2, and the third coordinate in the surface is the distance ζ from the middle surface, measured along the normal (see Figure 3.6).

General classical theory of thin elastic shells can be found in [32]. The theory of thin elastic isotropic shells involving transverse shear deformation and transverse normal stress is due to Reissner [42] and Naghdi [30], and for small deflection is based on the following assumptions:

- the thickness of shell h must be much smaller than the least radius of curvature R of the middle surface, i.e. $h/R \ll 1$ (a shell is regarded as thin if $\max(h/R) \leq 1/20$, otherwise it will be considered as a thick shell);

- the strains and displacements are small enough for changes in geometry to be neglected;

- the component of stress normal to the middle surface is small compared to other components of stress, and it may be neglected in the stress-strain relationships;

- straight lines normal to the undeformed middle surface would remain inextensible, straight and normal to the middle surface as the shell undergoes deformation, but would not necessarily remain normal to the middle surface. This condition implies that unlike the classical theory the transverse shear strains are not neglected.

The stresses are replaced by a system of stress resultants and stress couples in both classical and shear deformable shell theories.

A shell is considered to be shallow (i.e. slightly curved) if the middle surface is sufficiently smooth and all points on this surface are sufficiently close to a plane. According to Vlasov [56], a shell is shallow if the ratio of the rise to the shorter side (for a shell of rectangular plan) or to the diameter (for a shell of circular plan) (i.e. H/D) is less than $1/5$ (see Figure 3.7).

In shallow shell theory, all points on the middle surface are located by the Cartesian coordinates of their projection on the x_1x_2−plane, that is,

$$x_1 = \xi_1; \qquad x_2 = \xi_2; \qquad x_3 = f(\xi_1, \xi_2) \qquad (3.56)$$

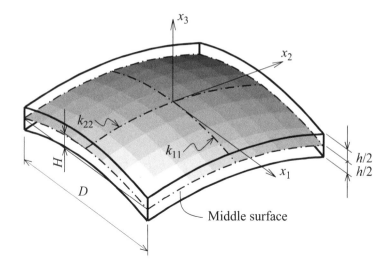

Figure 3.7: Shallow shell geometry.

where $\xi_1; \xi_2$ are the curvilinear coordinates and $f(\xi_1, \xi_2)$ is the equation of the middle surface of the shell. General theory of shallow shell is based on the following assumptions:

1. the squares of the derivatives df/dx_α and their products are negligible in comparison to unity, i.e. $(df/dx_\alpha)^2 \ll 1, (df/dx_\alpha)(df/dx_\beta) \ll 1$;

2. the transverse shear resultants Q_α in equilibrium equations of stress resultants in the x_1- and x_2-directions are negligible;

3. the influence of tangential displacements u_α in the transverse shear strain $\gamma_{\alpha 3}$ is negligible.

Consider an arbitrary shell of thickness h, as shown in Figure 3.7, with a quadratic middle surface given by

$$x_3 = -\frac{1}{2}(k_{11}x_1^2 + k_{22}x_2^2) \tag{3.57}$$

According to equation (3.57), the shell has $k_{11} = 1/R_1$ and $k_{22} = 1/R_2$ which are principal curvatures of the shell in the x_1-and x_2-directions, respectively, while $k_{12} = k_{21} = 0$. The lower surface of the shell is located at $(x_3 - h/2)$ and the upper surface at $(x_3 + h/2)$.

As in the shear deformable plate theory, described earlier, the generalized displacements are denoted as w_i and u_α, where w_α denotes rotations (ϕ_{x_1} and ϕ_{x_2}), w_3 denotes the out-of-plane displacement w, and u_α denotes in-plane displacements (u_1 and u_2). The generalized tractions are denoted as p_i and t_α, where p_α denotes tractions due to the stress couples (m_1 and m_2), p_3 denotes the traction due to shear stress resultant (t_3) and t_α denotes tractions due to membrane stress resultants (t_1 and t_2).

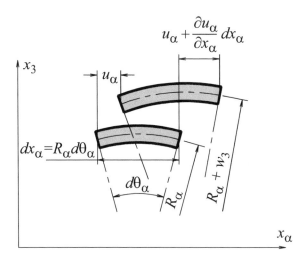

Figure 3.8: Basic definition of displacements and strains.

Strain-displacement Relationships

Figure 3.8 shows an element of a shallow shell. An increase of element length due to displacement along x_α is $(\partial u_\alpha/\partial x_\alpha)\, dx_\alpha$, and because of the radial displacement w_3, the element length increases by $w_3 d\theta_\alpha \approx w_3(dx_\alpha/R_\alpha)$. It follows that the in-plane normal strain is obtained as

$$\varepsilon_{\alpha\alpha} = u_{\alpha,\alpha} + \frac{w_3}{R_\alpha} \tag{3.58}$$

Similar to the elasticity theory of two-dimensional bodies, in-plane shear strain can be written (see Chapter 2) as

$$\gamma_{\alpha\beta} = 2\varepsilon_{\alpha\beta} = u_{\alpha,\beta} + u_{\beta,\alpha}; \qquad \alpha \neq \beta \tag{3.59}$$

Transverse shear strain is obtained from the distortion of angle of the face of the shell element perpendicular to x_1- or x_2- axis as

$$\gamma_{\alpha 3} = w_\alpha + w_{3,\alpha} \tag{3.60}$$

The curvature relationships with the derivatives of generalized displacements can be written as

$$\kappa_{\alpha\beta} = 2\chi_{\alpha\beta} = w_{\alpha,\beta} + w_{\beta,\alpha} \tag{3.61}$$

Stress Resultants and Stress Couples

The components of stress are shown in Figure 3.9, where e_1, e_2, and e_3 are unit vectors parallel to the x_1, x_2, and x_3-axes respectively.

To define stress resultants and stress couples, we now consider a face of the shell element shown in Figure 3.10a. The length of the arc in the middle surface is $dx_2 = R_2 d\theta$, and the length of the arc at a distance x_3 from the middle surface is $(R_2 + x_3)\, d\theta$. The forces (as a result of the stresses) acting on the entire surface of the shell are given as

$$F_1 = R_2 d\theta \int_{-h/2}^{+h/2} \left(1 + \frac{x_3}{R_2}\right) \sigma_{11} dx_3; \tag{3.62}$$

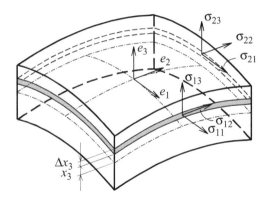

Figure 3.9: Component of stresses.

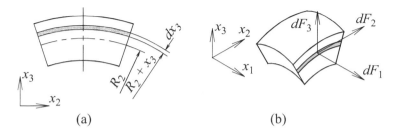

(a)   (b)

Figure 3.10: (a) A surface of an element perpendicular to x_1-axis; (b) components of a force acting on a face of an element.

$$F_2 = R_2 d\theta \int_{-h/2}^{+h/2} \left(1 + \frac{x_3}{R_2}\right) \sigma_{12} dx_3 \qquad (3.63)$$

$$F_3 = R_2 d\theta \int_{-h/2}^{+h/2} \left(1 + \frac{x_3}{R_2}\right) \sigma_{13} dx_3 \qquad (3.64)$$

The stress resultants (defined as forces per unit length along dx_2) are the membrane stress resultants $N_{\alpha\beta}$ and the shearing stress resultants Q_α, and are given as

$$N_{11} = \frac{F_1}{R_2 d\theta} = \int_{-h/2}^{+h/2} \left(1 + \frac{x_3}{R_2}\right) \sigma_{11} dx_3 \qquad (3.65)$$

$$N_{12} = \frac{F_2}{R_2 d\theta} = \int_{-h/2}^{+h/2} \left(1 + \frac{x_3}{R_2}\right) \sigma_{12} dx_3 \qquad (3.66)$$

$$Q_1 = \frac{F_3}{R_2 d\theta} = \int_{-h/2}^{+h/2} \left(1 + \frac{x_3}{R_2}\right) \sigma_{13} dx_3 \qquad (3.67)$$

and the bending stress couples $M_{\alpha\beta}$, defined as the bending and twisting moments per unit length along dx_2, are obtained in similar way to give

$$M_{11} = \int_{-h/2}^{+h/2} \left(1 + \frac{x_3}{R_2}\right) x_3 \sigma_{11} dx_3; \qquad (3.68)$$

$$M_{12} = \int_{-h/2}^{+h/2} \left(1 + \frac{x_3}{R_2}\right) x_3 \sigma_{12} dx_3 \tag{3.69}$$

Similarly, by considering the other face of the shell element which is perpendicular to the x_2-axis, other components of stress resultants and stress couples can be obtained as follows

$$N_{22} = \int_{-h/2}^{+h/2} \left(1 + \frac{x_3}{R_1}\right) \sigma_{22} dx_3 \tag{3.70}$$

$$N_{21} = \int_{-h/2}^{+h/2} \left(1 + \frac{x_3}{R_1}\right) \sigma_{21} dx_3 \tag{3.71}$$

$$Q_2 = \int_{-h/2}^{+h/2} \left(1 + \frac{x_3}{R_1}\right) \sigma_{23} dx_3 \tag{3.72}$$

$$M_{22} = \int_{-h/2}^{+h/2} \left(1 + \frac{x_3}{R_1}\right) x_3 \sigma_{22} dx_3 \tag{3.73}$$

$$M_{21} = \int_{-h/2}^{+h/2} \left(1 + \frac{x_3}{R_1}\right) x_3 \sigma_{21} dx_3 \tag{3.74}$$

For the sake of simplicity, throughout this chapter both the stress resultants and stress couples will be referred to as the generalized stress resultants.

Reissner [42] and Naghdi [30] assumed that the stresses due to membrane forces are uniform and the stresses due to bending and twisting moments vary linearly, that is

$$\left(1 + \frac{x_3}{R_\gamma}\right) \sigma_{\alpha\beta} = \frac{1}{h} N_{\alpha\beta} + \frac{12 x_3}{h^3} M_{\alpha\beta}; \quad \left\{ \begin{array}{l} \gamma = \beta \text{ if } \alpha \neq \beta \\ \gamma \neq \beta \text{ if } \alpha = \beta \end{array} \right. \tag{3.75}$$

and the transverse shear stresses vary parabolically over the thickness, so that

$$\left(1 + \frac{x_3}{R_\gamma}\right) \sigma_{\alpha 3} = \frac{3}{2h} \left[1 - \left(\frac{2 x_3}{h}\right)^2\right] Q_\alpha; \quad \gamma \neq \alpha \tag{3.76}$$

The generalized tractions at a boundary point can be defined as:

$$p_\alpha = M_{\alpha\beta} n_\beta \tag{3.77}$$

$$p_3 = Q_\alpha n_\alpha \tag{3.78}$$

$$t_\alpha = N_{\alpha\beta} n_\beta \tag{3.79}$$

where n_β are the components of the outward normal vector to the shell boundary (see Figure 3.2).

Equilibrium

The equilibrium equations for forces and moments taking into consideration the effect of the uniform load q_i acting over the entire of the middle surface area (see Figure 3.11), can be written as

$$\frac{\partial N_{11}}{\partial x_1} + \frac{\partial N_{21}}{\partial x_2} + q_1 = 0 \tag{3.80}$$

$$\frac{\partial N_{12}}{\partial x_1} + \frac{\partial N_{22}}{\partial x_2} + q_2 = 0 \tag{3.81}$$

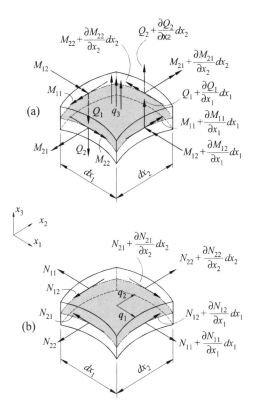

Figure 3.11: Stress resultant equilibrium of a shell element: (a) bending and shear stress resultants; (b) membrane stress resultants.

$$\frac{\partial M_{11}}{\partial x_1} + \frac{\partial M_{21}}{\partial x_2} - Q_1 = 0 \tag{3.82}$$

$$\frac{\partial M_{12}}{\partial x_1} + \frac{\partial M_{22}}{\partial x_2} - Q_2 = 0 \tag{3.83}$$

$$\frac{\partial Q_1}{\partial x_1} + \frac{\partial Q_2}{\partial x_2} - \left(\frac{N_{11}}{R_1} + \frac{N_{22}}{R_2}\right) + q_3 = 0 \tag{3.84}$$

$$N_{12} - N_{21} + \frac{M_{12}}{R_1} - \frac{M_{21}}{R_2} = 0 \tag{3.85}$$

where the effect of Q_α/R_α is ignored.

The equations of equilibrium (3.80)-(3.84) can be rewritten using indicial notation as

$$M_{\alpha\beta,\beta} - Q_\alpha = 0; \tag{3.86}$$

$$Q_{\alpha,\alpha} - k_{\alpha\beta}N_{\alpha\beta} + q_3 = 0; \tag{3.87}$$

$$N_{\alpha\beta,\beta} + q_\alpha = 0 \tag{3.88}$$

where $\alpha, \beta = 1, 2$ and $k_{12} = k_{21} = 0$.

Stress Resultant-displacement Relationships

The relationships between stress resultants, strains and displacements are as follows:

$$
\begin{aligned}
M_{\alpha\beta} &= D\frac{1-\nu}{2}\left(2\chi_{\alpha\beta} + \frac{2\nu}{1-\nu}\chi_{\gamma\gamma}\delta_{\alpha\beta}\right) \\
&= D\frac{1-\nu}{2}\left(w_{\alpha,\beta} + w_{\beta,\alpha} + \frac{2\nu}{1-\nu}w_{\gamma,\gamma}\delta_{\alpha\beta}\right)
\end{aligned} \tag{3.89}
$$

$$
\begin{aligned}
Q_\alpha &= D\frac{(1-\nu)}{2}\lambda^2\gamma_{\alpha3} \\
&= C(w_\alpha + w_{3,\alpha})
\end{aligned} \tag{3.90}
$$

$$
\begin{aligned}
N_{\alpha\beta} &= B\frac{1-\nu}{2}\left(\varepsilon_{\alpha\beta} + \varepsilon_{\beta\alpha} + \frac{2\nu}{1-\nu}\varepsilon_{\gamma\gamma}\delta_{\alpha\beta}\right) \\
&= B\frac{1-\nu}{2}\left(u_{\alpha,\beta} + u_{\beta,\alpha} + \frac{2\nu}{1-\nu}u_{\gamma,\gamma}\delta_{\alpha\beta}\right) + B\left[(1-\nu)k_{\alpha\beta} + \nu\delta_{\alpha\beta}k_{\phi\phi}\right]w_3 \\
&= N_{\alpha\beta}^{(i)} + N_{\alpha\beta}^{(ii)}
\end{aligned} \tag{3.91}
$$

where $B(= Eh/\left(1-\nu^2\right))$ is known as the tension stiffness; $D(= Eh^3/\left[12\left(1-\nu^2\right)\right])$ is the bending stiffness of the shell; $C(=\left[D\left(1-\nu\right)\lambda^2\right]/2)$ is the shear stiffness; $\lambda = \sqrt{10}/h$ is called the shear factor; and $\delta_{\alpha\beta}$ is the Kronecker delta function. The term $N_{\alpha\beta}$ is separated into $N_{\alpha\beta}^{(i)}$, which are due to in-plane displacements, and $N_{\alpha\beta}^{(ii)}$, which are due to curvature and out-of-plane displacements.

Equilibrium Equations in Terms of Displacements

By substituting (3.89)-(3.91) into (3.86)-(3.88), the equilibrium equations in term of displacements are obtained as follows:

$$
D\nabla^2 w_1 + \frac{D}{2}\left(1+\nu\right)\frac{\partial}{\partial x_2}\left(-\frac{\partial w_1}{\partial x_2} + \frac{\partial w_2}{\partial x_1}\right) - Cw_1 - C\frac{\partial w_3}{\partial x_1} = 0 \tag{3.92}
$$

$$
\frac{D}{2}\left(1+\nu\right)\frac{\partial}{\partial x_1}\left(\frac{\partial w_1}{\partial x_2} - \frac{\partial w_2}{\partial x_1}\right) + D\nabla^2 w_2 - Cw_2 - C\frac{\partial w_3}{\partial x_2} = 0 \tag{3.93}
$$

$$
C\nabla^2 w_3 + C\frac{\partial w_1}{\partial x_1} + C\frac{\partial w_2}{\partial x_2} + q_3 - B(k_{11} + \nu k_{22})\frac{\partial u_1}{\partial x_1}
$$
$$
- B(\nu k_{11} + k_{22})\frac{\partial u_2}{\partial x_2} - B(k_{11}^2 + k_{22}^2 + 2\nu k_{11}k_{22})w_3 = 0 \tag{3.94}
$$

$$
B\nabla^2 u_1 + \frac{B}{2}\left(1+\nu\right)\frac{\partial}{\partial x_2}\left(-\frac{\partial u_1}{\partial x_2} + \frac{\partial u_2}{\partial x_1}\right) + q_1 + B(k_{11} + \nu k_{22})\frac{\partial w_3}{\partial x_1} = 0 \tag{3.95}
$$

$$
\frac{B}{2}\left(1+\nu\right)\frac{\partial}{\partial x_1}\left(\frac{\partial u_1}{\partial x_2} - \frac{\partial u_2}{\partial x_1}\right) + B\nabla^2 u_2 + q_2 + B(\nu k_{11} + k_{22})\frac{\partial w_3}{\partial x_2} = 0 \tag{3.96}
$$

Equations (3.92)-(3.94) can be rewritten in a more compact form as

$$
L_{ik}^b w_k + f_i^b = 0 \tag{3.97}
$$

and similarly, equations (3.95) and (3.96) as

$$L_{\alpha\beta}^m u_\beta + f_\alpha^m = 0 \tag{3.98}$$

The Navier differential operator L_{ik}^b corresponds to shear deformable plate bending problems

$$L_{\alpha\beta}^b = \frac{D}{2}\left[(1-\nu)\left(\nabla^2 - \lambda^2\right)\delta_{\alpha\beta} + (1+\nu)\frac{\partial}{\partial x_\alpha}\frac{\partial}{\partial x_\beta}\right] \tag{3.99}$$

$$L_{\alpha 3}^b = -\frac{(1-\nu)D}{2}\lambda^2\frac{\partial}{\partial x_\alpha} \tag{3.100}$$

$$L_{3\alpha}^b = -L_{\alpha 3}^b \tag{3.101}$$

$$L_{33}^b = \frac{(1-\nu)D}{2}\lambda^2\nabla^2 \tag{3.102}$$

with $f_\alpha^b = 0$ and

$$f_3^b = q_3 - B[(1-\nu)k_{\alpha\beta} + \nu k_{\phi\phi}\delta_{\alpha\beta}]u_{\alpha,\beta} - B[k_{11}^2 + k_{22}^2 + 2\nu k_{11}k_{22})w_3 \tag{3.103}$$

The Navier differential operator $L_{\alpha\beta}^m$ corresponds to two-dimensional plane stress problems

$$L_{\alpha\beta}^m = B\nabla^2\delta_{\alpha\beta} + \frac{B(1+\nu)}{2}\frac{\partial}{\partial x_\alpha}\frac{\partial}{\partial x_\beta}(1-2\delta_{\alpha\beta}) \tag{3.104}$$

with

$$f_\alpha^m = q_\alpha + B[(1-\nu)k_{\alpha\beta} + \nu k_{\phi\phi}\delta_{\alpha\beta}]w_{3,\beta}$$

3.3.2 Integral Equation Formulations

Rotations and Out-of-plane Displacement

By utilizing the Betti's reciprocal theorem, the rotations and out-of-plane displacements for an internal source point $\mathbf{X}'$ can be written as [8]

$$w_i(\mathbf{X}') = \int_S U_{ik}^b(\mathbf{X}',\mathbf{x})p_k(\mathbf{x})\,dS - \int_S T_{ik}^b(\mathbf{X}',\mathbf{x})w_k(\mathbf{x})\,dS$$

$$+ \int_V U_{ik}^b(\mathbf{X}',\mathbf{X})f_k^b(\mathbf{X})dV \tag{3.105}$$

where $U_{ik}^b(\mathbf{X}',\mathbf{x})$ and $T_{ik}^b(\mathbf{X}',\mathbf{x})$ are the fundamental solutions for rotations and out-of-plane displacements and bending and shear tractions, respectively, as shown in Section 3.2.2. Substituting for f_k^b, we have

$$w_i(\mathbf{X}') = \int_S U_{ik}^b(\mathbf{X}',\mathbf{x})p_k(\mathbf{x})\,dS - \int_S T_{ik}^b(\mathbf{X}',\mathbf{x})w_k(\mathbf{x})\,dS$$

$$-B\int_V U_{i3}^b(\mathbf{X}',\mathbf{X})\left\{[(1-\nu)k_{\alpha\beta} + \nu k_{\phi\phi}\delta_{\alpha\beta}]u_{\alpha,\beta} - (k_{11}^2 + k_{22}^2 + 2\nu k_{11}k_{22})w_3\right\}dV$$

$$+ \int_V U_{i3}^b(\mathbf{X}',\mathbf{X})q_3(\mathbf{X})dV \tag{3.106}$$

In-plane Displacements

By choosing the $(\cdot)^*$ state to represent the fundamental state such as

$$N^{(i)}_{\theta\alpha\beta,\beta}(\mathbf{X}',\mathbf{X}) + \delta(\mathbf{X}',\mathbf{X})\delta_{\theta\alpha} = 0 \tag{3.107}$$

and making use of the Dirac delta property (3.33), integral equation for internal source point $\mathbf{X}'$ can be written as [8]

$$u_\alpha(\mathbf{X}') = \int_S U^m_{\alpha\beta}(\mathbf{X}',\mathbf{x})\, t_\beta(\mathbf{x})\, dS - \int_S T^m_{\alpha\beta}(\mathbf{X}',\mathbf{x})\, u_\beta(\mathbf{x})\, dS$$

$$+ \int_V U^m_{\alpha\beta}(\mathbf{X}',\mathbf{X})\, f^m_\beta\, dV \tag{3.108}$$

The above equation is equivalent to Somigliana's identity for and internal displacement described in Chapter 2. Substituting for f^m_α gives

$$u_\alpha(\mathbf{X}') = \int_S U^m_{\alpha\beta}(\mathbf{X}',\mathbf{x})\, t_\beta(\mathbf{x})\, dS - \int_S T^m_{\alpha\beta}(\mathbf{X}',\mathbf{x})\, u_\beta(\mathbf{x})\, dS$$

$$- B\int_V U^m_{\alpha\beta}(\mathbf{X}',\mathbf{X})\left[(1-\nu)k_{\beta\gamma} + \nu\delta_{\beta\gamma}k_{\phi\phi}\right] w_{3,\gamma}(\mathbf{X})\, dV$$

$$+ \int_V U^m_{\alpha\beta}(\mathbf{X}',\mathbf{X})\, q_\beta dV\mathbf{X} \tag{3.109}$$

where $U^m_{\alpha\beta}(\mathbf{X}',\mathbf{x})$ and $T^m_{\alpha\beta}(\mathbf{X}',\mathbf{x})$ are the fundamental solutions for in-plane displacements and membrane tractions, respectively (see Chapter 2).

3.3.3 Boundary Integral Equations

The above integrals are regular provided $r \neq 0$. If the point $\mathbf{X}'$ is taken to the boundary, that is $\mathbf{X}' \to \mathbf{x}' \in S$, the distance r tends to zero and, in the limit, the fundamental solutions exhibit singularities. A semi-circular domain with boundary S^+ is constructed around the point $\mathbf{x}'$ as shown in Figure 3.12.

By taking the point $\mathbf{X}'$ to the boundary, that is $\mathbf{X}' \to \mathbf{x}' \in S$, and assuming that the displacements w_i and u_θ satisfy Hölder continuity,

$$|w_i(\mathbf{x}) - w_i(\mathbf{x}')| < Ar^\alpha; \qquad A:\ \text{constant} > 1,\ \text{and}\ 0 < \alpha \leqslant 1 \tag{3.110}$$

equation (3.106) can be written as follows [8]:

$$w_i(\mathbf{x}') + \lim_{\varepsilon \to 0}\int_{S-S_\varepsilon+S^+} T^b_{ik}(\mathbf{x}',\mathbf{x})w_k(\mathbf{x})\, dS = \lim_{\varepsilon \to 0}\int_{S-S_\varepsilon+S^+} U^b_{ik}(\mathbf{x}',\mathbf{x})p_k(\mathbf{x})\, dS$$

$$- \lim_{\varepsilon \to 0}\int_V U^b_{i3}(\mathbf{x}',\mathbf{X})\left\{B[(1-\nu)k_{\alpha\beta} + \nu k_{\phi\phi}\delta_{\alpha\beta}]u_{\alpha,\beta} - B(k^2_{11} + k^2_{22} + 2\nu k_{11}k_{22})w_3\right\} dV$$

$$+ \lim_{\varepsilon \to 0}\int_V U^b_{i3}(\mathbf{x}',\mathbf{X})q_3(\mathbf{X})dV \tag{3.111}$$

All limiting process can be found in Appendix C. Taking into account all the limits and jump terms, the boundary integral equations can be written as

$$C^b_{ik}w_k(\mathbf{x}') + \fint_S T^b_{ik}(\mathbf{x}',\mathbf{x})w_k(\mathbf{x})\, dS = \int_S U^b_{ik}(\mathbf{x}',\mathbf{x})p_k(\mathbf{x})\, dS$$

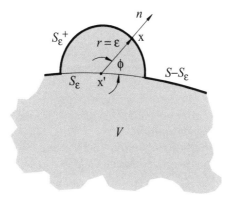

Figure 3.12: Semi-circular region around the source point when it approaches the boundary.

$$-\int_V U_{i3}^b(\mathbf{X}',\mathbf{X})\left\{B[(1-\nu)k_{\alpha\beta}+\nu k_{\phi\phi}\delta_{\alpha\beta}]u_{\alpha,\beta}-B(k_{11}^2+k_{22}^2+2\nu k_{11}k_{22})w_3\right\}dV$$

$$+\int_V U_{i3}^b(\mathbf{X}',\mathbf{X})q_3(\mathbf{X})dV \tag{3.112}$$

and equation (3.109) can be written as

$$u_\alpha(\mathbf{x}')+\lim_{\varepsilon\to0}\int_{S-S_\varepsilon+S^+}T_{\alpha\beta}^m(\mathbf{x}',\mathbf{x})\,u_\beta(\mathbf{x})\,dS$$

$$+B\lim_{\varepsilon\to0}\int_V U_{\alpha\beta}^m(\mathbf{x}',\mathbf{X})\,[(1-\nu)k_{\beta\gamma}+\nu\delta_{\beta\gamma}k_{\gamma\gamma}]\,w_{3,\gamma}(\mathbf{X})\,dV$$

$$=\lim_{\varepsilon\to0}\int_{S-S_\varepsilon+S^+}U_{\alpha\beta}^m(\mathbf{x}',\mathbf{x})\,t_\beta(\mathbf{x})\,dS+\lim_{\varepsilon\to0}\int_V U_{\alpha\beta}^m(\mathbf{x}',\mathbf{X})\,q_\beta dV \tag{3.113}$$

After the limiting process (see Appendix C), the boundary integral equation for membrane behaviour can be written as

$$C_{\alpha\beta}^m u_\beta(\mathbf{x}')+\oint_S T_{\alpha\beta}^m(\mathbf{x}',\mathbf{x})\,u_\beta(\mathbf{x})\,dS$$

$$+B\int_V U_{\alpha\beta}^m(\mathbf{x}',\mathbf{X})\,[(1-\nu)k_{\beta\gamma}+\nu\delta_{\beta\gamma}k_{\phi\phi}]\,w_{3,\gamma}(\mathbf{X})\,dV$$

$$=\int_S U_{\alpha\beta}^m(\mathbf{x}',\mathbf{x})\,t_\beta(\mathbf{x})\,dS+\int_V U_{\alpha\beta}^m(\mathbf{x}',\mathbf{X})\,q_\beta dV \tag{3.114}$$

where $C_{ik}^b(\mathbf{x}')$ and $C_{\alpha\beta}^m(\mathbf{x}')$ are the jump terms.

Equations (3.112) and (3.114) represent five boundary integral equations; the first two in (3.112) ($i=\alpha=1,2$) are for rotations, the third ($i=3$) is for the out-of-plane displacement and the two in (3.114) ($\alpha=1,2$) are for in-plane displacements, which can be used to solve shear deformable shallow shell bending problems.

3.3.4   Numerical Discretization

Discretization

Assume quadratic isoparametric boundary elements are used to describe the geometry and the function along the boundary, while for the domain, quadratic quadrilateral isoparametric elements are used to describe the geometry and constant interpolations for the function. Semi-discontinuous elements are used for the corners to avoid difficulties with traction discontinuities.

Assuming that q_3 is uniform, equation (3.112) can be rewritten (with slightly different arrangement) in a discretized forms as [8]

$$C_{ik}^b(\mathbf{x}')w_k(\mathbf{x}') + \sum_{n=1}^{N_e}\sum_{m=1}^{3} w_k^{nm} \int_{-1}^{1} T_{ik}^b(\mathbf{x}',\mathbf{x})N_m(\eta)J^n(\eta)d\eta$$

$$= \sum_{n=1}^{N_e}\sum_{m=1}^{3} p_k^{nm} \int_{-1}^{+1} U_{ik}^b(\mathbf{x}',\mathbf{x})N_m(\eta)J^n(\eta)d\eta$$

$$-B\frac{1-\nu}{2} \sum_{n=1}^{N_e}\sum_{m=1}^{3} k_{\alpha\beta}\left(u_\alpha^{nm}n_\beta^{nm} + u_\beta^{nm}n_\alpha^{nm} + \frac{2\nu}{1-\nu}u_\gamma^{nm}n_\gamma^{nm}\delta_{\alpha\beta}\right)$$

$$\times \int_{-1}^{+1} U_{i3}^b(\mathbf{x}',\mathbf{x})N_m(\eta)J^n(\eta)d\eta$$

$$+B\frac{1-\nu}{2} \sum_{k=1}^{N_c} k_{\alpha\beta}u_\alpha^k \int_{-1}^{1}\int_{-1}^{1} U_{i3,\beta}^b(\mathbf{x}',\mathbf{X})J^k(\eta_1,\eta_2)d\eta_1 d\eta_2$$

$$+B\frac{1-\nu}{2} \sum_{k=1}^{N_c} k_{\alpha\beta}u_\beta^k \int_{-1}^{1}\int_{-1}^{1} U_{i3,\alpha}^b(\mathbf{x}',\mathbf{X})J^k(\eta_1,\eta_2)d\eta_1 d\eta_2$$

$$+B\nu \sum_{k=1}^{N_c} k_{\alpha\beta}u_\gamma^k \int_{-1}^{1}\int_{-1}^{1} U_{i3,\gamma}^b(\mathbf{x}',\mathbf{X})J^k(\eta_1,\eta_2)d\eta_1 d\eta_2$$

$$-B \sum_{k=1}^{N_c} k_{\alpha\beta}[(1-\nu)k_{\alpha\beta} + \nu\delta_{\alpha\beta}k_{\gamma\gamma}]\, w_3^k \int_{-1}^{1}\int_{-1}^{1} U_{i3}^b(\mathbf{x}',\mathbf{X})J^k(\eta_1,\eta_2)d\eta_1 d\eta_2$$

$$+q \sum_{n=1}^{N_e} \int_{-1}^{1} V_{i,\alpha}^b(\mathbf{x}',\mathbf{x})n_\alpha(\eta)J^n(\eta)d\eta \qquad (3.115)$$

and assuming $q_\alpha = 0$, equation (3.114) can be rewritten with a slightly different arrangement of fundamental solutions and unknowns [8] as

$$C_{\alpha\beta}^m(\mathbf{x}')u_\beta(\mathbf{x}') + \sum_{n=1}^{N_e}\sum_{m=1}^{3} u_\beta^{nm} \int_{-1}^{1} T_{\alpha\beta}^m(\mathbf{x}',\mathbf{x})N_m(\eta)J^n(\eta)d\eta$$

$$+B \sum_{k=1}^{N_c} [k_{\beta\gamma}(1-\nu) + \nu\delta_{\beta\gamma}k_{\theta\theta}]\, w_3^k \int_{-1}^{1}\int_{-1}^{1} U_{\alpha\beta,\gamma}^m(\mathbf{x}',\mathbf{X})J^k(\eta_1,\eta_2)d\eta_1 d\eta_2$$

$$= \sum_{n=1}^{N_e} \sum_{m=1}^{3} t_\beta^{nm} \int_{-1}^{1} U_{\alpha\beta}^m(\mathbf{x}', \mathbf{x}) N_m J^n(\eta) d\eta \qquad (3.116)$$

where N_e and N_c are the number of boundary elements and internal cells, respectively, and N_m are the quadratic shape functions.

For every collocation node, equations (3.115)-(3.116) will give the following linear system of equations in a matrix form:

$$\begin{bmatrix} \mathbf{H}^b & \mathbf{H}^u \\ \mathbf{H}^w & \mathbf{H}^m \end{bmatrix}_{5\times5} \left\{ \begin{array}{c} \mathbf{w} \\ \mathbf{u} \end{array} \right\}_{5\times1} = \begin{bmatrix} \mathbf{G}^b & 0 \\ 0 & \mathbf{G}^m \end{bmatrix}_{5\times5} \left\{ \begin{array}{c} \mathbf{p} \\ \mathbf{t} \end{array} \right\}_{5\times1} + \left\{ \begin{array}{c} \mathbf{b} \\ \mathbf{0} \end{array} \right\}_{5\times1} \qquad (3.117)$$

where $\mathbf{w} = \{w_1, w_2, w_3\}^\mathsf{T}$, $\mathbf{u} = \{u_1, u_2\}^\mathsf{T}$, $\mathbf{p} = \{p_1, p_2, p_3\}^\mathsf{T}$, $\mathbf{t} = \{t_1, t_2\}^\mathsf{T}$ and $\mathbf{b} = \{0, 0, q_3\}^\mathsf{T}$; $\mathbf{H}^b$, $\mathbf{H}^m$, $\mathbf{G}^b$ and $\mathbf{G}^m$ are boundary element influence matrices for plate bending and plane stress formulations, respectively, $\mathbf{H}^u$, and $\mathbf{H}^w$ are matrices which contain coupled terms between plate bending and plane stress formulations. The matrices $\mathbf{H}^b$, $\mathbf{H}^m$, $\mathbf{H}^u$, $\mathbf{H}^w$, $\mathbf{G}^b$ and $\mathbf{G}^m$ then form shallow shell influence matrices. After performing all of the collocation process, equations (3.115)-(3.116) can be written as

$$[H]_{5M+3Nin\times5M+3Nin} \{u\}_{5M+3Nin\times1}$$
$$= [G]_{5M+3Nin\times15N_e} \{p\}_{15N_e\times1} + \{Q\}_{5M+3Nin\times1} \qquad (3.118)$$

where $[H]$ and $[G]$ are the well-known boundary element influence matrices, $\{u\}$ is the boundary and domain displacement vector, $\{p\}$ is the boundary traction vector, and $\{Q\}$ is the domain load vector. M, Nin and N_e are number of boundary nodes, internal nodes and boundary elements, respectively.

There are three possible boundary conditions, i.e. clamped, simply supported and free boundary. These boundary conditions can be summarized as follows:

Clamped boundary condition

$$w_t = 0, \quad w_n = 0, \quad w_3 = 0, \quad u_t = 0 \quad \text{and} \quad u_n = 0 \qquad (3.119)$$

Simply supported boundary condition

$$w_t = 0, \quad w_3 = 0, \quad M_n = 0 \quad \text{and} \quad (u_1 = 0 \quad \text{or} \quad u_2 = 0) \qquad (3.120)$$

Free boundary condition

$$M_t = 0, \quad M_n = 0, \quad p_3 = 0, \quad N_t = 0 \quad \text{and} \quad N_n = 0 \qquad (3.121)$$

After imposing boundary conditions, equation (3.118) can be written as

$$[A]_{5M+3Nin\times5M+3Nin} \{x\}_{5M+3Nin\times1} = \{b\}_{5M+3Nin\times1} \qquad (3.122)$$

where $[A]$ is the system matrix, $\{x\}$ is the unknown vector and $\{b\}$ is the vector of prescribed boundary values.

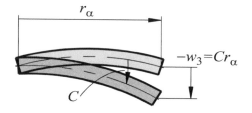

Figure 3.13: Rigid body rotations.

Treatment of Singularities

The boundary integral equations discussed above contain integrands with several different orders of singularities. These singular integrals are treated separately based on their order of singularity, as described in Chapter 11. The influence matrix $[G$ and the load vector matrix $\{Q\}$ contain weakly singular integrals, which are treated using a nonlinear coordinate transformation (see Chapter 11).

The matrix $[H$ contains strongly singular integrals, which can be computed indirectly by considering the generalized rigid body movements. This can be achieved as follows. If a traction-free problem is considered, five independent cases may be observed, that is, two rigid body rotations:

- $u_1 = 0$, $u_2 = 0$, $w_1 = C$ and $w_2 = 0$, then $w_3 = -Cr_1$,

- $u_1 = 0$, $u_2 = 0$, $w_2 = C$ and $w_1 = 0$, then $w_3 = -Cr_2$

as shown in Figure 3.13, and a rigid body out-of-plane translation:

- $u_1 = 0$, $u_2 = 0$, $w_3 = C$, $w_1 = 0$, and $w_2 = 0$

for the rotations and out-of-plane displacement integral equations, and two rigid body conditions for in-plane translations:

- $u_1 = C$, $u_2 = 0$, $w_1 = 0$, $w_2 = 0$, and $w_3 = 0$

- $u_2 = C$, $u_1 = 0$, $w_1 = 0$, $w_2 = 0$, and $w_3 = 0$

for the in-plane displacements integral equations. The term C is an arbitrary constant, and r_α denotes components of vector r in x_α coordinates.

By applying the above cases to the system of equations in (3.118), the following expressions can be written:

$$H^{i\alpha}(\mathbf{x}') = -\int_S [T^b_{i\alpha}(\mathbf{x}',\mathbf{x}) + (-r_\alpha)T^b_{i3}(\mathbf{x}',\mathbf{x})] dS$$

$$H^{i3}(\mathbf{x}') = -\int_S T^b_{i3}(\mathbf{x}',\mathbf{x}) dS$$

$$H^{(3+\theta)(3+\alpha)}(\mathbf{x}') = -\int_S T^b_{\theta\alpha}(\mathbf{x}',\mathbf{x}) dS \qquad (3.123)$$

where $H^{i\alpha}(\mathbf{x}')$, $H^{i3}(\mathbf{x}')$ and $H^{(3+\theta)(3+\alpha)}(\mathbf{x}')$ include the diagonal sub-matrix and the jump term C_{ij} in the influence matrix $[H$. All terms in the integrals in equation

(3.123) have already been computed except for the second term in the first integral. Fortunately, in the second term of the first integral, the distance r_α cancels the weak singularity in $T^b_{\alpha 3}$ and the strong singularity in T^b_{33} in the singular element under consideration.

There are also weak singular terms in the domain integrals. When these integrals are computed numerically using cell discretizations, the weak singular kernels are treated using a triangle to square transformation technique, as explained in Chapter 11.

Transformation of Domain Integrals

The domain integrals which appear in the boundary integral equations derived in the previous section can be transformed into boundary integrals with the use of the dual reciprocity technique [59].

If the membrane body forces $q_1 = q_2 = 0$, in the boundary integral equations (3.112) and (3.114) there are six domain integrals, as follows:

$$I_1 = \int_V U^b_{i3} w_3 dV, \qquad I_2 = \int_V U^b_{i3} \frac{\partial u_1}{\partial x_1} dV$$

$$I_3 = \int_V U^b_{i3} \frac{\partial u_2}{\partial x_2} dV, \qquad I_4 = \int_V U^b_{i3} q_3 dV$$

$$I_5 = \int_V U^m_{\alpha 1} \frac{\partial w_3}{\partial x_1} dV, \qquad I_6 = \int_V U^m_{\alpha 2} \frac{\partial w_3}{\partial x_2} dV \qquad (3.124)$$

As in section 2..6.5 the domain terms w_3, u_α and q_3 are approximated using radial basis functions. Here different $f^l(r)$ functions are used for each term as described next.

From the particular solution $\hat{w}_{lk}$ for plate bending which satisfy the differential equations

$$L^{b,adj}_{\alpha k} \hat{w}^l_k = 0 \qquad \text{and} \qquad L^{b,adj}_{3k} \hat{w}^l_k = f^l(r) \qquad (3.125)$$

where $L^{b,adj}_{\alpha k}$ and $L^{b,adj}_{3k}$ are adjoint operators of the original differential operator L^b_{ik} for plate bending problem in equation (3.97), and the boundary integral equations for the plate bending problem becomes

$$C^b_{ik}(\mathbf{x}')\hat{w}^l_k(\mathbf{x}') = \int_S U^b_{ik}(\mathbf{x}', \mathbf{x})\hat{p}^l_k dS - \int_S T^b_{ik}(\mathbf{x}', \mathbf{x})\hat{w}^l_k dS$$

$$+ \int_V U^b_{i3}(\mathbf{x}', \mathbf{X}) f^l dV \qquad (3.126)$$

which implies that

$$I_1 = \sum_{l=1}^{N+L} \left(C^b_{ik} \hat{w}^l_k - \int_S U^b_{ik}(\mathbf{x}', \mathbf{x})\hat{p}^l_k dS + \int_S U^b_{ik}(\mathbf{x}', \mathbf{x})\hat{w}^l_k dS \right) \mathbf{F}^{-1} w_3 \qquad (3.127)$$

The particular solutions $\hat{w}^l_k$ and $\hat{p}^l_k$ for the radial basis function $f^l(r) = 1 + r$ were derived in [59] and are given in Appendix D.

Similar to the previous procedure, if

$$L^{b,adj}_{\alpha k} \overline{\hat{w}}^l_k = 0 \qquad \text{and} \qquad L^{b,adj}_{3k} \overline{\hat{w}}^l_k = \frac{\partial f^l(r)}{\partial x_\alpha} = \frac{x_\alpha}{r} \qquad (3.128)$$

the domain integral

$$I_2 = \sum_{l=1}^{N+L} \left(C_{ik}^b \overline{\widehat{w}}_k^l - \int_S U_{ik}^b(\mathbf{x}',\mathbf{x}) \overline{\widehat{p}}_k^l dS + \oint_S T_{ik}^b(\mathbf{x}',\mathbf{x}) \overline{\widehat{w}}_k^l dS \right) \mathbf{F}^{-1} u_1 \qquad (3.129)$$

for $\alpha = 1$, and

$$I_3 = \sum_{l=1}^{N+L} \left(C_{ik}^b(\mathbf{x}') \overline{\widehat{w}}_k^l - \int_S U_{ik}^b(\mathbf{x}',\mathbf{x}) \overline{\widehat{p}}_k^l dS + \oint_S T_{ik}^b(\mathbf{x}',\mathbf{x}) \overline{\widehat{w}}_k^l dS \right) \mathbf{F}^{-1} u_2 \qquad (3.130)$$

for $\alpha = 2$.

Domain integral I_4 can be solved like equation (3.127) by replacing w_3 with q_3, to give

$$I_4 = \sum_{l=1}^{N+L} \left(C_{ik}^b \widehat{w}_k^l - \int_S U_{ik}^b(\mathbf{x}',\mathbf{x}) \widehat{p}_k^l dS + \oint_S T_{ik}^b(\mathbf{x}',\mathbf{x}) \widehat{w}_k^l dS \right) \mathbf{F}^{-1} q_3 \qquad (3.131)$$

The integral I_5 can be evaluated from the particular solution $\widehat{u}_\alpha^l$ for the two-dimensional plane stress elasticity problem, that satisfies the differential equation

$$L_{1\alpha}^{m,adj} \widehat{u}_\alpha^1 = \frac{\partial f^l(r)}{\partial x_1} = \frac{x_1}{r} \qquad \text{and} \qquad L_{2\alpha}^{m,adj} \widehat{u}_\alpha^l = 0 \qquad (3.132)$$

to give

$$I_5 = \left(\sum_{l=1}^{N+L} C_{\alpha\beta}^m \widehat{u}_\beta^l - \int_S U_{\alpha\beta}^m(\mathbf{x}',\mathbf{x}) \widehat{t}_\beta^l dS + \oint_S T_{\alpha\beta}^m(\mathbf{x}',\mathbf{x}) \widehat{u}_\beta^l dS \right) \mathbf{F}^{-1} w_3 \qquad (3.133)$$

The last domain integral I_6 can be obtained from the particular solution $\overline{\widehat{u}}_\alpha^l$ for the two-dimensional plane stress elasticity problem that the differential equation satisfies

$$L_{1\alpha}^{m,adj} \overline{\widehat{u}}_\alpha^l = 0 \qquad \text{and} \qquad L_{2\alpha}^{m,adj} \overline{\widehat{u}}_\alpha^l = \frac{\partial f^l(r)}{\partial x_2} = \frac{x_2}{r} \qquad (3.134)$$

to give

$$I_6 = \left(\sum_{l=1}^{N+L} C_{\alpha\beta}^m \overline{\widehat{u}}_\beta^l - \int_S U_{\alpha\beta}^m(\mathbf{x}',\mathbf{x}) \overline{\widehat{t}}_\beta^l dS + \oint_S S_{\alpha\beta}^m(\mathbf{x}',\mathbf{x}) \overline{\widehat{u}}_\beta^l dS \right) \mathbf{F}^{-1} w_3 \qquad (3.135)$$

3.3.5   Stress Resultants Integral Equations

The stress resultants at domain point $\mathbf{X}'$ can be evaluated from (3.106) and (3.109) by using relationships in equations (3.89)-(3.91), to give

$$M_{\alpha\beta}(\mathbf{X}') = \int_S D_{\alpha\beta k}^b(\mathbf{X}',\mathbf{x}) p_k(\mathbf{x}) dS - \int_S S_{\alpha\beta k}^b(\mathbf{X}',\mathbf{x}) w_k(\mathbf{x}) dS$$

$$+ \int_V D_{\alpha\beta k}^b(\mathbf{X}',\mathbf{X}) f_k^b dV \qquad (3.136)$$

$$Q_\beta(\mathbf{X'}) = \int_S D^b_{3\beta k}(\mathbf{X'}, \mathbf{x}) p_k(\mathbf{x}) dS - \int_S S^b_{3\beta k}(\mathbf{X'}, \mathbf{x}) w_k(\mathbf{x}) dS$$

$$+ \int_V D^b_{3\beta 3}(\mathbf{X'}, \mathbf{X}) f^b_k dV \tag{3.137}$$

and

$$N_{\alpha\beta}(\mathbf{X'}) = \int_S D^m_{\alpha\beta\gamma}(\mathbf{X'}, \mathbf{x}) t_\gamma(\mathbf{x}) dS - \int_S S^m_{\alpha\beta\gamma}(\mathbf{X'}, \mathbf{x}) u_\gamma(\mathbf{x}) dS$$

$$+ \int_V D^m_{\alpha\beta\gamma}(\mathbf{X'}, \mathbf{X}) f^b_\gamma dV \tag{3.138}$$

The kernels $D^b_{i\beta k}$ and $S^b_{i\beta k}$ are a linear combination of the first derivatives of U^b_{ij}, T^b_{ij}, and the kernels $D^m_{\alpha\beta\gamma}$ and $S^m_{\alpha\beta\gamma}$ are linear combination of the first derivatives of $U^m_{\alpha\beta}$ and $T^m_{\alpha\beta}$ (see Chapter 2). The expression of S^b_{ijk}, D^m_{ijk} were presented in Section 3.2.2.

3.3.6 Hypersingular Integral Equations

The stress resultant boundary integral equations are formed by considering the behaviour of equations (3.136)-(3.138) as $\mathbf{X'}$ approaches $\mathbf{x'}$ on boundary S. A semi-circular domain with boundary S^+ is constructed around the point $\mathbf{x'}$ as shown in Figure 3.12. Taking the limit as $\mathbf{X'} \to \mathbf{x'}$, equations (3.136)-(3.138) can be rewritten as follows:

$$M_{\alpha\beta}(\mathbf{x'}) + \lim_{\varepsilon \to 0} \int_{S-S_\varepsilon+S^+} S^b_{\alpha\beta k}(\mathbf{x'}, \mathbf{x}) w_k(\mathbf{x}) dS$$

$$= \lim_{\varepsilon \to 0} \int_{S-S_\varepsilon+S^+} D^b_{\alpha\beta k}(\mathbf{x'}, \mathbf{x}) p_k(\mathbf{x}) dS + \lim_{\varepsilon \to 0} \int_V D^b_{\alpha\beta k}(\mathbf{x'}, \mathbf{X}) f^b_k dV \tag{3.139}$$

$$Q_\beta(\mathbf{x'}) + \lim_{\varepsilon \to 0} \int_{S-S_\varepsilon+S^+} S^b_{3\beta k}(\mathbf{x'}, \mathbf{x}) w_k(\mathbf{x}) dS$$

$$= \lim_{\varepsilon \to 0} \int_{S-S_\varepsilon+S^+} D^b_{3\beta k}(\mathbf{x'}, \mathbf{x}) p_k(\mathbf{x}) dS + \lim_{\varepsilon \to 0} \int_V D^b_{3\beta k}(\mathbf{x'}, \mathbf{X}) f^b_k dV \tag{3.140}$$

$$N_{\alpha\beta}(\mathbf{x'}) + \lim_{\varepsilon \to 0} \int_{S-S_\varepsilon+S^+} S^m_{\alpha\beta\gamma}(\mathbf{x'}, \mathbf{x}) u_\gamma(\mathbf{x}) dS$$

$$= \lim_{\varepsilon \to 0} \int_{S-S_\varepsilon+S^+} D^m_{\alpha\beta\gamma}(\mathbf{x'}, \mathbf{x}) t_\gamma(\mathbf{x}) dS + \lim_{\varepsilon \to 0} \int_V D^m_{\alpha\beta\gamma}(\mathbf{x'}, \mathbf{X}) f^m_\gamma dV \tag{3.141}$$

Equations (3.139) and (3.140) represent the bending and shear stress resultant boundary integral equations, respectively, while equations (3.141) represent the membrane stress resultant boundary integral equations at the boundary point $\mathbf{x'}$.

The generalized displacements w_i and u_α are required to be $C^{1,\alpha}$, $(0 < \alpha < 1)$, and the generalized tractions p_i and t_α are required to be $C^{0,\alpha}$, $(0 < \alpha < 1)$, for the principal-value integrals to exist. To satisfy the continuity requirements, the point $\mathbf{x'}$ is assumed to be on a smooth boundary. The evaluation of the integrals in equations (3.139) and (3.141) is reported in Appendix C. In the limiting processes, some integrals in equations (3.139) and (3.141) lead to a jump on the stress resultants. Taking into account all the limits and the jump terms, as $\varepsilon \to 0$, for a source point on a smooth boundary, stress resultant integral equations are obtained as follows

$$\frac{1}{2}M_{\alpha\beta}(\mathbf{x}') + \oint_S S^b_{\alpha\beta\gamma}(\mathbf{x}',\mathbf{x})w_\gamma(\mathbf{x})dS + \oint_S S^b_{\alpha\beta3}(\mathbf{x}',\mathbf{x})w_3(\mathbf{x})dS$$

$$= \oint_S D^b_{\alpha\beta\gamma}(\mathbf{x}',\mathbf{x})p_\gamma(\mathbf{x})dS + \int_S D^b_{\alpha\beta3}(\mathbf{x}',\mathbf{x})p_3(\mathbf{x})dS$$

$$+ \int_V D^b_{\alpha\beta k}(\mathbf{x}',\mathbf{X})f^b_k dV \tag{3.142}$$

$$\frac{1}{2}Q_\beta(\mathbf{x}') + \oint_S S^b_{3\beta\gamma}(\mathbf{x}',\mathbf{x})w_\gamma(\mathbf{x})dS + \oint_S S^b_{3\beta3}(\mathbf{x}',\mathbf{x})w_3(\mathbf{x})dS$$

$$= \int_S D^b_{3\beta\gamma}(\mathbf{x}',\mathbf{x})p_\gamma(\mathbf{x})dS + \oint_S D^b_{3\beta3}(\mathbf{x}',\mathbf{x})p_3(\mathbf{x})dS$$

$$+ \int_V D^b_{3\beta k}(\mathbf{x}',\mathbf{X})f^b_k dV \tag{3.143}$$

$$\frac{1}{2}N_{\alpha\beta}(\mathbf{x}') + \oint_S S^m_{\alpha\beta\gamma}(\mathbf{x}',\mathbf{x})u_\gamma(\mathbf{x})dS$$

$$+ B\left[k_{\alpha\gamma}(1-\nu) + \nu\delta_{\alpha\gamma}k_{\phi\phi}\right]\oint_S S^m_{\alpha\beta\gamma}(\mathbf{X}',\mathbf{x})w_3(\mathbf{x})n_\gamma(\mathbf{x})dS$$

$$= \oint_S D^m_{\alpha\beta\gamma}(\mathbf{x}',\mathbf{x})t_\gamma(\mathbf{x})dS + \int_V D^b_{\alpha\beta\gamma}(\mathbf{x}',\mathbf{X})f^m_\gamma dV$$

$$+ \frac{1}{2}B\left[(1-\nu)k_{\alpha\beta} + \nu\delta_{\alpha\beta}k_{\phi\phi}\right]w_3(\mathbf{x}') \tag{3.144}$$

Equations (3.142), (3.143) and (3.144) represent five stress resultant integral equations for boundary point $\mathbf{x}'$ on a smooth boundary S.

Transformation of Domain Integrals

Domain integrals in the hypersingular integral equations can be transformed into boundary integrals using the dual reciprocity technique [60, 63]. To evaluate internal stress resultants using integral equations (3.142)-,(3.144) there are six domain integrals, as follows:

$$I_7 = \int_V D^b_{i\beta3}w_3 dV, \qquad I_8 = \int_V D^b_{i\beta3}\frac{\partial u_1}{\partial x_1}dV$$

$$I_9 = \int_V D^b_{i\beta3}\frac{\partial u_2}{\partial x_2}dV, \qquad I_{10} = \int_V D^b_{i\beta3}q_3 dV$$

$$I_{11} = \int_V D^m_{\alpha\beta1}\frac{\partial w_3}{\partial x_1}dV, \qquad I_{12} = \int_V D^m_{\alpha\beta2}\frac{\partial w_3}{\partial x_2}dV \tag{3.145}$$

The procedure is similar to that described in Section 3.3.4 using the particular solutions for two-dimensional plane-stress and plate bending given in Appendix D.

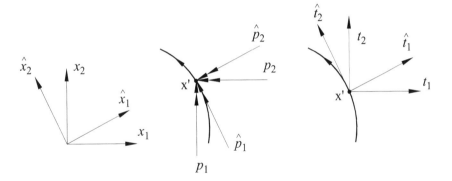

Figure 3.14: Local and global coordinate system at $\mathbf{x}'$.

3.3.7 Boundary Stress Resultants

The boundary stress resultants can be evaluated from the local tractions and by computing the generalized local strains on the boundary using displacement derivatives. Consider the local coordinate system shown in Figure 3.14. The generalized displacements ($\hat{w}_i$ and $\hat{u}_\theta$) and tractions ($\hat{p}_i$, $\hat{t}_\theta$) at point $\mathbf{x}'$ in the local system of coordinates are related to those in the global coordinate system, as follows:

$$\hat{w}_i = \hat{e}_{ij}w_j \qquad \text{and} \qquad \hat{u}_\theta = \hat{e}_{ij}u_\theta$$
$$\hat{p}_i = \hat{e}_{ij}p_j \qquad \text{and} \qquad \hat{t}_\theta = \hat{e}_{ij}t_\theta \tag{3.146}$$

and the tensor $\hat{e}_{ij}$ is the rotation matrix, and can be written in terms of the known normal components by

$$\hat{e}_{ij} = \begin{bmatrix} n_1 & n_2 & 0 \\ -n_2 & n_1 & 0 \\ 0 & 0 & 1 \end{bmatrix} \tag{3.147}$$

From the equilibrium of stress resultants in the local coordinate system, we have

$$\hat{M}_{1\alpha} = \hat{p}_\alpha \tag{3.148}$$
$$\hat{Q}_1 = \hat{p}_3 \tag{3.149}$$
$$\hat{N}_{1\alpha} = \hat{t}_\alpha \tag{3.150}$$

Other components of the local stress resultant tensor can be evaluated using equations (3.89)-(3.91) as follows:

$$\hat{M}_{22} = \nu\hat{p}_1 + D\left(1 - \nu^2\right)\hat{w}_{2,2} \tag{3.151}$$
$$\hat{Q}_2 = \frac{D(1-\nu)\lambda^2}{2}[\hat{w}_2 + \hat{w}_{3,2} \tag{3.152}$$
$$\hat{N}_{22} = \nu\hat{t}_1 + B\left(1 - \nu^2\right)\hat{u}_{2,2} + B\left(1 - \nu^2\right)\hat{k}_{22}\hat{w}_3 \tag{3.153}$$

The displacement approximation in terms of the element shape functions N_k may be written as

$$\hat{w}_j = N_k\hat{w}_j^k \tag{3.154}$$
$$\hat{u}_\theta = N_k\hat{u}_\theta^k \tag{3.155}$$

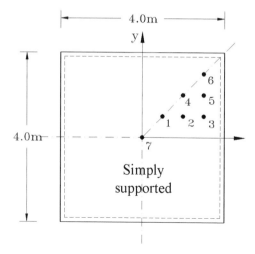

Figure 3.15: Simply supported thin square plate.

where $\hat{w}_j^k$ and $\hat{u}_\theta^k$ are the local generalized boundary displacements. The displacement derivatives in equation (3.151) can be rewritten in the following form:

$$\hat{w}_{j,2} = N_{k,2}\hat{w}_j^k = \frac{\partial N_k}{\partial\eta}\hat{e}_{ji}\hat{w}_i^k\frac{\partial\eta}{\partial\hat{x}_2(\mathbf{x}')} \tag{3.156}$$

$$\hat{u}_{\alpha,2} = N_{k,2}\hat{u}_\alpha^k = \frac{\partial N_k}{\partial\eta}\hat{e}_{\alpha i}\hat{u}_\alpha^k\frac{\partial\eta}{\partial\hat{x}_2(\mathbf{x}')} \tag{3.157}$$

The local curvature term $\hat{k}_{22}$ can be obtained from the following transformation:

$$\hat{k}_{22} = n_1^2 k_{11} + n_2^2 k_{22} \tag{3.158}$$

and hence the local boundary stress resultant tensor can be evaluated. The global stress resultant tensor can be evaluated via the following transformation:

$$M_{\alpha\beta} = \hat{e}_{\theta\alpha}\hat{e}_{S\beta}\hat{M}_{\theta S} \tag{3.159}$$

$$Q_\alpha = \hat{e}_{\beta\alpha}\hat{Q}_\beta \tag{3.160}$$

$$N_{\alpha\beta} = \hat{e}_{\theta\alpha}\hat{e}_{S\beta}\hat{N}_{\theta S} \tag{3.161}$$

3.4 Examples

In this section, several examples are presented for plate and shell structures. Application of the multi-domain BEM formulation to plate-structures is also presented.

3.4.1 Simply Supported Thin Square Plate

A square plate of 4m side length and simply supported from all sides, as shown in Figure 3.15, was analyzed by Rashed *et al.* [38] using displacement and traction

Table 3.1: Thin square plate, displacements at internal points.

Pt	$\phi_x \times -10^2 (rad)$			$\phi_y \times -10^2 (rad)$			$w \times -10^2 (m)$		
	Exact	DBIE	TBIE	Exact	DBIE	TBIE	Exact	DBIE	TBIE
1	0.184	0.184	0.185	0.184	0.184	0.185	0.613	0.615	0.618
2	0.355	0.355	0.356	0.142	0.142	0.143	0.477	0.478	0.481
3	0.490	0.490	0.492	0.078	0.078	0.078	0.264	0.265	0.266
4	0.275	0.275	0.276	0.275	0.275	0.276	0.373	0.373	0.375
5	0.382	0.382	0.383	0.151	0.151	0.152	0.207	0.207	0.208
6	0.212	0.212	0.213	0.212	0.212	0.213	0.115	0.115	0.116
7	0.000	0.000	0.000	0.000	0.000	0.000	0.709	0.711	0.712

Table 3.2: Thin square plate, bending moments at internal points.

Pt	$M_{xx}(tf.m/m)$			$M_{xy}(tf.m/m)$			$M_{yy}(tf.m/m)$		
	Exact	DBIE	TBIE	Exact	DBIE	TBIE	Exact	DBIE	TBIE
1	-0.440	-0.440	-0.440	0.038	0.038	0.038	-0.440	-0.440	-0.440
2	-0.374	-0.375	-0.375	0.072	0.072	0.072	-0.350	-0.350	-0.351
3	-0.241	-0.241	-0.241	0.096	0.096	0.096	-0.203	-0.203	-0.203
4	-0.301	-0.302	-0.302	0.137	0.137	0.137	-0.301	-0.302	-0.302
5	-0.198	-0.198	-0.198	0.186	0.186	0.186	-0.177	-0.177	-0.177
6	-0.121	-0.121	-0.121	0.258	0.258	0.258	-0.121	-0.121	-0.121
7	-0.491	-0.490	-0.490	0.000	0.000	0.000	-0.491	-0.490	-0.490

boundary integral equations for shear deformable plates. A uniform load -0.64 tf/m^2 is applied over the plate domain. Seven internal points were considered (see Figure 3.15). Due to the problem symmetry, only one-quarter of the plate was modelled in the analysis for points 1 to 6. The results for point 7, which is located at the centre of the plate are obtained by modelling the complete plate. Tables 3.1, 3.2 and 3.3 present the displacements, bending moments and shear stresses at the internal points shown in Figure 3.15. The exact results for this problem can be found in [50], and the corresponding numerical results for both the Displacement Boundary Integral Equation (DBIE) and the Traction Boundary Integral Equation (TBIE) are based on a model with 64 boundary elements. Referring to [20], accurate results for the DBIE can be achieved using only eight continuous elements; but it was found that by employing 32 elements the results for the TBIE have an error of 2% to 7%, whereas with 64 elements the error is within 0.6%.

The direct evaluation of stress resultants on the boundary is an important application of the hypersingular integral equations. To demonstrate this application, the same plate was considered and analyzed twice to compare the stress resultants on a boundary line (y-axis) for a quarter of the plate with the stress resultants when this line is considered as an internal line for the full plate analysis. In the first analysis, a quarter of the plate was considered by employing 16 boundary elements per side. The boundary stress resultants along the boundary along the y-axis are computed using the Shape Function Differentiation (SFD) method and using the hypersingular

Table 3.3: Thin square plate, shear forces at internal points.

Pt	$Q_x(tf/m)$			$Q_y(tf/m)$		
	Exact	DBIE	TBIE	Exact	DBIE	TBIE
1	0.1529	0.1527	0.1531	0.1529	0.1527	0.1531
2	0.3278	0.3275	0.3279	0.1200	0.1199	0.1203
3	0.5471	0.5467	0.5474	0.0668	0.0667	0.0671
4	0.2613	0.2610	0.2614	0.2613	0.2610	0.2614
5	0.4505	0.4502	0.4506	0.1485	0.1482	0.1485
6	0.2713	0.2710	0.2713	0.2713	0.2710	0.2713
7	0.0000	0.0000	0.0000	0.0000	0.0000	0.0000

integral equation (SIE). In the second analysis, the complete plate was reanalyzed by using 16 continuous boundary elements per side. The stress resultants were computed at the internal points, which are in the same place as the boundary points in the first analysis. Figures 3.16 and 3.17 show the shear and the bending stress resultants, respectively, along the line considered.

3.4.2 L-shaped Plate Structure

To demonstrate application of the multi-region BEM to Reissner plate theory, the problem of an L-shaped plate structure as shown in Figure 3.18 was analyzed by Dirgantara and Aliabadi [10]. The problem consists of two rectangular plates of the same size and thickness joined together to form an L-shape. The properties of the plates are: $L_1 = 1.0$; $L_2/L_1 = 1.0$; $L_3/L_1 = 2.0$; $t_1/L_1 = t_2/L_1 = 0.1$; $EL_1/q = 100\,000$ and $\nu = 0.0$. The plate is loaded by a distributed loading q along the tip edge of the horizontal plate as shown in Figure 3.18. Table 3.9 shows the BEM results for the tip deflection, together with exact solutions.

Table 3.4: Normalized deflection w/L_1 at the tip of an L-shape plate structure.

Deflection w/L_1	$\theta = 91^0$	$\theta = 95^0$	$\theta = 120^0$
BEM	0.16136	0.16493	0.16483
Bernoulli-Euler beam theory	0.16102	0.16460	0.16455
Exact	0.16142	0.16497	0.16476

3.4.3 Cantilever Beam

A cantilever beam with an H-shaped cross-section is subjected to a uniformly distributed load p, as shown in Figure 3.19. The beam is assembled with five flat plates and two interface lines. Material constants for each plate are the same and taken as Poisson's ratio $\nu = 0.3$ and Young's modulus E. The geometry parameters are chosen as: the width of middle plate $W = 0.5H$, where H denotes the height of the left-hand side and right-hand side plates, thickness for each plates $h = 0.01H$ and the length

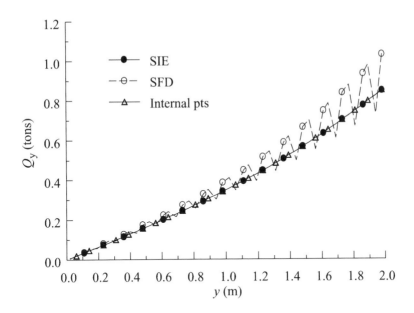

Figure 3.16: Shear stress resultant along the y-axis for the simply supported plate.

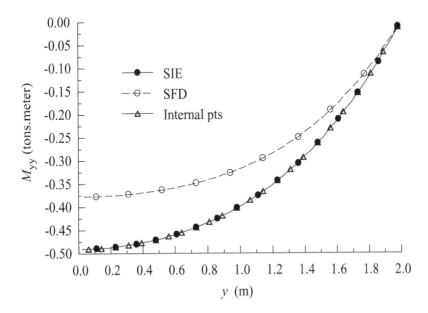

Figure 3.17: Bending stress resultant along the y-axis for the simply supported plate.

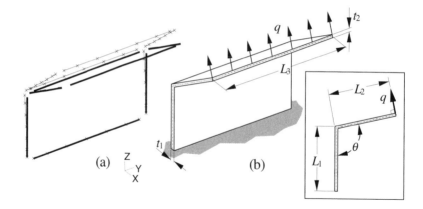

Figure 3.18: L-shaped plate subjected to distributed loading along the tip edge.

of beam $L = 2H$. The problem was analyzed by Wen, Aliabadi and Young [64] using 160 continuous quadratic elements. The deflection w_3 and membrane forces N_{11} and N_{12} in the left-hand side plate local coordinate system are plotted in Figures 3.20, 3.21 and 3.22. The results by the classical beam theory and elasticity are given in the same figures for comparison. As can be seen from these figures, the BEM results are in good agreement with analytical solutions.

3.4.4 X-core Structure

Consider an X-core structure loaded by uniform pressure q on the top plates, as shown in Figure 3.23. Sixteen plates are assembled with twelve joint lines, and are fixed at their ends as shown in Figure 10. Each plate is made of the same material with Young's modulus $E = 70$ GPa and Poisson's ratio $\nu = 0.3$. The problem was analyzed in [63] using 240 continuous quadratic elements. The boundary element mesh for BEM and FEM analysis are shown in Figure 3.24. BEM results for deflection at the free end on the top ($x_2 = 0$, $x_3 = 0.5$m) and bottom ($x_2 = 0$, $x_3 = 0$) of the structure are plotted in Figure 3.25. FEM results are presented in these figures for comparison, and good agreement is obtained.

3.4.5 Shallow Spherical Shell

In this example, a shallow spherical cap (see Figure 3.26) as analyzed by Dirgantara and Aliabadi [8] is presented. The geometric and material properties of the cap are as follows: $a = 5$; $h/a = 0.02$; $k_{11} = k_{22} = 1/R$, where $a/R = 0.05$. The cap is loaded with uniform pressure q_0 with the ratio of $E/q_0 = 210\,000$ and Poisson's ratio $\nu = 0.3$. Two types of boundary conditions employed are: clamped and simply supported. For the clamped edge problem, the boundary conditions are $w_i = 0$ and $u_\alpha = 0$ along the boundary, while for the simply supported edge, $M_n = 0$, $w_3 = 0$, and $u_2 = 0$ along the boundary.

Three different BEM meshes were used. Mesh A has 8 boundary elements and 9 cells, Mesh B has 12 boundary elements and 30 cells, Mesh C has 16 quadratic boundary elements and 81 constant cells. For comparison, three FEM meshes were

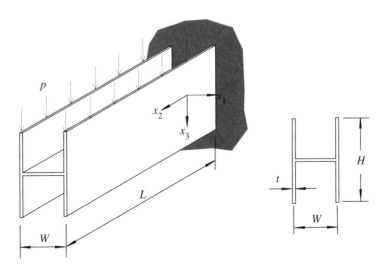

Figure 3.19: H-shape section beam under load p.

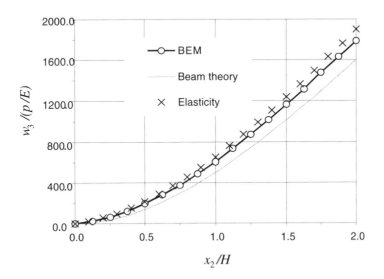

Figure 3.20: Deflection w_3 along the x_2-axis.

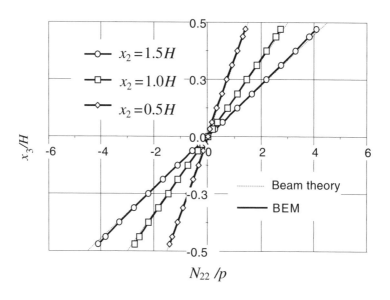

Figure 3.21: Internal stress N_{22}/p at different points.

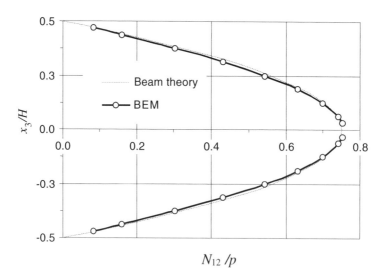

Figure 3.22: Internal stress N_{12}/p at different points.

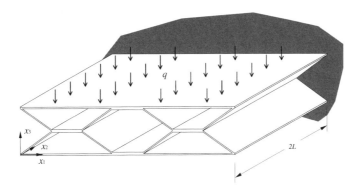

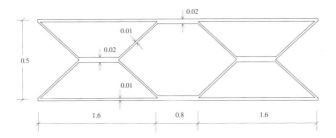

Figure 3.23: X-core section.

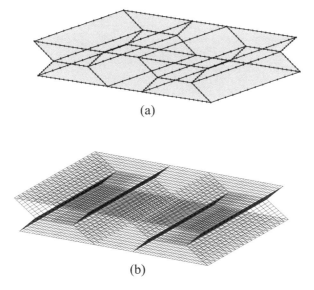

Figure 3.24: Boundary elements and finite elements meshes of X-core section.

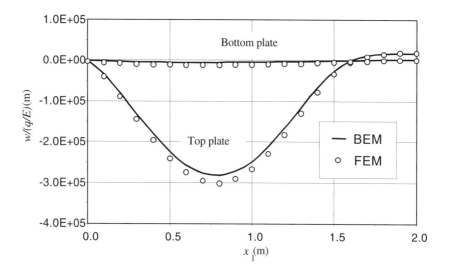

Figure 3.25: Deflection at the top and bottom plates.

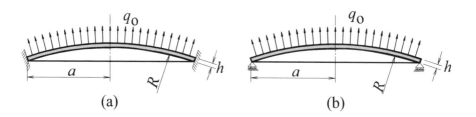

Figure 3.26: Circular shallow spherical shell: uniformly loaded. (a) Clamped edge; (b) simply supported edge.

also used: Mesh D has 120 linear shell elements, Mesh E has 240 elements, and Mesh F has 540 elements.

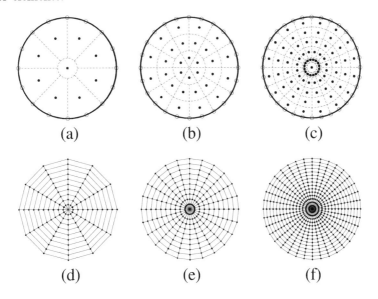

Figure 3.27: BEM and FEM meshes.

Table 3.5: Boundary results of the clamped circular shell.

Clamped	B E M			F E M		
	mesh A	mesh B	mesh C	mesh D	mesh E	mesh F
$M_n/\left(a^2 q_0\right)$	0.05658	0.06186	0.06028	0.05658	0.05911	0.05966
$Q_n/\left(a q_0\right)$	0.22300	0.29331	0.28588	--	--	--
$N_n/\left(a q_0\right)$	6.42208	4.63500	4.54572	4.5880	4.6358	4.5758

Tables 3.5-3.6 shows results obtained using the BEM and FEM. Results for the out-of-plane displacements are shown for both the BEM and FEM in Figure 3.28. All three BEM meshes are in excellent agreement with FEM mesh F for the clamped edge and FEM mesh E for the simply supported edge. It can be seen that, using the BEM formulation, convergence can be achieved with only a small number of elements and cells, while in the FEM, convergence was achieved after using a comparatively larger number of elements.

Comparison between the cell BEM formulation [8] and the dual reciprocity technique [59] were made by Dirgantara [9] by analyzing the clamped spherical shells for different ratios of a/R. The properties of these caps are the same as before, except the curvature of the shells are varied within the range of $ak_{11} = ak_{22} = a/R = 0.0 - 1.0$, where the ratio $a/R = 0.0$ represents a flat circular plate and $a/R = 1.0$ represents half of a sphere. The mesh used for the BEM with cells included 16 boundary elements and 81 constant cells, and the mesh for the BEM with the dual reciprocity technique included 16 boundary elements and 81 domain points. In the tests carried

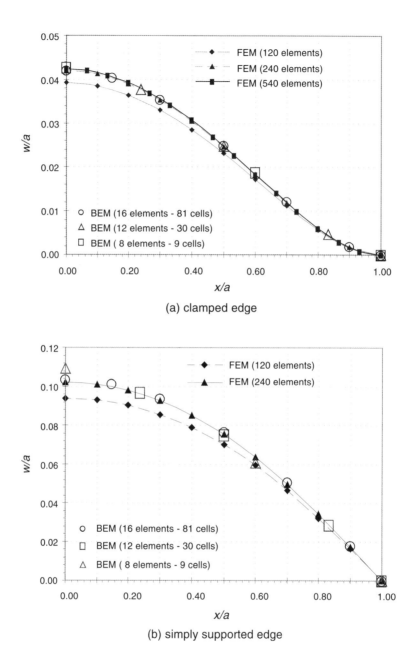

Figure 3.28: Out-of-plane displacement along the centre line of the shallow spherical shell.

Table 3.6: Boundary results of the simply supported circular shell.

Simply supported	B E M			F E M	
	mesh A	mesh B	mesh C	mesh D	mesh E
ϕ_n	0.11142	0.16491	0.16956	0.1618	0.1722
$Q_n/(aq_0)$	0.68831	0.68219	0.72014	$--$	$--$
$N_n/(aq_0)$	14.4539	11.8307	11.2759	10.3358	10.6640

out, both BEM models were found to be in excellent agreement ($< 1\%$) with each other and the FEM mesh F, as can be seen in Figure 3.29.

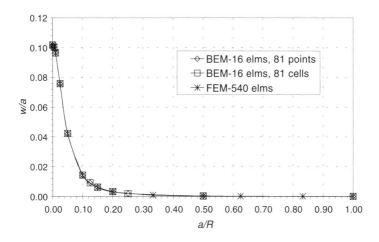

Figure 3.29: Out-of-plane displacement at the centre of clamped spherical shell for different ratio of a/R.

3.4.6 Simply Supported Spherical Shell

In this example, a square shallow spherical cap (see Figure 3.30) as analyzed by Dirgantara and Aliabadi [8], is presented. Several shells with different thickness h were analyzed. The properties of the caps are as follows: $a = 16$; $h/a = 0.02, 0.10$, 0.20 and 0.40; $ak_{11} = ak_{22} = a/R = 1/6$; $E/q = 10^5$ and $\nu = 0.3$. The boundary conditions for a simply supported edge are: $M_n = 0, w_3 = 0$, and $u_2 = 0$ along the boundary.

Three different BEM meshes are used. Mesh A has 20 boundary elements and 25 cells, Mesh B has 28 boundary elements and 49 cells, and Mesh C has 36 quadratic boundary elements and 81 constant cells.

This example was also analyzed using the dual reciprocity technique [9]. Three different BEM meshes were used. Mesh D has 20 boundary elements and 25 domain points, Mesh E has 20 boundary elements and 49 domain points, and Mesh F has 32 boundary elements and 49 domain points.

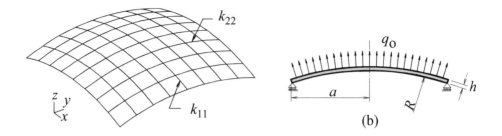

Figure 3.30: Simply supported square shallow shell: uniformly loaded.

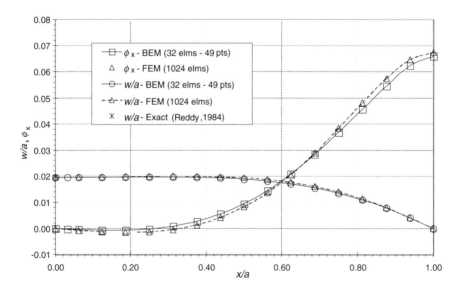

Figure 3.31: Dispacements along the x_1-axis for the simply supported shallow spherical shell.

Table 3.7: Boundary results of the clamped circular shells with a hole: vertical shear force stress resultant along the hole edge.

	BEM	*FEM*	Δ
— *outer boundary* —			
$M_n/\left(a^2 q_0\right)$	-0.006892	-0.006921	0.42 %
$Q_n/\left(a q_0\right)$	-0.008632	$--$	$--$
$N_n/\left(a q_0\right)$	0.474654	0.510060	6.94 %
— *inner boundary* —			
ϕ_n	0.0289094	0.029175	0.91 %
w/a	0.014777	0.014962	1.23 %

Table 3.8: Normalized deflection w/a (at $x = 0, y = 0$) of a simply supported square spherical shell: uniformly loaded ($w/a \times 10^3$).

h/a	0.0200	0.1000	0.2000	0.4000
BEM (20 elms-25 cells)	19.1731	2.9896	0.6816	0.1180
BEM (28 elms-49 cells)	19.3988	3.0320	0.6860	0.1183
BEM (36 elms-81 cells)	19.4850	3.0502	0.6879	0.1184
BEM (20 elms-25 points)	19.4331	3.0224	0.6893	0.1213
BEM (20 elms-49 points)	19.4569	3.0491	0.6924	0.1216
BEM (32 elms-49 points)	19.5850	3.0528	0.6926	0.1216
Exact - SDT (Reddy)	19.6163	3.1059	0.7041	0.1235
Exact - CT (Reddy)	19.6206	3.0952	0.6795	0.1042

SDT= Shear deformable theory; CT= Classical theory

Table 3.5 and Figure 3.31 show the results of this example. All the results from the BEM meshes show good agreement with exact solutions [40] and an FEM model with 1024 elements. Comparison of shear deformation theory and classical theory shows that the effect of shear deformation is significant for h/a bigger than 0.2. The differences between classical theory and shear deformation theory are 0.35% for $h/a = 0.1$, 3.6% for $h/a = 0.2$ and 18.6% for $h/a = 0.4$.

3.4.7 Cylindrical Shell

In this example, a square shallow cylindrical shell (see Figure 3.32) as analyzed in [8] is presented. The properties of the shell are the same as in the previous example, except $h/a = 0.02$ and the curvature $ak_{11} = 1/6$; $ak_{22} = 0$.

Three different BEM meshes were used. Mesh A has 20 boundary elements and 25 cells, Mesh B has 28 boundary elements and 49 cells, and Mesh C has 36 quadratic boundary elements and 81 constant cells. For comparison, four FEM meshes were also used, Mesh D has 256 linear shell elements, Mesh E has 576 elements, Mesh F has 1024 elements, and Mesh G has 1600 elements.

Table 3.9: Deflection w/a (at $x = 0, y = 0$) of a simply supported square cylindrical shell: uniformly loaded.

	w
BEM (20 elms - 5 × 5 cells)	0.176074
BEM (28 elms - 7 × 7 cells)	0.182253
BEM (36 elms - 9 × 9 cells)	0.184099
BEM (20 elms - 5 × 5 points)	0.159019
BEM (20 elms - 7 × 7 points)	0.170458
BEM (32 elms - 9 × 9 points)	0.177511
BEM (32 elms - 11 × 11 points)	0.179903
FEM (16 × 16 elms)	0.168244
FEM (24 × 24 elms)	0.173103
FEM (32 × 32 elms)	0.175734
FEM (40 × 40 elms)	0.177395

This example was also analyzed [9] using the dual reciprocity technique [59] using different meshes. Mesh H has 20 boundary elements and 25 domain points, Mesh J has 20 boundary elements and 49 domain points, Mesh K has 32 boundary elements and 81 domain points, and Mesh L has 32 boundary elements and 121 domain points.

Table 3.6 presents the results of this example. As can be seen, the two BEM results are in good agreement with each other and with the FEM results.

3.5   Kirchhoff Plate Theory

The direct boundary element formulation for Kirchhoff plate theory is formed in terms of four boundary quantities of deflection w_3, normal slop ϕ_n, boundary moment M_n and the equivalent shear force V. In this section, initially some basic concepts are reviewed and differences between the shear deformable plate bending theory and classical plate bending theory are highlighted. Next, the BEM formulation is derived using the Betti's reciprocal theorem.

3.5.1   Basic Concepts

In Kirchhoff plate theory the deflection w_3 is related to the rotation ϕ_α by assuming that transverse shear deformation can be neglected, that is

$$
\phi_1 = \phi_{x_1} = \frac{\partial w_3}{\partial x_1}
$$

$$
\phi_2 = \phi_{x_2} = \frac{\partial w_3}{\partial x_2}
$$

The moments are related to deformation via

$$
M_{11} = -D \left(\frac{\partial^2 w_3}{\partial x_1^2} + \nu \frac{\partial^2 w_3}{\partial x_2^2} \right)
$$

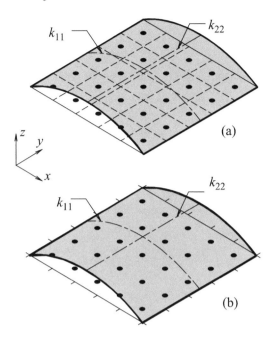

Figure 3.32: BEM model of a square shallow cylindrical shell: (a) cell discretization; (b) dual reciprocity technique.

$$M_{22} = -D\left(\frac{\partial^2 w_3}{\partial x_2^2} + \nu\frac{\partial^2 w_3}{\partial x_1^2}\right)$$

$$M_{12} = D(1-\nu)\frac{\partial^2 w_3}{\partial x_1\partial x_2} \tag{3.162}$$

Substituting for Q_α from (3.12) into (3.11) gives

$$\frac{\partial^2 M_{11}}{\partial x_1^2} - 2\frac{\partial^2 M_{12}}{\partial x_1\partial x_2} + \frac{\partial^2 M_{22}}{\partial x_2^2} = -q_3 \tag{3.163}$$

Substituting (3.162) into (3.163) gives

$$\frac{\partial^4 w_3}{\partial x_1^{24}} - 2\frac{\partial^4 w_3}{\partial x_1^2\partial x_2^2} + \frac{\partial^4 w_w}{\partial x_2^4} = \frac{q_3}{D} \tag{3.164}$$

or

$$(\nabla^2)^2 w_3 = \frac{q_3}{D} \tag{3.165}$$

where ∇^2 is the well known Laplace operator.

The generalized bending tractions p_α can be represented by their normal and tangential components, i.e.

$$p_\alpha = M_{\alpha\beta}n_\beta = -M^n n_\alpha + M^t s_\alpha \tag{3.166}$$

where s_α denotes the components of the unit tangent to the boundary S. The edge twisting moment M^t is combined with the shear traction p_3 to produce a resultant shear force V:

$$V = p_3 - \frac{\partial M^t}{\partial s} \tag{3.167}$$

At corners, a so-called *corner force* V_c is introduced such that

$$V_c = (M^s)_- - (M^s)_+ \tag{3.168}$$

where subscripts $-$ and $+$ relate to the boundary either side of a corner c.

3.5.2   Boundary Integral Formulation

The direct boundary integral equation can be obtained from the Betti's reciprocity relationship of two isotropic elastic states $(p_3, w_3, \phi_i, M_i, b)$ and $(p_3^*, w_3^*, \phi_i^*, M_i^*, b^*)$, that is

$$\int_S (p_3^* w_3 + M_i^* \phi_i) dS + \int_V b^* w_3 dV = \int_S (p_3 w_3^* + M_i \phi_i^*) dS + \int_V b w_3^* dV \tag{3.169}$$

Taking into account the boundary shear force V and the *corner forces*, equation (3.169) can be rewritten as

$$\int_S (V^* w_3 + M_i^* \phi_i) dS + \int_V b^* w_3 dV + V_c^* (w_3)_c$$

$$= \int_S (V w_3^* + M_i \phi_i^*) dS + \int_V b w_3^* dV + V_c (w_3^*)_c \tag{3.170}$$

where $(w_3)_c$ denotes the value of w_3 at a corner point c.

The state $(\,.\,)^*$ is taken to correspond to solution for a unit load, i.e.

$$-D(\nabla^2)^2 w_3 + \Delta(\mathbf{X}', \mathbf{X}) = 0 \tag{3.171}$$

The integral equation (3.170) can now be written as

$$
\begin{aligned}
w_3(\mathbf{X}') &= \int_S [V^k(\mathbf{X}', \mathbf{x}) w_3(\mathbf{x}) - M^k(\mathbf{X}', \mathbf{x}) \phi_n(\mathbf{x}) \\
&\quad + \phi_n^k(\mathbf{X}', \mathbf{x}) M_n(\mathbf{x}) - W^k(\mathbf{X}', \mathbf{x}) V(\mathbf{x})] dS \\
&\quad + \int_V W x^k(\mathbf{X}', \mathbf{X}) q_3 dV + W_c^k(\mathbf{X}', \mathbf{x}) V_c
\end{aligned}
\tag{3.172}
$$

where $\phi_n = \partial w_3 / \partial \mathbf{n}$, $W^k(\mathbf{X}', \mathbf{x})$, $M^K(\mathbf{X}', \mathbf{x})$, $V^k(\mathbf{X}', \mathbf{x})$ are fundamental solutions of (3.171), and are given as

$$W^k = \frac{1}{8\pi D} r^2 \ln(r)$$

$$\phi_n^k = -r_\alpha \frac{\ln r}{4\pi D} n_\alpha$$

$$M^k = \frac{-1}{4\pi} \left\{ \frac{(1-\nu) r_\alpha r_\alpha}{r^2} + \delta_{\alpha\beta} \left[(1+\nu) \ln(r) + \nu \right] \right\} n_\alpha n_\beta$$

$$V^k = -\frac{1}{2\pi} \frac{r_\alpha}{r^2} n_\alpha + \frac{1}{4\pi} \frac{\partial}{\partial s} \left[\left\{ \frac{(1-\nu) r_\alpha r_\alpha}{r^2} + \delta_{\alpha\beta} \left[(1+\nu) \ln(r) + \nu \right] \right\} n_\alpha s_\beta \right]$$

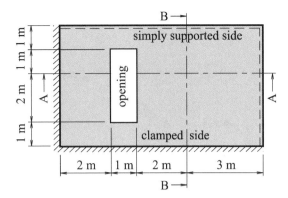

Figure 3.33: Rectangular plate with rectangular opening.

Considering the limit of (3.172) as $\mathbf{X}'$ tends to a boundary point $\mathbf{x}'$, we obtained the boundary integral equation for deflection, that is

$$C^k(\mathbf{x}')w_3(\mathbf{x}') = \int_S [V^k(\mathbf{x}',\mathbf{x})w_3(\mathbf{x}) - M^k(\mathbf{x}',\mathbf{x})\phi_n(\mathbf{x})$$

$$+\phi_n^k(\mathbf{x}',\mathbf{x})M_n(\mathbf{x}) - W^k(\mathbf{x}',\mathbf{x})V(\mathbf{x})]\,dS + \int_V W^k(\mathbf{x}',\mathbf{X})q_3 dV + W_c^k(\mathbf{x}',\mathbf{x})V_c \quad (3.173)$$

There are two unknowns in the above equation, that is the deflection w_3 or shear force V and rotation ϕ_n or moment M_n. An additional equation is obtained by differentiating (3.173) with respect to the normal at $\mathbf{x}'$ to give

$$C^k(\mathbf{x}')\phi_n(\mathbf{X}') = \int_S \left[V_{,n}^k(\mathbf{x}',\mathbf{x})w_3(\mathbf{x}) - M_{,n}^k(\mathbf{x}',\mathbf{x})\phi_n(\mathbf{x}) \right.$$

$$\left. +\phi_{n,n}^k(\mathbf{x}',\mathbf{x})M_n(\mathbf{x}) - \phi_n^k(\mathbf{x}',\mathbf{x})V(\mathbf{x})\right]dS + \int_V \phi_n^k(\mathbf{x}',\mathbf{X})q_3 dV + \phi_{nc}^k(\mathbf{x}',\mathbf{x})V_c \quad (3.174)$$

where the subscript n denotes a derivative with respect to the normal $\mathbf{n}$.

3.5.3   Example: Rectangular Plate with an Opening

In this example, a rectangular plate with a rectangular opening as shown in Figure 3.33 is considered. The plate is clamped on two sides and hinged on two other sides. The plate thickness is 0.16 m and the material properties are $E = 3 \times 10^7$ kN/m^2 and $\nu = 0.2$. A uniform load of 10 kN/m^2 is considered as the applied load. A mesh of 46 boundary elements was used [8] for the boundary element analysis, 8 and 5 elements for longer and shorter sides of the outer boundary, respectively, and 6 and 4 elements for longer and shorter sides of the opening, respectively.

Figures 3.34-3.37 show the out-of-plane displacement and bending moments along the lines $A - A$ and $B - B$ obtained using the BEM based on Reissner plate theory. Also presented are boundary elements results based on the Kirchhoff plate theory and finite element analysis given by Hartmann [17].

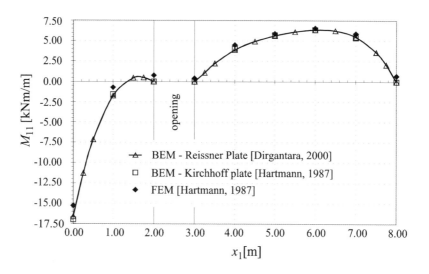

Figure 3.34: Bending stress resultant M_{11} along cross section A−A.

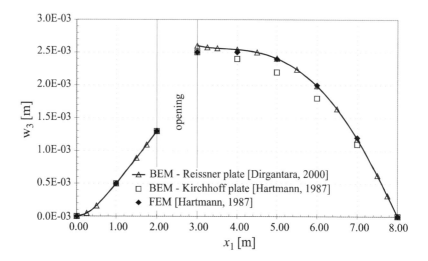

Figure 3.35: Out-of-plane displacement w_3 along cross-section $A − A$.

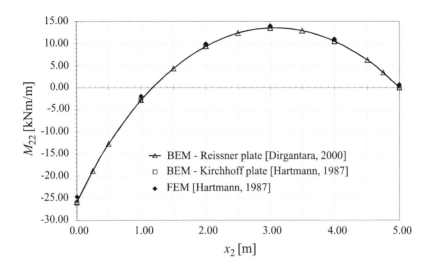

Figure 3.36: Bending stress resultant M_{22} along cross-section $B - B$.

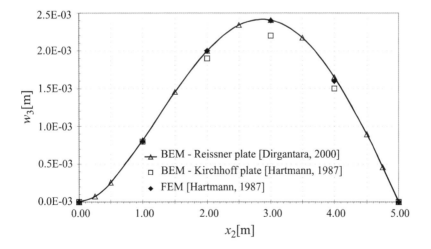

Figure 3.37: Out-of-plane displacement w_3 along cross-section $B - B$.

3.6 Summary

In this chapter the application of the boundary element method to plate and shell structures was described. The method was shown to yield highly accurate results in comparison to the finite element method. Furthermore, it was demonstrated that the BEM can be utilized to solve complex built up structures such as an aircraft X-core wing box section. The results presented demonstrated that the BEM is an efficient and robust method for solving thin structural problems.

Other recent developments of the BEM for plates and shells can be found in [13, 36, 45, 59, 65].

References

[1] Abdel-Akher, A. and Hartley, G.A., Evaluation of boundary integrals for plate bending, *International Journal for Numerical Methods in Engineering*, **28**, 75-93, 1989.

[2] Aliabadi, M.H., *Plate Bending Analysis with Boundary Elements,* Computational Mechanics Publications, Southampton, 1998.

[3] Atluri, S.N. and Pipkins, D.S. Large deformation analysis of plates and shells, *Boundary Element Analysis of Plates and Shells,* Springer-Verlag, Berlin, 142-165, 1991.

[4] Balas, J., Sladek, J. and Sladek, V., The boundary integral equation method for plates resting on a two-parameter foundation, *ZAMM*, **64,** 137-146, 1984.

[5] Barcellos, C. A. and Silva, L. H. M., A boundary element formulation for the Mindlin's plate model, in *Boundary Element Technology*, Computational Mechanics Publications, 123-130, 1989.

[6] Beskos, D.E., Static and dynamic analysis of shells, in *Boundary Element Analysis of Plate and Shells*, Spinger-Verlag, Berlin, 93-140, 1991

[7] Bézine, G., Boundary integral formulation for plate flexure with arbitrary boundary conditions, *Mech. Research Communications*, **5**(4), 197-206, 1978.

[8] Dirgantara, T. and Aliabadi, M. H., A new boundary element formulation for shear deformable shells analysis, *International Journal for Numerical Methods in Engineering*, **45**, 1257-1275, 1999.

[9] Dirgantara, T., Boundary element analysis of cracks in shear deformable plates and shells, PhD Thesis, Department of Engineering, Queen Mary, University of London, 2000.

[10] Dirgantara, T. and Aliabad, M. H., Boundary element method analysis of assembled plate-structure, *Communications in Numerical Methods in Engineering*, **17**, 749-760, 2001.

[11] Dirgantara, T. and Aliabadi, M.H., Dual boundary element formulation for fracture analysis of shear deformable shells, *International Journal of Solids and Structures*, **38**, 7769-7800, 2001.

[12] El-Zafrany, A., Debbih, M. and Fadhil, S., An efficient approach for boundary element bending analysis of thin and thick plates, *Computers and Structures,* **56**, 565-576, 1995.

[13] El-Zafrany, A., Al-Hosani, K. and Razzaq, R.J., A generalized fundamental solution for thick plates on two parameter elastic foundation, *Advances in Boundary Element Techniques II,* Hoggar, Geneva, 425-436, 2001.

[14] Fadhil, S. and El-Zafrany, A., Boundary element analysis of thick Reissner plates on two-parameter foundation, *International Journal of Solids and Structures,* **31**, 2901-2917, 1994.

[15] Forbes, D.J. and Robinson, A.R., *Numerical Analysis of Elastic Plates and Shallow Shells by an Integral Equation Method,* Structural research series report no 345, University of Illinois, Urbana, 1969.

[16] Hartmann, F. and Zormantel, R., The direct boundary element method in plate bending, *International Journal for Numerical Methods in Engineering,* **23**, 2049-2069, 1986

[17] Hartmann, F., Boundary elements on personal computers, *Microsoftware for Engineering,* **3**, 213-215, 1987.

[18] Jaswon, M.A., Maiti, M. and Symm, G.T., Numerical biharmonic analysis and some applications, *International Journal of Solids and Structures,* **3**, 309-332, 1967.

[19] Kamiya, N. and Sawaki, Y., An integral equation approach to finite deflection of elastic plates, *International Journal of Non-linear Mechanics,* **17**, 187-194, 1988.

[20] Karam, V.J. and Telles, J.C.F., On boundary elements for Reissner's plate theory, *Engineering Analysis*, **5**, 21-27, 1988

[21] Karami, G., Zarrinchang, J. and Foroughi, B., Analytical treatment of boundary integrals in direct boundary element analysis of plate bending problems, *International Journal for Numerical Methods in Engineering,* **37**, 2409-2427, 1994

[22] Katsikadelis, J.T. and Nerantzaki, M.S., The boundary element method for nonlinear problems, *Engineering Analysis with Boundary Elements,* **23**, 365-373, 1999.

[23] Kawabe, H., Plate buckling analysis by the boundary element method, *Theory and Applications of Boundary Element Methods,* Pergamon, Oxford, 267-374, 1987.

[24] Kirchhoff, G., Uber das gleichgewicht und die bewegung einer elastischen scheibe, *J.Rein Angew Math.,* **40**, 51-88, 1850.

[25] Lei, X.Y., Haung, M.K. and Wang, X.X., Geometrically nonlinear analysis of a Reissner type plate by the boundary element method, *Computers and Structures,* **37**, 911-916, 1990.

[26] Jinmu, L and Shuyao, L., Geometrically nonlinear analysis of the shallow shell by the displacement-based boundary element formulation, *Engineering Analysis with Boundary Elements*, **18**, 63-70, 1996.

[27] Lu, P. and Haung, M., Boundary element analysis of shallow shells involving shear deformation, *International Journal of Solids and Structures*, **29**, 1273-1282, 1992.

[28] Mindilin, R.D., Influence of rotatory inertia and shear on flexural motions of isotropic, elastic plates, *Journal of Applied Mechanics*, **18**, 31-38, 1951.

[29] Naghdi, P.M., Note on the equations of shallow elastic shells, *Quarterly Journal of Applied Mathematics*, **14**, 331-333, 1956

[30] Naghdi, P.M., On the theory of thin elastic shells, *Quarterly Journal of Applied Mathematics*, **14**, 369-380, 1956

[31] Nerantzaki, M.S. and Katsikadelis, J.T., Analysis of plates with variable thickness. An analog equation solution, *Plate Bending Analysis with Boundary Elements*, Computational Mechanics Publications, Southampton, 275-308, 1998.

[32] Novozhilov, V.V., *The Theory of Thin Shells*, translated by P. G. Lowe, P. Noordhoff Ltd., Gronigen, The Netherland, 1959

[33] Newton D.A. and Tottenham, H., Boundary value problems in thin shallow shells of arbitrary plan form, *Journal of Engineering Mathematics*, **2**, 211-223, 1968

[34] Paiva, J.B. and Aliabadi, M.H., Boundary element analysis of zoned plates in bending, *Computational Mechanics*, **25**, 560-566, 2000.

[35] Providakis, C.P. and Beskos, D.E., Free and forced vibration of shallow shells by boundary and interior elements, *Computer Methods in Applied Mechanics and Engineering*, **92**(1), 55-74, 1991

[36] Providakis, C.P., The effect of internal supports on the D/BEM transient dynamic analysis of elastoplastic Reissner-Mindlin Plates, *Advances in Boundary Element Techniques II*, Hoggar, Geneva, 467-474, 2001.

[37] Rashed, Y.F., Aliabadi, M.H. and Brebbia, C.A., The boundary element method for thick plates on a Winkler foundation, *International Journal of Numerical Methods for Engineering*, **41**, 1435-1462, 1998.

[38] Rashed, Y.F., Aliabadi, M.H and Brebbia, C.A., Hypersingular boundary element formulation for Reissner plates, *International J. Solids Structures*, **35(18)**, 2229-2249, 1998

[39] Rashed, Y.F., Aliabadi, M.H. and Brebbia, C.A., A boundary element formulation for a Reissner plate on a Pasternak foundation, *Computers and Structures*, **70**, 515-532, 1999.

[40] Reddy, J.N., Exact solutions of moderately thick laminated shells, *Journal of Engineering Mechanics*, ASCE, **110**, 794-809, 1984.

[41] Reissner, E., On bending of elastic plates, *Quarterly Journal of Applied Mathematics*, **5**, 55-68, 1947

[42] Reissner, E., Stress strain relations in the theory of thin elastic shells, *Journal of Mathematics and Physics*, **31**, 109-119, 1952.

[43] Shi, G. and Bezine, G., A general boundary integral formulation for the anisotropic plate bending problems, *Journal of Composite Materials*, **22**, 694-716, 1988.

[44] Shi, G., Flexural vibration and buckling analysis of orthotropic plates by the boundary element method, *International Journal of Solids and Structures*, **26**, 1351-1370, 1990.

[45] Sladek, J., Sladek, V. and Mang, H.A., Meshless local boundary integral equation method for plate bending problems, *Advances in Boundary Element Techniques II*, Hoggar, Geneva, 415-424, 2001.

[46] Stern, M., A General boundary integral formulation for plate bending problems, *International Journal of Solids and Structures*, **15**, 769-782, 1979.

[47] Syngellakis, S., Elzein, A., Plate buckling loads by the boundary element method, *International Journal for Numerical Methods in Engineering*, **37**, 1736-1778, 1994.

[48] Tanaka, M. and Miyazaki, K., A direct BEM for elastic plate-structures subjected to arbitrary loadings, in *Boundary Elements VII*, Springer-Verlag, Berlin, 4-3, 4-16, 1985.

[49] Telles, J.C.F. and Karam, V.J., Nonlinear material analysis of Reissner's plates, In *Plate Bending with Boundary Elements*, Computational Mechanics Publications, Southampton, 127-164, 1998.

[50] Timoshenko, S. and Woinowsky-Krieger, S., *Theory of Plates and Shells.* MacGraw-Hill, New York, 1959.

[51] Tosaka, N. and Miyake, S., A boundary integral equation formulation for elastic shallow shell bending problems, in *Boundary Elements*, Springer Verlag, Berlin, 527-538, 1983.

[52] Tottenham, H., The boundary element method for plates and shells, in *Developments in Boundary Element Methods -I*, Applied Science, London, 173-205, 1979.

[53] Valliapan, S. and Murti, V., Automatic remeshing technique in quasi static and dynamic crack propagation, in *Proceedings of NUMETA '85 Conference*, Swansea, 107-116, 1985.

[54] Vander Weeën, F., Application of the boundary integral equation method to Reissner's plate model, *International Journal for Numerical Methods in Engineering*, **18**, 1-10, 1982.

[55] Venturini, W.S. and Paiva, J.B., Plate bending analysis by the boundary element method considering zoned thickness domain, *Software for Engineering Workstations*, **4**, 183-185, 1988.

[56] Vlasov, V.Z., *General Theory of Shells and Its Application in Engineering*, National Technical Information Services, U.S. Department of Commerce, 1964.

[57] Wang, X.Ji and Tanaka, M., A dual reciprocity boundary element approach for the problem of large deflection of thin elastic plates, *Computational Mechanics*, **26**, 58-65, 2000.

[58] Wen, P.H., Aliabadi, M.H. and Young, A., Transformation of domain integrals to boundary integrals in BEM analysis of shear deformable plate bending problems, *Computational Mechanics*, **24**, 304-309, 1999.

[59] Wen, P.H., Aliabadi, M. H. and Young, A., Plane stress and plate bending coupling in BEM analysis of Shallow shells, *International Journal for Numerical Methods in Engineering*, **48**, 1107-1125, 2000.

[60] Wen,P.H., Aliabadi,M.H. and Young,A., Stiffened cracked plates analysis by dual boundary element method, *International Journal of Fracture*, **106,** 245-258, 2000.

[61] Wen, P.H. and Aliabadi, M.H., Boundary element method for dynamic plate bending problems, *International Journal of Solids and Structures*, **37**, 5177-5188, 2000.

[62] Wen, P.H., Aliabadi, M .H. and Young, A., Application of dual reciprocity method to plates and shells, *Engineering Analysis with Boundary Elements*, **24**, 583-590, 2000.

[63] Wen, P.H. Aliabadi,M.H. and Young, A., Boundary element analysis of assembled multilayered structures, *Advances in Boundary Element Techniques II*, Hoggar, Geneva, 555-562, 2001.

[64] Wen, P.H. Aliabadi, M.H. and Young, A., Crack growth analysis for airframe structures by the boundary element method, (to appear).

[65] Wen, P.H., Aliabadi, M.H. and Young, A., Large deflection analysis of Reissner plate by boundary element method, (to appear).

[66] Westphal, T. Jr. and Barcellos, C.A., Application of the boundary element method to Reissner's and Mindlin's plate model, in *Boundary Elements XII*, Computational Mechanics Publication, Southampton, **1**, 467-477, 1990.

[67] Xiao-Yan, L, Mao-Kuang, H and Xiuxi, W., Geometrically nonlinear analysis of a Reissner type plate by the boundary element method, *Computers and Structures*, **37**, 911-916, 1990.

[68] Yu, Y.Y., Axisymmetrical bending of circular plates under simultaneous action of lateral load, force in the middle plane, and elastic foundation. *Journal of Applied Mechanics*, **24,** 141-143, 1957.

[69] Zhang, J. D. and Atluri, S. N., A boundary/ interior element method for quasi static and transient response analysis of shallow shells, *Computer and Structures*, **24**, 213-223, 1986.

Chapter 4

Thermoelasticity

4.1 Introduction

In many engineering structures such as nuclear power plants, engines and electronic devices, the effects of thermo-mechanical loads on the parts must be studied. An early application of the BEM to steady-state thermoelasticity can be found in Rizzo and Shippy [25] and Cruse, Snow and Wilson [6] for three-dimensional and axisymmetric problems, respectively. For transient problems three different approached have been reported. They included volume-based formulations [30], Laplace transform domain formulation [27] and time domain formulations [4, 7, 28]. Rudolphi [26] presented a procedure for coupling the BEM and FEM for axisymmetric thermoelastic problems. Adaptive formulation has be used by Kamiya *et al.* [14] for steady thermoelastic analysis. Hypersingular boundary integral equations for flux and stress equations were presented by Prasad, Aliabadi and Rooke [19, 20] for steady state and transient two-dimensional thermoelastic problems, and dell'Erba and Aliabadi [11] for three-dimensional problems. Martin and Dulikravich [17] applied the BEM to aero-thermo-elastic concurrent design optimization of internally cooled turbine blades. Raveendra [24] used the BEM to study thermal stresses distribution on a turbine blade. Some recent applications of the BEM to thermoelastic problems can be found in Kassab and Aliabadi [15].

It is worth noting that a mathematical analogy exists between the equations of coupled thermoelasticity (see Nowacki [16]) and the equations of poroelasticity [2]. The Green's function for fully coupled thermoelasticity may, with simple substitutions, be used for poroelasticity, as demonstrated by Cheng and Detournay [5]. Application of the BEM to dynamic poroelastic and thermoelastic consolidation problems can be found in papers by Dominguez [10] and Smith and Booker [29], respectively.

In this chapter, the temperature and displacement boundary integral equations for thermoelastic problems are initially derived. The boundary flux and stress integral equations are also described. The formulations are presented for steady state and transient problems. For transient problems, the time domain formulation is derived using the reciprocal theorem (see, for example, Dargush and Banerjee [7, 8]). Time integration schemes using constant and linear approximations are discussed. Also described are effective numerical implementations of the BEM for both steady state and transient thermoelastic problems.

A different notation is used here for the temperature and flux boundary integral equations compared to Volume I, to avoid confusion with elasticity terms.

4.2 Basic Concepts

The material under consideration is assumed to be isotropic and homogeneous, and material properties are assumed to be independent of the temperature.

Thermal stresses may arise in a heated body either because of a non-uniform temperature distribution, external constraints or a combination of the two conditions. The total strains at each point of the heated body are made up of two parts. The first part is a uniform expansion proportional to the temperature rise θ. The second part comprises the strains required to maintain the continuity of the body, as well as those arising because of external loads. These strains are related to the stresses by means of the Hooke's law of isothermal elasticity. The total strains are the sum of the two components, and are related to the stresses as follows:

$$\varepsilon_{ij} = \frac{1+\nu}{E}\sigma_{ij} - \frac{\nu}{E}\sigma_{kk}\delta_{ij} + h\theta\delta_{ij} \tag{4.1}$$

where E is the Young's modulus, ν is the Poisson's ratio and h is the coefficient of linear thermal expansion. It can be seen from (4.1) that shear strains are not affected by the temperature as free thermal expansion does not produce angular distortion in an isotropic material.

Equation (4.1) can be conveniently expressed in terms of stresses:

$$\sigma_{ij} = 2\mu\left[\varepsilon_{ij} + \frac{\nu}{1-2\nu}\varepsilon_{kk}\delta_{ij}\right] - \frac{E}{1-2\nu}h\theta\delta_{ij} \tag{4.2}$$

The equations of equilibrium are the same as those of isothermal elasticity, since they are based on purely mechanical considerations. The stress tensor can be expressed in terms of displacement derivatives by combining Hooke's Law in equation (4.2) with the strain-displacement equation to give the Navier's equations:

$$\mu u_{i,kk} + \frac{\mu}{(1-2\nu)}u_{k,ik} - \frac{E}{(1-2\nu)}h\theta_{,i} + b_i = 0 \tag{4.3}$$

4.2.1 Heat Conduction Equation

The heat conduction equation can be written as follows:

$$k\theta_{,ii} - \bar{\theta}\frac{E}{(1-2\nu)}h\delta_{ij}\varepsilon_{ij} - \rho c\dot{\theta} = 0 \tag{4.4}$$

where $\bar{\theta}$ denotes the reference absolute temperature for the state of zero stress and strain, k is the conductivity, ρ is the density and c is the heat capacity per unit volume at constant strain. As expected, because of the cooling and heating associated with a change of volume, the temperature and strain fields are coupled in equation (4.4).

4.2.2 Thermoelasticity Equations

In this section governing equations for coupled and uncoupled transient thermoelasticity, as well as steady state thermoelasticity, are briefly reviewed.

Coupled Thermoelasticity

The equation of motion (6.56) and the heat conduction equation (4.4) recalled below represent the thermoelastic equations:

$$k\theta_{,ii} - \bar{\theta}\frac{E}{(1-2\nu)}h\delta_{ij}\varepsilon_{ij} - \rho c\dot{\theta} = 0 \tag{4.5}$$

$$\mu u_{i,kk} + \frac{\mu}{(1-2\nu)}u_{k,ik} - \frac{E}{(1-2\nu)}h\theta_{,i} + b_i = 0 \tag{4.6}$$

Depending on the temperature and loading conditions, some terms in the above equations can be neglected. Three types of thermoelasticity can be obtained by neglecting some terms in (4.5) and (4.6): coupled thermoelasticity, uncoupled thermoelasticity and steady-state thermoelasticity.

Uncoupled Transient Thermoelasticity

Neglecting the strains from equation (4.5) leads to the uncoupled equations being obtained, where the heat conduction equation is still time dependent. The governing equations are given by

$$k\theta_{,ii} - \rho c\dot{\theta} = 0 \tag{4.7}$$

$$\mu u_{i,kk} + \frac{\mu}{(1-2\nu)}u_{k,ik} - \frac{E}{(1-2\nu)}h\theta_{,i} + b_i = 0 \tag{4.8}$$

Steady State Thermoelasticity

If the loads are applied slowly and all the resulting diffusive processes are completed, the time dependency vanishes. In this case, the uncoupled time independent equations are obtained as

$$\theta_{,ii} = 0 \tag{4.9}$$

$$2\mu u_{i,kk} + \frac{2\mu}{(1-2\nu)}u_{k,ik} - \frac{E}{(1-2\nu)}h\theta_{,i} + b_i = 0 \tag{4.10}$$

4.3   Boundary Integral Formulations

In this section boundary integral equations for the steady-state thermoelasticity and transient thermoelasticity equations are presented. The boundary integral equations are first derived for interior points and then modified for boundary points.

4.3.1   Steady State Thermoelasticity

Boundary integral equations for steady-state thermoelasticity can be obtained from the Laplace equation and the static equilibrium equation. The two equations of steady-state thermoelasticity for a linear, elastic, isotropic and homogeneous body occupying a domain V enclosed by a boundary S can be expressed as follows:

$$\theta_{,ii}(\mathbf{X}) = 0 \quad \text{for} \quad \mathbf{X} \in V \tag{4.11}$$

and

$$\sigma_{ij,j}(\mathbf{X}) + b_i(\mathbf{X}) = 0 \quad \text{for} \quad \mathbf{X} \in V \tag{4.12}$$

Equation (4.12) is for static elastic response. If the relationship between stress and displacement is applied to equation (4.12), the steady state thermoelastic equations (4.9) and (4.10) can be obtained.

In general, equations (4.11) and (4.12) are solved subject to temperature boundary conditions on the boundary S_θ and flux boundary conditions on the boundary S_q, displacement boundary conditions on the boundary S_u and traction boundary conditions on the boundary S_t:

$$\theta(\mathbf{x}) = \bar{\theta}(\mathbf{x}) \text{ on } \mathbf{x} \in S_\theta$$

$$q(\mathbf{x}) = \bar{q}(\mathbf{x}) \text{ on } \mathbf{x} \in S_q$$

$$u_i(\mathbf{x}) = \bar{u}_i(\mathbf{x}) \text{ on } \mathbf{x} \in S_u$$

$$t_i(\mathbf{x}) = \bar{t}_i(\mathbf{x}) \text{ on } \mathbf{x} \in S_t \tag{4.13}$$

where $\mathbf{x}$ represents a point on body S and $S_\theta + S_q = S_u + S_t = S$. The flux is defined as

$$q = -k\theta_{,n}$$

where n preceded by a comma represents differentiation with respect to the outward normal, and k is the thermal conductivity.

The boundary integral equations for (4.11) and (4.12) are referred to as the 'temperature equation' and the 'displacement equation' in the following sections.

Temperature Integral Equation

The integral equation formulation for the temperature equation of an internal point $\mathbf{X}'$ can be derived from equations (4.11) and (4.13) to give (see Chapter 2, Volume I)

$$\theta(\mathbf{X}') - \int_S \theta(\mathbf{x})q^*(\mathbf{X}',\mathbf{x})dS = -\int_S q(\mathbf{x})\theta^*(\mathbf{X}',\mathbf{x})dS \tag{4.14}$$

where

$$\theta^*(\mathbf{X}',\mathbf{x}) = \frac{1}{4\pi kr}$$

$$q^*(\mathbf{X}',\mathbf{x}) = \frac{r_{,i}n_i}{4\pi r^2} \qquad i = 1,2,3$$

for three-dimensional problems, and

$$\theta^*(\mathbf{X}',\mathbf{x}) = \frac{1}{2\pi k}\ln\frac{1}{r}$$

$$q^*(\mathbf{X}',\mathbf{x}) = \frac{r_{,i}n_i}{2\pi r} \qquad i = 1,2$$

for two-dimensional problems.

The temperature equation for a boundary point $\mathbf{x}'$ can be written

$$C(\mathbf{x}')\theta(\mathbf{x}') - \int_S q^*(\mathbf{x}',\mathbf{x})\theta(\mathbf{x})dS = -\int_S \theta^*(\mathbf{x}',\mathbf{x})q(\mathbf{x})dS \tag{4.15}$$

where $C(\mathbf{x}') = \gamma/2\pi$ and γ is the internal angle at $\mathbf{x}'$.

Flux Boundary Integral Equation

The flux boundary integral equation is useful for boundary element formulation for cracks problems. It has been derived and implemented by Prasad, Aliabadi and Rooke [22] and dell'Erba and Aliabadi [11] for two- and three-dimensional problems, respectively.

By differentiating the temperature integral equation (4.14), the temperature derivative equation for an internal point $\mathbf{X}'$ can be obtained as [11, 22]

$$\frac{\partial \theta(\mathbf{X}')}{\partial X_i'} - \int_S \frac{\partial q^*(\mathbf{X}',\mathbf{x})}{\partial X_i'}\theta(\mathbf{x})dS = -\int_S \frac{\partial \theta^*(\mathbf{X}',\mathbf{x})}{\partial X_i'}q(\mathbf{x})dS \qquad (4.16)$$

where

$$\frac{\theta^*(\mathbf{X}',\mathbf{x})}{\partial X_i'} = -\frac{r_{,i}}{4\pi k r^2}$$

$$\frac{\partial q^*(\mathbf{X}',\mathbf{x})}{\partial X_i'} = \frac{1}{4\pi r^3}[3r_{,i}r_{,j}n_j - n_i]$$

for three-dimensional problems, and

$$\frac{\theta^*(\mathbf{X}',\mathbf{x})}{\partial X_i'} = -\frac{r_{,i}}{2\pi k r}$$

$$\frac{\partial q^*(\mathbf{X}',\mathbf{x})}{\partial X_i'} = \frac{1}{4\pi r^2}[2r_{,i}r_{,j}n_j - n_i]$$

for two-dimensional problems.

In equation (4.16) the integrands $\partial \theta^*/\partial X_i'$ and $\partial q^*/\partial X_i'$ are strongly singular and hypersingular, respectively.

Let us assume $\mathbf{x}'$ is on a smooth boundary and q and θ are $C^{0,\alpha}$ and $C^{1,\alpha}$ continuous, respectively. Following a similar limiting procedure as that described in Chapter 2, the derivative of temperature equation for a boundary point can be written as

$$\frac{\partial \theta(\mathbf{x}')}{\partial x_i'} - \lim_{\varepsilon \to 0}\int_{S-S_\varepsilon+S_\varepsilon^+} \frac{\partial q^*(\mathbf{x}',\mathbf{x})}{\partial x_i'}\theta(\mathbf{x})dS = -\lim_{\varepsilon \to 0}\int_{S-S_\varepsilon+S_\varepsilon^+} \frac{\partial \theta^*(\mathbf{x}',\mathbf{x})}{\partial x_i'}q(\mathbf{x})dS$$

$$(4.17)$$

By using the first term of Taylor's expansion, the integral on the right-hand side of equation (4.17) can be presented as

$$\begin{aligned}
\lim_{\varepsilon \to 0}\int_{S-S_\varepsilon+S_\varepsilon^+} \frac{\partial \theta^*(\mathbf{x}',\mathbf{x})}{\partial x_i'}q(\mathbf{x})dS &= \lim_{\varepsilon \to 0}\int_{S-S_\varepsilon} \frac{\partial \theta^*(\mathbf{x}',\mathbf{x})}{\partial x_i'}q(\mathbf{x})dS \\
&\quad + \lim_{\varepsilon \to 0}\int_{S_\varepsilon^+} \frac{\partial \theta^*(\mathbf{x}',\mathbf{x})}{\partial x_i'}[-k\theta_{,k}(\mathbf{x})n_k(\mathbf{x})]dS \\
&= \lim_{\varepsilon \to 0}\int_{S-S_\varepsilon} \frac{\partial \theta^*(\mathbf{x}',\mathbf{x})}{\partial x_i'}q(\mathbf{x})dS \\
&\quad -k\lim_{\varepsilon \to 0}\int_{S_\varepsilon^+} \frac{\partial \theta^*(\mathbf{x}',\mathbf{x})}{\partial x_i'}n_k(\mathbf{x})[\theta_{,k}(\mathbf{x}) - \theta_{,k}(\mathbf{x}')]dS \\
&\quad -k\theta_{,k}(\mathbf{x}')\lim_{\varepsilon \to 0}\int_{S_\varepsilon^+} \frac{\partial \theta^*(\mathbf{x}',\mathbf{x})}{\partial x_i'}n_k(\mathbf{x})dS \\
&= I_1 + I_2 + I_3 \qquad (4.18)
\end{aligned}$$

In equation (4.18), $I_2 \to 0$ as $\varepsilon \to 0$. The integral I_1 can be written in a Cauchy principal value sense as

$$I_1 = \lim_{\varepsilon \to 0} \int_{S-S_\varepsilon} \frac{\partial \theta^*(\mathbf{x}', \mathbf{x})}{\partial x_i'} q(\mathbf{x}) dS = \int_S \frac{\partial \theta^*(\mathbf{x}', \mathbf{x})}{\partial x_i'} q(\mathbf{x}) dS \qquad (4.19)$$

The integral I_3 for three-dimensional problems can be shown [9] to be

$$
\begin{aligned}
I_3 &= -k\theta_{,k}(\mathbf{x}') \lim_{\varepsilon \to 0} \int_{S_\varepsilon^+} \frac{\partial \theta^*(\mathbf{x}', \mathbf{x})}{\partial x_i'} n_k(\mathbf{x}) dS \\
&= -k\theta_{,k}(\mathbf{x}') \lim_{\varepsilon \to 0} \int_0^{2\pi} \int_0^{\pi/2} \left(-\frac{r_{,i} n_k}{4\pi k \varepsilon^2} \right) \varepsilon^2 \cos\phi \, d\phi \, d\varphi \\
&= \frac{\theta_{,i}(\mathbf{x}')}{2} \qquad (4.20)
\end{aligned}
$$

Equation (4.18) can be obtained from equations (4.19) and (4.20) as

$$\lim_{\varepsilon \to 0} \int_{S-S_\varepsilon+S_\varepsilon^+} \frac{\partial \theta^*(\mathbf{x}', \mathbf{x})}{\partial x_i'} q(\mathbf{x}) dS = \int_S \frac{\partial \theta^*(\mathbf{x}', \mathbf{x})}{\partial x_i'} q(\mathbf{x}) dS + \frac{\theta_{,i}(\mathbf{x}')}{2} \qquad (4.21)$$

As the integral on the left-hand side of equation (4.17) is hypersingular, two terms in the Taylor expansion are used for the regularization process, so that

$$
\begin{aligned}
\lim_{\varepsilon \to 0} \int_{S-S_\varepsilon+S_\varepsilon^+} \frac{\partial q^*(\mathbf{x}', \mathbf{x})}{\partial x_i'} \theta(\mathbf{x}) dS &= \lim_{\varepsilon \to 0} \int_{S-S_\varepsilon} \frac{\partial q^*(\mathbf{x}', \mathbf{x})}{\partial x_i'} \theta(\mathbf{x}) dS \\
&\quad + \lim_{\varepsilon \to 0} \int_{S_\varepsilon^+} \frac{\partial q^*(\mathbf{x}', \mathbf{x})}{\partial x_i'} [\theta(\mathbf{x}) - \theta(\mathbf{x}') \\
&\quad - \theta_{,k}(\mathbf{x}')(x_k - x_k')] \, dS \\
&\quad + \theta(\mathbf{x}') \lim_{\varepsilon \to 0} \int_{S_\varepsilon^+} \frac{\partial q^*(\mathbf{x}', \mathbf{x})}{\partial x_i'} dS \\
&\quad + \theta_{,k}(\mathbf{x}') \lim_{\varepsilon \to 0} \int_{S_\varepsilon^+} \frac{\partial q^*(\mathbf{x}', \mathbf{x})}{\partial x_i'} (x_k - x_k') dS \\
&= I_1' + I_2' + I_3' + I_4' \qquad (4.22)
\end{aligned}
$$

In equation (4.22), the integrals I_1' and I_3' together form a Hadamard principal value integral as follows:

$$
\begin{aligned}
I_1' + I_3' &= \lim_{\varepsilon \to 0} \int_{S-S_\varepsilon} \frac{\partial q^*(\mathbf{x}', \mathbf{x})}{\partial x_i'} \theta(\mathbf{x}) dS + \theta(\mathbf{x}') \lim_{\varepsilon \to 0} \int_{S_\varepsilon^+} \frac{\partial q^*(\mathbf{x}', \mathbf{x})}{\partial x_i'} dS \\
&= \fint_S \frac{\partial q^*(\mathbf{x}', \mathbf{x})}{\partial x_i'} \theta(\mathbf{x}) dS \qquad (4.23)
\end{aligned}
$$

The integral $I_2' \to 0$ as $\varepsilon \to 0$. The integral I_4' can be presented for three-dimensional problems as

$$
\begin{aligned}
I_4' &= \theta_{,k}(\mathbf{x}') \lim_{\varepsilon \to 0} \int_{S_\varepsilon^+} \frac{\partial q^*(\mathbf{x}', \mathbf{x})}{\partial x_i'} (x_k - x_k') dS \\
&= \theta_{,k}(\mathbf{x}') \lim_{\varepsilon \to 0} \int_{S_\varepsilon^+} \frac{1}{4\pi r^3} [3r_{,i} r_{,m} n_m - n_i] (x_k - x_k') dS
\end{aligned}
$$

$$= \frac{\theta_{,k}(\mathbf{x}')}{4\pi} \lim_{\varepsilon \to 0} \int_0^{2\pi} \int_0^{\pi/2} \frac{(3r_{,i} - n_i)}{\varepsilon^3} \varepsilon n_k(\mathbf{x}) \varepsilon^2 \cos \phi d\phi d\varphi$$

$$= \theta_{,i}(\mathbf{x}') \tag{4.24}$$

Rewriting equation (4.22) yields

$$\lim_{\varepsilon \to 0} \int_{S - S_\varepsilon + S_\varepsilon^+} \frac{\partial q^*(\mathbf{x}', \mathbf{x})}{\partial x_i'} \theta(\mathbf{x}) dS = \theta_{,i}(\mathbf{x}') + \oint_S \frac{\partial q^*(\mathbf{x}', \mathbf{x})}{\partial x_i'} \theta(\mathbf{x}) dS \tag{4.25}$$

The temperature derivative equation for a smooth boundary point $\mathbf{x}'$ can be obtained by substituting equations (4.21) and (4.25) into equation (4.17). It can be shown to be

$$\frac{1}{2} \frac{\partial \theta(\mathbf{x}')}{\partial x_i'} - \oint_S \frac{\partial q^*(\mathbf{x}', \mathbf{x})}{\partial x_i'} \theta(\mathbf{x}) dS = - \int_S \frac{\partial \theta^*(\mathbf{x}', \mathbf{x})}{\partial x_i'} q(\mathbf{x}) dS \tag{4.26}$$

Using the definition of flux, the final form of the flux equation for a boundary point can be obtained from equation (4.26) as

$$\frac{1}{2} q(\mathbf{x}') - n_i(\mathbf{x}') \oint_S \theta_i^{**}(\mathbf{x}', \mathbf{x}) q(\mathbf{x}) dS = -n_i(\mathbf{x}') \oint_S q_i^{**}(\mathbf{x}', \mathbf{x}) \theta(\mathbf{x}) dS \tag{4.27}$$

where

$$\theta^{**}(\mathbf{x}', \mathbf{x}) = \frac{r_{,i}}{4\pi r^2}$$

$$q^{**}(\mathbf{X}', \mathbf{x}) = -\frac{k}{4\pi r^3} [3r_{,i} r_{,j} n_j - n_i]$$

for three-dimensional problems, and

$$\theta^{**}(\mathbf{x}', \mathbf{x}) = \frac{r_{,i}}{2\pi r}$$

$$q^{**}(\mathbf{X}', \mathbf{x}) = -\frac{k}{4\pi r^2} [2r_{,i} r_{,j} n_j - n_i]$$

for two-dimensional problems.

A similar procedure may be followed for two-dimensional problems [21]. It is interesting to note that for two-dimensional problems, $I_3 = -\theta_{,i}/4$ and $I_4' = \theta_{,i}/4$. They do, however, add up with the free term $\theta_{,i}$ in (4.17) to give $\theta_{,i}/2$, as in (4.26).

Displacement Integral Equation

In this section the displacement integral equation for thermoelastic problems is derived. The derivation is similar to that presented in Chapter 2, with the thermal body force term now included.

The Betti's reciprocal theorem for thermoelastic problems can be written as

$$\int_V \varepsilon_{ij}^* [\sigma_{ij} + \frac{Eh}{1 - 2\nu} \theta \delta_{ij}] dV = \int_V \varepsilon_{ij} [\sigma_{ij}^* + \frac{Eh}{1 - 2\nu} \theta^* \delta_{ij}] dV \tag{4.28}$$

After some manipulation, it can be shown that (4.28) can be written as

$$\int_V u_k(\mathbf{X}) \sigma_{kj,j}^*(\mathbf{X}', \mathbf{X}) dV + \frac{Eh}{1 - 2\nu} \int_V \theta(\mathbf{X}) \varepsilon_{kk}^*(\mathbf{X}', \mathbf{X}) dV$$

$$= - \int_S t_k(\mathbf{X}) u_k^*(\mathbf{X}', \mathbf{x}) dS + \int_S u_k(\mathbf{x}) t_k^*(\mathbf{X}', \mathbf{x}) dS \tag{4.29}$$

Taking the stresses σ^*_{kj} to satisfy

$$\sigma^*_{kj,j}(\mathbf{X}',\mathbf{X})+\Delta(\mathbf{X}',\mathbf{X})e_k = 0 \tag{4.30}$$

where e_k is the unit vector of the directions of point forces. Substituting (4.30) into (4.29) and using the property of the Dirac delta function results in

$$u_i(\mathbf{X}')e_i + \int_S u_j(\mathbf{x})t^*_j(\mathbf{X}',\mathbf{x})dS$$
$$= \int_S t_j(\mathbf{x})u^*_j(\mathbf{X}',\mathbf{x})dS + \frac{Eh}{1-2\nu}\int_V \theta(\mathbf{X})\varepsilon^*_{kk}(\mathbf{X}',\mathbf{X})dV \tag{4.31}$$

Taking the components of point load separately, the fundamental traction and displacement fields can be rewritten as

$$u^*_j(\mathbf{X}',\mathbf{x}) = U_{ij}(\mathbf{X}',\mathbf{x})e_i \tag{4.32}$$
$$t^*_j(\mathbf{X}',\mathbf{x}) = T_{ij}(\mathbf{X}',\mathbf{x})e_i \tag{4.33}$$

The boundary integral equation for an internal displacement field can be obtained by substituting (4.32) and (4.33) into (4.31):

$$u_i(\mathbf{X}') + \int_S T_{ij}(\mathbf{X}',\mathbf{x})u_j(\mathbf{x})dS$$
$$= \int_S U_{ij}(\mathbf{X}',\mathbf{x})t_j(\mathbf{x})dS + \frac{Eh}{1-2\nu}\int_V U_{ik.k}(\mathbf{X}',\mathbf{X})\theta(\mathbf{X})dV \tag{4.34}$$

where U_{ij} and T_{ij} are the Kelvin fundamental solutions of the elasticity problem presented in Chapter 2.

The domain integral in (4.34) can be transformed to the boundary with the aid of the Galerkin vector, provided the temperature field satisfies the Laplace equation (i.e. $\theta_{,kk}$). Substituting the Galerkin vector (2.70) into the domain integral, we have

$$I = \frac{Eh}{1-2\nu}\int_V U_{ik.k}(\mathbf{X}',\mathbf{X})\theta(\mathbf{X})dV$$
$$\frac{Eh}{1-2\nu}\int_V G_{ik.kjj}(\mathbf{X}',\mathbf{X})\theta(\mathbf{X})dV \tag{4.35}$$

Since $\theta_{,jj} = 0$, equation (4.35) can be written as

$$I = \frac{Eh}{1-2\nu}\int_V \{G_{ik.kjj}(\mathbf{X}',\mathbf{X})\theta(\mathbf{X})-G_{ik,k}(\mathbf{X}',\mathbf{X})\theta_{,jj}\} dV$$
$$= \frac{Eh}{1-2\nu}\int_V \frac{\partial}{\partial X_j}(G_{ik.kj}(\mathbf{X}',\mathbf{X})\theta(\mathbf{X})-G_{ik,k}(\mathbf{X}',\mathbf{X})\theta_{,j}) dV \tag{4.36}$$

which can be transformed into the boundary using the divergent theorem, that is

$$I = \frac{Eh}{1-2\nu}\int_S (G_{ik.kj}(\mathbf{X}',\mathbf{X})\theta(\mathbf{X})-G_{ik,k}(\mathbf{X}',\mathbf{X})\theta_{,j})\,n_j(\mathbf{x})dS \tag{4.37}$$

Substituting for Galerkin vectors (2.76) and (2.86) into (4.37) gives

$$I = \int_S \overline{P}_i(\mathbf{X}',\mathbf{x})\theta(\mathbf{x})dS - \int_S \overline{Q}_i(\mathbf{X}',\mathbf{x})q(\mathbf{x})dS \tag{4.38}$$

where the fundamental fields for three-dimensional problems are given as

$$\overline{P}_i(\mathbf{X}',\mathbf{x}) = \frac{h(1+\nu)}{8\pi(1-\nu)r}\left(n_i - \frac{\partial r}{\partial n}r_{,i}\right)$$

$$\overline{Q}_i(\mathbf{X}',\mathbf{x}) = \frac{h(1+\nu)}{8\pi k(1-\nu)}r_{,i}$$

and for two-dimensional space are given by

$$\bar{P}_i(\mathbf{X}',\mathbf{x}) = \frac{(1+\nu)h}{4\pi(1-\nu)}\left\{\left[\ln\left(\frac{1}{r}\right) - \frac{1}{2}\right]n_i - r_{,i}r_{,k}n_k\right\} \tag{4.39}$$

$$\bar{Q}_i(\mathbf{X}',\mathbf{x}) = -\frac{(1+\nu)}{4\pi k(1-\nu)}hrr_{,i}\left[\ln\left(\frac{1}{r}\right) - \frac{1}{2}\right]$$

Finally, substituting the domain term in Somigliana's identity (4.34) for equation (4.38), and using the definition of flux, the thermoelastic displacement equation for an internal point $\mathbf{X}'$ can be written as

$$u_i(\mathbf{X}') + \int_S T_{ij}(\mathbf{X}',\mathbf{x})u_j(\mathbf{x})dS - \int_S \overline{P}_i(\mathbf{X}',\mathbf{x})\theta(\mathbf{x})dS$$

$$= \int_S U_{ij}(\mathbf{X}',\mathbf{x})t_j(\mathbf{x})dS - \int_S \overline{Q}_i(\mathbf{X}',\mathbf{x})q(\mathbf{x})dS \tag{4.40}$$

The displacement boundary integral equation for a boundary point $\mathbf{x}'$ can be obtained from equation (4.40) by taking the internal point $\mathbf{X}'$ to the boundary point $\mathbf{x}'$ to give

$$u_i(\mathbf{x}') + \lim_{\varepsilon\to0}\int_{S-S_\varepsilon+S_\varepsilon^+} T_{ij}(\mathbf{x}',\mathbf{x})u_j(\mathbf{x})dS - \lim_{\varepsilon\to0}\int_{S-S_\varepsilon+S_\varepsilon^+} \overline{P}_i(\mathbf{x}',\mathbf{x})\theta(\mathbf{x})dS$$

$$= \lim_{\varepsilon\to0}\int_{S-S_\varepsilon+S_\varepsilon^+} U_{ij}(\mathbf{x}',\mathbf{x})t_j(\mathbf{x})dS - \lim_{\varepsilon\to0}\int_{S-S_\varepsilon+S_\varepsilon^+} \overline{Q}_i(\mathbf{x}',\mathbf{x})q(\mathbf{x})dS \tag{4.41}$$

where U_{ij}, $\overline{P}_i$ are weakly singular and $\overline{Q}_i$ is not singular. The weakly singular terms do not give any free term, and the boundary integral representation of (4.41) can be written as

$$C_{ij}(\mathbf{x}')u_i(\mathbf{x}') + \fint_S T_{ij}(\mathbf{x}',\mathbf{x})u_j(\mathbf{x})dS - \int_S \overline{P}_i(\mathbf{x}',\mathbf{x})\theta(\mathbf{x})dS$$

$$= \int_S U_{ij}(\mathbf{x}',\mathbf{x})t_j(\mathbf{x})dS - \int_S \overline{Q}_i(\mathbf{x}',\mathbf{x})q(\mathbf{x})dS \tag{4.42}$$

where $C_{ij}(\mathbf{x}') = \delta_{ij}/2$ for smooth boundaries.

Stress Integral Equation

In this section the stress integral equation for thermoelastic problems is derived. The derivation is similar to that presented in Chapter 2, with the thermal body force term now included.

The stress equation for an internal point $\mathbf{X}'$ will be obtained by using the differentials of the displacement equation (4.40), with respect to the source point $\mathbf{X}'$, to give

$$\frac{\partial u_i(\mathbf{X}')}{\partial X'_j} + \int_S \frac{\partial T_{ik}(\mathbf{X}', \mathbf{x})}{\partial X'_j} u_k(\mathbf{x}) dS - \int_S \frac{\partial \overline{P}_i(\mathbf{X}', \mathbf{x})}{\partial X'_j} \theta(\mathbf{x}) dS$$
$$= \int_S \frac{\partial U_{ik}(\mathbf{X}', \mathbf{x})}{\partial X'_j} t_k(\mathbf{x}) dS - \int_S \frac{\partial \overline{Q}_i(\mathbf{X}', \mathbf{x})}{\partial X'_j} q(\mathbf{x}) dS$$

The displacement derivatives are combined through Hooke's law for thermoelasticity, equation (4.2), by using the displacement-strain relationships. This combination gives the following stress equation for an internal point $\mathbf{X}'$:

$$\sigma_{ij}(\mathbf{X}') + \int_S S_{kij}(\mathbf{X}', \mathbf{x}) u_k(\mathbf{x}) dS - \int_S \overline{P}_{ij}(\mathbf{X}', \mathbf{x}) \theta(\mathbf{x}) dS + \frac{E}{(1 - 2\nu)} h\theta(\mathbf{X}') \delta_{ij}$$
$$= \int_S D_{kij}(\mathbf{X}', \mathbf{x}) t_k(\mathbf{x}) dS - \int_S \overline{Q}_{ij}(\mathbf{X}', \mathbf{x}) q(\mathbf{x}) dS \tag{4.43}$$

where S_{kij} and D_{kij} are listed in Section 2.3.3. The fundamental fields $\overline{P}_{ij}$ and $\overline{Q}_{ij}$ have the following form:

$$\overline{P}_{ij}(\mathbf{x}', \mathbf{x}) = \frac{Eh}{8\pi(1 - \nu)r^2} \left\{ n_k r_{,k} \left[\frac{\delta_{ij}}{1 - 2\nu} - 3r_{,i} r_{,j} \right] + n_i r_{,j} + n_j r_{,i} \right\} \tag{4.44}$$

$$\overline{Q}_{ij}(\mathbf{x}', \mathbf{x}) = \frac{Eh}{8\pi k(1 - \nu)r} \left(r_{,i} r_{,j} - \frac{\delta_{ij}}{1 - 2\nu} \right) \tag{4.45}$$

for three-dimensional problems, and

$$\bar{P}_{ij}(\mathbf{X}', \mathbf{x}) = \frac{h\mu(1 + \nu)}{2\pi(1 - \nu)r} \left\{ n_k r_{,k} \left[\frac{\delta_{ij}}{(1 - 2\nu)} - 2r_{,i} r_{,j} \right] + n_i r_{,j} + n_j r_{,i} \right\} \tag{4.46}$$

$$\bar{Q}_{ij}(\mathbf{X}', \mathbf{x}) = -\frac{h\mu(1 + \nu)}{2\pi k(1 - \nu)} \left\{ r_{,i} r_{,j} + \frac{\delta_{ij}}{(1 - 2\nu)} \left[\frac{(1 + 2\nu)}{2} - \ln\left(\frac{1}{r}\right) \right] \right\} \tag{4.47}$$

for two-dimensional problems.

The stress boundary integral equation for a smooth boundary point $\mathbf{x}'$ can be obtained by performing a limiting process, in a similar way as for the displacement equation. The stress equation can be written as

$$\sigma_{ij}(\mathbf{x}') + \lim_{\varepsilon \to 0} \int_{S - S_\varepsilon + S_\varepsilon^+} S_{kij}(\mathbf{x}', \mathbf{x}) u_k(\mathbf{x}) dS$$
$$- \lim_{\varepsilon \to 0} \int_{S - S_\varepsilon + S_\varepsilon^+} \overline{P}_{ij}(\mathbf{x}', \mathbf{x}) \theta(\mathbf{x}) dS + \frac{E}{(1 - 2\nu)} h\theta(\mathbf{x}') \delta_{ij} \tag{4.48}$$
$$= \lim_{\varepsilon \to 0} \int_{S - S_\varepsilon + S_\varepsilon^+} D_{kij}(\mathbf{x}', \mathbf{x}) t_k(\mathbf{x}) dS - \lim_{\varepsilon \to 0} \int_{S - S_\varepsilon + S_\varepsilon^+} \overline{Q}_{ij}(\mathbf{x}', \mathbf{x}) q(\mathbf{x}) dS$$

In equation (4.48) $\overline{Q}_{ij}$ is weakly singular, so this term does not give a free term. The integrals involving S_{kij} and D_{kij} which are hypersingular and strongly singular, respectively, were dealt with in Chapter 2. The integral containing $\overline{P}_{ij}$ in equation

(4.48) can be regularized using the first term of the Taylor's expansion. Assuming the $C^{0,\alpha}$ continuity of $\theta(\mathbf{x})$, the resulting equation is

$$
\lim_{\varepsilon \to 0} \int_{S-S_\varepsilon+S_\varepsilon^+} \overline{P}_{ij}(\mathbf{x}',\mathbf{x})\theta(\mathbf{x})dS = \lim_{\varepsilon \to 0} \int_{S-S_\varepsilon} \overline{P}_{ij}(\mathbf{x}',\mathbf{x})\theta(\mathbf{x})dS
$$

$$
+ \lim_{\varepsilon \to 0} \int_{S_\varepsilon^+} \overline{P}_{ij}(\mathbf{x}',\mathbf{x})\left[\theta(\mathbf{x}) - \theta(\mathbf{x}')\right]dS
$$

$$
+\theta(\mathbf{x}') \lim_{\varepsilon \to 0} \int_{S_\varepsilon^+} \overline{P}_{ij}(\mathbf{x}',\mathbf{x})dS
$$

$$
= I_1'' + I_2'' + I_3'' \tag{4.49}
$$

In equation (4.49), $I_2'' \to \mathbf{0}$ as $\varepsilon \to \mathbf{0}$. The integral I_1'' exist in the Cauchy principal value sense, in other words

$$
\lim_{\varepsilon \to 0} \int_{S-S_\varepsilon} \overline{P}_{ij}(\mathbf{x}',\mathbf{x})\theta(\mathbf{x})dS = \int_S \overline{P}_{ij}(\mathbf{x}',\mathbf{x})\theta(\mathbf{x})dS \tag{4.50}
$$

The integral I_3'' leads to a free term

$$
\theta(\mathbf{x}') \lim_{\varepsilon \to 0} \int_{S_\varepsilon^+} \overline{P}_{ij}(\mathbf{x}',\mathbf{x})dS = \frac{E(1+\nu)}{6(1-\nu)(1-2\nu)}h\theta(\mathbf{x}')\delta_{ij} \tag{4.51}
$$

Substituting the limiting forms of the integrals involving S_{kij} and D_{kij}, as well as (4.51) into equation (4.48), gives the following stress boundary integral equation for a boundary point:

$$
\frac{1}{2}\sigma_{ij}(\mathbf{x}')+ \int_S S_{kij}(\mathbf{x}',\mathbf{x})u_k(\mathbf{x})dS + \frac{E}{2(1-2\nu)}h\theta(\mathbf{x}')\delta_{ij} \tag{4.52}
$$

$$
= - \int_S \overline{P}_{ij}(\mathbf{x}',\mathbf{x})\theta(\mathbf{x})dS + \int_S D_{kij}(\mathbf{x}',\mathbf{x})t_k(\mathbf{x})dS - \int_S \overline{Q}_{ij}(\mathbf{x}',\mathbf{x})q(\mathbf{x})dS
$$

Numerical Discretization

The solution of the thermoelasticity problem requires the thermoelastic fields θ, q, u_i and t_i which are linked by equations (4.15) and (4.42). The point collocation procedure, with the standard boundary element discretization, is summarized below.

The thermoelastic fields can be represented using isoparametric representation as

$$
\theta = \sum_{\alpha=1}^{m} N_\alpha \theta^\alpha \qquad q = \sum_{\alpha=1}^{m} N_\alpha q^\alpha \tag{4.53}
$$

$$
u_i = \sum_{\alpha=1}^{m} N_\alpha u_i^\alpha \qquad t_i = \sum_{\alpha=1}^{m} N_\alpha t_i^\alpha \tag{4.54}
$$

where N_α denotes the shape functions, and α refers to nodal values.

The discretized form of the temperature boundary integral equation, in a boundary divided into N_e elements, after collocating at node c can be written as

$$
C(\mathbf{x}^c)\theta(\mathbf{x}^c) = \sum_{n=1}^{N_e}\sum_{\alpha=1}^{m} \theta^{n\alpha}V^{n\alpha} - \sum_{n=1}^{N_e}\sum_{\alpha=1}^{m} q^{n\alpha}W^{n\alpha} \tag{4.55}
$$

where

$$V^{n\alpha} = \int_{-1}^{1}\int_{-1}^{1} q^* N_\alpha J^n d\eta_1 d\eta_2 \qquad W^{n\alpha} = \int_{-1}^{1}\int_{-1}^{1} \theta^* N_\alpha J^n d\eta_1 d\eta_2$$

for three-dimensional problems, and

$$V^{n\alpha} = \int_{-1}^{1} q^* N_\alpha J^n d\eta \qquad W^{n\alpha} = \int_{-1}^{1} \theta^* N_\alpha J^n d\eta$$

for two-dimensional problems. Similarly, for the displacement integral equation, we have

$$C_{ij}(\mathbf{x}^c)u_i(\mathbf{x}^c) = \sum_{n=1}^{N_e}\sum_{\alpha=1}^{m} t_j^{n\alpha} P_{ij}^{n\alpha} - \sum_{n=1}^{N_e}\sum_{\alpha=1}^{m} u_j^{n\alpha} Q_{ij}^{n\alpha} + \sum_{n=1}^{N_e}\sum_{\alpha=1}^{m} \theta^{n\alpha} P_i^{n\alpha} - \sum_{n=1}^{N_e}\sum_{\alpha=1}^{m} q^{n\alpha} Q_i^{n\alpha}$$

$$(4.56)$$

where

$$P_{ij}^{n\alpha} = \int_{-1}^{1}\int_{-1}^{1} U_{ij} N_\alpha J^n d\eta_1 d\eta_2 \qquad Q_{ij}^{n\alpha} = \int_{-1}^{1}\int_{-1}^{1} T_{ij} N_\alpha J^n d\eta_1 d\eta_2$$

$$P_i^{n\alpha} = \int_{-1}^{1}\int_{-1}^{1} P_i N_\alpha J^n d\eta_1 d\eta_2 \qquad Q_i^{n\alpha} = \int_{-1}^{1}\int_{-1}^{1} Q_i N_\alpha J^n d\eta_1 d\eta_2$$

for three-dimensional problems, and

$$P_{ij}^{n\alpha} = \int_{-1}^{1} U_{ij} N_\alpha J^n d\eta \qquad Q_{ij}^{n\alpha} = \int_{-1}^{1} T_{ij} N_\alpha J^n d\eta$$

$$P_i^{n\alpha} = \int_{-1}^{1} P_i N_\alpha J^n d\eta \qquad Q_i^{n\alpha} = \int_{-1}^{1} Q_i N_\alpha J^n d\eta$$

The System Matrices

If M is the total number of nodes on the discretized geometry, the collocation of equations (4.55) and (4.56) throughout all the nodes ($c = 1, M$) leads to a system of algebraic equations, which can be expressed in matrix form as

$$\mathbf{Hu} = \mathbf{Gt} \qquad (4.57)$$

The usual method of analysis of uncoupled thermoelasticity problems has been to solve the potential equation first and then to solve the elasticity equation, since in uncoupled thermoelasticity the thermal field does not depend upon the stresses. Both equations can be coupled to form a single set and they can be solved together. In this scheme, less computational effort is required for the numerical integration, as parameters are calculated once for the thermoelastic and thermal problems. Also, a simpler update of the formulation to deal with coupled and time dependant thermoelasticity is possible. Hence, the system of equation (4.57) can be written as

$$\mathbf{H}^\alpha \mathbf{u}_\alpha = \mathbf{G}^\alpha \mathbf{t}_\alpha \tag{4.58}$$

where the vectors $\mathbf{u}_\alpha$, $\mathbf{t}_\alpha$ and the matrices $\mathbf{G}^\alpha$, $\mathbf{H}^\alpha$ are given by

$$\mathbf{u}_\alpha = \begin{bmatrix} u_1 \\ u_2 \\ u_3 \\ \theta \end{bmatrix} \quad \text{and} \quad \mathbf{t}_\alpha = \begin{bmatrix} t_1 \\ t_2 \\ t_3 \\ q \end{bmatrix} \tag{4.59}$$

and

$$\mathbf{G}^\alpha = \begin{bmatrix} G_{11} & G_{12} & G_{13} & G_{1\theta} \\ G_{21} & G_{22} & G_{23} & G_{2\theta} \\ G_{31} & G_{32} & G_{33} & G_{3\theta} \\ 0 & 0 & 0 & G_{\theta\theta} \end{bmatrix} \quad \text{and} \quad \mathbf{H}^\alpha = \begin{bmatrix} H_{11} & H_{12} & H_{13} & H_{1\theta} \\ H_{21} & H_{22} & H_{23} & H_{2\theta} \\ H_{31} & H_{32} & H_{33} & H_{3\theta} \\ 0 & 0 & 0 & H_{\theta\theta} \end{bmatrix} \tag{4.60}$$

The coefficients of the matrices $\mathbf{G}^\alpha$ and $\mathbf{H}^\alpha$ are obtained from the integration of equation (4.56) in the following way: the terms containing U_{ij} (G_{11} to G_{33}) and T_{ij} (H_{11} to H_{33}) and the terms containing Q_i ($G_{1\theta}$ to $G_{3\theta}$) and P_i ($H_{1\theta}$ to $H_{3\theta}$); and from equation (4.55) in the following way: the terms containing θ^* ($G_{\theta\theta}$) and q^* ($H_{\theta\theta}$).

Since the problems considered above are three-dimensional, and because of the coupling of the potential and thermoelasticity equations, the dimensions of $\mathbf{G}^\alpha$ and $\mathbf{H}^\alpha$ are $4M \times 4M$, which clearly gives the dimensions of the vectors $\mathbf{u}_\alpha$ and $\mathbf{t}_\alpha$ as $4M$.

Similarly, for two-dimensional problems, we have

$$\mathbf{u}_\alpha = \begin{bmatrix} u_1 \\ u_2 \\ \theta \end{bmatrix} \quad \text{and} \quad \mathbf{t}_\alpha = \begin{bmatrix} t_1 \\ t_2 \\ q \end{bmatrix}$$

and

$$\mathbf{G}^\alpha = \begin{bmatrix} G_{11} & G_{12} & G_{1\theta} \\ G_{21} & G_{22} & G_{2\theta} \\ 0 & 0 & G_{\theta\theta} \end{bmatrix} \quad \text{and} \quad \mathbf{H}^\alpha = \begin{bmatrix} H_{11} & H_{12} & H_{1\theta} \\ H_{21} & H_{22} & H_{2\theta} \\ 0 & 0 & H_{\theta\theta} \end{bmatrix} \tag{4.61}$$

By applying the boundary conditions and re-arranging equation (4.58) such that the unknowns are on the left-hand side and the known on the right-hand side, the equation is

$$\mathbf{A}\mathbf{x} = \mathbf{f} \tag{4.62}$$

where the matrix $\mathbf{A}$ contains coefficients from $\mathbf{G}^\alpha$ and $\mathbf{H}^\alpha$.

4.3.2 Transient Thermoelasticity

Boundary integral equations for transient thermoelasticity can be obtained from the diffusion equation and an equilibrium equation. The two equations of uncoupled transient thermoelasticity for a linear, elastic, isotropic and homogeneous body occupying a domain V enclosed by a boundary S can be expressed as follows:

$$\theta_{,ii} - \frac{1}{\kappa}\dot{\theta} = 0 \tag{4.63}$$

$$\sigma_{ij,j} + b_i = 0 \tag{4.64}$$

where κ is the thermal diffusivity which is equal to $k/\rho c$. If the relationship between stress and displacement is applied to equation (4.64), the uncoupled transient thermoelastic equations (4.7) and (4.8) can be obtained.

The differential equation (4.63) is solved subject to initial temperature conditions in V and temperature and flux boundary conditions on S_θ and S_q, respectively. For equation (4.63), with initial conditions ($\tau = \tau_o$):

$$\theta(\mathbf{x},\tau) = \bar{\theta}(\mathbf{x},\tau_o) \text{ on } \mathbf{x} \in S_\theta,$$

$$q(\mathbf{x},\tau) = \bar{q}(\mathbf{x},,\tau_o) \text{ on } \mathbf{x} \in S_q,$$

$$u_i(\mathbf{x},\tau) = \bar{u}_i(\mathbf{x},\tau_o) \text{ on } \mathbf{x} \in S_u$$

$$t_i(\mathbf{x},\tau) = \bar{t}_i(\mathbf{x},\tau_o) \text{ on } \mathbf{x} \in S_t, \tag{4.65}$$

where τ_o represents the initial and τ_F represents the final time.

Temperature Integral Equation

The temperature boundary integral formulation of the diffusion equation for an internal point $\mathbf{X}'$ is as follows:

$$\theta(\mathbf{X}',\tau_F) - \theta(\mathbf{X}',\tau_o) - \int_{\tau_o}^{\tau_F}\int_S \theta(\mathbf{x},\tau)Q(\mathbf{X}',\mathbf{x},\tau_F,\tau)dSd\tau$$

$$= -\int_{\tau_o}^{\tau_F}\int_S q(\mathbf{x},\tau)\Theta(\mathbf{X}',\mathbf{x},\tau_F,\tau)dSd\tau$$

$$+ \int_S q(\mathbf{x},\tau_o)\bar{\Theta}(\mathbf{X}',\mathbf{x},\tau_F,\tau_o)dS - \int_S \theta(\mathbf{x},\tau_o)\bar{Q}(\mathbf{X}',\mathbf{x},\tau_F,\tau_o)dS \tag{4.66}$$

The temperature boundary integral equation is obtained by letting $\mathbf{X}' \to \mathbf{x}'$ in equation (4.66) as

$$C(\mathbf{x}')\theta(\mathbf{x}',\tau_F) - C(\mathbf{x}')\theta(\mathbf{x}',\tau_o) - \int_S\int_{\tau_o}^{\tau_F} \theta(\mathbf{x},\tau)Q(\mathbf{x}',\mathbf{x},\tau_F,\tau)d\tau dS$$

$$= -\int_S\int_{\tau_o}^{\tau_F} q(\mathbf{x},\tau)\Theta(\mathbf{x}',\mathbf{x},\tau_F,\tau)d\tau dS$$

$$+ \int_S q(\mathbf{x},\tau_o)\bar{\Theta}(\mathbf{x}',\mathbf{x},\tau_F,\tau_o)dS - \int_S \theta(\mathbf{x},\tau_o)\bar{Q}(\mathbf{x}',\mathbf{x},\tau_F,\tau_o)dS \tag{4.67}$$

where $C(\mathbf{x}')$ depends upon the boundary. The fundamental solutions can be found in [1, 3, 12, 18, 22] as

$$\Theta(\mathbf{x}', \mathbf{x}, \tau_F, \tau) = \frac{1}{4\pi k(\tau_F - \tau)^{3/2}} e^{-s} H(\tau_F - \tau) \qquad \text{for } \tau \le \tau_F \qquad (4.68)$$

$$Q(\mathbf{x}', \mathbf{x}, \tau_F, \tau) = \frac{r r_{,i} n_i}{16\pi^{3/2}[\kappa(\tau_F - \tau)]^{5/2}} e^{-s} \qquad (4.69)$$

$$\bar{\Theta}(\mathbf{x}', \mathbf{x}, \tau_F, \tau_o) = \frac{1}{4\pi k r}\left[1 - \frac{2}{\sqrt{\pi}} \int_0^{\zeta/2} e^{-x^2} dx\right] \qquad (4.70)$$

$$\bar{Q}(\mathbf{x}', \mathbf{x}, \tau_F, \tau_o) = \frac{r_{,i} n_i}{4\pi r^2}\left[1 - \frac{2}{\sqrt{\pi}} \int_0^{\zeta/2} e^{-x^2} dx - \frac{\zeta}{\sqrt{\pi}} e^{-\zeta^2/4}\right] \qquad (4.71)$$

where $\zeta = \frac{r}{\kappa(\tau_F - \tau_o)}$, for three-dimensional problems, and

$$\Theta(\mathbf{x}', \mathbf{x}, \tau_F, \tau) = \frac{1}{4\pi k(\tau_F - \tau)} e^{-s} H(\tau_F - \tau) \qquad \text{for } \tau \le \tau_F \qquad (4.72)$$

$$Q(\mathbf{x}', \mathbf{x}, \tau_F, \tau) = \frac{r r_{,i} n_i}{8\pi\left[\kappa(\tau_F - \tau)\right]^2} e^{-s} \qquad (4.73)$$

$$\bar{\Theta}(\mathbf{x}', \mathbf{x}, \tau_F, \tau_o) = \frac{1}{4\pi k} E_1(s_o)$$

$$\bar{Q}(\mathbf{x}', \mathbf{x}, \tau_F, \tau_o) = \frac{r_{,i} n_i}{2\pi r} e^{-s_o} \qquad (4.74)$$

where

$$s = \frac{r^2}{4\kappa(\tau_F - \tau)}$$

$$s_o = \frac{r^2}{4\kappa(\tau_F - \tau_o)}$$

$$E_1(s_o) = \int_{s_o}^{\infty} \frac{e^{-x}}{x} dx$$

for two-dimensional problems.

Flux Boundary Integral Equation

The flux equation across any smooth boundary point $\mathbf{x}'$ requires temperature derivative equations at that boundary point. The temperature derivative at an internal point $\mathbf{X}'$ is

$$\frac{\partial \theta(\mathbf{X}', \tau_F)}{\partial \mathbf{X}'_i} - \frac{\partial \theta(\mathbf{X}', \tau_o)}{\partial \mathbf{X}'_i} - \int_{\tau_o}^{\tau_F} \int_S \theta(\mathbf{x}, \tau) \frac{\partial Q(\mathbf{X}', \mathbf{x}, \tau_F, \tau)}{\partial \mathbf{X}'_i} dS d\tau$$

$$= -\int_{\tau_o}^{\tau_F} \int_S q(\mathbf{x}, \tau) \frac{\partial \Theta(\mathbf{X}', \mathbf{x}, \tau_F, \tau)}{\partial \mathbf{X}'_i} dS d\tau$$

$$+ \int_S q(\mathbf{x}, \tau_o) \frac{\partial \bar{\Theta}(\mathbf{X}', \mathbf{x}, \tau_F, \tau_o)}{\partial \mathbf{X}'_i} dS - \int_S \theta(\mathbf{x}, \tau_o) \frac{\partial \bar{Q}(\mathbf{X}', \mathbf{x}, \tau_F, \tau_o)}{\partial \mathbf{X}'_i} dS \qquad (4.75)$$

The two-dimensional derivatives of fundamental fields Θ, Q, $\bar{\Theta}$ and $\bar{q}$ in the temperature derivative equation (4.75) are given as follows:

$$\frac{\partial \Theta(\mathbf{X}', \mathbf{x}, \tau_F, \tau)}{\partial X'_i} = \frac{rr_{,i}}{8\pi\kappa k(\tau_F - \tau)^2} e^{-s} \tag{4.76}$$

$$\frac{\partial Q(\mathbf{X}', \mathbf{x}, \tau_F, \tau)}{\partial X'_i} = \frac{1}{\kappa}\left[-\frac{n_i}{8\pi(\tau_F - \tau)^2} + \frac{r_{,k}n_k r_{,i} r^2}{16\pi\kappa(\tau_F - \tau)^3}\right] e^{-s} \tag{4.77}$$

$$\frac{\partial \bar{\Theta}(\mathbf{X}', \mathbf{x}, \tau_F, \tau_o)}{\partial X'_i} = \frac{r_{,i}}{2\pi kr} e^{-s_o} \tag{4.78}$$

$$\frac{\partial \bar{\Theta}(\mathbf{X}', \mathbf{x}, \tau_F, \tau_o)}{\partial X'_i} = \frac{r_{,i}}{2\pi kr} e^{-s_o} \tag{4.79}$$

$$\frac{\partial \bar{Q}(\mathbf{X}', \mathbf{x}, \tau_F, \tau_o)}{\partial X'_i} = \frac{1}{2\pi}\left(-\frac{n_i}{r^2} + \frac{2r_{,k}n_k r_{,i}}{r^2} + \frac{r_{,k}n_k r_{,i}}{2\kappa(\tau_F - \tau_o)}\right) e^{-s_o} \tag{4.80}$$

The temperature derivative equation for a point on a smooth boundary $\mathbf{x}'$ can be obtained by letting $\mathbf{X}' \to \mathbf{x}'$ in equation (4.75). The time integration for derivatives of Θ and Q integrands in equation (4.75) is carried out by dividing the time range τ_o to τ_F into equal time steps of $\Delta\tau$. The integration over the last time step can be written as

$$\int_S \int_{\tau_{(F-1)}}^{\tau_F} \frac{\partial Q(\mathbf{X}', \mathbf{x}, \tau_F, \tau)}{\partial X'_i} \theta(\mathbf{x}, \tau) d\tau dS \quad \text{and} \quad \int_S \int_{\tau_{(F-1)}}^{\tau_F} \frac{\partial \Theta(\mathbf{X}', \mathbf{x}, \tau_F, \tau)}{\partial X'_i} q(\mathbf{x}, \tau) d\tau dS$$

For constant interpolation over time, and noting that as $\tau \to \tau_F$, we have

$$s \to \infty \quad \text{and} \quad e^{-s}, \quad \frac{e^{-s}}{(\tau_F - \tau)} \to 0$$

the two integrals (for 2D problems) can be written as follows:

$$\int_S \left\{ \frac{r_{,k}n_k r_{,i}}{4\pi\Delta\tau} e^{-s_{F-1}} - \left(\frac{n_i}{2} - r_{,k}n_k r_{,i}\right) \frac{1}{\pi r^2} e^{-s_{F-1}} \right\} \theta(\mathbf{x}, \tau_F) dS$$

$$= \int_S A_i^F \theta(\mathbf{x}, \tau_F) dS \tag{4.81}$$

$$\int_S \left\{ \frac{r_{,i}}{2\pi kr} e^{-s_{F-1}} \right\} q(\mathbf{x}, \tau_F) dS = \int_S B_i^F q(\mathbf{x}, \tau_F) dS \tag{4.82}$$

where A_i^F and B_i^F represent the integrands. The flux q is assumed to be $C^{0,\alpha}$ continuous and the temperature θ to be $C^{1,\alpha}$ continuous. The regularized form of equation (4.81) can be written as follows:

$$\lim_{\mathbf{X}' \to \mathbf{x}'} \int_S A_i^F \theta(\mathbf{x}, \tau_F) dS = \lim_{\epsilon \to 0} \int_{S-S_\epsilon + S_\epsilon^+} A_i^F \theta(\mathbf{x}, \tau_F) dS = \lim_{\epsilon \to 0} \int_{S-S_\epsilon} A_i^F \theta(\mathbf{x}, \tau_F) dS$$

$$+ \lim_{\epsilon \to 0} \int_{S_\epsilon^+} A_i^F \left\{ \theta(\mathbf{x}, \tau_F) - \theta(\mathbf{x}', \tau_F) - \theta_{,i}(\mathbf{x}', \tau_F)(\mathbf{x}_i - \mathbf{x}'_i) \right\} dS$$

$$+ \theta(\mathbf{x}', \tau_F) \lim_{\epsilon \to 0} \int_{S_\epsilon^+} A_i^F dS + \theta_{,i}(\mathbf{x}', \tau_F) \lim_{\epsilon \to 0} \int_{S_\epsilon^+} A_i^F (\mathbf{x}_i - \mathbf{x}'_i) dS \tag{4.83}$$

The first and third integrals together form the Hadamard principal value integral, the second integral goes to zero as $\epsilon \to 0$, and the fourth integral gives a bounded term. The final integral is written as

$$\lim_{\mathbf{x}' \to \mathbf{x}'} \int_S A_i^F \theta(\mathbf{x}, \tau_F) dS = \lim_{\epsilon \to 0} \oint_S A_i^F \theta(\mathbf{x}, \tau_F) dS + \theta_{,i}(\mathbf{x}', \tau_F) \lim_{\epsilon \to 0} \int_{S_\epsilon^+} A_i^F (\mathbf{x}_i - \mathbf{x}'_i) dS \tag{4.84}$$

Similarly, equation (4.82) is as follows:

$$\lim_{\mathbf{x}' \to \mathbf{x}'} \int_S B_i^F q(\mathbf{x}, \tau_F) dS = \lim_{\epsilon \to 0} \int_{S - S_\epsilon + S_\epsilon^+} B_i^F q(\mathbf{x}, \tau_F) dS = \lim_{\epsilon \to 0} \int_{S - S_\epsilon} B_i^F q(\mathbf{x}, \tau_F) dS$$

$$+ \lim_{\epsilon \to 0} \int_{S_\epsilon^+} B_i^F \left(-k n_i(\mathbf{x}) \right) \left(\theta_{,i}(\mathbf{x}, \tau_F) - \theta_{,i}(\mathbf{x}', \tau_F) \right) dS$$

$$- k\theta_{,i}(\mathbf{x}', \tau_F) \lim_{\epsilon \to 0} \int_{S_\epsilon^+} B_i^F n_i(\mathbf{x}) dS \tag{4.85}$$

where the relation $q = -k\theta_{,i} n_i$ is used. Of the three terms on the right-hand side of equation (4.85), the first term is a Cauchy principal value integral and the second term tends to zero as $\epsilon \to 0$. The third integral gives a bounded term similar to that in equation (4.84). Then equation (4.85) can be written as follows:

$$\lim_{\mathbf{x}' \to \mathbf{x}'} \int_S B_i^F q(\mathbf{x}, \tau_F) dS = \lim_{\epsilon \to 0} \oint_S B_i^F q(\mathbf{x}, \tau_F) dS - k\theta_{,i}(\mathbf{x}', \tau_F) \lim_{\epsilon \to 0} \int_{S_\epsilon^+} B_i^F n_i(\mathbf{x}) dS. \tag{4.86}$$

After substituting the parameters for a semi-circular region (see Chapter 2), in equations (4.84) and (4.85) the following equalities are obtained:

$$\lim_{\mathbf{x}' \to \mathbf{x}'} \int_S A_i^F \theta(\mathbf{x}, \tau_F) dS = \lim_{\epsilon \to 0} \oint_S A_i^F \theta(\mathbf{x}, \tau_F) dS + \frac{\theta_{,i}(\mathbf{x}', \tau_F)}{4} \tag{4.87}$$

and

$$\lim_{\mathbf{x}' \to \mathbf{x}'} \int_S B_i^F q(\mathbf{x}, \tau_F) dS = \lim_{\epsilon \to 0} \oint_S B_i^F q(\mathbf{x}, \tau_F) dS - \frac{\theta_{,i}(\mathbf{x}', \tau_F)}{4} \tag{4.88}$$

Similar relations can be obtained for the terms of initial conditions of equation (4.75). The final temperature derivative boundary integral equation for point $\mathbf{x}'$ on a smooth boundary is as follows:

$$\frac{1}{2} \frac{\partial \theta(\mathbf{x}', \tau_F)}{\partial \mathbf{x}'_i} - \frac{1}{2} \frac{\partial \theta(\mathbf{x}', \tau_o)}{\partial \mathbf{x}'_i} - \int_{\tau_o}^{\tau_F}\!\!\!\int_S \theta(\mathbf{x}, \tau) \frac{\partial Q(\mathbf{x}', \mathbf{x}, \tau_F, \tau)}{\partial \mathbf{x}'_i} dS d\tau$$

$$= - \int_{\tau_o}^{\tau_F}\!\!\!\int_S q(\mathbf{x}, \tau) \frac{\partial \Theta(\mathbf{x}', \mathbf{x}, \tau_F, \tau)}{\partial \mathbf{x}'_i} dS d\tau$$

$$+ \int_S q(\mathbf{x}, \tau_o) \frac{\partial \bar{\Theta}(\mathbf{x}', \mathbf{x}, \tau_F, \tau_o)}{\partial \mathbf{x}'_i} dS - \oint_S \theta(\mathbf{x}, \tau_o) \frac{\partial \bar{Q}(\mathbf{x}', \mathbf{x}, \tau_F, \tau_o)}{\partial \mathbf{x}'_i} dS \tag{4.89}$$

The boundary flux integral equation is obtained by multiplying equation (4.89) with $-k n_i(\mathbf{x}')$, giving

$$\frac{q(\mathbf{x}', \tau_F)}{2} - \frac{q(\mathbf{x}', \tau_o)}{2} + n_i(\mathbf{x}') \oint_S \int_{\tau_o}^{\tau_F} \theta(\mathbf{x}, \tau) Q_i(\mathbf{x}', \mathbf{x}, \tau_F, \tau) d\tau dS$$

$$= n_i(\mathbf{x}') \oint_S \int_{\tau_o}^{\tau_F} q(\mathbf{x}, \tau) \Theta_i(\mathbf{x}', \mathbf{x}, \tau_F, \tau) d\tau dS$$

$$-n_i(\mathbf{x}') \oint_S q(\mathbf{x}, \tau_o) \bar{\Theta}_i(\mathbf{x}', \mathbf{x}, \tau_F, \tau_o) dS + n_i(\mathbf{x}') \oint_S \theta(\mathbf{x}, \tau_o) \bar{Q}_i(\mathbf{x}', \mathbf{x}, \tau_F, \tau_o) dS \quad (4.90)$$

The fundamental fields Θ_i, Q_i, $\bar{\Theta}_i$ and $\bar{q}_i$ in the flux equation (4.89) are given as follows:

$$\Theta_i(\mathbf{X}', \mathbf{x}, \tau_F, \tau) = k \frac{\partial \Theta(\mathbf{X}', \mathbf{x}, \tau_F, \tau)}{\partial \mathbf{X}'_i} = \frac{r r_{,i}}{8\pi\kappa(\tau_F - \tau)^2} e^{-s} \quad (4.91)$$

$$Q_i(\mathbf{X}', \mathbf{x}, \tau_F, \tau) = k \frac{\partial Q(\mathbf{X}', \mathbf{x}, \tau_F, \tau)}{\partial \mathbf{X}'_i} = c \left[-\frac{n_i}{8\pi(\tau_F - \tau)^2} + \frac{r_{,k} n_k r_{,i} r^2}{16\pi\kappa(\tau_F - \tau)^3} \right] e^{-s} \quad (4.92)$$

$$\bar{\Theta}_i(\mathbf{X}', \mathbf{x}, \tau_F, \tau_o) = k \frac{\partial \bar{\Theta}(\mathbf{X}', \mathbf{x}, \tau_F, \tau)}{\partial \mathbf{X}'_i} = \frac{r_{,i}}{2\pi r} e^{-s_o} \quad (4.93)$$

$$\bar{Q}_i(\mathbf{X}', \mathbf{x}, \tau_F, \tau_o) = k \frac{\partial \bar{Q}(\mathbf{X}', \mathbf{x}, \tau_F, \tau)}{\partial \mathbf{X}'_i}$$

$$= \frac{k}{2\pi} \left(-\frac{n_i}{r^2} + \frac{2 r_{,k} n_k r_{,i}}{r^2} + \frac{r_{,k} n_k r_{,i}}{2\kappa(\tau_F - \tau_o)} \right) e^{-s_o} \quad (4.94)$$

Displacement Integral Equation

In this section the displacement integral equation is presented for transient thermoelasticity using time domain formulation. The reciprocal theorem that was used by Dargush et al. [7] was for coupled thermoelasticity. Following Prasad et al. [22, 21, 23] the same theorem is used in this section and later simplified to uncoupled thermoelasticity. Consider two independent states of a body, one state represented by u_i, t_j, θ, q with variables as $\mathbf{X}$ and τ in a domain V with no body forces, and another state represented by u_j^*, t_j^*, θ^*, q^* with variables as $\mathbf{X}$ and $\tau_F - \tau$ in domain a V^* with a body force b_i^*. These can be written as

$$\int_S \int_0^{\tau_F} t_j(\mathbf{x}, \tau) u_j^*(\mathbf{x}, \tau_F, \tau) d\tau dS - \int_S \int_0^{\tau_F} u_j(\mathbf{x}, \tau) t_j^*(\mathbf{x}, \tau_F, \tau) d\tau dS$$

$$+ \int_S \int_0^{\tau_F} q(\mathbf{x}, \tau) \theta^*(\mathbf{x}, \tau_F, \tau) d\tau dS - \int_S \int_0^{\tau_F} \theta(\mathbf{x}, \tau) q^*(\mathbf{x}, \tau_F, \tau) d\tau dS$$

$$- \int_V \int_0^{\tau_F} u_j(\mathbf{X}, \tau) b_j^*(\mathbf{X}, \tau) d\tau dV \quad (4.95)$$

Consider a body force b_j^* at an internal point $\mathbf{X}'$ of V^*, an infinite domain. Let the body force be

$$b_j^* = \Delta(\mathbf{X}', \mathbf{X}) \Delta(\tau_F, \tau) e_j \quad (4.96)$$

where e_j is the unit directional vector. The components of the displacement, traction, temperature and flux responses can be separated as

$$u_j^*(\mathbf{x}, \tau_F, \tau) = U_{ij}(\mathbf{X}', \mathbf{x}, \tau_F, \tau) e_i(\mathbf{X}')$$

$$t_j^*(\mathbf{x}, \tau_F, \tau) = T_{ij}(\mathbf{X}', \mathbf{x}, \tau_F, \tau) e_i(\mathbf{X}')$$

$$\theta^*(\mathbf{x}, \tau_F, \tau) = G_i(\mathbf{X}', \mathbf{x}, \tau_F, \tau) e_i(\mathbf{X}')$$

$$Q^*(\mathbf{x}, \tau_F, \tau) = F_i(\mathbf{X}', \mathbf{x}, \tau_F, \tau)e_i(\mathbf{X}').\tag{4.97}$$

The displacement integral equation for an internal point is obtained by substituting equations (4.97) and (4.96) into the reciprocity theorem of equation (4.95), to give

$$u_i(\mathbf{X}', \tau_F) = \int_S \int_0^{\tau_F} T_{ij}(\mathbf{X}', \mathbf{x}, \tau_F, \tau)u_j(\mathbf{x}, \tau)d\tau dS$$

$$+ \int_S \int_0^{\tau_F} F_i(\mathbf{X}', \mathbf{x}, \tau_F, \tau)\theta(\mathbf{x}, \tau)d\tau dS$$

$$= \int_S \int_0^{\tau_F} U_{ij}(\mathbf{X}', \mathbf{x}, \tau_F, \tau)t_j(\mathbf{x}, \tau)d\tau dS + \int_S \int_0^{\tau_F} G_i(\mathbf{X}', \mathbf{x}, \tau_F, \tau)q(\mathbf{x}, \tau)d\tau dS \tag{4.98}$$

Equation (4.98) is for coupled thermoelasticity. Fundamental fields U_{ij} and T_{ij} for uncoupled thermoelasticity can be derived directly by taking the limiting form of the fundamental fields of coupled thermoelasticity. As temperature changes have no effect on displacements and the unit force has an instantaneous response for displacements, there is no time dependence in the kernels U_{ij} and T_{ij}.

Considering the initial time as τ_o instead of τ, the displacement boundary integral equation for uncoupled thermoelasticity can be written as follows:

$$u_i(\mathbf{X}', \tau_F) + \int_S T_{ij}(\mathbf{X}', \mathbf{x})u_j(\mathbf{x}, \tau_F)dS + \int_S \int_{\tau_o}^{\tau_F} F_i(\mathbf{X}', \mathbf{x}, \tau_F, \tau)\theta(\mathbf{x}, \tau)d\tau dS$$

$$= \int_S U_{ij}(\mathbf{X}', \mathbf{x})t_j(\mathbf{x}, \tau_F)dS + \int_S \int_{\tau_o}^{\tau_F} G_i(\mathbf{X}', \mathbf{x}, \tau_F, \tau)q(\mathbf{x}, \tau)d\tau dS \tag{4.99}$$

The fundamental fields for two-dimensional problems [7, 19] can be written as

$$F_i(\mathbf{X}', \mathbf{x}, \tau_F, \tau) = \frac{(1+\nu)h}{4\pi(1-\nu)}\left[\frac{2\kappa}{r^2}(2r_{,i}r_{,k}n_k - n_i)(1 - e^{-s}) - \frac{r_{,i}r_{,k}n_k}{(\tau_F - \tau)}e^{-s}\right] \tag{4.100}$$

$$G_i(\mathbf{X}', \mathbf{x}, \tau_F, \tau) = \frac{(1+\nu)h}{2\pi c(1-\nu)}\frac{r_{,i}}{r}(1 - e^{-s}). \tag{4.101}$$

The order of singularities of F_i and G_i becomes apparent after time integration, that F_i is $O(\ln r)$ and G_i is not singular for two-dimensional problems.

The displacement boundary integral equation can be obtained from equation (4.99) by letting $\mathbf{X}' \to \mathbf{x}'$. The fundamental fields U_{ij} and T_{ij} for the displacement equation in transient thermoelasticity are the same as the fundamental fields of the displacement equation in steady-state thermoelasticity. The order of singularity of F_i and G_i after time integration is the same as $\bar{P}_i$ and $\bar{Q}_i$ of equation (4.42) in steady-state thermoelasticity. The free term coming from the singularity in T_{ij} is the same as in equation (4.42). Finally, the displacement equation for a boundary point $\mathbf{x}'$ is as follows:

$$C_{ij}(\mathbf{x}')u_j(\mathbf{x}', \tau_F) + \int_S T_{ij}(\mathbf{x}', \mathbf{x})u_j(\mathbf{x}, \tau_F)dS$$

$$+ \int_S \int_{\tau_o}^{\tau_F} F_i(\mathbf{x}', \mathbf{x}, \tau_F, \tau)\theta(\mathbf{x}, \tau)d\tau dS$$

$$= \int_S U_{ij}(\mathbf{x}', \mathbf{x})t_j(\mathbf{x}, \tau_F)dS + \int_S \int_{\tau_o}^{\tau_F} G_i(\mathbf{x}', \mathbf{x}, \tau_F, \tau)q(\mathbf{x}, \tau)d\tau dS \tag{4.102}$$

Stress Integral Equation

In this section the stress integral equation is presented for transient thermoelasticity using time domain formulation. The stress equations are obtained by substituting the derivatives, with respect to $\mathbf{X}'$, of the displacement equation (4.99), in, the constitutive law. Derivatives of the displacement are given by the following:

$$\frac{\partial u_i(\mathbf{X}', \tau_F)}{\partial \mathbf{X}'_j} + \int_S \frac{\partial T_{ik}(\mathbf{X}', \mathbf{x})}{\partial \mathbf{X}'_j} u_k(\mathbf{x}, \tau_F) dS$$

$$+ \int_S \int_{\tau_o}^{\tau_F} \frac{\partial F_i(\mathbf{X}', \mathbf{x}, \tau_F, \tau)}{\partial \mathbf{X}'_j} \theta(\mathbf{x}, \tau) d\tau dS$$

$$= \int_S \frac{\partial U_{ik}(\mathbf{X}', \mathbf{x})}{\partial \mathbf{X}'_j} t_k(\mathbf{x}, \tau_F) dS + \int_S \int_{\tau_o}^{\tau_F} \frac{\partial G_i(\mathbf{X}', \mathbf{x}, \tau_F, \tau)}{\partial \mathbf{X}'_j} q(\mathbf{x}, \tau) d\tau dS \qquad (4.103)$$

The stress equation for an interior point $\mathbf{X}'$ can be written as follows:

$$\sigma_{ij}(\mathbf{X}', \tau_F) + \int_S S_{kij}(\mathbf{X}', \mathbf{x}) u_k(\mathbf{x}, \tau_F) dS$$

$$+ \int_S \int_{\tau_o}^{\tau_F} F_{ij}(\mathbf{X}', \mathbf{x}, \tau_F, \tau) \theta(\mathbf{x}, \tau) d\tau dS + \frac{2\mu(1+\nu)}{(1-2\nu)} h\theta(\mathbf{X}', \tau_F)\delta_{ij}$$

$$= \int_S D_{kij}(\mathbf{X}', \mathbf{x}) t_k(\mathbf{x}, \tau_F) dS$$

$$+ \int_S \int_{\tau_o}^{\tau_F} G_{ij}(\mathbf{X}', \mathbf{x}, \tau_F, \tau) q(\mathbf{x}, \tau) d\tau dS \qquad (4.104)$$

The fundamental solutions for two-dimensional problems are

$$F_{ij}(\mathbf{X}', \mathbf{x}, \tau_F, \tau) = -\frac{Eh}{4\pi(1-\nu)} \frac{1}{r} \Big[(4r_{,l}n_l r_{,i} r_{,j} - r_{,j}n_i - r_{,i}n_j - \delta_{ij} r_{,l} n_l)$$

$$\times \left\{ \frac{e^{-s}}{(\tau_F - \tau)} - \frac{4k}{r^2}(1 - e^{-s}) \right\} + \frac{r_{,l}n_l r^2}{2k(\tau_F - \tau)^2} \left\{ r_{,i} r_{,j} + \frac{\nu}{(1-2\nu)} \delta_{ij} \right\} e^{-s} \Big] \qquad (4.105)$$

$$G_{ij}(\mathbf{X}', \mathbf{x}, \tau_F, \tau) = -\frac{Eh}{4\pi k(1-\nu)} \Big[\frac{e^{-s}}{(\tau_F - \tau)} (r_{,i} r_{,j} + \frac{\nu}{(1-2\nu)} \delta_{ij})$$

$$- \frac{4\kappa}{r^2} (r_{,i} r_{,j} - \frac{\delta_{ij}}{2})(1 - e^{-s}) \Big] \qquad (4.106)$$

The boundary stress integral equation can be obtained from equation (4.104) by letting $\mathbf{X}' \to \mathbf{x}$. As the fundamental fields S_{kij} and D_{kij} are the same as in steady-state thermoelasticity, equation (4.104) is valid in transient thermoelasticity. The fundamental fields G_{ij} and F_{ij} are dependent on time. By dividing the time range into equal time steps, and carrying out the time integration on the final time step assuming that θ and q remain constant, gives the following:

$$\int_{\tau_{F-1}}^{\tau_F} F_{ij} d\tau = f_{ij} = -\frac{Eh}{4\pi(1-\nu)r} \Big[(r_{,j}n_i + r_{,i}n_j + r_{,l}n_l(\delta_{ij} - 4r_{,i}r_{,j})) \frac{(1 - e^{-s_{F-1}})}{s_{F-1}}$$

$$+ 2r_{,l}n_l \left(r_{,i}r_{,j} + \frac{\nu\delta_{ij}}{(1-2\nu)} \right) e^{-s_{F-1}} \Big] \qquad (4.107)$$

and

$$\int_{\tau_{F-1}}^{\tau_F} G_{ij}d\tau = g_{ij} = -\frac{Eh}{4\pi k(1-\nu)}\left[\left(\frac{\delta_{ij}}{2} - r_{,i}r_{,j}\right)\frac{(1-e^{-s_{F-1}})}{s_{F-1}} + \frac{\delta_{ij}}{2(1-2\nu)}E_1(s_{F-1})\right] \tag{4.108}$$

where $s_{F-1} = r^2/(4\kappa\Delta\tau)$ and as $\tau \to \tau_F$, $e^{-s_F} = 0$. The order of singularities of f_{ij} and g_{ij} are $O(1/r)$ and $O(\ln r)$, respectively. Equation (4.108) is weakly singular, and the singularity and does not give any free term. The free term from equation (4.107) is as follows:

$$\lim_{\mathbf{X}'\to\mathbf{x}'}\int_S\left\{\int_{\tau_o}^{\tau_F}F_{ij}(\mathbf{X}',\mathbf{x},\tau_F,\tau)\theta(\mathbf{x},\tau)d\tau\right\}dS = \lim_{\mathbf{X}'\to\mathbf{x}'}\int_S f_{ij}\theta(\mathbf{X}',\tau_F)dS$$

$$= \fint_S\left\{\int_{\tau_o}^{\tau_F}F_{ij}(\mathbf{x}',\mathbf{x},\tau_F,\tau)\theta(\mathbf{x},\tau)d\tau\right\}dS - \frac{Eh}{4(1-\nu)(1-2\nu)}\theta(\mathbf{x}',\tau_F)\delta_{ij} \tag{4.109}$$

Using equation (4.109), and other limiting forms of equations involving the other kernels derived previously with equation (4.104), the stress boundary integral equation in transient thermoelasticity can be written as follows:

$$\frac{\sigma_{ij}(\mathbf{x}',\tau_F)}{2} + \fint_S S_{kij}(\mathbf{x}',\mathbf{x})u_k(\mathbf{x},\tau_F)dS$$

$$+ \fint_S\int_{\tau_o}^{\tau_F}F_{ij}(\mathbf{x}',\mathbf{x},\tau_F,\tau)\theta(\mathbf{x},\tau)d\tau dS + \frac{Eh}{2(1-2\nu)}\theta(\mathbf{x}',\tau_F)\delta_{ij}$$

$$= \fint_S D_{kij}(\mathbf{x}',\mathbf{x})t_k(\mathbf{x},\tau_F)dS + \int_S\int_{\tau_o}^{\tau_F}G_{ij}(\mathbf{x}',\mathbf{x},\tau_F,\tau)q(\mathbf{x},\tau)d\tau dS \tag{4.110}$$

The boundary traction equation can be obtained by multiplying the above equation with the components of the outward normal to the boundary at the source point.

Numerical Implementation

The boundary integral equations of transient thermoelasticity require spatial and temporal discretization. Temporal integration is carried out using constant and linear time elements. Time integration is carried out analytically first, and the spatial integration can be carried out either analytically or numerically, depending, on the type of element.

Consider that the time range τ_o to τ_F is divided into F time steps of equal size $\Delta\tau$. Assuming that M^b are the temporal shape functions for the fth time step and τ^b are the temporal nodes, then the time can be written as

$$\tau = M^b\tau^b \qquad \text{for } \tau_{f-1} \leq \tau \leq \tau_f; \tag{4.111}$$

where

$$[M^b] = 1$$

for constant time interpolation, and

$$[M^b] = \left[\frac{\tau_f - \tau}{\Delta\tau}, \frac{\tau - \tau_{f-1}}{\Delta\tau}\right] \tag{4.112}$$

for linear time interpolation.

For isoparametric elements, the primary variables, required for equations (4.102) and (4.67), at any point on the element S_n at any time in the fth time step can be written as follows:

$$\theta(\mathbf{x}, \tau) = N_a M^b \theta^{abnf}, \qquad\qquad q(\mathbf{x}, \tau) = N_a M^b q^{abnf},$$

$$u_i(\mathbf{x}, \tau_F) = N_a u_i^{an} \qquad \text{and} \qquad t_i(\mathbf{x}, \tau_F) = N_a t_i^{an} \qquad (4.113)$$

where θ^{abnf}, q^{abnf}, u_i^{an} and t_i^{an} are the nodal temperature, flux, displacement and traction values, respectively. Superscripts a and b represent the spatial and temporal node numbers in each element, and n and f represent the spatial and temporal element numbers.

Discretization   The temperature boundary integral equation (4.67) can be written in a discretized form as follows:

$$C(\mathbf{x}')(\theta(\mathbf{x}', \tau_F) - \theta(\mathbf{x}', \tau_o)) - \sum_{n=1}^{N_e} \left[\int_{S_n} N_a \left\{ \sum_{f=1}^{F} \int_{\tau_{f-1}}^{\tau_f} M^b Q d\tau \right\} dS \right] \theta^{abnf}$$

$$= -\sum_{n=1}^{N_e} \left[\int_{S_n} N_a \left\{ \sum_{f=1}^{F} \int_{\tau_{f-1}}^{\tau_f} M^b \Theta d\tau \right\} dS \right] q^{abnf} \qquad (4.114)$$

The discretization of the displacement boundary integral equation (4.102) is as follows:

$$C_{ij}(\mathbf{x}') u_j(\mathbf{x}', \tau_F) + \sum_{n=1}^{N_e} \left[\int_{S_n} N_a T_{ij} dS \right] u_j^{an}$$

$$+ \sum_{n=1}^{N_e} \left[\int_{S_n} N^a \left\{ \sum_{f=1}^{F} \int_{\tau_{f-1}}^{\tau_f} M^b F_i d\tau \right\} dS \right] \theta^{abnf}$$

$$= \sum_{n=1}^{N_e} \left[\int_{S_n} N_a U_{ij} dS \right] t_j^{an} + \sum_{n=1}^{N_e} \left[\int_{S_n} N_a \left\{ \sum_{f=1}^{F} \int_{\tau_{f-1}}^{\tau_f} M^b G_i d\tau \right\} dS \right] q^{abnf} \quad (4.115)$$

In the above discretization of the boundary integral equations, different integrals can have different numbers of spatial and temporal points (nodes).

Time Integration   The time integrations can be evaluated analytically. It can be seen from the fundamental solutions that the primary time variable is $(\tau_F - \tau)$, where τ_F is the time at which the results are required, and τ is the integration variable which varies from τ_o to τ_F. For the temporal integration $(\tau_F - \tau_o)$ is divided into F time steps of fixed size $\Delta\tau$ each $(\tau_F = \tau_o + F\Delta\tau)$.

The unknown boundary values at τ_F, are calculated from the boundary values at all the previous times $(\tau_f, f = 1, F - 1)$. Since the time is divided into equal time steps of $\Delta\tau$, only one new set of matrices needs to be calculated in order to obtain results at τ_F. The matrices calculated at all the previous time steps are used again. Furthermore, by using a LU decomposition solver, the decomposition needs be done only once, and the same matrix can be used at all the time steps.

The time progression scheme for the temperature (4.114) at time t_F can be written as:

$$C(\mathbf{x}')(\theta(\mathbf{x}',\tau_F) - \theta(\mathbf{x}',\tau_o)) - \sum_{n=1}^{N_e}\left[\int_{S_n} N_a \int_{\tau_{F-1}}^{\tau_F} M^b Q(\mathbf{x}',\mathbf{x},\tau_F,\tau)d\tau dS\right]\theta^{abnF}$$

$$= -\sum_{n=1}^{N_e}\left[\int_{S_n} N_a \int_{\tau_{F-1}}^{\tau_F} M^b \Theta(\mathbf{x}',\mathbf{x},\tau_F,\tau)d\tau dS\right]q^{abnF}$$

$$+\sum_{n=1}^{N_e}\sum_{f=1}^{F-1}\left[\int_{S_n} N_a \int_{\tau_{f-1}}^{\tau_f} M^b Q(\mathbf{x}',\mathbf{x},\tau_F,\tau)d\tau dS\right]\theta^{abnf}$$

$$-\sum_{n=1}^{N_e}\sum_{f=1}^{F-1}\left[\int_{S_n} N_a \int_{\tau_{f-1}}^{\tau_f} M^b \Theta(\mathbf{x}',\mathbf{x},\tau_F,\tau)d\tau dS\right]q^{abnf} \qquad (4.116)$$

The unknowns at time τ_F are in the first two lines of equation (4.116), and the integrals in the last two lines include the results for time steps 1 to $F-1$. When $F = 1$, the results at the first step ($\tau_1 = \tau_o + \Delta\tau$) are calculated. The matrices evaluated for the first step (i.e. for the integrals in the first two lines of equation (4.116)) can be used repeatedly for all the subsequent time steps. For each new time step, two new matrices for the integrals in the last two lines of the equation have to be calculated; they can be stored and subsequently used for the next time step.

The time progression scheme for the displacement boundary integral equation can be written from equation (4.115) as follows:

$$C_{ij}(\mathbf{x}')u_j(\mathbf{x}',\tau_F) + \sum_{n=1}^{N_e}\left[\int_{S_n} N_a T_{ij}dS\right]u_j^{an}$$

$$+\sum_{n=1}^{N_e}\left[\int_{S_n} N^a \int_{\tau_{F-1}}^{\tau_F} M^b F_i d\tau dS\right]\theta^{abnF}$$

$$= \sum_{n=1}^{N_e}\left[\int_{S_n} N_a U_{ij}dS\right]t_j^{an} + \sum_{n=1}^{N_e}\left[\int_{S_n} N_a \int_{\tau_{F-1}}^{\tau_F} M^b G_i d\tau dS\right]q^{abnF}$$

$$-\sum_{n=1}^{N_e}\sum_{f=1}^{F-1}\left[\int_{S_n} N_a \int_{\tau_{f-1}}^{\tau_f} M^b F_i d\tau dS\right]\theta^{abnf}$$

$$+\sum_{n=1}^{N_e}\sum_{f=1}^{F-1}\left[\int_{S_n} N_a \int_{\tau_{f-1}}^{\tau_f} M^b G_i d\tau dS\right]q^{abnf} \qquad (4.117)$$

The above kernels and the kernels of other boundary integral equations, after time integration, are given in the following subsection.

Time Integration of Fundamental Fields   The linear and constant time-integration of the two-dimensional kernels are presented in this section. The following definitions are used to simplify the equations

$$s_f = \frac{r^2}{4\kappa(\tau_F - \tau_f)} \quad \text{and} \quad s_{f-1} = \frac{r^2}{4\kappa(\tau_F - \tau_{f-1})} \tag{4.118}$$

The constant time interpolation of the temperature equation gives

$$\int_{\tau_{f-1}}^{\tau_f} Q d\tau = Q^f = -\frac{r_{,k} n_k}{2\pi r}\left[e^{-s_f} - e^{-s_{f-1}}\right] \tag{4.119}$$

$$\int_{\tau_{f-1}}^{\tau_f} \Theta d\tau = \Theta^f = \frac{1}{2\pi k}\left[E_1(s_f) - E_1(s_{f-1})\right] \tag{4.120}$$

For linear time interpolation, we have

$$\int_{\tau_{f-1}}^{\tau_f} QM^1 d\tau = -(F-f)Q^f - \frac{r_{,k} n_k r}{8\pi\kappa\Delta\tau}[E_1(s_f) - E_1(s_{f-1})] \tag{4.121}$$

$$\int_{\tau_{f-1}}^{\tau_f} QM^2 d\tau = (F-f+1)Q^f + \frac{r_{,k} n_k r}{8\pi\kappa\Delta\tau}[E_1(s_f) - E_1(s_{f-1})] \tag{4.122}$$

and

$$\int_{\tau_{f-1}}^{\tau_f} \Theta M^1 d\tau = -(F-f)\Theta^f - \frac{r^2}{16\pi\kappa\Delta\tau}\left[\frac{e^{-s_f}}{s_f} - \frac{e^{-s_{f-1}}}{s_{f-1}}\right] \tag{4.123}$$

$$+ \frac{r^2}{16\pi\kappa k\Delta\tau}[E_1(s_f) - E_1(s_{f-1})]$$

$$\int_{\tau_{f-1}}^{\tau_f} \Theta M^2 d\tau = (F-f+1)\Theta^f + \frac{r^2}{16\pi\kappa\Delta\tau}\left[\frac{e^{-s_f}}{s_f} - \frac{e^{-s_{f-1}}}{s_{f-1}}\right] \tag{4.124}$$

$$- \frac{r^2}{16\pi\kappa k\Delta\tau}[E_1(s_f) - E_1(s_{f-1})]$$

The constant time interpolation of the displacement equation gives

$$\int_{\tau_{f-1}}^{\tau_f} F_i d\tau = f_i$$

$$= \frac{(1+\nu)h}{4\pi(1-\nu)}\left[(2r_{,i} r_{,k} n_k - n_i)\left(-\frac{(1-e^{-s_f})}{s_f} + \frac{(1-e^{-s_{f-1}})}{s_{f-1}}\right)\right.$$

$$\left. + \frac{n_i}{2}[E_1(s_f) - E_1(s_{f-1})]\right] \tag{4.125}$$

and

$$\int_{\tau_{f-1}}^{\tau_f} G_i d\tau = g_i$$

$$= \frac{(1+\nu)hr_{,i} r}{4\pi\kappa(1-\nu)}\left[\left(-\frac{(1-e^{-s_f})}{s_f} + \frac{(1-e^{-s_{f-1}})}{s_{f-1}}\right)\right.$$

$$-\frac{1}{2}\left[E_1(s_f) - E_1(s_{f-1})\right]$$ (4.126)

The linear time interpolation of the displacement equation gives

$$\int_{\tau_{f-1}}^{\tau_f} F_i M^1 d\tau = -(F-f)f_i$$ (4.127)

$$-\frac{(1+\nu)hr^2}{32\pi(1-\nu)\kappa\Delta\tau}\left[(r_{,i}r_{,k}n_k - \frac{n_i}{2})\left(\frac{(1-e^{-s_f})}{s_f^2} - \frac{(1-e^{-s_{f-1}})}{s_{f-1}^2}\right)\right.$$
$$\left. + (r_{,i}r_{,k}n_k + \frac{n_i}{2})\left\{(E_1(s_f) - E_1(s_{f-1})) - \left(\frac{e^{-s_f}}{s_f} - \frac{e^{-s_{f-1}}}{s_{f-1}}\right)\right\}\right]$$

$$\int_{\tau_{f-1}}^{\tau_f} F_i M^2 d\tau = (F-f+1)f_i$$ (4.128)

$$+\frac{(1+\nu)hr^2}{32\pi(1-\nu)\kappa\Delta\tau}\left[(r_{,i}r_{,k}n_k - \frac{n_i}{2})\left(\frac{(1-e^{-s_f})}{s_f^2} - \frac{(1-e^{-s_{f-1}})}{s_{f-1}^2}\right)\right.$$
$$\left. + (r_{,i}r_{,k}n_k + \frac{n_i}{2})\left\{(E_1(s_f) - E_1(s_{f-1})) - \left(\frac{e^{-s_f}}{s_f} - \frac{e^{-s_{f-1}}}{s_{f-1}}\right)\right\}\right]$$

$$\int_{\tau_{f-1}}^{\tau_f} G_i M^1 d\tau = -(F-f)g_i$$ (4.129)

$$-\frac{(1+\nu)hr_{,i}r^3}{64\pi(1-\nu)\kappa\Delta\tau}\left[\left(\frac{(1-e^{-s_f})}{s_f^2} - \frac{(1-e^{-s_{f-1}})}{s_{f-1}^2}\right)\right.$$
$$\left. \left(\frac{e^{-s_f}}{s_f} - \frac{e^{-s_{f-1}}}{s_{f-1}}\right) - (E_1(s_f) - E_1(s_{f-1}))\right]$$

$$\int_{\tau_{f-1}}^{\tau_f} G_i M^1 d\tau = (F-f+1)g_i$$ (4.130)

$$+\frac{(1+\nu)hr_{,i}r^3}{64\pi(1-\nu)kk\Delta\tau}\left[\left(\frac{(1-e^{-s_f})}{s_f^2} - \frac{(1-e^{-s_{f-1}})}{s_{f-1}^2}\right)\right.$$
$$\left. \left(\frac{e^{-s_f}}{s_f} - \frac{e^{-s_{f-1}}}{s_{f-1}}\right) - (E_1(s_f) - E_1(s_{f-1}))\right]$$

In the above equations, the exponential integral $E_1(s) = \int_s^\infty e^{-x}/x \, dx$. The extra terms due to linear interpolation are nonsingular, i.e.

$$\left.\begin{array}{l} e^{-s_f} \to 1 \text{ as } r \to 0 \\ e^{-s_{f-1}} \to 1 \text{ as } r \to 0 \end{array}\right\} \text{ when } \tau_f \neq \tau_F$$ (4.131)

If $\tau_f = \tau_F$ then $e^{-s_F} = 0$.

From equation (4.131) it can be seen that, for time steps when $f \neq F$, there is no spatial singularity. In the evaluation of the singularities in the kernels, it should be observed that

$$\lim_{r \to 0} \left\{ 1 - e^{-\frac{r^2}{4k\tau}} \right\} = O(r^2)$$

$$\lim_{x \to 0} \{ E_1(x) \} = O(\ln x)$$

The order of singularity in equation (4.119) when $f = F$ is $O(r_{,k} n_k / r)$. Similarly, the exponential integral in equation (4.120) is $O(\ln r)$. Hence, the order of singularities of boundary integral equations in transient thermoelasticity are the same as in steady-state thermoelasticity.

4.4   Examples

4.4.1   Thick Hollow Cylinder

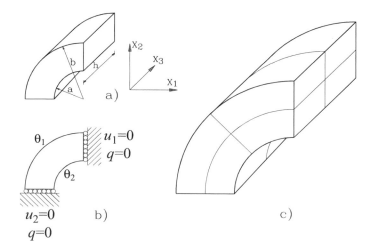

Figure 4.1: Hollow cylinder. (a) Geometry, (b) boundary conditions, (c)BEM discretization.

Consider a thick hollow cylinder under a radial thermal gradient, as shown in Figure 4.1. Due to the symmetry of the problem, only one quarter of the hollow cylinder was modelled. The BEM model consists of 24 eight-noded continuous elements and 74 nodes [9]. The physical dimensions of the cylinder are $a/b = 0.4$ and $b/h = 1.0$. The coefficient of linear thermal expansion was taken as $h = 1 \cdot 10^{-5}/°\mathrm{C}$ and results are independent of the thermal conductivity k. The normal displacement on the top and bottom surfaces of the cylinder are constrained, while the lateral surfaces are traction free. The thermal boundary conditions are $\theta_1 = 1°\mathrm{C}$ on the external cylindrical surface and $\theta_2 = 0°\mathrm{C}$ on the internal one; $q = 0$ was prescribed on the remaining surfaces. Numerical results for values of radial displacement and thermal stresses, compared with analytical values, are presented in Tables 4.1 to 4.3, where $\rho = r/b$ represents the radial position r divided by the external radius b. For comparison, the results have been normalized by dividing by a factor $F = hb\theta_1$ for displacements and

Table 4.1: Radial displacements on a hollow cylinder subjected to radial temperature variation.

ρ	BEM	Exact
0.4	0.325	0.322
0.55	0.347	0.346
0.7	0.454	0.453
0.85	0.615	0.611
1.00	0.807	0.806

Table 4.2: Circumferential stress on a hollow cylinder subjected to radial temperature variation.

ρ	BEM	Exact
0.4	0.827	0.857
0.55	0.262	0.254
0.7	-0.090	-0.082
0.85	-0.305	-0.307
1.00	-0.474	-0.474

$F = hE\theta_1$ for stresses. The circumferential and axial stresses were obtained from boundary tractions by applying the relation between the stress tensor. The results presented demonstrate that the BEM formulation accurately predicts displacements under thermal loads, as can be seen in Table 4.1, where the results have a maximum difference of 0.2% with respect to the exact solution. The results obtained for stresses are less accurate. They have a difference of 3.5% and 0.3% for circumferential and axial stress, respectively, at the maximum values, compared to the exact solution.

4.4.2 Transient Thermal Problems

The temperature boundary integral equation yields reasonably accurate solutions when using constant interpolation for time approximation. However, as shown by Wrobel [31] accuracy can be much improved by using linear time interpolation.

Prasad, Aliabadi and Rooke [20] have shown the results from constant time interpolation of the hypersingular diffusion (flux) equations are generally inaccurate, but the accuracy can be dramatically improved by the use of linear time interpolation. To demonstrate this behaviour, two examples presented, Prasad et $al.$ [20] are presented next. The thermal conductivity $k = 1.0$ and the initial boundary conditions (θ_o and q_o) are set to zero for both problems.

It is worth noting that, if a problem being modelled is a thermal shock problem (a sudden rise in temperature), the flux is infinite at the boundary where the thermal shock is applied. This causes difficulties if the time interpolation is linear. To overcome the difficulty, Prasad et $al.$ [20] approximated the first step as a linear increase in temperature, as shown in Figure 4.2

Table 4.3: Axial stress on a hollow cylinder subjected to radial temperature variation.

ρ	BEM	Exact
0.4	0.198	0.215
0.55	-0.239	-0.248
0.7	-0.608	-0.599
0.85	-0.880	-0.882
1.00	-1.121	-1.118

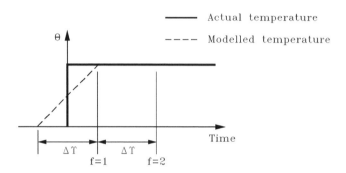

Figure 4.2: Thermal shock modelling for linear time interpolation.

Rectangular Plate

Consider the rectangular plate shown in Figure 4.3. The problem was solved [20] using 12 discontinuous quadratic elements. Flux equation results and temperature equation results for points at $x = 0.0$ and $x = 4.0$ with different time interpolation methods and time steps ($\Delta\tau$) are compared in Figures 4.4 and 4.5, respectively. It has been shown [31] that the results from the temperature equation with linear interpolation are more accurate than with constant time interpolation.

In Figure 4.4 the temperature on the surface $x = 4.0$ is plotted as a function of time. It can be seen from the figure that the flux results, obtained with linear time interpolation, are very similar to the results from the temperature equation with linear time interpolation. Of the two linear time interpolation results for the flux

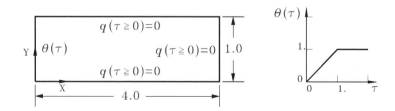

Figure 4.3: Rectangular plate with boundary conditions.

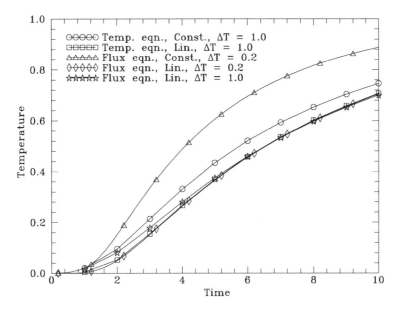

Figure 4.4: Temperature at $x = 4.0$ surface.

equation, the one with the smaller time step conforms more closely to those from the temperature equation with linear interpolation, as expected. The temperature equation with constant time interpolation differs by around 5-10% from the above three linear interpolation curves. Results from the flux equation with constant time interpolation are the most inaccurate of all, with differences in the order of 20%.

In Figure 4.5, the flux at $x = 0.0$ is plotted as a function of time. It can be seen that the results from all the different methods, expect those from the flux equation with constant time interpolation, agree with each other. It should be observed that the graphs with time steps of 1.0 and 0.2 start at 1.0 and 0.2, respectively. In this case, both constant and linear interpolation of the temperature equation gives accurate results.

The solutions from the flux equation with constant time interpolation and a time step of 0.2 give different results from those with linear time interpolation and a time step of 1.0; the difference is more than 20%.

Semi-circular Plate

A semi-circular plate, as shown in Figure 4.6, was analyzed by Prasad *et al.* [20] with six elements on the straight edge and eight elements on the curved edge. All the elements are discontinuous quadratic. The elements on the curved edge for the flux equation are straight and for the temperature equation curved. Results for the temperature on the straight edge at $R = 0.6$ are shown as a function of time in Figure 4.7. The results from both the temperature equation and the flux equation with linear time interpolation are shown to agree well with the analytical results [3], whereas results from the flux equation with constant time interpolation diverge from the analytical results by up to 40%.

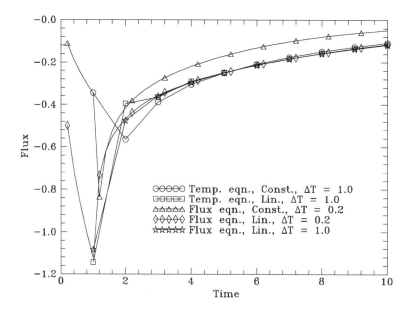

Figure 4.5: Flux at $x = 0.0$ surface.

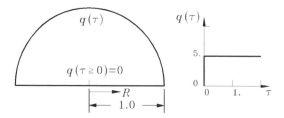

Figure 4.6: Semicircular plate with boundary conditions.

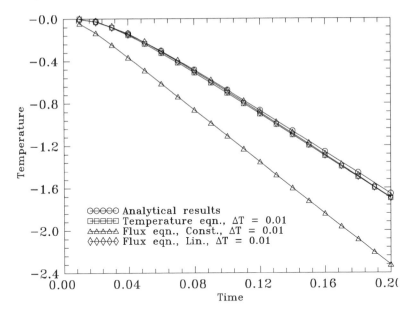

Figure 4.7: Temperature at $R = 0.6$ from the centre.

4.5 Summary

In this chapter the boundary integral equations for steady state and transient ther-moelasticity have been derived using the reciprocal theorem. Efficient methods for effective numerical implementation of the method were discussed. Also presented were hypersingular traction and flux equations, which can be utilized in the dual boundary element formulation for crack problems, described in Chapter 8.

References

[1] Banerjee, P.K., *The Boundary Element Methods in Engineering*, MacGraw-Hill, New York, 1994.

[2] Biot, M.A., General theory of three dimensional consolidation, *Journal of Applied Physics*, **12**, 155-164, 1941.

[3] Carslaw, H.S. and Jaeger, J.C., *Conduction of Heat in Dolids*, 2nd ed., Clarendon Press, Oxford, 1959.

[4] Chaudouet, A., Three-dimensional transient thermo-elastic analyses by the BIE method. *International Journal for Numerical Methods in Engineering*, **24**, 25-45, 1987.

[5] Cheng, A.H.D. and Detournay, E., A direct boundary element method for plane strain poroelasticity, *International Journal for Numerical Methods in Engineering*, **12**, 551-572, 1988.

[6] Cruse, T.A., Snow, D.W. and Wilson, R.B., Numerical solutions in axisymmetric elasticity, *Computers and Structures*, **7**, 445-451, 1977.

[7] Dargush, G.F. and Banerjee, P.K., Development of a boundary element method for time-dependent planar thermoelasticity, *International Journal of Solids and Structures*, **25**, 999-1021, 1989.

[8] Dargush, G.F. and Banerjee, P.K., BEM analysis for three-dimensional problems of transient thermoelasticity, *International Journal of Solids and Structures*, **26**, 199-216, 1990.

[9] dell'Erba, D.N., Dual boundary element formulation for three-dimensional thermoelastic fracture mechanics, PhD thesis, Wessex Institute of Technology, University of Wales, 1999.

[10] Dominguez, J., Boundary element approach for dynamic poroelastic problems, *International Journal for Numerical Methods in Engineering*, **35**, 307-324, 1992.

[11] dell'Erba, D.N., Aliabadi, M.H. and Rooke, D.P,. Dual boundary element method for three-dimensional thermoelastic crack problems, *International Journal of Fracture*, **94**, 89-101, 1998.

[12] El-Zafrany, A., *Techniques of The Boundary Element Method*, Ellis Horwood Series in Mathematics and Its Applications, 1993.

[13] Emmel, E and Stamm, H., Calculation of stress intensity factors of thermally loaded cracks using the finite element method. *International Journal of Pressure Vessels and Piping*, **19**, 1-17, 1985.

[14] Kamiya,N., Aikawa,Y. and Kawaguchi,K., An adaptive boundary element scheme for steady thermoelastic analysis, *Computer Methods in Applied Mechanics and Engineering*, **119**, 311-324, 1994.

[15] Kassab, A. and Aliabadi, M.H., *Coupled Field Problems*, WIT Press, Southampton, 2001.

[16] Nowacki, W., *Thermoelasticity*, Pergamon Press, 1986.

[17] Martin, T.J. and Dulikravich, G.S., Aero-thermo-elastic concurrent design optimization of internally cooled turbine blades, Chapter 5, in *Coupled Field Problems*, WIT Press, Southampton, 137-184, 2001.

[18] Morse, P.M. and Feshback, H., *Methods of Theoretical Physics*, MacGraw-Hill, New York, 1953.

[19] Prasad, N.N.V., Aliabadi, M.H. and Rooke, D.P., The dual boundary element method for thermoelastic crack problems, *International Journal of Fracture*, **66**, 255-272, 1994.

[20] Prasad, N.N.V., Aliabadi, M.H. and Rooke, D.P., On the use of the hypersingular boundary element diffusion equation, *Boundary Element Communications*, **5**, 215-217, 1994.

[21] Prasad, N.N.V. and Aliabadi, M.H. and Rooke, D.P., The dual boundary element method for transient thermoelastic crack problems, *International Journal of Solids and Structures,* **33**, 2695-2718, 1996.

[22] Prasad, N.N.V. Thermo-mechanical crack growth: boundary element analysis., PhD thesis, Wessex Institute of Technology, University of Portsmouth, 1995

[23] Prasad, N.N.V., *Thermomechanical Crack Growth using Boundary Elements*, Topics in Engineering 24, Computational Mechanics Publications, Southampton, 1998.

[24] Raveendra, S.T., The use of a piecewise continuous polynomial basis function for the surface reduction of integral equations in thermoelastic analysis, *Computers and Structures,* **77,** 601-614, 2000.

[25] Rizzo, F.J. and Shippy, D.J., An advanced boundary integral equation method for three-dimensional thermoelasticity, *International Journal for Numerical Methods in Engineering,* **11**, 1753-1768, 1977.

[26] Rudolphi, T.J., Axisymmetric thermoelasticity by combined boundary and finite elements, *Engineering Computations,* **5**, 59-64, 1988.

[27] Sladek, V. and Sladek, J., Boundary integral equation method in thermoelasticity Part I: General analysis. *Applied Mathematical Modelling,* **7,** 241-253, 1983.

[28] Sladek, V. and Sladek, J., Boundary integral equation method in thermoelasticity part III, uncoupled thermoelasticity, *Applied Mathematical Modelling,* **8,** 413-418, 1984.

[29] Smith, D.W. and Booker, J.R., Boundary element analysis of linear thermoelastic consolidation, *International Journal for Numerical and Analytical Methods in Geomechanics,* **20**, 457-488, 1996.

[30] Tanaka, M., Togoh, H. and Kikuta, M., Boundary element method applied to 2-D thermoelastic problems in steady and non-steady states, *Engineering Analysis,* **1,** 13-19, 1984.

[31] Wrobel, L.C., Potential and viscous flow problems using the boundary element method, PhD. thesis, University of Southampton, 1981.

Chapter 5

Elastodynamics

5.1 Introduction

Solutions in elastodynamics using the boundary element method are usually obtained by either the Time Domain method (TDM), Laplace (or Fourier) Transform Method (LTM) or the Dual Reciprocity Method (DRM).

Early formulations of the boundary elements related to elastodynamics problems can be traced back to works of Friedman and Shaw [22], Banaugh and Goldsmith [5] and Cruse and Rizzo [12]. The first direct BEM formulation for transient elastodynamic problems in conjunction with the Laplace transform were derived in [12, 13]. Manolis and Beskos [37] improved certain aspects of the formulation.

The time domain boundary element formulation was originally presented by Cole, Kosloff and Minster [11] for anti-plane problems, followed by a more general formulation by Niwa, Kobayashi and Kitahara [41]. Further improvements were reported [3, 30, 32]. Nardini and Brebbia [40] presented a formulation where mass and stiffness matrices were obtained from a static formulation. They subsequently developed the Dual Reciprocity Method (see Chapter 2) to transform the domain integral containing the inertial term to the boundary.

The application of the BEM to elastodynamic problems has been extensively studied due to its importance in many fields, including soil structure interactions [1, 2] and fracture mechanics [18, 23, 54]. Readers are recommended to consult specialized textbooks on the subject [16, 38] as well as review articles [7, 8] for historical overviews. Other interesting contributions can be found in [4, 9, 14].

More recently, attention has been paid to the boundary stress integral equation due to its importance to the dual boundary element formulation in fracture mechanics (see Chapter 8). Fedelinski, Aliabadi and Rooke [18, 19, 20] presented the derivations and effective implementation of the hypersingular stress equations for the time domain, Laplace transform and dual reciprocity method for two-dimensional problems. Later, Wen, Aliabadi and Rooke [54, 55, 56] presented the same three formulations for three-dimensional problems. Symmetric Galerkin boundary element formulation in elastodynamics can be found in [44] using the dual reciprocity method, and [45] using the Laplace transform method.

In this chapter, formulations for displacement and stress integral equations are presented for the time domain, Laplace transform and dual reciprocity methods.

5.2　Transient Elastodynamics

5.2.1　Basic Concepts

The governing equation of the dynamic Naviers's equation for an elastic body can be written as

$$(\lambda + \mu)u_{i,ij}(\mathbf{x},t) + \mu u_{j,ii}(\mathbf{x},t) + \rho\,[b_j(\mathbf{x},t) - \ddot{u}_j(\mathbf{x},t)] = 0 \tag{5.1}$$

where λ and μ are constants denoting the Lame' and shear modulus of elasticity, ρ is the mass density, b_j is the body force (per unit mass), t denotes the time and $\ddot{u}_j = \partial^2 u_j/\partial t^2$ is the acceleration. Equation (5.1) can be rewritten in a form

$$(c_1^2 - c_2^2)u_{i,ij}(\mathbf{x},t) + c_2^2 u_{j,ii}(\mathbf{x},t) + b_j(\mathbf{x},t) - \ddot{u}_j(\mathbf{x},t) = 0 \tag{5.2}$$

where the dilatational and shear wave velocities are given as $c_1^2 = (\lambda + 2\mu)/\rho$ and $c_2^2 = \mu/\rho$, respectively.

Considering a domain V with a bounded surface S, the boundary condition can be written as

$$u_i(\mathbf{x}, t) = u_i^o \qquad t = 0 \tag{5.3}$$

$$\dot{u}_i(\mathbf{x}, t) = \dot{u}_i^o \qquad t = 0 \tag{5.4}$$

for the initial condition, and

$$u_i(\mathbf{x}, t) = \hat{u}_i(\mathbf{x},t) \qquad on\ S_u \tag{5.5}$$

$$t_i(\mathbf{x}, t) = \hat{t}_i(\mathbf{x}, t) \qquad on\ S_t \tag{5.6}$$

where $t_i = \sigma_{ij}n_j$, with n_j denoting the components of the outward normal to the boundary S at $\mathbf{x}$, and S_u and S_t denote parts of the boundary with either displacement or traction boundary conditions. The stress tensors are given by

$$\sigma_{ij}(\mathbf{x},t) = \rho(c_1^2 - 2c_2^2)\delta_{ij}u_{m,m}(\mathbf{x},t) + \rho c_2^2(u_{i,j}(\mathbf{x}, t) + u_{j,i}(\mathbf{x}, t)) \tag{5.7}$$

5.2.2　Time Domain Integral Equations

The reciprocal theorem in dynamics is due to Graf [24], and is the extension of the well known Betti's reciprocal theorem in elastostatics. Considering two independent elastodynamic states with parameters $(u_i, t_i, u_i^o, \dot{u}_i^o, b_i)$ and $(u_i^*, t_i^*, u_i^{*o}, \dot{u}_i^{*o}, b_i^*)$ defined in a same domain V bounded by a surface S, the reciprocal theorem for $t \geq 0$ can be written as

$$\int_S t_i * u_i^* dS + \int_V \rho\,[b_i * u_i^* + \dot{u}_i^o u_i^* + \dot{u}_i^o \ddot{u}_i^*]\,dV$$

$$= \int_S t_i^* * u_i dS + \int_V \rho\,[b_i^* * u_i + \dot{u}_i^{*o} u_i + \dot{u}_i^{*o} \ddot{u}_i]\,dV \tag{5.8}$$

where $*$ denotes Reimann convolution.

The elastic state $(u_i^*, t_i^*, u_i^{*o}, \dot{u}_i^{*o}, b_i^*)$ is chosen to correspond to a unit impulse applied at a time τ at point $\mathbf{X}'$, in a direction e_i, that is

$$\rho b_i^* = \Delta(t - \tau)\Delta(\mathbf{X} - \mathbf{X}')e_i \tag{5.9}$$

where Δ is the Dirac delta function. The displacement at point $\mathbf{x}$ at time t due to the unit impulse is given as

$$u_i(\mathbf{X}, t) = U_{ij}(\mathbf{X}, \mathbf{X}', t')e_j \tag{5.10}$$

and the traction as

$$t_i(\mathbf{X}, t) = T_{ij}(\mathbf{X}, \mathbf{X}', t')e_j \tag{5.11}$$

where $t' = t - \tau$ is the retarded time. Substituting expressions (5.9),(5.10) and (5.11) into (5.8) gives

$$
\begin{aligned}
u_i(\mathbf{X}', t) \;=\; & \int_S \int_0^t \left(U_{ij}(\mathbf{x}, \mathbf{X}', t')t_j(\mathbf{x}, t) - T_{ij}(\mathbf{x}, \mathbf{X}', t')u_j(\mathbf{x}, t) \right) d\tau dS \\
& + \rho \int_V \int_0^t U_{ij}(\mathbf{X}, \mathbf{X}', t')b_j(\mathbf{X}, t)d\tau dV \\
& + \rho \int_V \left(U_{ij}(\mathbf{X}, \mathbf{X}', t - 0)\frac{\partial u_j(\mathbf{X}, 0)}{\partial t} + \frac{\partial U_{ij}(\mathbf{X}, \mathbf{X}', t - 0)}{\partial t}u_j(\mathbf{X}, 0) \right) dV
\end{aligned}
\tag{5.12}
$$

Equation (5.12) relates the interior displacements to the boundary values of displacement and tractions. The strain field throughout the body may be obtained by differentiating equation (5.12), which leads to the equation

$$
\begin{aligned}
\frac{\partial u_i(\mathbf{X}', t)}{\partial X_k'} \;=\; & \int_S \int_0^t \left(\frac{\partial U_{ij}(\mathbf{x}, \mathbf{X}', t')}{\partial X_k'}t_j(\mathbf{x}, t) - \frac{\partial T_{ij}(\mathbf{x}, \mathbf{X}', t')}{\partial X_k'}u_j(\mathbf{x}, t) \right) d\tau dS \\
& + \rho \int_V \int_0^t \frac{\partial U_{ij}(\mathbf{X}, \mathbf{X}', t')}{\partial X_k'}b_i(\mathbf{X}, t)d\tau dV \\
& + \rho \int_V \left(\frac{\partial U_{ij}(\mathbf{X}, \mathbf{X}', t - 0)}{\partial X_k'}\frac{\partial u_j(\mathbf{X}, 0)}{\partial t} + \frac{\partial \left(\dot{U}_{ij}(\mathbf{X}, \mathbf{X}', t - 0) \right)}{\partial X_k'}u_j(\mathbf{X}, 0) \right) dV
\end{aligned}
\tag{5.13}
$$

substituting equation (5.13) into Hooke's law (5.7) yields the stresses at an interior point $\mathbf{X}'$:

$$
\begin{aligned}
\sigma_{ij}(\mathbf{X}', t) \;=\; & \int_S \int_0^t \left(D_{kij}(\mathbf{x}, \mathbf{X}', t')t_k(\mathbf{x}, t) - S_{kij}(\mathbf{x}, \mathbf{X}', t')u_k(\mathbf{x}, t) \right) d\tau dS \\
& + \rho \int_V \int_0^t D_{kij}(\mathbf{X}, \mathbf{X}', t')b_k(\mathbf{X}, t)d\tau dV \\
& + \rho \int_V \left(D_{kij}(\mathbf{X}, \mathbf{X}', t - 0)\frac{\partial u_k(\mathbf{X}, 0)}{\partial t} + \frac{\partial D_{kij}(\mathbf{X}, \mathbf{X}', t - 0)}{\partial t}u_k(\mathbf{X}, 0) \right) dV
\end{aligned}
\tag{5.14}
$$

where fundamental solutions U_{ij}, T_{ij}, D_{kij} and S_{kij} are given in Appendix E for two- and three-dimensional problems.

The boundary displacement integral equation can be obtained by taking the limiting process as the internal point goes to the boundary that is as $\mathbf{X}' \to \mathbf{x}'$, to give

$$C_{ij}(\mathbf{x}')u_j(\mathbf{x}', t) = \int_S \int_0^t U_{ij}(\mathbf{x}, \mathbf{x}', t')t_j(\mathbf{x}, t)d\tau dS - \oint_S \int_0^t T_{ij}(\mathbf{x}, \mathbf{x}', t')u_j(\mathbf{x}, t)d\tau dS$$

$$+\rho \int_V \int_0^t U_{ij}(\mathbf{X}, \mathbf{x}', t') b_i(\mathbf{X}, t) d\tau dV$$

$$+\rho \int_V \left(U_{ij}(\mathbf{X}, \mathbf{x}', t-0) \frac{\partial u_j(\mathbf{X}, 0)}{\partial t} + \frac{\partial U_{ij}(\mathbf{X}, \mathbf{x}', t-0)}{\partial t} u_j(\mathbf{X}, 0) \right) dV \qquad (5.15)$$

where C_{ij} is a constant that depends upon the position of the source point $\mathbf{x}'$ and can be obtained from the consideration of the rigid body condition.

The boundary stress integral equation can be derived in a similar way by considering the limiting form of equation (7.40) as $\mathbf{x}'$ tends to a smooth boundary, to give

$$\frac{1}{2}\sigma_{ij}(\mathbf{x}', \tau) = \fint_S \int_0^t D_{kij}(\mathbf{x}, \mathbf{x}', t') t_k(\mathbf{x}, t) d\tau dS - \fint_S \int_0^t S_{kij}(\mathbf{x}, \mathbf{x}', t') u_k(\mathbf{x}, t) d\tau dS$$

$$+\rho \int_V \int_0^t D_{kij}(\mathbf{X}, \mathbf{x}', t') b_k(\mathbf{X}, t) d\tau dV$$

$$+\rho \int_V \left(D_{kij}(\mathbf{X}, \mathbf{x}', t-0) \frac{\partial u_k(\mathbf{X}, 0)}{\partial t} + \frac{\partial D_{kij}(\mathbf{X}, \mathbf{x}', t-0)}{\partial t} u_k(\mathbf{X}, 0) \right) dV \qquad (5.16)$$

The traction boundary integral equation is obtained by multiplying the boundary stress integral equation (5.16) by the outward normal at the source point, i.e. $n_j(\mathbf{x}')$, to give

$$\frac{1}{2} t_j(\mathbf{x}', \tau) = n_j(\mathbf{x}') \fint_S \int_0^t D_{kij}(\mathbf{x}, \mathbf{x}', t') t_k(\mathbf{x}, t) d\tau dS$$

$$-n_j(\mathbf{x}') \fint_S \int_0^t S_{kij}(\mathbf{x}, \mathbf{x}', t') u_k(\mathbf{x}, t) d\tau dS + \rho n_j(\mathbf{x}') \int_V \int_0^t D_{kij}(\mathbf{X}, \mathbf{x}', t') b_k(\mathbf{X}, t) d\tau dV$$

$$+\rho n_j(\mathbf{x}') \int_V \left(D_{kij}(\mathbf{X}, \mathbf{X}', t-0) \frac{\partial u_k(\mathbf{X}, 0)}{\partial t} + \frac{\partial D_{kij}(\mathbf{X}, \mathbf{X}', t-0)}{\partial t} u_k(\mathbf{X}, 0) \right) dV$$

$$(5.17)$$

5.2.3   Numerical Discretization

The numerical solution of (5.15) is obtained after discretizing both space and time variations. The boundary S is divided into N_e elements, and observation time t is divided into M time steps ($M\Delta t$). For the spatial distributions, the same shape functions are used for both $u_j(\mathbf{x}, \tau)$ and $t_j(\mathbf{x}, \tau)$, and the values of displacement and traction at time step $m\Delta t (m = 1, 2, ...M)$ can be written as:

$$t_i(\mathbf{x}, \tau) = \sum_{m=1}^M \phi_m(\tau) t_i^m(\mathbf{x})$$

$$u_i(\mathbf{x}, \tau) = \sum_{m=1}^M \left(\Xi_1(\tau) u_i^n(\mathbf{x}) + \Xi_2(\tau) u_i^{n-1}(\mathbf{x}) \right) \qquad (5.18)$$

where

$$\phi_n(\tau) = H(\tau - [n-1\Delta t) - H(\tau - n\Delta t)$$

$\Xi_1(\tau)$ and $\Xi_2(\tau)$ are temporal linear interpolation functions

$$\Xi_1(\tau) = \frac{\tau - (m-1)\Delta t}{\Delta t}\phi_m(\tau); \quad \Xi_2(\tau) = \frac{m\Delta t - \tau}{\Delta t}\phi_m(\tau)$$

$H(\tau)$ stands for the Heaviside step function, and $t_i^m(\mathbf{x})$ and $u_i^m(\mathbf{x})$ are the traction and displacement at time $t = m\Delta t$ at boundary point $\mathbf{x}$. The time functions in the fundamental solution are simple enough to carry out the time integration analytically. Considering the above approximation, the equations at observation time $t(= M\Delta t)$, in the absence of body forces for initial conditions zero, are

$$C_{ij}u_j^M(\mathbf{x}') = \sum_{m=1}^{M} \int_S \int_{(m-1)\Delta t}^{m\Delta t} U_{ij}(\mathbf{x}', \mathbf{x}, t')t_j^m(\mathbf{x})d\tau dS$$

$$-\sum_{m=1}^{M} \fint_S \int_{(m-1)\Delta t}^{m\Delta t} T_{ij}(\mathbf{x}', \mathbf{x}, t') \left(\Xi_1(\tau)u_i^m(\mathbf{x}) + \Xi_2(\tau)u_i^{m-1}(\mathbf{x})\right) d\tau dS \qquad (5.19)$$

and similarly, for the traction integral equation, we have

$$\frac{1}{2}t_j^M(\mathbf{x}') = n_i(\mathbf{x}') \sum_{m=1}^{M} \fint_S \int_{(m-1)\Delta t}^{m\Delta t} D_{kij}(\mathbf{x}', \mathbf{x}, t')t_k^m(\mathbf{x})d\tau dS$$

$$-n_i(\mathbf{x}') \sum_{m=1}^{M} \fint_S \int_{(m-1)\Delta t}^{m\Delta t} S_{kij}(\mathbf{x}', \mathbf{x}, t') \left(M_1(\tau)u_k^m(\mathbf{x}) + M_2(\tau)u_k^{m-1}(\mathbf{x})\right) d\tau dS \quad (5.20)$$

Displacement, traction and geometry on the element n are represented in terms of isoparametric shape functions, where $u_i^{nm\alpha}, t_i^{nm\alpha}$ are values of displacement and traction at node α of element n at time $m\Delta t$, and $x_i^{n\alpha}$ denotes the node coordinate. Because of the presence of the $\dot{\delta}(t - \tau - r/c)$ function and $\ddot{\delta}(t - \tau - r/c)$ in fundamental solutions T_{ij} and D_{kij}, the following temporal integrations can be approximated as:

$$\int_{(n-1)\Delta t}^{n\Delta t} \dot{\delta}\left(t - \tau - \frac{r}{c}\right)f(\mathbf{x}, \tau)d\tau \simeq \frac{f^n(\mathbf{x}) - f^{n-1}(\mathbf{x})}{\Delta t}\phi_n(t) \qquad (5.21)$$

$$\int_{(n-1)\Delta t}^{n\Delta t} \ddot{\delta}\left(t - \tau - \frac{r}{c}\right)f(\mathbf{x}, \tau)d\tau \simeq \frac{f^n(\mathbf{x}) - 2f^{n-1}(\mathbf{x}) + f^{n-2}(\mathbf{x})}{(\Delta t)^2}\phi_n(t) \qquad (5.22)$$

The use of these approximations enables the discretized form of the displacement boundary integral equations for three-dimensional problems to be written as

$$C_{ij}u_j^M(\mathbf{x}^c) = \sum_{m=1}^{M}\sum_{n=1}^{N_e}\sum_{\alpha=1}^{} t_j^{nm\alpha} \int_{-1}^{1}\int_{-1}^{1} U_{ij1}^{M-m+1}N_\alpha J^n d\eta_1 d\eta_2$$

$$-\sum_{m=1}^{M}\sum_{n=1}^{N_e}\sum_{\alpha=1}^{} u_j^{nm\alpha} \int_{-1}^{1}\int_{-1}^{1} \left(T_{ij1}^{M-m+1} + T_{ij2}^{M-m}\right)N_\alpha J^m d\eta_1 d\eta_2 \qquad (5.23)$$

where $U_{ij\gamma}^{M-m+1}$ and $T_{ij\gamma}^{M-m+1}$ can be found in Appendix E. A similar set of integral equations can be written for two-dimensional problems in terms of the local coordinates η. The set of discretized boundary equations can be written in matrix form at time step M as

$$\tilde{\mathbf{H}}^{MM}\mathbf{u}^M = \tilde{\mathbf{G}}^{MM}\mathbf{t}^M + \sum_{m=1}^{M-1}(\tilde{\mathbf{G}}^{Mm}\mathbf{t}^m - \tilde{\mathbf{H}}^{Mm}\mathbf{u}^m) \tag{5.24}$$

where $\mathbf{u}^m$, $\mathbf{t}^m$ contain nodal values of displacements and tractions at the time step m; $\tilde{\mathbf{H}}^{Mm}$ and $\tilde{\mathbf{G}}^{Mm}$ depend upon the integrals of the fundamental solutions and interpolating functions. The superscripts Mm emphasize that the matrix depends upon the difference between the time steps M and m. The columns of matrices $\tilde{\mathbf{H}}^{MM}$, $\tilde{\mathbf{G}}^{MM}$ are reordered according to the boundary conditions, giving new matrices $\tilde{\mathbf{A}}^{MM}$ and $\tilde{\mathbf{B}}^{MM}$. The matrix $\tilde{\mathbf{A}}^{MM}$ is multiplied by the vector $\mathbf{x}^M$ of unknown displacements and tractions, and the matrix $\tilde{\mathbf{B}}^{MM}$ by the vector $\mathbf{y}^M$ of known boundary conditions, as follows:

$$\tilde{\mathbf{A}}^{MM}\mathbf{x}^M = \tilde{\mathbf{B}}^{MM}\mathbf{y}^M + \sum_{m=1}^{M-1}(\tilde{\mathbf{G}}^{Mm}\mathbf{t}^m - \tilde{\mathbf{H}}^{Mm}\mathbf{u}^m) \tag{5.25}$$

In each time step, only the matrices which correspond to the maximum difference $M - m$ are computed. The rest of the matrices are known from the previous steps. The matrices $\tilde{\mathbf{A}}^{MM}$ and $\tilde{\mathbf{B}}^{MM}$ are only calculated in the first step since they are the same at each time step; $\tilde{\mathbf{A}}^{MM} = \tilde{\mathbf{A}}$ and $\tilde{\mathbf{B}}^{MM} = \tilde{\mathbf{B}}$. The matrix equation (5.25) can be written in a simpler form as

$$\tilde{\mathbf{A}}\mathbf{x}^M = \mathbf{f}^M \tag{5.26}$$

where

$$\mathbf{f}^M = \tilde{\mathbf{B}}\mathbf{y}^M + \sum_{m=1}^{M-1}(\tilde{\mathbf{G}}^{Mm}\mathbf{t}^m - \tilde{\mathbf{H}}^{Mm}\mathbf{u}^m) \tag{5.27}$$

is a known vector. The matrix equation is solved step-by-step giving the unknown displacements and tractions at each time step. During the initial steps the fundamental solutions are only non-zero in the neighbourhood of the collocation point; they are therefore integrated only over that part of the boundary. The solution process becomes progressively slower at later times because the vector $\mathbf{f}^M$ depends upon all the matrices from the previous steps.

Similarly, the traction equation can be written in a discretized form [20, 55] as

$$\frac{1}{2}t_j^M(\mathbf{x}^c) = n_i(\mathbf{x}^c)\sum_{m=1}^{M}\sum_{n=1}^{N_e}\sum_{\alpha=1}t_k^{nm\alpha}\int_{-1}^{1}\int_{-1}^{1}\left(D_{kij1}^{M-m+1} + D_{kij2}^{M-m}\right)N_\alpha J^n d\eta_1 d\eta_2$$

$$-n_i(\mathbf{x}^c)\sum_{m=1}^{M}\sum_{n=1}^{N_e}\sum_{\alpha=1}u_k^{nm\alpha}\int_{-1}^{1}\int_{-1}^{1}\left(S_{kij1}^{M-m+1} + S_{kij2}^{M-m} + S_{kij3}^{M-m-1}\right)N_\alpha J^n d\eta_1 d\eta_2 \tag{5.28}$$

where $D_{kij\gamma}^{M-m+1}$ and $S_{kij\gamma}^{M-m+1}$ can be found in Appendix E.

Evaluation of the Space Integrals

The orders of singularity in U_{ij}, T_{ij}, D_{kij} and S_{kij} are the same as in the static fundamental solutions (see Chapter 2), and can be dealt with in the following manner:

$$\int()^{transient}ds = \int()^{static}ds + \int[()^{transient} - ()^{static}]ds \tag{5.29}$$

where the second integral on the right-hand side of (5.29) is non-singular and can be evaluated using the standard Gaussian quadrature. The first integral on the right-hand side of (5.29) can be evaluated using one of the methods described in Chapter 11.

It is worth noting that elements which the wave front has not reached need not be computed, as they contribute nothing to the system. In this way, substantial savings are possible as far as the numerical integrations are carried out.

5.3   Laplace Transform Method

5.3.1   Basic Concepts

Application of the Laplace transform to the equation of motion (5.2) will give

$$(c_1^2 - c_2^2)\bar{u}_{i,jj}(\mathbf{x}, s) + c_2^2 \bar{u}_{j,ii}(\mathbf{x}, s) + \bar{b}_j(\mathbf{x}, s) - s^2 \bar{u}_j(\mathbf{x}, s) = 0 \qquad (5.30)$$

where $\bar{u}(\mathbf{x}, s)$ is a component of the transformed displacement of a point $\mathbf{x}$ and s is a Laplace parameter. The Laplace transform of function $f(\mathbf{x},t)$ is defined as

$$\mathcal{L}[f(\mathbf{x},t)] = \bar{f}(\mathbf{x}, s) = \int_0^\infty f(\mathbf{x}, t)e^{-st}dt \qquad (5.31)$$

The transformed displacements and tractions satisfy the following boundary conditions:

$$\bar{u}_i(\mathbf{x}, t) = \hat{u}_i(\mathbf{x},t) \qquad \text{on } S_u$$
$$\bar{t}_i(\mathbf{x}, t) = \hat{t}_i(\mathbf{x},t) \qquad \text{on } S_t \qquad (5.32)$$

where S_u and S_t denote parts of the boundary with either displacement or traction boundary conditions are prescribed. The Laplace transform of the stress tensors are given by

$$\bar{\sigma}_{ij}(\mathbf{x}, s) = \rho(c_1^2 - 2c_2^2)\delta_{ij}\bar{u}_{m,m}(\mathbf{x}, s) + \rho c_2^2(\bar{u}_{i,j}(\mathbf{x}, s) + \bar{u}_{j,i}(\mathbf{x}, s)) \qquad (5.33)$$

5.3.2   Integral Equation Formulations

Assuming zero initial displacements and velocities, the reciprocal identity for two solutions $\bar{u}_i(\mathbf{X}, s)$ and $\bar{u}_i^*(\mathbf{X}, s)$ can be written as

$$\int_S \bar{t}_i(\mathbf{X}, s)\bar{u}_i^*(\mathbf{X}, s)dS + \int_V \bar{b}_i(\mathbf{X}, s)\bar{u}_i^*(\mathbf{X}, s)dV$$

$$= \int_S \bar{t}_i^*(\mathbf{X}, s)\bar{u}_i(\mathbf{X}, s)dS + \int_V \bar{b}_i^*(\mathbf{X}, s)\bar{u}_i(\mathbf{X}, s)dV \qquad (5.34)$$

The elastic state $(\bar{u}_i^*, \bar{t}_i^*, \bar{b}_i^*)$ is chosen to correspond to a unit harmonic body force. The Somigliana's identity for displacement at an internal point $\mathbf{X}'$ can be written as

$$\bar{u}_i(\mathbf{X}', s) = \int_S \left(\bar{U}_{ij}(\mathbf{x}, \mathbf{X}', s)\bar{t}_i(\mathbf{x}, s) - \bar{T}_{ij}(\mathbf{x}, \mathbf{X}', s)\bar{u}_i(\mathbf{x}, s) \right) dS$$

$$+ \int_V \bar{U}_{ij}(\mathbf{X}, \mathbf{X}', s)\bar{b}_i(\mathbf{X}, s)dV \qquad (5.35)$$

The stresses at an interior point $\mathbf{X}'$ can be obtained by differentiating the displacement integral equation and the application of the Hooke's law to give

$$\sigma_{ij}(\mathbf{X}',s) = \int_S \left(\bar{D}_{kij}(\mathbf{x},\mathbf{X}',s)\bar{t}_i(\mathbf{x},s) - \bar{S}_{kij}(\mathbf{x},\mathbf{X}',s)\bar{u}_i(\mathbf{x},s) \right) dS$$

$$+ \int_S \bar{D}_{kij}(\mathbf{X},\mathbf{X}',s)\bar{b}_i(\mathbf{X},s)dV \tag{5.36}$$

The displacement boundary integral equation can be obtained by considering the limiting form of the equation (5.35) as $\mathbf{X}'$ tends to the boundary, to give

$$C_{ij}(\mathbf{x}')\bar{u}_j(\mathbf{x}',s) = \int_S \bar{U}_{ij}(\mathbf{x}',\mathbf{x},s)\bar{t}_j(\mathbf{x},s)dS + \int_S \bar{T}_{ij}(\mathbf{x}',\mathbf{x},s)\bar{u}_j(\mathbf{x},s)dS$$

$$+ \int_V \bar{U}_{ij}(\mathbf{x}',\mathbf{X},s)\bar{b}_j(\mathbf{x},s)dV$$

Similarly, the boundary stress integral equation can be written [20, 57] as

$$\frac{1}{2}\bar{\sigma}_{ij}(\mathbf{x}',s) = \fint_S \bar{D}_{kij}(\mathbf{x}',\mathbf{x},s)\bar{t}_i(\mathbf{x},s) - \fint_S \bar{S}_{kij}(\mathbf{x}',\mathbf{x},s)\bar{u}_i(\mathbf{x},s)dS$$

$$+ \int_V \bar{D}_{kij}(\mathbf{x}',\mathbf{X},s)\bar{b}_i(\mathbf{X},s)dV \tag{5.37}$$

and the boundary traction integral equation is obtained by multiplying (5.37) by the normal at the collocation point, that is

$$\frac{1}{2}\bar{t}_j(\mathbf{x}',s) = n_i(\mathbf{x}')\left\{ \fint_S \bar{D}_{kij}(\mathbf{x}',\mathbf{x},s)\bar{t}_i(\mathbf{x},s)dS - \fint_S \bar{S}_{kij}(\mathbf{x}',\mathbf{x},s)\bar{u}_i(\mathbf{x},s)dS \right\}$$

$$+ n_i(\mathbf{x}')\int_S \bar{D}_{kij}(\mathbf{x}',\mathbf{X},s)\bar{b}_i(\mathbf{X},s)dV \tag{5.38}$$

where fundamental solutions $\bar{U}_{ij}$, $\bar{T}_{ij}$, $\bar{D}_{kij}$ and $\bar{S}_{kij}$ for two- and three-dimensional problems are given in Appendix E. The integration of the transformed fundamental solutions requires an understanding of their behaviour in the neighbourhood of the collocation point. A detailed explanation is presented in Appendix E.

It is also possible to arrive at the solution of $\bar{u}_i$ via the Fourier transform. In this case, the Laplace parameter s should be replaced with $-i\omega$, where ω is the angular frequency. The above formulation results in the reduction of the transient problem to a steady state one. To obtain the transient solution, it is necessary to invert back to the real space by some efficient numerical technique.

5.3.3 Numerical Discretization

The boundary is discretized in the same way as for the time domain method. As a result of the approximation of spatial variations of transformed boundary displacements and tractions, the displacement equation, are

$$C_{ij}\bar{u}_j(s) = \sum_{n=1}^{N_e}\sum_{\alpha=1}^{} \left[\bar{t}_j^{n\alpha}(s) \int_{-1}^1 \bar{U}_{ij}(\eta,s)N_\alpha(\eta)J^n(\eta)d\eta \right.$$

$$-\bar{u}_j^{n\alpha}(s)\int_{-1}^{1}\bar{T}_{ij}(\eta,s)N_\alpha(\eta)J^n(\eta)d\eta\bigg] \tag{5.39}$$

The boundary equations (5.39) and (5.43) are applied for the boundary nodes. The set of discretized boundary integral equations can be written in matrix form as

$$\bar{\mathbf{H}}\bar{\mathbf{u}} = \bar{\mathbf{G}}\bar{\mathbf{t}}, \tag{5.40}$$

where $\bar{\mathbf{u}}$ and $\bar{\mathbf{t}}$ contain nodal values of the transformed displacements and tractions respectively, and $\bar{\mathbf{H}}$ and $\bar{\mathbf{G}}$ depend upon integrals of the transformed fundamental solutions and the interpolating functions. The matrices $\bar{\mathbf{H}}$ and $\bar{\mathbf{G}}$ are reordered according to the boundary conditions, in the same way as in elastostatics, to give new matrices $\bar{\mathbf{A}}$ and $\bar{\mathbf{B}}$. The matrix $\bar{\mathbf{A}}$ is multiplied by the vector $\bar{\mathbf{x}}$ of unknown transformed displacements and tractions, and $\bar{\mathbf{B}}$ by the vector $\bar{\mathbf{y}}$ of the known transformed boundary conditions, as follows:

$$\bar{\mathbf{A}}\bar{\mathbf{x}} = \bar{\mathbf{B}}\bar{\mathbf{y}}, \tag{5.41}$$

or

$$\bar{\mathbf{A}}\bar{\mathbf{x}} = \bar{\mathbf{f}}, \tag{5.42}$$

where $\bar{\mathbf{f}} = \bar{\mathbf{B}}\bar{\mathbf{y}}$ is a known vector.

The matrix equation (5.42) is solved giving the unknown transformed displacements and tractions for a particular Laplace parameter. For a simple temporal variation of the prescribed boundary conditions, their Laplace transforms can be calculated analytically. In order to obtain the unknown displacements and tractions as functions of time, the unknown transformed variables must be computed for a series of Laplace parameters.

Similarly, the discretized form of the traction integral equation can be written as

$$\frac{1}{2}\bar{t}_j(s) = n_i\sum_{n=1}^{N_e}\sum_{\alpha=1}\bigg[\bar{t}_k^{n\alpha}(s)\int_{-1}^{1}\bar{D}_{kij}(\eta,s)N_\alpha(\eta)J^n(\eta)d\eta$$

$$-\bar{u}_k^{n\alpha}(s)\rlap{\Big/}\int_{-1}^{1}\bar{S}_{kij}(\eta,s)N_\alpha(\eta)J^n(\eta)d\eta\bigg] \tag{5.43}$$

5.3.4 Laplace Inversion Method

The numerical inversion of the Laplace can be obtained by using the Durbin method, which is based on a sine and cosine Fast Fourier Transform (FFT) [17].

The values of a transformed function $\bar{f}(s)$ are calculated for a series of Laplace parameters $s_k = a + i2k\pi/T$, where a is a constant, $i = \sqrt{-1}$ and T is a time interval of interest. The values of the original function $f(\tau)$ are obtained at periodic time $\tau_m = m\Delta t$ ($m = 1, 2, ...M - 1$) from the following equation:

$$f(\tau_m) = \frac{2}{T}\exp(am\Delta\tau)\left\{-\frac{1}{2}\Re[\bar{f}(a)] + \Re\left[\sum_{k=0}^{M-1}(A(k)+iB(k))W^{mk}\right]\right\}$$

where

$$A(k) = \sum_{l=0}^{L}\Re\left[\bar{f}\left(a + i(k+lM)\frac{2\pi}{T}\right)\right]$$

$$B(k) = \sum_{l=0}^{L} \Im\left[\bar{f}\left(a + i(k + lM)\frac{2\pi}{T}\right)\right]$$

and

$$W = \exp\left(i\frac{2\pi}{M}\right)$$

where $\Re[$ denotes the real part and $\Im[$ denotes the imaginary part. The usual range of parameters is aT between 5 and 10 and the product LM between 50 and 5000.

5.4   Dual Reciprocity Method

The differential equation of motion can be expressed as follows:

$$\sigma_{jk,k} = \rho\ddot{u}_j \tag{5.44}$$

The differential equation (5.44) is solved subjected to the boundary conditions given in (5.3) to (5.6). By utilizing Hooke's law, we have

$$\mu u_{k,jj} + \frac{\mu}{1 - 2\nu}u_{j,jk} = \rho\ddot{u}_k \tag{5.45}$$

The boundary element formulation for elastodynamic problems can be derived from the Maxwell-Betti reciprocal theorem for two independent states (u_i, t_i) and (u_i^*, t_i^*), and expressed as

$$\int_V u_j^*(\mathbf{X}, \mathbf{X}')\sigma_{jk,k}(\mathbf{X}, \mathbf{X}')dV - \int_V u_j(\mathbf{X}, \mathbf{X}')\sigma_{jk,k}^*(\mathbf{X}, \mathbf{X}')dV$$

$$= \int_S u_j^*(\mathbf{x}, \mathbf{X}')t_j(\mathbf{x}, \mathbf{X}')dS - \int_S u_j(\mathbf{x}, \mathbf{X}')t_j^*(\mathbf{x}, \mathbf{X}')dS \tag{5.46}$$

If the state (u^*, t^*) is chosen to correspond to the solution of a static problem due to a unit force, equation (5.46) can be rewritten as

$$u_i(\mathbf{X}') = \int_S \left[U_{ij}(\mathbf{x}, \mathbf{X}')t_j(\mathbf{x}) - T_{ij}(\mathbf{x}, \mathbf{X}')u_j(\mathbf{x})\right]dS - \int_V U_{ij}(\mathbf{X}, \mathbf{X}')\sigma_{jk,k}(\mathbf{X})dV \tag{5.47}$$

where U_{ij} and T_{ij} are static fundamental solutions (see Chapter 2). Substituting $\sigma_{jk,k}$ from equation (5.47), results in

$$u_i(\mathbf{X}') = \int_S \left[U_{ij}(\mathbf{x}, \mathbf{X}')t_j(\mathbf{x}) - T_{ij}(\mathbf{x}, \mathbf{X}')u_j(\mathbf{x})\right]dS - \rho\int_V U_{ij}(\mathbf{X}, \mathbf{X}')\ddot{u}_j(\mathbf{X}, \tau)dV \tag{5.48}$$

Similarly, stresses at an internal point can be written as

$$\sigma_{ij}(\mathbf{X}') = \int_S \left[D_{kij}(\mathbf{x}, \mathbf{X}')t_k(\mathbf{x}) - D_{kij}(\mathbf{x}, \mathbf{X}')u_k(\mathbf{x})\right]dS - \rho\int_V S_{kij}(\mathbf{x}, \mathbf{X}')\ddot{u}_k(\mathbf{x}, \tau)dV \tag{5.49}$$

The displacement boundary integral equations are obtained through the usual limiting process, and can be written as

$$C_{ij}(\mathbf{x}')u_j(\mathbf{x}') + \int_S T_{ij}(\mathbf{x}, \mathbf{x}')u_j(\mathbf{x})dS = \int_S U_{ij}(\mathbf{x}, \mathbf{x}')t_j(\mathbf{x})dS$$

$$+\rho \int_V U_{ij}(\mathbf{X}, \mathbf{x}')\ddot{u}_j(\mathbf{X}, \tau)dV \tag{5.50}$$

and the boundary stress integral equation as

$$\frac{1}{2}\sigma_{ij}(\mathbf{x}') + \int_S S_{kij}(\mathbf{x}, \mathbf{x}')u_k(\mathbf{x})dS = \int_S D_{kij}(\mathbf{x}, \mathbf{x}')t_k(\mathbf{x})dS$$

$$+\rho \int_V S_{kij}(\mathbf{X}, \mathbf{x}')\ddot{u}_{kj}(\mathbf{X}, \tau)dV \tag{5.51}$$

where D_{kij} and S_{kij} are the Kelvin fundamental solutions given in Chapter 2.

To transform the domain integrals in (5.50) into boundary integrals, the DRM was developed by Nardini and Brebbia [40] for the displacement integral equation. The method is essentially a generalized way of constructing particular solutions as described in Section 2.6.5 that can be used to solve time-dependent problems. In this method the equations of motion are expressed in a boundary integral form using the fundamental solutions of elastostatics. This can be achieved by approximating the acceleration of a point $\mathbf{x}$ of the body by a sum of N coordinate functions $f(\mathbf{x^s}, \mathbf{x})$ multiplied by unknown time-dependent coefficients $\ddot{\alpha}_l^s(\tau)$:

$$\ddot{u}_l(\mathbf{x}, \tau) = \sum_{s=1}^{N+L} \ddot{\alpha}_l^s(\tau)f(\mathbf{x^s}, \mathbf{x}), \tag{5.52}$$

The approximation function $f(\mathbf{x^s}, \mathbf{x}) = c + r(\mathbf{x^s}, \mathbf{x})$ is chosen, where c is a constant and $r(\mathbf{x^s}, \mathbf{x})$ is the distance between a defining point $\mathbf{x}^n$ and the point $\mathbf{x}$. The defining point can be a boundary or a domain point s.

Using this assumption, the displacement boundary equation of motion, for a homogeneous and isotropic linear elastic body, can be written as

$$C_{ij}(\mathbf{x}')u_j(\mathbf{x}', \tau) - \int_S U_{ij}(\mathbf{x}', \mathbf{x})t_j(\mathbf{x}, \tau)dS + \int_S T_{ij}(\mathbf{x}', \mathbf{x})u_j(\mathbf{x}, \tau)dS$$

$$= \sum_{s=1}^{N+L} \rho\ddot{\alpha}_l^s(\tau)\left[C_{ij}\hat{u}_{lj}^s - \int_S U_{ij}(\mathbf{x}', \mathbf{x})\hat{t}_{lj}^s dS + \int_S T_{ij}(\mathbf{x}', \mathbf{x})\hat{u}_{lj}^s dS\right] \tag{5.53}$$

The stress integral equation for a point which belongs to a smooth boundary has the form [18, 54]

$$\frac{1}{2}\sigma_{ij}(\mathbf{x}', \tau) - \left[\int_S D_{kij}(\mathbf{x}', \mathbf{x})t_k(\mathbf{x}, \tau)dS - \int_S S_{kij}(\mathbf{x}', \mathbf{x})u_k(\mathbf{x}, \tau)dS\right]$$

$$= \sum_{s=1}^{N+L} \rho\ddot{\alpha}_l^s(\tau)\left\{\frac{1}{2}\hat{\sigma}_{lj}^s - \left[\int_S D_{kij}(\mathbf{x}', \mathbf{x})\hat{t}_{lk}^s dS - \int_S S_{kij}(\mathbf{x}', \mathbf{x})\hat{u}_{lk}^s dS\right]\right\} \tag{5.54}$$

where $i, j, k, l, = 1, 2$ for two-dimensional problems and $i, j, k, l, = 1, 2, 3$ for three-dimensional problems, ρ is the mass density, and $\hat{u}_{lj}^s$ and $\hat{t}_{lj}^s$ are particular displacements and tractions, which correspond to the function f^s; they are given in Section 2.6.5.

5.4.1 Numerical Discretization

The boundary of the body is discretized as in the previous approaches. The displacements and tractions, $u_j(\mathbf{x}, \tau)$, $t_j(\mathbf{x}, \tau)$ and $\hat{u}_{lj}^n(\mathbf{x}'', \mathbf{x})$, $\hat{t}_{lj}^n(\mathbf{x}'', \mathbf{x})$, within each element are approximated using the same interpolation functions. As a result of the approximation, the following displacement and traction equations for two-dimensional problems are obtained:

$$C_{ij}u_j(\mathbf{x}^c, \tau) - \sum_{n=1}^{N_e}\sum_{\alpha=1} \left[t_j^{n\alpha}(\tau) \int_{-1}^{1} U_{ij}(\mathbf{x}^c, \eta)N_\alpha(\eta)J^n(\eta)d\eta \right.$$

$$\left. - u_j^{n\alpha}(\tau) \int_{-1}^{1} T_{ij}(\mathbf{x}^c, \eta)N_\alpha(\eta)J^n(\eta)d\eta \right] =$$

$$\sum_{s=1}^{N+L} \rho\ddot{\alpha}_l^s(\tau) \left\{ C_{ij}\hat{u}_{lj}^n - \sum_{n=1}^{N_e}\sum_{\alpha=1} \left[\hat{t}_{lj}^{sn\alpha} \int_{-1}^{1} U_{ij}(\mathbf{x}^c, \eta)N_\alpha(\eta)J^n(\eta)d\eta \right. \right.$$

$$\left. \left. - \hat{u}_{lj}^{sn\alpha} \int_{-1}^{1} T_{ij}(\mathbf{x}^c, \eta)N_\alpha(\eta)J^n(\eta)d\eta \right] \right\} \tag{5.55}$$

The boundary equations are applied at the boundary nodes as in the other approaches. The displacement equations are applied at the domain points, when they are used to improve the approximation of accelerations. The set of equations can be written in matrix form as

$$\mathbf{Hu} - \mathbf{Gt} - \rho(\mathbf{H}\hat{\mathbf{u}} - \mathbf{G}\hat{\mathbf{t}})\ddot{\boldsymbol{\alpha}} = 0 \tag{5.56}$$

where $\mathbf{H}$ and $\mathbf{G}$ depends upon integrals of fundamental solutions and interpolating functions; they are the same as in elastostatics. The vectors $\mathbf{u}$, $\mathbf{t}$, $\hat{\mathbf{u}}$ and $\hat{\mathbf{t}}$ contain nodal values of real and particular displacements and tractions. The relationship between $\ddot{\mathbf{u}}$ and $\ddot{\boldsymbol{\alpha}}$ is established by applying equation (5.52) to every boundary and domain node. The resulting set of equations can be written in matrix form:

$$\ddot{\mathbf{u}} = \mathbf{F}\ddot{\boldsymbol{\alpha}} \tag{5.57}$$

where the elements of the matrix $\mathbf{F}$ are the values of the function $f^n(\mathbf{x}^s, \mathbf{x})$ at all S nodes. The unknown coefficients $\ddot{\boldsymbol{\alpha}}$ can be expressed in terms of the accelerations $\ddot{\mathbf{u}}$ as follows:

$$\ddot{\boldsymbol{\alpha}} = \mathbf{E}\ddot{\mathbf{u}} \tag{5.58}$$

Substitution of equation (5.58) into equation (5.56) gives [18]

$$\mathbf{Hu} - \mathbf{Gt} - \rho(\mathbf{H}\hat{\mathbf{u}} - \mathbf{G}\hat{\mathbf{t}})\mathbf{E}\ddot{\mathbf{u}} = 0 \tag{5.59}$$

or

$$\mathbf{Hu} - \mathbf{Gt} + \mathbf{M}\ddot{\mathbf{u}} = 0 \tag{5.60}$$

where $\mathbf{M} = -\rho(\mathbf{H}\hat{\mathbf{u}} - \mathbf{G}\hat{\mathbf{t}})\mathbf{E}$ and is the mass matrix of the structure.

The discretized form of the traction equation used by [18] and [54] is given as

$$\frac{1}{2}t_j(\mathbf{x}^c, \tau) - n_i \sum_{n=1}^{N_e}\sum_{\alpha=1}^{m} \left[t_k^{n\alpha}(\tau) \int_{-1}^{1} D_{kij}(\mathbf{x}^c\eta)N_\alpha(\eta)J^n(\eta)d\eta \right.$$

$$-u_k^{n\alpha}(\tau)\fint_{-1}^{1}S_{kij}(\mathbf{x}^c,\eta)N_\alpha(\eta)J^n(\eta)d\eta\Bigg] =$$

$$\sum_{s=1}^{N+L}\rho\ddot{\alpha}_l^s(\tau)\Bigg\{\frac{1}{2}\hat{t}_{lj}^s - n_i\sum_{n=1}^{N_e}\sum_{\alpha=1}^{m}\Bigg[\hat{t}_{lk}^{sn\alpha}\fint_{-1}^{1}D_{kij}(\mathbf{x}^c,\eta)N_\alpha(\eta)J^n(\eta)d\eta \tag{5.61}$$

$$-\hat{u}_{lk}^{sn\alpha}\fint_{-1}^{1}S_{kij}(\mathbf{x}^c,\eta)N_\alpha(\eta)J^n(\eta)d\eta\Bigg]\Bigg\}.$$

For three-dimensional problems, the discretized equations can be obtained in a similar way to that described for elastostatic problems in Chapter 2.

5.4.2   Dual Reciprocity in the Laplace Domain

An alternative formulation for the dual reciprocity in elastodynamic problems was developed by Wen, Aliabadi and Rooke [57] in the Laplace domain (rather than time domain described above). Taking the Laplace of (5.45), we have

$$\mu\bar{u}_{k,jj} + \frac{\mu}{1-2\nu}\bar{u}_{j,jk} = \rho s^2\bar{u}_k - \bar{w}_k \tag{5.62}$$

where

$$\bar{w}_k = su_k^o(\mathbf{x},0) - \dot{u}_k^o(\mathbf{x},0)$$

In the transformed domain the solution of (5.62) should satisfy the boundary conditions (5.32). Using Betti's reciprocal theorem for two independent states, the displacement at an interior point $\mathbf{X}'$ can be written as

$$\bar{u}_i(\mathbf{X}') = \int_S \left[U_{ij}(\mathbf{x},\mathbf{X}')\bar{t}_j(\mathbf{x}) - T_{ij}(\mathbf{x},\mathbf{X}')\bar{u}_j(\mathbf{x})\right]dS - \rho\int_V U_{ij}(\mathbf{X},\mathbf{X}')\left[s^2\bar{u}_k - \bar{w}_k\right]dV \tag{5.63}$$

Similarly, stresses at an internal point can be written as

$$\bar{\sigma}_{ij}(\mathbf{X}') = \int_S \left[U_{kij}(\mathbf{x},\mathbf{X}')\bar{t}_k(\mathbf{x}) - T_{kij}(\mathbf{x},\mathbf{X}')\bar{u}_k(\mathbf{x})\right]dS$$

$$-\rho\int_V U_{kij}(\mathbf{X},\mathbf{X}')\left[s^2\bar{u}_k - \bar{w}_k\right]dV \tag{5.64}$$

The displacement boundary integral equations are obtained through the usual limiting process, and can be written as

$$C_{ij}(\mathbf{x}')\bar{u}_j(\mathbf{x}') + \int_S T_{ij}(\mathbf{x},\mathbf{x}')\bar{u}_j(\mathbf{x})dS = \int_S U_{ij}(\mathbf{x},\mathbf{x}')\bar{t}_j(\mathbf{x})dS$$

$$+\rho\int_V U_{ij}(\mathbf{X},\mathbf{x}')\left[s^2\bar{u}_k - \bar{w}_k\right]dV \tag{5.65}$$

The approximation now takes the following form:

$$s^2\bar{u}_k(\mathbf{X},\mathbf{s}) - \bar{w}_k(\mathbf{X},\mathbf{s}) = \sum_{n=1}^{N}f^n(\mathbf{X})\alpha_k^n(s) \tag{5.66}$$

where $f^n(\mathbf{X}) = c + r$ and unknown α_k^n can be determined by the values of the accelerations at N collocation points, and written in matrix form as

$$\boldsymbol{\alpha} = \mathbf{F}^{-1}[s^2 \bar{u}_k - \bar{w}_k] \tag{5.67}$$

The domain integral on the right-hand side of (5.65) can be written as

$$\rho \int_V U_{ij}(\mathbf{X}, \mathbf{x}') \left[s^2 \bar{u}_k - \bar{w}_k \right] dV$$

$$= \rho \sum_{n=1}^{N+L} \left(C_{ij}(\mathbf{x}') \hat{u}_{lj}^n - \int_S U_{ij}(\mathbf{x}', \mathbf{x}) \hat{t}_{lj}^n dS + \oint_S T_{ij}(\mathbf{x}', \mathbf{x}) \hat{u}_{lj}^n(\mathbf{x}^n, \mathbf{x}) dS \right) \alpha_k^n \tag{5.68}$$

where particular solutions $\hat{u}_{lj}^n$ and $\hat{t}_{lj}^n$ are given in Section 2.6.5.

The discretized form of (5.68) can be written in a matrix form as

$$\mathbf{H}\bar{u} - \mathbf{G}\bar{t} - \rho(\mathbf{H}\hat{u} - \mathbf{G}\hat{t})\boldsymbol{\alpha} = 0 \tag{5.69}$$

Substituting (5.67) into (5.69) gives

$$\mathbf{H}\bar{u} - \mathbf{G}\bar{t} - \rho(\mathbf{H}\hat{u} - \mathbf{G}\hat{t})\mathbf{F}^{-1}[s^2 \bar{u}_k - \bar{w}_k] = 0 \tag{5.70}$$

which can be rewritten as

$$\mathbf{M}\bar{u} = \mathbf{G}\bar{t} - \mathbf{M}^0 \bar{w} \tag{5.71}$$

where $\mathbf{M}$ is complex matrix and a function of the Laplace transform parameter s, $\mathbf{M} = \mathbf{H} - s^2 \mathbf{M}^0$ and $\mathbf{M}^0 = \rho(\mathbf{H}\hat{U} - \mathbf{G}\hat{T})\mathbf{F}^{-1}$. For the zero initial condition at $\tau = 0$, $\bar{w} = 0$ and the unknowns are evaluated for a set of Laplace transform parameters s. It is sufficient to calculate all matrices $\mathbf{H}, \mathbf{G}, ..\mathbf{M}^0$ once in the Laplace transform domain.

5.4.3 Time Integration

Several time-integration schemes have been proposed [6] to deal with equations of (5.60). In all of them, some type of finite difference equation is used to calculate the accelerations using the displacements at predefined intervals of time assuming a form for the variation within the intervals. Most methods differentiate from each other by the type of variation assumed, i.e. linear, quadratic cubic, etc., and by whether the equilibrium conditions are considered at time t (explicit methods). Hubolt [26] has been used extensively in combination with DRM, because it is unconditionally stable and does not require parameters to be specified:

$$\ddot{u}_{\tau+\Delta\tau} = \frac{1}{\Delta\tau^2} [2\mathbf{u}_{\tau+\Delta\tau} - 5\mathbf{u}_\tau + 4\mathbf{u}_{\tau-\Delta\tau} - \mathbf{u}_{\tau-2\Delta\tau}] \tag{5.72}$$

Considering equation (5.71) at time interval $t + \Delta t$ and substituting the above expression leads to

$$\left(\frac{2}{\Delta t^2}\mathbf{M} + \mathbf{H} \right) \mathbf{u}_{t+\Delta t} = \mathbf{G}t_{t+\Delta t} + \frac{1}{\Delta t^2}\mathbf{M}(5\mathbf{u}_t - 4\mathbf{u}_{t-\Delta t} + \mathbf{u}_{t-2\Delta t}) \tag{5.73}$$

Substituting the boundary conditions at time $t + \Delta t$ and rearranging the expression leads to

$$\mathbf{A}\mathbf{X}_{t+\Delta t} = \mathbf{f}_{t+\Delta t}$$

Once the system of equations is solved for $\mathbf{u}_{t+\Delta t}$, the process can continue to the next step.

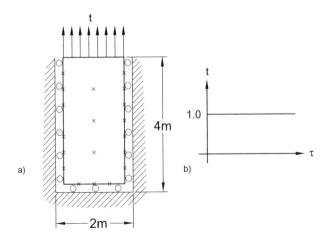

Figure 5.1: Strip subjected to tensile load ($E = 100\,000$, $\nu = 0.25$ and $\rho = 1.0$).

5.5 Examples

In this section several examples are presented to demonstrate the application of the time-domain, Laplace transform and dual reciprocity method for the solution of elastodynamic problems.

5.5.1 A Strip Subjected to Pure Tension and Shear Loads

A rectangular strip, 2 m wide and 4 m high, is initially studied under pure tension Heaviside load (see Figure 5.1), and then with shear load having a triangular variation in time (see Figure 5.4). The strip is modelled with 12 quadratic boundary elements.

Tensile load

Perez-Gavilan and Aliabadi [44] solved the problem using both collocation and Galerkin methods. The computed values of the displacements at the mid-point of the free edge and tractions at the mid-point of the base are presented in Figures 5.2 and 5.3, respectively.

For the dual reciprocity method two sets of results are presented using 0 and 3 internal points. The time domain solutions presented correspond to the collocation method using time interval of 0.001 sec . There is good agreement between the different BEM formulations. Dominguez and Gallego [15] used the above example to study the sensitivity of the time-domain formulation to the size of the time step. They recommended the time step should be chosen so that $c_p \Delta t / L$ (L being the distance between two nodes on an element) is close to unity. Large time steps were found to be unable to capture rapid changes in the solution while smaller time steps were shown to yield unstable solutions.

Shear load

For the shear load the peak impulse is at 0.01 sec and duration is 0.02 sec . The maximum peak is at 0.01 sec . Here, the problem is analyzed using the Dual Reciprocity

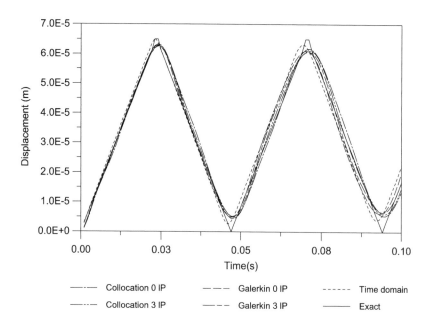

Figure 5.2: Displacements vs. time for the strip subjected to tensile load.

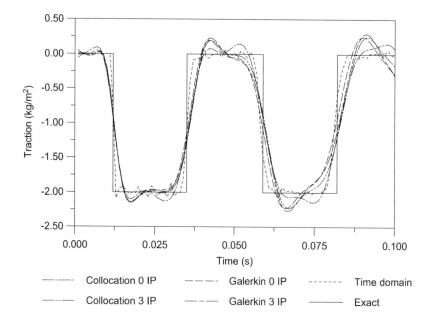

Figure 5.3: Tractions vs. time for stip subjected to tensile load.

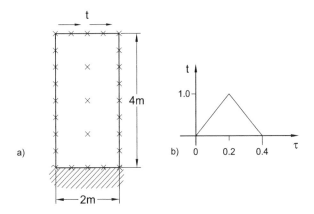

Figure 5.4: Strip subjected to shear load ($E = 100\,000$, $\nu = 0.25$ and $\rho = 1.0$).

Method (DRM), Laplace Transform Method (LTM) and Time Domain Method (TDM). For dual reciprocity, three internal points uniformly distributed are used, as shown in Figure 5.4. For the Laplace transform 25 Laplace parameters were used. For the time domain, the time step used is 0.003 sec, while for dual reciprocity a time step of 0.001 sec was used.

Figure 5.5 presents the solutions obtained for the horizontal displacements u_1 at the mid-pint of the free edge where the load is applied. The agreement between different BEM methods is found to be good.

5.5.2 Plate with a Hole

A plate with a hole, as shown in Figure 6.6, is subjected to tensile stress at one end. Only a quarter of the plate was modelled [44] due to symmetry. The material properties used are as follows: shear modulus $\mu = 2.03 \times 10^6$ N/m^2, mass density $\rho = 1000$ kg/m^3, Poisson ratio $\nu = 0.3$. The longitudinal wave velocity $c_1 = 84.2911$ m/s and the first four natural frequencies are $\omega_1 = 8.6$ rad, $\omega_2 = 15.5$ rad/s, $\omega_3 = 14.39$ rad/s and $\omega_4 = 50.0$ rad/s. The results in Figures 5.7 present the solutions obtained using both collocation and Galerkin formulations using 14 quadratic elements to discretize the boundary.

5.5.3 Rectangular Bar Subjected to Tensile Load

Consider a bar with, square section of width $2w$ and height $2h$, as shown in Figure 5.8, loaded by an impact load $\sigma_0 H(\tau)$ at the top of the bar with the bottom fixed, where $H(\tau)$ is the step function. The bar geometry is defined by the ratio $w/h = 0.5$ with zero Poisson's ratio. There are 40 continuous elements on the boundary with 122 nodes. The numerical results for displacements $u_3(0, w, 2h)$ and $u_3(0, w, h)$ are shown in Figure 5.9. For the time domain method, the normalized time increment $\Delta \bar{t}$ is chosen as 0.2 ($\bar{t} = c_1 t/h$, c_1 is the velocity of dilatational waves) and 80 time steps are calculated. The total number of Laplace transform parameters s is chosen as 25. For the dual reciprocity method, the total number of collocation points is 244 on the boundary and in the domain (to evaluate acceleration). The normalized time

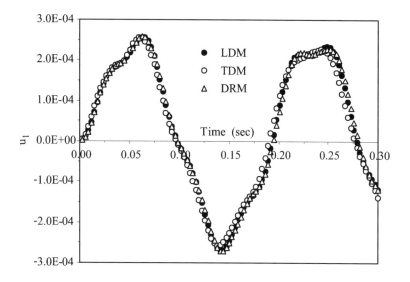

Figure 5.5: Horizontal displacements vs. time for the rectangular strip subjected to shear load.

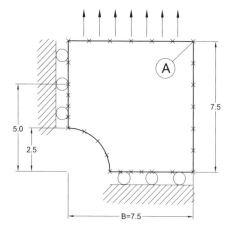

Figure 5.6: Plate with a hole.

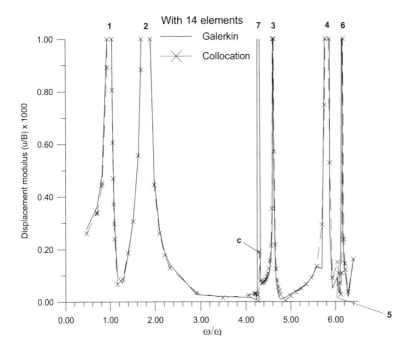

Figure 5.7: Vertical displacement modulus of point A by load frequency ratio.

step $\Delta\bar{\tau} = 0.1$. The exact solutions for the dynamic stress are presented in the same figure for comparison, and reasonable agreement is observed.

Comparisons were also made between the two DRM approaches presented in Section 5.4 by Wen and Aliabadi [57]. Figures 5.10 and 5.11 show the normalized values of displacements and tractions obtained.

5.5.4 Comparison of DRM, TDM and LTM

In this section the three formulations presented in previous sections are compared. As has been described, the formulation and numerical implementation for each method is very different. Fedelinski, Aliabadi and Rooke [20] and Wen, Aliabadi and Young [57] carried out a comparison of these methods for two- and three-dimensional problems, respectively. Here, their findings are reported again.

Fundamental Solutions

The TDM uses fundamental solutions of elastodynamics, which are time-dependent. They require time and space integration. For a simple variation of the boundary displacements and tractions, i.e. piecewise constant or linear, the time integrals can calculated analytically. As a result of time integration, the convoluted fundamental solutions are obtained. They have a more complicated form than the static fundamental solutions, therefore their space integration is more difficult. At short times the time-dependent fundamental solutions are non-zero in the neighbourhood of the collocation point only; therefore, initially integrations are required along part of the

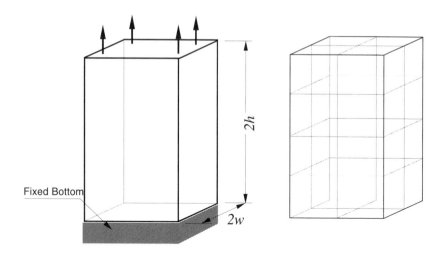

Figure 5.8: (a) Rectangular bar and (b) element mesh.

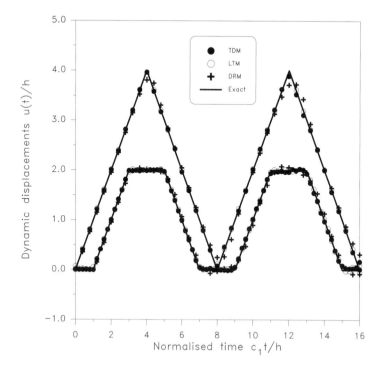

Figure 5.9: Normalized dynamic displacement on the boundary.

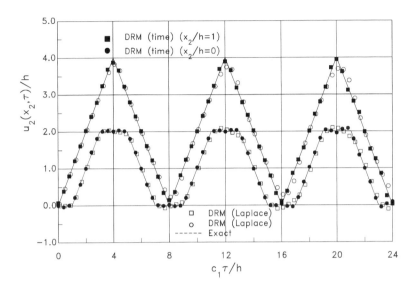

Figure 5.10: Displacements vs. time for the rectangular bar subjected to impact load.

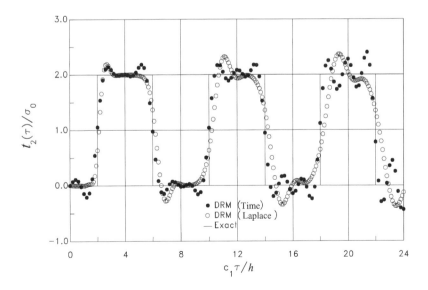

Figure 5.11: Tractions on the lower surface of the bar.

boundary only.

In the LTM the Laplace transform of the fundamental solutions are applied. They require the space integration only. However, their forms are complicated, as seen in Section 5.3.2.

The DRM uses the fundamental solutions of elastostatics, and particular solutions, which depends upon an approximation function. The fundamental solutions of elastostatics are time independent, therefore only space integration is required. The particular solutions do not require integration.

The convoluted fundamental solutions and the Laplace transforms of the fundamental solutions of elastodynamics converge to the static ones if a small mass density or a long time is assumed. The order of spatial singularity of fundamental solutions used in each method is the same.

Matrix formulation of Equations of Motion

In the TDM, for each time step, only two matrices $\mathbf{H}^{Nn}$ and $\mathbf{G}^{Nn}$, which correspond to the maximum difference between time steps $N - n$, need to be computed. Both matrices require time and space integration. For the LTM two matrices, $\bar{\mathbf{H}}$ and $\bar{\mathbf{G}}$, are calculated for each Laplace parameter. In the DRM two matrices, $\mathbf{H}$ and $\mathbf{G}$, need to be formed; they are obtained by boundary integration, and are the same as the matrices used in elastostatics. An additional matrix $\mathbf{F}$, which contains values of the approximation functions, and two matrices $\hat{\mathbf{u}}$ and $\hat{\mathbf{t}}$, which contain the values of the particular solutions, are required in order to calculate the mass matrix $\mathbf{M}$ of the structure. The calculation of $\mathbf{F}$, $\hat{\mathbf{u}}$ and $\hat{\mathbf{t}}$ is straightforward, but the inverse of the matrix $\mathbf{F}$ and double multiplication required to obtain $\mathbf{M}$ is very time consuming, although it needs to be done only once.

Modification of the System of Equations

The matrix system of equations in all approaches is modified according to the boundary conditions. In the TDM the matrices $\mathbf{H}^{NN}$ and $\mathbf{G}^{NN}$ are divided into two submatrices each, and arranged giving new matrices $\mathbf{A}$ and $\mathbf{B}$. The matrices are modified only once during the first time step. For the LTM the matrices are rearranged in a similar way as for the time domain method. This modification is required for each Laplace parameter. In the DRM each of the matrices $\mathbf{H}, \mathbf{G}$ and $\mathbf{M}$ is divided into four submatrices, which correspond to the nodes with specified displacements and tractions, respectively. The reduced system of equations is obtained by matrix operations applied to these matrices.

The above described modifications applied for the TDM and LTM are similar to those used in matrices, and are much simpler than in the DRM.

Solution of the System of Equations

The size of the system of equations generated by the TDM and LTM is equal to the number of nodes times the number of degrees of freedom (d.o.f) per node. The number of equations which needs to be solved in each time step in the DRM equals the number of unknown displacements of nodes and domain points. The ratio of the sizes of the systems obtained by the methods depends upon the problem. Usually, the system given by the DRM is the largest because the method requires additional domain points to obtain an accurate solution. For the TDM and DRM the matrix of coefficients of the system of equations is the same in each time step. Therefore, efficient methods for the systems with a constant matrix of coefficients and varying right-hand side vectors can be used. Both systems can be solved using the LU-decomposition method. The computation of the RHS vector in the TDM, which depends on matrices $\mathbf{H}^{Nn}$ and $\mathbf{G}^{Nn}$ from all previous steps, is time consuming, particularly if the solution for many time steps is needed. The system of equations obtained by the LTM for each Laplace parameter is different.

Memory Requirements, Speed, Accuracy

In the TDM only matrices $\mathbf{H}^{Nn}$ and $\mathbf{G}^{Nn}$ are calculated. They are reordered only once at the first step, giving the matrices $\mathbf{A}$ and $\mathbf{B}$. The RHS vector in this approach requires matrices $\mathbf{H}^{Nn}$ and $\mathbf{G}^{Nn}$ from all previous steps. These matrices can be stored using one file. For the LTM two matrices $\bar{\mathbf{H}}$ and $\bar{\mathbf{G}}$ are rearranged to give new matrices $\bar{\mathbf{A}}$ and $\bar{\mathbf{B}}$. No other matrices are needed in this approach. The DRM requires more matrices to formulate the system of equations of motion ($\mathbf{H}, \mathbf{G}, \hat{\mathbf{u}}, \hat{\mathbf{t}}, \mathbf{F}, \mathbf{F}^{-1}, \mathbf{M}$ and submatrices obtained by the partition of $\mathbf{H}, \mathbf{G}$ and $\mathbf{M}$). These matrices are not all required at the same time. However, in order to improve the speed it is better to store all the matrices for future use.

The formulation of the system of equations in TDM is simple, but the RHS vector requires integration in each time step, and depends on the history of the process. The generation of equations in the LTM is simple but time consuming because of the complexity of the transformed fundamental solutions. The total time is approximately proportional to the number of Laplace parameters. This method requires a solution of a new system of equations for each new parameter. On the other hand, although the formulation of the system of equations in the DRM takes more time because of the larger number of matrices that calculated and the modification of the system is more complicated, the calculation of the RHS is very simple. In most cases, the DRM is several times faster than TDM and LTM.

The TDM method requires the approximation of boundary displacements and tractions and their variation in time. In this method the time step should be related to the size of the boundary elements, in order to obtain stable solutions, as discussed in the previous section.

For the DRM the approximation of boundary displacements and tractions and their variation in time, and an additional approximation of accelerations using the radial basis function, is used. If the acceleration field has a steep gradient, as in the case of a structure subjected to impact loading, it is expressed in terms of boundary nodal values and additional domain points. The use of the domain points is not convenient. Choosing the position of points requires experienced users or special automatic adaptive methods. This additional approximation decreases the accuracy of the solution. The solution of the system of equations of motion is usually obtained by the Houbolt method, which imposes strong numerical damping. This damping can be reduced by using smaller time steps.

Test Results

In order to compare the memory and time requirements for three methods, Fedelinski *et al.* [20] solved an example of a rectangular plate with an inclined central crack.

Table 5.1: Size of major matrices created by the methods.

Matrices	TDM	LTM	DRM
H	192×192	192×192	272×192
G	192×240	192×240	272×240
F			272×272
û			192×272
t̂			240×272
$\sum$	2073600	82994	308992
Comments	Matrices are stored for all 25 steps	Elements of matrices are complex numbers	

The boundary of the plate was modelled using quadratic elements. The total number of nodes was 96 and the system has 96 degrees of freedom (d.o.f.). An additional 40 domain points were used for the DRM, increasing the d.o.f. to 272. The solutions were obtained using 25 steps for the TDM, 25 Laplace parameters for the LTM and 100 steps for the DRM. The number of elements of the major matrices created is given in Table 5.1. The computer times obtained are presented in Figures 5.12 and 5.13.

The actual CPU times will be several times smaller with present day computers, but the trends still remain the same.

The TDM is fast during the initial steps because the time-dependent fundamental solutions are integrated along part of the boundary only. For this problem, the integration of kernels along the whole boundary is required by the TDM after 11 steps. From then on, the solution needs about 2 min per time step, as shown in Figure 5.13; this time slowly increases, approximately linearly, with the number of steps because the solution depends upon all the previous time steps.

The time required to find the solution for each Laplace parameter is approximately constant, and equals about 5 min, as we can see from Figure 5.13. The DRM requires about 11 min to formulate the system of equations and to decompose the matrix of coefficients, but the solution in each time step takes about 2 sec only.

The system of equations of motion is solved faster by the TDM than the DRM during the first 10 steps. The total time for the TDM is a similar function of the number of steps as the time for the LTM is a function of the number of Laplace parameters. The LTM is the slowest method because of the complexity of the fundamental solutions. The time of computation for the TDM and the LTM is mainly determined by the integration of fundamental solutions, which may have complicated forms. Any improvements in the efficiency of integration is unlikely to reduce the time to that required by the DRM. Therefore, the ratio of the time required by the TDM and LTM to the DRM increases as the number of the steps increases. The solutions of this example were obtained using the same number of boundary elements in order to compare memory and time requirements. However, similar accuracy was obtained using a different number of boundary elements for each method, namely 32 for LTM, 40 for DRM and 64 for TDM. For this new discretization the solution times are shown in Figure 5.14

For three-dimensional problems Wen, Aliabadi and Young [57] reported similar trends for computational times of three methods.

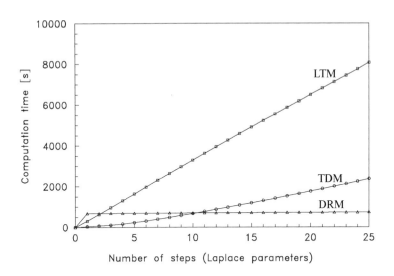

Figure 5.12: Computation time vs. number of steps (number of Laplace parameter).

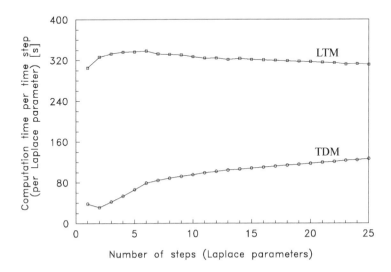

Figure 5.13: Number of steps (Laplace parameter).

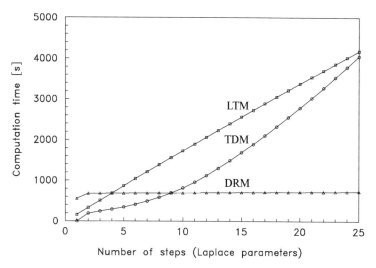

Figure 5.14: Computation time vs. number of steps (number of Laplace parameters).

5.6 Acoustic Scattering in Fluid-Solid Problems

Scattering of acoustic and electromagnetic waves is important in many engineering fields such as mechanics and aeronautics, among others. Contributions to the acoustic scattering problem in the case of an impenetrable obstacle can be found in Seybert et al. [50, 51] and Goswami et al. [25] for three-dimensional problems, and Mallardo and Aliabadi [36] for two-dimensional problems. The formulation consists of the coupling of two sets of integral equations which respectively represent the response of an obstacle and the acoustic behaviour of the fluid in presence of an incident beam. The two equations are coupled by enforcing continuity of normal components of velocities and equilibrium of pressures along the interface. The obstacle response is characterized by the internal pressure p_i, while the external behaviour is characterized by the total pressure p_e. The total pressure is taken as the sum of the incident p_{inc} and scattered p_{sc} pressure, i.e.

$$p_e(\mathbf{x}) = p_{inc}(\mathbf{x}) + p_{sc}(\mathbf{x})$$

The scattered pressure p_{sc} satisfies the Sommerfield radiation condition at infinity [42].

The boundary integral equation for the external problem is governed by the Helmholz equation (see Volume I) and the internal elastic region by equation (5.30). The compatibility and equilibrium conditions at the fluid-solid interface, and the traction free condition on the boundary of the cavity, can be written as

$$\begin{aligned}
q(\mathbf{x}) &= \rho_e \omega^2 \mathbf{u}(\mathbf{x}).\mathbf{n}(\mathbf{x}) & \mathbf{x} \in S_e \\
\mathbf{t}(\mathbf{x}) &= -p(\mathbf{x})\mathbf{n}(\mathbf{x}) & \mathbf{x} \in S_e \\
\mathbf{t}(\mathbf{x}) &= 0 & \mathbf{x} \in S_i
\end{aligned}$$ (5.74)

where $\mathbf{n}$ denotes the inward normal on the external boundary S_e, the flux $q(\mathbf{x})$ is the normal derivative of $p(\mathbf{x})$, ρ_e is the external density and ω is the circular frequency.

Table 5.2: Material properties.

Brass cylinder	
Mass density	$\rho = 8500 kg/m^3$
Young's modulus	$E = 10.5 \times 10^{11} Pa$
Poisson's ratio	$\nu = 1/3$
Radius	$R = 1.0m$
Acoustic medium: water	
Mass density	$\rho = 998 kg/m^3$
Sound speed	$c = 1486 m/s$

The coupled integral equations are discretized in the usual manner, and can be written in a matrix form as [36]

$$
\begin{bmatrix}
H^e & 0 & 0 & -G^e & 0 \\
0 & H^i_{S_e S_e} & H^i_{S_e S_i} & 0 & -G^i_{S_e S_e} \\
0 & H^i_{S_i S_e} & H^i_{S_i S_i} & 0 & -G^i_{S_i S_e} \\
0 & \rho_e \omega^2 N^t & 0 & -I & 0 \\
\mathbf{n} & 0 & 0 & 0 & I
\end{bmatrix}
\begin{Bmatrix}
\mathbf{p} \\
\mathbf{u}_{S_e} \\
\mathbf{u}_{S_i} \\
\mathbf{q} \\
\mathbf{t}_{S_e}
\end{Bmatrix}
=
\begin{Bmatrix}
\mathbf{p}_{inc} \\
0 \\
0 \\
0 \\
0
\end{Bmatrix}
\tag{5.75}
$$

where H^e and G^e correspond to the boundary integral equations for Helmholtz problem (see Volume I) and H^i and G^i correspond to equation (5.40), $\mathbf{n}$ contains the components of the inward normal at every node on S_e and I is the identity matrix.

5.6.1 Example : Cylinder Immersed in Fluid

The elastic scatterer has a cylindrical shape and is immersed in an infinite fluid carrying an incoming incident wave (see Figure 5.15). The properties of the fluid and solid are given in Table 5.2.

The problem was studied by Mallardo and Aliabadi[36] for the case of the central cavity (i.e. $D = 0$) with $\alpha_{inc} = 0$ and $R_{cav} = 0.05R$. Figures 5.16 to 5.18 compare the analytical and boundary element far field coefficients in the cases $k_e R = 1$, 5 and 10 (k_e is the external wave number $k_e = \omega/c_e$, with c_e being the external wave speed), as presented in Mallardo and Aliabadi [36].

For this problem, an analytical solution was obtained in the presence of a central cavity. Following Lin and Raptis [31], but modifying the compressional and transverse waves inside the solid in order to keep the internal flaw in count, i.e.

$$\mathbf{u} = \nabla \phi + \nabla \times \psi$$

$$\phi = \sum_{m=0}^{\infty} \epsilon_m i^m a_m J_m(k_1 r) Cos(m\theta) + \sum_{m=0}^{\infty} \epsilon_m i^m c_m H_m(k_1 r) Cos(m\theta)$$

$$\psi = \sum_{m=0}^{\infty} \epsilon_m i^m b_m J_m(k_2 r) Sin(m\theta) + \sum_{m=0}^{\infty} \epsilon_m i^m d_m H_m(k_2 r) Sin(m\theta)$$

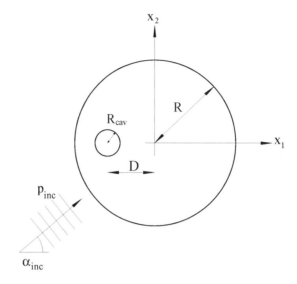

Figure 5.15: Cylinder with an internal circular cavity.

it is possible to obtain the scattering pressure on the external circle as the sum of a series in which the mth term involves the resolution of a 5×5 (and not 3×3) system of linear equations: the matrix coefficients are in terms of the Bessel functions of the first and third kind, of order m. The first 25 terms of the series solution and 20 terms of the Bessel function expansions were taken into account in order to obtain a satisfactory stable analytical solution [36].

The BEM results show good agreement with the analytical solutions. In the case $k_e R = 1$, 8+8 elements on $S_e \cup S_i$ was sufficient to achieve errors of less than 1%, while 16+8 and 28+8 elements were necessary to achieve the same precision for $k_e R = 5$ and 10, respectively. The reason for increasing the number of elements is an increased variation in the solution requiring finer meshes.

5.7   Summary

In this chapter displacement and stress integral equations were presented for elasto-dynamic problems. Three solution methods based upon the time-domain, frequency domain and mass matrix were presented. Each method was shown to have certain advantages and disadvantages. The time domain TDM is the most time consuming of the three methods. TDM is, however, the most general, allowing the solution of more complicated problems such as contact and crack propagation. Nonlinear or moving boundary problems would be difficult and inefficient to solve using the frequency do-main approach. The Laplace transform, however, is the most efficient is reproducing rapid changes in the solution, and is usually found to be the most accurate of the three methods. The dual reciprocity method DRM is perhaps the least accurate of the three methods, usually introducing some damping into the solution. The method is, nevertheless, the simplest to implement and fastest of the three.

Other recent developments can be found in [10, 21, 48, 59, 60] and applied to crack

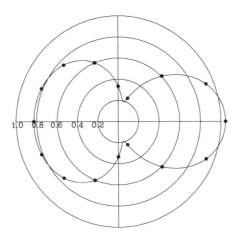

Figure 5.16: Angular distribution $| \, p_{sc}/p_{inc} \, |$ for brass in water with a central cavity. Elastic scatter; $k_e R = 1$, __ Analytical solutions, • BEM solutions.

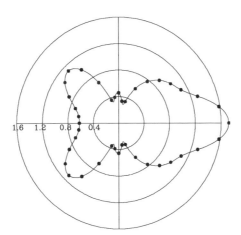

Figure 5.17: Angular distribution $| \, p_{sc}/p_{inc} \, |$ for brass in water with a central cavity. Elastic scatter; $k_e R = 5$, __ Analytical solutions, • BEM solutions.

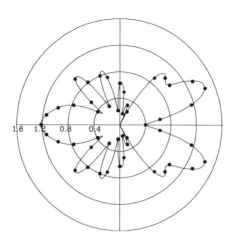

Figure 5.18: Angular distribution $| p_{sc}/p_{inc} |$ for brass in water with a central cavity. Elastic scatter; $k_e R = 10$, _ Analytical solutions, • BEM solutions.

problems.

References

[1] Abascal, R. and Dominguez, J., Vibrations of footings on zoned viscoelastic soil. *Journal of Engineering Mechanics, ASCE,* **112**, 433-447, 1986.

[2] Abascal, R. and Dominguez, J., Dynamic response of two-dimensional foundations allowed to uplift, *Computers and Geomechanics,* **9**, 113-129, 1990.

[3] Antes, H., A boundary element procedure for transient wave propagations in two-dimensional isotropic elastic media, *Finite Elements Analysis and Design,* **1**, 313-322, 1985.

[4] Antes, H. and Von Estorff, O., Analysis of absorption effects on the dynamic response of dam reservoir systems by element methods, *Earthquake Engineering and Structural Dynamics,* **15**, 1023-1036, 1987.

[5] Banaugh, R.P. and Goldsmith, W., Diffraction of steady elastic waves by surfaces of arbitrary shape, *Journal of Applied Mechanics,* **30**, 589-597, 1963.

[6] Bathe, K.J. and Wilson, E.L., *Numerical Methods in Finite Element Analysis,* Prentice-Hall: Englewood Cliffs, New Jersey, 1976.

[7] Beskos, D.E., Boundary element methods in dynamic analysis, *Applied Mechanics Review,* **40**, 1-23, 1987.

[8] Beskos, D.E., Boundary element methods in dynamic analysis, part II (1986-1996), *Applied Mechanics Review,* **50**, 149-197, 1997.

[9] Coda, H.R. and Venturini, W.S., Three-dimensional transient BEM analysis, *Computers & Structures*, 751-768, 1995.

[10] Coda, H., Venturini, W.S. and Aliabadi, M.H., A general 3D BEM/FEM coupling applied to elastodynamic continua/frame structures interaction analysis, *International Journal for Numerical Methods in Engineering*, **46**, 695-712, *1999*.

[11] Cole, D.M. Kosloff and Minster, J.B., A numerical boundary integral equation method for elastodynamics-I, *Bulletin of Seismological Society of America*, **68**, 1331-1357, 1978.

[12] Cruse, T.A. and Rizzo, F.J., A direct formulation and numerical solution for the general transient elastodynamics problem-I, *Journal of Math. Anal. Appl.*, **22**, 244-259, 1968.

[13] Cruse, T.A., A direct formulation and numerical solution of the general transient elastodynamic problem-II, *Journal of Math. Anal. Appl.*, **22,** 341-355, 1968**.**

[14] Davi, G. and Milazzo, A., A symmetric and positive definite variational BEM for 2-D free vibration analysis. *Engineering Analysis with Boundary Elements*, **14**, 357-362, 1994.

[15] Dominguez, J. and Gallego, R., The time-domain boundary element method for elastodynamic problems. *Mathemetical and Computer Modelling*, **15**, 119-129, 1991.

[16] Dominguez, J., *Boundary Elements in Dynamics*, Computational Mechanics Publications, Southampton 1993.

[17] Durbin, F., Numerical inversion of Laplace transforms: an efficient improvement to Dubner and Abate's method, *Computing Journal*, **17**, 1974. **17,**

[18] Fedelinski, P., Aliabadi, M.H. and Rooke, D.P., The dual boundary element method in dynamic fracture mechanics, *Engineering Analysis with Boundary Elements*, **12**, 203-210, 1993.

[19] Fedelinski, P., Aliabadi, M.H. and Rooke, D.P., A single region time domain BEM for dynamic crack problems, *International Journal Solids and Structures*, **32,** 3555-3571, 1995.

[20] Fedelinski, P., Aliabadi, M.H. and Rooke, D.P., The Laplace transform DBEM method for mixed-mode dynamic crack analysis, *Computers and Structures*, **59,** 1021-1031, 1996.

[21] Frangi, A. and Novati, G., On the numerical stability of time domain elastodynamic analyses by BEM, *Computer Methods in Applied Mechanics and Engineering*, **173,** 403-417, 1999.

[22] Friedman,M.B. and Shaw,R.P., Diffraction of pulses by cylindrical obstacles of arbitrary cross section, *Journal of Applied Mechanics*, **29**, 40-46, 1962.

[23] Gallego, R.. and Dominguez, J., Dynamic crack propagation analysis by moving singular boundary elements, *Journal of Applied Mechanics, ASME*, **59,** S158-S162, 1992.

[24] Graff, K.F., *Wave Motion in Elastic Solids,* Columbus, University of Ohio Press, 1975.

[25] Goswami, P.P., Rudulphi, T.J., Rizzo, F.J. and Shippy, D.J., A boundary element model for acoustic-elastic interaction with applications in ultrasonic NDE, *Journal of Nondestructive Evaluation,* **9**, 101-112, 1990.

[26] Houbolt, J.C. A recurrence matrix solution for the dynamic response of elastic aircraft. *Journal of Aeronautical Science,* **17**, 540-550, 1950.

[27] Guoyou Yu, Mansur, W.J., Carrer, J.A.M. and Gong, L., Time weighting in time domain BEM, *Engineering Analysis with Boundary Elements,* **22**, 175-181, 1998.

[28] Israil, A.S.M. and Banerjee, P.K., Two-dimensional transient wave- propagation problems by time-domain BEM, *International Journal of Solids and Structures,* **26** , 851-864, 1990

[29] Israil, A.S.M. & Banerjee, P.K., Interior stress calculations in 2-D time-domain transient BEM analysis, *International Journal of Solids and Structures,* **27**, 915-927, 1991.

[30] Karabalis, D.L. and Beskos, D.E., Dynamic response of 3-D rigid surface foundations by time domain boundary element method, *Earthquake Engineering Structures and Dynamics,* **12,** 73-93, 1984.

[31] Lin, W.H. and Raptis, A.C., Acoustic scattering by elastic solid cylinder and spheres in viscous fluids, *Journal of Acoustical Society of America,* **73,** 736-748, 1983.

[32] Mansur, W.J., A time stepping technique to solve wave propagation problems using the boundary element method, PhD thesis, University of Southampton, 1983.

[33] Mansur, W.J. , Carrer, J.A.M. and Siqueira, F.N., Time discontinuous linear traction approximation in time domain BEM scalar wave approximation, *International Journal for Numerical Methods in Engineering,* **42**, 667-683, 1988.

[34] Mansur, W.J., Carrer, J.A.M. and Siqueira, F.N., Time discontinuous linear traction approximation in time-domain BEM scalar wave approximation. *Communications in Numerical Methods in Engineering, 2001.*

[35] Maier, G. Diligenti, M. and Carini, A., A variational approach to boundary element elastodynamic analysis and extension to multidomain problems. *Computer Methods in Applied Mechanics and Engineering,* **92**, 192-213, 1991.

[36] Mallardo, V. and Aliabadi, M.H., Boundary element method for acoustic scattering in fluid-fluidlike and fluid-solid problems, *Journal of Sounds and Vibration,* **216**, 413-434, 1998.

[37] Manolis, D.E. and Beskos, D.E., Dynamic stress concentration studies by the boundary integral equation method. *Innovative numerical analysis for the engineering sciences,* University of Virginia Press, Charlottesville, 459-463, 1980.

[38] Manolis, G.D. and Beskos, D.E., *Boundary Element Methods in Elastodynamics*, Unwin Hyman, London 1988.

[39] Manolis, G.D. and Shaw, R.P. Fundamental solutions for variable density two-dimensional elastodynamic problems, *Engineering Analysis with Boundary Elements,* **24**, 739-750, 2000.

[40] Nardini, D.. and Brebbia, C.A., A new approach for free vibration analysis using boundary elements, *Boundary element methods in engineering*, 312-326, Springer-Verlag, Berlin, 1982.

[41] Niwa, Y. and Hirose, S., Three-dimensional analysis of ground motion by integral equation method in wave number domain. *Numerical Methods in Geomechanics,* A.A.Balkema, Rotterdam, 143-149, 1995.

[42] Pierce, A.D., *Acoustics: An Introduction to its Physical Principles and Applications,* 177-178, McGraw-Hill, New York, 1981.

[43] Peirce, A. and Siebrits, E., Stability analysis and design of time-stepping schemes for general elastodynamics boundary element models, *International Journal for Numerical Methods in Engineering,* **40**, 319-342, 1997.

[44] Perez-Gavilan, J.J. and Aliabadi, M.H., A Galerkin boundary element formulation with dual reciprocity for elastodynamics, *International Journal for Numerical Methods in Engineering,* **48**, 1331-1344, 2000.

[45] Perez-Gavilan, J.J. and Aliabadi, M.H., A symmetric Galerkin boundary element formulation for dynamic frequency viscoelastic problems. *Computers and Structures,* **79,** 2621-2633, 2001.

[46] Rizos, D.C. and Karabalis, D.L., An advanced direct time-domain BEM formulation for general 3D elastodynamic problems, *Computational Mechanics,* **15**, 249-269, 1994.

[47] Schanz, M. and Antes, H., A new visco- and elastodynamic time domain boundary element formulation. *Computational Mechanics,* **20**, 452-459, 1997.

[48] Schanz, M. and Antes, H., A boundary integral formulation for the dynamic behavior of a Timoshenko beam, *Advances in Boundary Element Techniques II,* Hoggar, Geneva, 475-482, 2001.

[49] Spyrakos, C.C. and Beskos, D.E., Dynamic response of flexible strip foundations by boundary and finite elements, *Soil Dynamics Earthquake Engineering,* **5**, 84-96, 1986.

[50] Seybert, A.F., Wu, T.W. and Wu, X.F., Radiation and scattering of acoustic waves from elastic solids and shells using the boundary element method. *Journal of American Acoustical Society of America,* **84**, 1906-1912, 1988.

[51] Seybert, A.F., Cheng, C.Y.R. and Wu, T.W., The solution of coupled interior/exterior acoustic problems using boundary element method, *Journal of Acoustic Society of America,* **88**, 1612-1618, 1990

[52] Wang, H.C. and Banerjee, P.K., Axisymmetric transient elastodynamic analysis by boundary element method, *International Journal of Solids and Structures* **26**, 401-415, 1990.

[53] Wen, P.H., Aliabadi,M.H. and Rooke,D.P., The influence of elastic waves on dynamic stress intensity factors (three-dimensional problems), *Archive of Applied Mechanics,* **66,** 385-394, 1996.

[54] Wen, P.H., Aliabadi, M.H. and Rooke, D.P., Cracks in three dimensions: a dynamic dual boundary element analysis, *Computer Methods in Applied Mechanics and Engineering,* **167**, 139-151, 1998.

[55] Wen, P.H., Aliabadi, M.H. and Young, A., A time-dependent formulation of dual boundary element method for 3D dynamic crack problems, *International Journal of Numerical Methods in Engineering,* **45**, 1887-1905, 1999.

[56] Wen, P.H, Aliabadi, M.H. and Rooke, D.P., A mass matrix formulation for three-dimensional dynamic fracture mechanics, *Computer Methods in Applied Mechanics and Engineering,* **173,** 365-374, 1999.

[57] Wen, P.H., Aliabadi, M.H. and Rooke, D.P., Three-dimensional dynamic fracture analysis with dual reciprocity method in Laplace domain, *Engineering Analysis with Boundary Elements,* **23**, 51-58, 1999.

[58] Wu, T.W. and Lee, L., A choice of interpolation schemes for the boundary element method in elastodynamics. *Communications in Numerical Methods in Engineering,* **9**, 375-385, 1993.

[59] Zhange, Ch., Transient elastodynamic antiplane crack analysis in anisotropic solids. *International Journal of Solids and Structures,* **37,** 6107-6130, 2000.

[60] Zhang,Ch., Transient dynamic analysis of an antiplane interface crack in anisotropic solids, *Advances in Boundary Element Techniques II,* Hoggar, Geneva, 459-466, 2001.

Chapter 6

Elastoplasticity

6.1 Introduction

Engineering design is mainly concerned with maintaining structures operating within their elastic range. There are, however, circumstances in which prior plastic deformations resulting in compressive residual stresses are desirable. These include surface treatment techniques such as cold expansion, shot peening, cold rolling, etc. Other cases in which plastic deformation occurs are the gross overload of parts or stress concentration causing material to exceed the elastic range locally.

The first application of the boundary element method to elastoplastic problems is due to Riccardella [37]. During the same period, Mendelson [29] presented indirect and direct formulations. The direct formulation was presented including the integral expression for the internal stresses (two- and three-dimensional problems). These expressions however, were later shown to be incorrect due to the way in which the plastic strain term was considered. In 1975 an extension of the above work was presented by Mendelson and Albers [30].

Two years later, a theoretical paper concerned with reducing the three-dimensional direct boundary element formulation to the plane stress case was presented by Mukherjee [31]. In this paper, the equations presented in [29] and [30] were partially corrected, and modified versions of the kernels of the plastic strain integrals presented. An important contribution towards the proper inelastic boundary element formulation was published by Bui [3]. Here, the appropriate concept for the derivative of the singular integral of the inelastic term was discussed, and the three-dimensional integral expressions were presented. The author also indicates the existence of a free term. A year later, the complete boundary element formulation for two-dimensional plasticity problems was published by Telles and Brebbia [39]. The correct expressions for internal stresses were given, including the proper derivatives of the singular domain integrals. The initial stress approach, which is similar to the technique normally used in finite elements, was presented by Chaudonneret [6] and Cathie [4].

One of the main difficulties in the early application of the BEM to nonlinear problems such as plasticity was the availability of accurate integration techniques for higher order discretization. The principal difficulty was the accurate evaluation of the Cauchy principal value volume integrals when computing interior stresses. The evaluation of these integrals mostly relied on analytical (see [27]) or semi-analytical

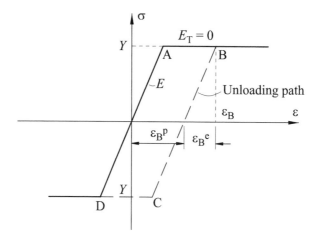

Figure 6.1: Elastic-perfectly plastic.

integration (see [40]). Unfortunately, these techniques are only applicable to plane problems using piecewise constant or, at most, linear cells, since higher order interpolations produce very complex integrals. The most general procedure is the direct evaluation of the Cauchy principal value integrals. This technique was first implemented for two-dimensional elastoplasticity problems by Leitão et al. [17, 18] and later by Cisilino and Aliabadi [7, 8] for three-dimensional problems. Other contributions to inelastic problems can be found in [1, 2, 3, 5, 6, 12, 15, 16, 38, 25, 34, 35].

In this chapter, a BEM implementation for two- and three-dimensional elastoplasticity is presented. It closely follows the advanced numerical implementations of Leitão et al. [19, 20, 21] and Cisilino and Aliabadi [9, 10, 11].

6.2   Basic Concepts

All materials possess an elastic limit, beyond which they start to yield. The elastic limit can be described in terms of yield stress, σ_Y (or abbreviated to Y); see Figures 6.2 and 6.1. Plastic deformations depend not only on the final stresses, but also on the stress history determined by the loading path. Due to the nonlinearity and irreversibility of plastic deformations, and also to other phenomena occurring only after the material becomes plastic such as viscosity, creep, hardening and softening, the physical relations describing the process are more complicated than those for purely elastic behaviour.

Consider a body subjected to uniaxial loading, where $\sigma_{ii} \neq 0$, $\sigma_{ij} = \sigma_{jj} = 0$, $j \neq i$ (here the summation is not implied). This stress state may be the result of the simple tension and compression test, used for estimation of the yield stress of the material. Two possible material behaviours are: elastic-perfectly plastic (no hardening), and elastoplastic (with hardening), as shown in Figures 6.1 and 6.2, respectively.

Figure 6.2 presents the hardening with the segment of positive slope E_T (with E representing the elastic behaviour) being the ability of the material to withstand a greater stress after plastic deformation occurs. Conversely, softening is said to exist when that slope is negative, which means that after the material has yielded, the

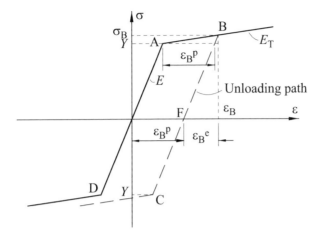

Figure 6.2: Elastoplastic behaviour (linear hardening).

plastic strain may increase even if the load is decreased. It is common to linearize the hardening behaviour, as shown in Figure 6.2, but if necessary more general curves can be considered for specific materials.

The material behaviour during loading and unloading is significantly different, as shown in Figures 6.1 and 6.2. For the elastic-perfectly plastic model, loading the sample past the yield point, Y, up to point B will cause some plastic (irrecoverable) strain, ε_B^p, together with the elastic strain the material has undergone to reach yielding, $\varepsilon_B^e = \varepsilon_A^e$. If the load is completely removed, the point F on the positive region of the strain axis will be reached, and the plastic strain is given by $\varepsilon_F^p = \varepsilon_B^p$, and the elastic strain is now equal to zero. The unloading is done in an elastic way, and the slope of the unloading segment is the Young's modulus. Reversing the load leads to a yield in compression being attained at point C (if the yield stress in compression is the same as in tension); at this point the elastic strain is again the yield strain. Point D is reached, for which the total strain is the yield strain.

For the case of linear hardening, as the material is loaded past the yield stress in tension and reaches point B in Figure 6.2, the plastic component of the strain is denoted by ε_B^p and the elastic part by ε_B^e. It can be seen from Figure 6.1 that ε_B^e is different to ε_A^e. If the material is subsequently unloaded completely, point F on the strain axis is reached; at this point, the elastic strain is again zero, and the plastic strain is $\varepsilon_F^p = \varepsilon_B^p$. Reloading at this stage will cause an increase in the plastic strain value only if the stress goes beyond its previous value, σ_B; this stress value is called the subsequent yield stress, and it changes as the plastic strain changes. Reversing the load will cause the material to yield in compression; at which point this yielding in compression occurs depends on the type of hardening behaviour. It is known from experiments that the compressive yield stress will vary depending on the previous deformation history (*Bauschinger effect*).

The most popular models used for the simulation of hardening are:

- **isotropic hardening**: both compressive and tensile yield stresses are equal, $\sigma_A = \sigma_D$. The same behaviour happens for subsequent yield points. The isotropic model neglects the Bauschinger effect;

- **kinematic hardening:** the elastic range is assumed not to change during hardening. This implies that the difference between yielding in tension and in compression is, at all stages, equal to the initial yield stress difference;

- **independent or mixed hardening:** is the more general rule. The material hardens independently in tension and in compression.

6.2.1 Elastoplastic Relationships

One of the simplest relationships between plastic strains and stresses occurs when a material with linear hardening behaviour is loaded uniaxially.

The total strain ε is expressed as a sum of the elastic and plastic components as follows:

$$\varepsilon = \varepsilon^e + \varepsilon^p \tag{6.1}$$

where $\varepsilon^e = \frac{\sigma}{E}$ is the elastic part of the strain tensor and ε^p represents the plastic part.

For initial loading, elastic behaviour is obtained for

$$\sigma - Y < 0 \tag{6.2}$$

Once the stress has exceeded the original yield stress Y, we have

$$\sigma - \sigma_o < 0 \tag{6.3}$$

where the value of σ_o depends on plastic flow rule. The relationship between stress and strain for monotonic loading in tension has the form

$$\sigma_o = Y + E\varepsilon^e \ \ \text{if } \sigma > Y \tag{6.4}$$

which may be written as

$$\sigma_o = Y + E_T\varepsilon \tag{6.5}$$

The strain decomposition in equation (6.1) allows for (6.5) to be expressed as

$$\sigma_o = Y + E_T(\varepsilon^e + \varepsilon^p) \tag{6.6}$$

Substituting equation (6.4) into equation (6.6) yields

$$\sigma_o = Y + E_T\varepsilon^p + E_T\frac{\sigma - Y}{E} \tag{6.7}$$

Finally, the following equation is obtained:

$$\sigma_o = Y + \frac{E_T}{1 - E_T/E}\varepsilon^p = Y + H'\varepsilon^p \tag{6.8}$$

where H' is the plastic modulus defined as the slope of the hardening curve now expressed in terms of stress versus plastic strain. For the case of linear hardening, H' is a constant, but in general, it changes continuously along the hardening curve, i.e.

$$H' = \frac{d\sigma}{d\varepsilon^p} \tag{6.9}$$

where $d\sigma$ is the stress increment that causes the plastic strain increment $d\varepsilon^p$.

In general, the yield stress is a function of a hardening parameter κ, which represents the total plastic work, that is

$$\kappa = W_p = \int \sigma d\varepsilon^p \tag{6.10}$$

Hence

$$\sigma_o(\kappa) = \sigma_o \int \sigma d\varepsilon^p \tag{6.11}$$

Plastic behaviour is possible if the following condition is satisfied:

$$f(\sigma, \kappa) = \sigma - \sigma_o(\kappa) = 0 \tag{6.12}$$

where $f(\sigma, \kappa)$ is a yield function subjected to the condition

$$f(\sigma, \kappa) \leq 0 \tag{6.13}$$

6.2.2 Multiaxial Loading

The uniaxial plastic behaviour is only possible if condition (6.12) is satisfied. The elastoplastic model described above cannot be used to represent the complex and general case of multiaxial loading. The generalization above models to the more complex multiaxial stress states, requires the definition of more general yield criteria. Some of these are:

- **Rankine**: assumes that yielding occurs when the maximum principal stress attains a value equal to the current yield stress, tensile or compressive;

- **Saint-Venant**: assumes yielding occurs when the maximum principal strain attains a value equal to the yield strain, both in simple tension or compression;

- **Tresca**: assumes the maximum shear stress attains a value equal to the maximum shear stress under simple tension;

- **von Mises**: assumes the maximum shear or distortion strain energy becomes equal to that at yield in simple tension or compression.

The maximum values referred to above correspond to the current yield stress values when hardening exists.

The choice of a suitable criterion depends upon the material properties. For example, the von Mises and Tresca criteria have been shown to give good agreement with experimental data when applied to metals. This is reasonable since, for these materials, only the shear stress determines the onset of yielding. It is thus convenient to separate the stress tensor into its spherical component as follows:

$$\overline{\sigma}_{ij} = \frac{1}{3}\sigma_{kk}\delta_{ij} \tag{6.14}$$

and its shear or deviatoric component as

$$S_{ij} = \sigma_{ij} - \overline{\sigma}_{ij} \tag{6.15}$$

which determines the plastic behaviour. Since only the deviatoric stress tensor contributes to plasticity, the yield function can be written in terms of the invariants J_2 and J_3 which are, respectively,

$$J_2 = \frac{1}{2} S_{ij} S_{ij} \tag{6.16}$$

and

$$J_3 = \frac{1}{3} S_{ij} S_{jk} S_{ki} \tag{6.17}$$

A yield function which describes the plastic behaviour of metals takes the following general form:

$$f(J_2, J_3, \kappa) = F(\sigma_{ij}) - \sigma_o(\kappa) = 0 \tag{6.18}$$

where the yield stress $\sigma_o(\kappa)$ depends on the stress and strain history of the material and its strain hardening properties. The function f is called a yield or loading surface. It defines a surface on the stress space with points lying on this surface representing possible stress states foe which yielding occurs. Expression (6.10) for the plastic work κ must now be substituted by

$$\kappa = W_p = \int \sigma_{ij} d\varepsilon_{ij}^p \tag{6.19}$$

The von Mises yield criterion does not depend, upon J_3, and can be therefore written as

$$f(J_2, \kappa) = \sqrt{3J_2} - \sigma_o(\kappa) = 0 \tag{6.20}$$

In order to describe the plastic deformation of a strain hardening material, three cases can be distinguished:

$$F(\sigma_{ij}) = \sigma_o(\kappa), \qquad dF = \frac{\partial F}{\partial \sigma_{ij}} d\sigma_{ij} > 0 \tag{6.21}$$

which constitutes *loading*,

$$F(\sigma_{ij}) = \sigma_o(\kappa), \qquad dF = \frac{\partial F}{\partial \sigma_{ij}} d\sigma_{ij} = 0 \tag{6.22}$$

which constitutes *neutral loading*, and

$$F(\sigma_{ij}) = \sigma_o(\kappa), \qquad dF = \frac{\partial F}{\partial \sigma_{ij}} d\sigma_{ij} < 0 \tag{6.23}$$

which constitutes *unloading*. For perfectly plastic materials, flow occurs for

$$F(\sigma_{ij}) = \sigma_o, \qquad dF = 0 \tag{6.24}$$

where it should be noticed that σ_o is a constant, and the case $dF > 0$ does not exist.

The von Mises yield surface represents a cylinder with an elliptical section, whose axis is the so-called 'hydrostatic' axis in which the principal stresses have the same value $\sigma_1 = \sigma_2 = \sigma_3$, as shown in Figure 6.3.

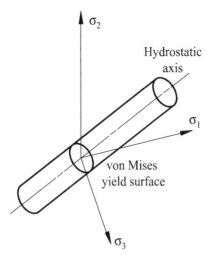

Figure 6.3: von Mises yield surface.

6.2.3 Plastic Stress-Strain Relationships

As described previously, plasticity is a path-dependent phenomenon; therefore, it becomes necessary to compute the differentials or increments of plastic strain throughout the loading history, and then obtain the accumulated strains by integration or summation. A suitable relationship for evaluation of the plastic strain increments is given by the Prandtl-Reuss equations

$$d\varepsilon_{ij}^p = S_{ij}d\lambda \qquad (6.25)$$

where $d\lambda$ is a proportionality factor which may vary throughout the loading history, but remians always positive.

We define an equivalent or effective stress and an equivalent or effective plastic strain increment as

$$\sigma_e = \sqrt{3J_2} \qquad (6.26)$$

and

$$d\varepsilon_e^p = \sqrt{\frac{2}{3}d\varepsilon_{ij}^p d\varepsilon_{ij}^p} \qquad (6.27)$$

Expressing the proportionality factor $d\lambda$ in terms of the above variables

$$d\varepsilon_{ij}^p d\varepsilon_{ij}^p = S_{ij}S_{ij}d\lambda^2 \qquad (6.28)$$

leads to

$$\frac{3}{2}(d\varepsilon_e^p)^2 = (\sigma_e d\lambda)^2 \qquad (6.29)$$

or

$$d\lambda = \frac{3}{2}\frac{d\varepsilon_e^p}{\sigma_e} \qquad (6.30)$$

The Prandtl-Reuss equations (6.25) can also be written in terms of the plastic strain increment and the total strain only. The total strain can be written as

$$\varepsilon_{ij} = \varepsilon_{ij}^e + \varepsilon_{ij}^p + \Delta\varepsilon_{ij}^p \qquad (6.31)$$

where ε^e_{ij} is the elastic component of the total strain, already including the current load increment, ε^p_{ij} is the plastic component up to but not including the current load increment, and $\Delta\varepsilon^p_{ij}$ is the increment of plastic strain due to the load increment.

Consider now a new variable, the modified total strain, as follows:

$$\varepsilon'_{ij} = \varepsilon_{ij} - \varepsilon^p_{ij} \tag{6.32}$$

or

$$\varepsilon'_{ij} = \varepsilon^e_{ij} + \Delta\varepsilon^p_{ij} \tag{6.33}$$

which using Hooke's law can be written as

$$\varepsilon'_{ij} = \frac{1}{2\mu}\left(\sigma_{ij} - \frac{\nu}{1+\nu}\sigma_{kk}\delta_{ij}\right) + \Delta\varepsilon^p_{ij} \tag{6.34}$$

Expression (6.34) can also be written in deviatoric form as (note that plastic strains are incompressible, i.e., $\Delta\varepsilon^p_{kk} = 0$)

$$e'_{ij} = \frac{S_{ij}}{2\mu} + \Delta\varepsilon^p_{ij} \tag{6.35}$$

in which

$$e'_{ij} = \varepsilon'_{ij} - \frac{\delta_{ij}}{3}\varepsilon'_{kk} \tag{6.36}$$

Recalling the Prandtl-Reuss equations given by (6.25), expression (6.35) yields

$$e'_{ij} = \left(1 + \frac{1}{2\mu\Delta\lambda}\right)\Delta\varepsilon^p_{ij} \tag{6.37}$$

An equivalent modified total strain can be defined as

$$\varepsilon'_e = \sqrt{\frac{2}{3}e'_{ij}e'_{ij}} \tag{6.38}$$

and an equivalent plastic strain increment as

$$\Delta\varepsilon^p_e = \sqrt{\frac{2}{3}\Delta\varepsilon^p_{ij}\Delta\varepsilon^p_{ij}} \tag{6.39}$$

A new expression for equation (6.37) in terms of the equivalent quantities introduced above can be obtained as

$$1 + \frac{1}{2\mu\Delta\lambda} = \frac{\varepsilon'_e}{\Delta\varepsilon^p_e} \tag{6.40}$$

Substitution of (6.40) into (6.37) gives

$$\Delta\varepsilon^p_{ij} = \frac{\Delta\varepsilon^p_e}{\varepsilon'_e}e'_{ij} \tag{6.41}$$

which are the Prandtl-Reuss equations written in terms of strains only.

From equation (6.41) it can be seen that in order to determine the actual magnitude of the plastic strain increments $\Delta\varepsilon^p_{ij}$, the equivalent plastic strain increment must be known. Therefore, substituting the proportionality factor $\Delta\lambda$ given in equation (6.30) into (6.40) results in

$$1 + \frac{\sigma_e}{3\mu\Delta\varepsilon_e^p} = \frac{\varepsilon_e'}{\Delta\varepsilon_e^p} \tag{6.42}$$

which, if solved for the equivalent plastic strain increment, gives

$$\Delta\varepsilon_e^p = \varepsilon_e' - \frac{1}{3\mu}\sigma_e \tag{6.43}$$

Since the condition expressed in expression (6.26) must be satisfied throughout the plastic process, σ_o can be substituted for σ_e in equation (6.43)

$$\Delta\varepsilon_e^p = \varepsilon_e' - \frac{1}{3\mu}\sigma_o \tag{6.44}$$

where σ_0 corresponds to the uniaxial yield stress after application of the current load increment; consequently it is still unknown. This term, however, can by approximated by a Taylor series about the preceding value of σ_o, as

$$\sigma_{o,i} = \sigma_{o,i-1} + \left(\frac{d\sigma_o}{d\sigma_e^p}\right)_{i-1}\Delta\varepsilon_e^p + \ldots \tag{6.45}$$

The basic elastoplastic relationship between total strain and the increment of plastic strain is obtained from (6.45) as

$$\Delta\varepsilon_e^p = \frac{3\mu\varepsilon_e' - \sigma_{o,i-1}}{3\mu + H_{i-1}'} \tag{6.46}$$

where $H_{i-1}' = \left(\frac{d\sigma_o}{d\sigma_e^p}\right)_{i-1}$ is the plastic modulus computed before the load increment.

6.3 The Governing Elastoplastic Equations

The basic differential equations for continuum elastoplastic problems are introduced in this section. The rate for the equation is presented, in order to maintain a general notation with respect to other time-dependent inelastic phenomena, such as viscoplasticity and creep. Classical plasticity is a time-independent phenomenon, and pure incremental quantities could be used instead. For example, the stress for the load step k can be represented by

$$\dot{\sigma}_{ij}^k = \dot{\sigma}_{ij}^{k-1} + \Delta\sigma_{ij} \tag{6.47}$$

The superscript k is normally omitted for the sake of simplifying the notation.

The rate form of the equilibrium equations presented in Chapter 2 can be written as

$$\dot{\sigma}_{ij,i} + \dot{b}_j = 0 \tag{6.48}$$

where $\dot{b}_j$ are body force rates.

Under the assumption of small strains, the relationship between strains and displacements can be written in terms of rate quantities as follows:

$$\dot{\varepsilon}_{ij} = \frac{1}{2}(\dot{u}_{i,j} + \dot{u}_{j,i}) \tag{6.49}$$

Recalling the decomposition of the strain tensor into elastic and inelastic components, the following identity is obtained

$$\dot{\varepsilon}_{ij} = \frac{1}{2}(\dot{u}_{i,j} + \dot{u}_{j,i}) = \dot{\varepsilon}^e_{ij} + \dot{\varepsilon}^p_{ij} \tag{6.50}$$

The inelastic part of the strain rate can include any kind of inelastic strain such as plastic, creep, thermal or others.

The application of the generalized Hooke's law to the elastic component of the strain tensor $\dot{\varepsilon}^e_{ij}$, and the consideration of the plastic component $\dot{\varepsilon}^p_{ij}$ as an initial strain, allows for the relationship between the stress and the strain rates to be expressed by

$$\dot{\sigma}_{ij} = 2\mu(\dot{\varepsilon}_{ij} - \dot{\varepsilon}^p_{ij}) + \frac{2\mu\nu}{1 - 2\nu}(\dot{\varepsilon}_{ll} - \dot{e})\delta_{ij} \tag{6.51}$$

where $\dot{e} = \dot{\varepsilon}^p_{kk}$, i.e. the inelastic dilatational strain rate.

Equation (6.51) may be rewritten for initial stresses,

$$\dot{\sigma}_{ij} = 2\mu\dot{\varepsilon}_{ij} + \frac{2\mu\nu}{1 - 2\nu}\dot{\varepsilon}_{kk}\delta_{ij} - \dot{\sigma}^p_{ij} \tag{6.52}$$

where $\dot{\sigma}^p_{ij}$ stands for the components of the initial stress given by

$$\dot{\sigma}^p_{ij} = 2\mu\dot{\varepsilon}^p_{ij} + \frac{2\mu\nu}{1 - 2\nu}\dot{e}\delta_{ij} \tag{6.53}$$

Recalling the plastic strain rates

$$\Delta\varepsilon^p_e = \frac{3\mu\varepsilon_e - \sigma_o}{3\mu + H'} \tag{6.54}$$

The elastoplastic flow rule may be written in different ways, for example, expressing the relationship between the increments of plastic strain and increments of stress.

The governing Navier rate differential equations of the problem are obtained in a similar way to the procedure described in Chapter 2, but now using the rate form of the equations instead, we have

$$\dot{u}_{j,ll} + \frac{1}{1 - 2\nu}\dot{u}_{l,lj} = 2(\dot{\varepsilon}^p_{ij,i} + \frac{\nu}{1 - 2\nu}\dot{e}_{,j}) - \frac{\dot{b}_j}{\mu} \tag{6.55}$$

The equation representing the traction boundary conditions of the problem can also be obtained in rate form,

$$\dot{t}_i + 2\mu(\dot{\varepsilon}^p_{ij}n_j + \frac{\nu}{1 - 2\nu}\dot{e}n_i) = \frac{2\mu\nu}{1 - 2\nu}\dot{u}_{l,l}n_i + \mu(\dot{u}_{i,j} + \dot{u}_{j,i})n_j \tag{6.56}$$

Notice that the above equations are valid for three-dimensional problems. For planar problems the indices now vary from 1 to 2, and $\dot{\varepsilon}^p_{33} = 0$ for the plane stress state.

The resulting governing equations are not linear, and so a direct solution of the problem is no longer possible. To solve the nonlinear system of equations thus obtained, the method of successive elastic solutions [28] is used; the method takes its name because each iteration essentially involves the solution of an elastic problem. This can be summarized as: starting from any position on the stress-strain curve for which some plastic as well as elastic strain exists, the next load increment to be applied to the body will follow an elastic path. Since the phenomenon is non-linear, errors will arise between that *elastic* and the *real* position obtained by using the real constitutive relationship. This error must be eliminated or diminished, and several procedures to accomplish this will be discussed later in this chapter. For the moment, what is important is that a succession of elastic solutions is used to solve a non-linear elastoplastic problem. The inelastic components of each variable are accounted for as an initial term corresponding to the previous accumulated load state. The way in which these initial terms are considered defines the approach being used. Two commonly used approaches are the initial strain and the initial stress approaches. The initial strain formulation is described in detail next, and other procedures are described later in the chapter.

6.4   Initial Strain Boundary Integral Formulation

The Navier rate equations (6.56) together with Betti's reciprocal theorem are used in this section to derive the displacement boundary integral equations for elastoplasticity. The procedure is similar to that employed in Chapter 2, where the boundary integral equation for elasticity was presented.

Consider two domains, V and V^*, as defined in section 2.3. In order to obtain the boundary integral representation of the problem, Betti's theorem may be used. Due to the symmetry of Hooke's law, the theorem proves the validity of the following identity:

$$\int_V \dot{\sigma}_{ij}^*(\mathbf{X})\dot{\varepsilon}_{ij}^e(\mathbf{X})dV = \int_V \dot{\sigma}_{ij}(\mathbf{X})\dot{\varepsilon}_{ij}^*(\mathbf{X})dV \qquad (6.57)$$

where the starred and unstarred fields are independent fields that satisfy equilibrium, compatibility and constitutive relationships.

Expressing the elastic component of the strain rate tensor in terms of the total plastic strain rates, equation (6.57) can be written as

$$\int_V \dot{\sigma}_{ij}^*(\mathbf{X})[\dot{\varepsilon}_{ij}(\mathbf{X}) - \dot{\varepsilon}_{ij}^p(\mathbf{X})]dV = \int_V \dot{\sigma}_{ij}(\mathbf{X})\dot{\varepsilon}_{ij}^*(\mathbf{X})dV \qquad (6.58)$$

By virtue of repeated subscripts, i and j, the following relationship holds:

$$\dot{\sigma}_{ij}^*\dot{u}_{j,i} = \dot{\sigma}_{ji}^*\dot{u}_{i,j} \qquad (6.59)$$

Hence, because of the symmetry of the stress tensor and the geometrical linearity, the following identity holds:

$$\dot{\sigma}_{ij}^*\dot{\varepsilon}_{ij} = \dot{\sigma}_{ij}^*\frac{1}{2}(\dot{u}_{i,j} + \dot{u}_{j,i}) = \frac{1}{2}(\dot{\sigma}_{ij}^* + \dot{\sigma}_{ji}^*)\dot{u}_{i,j} = \dot{\sigma}_{ij}^*\dot{u}_{i,j} \qquad (6.60)$$

Using this result, equation (6.58) can be rewritten as follows:

$$\int_V \dot{\sigma}_{ij}^*(\mathbf{X})\dot{u}_{i,j}(\mathbf{X})dV - \int_V \dot{\sigma}_{ij}^*(\mathbf{X})\dot{\varepsilon}_{ij}^p(\mathbf{X})dV = \int_V \dot{\sigma}_{ij}(\mathbf{X})\dot{u}_{i,j}^*(\mathbf{X})dV \qquad (6.61)$$

Integrating by parts those terms not containing the initial strains leads to

$$\int_V \dot{\sigma}_{ij,j}^*(\mathbf{X})\dot{u}_i(\mathbf{X})dV + \int_S \dot{\sigma}_{ij}^*(\mathbf{x})n_j(\mathbf{x})\dot{u}_i(\mathbf{x})dS$$

$$-\int_V \dot{\sigma}_{ij}^*(\mathbf{X})\dot{\varepsilon}_{ij}^p(\mathbf{X})dV = \int_V \dot{\sigma}_{ij,j}(\mathbf{X})\dot{u}_i^*(\mathbf{X})dV$$

$$+\int_S \dot{\sigma}_{ij}(\mathbf{x})n_j(\mathbf{x})\dot{u}_i^*(\mathbf{x})dS \tag{6.62}$$

where $\mathbf{x} \in S$. Upon substitution of the equilibrium equations (6.48) and the traction definition ($\dot{t}_i = \dot{\sigma}_{ij}n_j$), and in the absence of body forces, equation (6.62) yields

$$\int_V \dot{b}_i^*(\mathbf{X})\dot{u}_i(\mathbf{X})dV + \int_S \dot{t}_i^*(\mathbf{x})\dot{u}_i(\mathbf{x})dS - \int_V \dot{\sigma}_{ij}^*(\mathbf{X})\dot{\varepsilon}_{ij}^p(\mathbf{X})dV$$

$$= \int_S \dot{t}_i(\mathbf{x})\dot{u}_i^*(\mathbf{x})dS \tag{6.63}$$

As in Section 2.3, the state of displacement u^*, traction t^*, and stress σ^* are chosen to correspond to a known solution of the governing rate Navier's equation (6.56). The displacement, traction and stress fundamental fields, representing the displacement, traction and stress at point $\mathbf{X}$ along direction j when a unit point load is applied at $\mathbf{X}'$ in direction i, can be expressed as

$$\dot{u}_i^*(\mathbf{X}) = \dot{u}_j^*(\mathbf{X})\delta_{ij} = U_{ij}(\mathbf{X}',\mathbf{X})\delta_{ij}e_i$$

$$\dot{t}_i^*(\mathbf{X}) = \dot{t}_j^*(\mathbf{X})\delta_{ij} = T_{ij}(\mathbf{X}',\mathbf{X})\delta_{ij}e_i$$

and

$$\dot{\sigma}_{jk}^*(\mathbf{X}) = \sigma_{ijk}(\mathbf{X}',\mathbf{X})e_i \tag{6.64}$$

The expressions for the fundamental displacement and traction fields $u_{ij}^*(\mathbf{X}',\mathbf{X})$ and $t_{ij}^*(\mathbf{X}',\mathbf{X})$ are the same to those given in Section 2.3. The fundamental stress field $\sigma_{ijk}(\mathbf{X}',\mathbf{X})$ for 3D is given as

$$\sigma_{ijk}(\mathbf{X}',\mathbf{X}) = \frac{-1}{8\pi(1-\nu)r^2}\{(1-2\nu)(r_{,j}\delta_{ik} + r_{,i}\delta_{jk} - r_{,k}\delta_{ij}) + 3r_{,i}r_{,j}r_{,k}\} \tag{6.65}$$

For the plane stress case,

$$\sigma_{ijk}(\mathbf{X}',\mathbf{X}) = \frac{-1}{4\pi(1-\nu')r}\{(1-2\nu')(r_{,j}\delta_{ki} + r_{,i}\delta_{jk} - r_{,k}\delta_{ij}) + 2r_{,i}r_{,j}r_{,k}\} \tag{6.66}$$

where r_i are the components of the distance $r = \sqrt{r_i r_i}$ between the source point and the field point, and n_j is the component of the outward normal.

For the plane strain case, σ_{ijk} is given by

$$\hat{\sigma}_{ijk}(\mathbf{X}',\mathbf{X}) = \sigma_{ijk}(\mathbf{X}',\mathbf{X}) + \frac{2\nu\delta_{ij}r_{,k}}{4\pi(1-\nu)r} \tag{6.67}$$

with ν' in σ_{ijk} being replaced by ν.

Substituting the fundamental fields given above into equation (6.63), the rate version of the Somigliana equation is obtained:

$$\dot{u}_i(\mathbf{X}') = \int_S U_{ij}(\mathbf{X}',\mathbf{x})\dot{t}_j(\mathbf{x})dS - \int_S T_{ij}(\mathbf{X}',\mathbf{x})\dot{u}_j(\mathbf{x})dS$$

$$\int_V \sigma_{ijk}(\mathbf{X}',\mathbf{X})\dot{\varepsilon}_{jk}^p(\mathbf{X})dV \tag{6.68}$$

Equation (6.68) is the non-linear counterpart of equation (2.65), and gives the displacements $\dot{u}_i(\mathbf{X}')$ at any point $\mathbf{X}'$ of the domain V, when the value of the boundary displacements and tractions as well as the plastic deformations $\dot{\varepsilon}^p_{jk}(\mathbf{x})$ within the domain are known.

The direct boundary element formulation relating the boundary displacements to the boundary tractions and the domain plastic strains can be obtained from equation (6.68) by considering the limiting process as an internal point goes to the boundary, that is as $\mathbf{X}' \to \mathbf{x}'$ (see Section 2.3.4). This leads to the following boundary integral representation of the boundary displacements when the initial strain approach for the solution of elastoplastic problems is used:

$$C_{ij}(\mathbf{x}')\dot{u}_j(\mathbf{x}') + \fint_S T_{ij}(\mathbf{x}',\mathbf{x})\dot{u}_j(\mathbf{x})dS = \int_S U_{ij}(\mathbf{x}',\mathbf{x})\dot{t}_j(\mathbf{x})dS$$

$$+ \int_V \sigma_{ijk}(\mathbf{X}',\mathbf{X})\dot{\varepsilon}^p_{jk}(\mathbf{X})dV \tag{6.69}$$

where $\fint$ stands for the Cauchy principal value integral, and the term $C_{ij}(\mathbf{x}')$ is the same as that described in Chapter 2.

6.4.1   Internal Stresses

The boundary integral representation of stresses at internal points is the result of applying the generalized Hooke's law to the elastic part of the total strain rate tensor. This is represented by the following equation:

$$\dot{\sigma}_{ij} = \mu\left(\frac{\partial \dot{u}_i}{\partial x_j} + \frac{\partial \dot{u}_j}{\partial x_i}\right) + \frac{2\mu\nu}{1-2\nu}\frac{\partial \dot{u}_k}{\partial x_k}\delta_{ij} - 2\mu\dot{\varepsilon}^p_{ij} + \frac{2\mu\nu}{1-2\nu}\dot{e}\delta_{ij} \tag{6.70}$$

which can be computed by substituting equation (6.68) into equation (6.70) on condition that the space derivatives be taken with respect to the coordinates of the collocation point. This takes the form

$$\frac{\partial \dot{u}_i}{\partial x_m}(\mathbf{X}') = \int_S \frac{\partial U_{ij}(\mathbf{X}',\mathbf{x})}{\partial x_m}\dot{t}_j dS - \int_S \frac{\partial T_{ij}(\mathbf{X}',\mathbf{x})}{\partial x_m}\dot{u}_j dS + \frac{\partial}{\partial x_m}\int_V \sigma^*_{ijk}(\mathbf{X}',\mathbf{X})\dot{\varepsilon}^p_{jk}(\mathbf{X})dV \tag{6.71}$$

where the differentiation can be applied directly to the fundamental solution tensors for the first two integrals. However, the last one must be carried out using the Leibnitz[1] formula, so that

$$\frac{\partial}{\partial x_m}\int_V \sigma^*_{ijk}(\mathbf{X}',\mathbf{X})\dot{\varepsilon}^p_{jk}(\mathbf{X}')dV = \int_V \frac{\partial \sigma^*_{ijk}(\mathbf{X}',\mathbf{X})}{\partial x_m}\dot{\varepsilon}^p_{jk}(\mathbf{X})dV - \dot{\varepsilon}^p_{jk}\int_{S'} \sigma^*_{ijk}(\mathbf{X}',\mathbf{x})r_{,m}dS \tag{6.72}$$

The singularity of $O(r^{d-1})$ ($d = 2$ for 2D and $d = 3$ for 3D) in the term σ^*_{ijk} requires the derivative of the domain integral in equation (6.71) to be considered in the Cauchy principal value sense, and that is the reason why the region denoted by S', corresponding to a unit sphere around the source point, has to be considered.

---

[1] The Leibnitz formula states that

$$\frac{d}{d\alpha}\int_{\varphi_1(\alpha)}^{\varphi_2(\alpha)} F(x,\alpha)dx = \int_{\varphi_1(\alpha)}^{\varphi_2(\alpha)} \frac{F(x,\alpha)}{d\alpha}dx - F(\varphi_1(\alpha),\alpha)\frac{d\varphi_1(\alpha)}{d\alpha} + F(\varphi_2(\alpha),\alpha)\frac{d\varphi_2(\alpha)}{d\alpha}$$

By collecting all the terms in equation (6.70), the boundary integral representation of the internal stresses can be expressed as

$$\dot{\sigma}_{ij}(\mathbf{X}') = \int_S D_{ijk}(\mathbf{X}',\mathbf{x})\dot{t}_k(\mathbf{x})dS - \int_S S_{ijk}(\mathbf{X}',\mathbf{x})\dot{u}_k(\mathbf{x})dS$$

$$+ \int_V \Sigma_{ijkl}(\mathbf{X}',\mathbf{X})\dot{\varepsilon}^p_{kl}(\mathbf{X})dV + f_{ij}(\dot{\varepsilon}^p_{kl}) \tag{6.73}$$

where D_{ijk} and S_{ijk} are, respectively, the generalized displacement and traction at a point $\mathbf{x}$ along the direction defined by k when an appropriate load is applied at the load point $\mathbf{x}'$. The term $f_{ij}(\dot{\varepsilon}^p_{kl})$ arises from the integration of the stress kernel Σ^*_{ijkl} over the surface S' centred at the load point $\mathbf{x}'$.

The expressions for D_{ijk} and S_{ijk} were given in Section 2.3.3; while Σ_{ijkl} and $f_{ij}(\dot{\varepsilon}^p_{kl})$ for a three-dimensional domain are given by

$$\Sigma_{ijkl}(\mathbf{X}',\mathbf{X}) = \frac{\mu}{4\pi(1-\nu)r^3}\{3(1-2\nu)(\delta_{ij}r_{,k}r_{,l} + \delta_{kl}r_{,i}r_{,j})$$

$$+3\nu(\delta_{li}r_{,j}r_{,k} + \delta_{jk}r_{,l}r_{,i} + \delta_{ik}r_{,l}r_{,j} + \delta_{jl}r_{,i}r_{,k}) - 15r_{,i}r_{,j}r_{,k}r_{,l}$$

$$+(1-2\nu)(\delta_{ik}\delta_{lj} + \delta_{jk}\delta_{li}) - (1-4\nu)\delta_{ij}\delta_{kl}\} \tag{6.74}$$

and the free term is given by

$$f_{ij}(\dot{\varepsilon}^p_{kl}) = -\frac{2\mu}{15(1-\nu)}\left[(7-5\nu)\dot{\varepsilon}^p_{ij} + (1+5\nu)\dot{\varepsilon}^p_{ll}\delta_{ij}\right] \tag{6.75}$$

The Σ_{ijkl} kernel for the plane stress case is given by

$$\Sigma_{ijkl}(\mathbf{X}',\mathbf{X}) = \frac{\mu}{2\pi(1-\nu')r^2}\{2(1-2\nu')\delta_{ij}r_{,k}r_{,l} + \delta_{kl}r_{,i}r_{,j})$$

$$+2\nu'(\delta_{li}r_{,j}r_{,k} + \delta_{jk}r_{,l}r_{,i} + \delta_{ik}r_{,l}r_{,j} + \delta_{jl}r_{,i}r_{,k}) - 8r_{,i}r_{,j}r_{,k}r_{,l}$$

$$+(1-2\nu')(\delta_{ik}\delta_{lj} + \delta_{jk}\delta_{li}) - (1-4\nu')\delta_{ij}\delta_{kl}\} \tag{6.76}$$

and the free term is given by

$$f_{ij}(\dot{\varepsilon}^p_{kl}) = -\frac{\mu}{4(1-\nu')}(2\dot{\varepsilon}^p_{ij} + \dot{\varepsilon}^p_{ll}\delta_{ij}) \tag{6.77}$$

For the plane strain case, Σ_{ijkl} is replaced by $\hat{\Sigma}_{ijkl}$ with

$$\hat{\Sigma}_{ijkl} = \Sigma_{ijkl} + \frac{\mu}{2\pi(1-\nu)r^2}(4\nu r_{,i}r_{,j}\delta_{kl} - 2\nu\delta_{ij}\delta_{kl}) \tag{6.78}$$

with ν' in Σ_{ijkl} being replaced by ν. The free term takes the form

$$f_{ij} = -\frac{\mu}{4\pi(1-\nu)}[2\dot{\varepsilon}^p_{ij} + (1-4\nu)\dot{\varepsilon}^p_{ll}\delta_{ij}. \tag{6.79}$$

6.4.2 Stresses at Boundary Points

The boundary stresses can be evaluated using the boundary tractions and tangential strains in a manner similar to elastostatics:

$$\hat{\sigma}_{11} = \frac{1}{1-\nu}(2\mu\hat{\varepsilon}_{11} + \nu\hat{\sigma}_{22}) + 2\mu\left(\frac{\nu}{1-\nu}\hat{\varepsilon}_{22}^p - \hat{\varepsilon}_{11}^p\right)$$

$$\hat{\sigma}_{12} = \hat{t}_1$$

$$\hat{\sigma}_{22} = \hat{t}_2 \tag{6.80}$$

for plane strain, and

$$\hat{\sigma}_{11} = \frac{1}{1-\nu}(2\mu\hat{\varepsilon}_{11} + \nu\hat{\sigma}_{22}) - 2\mu\frac{\nu}{1-\nu}\hat{\varepsilon}_{11}^p$$

$$\hat{\sigma}_{12} = \hat{t}_1$$

$$\hat{\sigma}_{22} = \hat{t}_2 \tag{6.81}$$

for plane stress, where the ^above the symbol denotes the local coordinate system (see Figure 2.22).

Similarly, for three-dimensional problems, we have

$$\hat{\sigma}_{3i} = \hat{t}_i \qquad i = 1, 2, 3$$

$$\hat{\sigma}_{11} = 2\mu(\hat{\varepsilon}_{11} - \hat{\varepsilon}_{11}^p) + \frac{2\mu}{1-2\nu}\hat{\varepsilon}_{ll}$$

$$\hat{\sigma}_{22} = 2\mu(\hat{\varepsilon}_{22} - \hat{\varepsilon}_{22}^p) + \frac{2\mu}{1-2\nu}\hat{\varepsilon}_{ll}$$

$$\hat{\sigma}_{12} = 2\mu(\hat{\varepsilon}_{12} - \hat{\varepsilon}_{12}^p) \tag{6.82}$$

where

$$\hat{\varepsilon}_{33} = \left(\frac{1-2\nu}{1-\nu}\right)\left[\frac{\hat{\sigma}_{33}}{2\mu} + \hat{\varepsilon}_{33}^p - \frac{\nu}{1-2\nu}(\hat{\varepsilon}_{11} + \hat{\varepsilon}_{22})\right] \tag{6.83}$$

Alternatively, the boundary stresses may be evaluated using the boundary integral representation of stresses [8, 19]. To obtain the integral representation, it is necessary to take the limit of equation (6.72), as the source point $\mathbf{x}'$ tends to the boundary S. The procedures for the boundary integrals in (6.73) have already been illustrated in Chapter 2. It can be shown [19] that when the point is on the boundary, the differentiation of the domain integral gives rise to a contribution from the second integral on the right-hand side of (6.72) that is half the integral over a unit circle,

$$\dot{\varepsilon}_{lk}^p \int_0^\pi \sigma_{jki}^* r_{,m} dS = \frac{1}{2}\dot{\varepsilon}_{lk}^p \int_0^{2\pi} \sigma_{jki}^* r_{,m} dS$$

Taking the limit of (6.72) as the point approaches the boundary and applying Hooke's law gives

$$\frac{1}{2}\dot{\sigma}_{ij}(\mathbf{x}') = \int_S D_{ijk}(\mathbf{x}',\mathbf{x})\dot{t}_k(\mathbf{x})dS - \oint_S S_{ijk}(\mathbf{x}',\mathbf{x})\dot{u}_k(\mathbf{x})dS$$

$$+ \int_V \Sigma_{ijkl}(\mathbf{x}',\mathbf{X})\dot{\varepsilon}_{kl}^p(\mathbf{X})dV + \frac{1}{2}f_{ij}(\dot{\varepsilon}_{kl}^p(\mathbf{x}')) \tag{6.84}$$

6.4.3   Alternative Approaches

Two alternative approaches to the initial strain formulation presented above, namely the initial stress and the pseudo body force formulations, are available in the context of elastoplastic BEM formulations.

Initial Stress Approach

The initial stress approach is similar to the initial strain approach, differing only for the primary unknown in the domain integrals. In general, it is only necessary to replace the domain integral in the initial strain formulation

$$\int_V \sigma^\star_{ijk}\dot{\varepsilon}^p_{jk}dV$$

by the equivalent quantity

$$\int_V \varepsilon^\star_{ijk}\dot{\sigma}^p_{jk}dV$$

The choice between the initial strain and initial stress is not significant, as it does not determine the path followed by the iterative process, since reciprocity is verified throughout the loading path, that is

$$\int_V \sigma^\star_{ijk}\dot{\varepsilon}^p_{jk}dV = \int_V \varepsilon^\star_{ijk}\dot{\sigma}^p_{jk}dV$$

Initial strain has been shown to work well when applied to metals [40].

Pseudo Body Force Approach

The pseudo body force approach leads to the simplest algebraic equations, although not necessarily the most efficient or easiest to deal with. This approach treats the terms involving the derivatives of the plastic strain rate as a type of body forces, allowing the Navier equation to be expressed in a form which bears a close resemblance to the Navier equation for elastostatics [40], i.e.

$$C_{ij}(\mathbf{x}')\tilde{u}_j(\mathbf{x}') + \int_S T_{ij}(\mathbf{x}',\mathbf{x})\tilde{u}_j(\mathbf{x})dS = \int_S U_{ij}(\mathbf{x}',\mathbf{x})\tilde{t}_j(\mathbf{x})dS + \int_V U_{ij}(\mathbf{x}',\mathbf{X})\tilde{b}_j(\mathbf{X})dV \tag{6.85}$$

with

$$\tilde{b}_j = \dot{b}_j - 2\mu\left(\dot{\varepsilon}^p_{ij,i} + \frac{\nu}{1-2\nu}\dot{e}_{,j}\right) = \dot{b}_j - \dot{\sigma}^p_{ij,i}$$

and

$$\tilde{t}_i = \dot{t}_i - 2\mu\left(\dot{\varepsilon}^p_{ij}n_j + \frac{\nu}{1-2\nu}\dot{e}n_i\right) = \dot{t}_i - \dot{\sigma}^p_{ij}n_j$$

The use of pseudo traction rates instead of traction rates further complicates the numerical formulation because the boundary conditions for the physical problem are defined in terms of displacement and traction rates. So, the advantage of having a close resemblance to the Navier equation of elastostatics may be overridden by the more complex boundary conditions.

6.5   Numerical Discretization

In order to solve the integral equations presented in the previous sections, the boundary S and the part of the domain V that is likely to yield must be discretized. This is an advantage of the BEM over other domain methods, such as the Finite Element Method, for which the complete domain has to be discretized. In what follows, the characteristics of the boundary and domain discretizations are discussed.

For the domain, only those regions susceptible to yielding, such as stress concentration sites or areas of higher stresses, are discretized with cells. The cells for two-dimensional boundary elements can be triangular or quadrilateral, similar to surface elements used for 3D problems (see Chapter 2). Three-dimensional cells are used to discretize the volume for 3D problems. Shape functions for 27-node brick elements are given in Appendix F.

The domain, V_Y, is divided in M_c cells:

$$V_Y = \bigcup_{m=1}^{M_c} V_m$$

The strain and stress rate tensors are given, locally at the cell V_m, by

$$\dot{\varepsilon}_{ij}^p(\mathbf{x}) = \sum_{\alpha=1}^{m_c} \Phi_\alpha(\mathbf{x})\dot{\varepsilon}_{ij}^{p,\alpha} \tag{6.86}$$

and

$$\dot{\sigma}_{ij}^p(\mathbf{x}) = \sum_{\alpha=1}^{m_c} \Phi_\alpha(\mathbf{x})\dot{\sigma}_{ij}^{p,\alpha} \tag{6.87}$$

where m_c is the number of nodes in the cell and Φ_α are the shape functions.

The discretized rate boundary integral equation can be written as

$$\mathbf{C}(\mathbf{x}')\dot{\mathbf{u}}(\mathbf{x}') + \sum_{n=1}^{N_e}\left(\int_{S_n}\mathbf{T}\mathbf{N}dS\right)\dot{\mathbf{u}}^n$$

$$= \sum_{n=1}^{N_e}\left(\int_{S_n}\mathbf{U}\mathbf{N}dS\right)\dot{\mathbf{t}}^n + \sum_{m=1}^{M_c}\left(\int_{V_m}\boldsymbol{\sigma}\boldsymbol{\Phi}dS\right)\dot{\boldsymbol{\varepsilon}}^{p,m} \tag{6.88}$$

Similarly, the discretized expression for the domain stresses can be obtained

$$\dot{\boldsymbol{\sigma}}(\mathbf{x}') = \sum_{n=1}^{N_e}\left(\int_{S_n}\mathbf{S}\mathbf{N}dS\right)\dot{\mathbf{u}}^n$$

$$- \sum_{n=1}^{N_e}\left(\int_{S_n}\mathbf{D}\mathbf{N}^TdS\right)\dot{\mathbf{t}}^n + \sum_{m=1}^{M_c}\left(\int_{V_m}\boldsymbol{\Sigma}\boldsymbol{\Phi}dS\right)\dot{\boldsymbol{\varepsilon}}^{p,m} + \mathbf{C}'(\mathbf{x}')\dot{\boldsymbol{\varepsilon}}^p(\mathbf{x}') \tag{6.89}$$

where N_e and M_c are the number of elements and cells, respectively, used for the model discretization; the quantities $\mathbf{U}, \mathbf{T}, \boldsymbol{\sigma}, \mathbf{D}, \mathbf{S}, \boldsymbol{\Sigma}$ are submatrices containing the fundamental solutions. Finally, $\mathbf{N}$ and $\boldsymbol{\Phi}$ are vectors containing the boundary element and cell shape functions respectively.

6.5.1   Treatment of the Integrals

Techniques for the treatment of boundary integrals are the same as in elasticity, and are described in Chapter 11. Elastoplastic BEM problems also require the evaluation of singular boundary and domain integrals. Next, integration techniques required for the evaluation of integrals in elastoplastic BEM are discussed.

Boundary Integrals

Boundary integrals have to be evaluated for both the rate boundary integral equation (6.88) and the integral representation of the internal stress (6.89). Integrals arising in equation (6.88) are the same as those in elasticity. Their evaluation is carried out here using the same techniques as in Chapter 11.

Boundary integrals arising in equation (6.89) present kernels with strong and hypersingular characteristics (kernels D_{ijk} and S_{ijk}, respectively). However, since equation (6.89) is not used to compute boundary stresses, the collocation point never lies on the integration element, and the integrals always present a near singular behaviour. Therefore, they can be evaluated using the standard Gauss quadrature with an element subdivision technique (see Section 11.2.1).

Volume Integration

Different types of integrals can be identified according to the nature of the kernel and the relative position of the collocation point with respect to the integration domain:

- *Nearly singular:* this is the case for all kernels when the collocation node does not lie on the integration cell. These integrals have integrands which vary sharply as the source point approaches the integration cell. Near singular integrals are evaluated using standard Gauss quadrature formulae with a cell subdivision technique. Cells are subdivided according to their relative distance to the collocation node (see Section 11.2.1).

- *Weakly singular:* this term is used to define integrals which present a singular kernel of order $1/r$ and $1/r^2$ for 2D and 3D problems, respectively. This is the case of the stress kernel σ_{ijk} given in equation (6.65) and arising in equation (6.89). These kernels have their singularity cancelled by employing a transformation of variable technique analogous to that used for boundary integration. For two-dimensional cells different techniques are described in Section 11.3.2. Leitão *et al.* [19] adopted the transformation of variable technique. For three-dimensional cells, analogous techniques can be used. Cisilino and Aliabadi [9] developed a technique based on the transformation of variable technique. In this method, cells are divided into pyramid shaped sub-cells, with their vertices at the collocation node and the cell faces as the basis. Then, each pyramid is mapped into cuboids, resulting in a Jacobian of transformation proportional to r^2, which exactly cancels out the $1/r^2$ singularity.

- *Strongly singular:* this term is used to define integrals involving the kernel Σ_{ijkl} (equation (6.74)) arising in equation (6.89), which present a singularity or order $1/r^2$ and $1/r^3$ for 2D and 3D problems, respectively. For two-dimensional cells, the techniques to deal with strongly singular integrals are similar to those described in section 11.4.3. For three-dimensional cells, a similar method was developed by Cisilino and Aliabadi [7].

Strongly Singular Integrals Strongly singular integrals of $O(1/r^3)$ arising from integrals containing the Σ_{ijkl} kernel are integrated using a singularity subtraction technique, similar to that outlined for hypersingular boundary integral evaluation (see Chapter 11). The technique was implemented in two- and three-dimensional elastoplastic BEM by [17] and [8].

The original form of the integral is after using the limit notation:

$$I = \int_V \Sigma^*_{ijkl}\dot{\varepsilon}^p_{kl}dV = \lim_{s\to 0}\int_{V-V_s}\Sigma^*_{ijkl}\dot{\varepsilon}^p_{kl}dV \qquad (6.90)$$

where V_s is a neighbourhood of radius s of the collocation point, as depicted in Figures 6.4 and [14] for two-dimensional and three-dimensional problems, respectively. Discretizing in the usual way, the following expression in local coordinates is obtained:

$$I = \lim_{s\to 0}\sum_{m=1}^{M_c}\left(\int_{R-R_s}\Sigma^*_{ijkl}\dot{\varepsilon}^p_{kl}\Phi J dV\right) \qquad (6.91)$$

where M_c is the number of cells containing the collocation point; J is the Jacobian of the transformation to the local coordinate system; and R and R_s are the regions defined by the mapping of V and V_s, respectively, to the local system.

Two-dimensional problems To achieve the desired regularization two steps are needed. The first is the introduction of a polar coordinate transformation as in equations (11.39) and (11.40). The next step is the introduction of a Taylor series expansion about the collocation point for the components of the distance r and its derivatives, as follows:

$$r_i = \rho\left(\left(\frac{\partial x_i}{\partial \eta_1}\right)_{\eta_1=\eta'_1}\cos\theta + \left(\frac{\partial x_i}{\partial \eta_2}\right)_{\eta_2=\eta'_2}\sin\theta\right) + O(\rho^2) \qquad i=1,2 \qquad (6.92)$$

and

$$r_{,i} = \frac{A_i(\theta)}{\left(\sum_{k=1}^2 A_k(\theta)^2\right)^{\frac{1}{2}}} + O(\rho) \qquad i=1,2 \qquad (6.93)$$

where

$$A_i(\theta) = \left(\frac{\partial x_i}{\partial \eta_1}\right)_{\eta_1=\eta'_1}\cos\theta + \left(\frac{\partial x_i}{\partial \eta_2}\right)_{\eta_2=\eta'_2}\sin\theta \qquad i=1,2 \qquad (6.94)$$

The coordinate transformation leads to the following form for expression (6.91)

$$\lim_{s\to 0}\sum_{m=1}^{M_c}\int_{\theta_1^m}^{\theta_2^m}\int_{\alpha^m(s,\theta)}^{\bar{\rho}^m(\theta)}F^m_{ijkl}(\rho,\theta)d\rho d\theta \qquad (6.95)$$

The integrand function F_{ijkl} in (6.95) denotes the product of the stress kernel, shape function and Jacobians of both transformations and the initial strain tensor. The terms θ_1 and θ_2 are the initial and final angles, and α and $\bar{\rho}$ are the initial and final radii, which depends on θ, as shown in Figure 6.4.

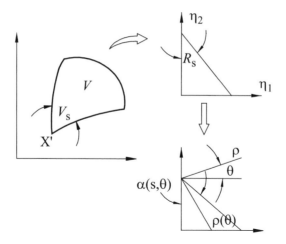

Figure 6.4: Definition of the integration limits.

The use of the Taylor expansion leads to the definition of a function $f_{ijkl}(\theta)$, which when divided by ρ has the same asymptote as the integrand function $F_{ijkl}(\rho, \theta)$ when $\rho \to 0$. In this case, the singularity of order $\frac{1}{\rho}$ is completely separated from the rest of the integral:

$$F_{ijkl}(\rho, \theta) = \frac{1}{\rho} \left[f_{ijkl}(\theta) + O(\rho) \right] \tag{6.96}$$

It can be immediately seen that

$$F_{ijkl}(\rho, \theta) - \frac{1}{\rho} f_{ijkl}(\theta) = O(1) \text{ as } \rho \to 0 \tag{6.97}$$

Consequently, the original integral in polar coordinates, (6.95), can be represented by

$$\lim_{s \to 0} \sum_{m=1}^{M_c} \left[\int_{\theta_1^m}^{\theta_2^m} \int_{\alpha^m(s,\theta)}^{\bar{\rho}^m(\theta)} \left(F_{ijkl}^m(\rho, \theta) - \frac{f_{ijkl}^m(\theta)}{\rho} \right) d\rho d\theta \right.$$

$$\left. + \int_{\theta_1^m}^{\theta_2^m} \int_{\alpha^m(s,\theta)}^{\bar{\rho}^m(\theta)} \frac{f_{ijkl}^m(\theta)}{\rho} d\rho d\theta \right] \tag{6.98}$$

where the asymptotic function $f_{ijkl}(\theta)$ was subtracted and added back.

The first of these integrals is, considering equation (6.97), regular, hence the integral can be evaluated using a standard Gaussian integration. The second term can be simply integrated with respect to ρ, since the asymptotic function only depends on θ. It is important to appropriately define the vanishing neighbourhood R_s^m represented by the lower limit of the integral (6.95), $\alpha^m(\rho, \theta)$ [14]. That is achieved by expressing it in terms of the Taylor expansion previously introduced:

$$\alpha^m(s, \theta) = \frac{s}{\left[\sum_{k=1}^{2} A_k(\theta)^2 \right]^{\frac{1}{2}}} + O(\rho^2) = s\beta(\theta) + O(\rho^2) \tag{6.99}$$

Therefore, the second term in equation (6.98) becomes

$$\lim_{s \to 0} \sum_{m=1}^{M_c} \int_{\theta_1^m}^{\theta_2^m} f_{ijkl}^m(\theta) \ln \frac{\bar{\rho}^m(\theta)}{s\beta(\theta)} d\theta \tag{6.100}$$

This results in the following expression:

$$\sum_{m=1}^{M_c} \left[\int_{\theta_1^m}^{\theta_2^m} f_{ijkl}^m(\theta) \ln\left(\frac{\overline{\rho}^m(\theta)}{\beta(\theta)}\right) d\theta - \lim_{s\to 0}\left(\ln s \int_{\theta_1^m}^{\theta_2^m} f_{ijkl}^m(\theta) d\theta\right) \right] \tag{6.101}$$

This second term, the only one that is still affected by the limit process, can be shown to vanish for an internal singular point.

The final expression for the Cauchy principal value is then

$$\sum_{m=1}^{M_c} \left[\int_{\theta_1^m}^{\theta_2^m} \int_0^{\overline{\rho}^m(\theta)} \left(F_{ijkl}^m(\rho,\theta) - \frac{f_{ijkl}^m(\theta)}{\rho}\right) d\rho d\theta \right.$$

$$\left. + \int_{\theta_1^m}^{\theta_2^m} f_{ijkl}^m(\theta) \ln\left(\frac{\overline{\rho}^m(\theta)}{\beta^m(\theta)}\right) d\theta \right] \tag{6.102}$$

Three-dimensional problems To achieve the desired regularization, two steps are required. The first is the introduction of a spherical coordinates system (ρ, θ, φ) centred at the collocation node, and the second one is the introduction of a Taylor series expansion about the collocation node for the components of the distance r and its derivatives (see Chapter 11). The coordinate transformation leads to the following form for expression (6.91):

$$I = \lim_{s\to 0} \sum_{m=1}^{M_c} \int_{\theta_1^n}^{\theta_2^m} \int_{\varphi_1^n}^{\varphi_2^m} \int_{\alpha^n(s,\theta,\varphi)}^{\overline{\rho}^m(\theta,\varphi)} F_{ijkl}^m(\rho,\theta,\varphi) d\rho d\varphi d\theta \tag{6.103}$$

where the integrand function F_{ijkl}^m stands for the product of the kernel by all the other terms, namely shape functions, Jacobians of both transformations and the initial strain tensor. The terms θ_1^m, θ_2^m, φ_1^m and φ_2^m denote the initial and final angles and α and $\overline{\rho}$ the initial and final radii, which depend on θ and φ (see Figure 6.5(a)).

It is worth noting that the integrand function $F_{ijkl}^m(\rho,\theta,\varphi)$ is of $O(r^{-1})$, since the original singularity of $O(r^{-3})$ has been weakened by spherical-coordinates transformation Jacobian, which is proportional to r^2. The use of the Taylor series expansion about the collocation node for the components of the distance r and its derivatives leads to the definition of a function $f_{ijkl}^m(\theta,\varphi)$, which when divided by ρ has the same asymptote as the integrand function $F_{ijkl}^m(\rho,\theta,\varphi)$ when $\rho \to 0$. In this case, the remaining singularity of $O(r^{-1})$ is completely separated from the rest of the integral. Consequently, the original integral in polar coordinates (6.103) can be represented by

$$I = \lim_{s\to 0} \sum_{m=1}^{M_c} \left\{ \int_{\theta_1^m}^{\theta_2^m} \int_{\varphi_1^m}^{\varphi_2^m} \int_{\alpha^m(s,\theta,\varphi)}^{\overline{\rho}^m(\theta,\varphi)} \left[F_{ijkl}^m(\rho,\theta,\varphi) - \frac{f_{ijkl}^m(\theta,\varphi)}{\rho}\right] d\rho d\varphi d\theta \right.$$

$$\left. + \int_{\theta_1^m}^{\theta_2^m} \int_{\varphi_1^m}^{\varphi_2^m} \int_{\alpha^n(s,\theta,\varphi)}^{\overline{\rho}^m(\theta,\varphi)} \frac{f_{ijkl}^m(\theta,\varphi)}{\rho} d\rho d\varphi d\theta \right\} \tag{6.104}$$

where the asymptotic function $f_{ijkl}^m(\theta,\varphi)$ was subtracted and added back. After performing the limiting process (see Chapter 11), the final expression for the integral

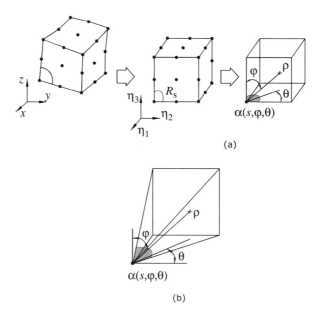

(a)

(b)

Figure 6.5: Evaluation of strongly-singular integrals. Coordinate transformation and integration limits.

is given as

$$
I = \sum_{m=1}^{M_c} \left\{ \int_{\theta_1^m}^{\theta_2^m} \int_{\varphi_1^m}^{\varphi_2^m} \int_{\alpha^m(s,\theta,\varphi)}^{\bar{\rho}^m(\theta,\varphi)} \left(F_{ijkl}^m(\rho,\theta,\varphi) - \frac{f_{ijkl}^m(\theta,\varphi)}{\rho} \right) d\rho d\varphi d\theta \right.
$$
$$
\left. + \int_{\theta_1^m}^{\theta_2^m} \int_{\varphi_1^m}^{\varphi_2^m} f_{ijkl}^m(\theta,\varphi) \ln\left(\frac{\bar{\rho}^m(\theta)}{\beta^m(\theta)} \right) d\varphi d\theta \right\} \qquad (6.105)
$$

where $\beta^n(\theta)$ is the asymptotic expression of $\alpha^m(s,\theta,\varphi)$. Both integrals are now regular and can be evaluated using standard Gauss quadrature formulae. For the sake of simplicity and accuracy when determining integration limits, the first integrand of equation (6.105) is carried out after dividing each volume cell into pyramid shaped subcells. These subcells are defined according the position of the collocation node within the cell domain (see Figure 6.5(b)).

6.5.2   System Matrices Assembly

The matrix representation of the boundary displacement integral equation (6.88), when applied to all boundary nodes, can be written as

$$
\mathbf{H}\dot{\mathbf{u}} = \mathbf{G}\dot{\mathbf{t}} + \mathbf{D}\dot{\boldsymbol{\varepsilon}}^p \qquad (6.106)
$$

where matrices $\mathbf{H}$ and $\mathbf{G}$ are the same as those obtained for the purely elastic case, and matrix $\mathbf{D}$ is due to the inelastic strain integral.

A matrix representation can be also formed for (6.89) as

$$
\dot{\boldsymbol{\sigma}} = \mathbf{G}'\dot{\mathbf{t}} - \mathbf{H}'\dot{\mathbf{u}} + (\mathbf{D}' + \mathbf{C}')\dot{\boldsymbol{\varepsilon}}^p \qquad (6.107)
$$

where matrix $\mathbf{C}'$ contains the free term in equation (6.89) and $\mathbf{D}'$ accounts for the inelastic strain influence over the discretized domain.

Reordering equation (6.106) leads to

$$\mathbf{A}\dot{\mathbf{y}} = \dot{\mathbf{f}} + \mathbf{D}\dot{\varepsilon}^p \tag{6.108}$$

where vector $\dot{\mathbf{y}}$ contains boundary unknowns and the vector $\dot{\mathbf{f}}$ contributions of the prescribed values on the boundary, and $\mathbf{A}$ corresponds to the system matrix derived from $\mathbf{G}$ and $\mathbf{H}$. Further rearrangement gives

$$\dot{\mathbf{y}} = \mathbf{K}\dot{\varepsilon}^p + \dot{\mathbf{m}} \tag{6.109}$$

where

$$\mathbf{K} = \mathbf{A}^{-1}\mathbf{D} \tag{6.110}$$

and

$$\dot{\mathbf{m}} = \mathbf{A}^{-1}\dot{\mathbf{f}} \tag{6.111}$$

is the vector of the purely elastic solution for the boundary tractions and displacements.

In a similar way, equation (6.107) can be rewritten as

$$\dot{\sigma} = -\mathbf{A}'\dot{\mathbf{y}} + \dot{\mathbf{f}}' + \mathbf{E}\dot{\varepsilon}^p \tag{6.112}$$

where $\mathbf{E} = \mathbf{D}' + \mathbf{C}'$ accounts for the inelastic strain influence, and $\mathbf{A}'$, $\dot{\mathbf{f}}'$ and $\dot{\mathbf{y}}$ correspond to the system matrix, the elastic part of the independent term and the vector of the unknowns, respectively.

The system is reduced to one equation by substituting (6.109) into (6.112). The elastic stresses at internal points are given by

$$\dot{\sigma} = \mathbf{B}\dot{\varepsilon}^p + \dot{\mathbf{n}} \tag{6.113}$$

where

$$\mathbf{B} = \mathbf{E} - \mathbf{A}'\mathbf{K} \tag{6.114}$$

and

$$\dot{\mathbf{n}} = \dot{\mathbf{f}}' - \mathbf{A}'\dot{\mathbf{m}} \tag{6.115}$$

According to what has been shown above, two methods can be followed. In the first, equations (6.108) and (6.112) are used in succession, while the second only uses equation (6.113). The first method leads to a more complex algorithm, but avoids the need for costly matrix manipulations. However, the second method under certain circumstances is more efficient. These circumstances are related to the number of unknowns of the problem, as well as the number of iterations to be performed. Leitão [20, 21], analyzed this problem in the case of a two-dimensional algorithm. He compared the number of multiplications required by both methods, and found that in general the first method is more efficient for a low number of iterations, while the second is more convenient as the number of iterations increases. Obviously, the number of iterations is problem dependant, and therefore there is no optimum method.

6.6   Non-linear Solution Algorithm

In continuum mechanics, nonlinear problems are commonly solved by adopting a load incremental procedure and by iterating on the strain-stress constitutive equation. When primary unknowns of the problem (stresses or plastic strains) appear explicitly in the governing equations the algorithm is said to be of the *explicit* kind. This is the case for the initial strain formulation presented in this chapter. As these initial strains are not known *a priori*, the only way to solve the equations is by the use of iterative procedures.

Alternative solution algorithms are given by the so-called *implicit* procedures. Implicit procedures allow for the reduction of the number of iterations, or even the suppression of iterations, which means a direct procedure. Their main disadvantage is the increase in the computer effort for each load increment [43].

In what follows, a detailed description of the explicit procedure adopted in this book is presented, together with a brief description of the alternative implicit approach.

6.6.1   Implicit Procedures

The starting equation of implicit procedures can be obtained by substituting the relationship between the plastic strain increments and the stresses, given by [43] as

$$\dot{\varepsilon}^p_{ij} = D^p_{ijmn}\dot{\sigma}_{mn} \tag{6.116}$$

into the expression of the elastic internal stresses (6.113). For an incremental scheme, the solution for a given step k yields

$$\sigma_{k-1} + \Delta_k\sigma = B(\varepsilon^p_{k-1} + D^p_k\Delta_k\sigma) + N_{k-1} + \Delta_kN_k \tag{6.117}$$

where N_{k-1} corresponds to the elastic response at the load step $k-1$, and B is the same as that in equation (6.113). A residual vector can be defined as follows:

$$R_k = D\varepsilon^p_{k-1} + N_{k-1} - \sigma_{k-1} \tag{6.118}$$

and substituting into equation (6.117) becomes

$$B^p_k\Delta_k\sigma = \Delta_kN + R_k \tag{6.119}$$

in which

$$B^p_k = I - B_kD^p \tag{6.120}$$

Once this expression is obtained, several schemes are available, depending on how the matrix B^p is updated.

A direct solution can be obtained (i.e. no iterations are required) if the matrix B^p is updated for each load increment. Different schemes to obtain direct solutions were proposed by Banerjee and Raveendra [1] and Telles and Carrer [43]. The drawback of the approach is the large computational effort required to assemble the B^p matrix at each load step. However, the advantage of using this approach is still remarkable when high degree of nonlinearities are involved, i.e., when the classical iterative algorithms take an unduly large number of iterations to converge.

Iterative implicit algorithms, on the other hand, keep the matrix B^p constant for one or a small number of load increments, and perform some iterations in each of them.

In this way they produce accurate solutions with a minimum number of iterations. However, the complete system of equations has to be solved for each solution step. Evaluation of different implicit iterative solution schemes can be found in [43].

6.6.2 Explicit Procedure

Explicit techniques are characterized by the use of a recursive relationship between the stresses, the boundary unknowns and the plastic strains. The advantage of these techniques is that the matrices are kept constant, but the increase in the number of iterations is a drawback.

After computation of all the matrices and known vectors, the following step consists of solving equations:

$$\mathbf{A}\dot{\mathbf{y}} = \mathbf{f} + \mathbf{D}(\dot{\boldsymbol{\varepsilon}}^p + \Delta\dot{\boldsymbol{\varepsilon}}^p) \tag{6.121}$$

$$\dot{\boldsymbol{\sigma}} = -\mathbf{A}'\dot{\mathbf{y}} + \mathbf{f}' + \mathbf{E}(\dot{\boldsymbol{\varepsilon}}^p + \Delta\dot{\boldsymbol{\varepsilon}}^p) \tag{6.122}$$

for purely elastic behaviour, i.e. $\dot{\boldsymbol{\varepsilon}}^p = \Delta\dot{\boldsymbol{\varepsilon}}^p = 0$ ($\Delta\dot{\boldsymbol{\varepsilon}}^p$ denotes the increment of plastic strains) for the total value of external load. The *load at first yield* can be calculated by taking the most highly stressed (boundary or internal) node, and comparing its equivalent stress, $\sigma_e^{\max}$, with the uniaxial yield stress of the material. The incremental process starts by reducing this stress value with a load factor, defined as

$$\lambda_o = \frac{\sigma_Y}{\sigma_e^{\max}} \tag{6.123}$$

Then $\sigma_e^{\max}$ is put equal to the uniaxial yield stress σ_Y, and the elastic response is scaled by the initial load factor. The new load increment is then calculated, and further values of the load factor are given by

$$\lambda_i = \lambda_{i-1} + \Delta\lambda_o \qquad \lambda_i \le 1 \tag{6.124}$$

where $\Delta\lambda_0$ is the given value of the load increment with reference to the load at first yield. The stresses associated with this load level are obtained using equations (6.108) and (6.112) in their incremental form:

$$\mathbf{A}\dot{\mathbf{y}} = \lambda_i\mathbf{f} + \mathbf{D}(\dot{\boldsymbol{\varepsilon}}^p + \Delta\dot{\boldsymbol{\varepsilon}}^p) \tag{6.125}$$

$$\dot{\boldsymbol{\sigma}} = -\mathbf{A}'\dot{\mathbf{y}} + \lambda_i\mathbf{f}' + \mathbf{E}(\dot{\boldsymbol{\varepsilon}}^p + \Delta\dot{\boldsymbol{\varepsilon}}^p) \tag{6.126}$$

where $\dot{\boldsymbol{\varepsilon}}^p$ stores the plastic strains up to but not including the current load increment.

For a given value λ_i, the plastic strain increment, $\Delta\dot{\boldsymbol{\varepsilon}}^p$, is determined iteratively at each selected node as follows:

a) Estimate $\Delta\dot{\varepsilon}_{ij}^p$.

- $\Delta\dot{\varepsilon}_{ij}^p = 0$ for the first iteration.
- $\left(\Delta\dot{\varepsilon}_{ij}^p\right)_k = \left(\Delta\dot{\varepsilon}_{ij}^p\right)_{k-1}$ for subsequent iterations.

b) Compute $\mathbf{y}$ and $\boldsymbol{\sigma}$ (Equations (6.125) and (6.126))

c) Calculate:

From (6.34) $\dot{\varepsilon}'_{ij} = \frac{1}{2\mu}(\sigma_{ij} - \frac{\nu}{1-\nu}\sigma_{kk}\delta_{ij}) + \Delta\varepsilon^p_{ij}$

From (6.38) $\dot{\varepsilon}'_{eq} = \sqrt{\frac{2}{3}e'_{ij}e'_{ij}}$

From (6.46) $\Delta\dot{\varepsilon}^p_{eq} = \frac{3\mu\varepsilon'_e - \sigma_{o,i-1}}{3\mu + H'_{i-1}} \geq 0$

d) Verify convergence. Process is complete if the value of $\Delta\dot{\varepsilon}^p_{eq}$ is equal to its previous value, within a prescribed tolerance.

e) Compute a new estimate of $\Delta\dot{\varepsilon}^p_{ij}$ (Equation (6.41))

$$\Delta\dot{\varepsilon}^p_{ij} = \frac{\Delta\dot{\varepsilon}^p_e}{\dot{\varepsilon}'_e}e'_{ij}$$

f) Go to **a)** for a new iteration.

Once convergence is obtained at all control points, $\Delta\dot{\varepsilon}^p$ is added to $\dot{\varepsilon}^p$, and its value is used as an initial guess for the next load increment.

There are some variants to the above general method, for example the variable stiffness method [12, 16], in which the elastic constitutive relationship is changed according to the elastoplastic relationship; but the matrices are now no longer constant.

6.7   Examples

In this section, several two-dimensional and three-dimensional examples are analysed and results have been compared with analytical solutions or experimental data where available.

6.7.1   Two-dimensional Benchmarks

Perforated Aluminium Plate

The geometry and discretization are shown in Figure 6.6. The boundary of the problem is discretized with quadratic elements, while for the interior cells both linear and quadratic have been considered. The problem has been analyzed experimentally by Theocaris and Marketos [44], and for the numerical results plane stress state has been considered. The work hardening material considered has the following properties: Young's modulus, $E = 70$ GPa; Poisson's ratio, $\nu = 0.2$; plastic modulus, $H' = 0.032E$; yield stress, $\sigma_Y = 243$ MPa. The von Mises yield criterion was considered.

Figure 6.7 shows a comparison between experimental stresses at the root of the perforated plate and the values obtained by the BEM [26], for a load $\sigma_a = 0.47\sigma_y$. Results differ slightly near the hole, but are considered acceptable (see, for example, [21, 40]). Figure 6.8 shows the extension of the plastic zone for a final load of $\sigma_a = 0.53\sigma_Y$. Results agree with those presented by Zienkiewicz and Cormeau [45].

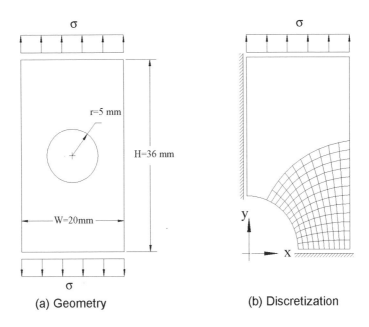

Figure 6.6: Perforated aluminium plate.

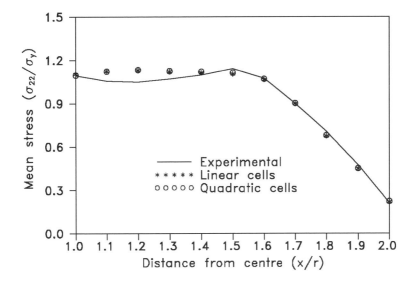

Figure 6.7: Computed and experimental stresses on net section of the plate.

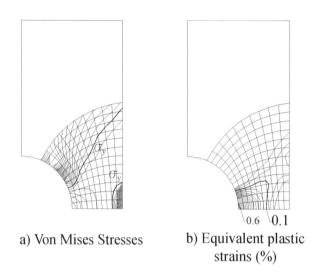

a) Von Mises Stresses   b) Equivalent plastic
                             strains (%)

Figure 6.8: Plastic zones obtained by BEM linear cells ($\sigma_a = 0.53\sigma_Y$).

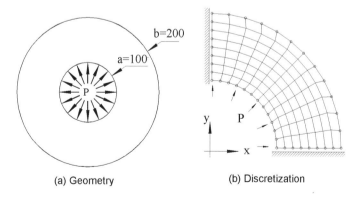

(a) Geometry   (b) Discretization

Figure 6.9: Thick cylinder under internal pressure P.

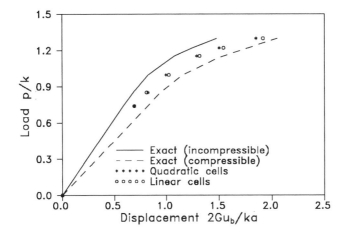

Figure 6.10: Outer surface displacements.

Thick Cylinder under Internal Pressure

Figure 6.9 shows the geometry and discretization of a thick cylinder subjected to internal pressure. The problem is considered under plain strain conditions. The example is discretized into 20 quadratic boundary elements, while the domain is discretized into 24 quadratic cells or 96 linear cells. Internal points are the same in both cases. The material properties are as follows: $E = 120$ GPa, $\nu = 0.3$, $H' = 0$ (perfectly plastic material), $\sigma_Y = 240$ MPa and yield stress in pure shear, $k = \frac{\sigma_y}{\sqrt{3}}$. The von Mises yield criterion was used.

Results are compared with analytical results obtained by Prager and Hodge [36], who presented results assuming compressibility and incompressibility of the material. Results obtained by the BEM [26] for displacements of the external surface fall between both curves, as shown in Figure (6.10). This is explained by the fact that the BEM considers compressibility in the elastic regime, and incompressibility of plastic strains. Circumferential stresses are shown in Figure 6.11. Results agree with the analytical curves, as stresses depend upon equilibrium rather than volume change.

Prestress and Cold Expansion Specimen

In the aerospace industry, cold expansion is a widely used mechanical treatment for fatigue life enhancement of components with circular cut-outs, such as fastener holes. The cold expansion process consists of pulling an oversized mandrel through a hole, thus deforming the surrounding material (see Figure 6.12). If the degree of expansion is sufficient, plastic deformation occurs and compressive residual stresses are introduced around the rim of the hole. These residual stresses are reputed to improve the fatigue life of components by as much as 10 times. Leitão, Aliabadi, Rooke and Cook [22, 23] analyzed the fatigue test specimen shown in Figure 6.13 subjected to prestressing technique and cold-expansion. The prestressing technique consists of pulling the extremities of specimens until the material undergoes deformation in the region of interest followed by the removal of the applied load. This is a very simple technique used under experimental conditions, but its use on aircraft components is

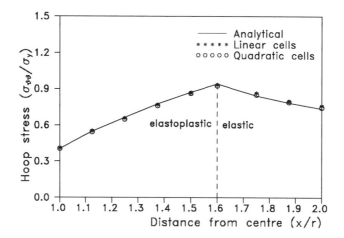

Figure 6.11: Circumferential stress distribtuion.

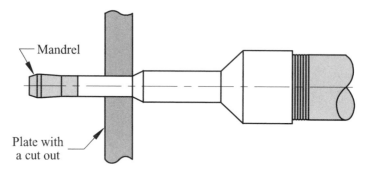

Figure 6.12: Cold expansion mandrel.

not practical.

For the analysis, a specimen made of BS2L65 aluminium alloy was considered. Plane stress conditions are assumed. The specimen geometry and material properties are shown in Figure 6.13. It is important to note that this part has a low edge margin; the diameter of the central hole is 12.7 mm and the width of the central section of the component is only 35.5 mm.

Prestressing is modelled numerically by the application of a single uniaxial tensile load at both ends of the specimen. This causes local plasticity in the region around the central hole. On removal of the load, a residual stress field is formed. The overloads used were 237, 277 and 317 KN, which resulted in average net section stresses equivalent to 60%, 70% and 80% of the 0.1% proof stress. The residual stress distributions obtained [17] using the BEM initial strain approach are shown in Figure 6.14. As can be seen from the Figure, in all cases the residual stress distribution was compressive at the edge of the hole, becoming tensile and tending to zero further away.

Unlike the prestressing, the cold expansion of holes is a widely used mechanical treatment for fatigue life improvements. The cold expansion effect is numerically

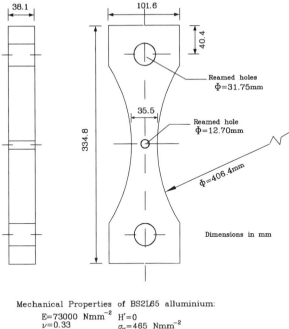

Mechanical Properties of BS2L65 alluminium:
E=73000 Nmm^{-2} H'=0
ν=0.33 σ_Y=465 Nmm^{-2}

Figure 6.13: Fatigue test specimen.

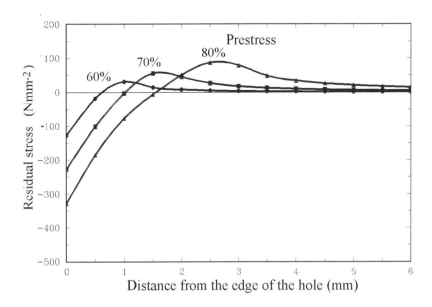

Figure 6.14: Residual stress distribution due to the prestress.

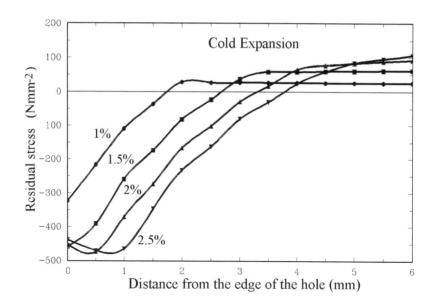

Figure 6.15: Residual stress distribution due to the cold-expansion.

simulated by the application of a radial displacement field around the hole. On removal of this radial expansion (i.e. by removal of the equivalent tractions on the edge of the hole) at the end of the expansion process, a residual stress field is formed. The BEM results obtained [22] for four different maximum expansion levels, namely 1%, 1.5%, 2% and 2.5% of the initial radius of the hole, are shown in Figure 6.15.

6.7.2   Three-dimensional Benchmarks

Thick Pressurized Cylinder

Consider a thick pressurized cylinder of inner radius $a = 100$ mm and outer radius $b = 200$ mm as shown in Figure 6.16. Due to the symmetry of the problem, only one quarter is modelled [9]. Appropriate boundary conditions are also applied in order to make the analysis under a plane strain condition in the z-direction. The boundary surface is discretized using 46, 9-noded elements, while the volume is only partially discretized up to 60% of the cylinder thickness, using 10 internal cells (hatched volume in Figure 6.16)

The constitutive model is considered as elastic-perfectly plastic. The material data is: modulus of elasticity, $E = 120$ GPa, Poisson's ratio, $\nu = 0.3$, yield stress, $\sigma_{yield} = 240$ MPa, and plastic strain hardening parameter, $H = 0$.

The internal pressure, $p = 181.5$ MPa, is applied in 15 load increments after load to first yield is determined. This final value of pressure corresponds to that makes 60% of the cylinder thickness go under the elastoplastic regime.

Numerical results are compared with those obtained analytically by Prager and Hodge [36]. In Figure 6.17, the internal and external radial displacements, together with the pressure, are plotted as a function of the radius of elastic-plastic boundary ρ, while in Figure 6.18, pressure is plotted as a function of the external radial displace-

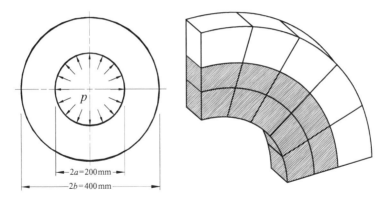

Figure 6.16: Thick pressurized cylinder. Model geometry and discretization.

ment. The stress distribution for a position $\rho/a = 1.6$ of the elastic-plastic boundary is shown in Figures 6.19 and 6.20 for internal and boundary nodes. Although the general agreement between the results is very good, some deviations are found (note, for example, the $\sigma_{\theta\theta}$ stress component at boundary nodes in Figure 6.20). This deviations can be attributed to the inaccuracy when determining the exact position of the elastic-plastic boundary, since its position as well as the applied load are given by discrete increments [9].

Perforated Strip

In this example the problem of a perforated strip of strain-hardening material is analyzed. The geometry of the problem as well as its discretization are illustrated in Figure 6.21. Due to the symmetry, only a quarter of the problem is discretized using 122, 9-noded boundary elements and 25 internal cells (hatched volume in Figure 6.21). A uniaxial tension $\sigma = 140$ MPa is applied in 13 equal load increments after first yield. The material properties are: elastic modulus, $E = 70$ GPa, Poisson's ratio, $\nu = 0.2$, yield stress, $\sigma_{yield} = 243$ MPa and plastic strain hardening modulus, $H' = 2240$ MPa. In Figure 6.22 stress-displacement plots for some key positions on the strip (see Figure 6.21) are shown. BEM results are compared with FEM results obtained using the LUSAS package [24]. The FEM discretization was the same as for the BEM, but using 20-noded brick elements. In Fig. 6.23 stress component σ_{xx} at the net section of the strip at a load level $\sigma = 0.47\sigma_{yield}$ is plotted. Results are compared to those obtained by the FEM analysis and experimentally by Theocaris and Marketos [44].

6.8   Summary

In this chapter a direct formulation for three-dimensional elastoplastic boundary elements has been presented. The boundary element solution of elastoplastic stress were shown to be accurate and in agreement with the experimental results. It was also shown that for elastoplastic problems, only the region expected yield needs to be discretized. In most practical situation the plastic region is very small compared

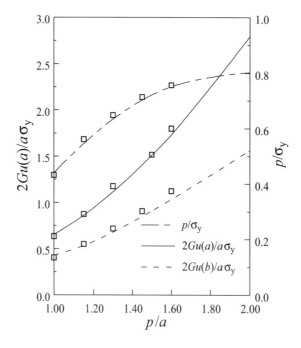

Figure 6.17: Internal and external radial displacements and pressure vs. the radius of the elastic boundary.

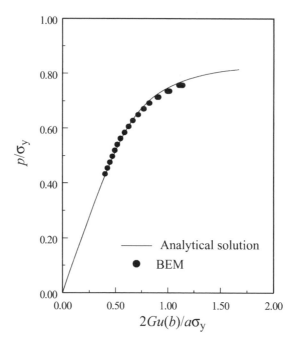

Figure 6.18: Internal pressure vs. external radial displacement for the thick pressurized cylinder.

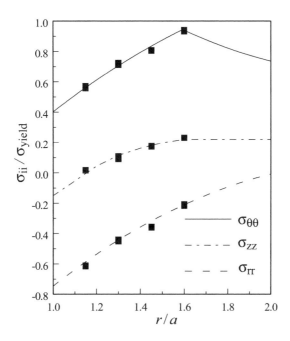

Figure 6.19: Stress distribution for internal nodes ($\rho/a = 1.6$).

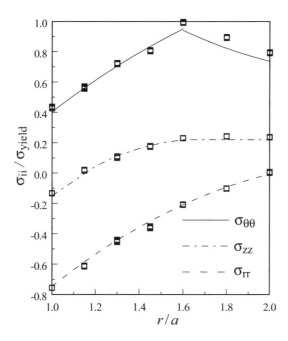

Figure 6.20: Stress distribution for boundary nodes ($\rho/a = 1.6$).

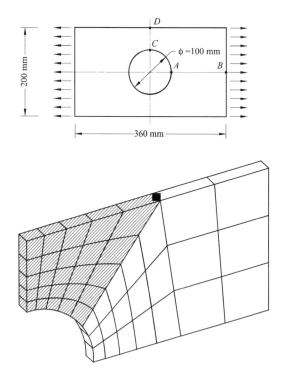

Figure 6.21: Perforated strip, model geometry and discretization.

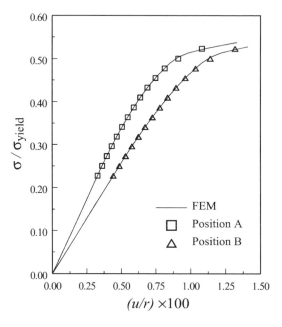

Figure 6.22: Perforated strip: displacements at positions A and B.

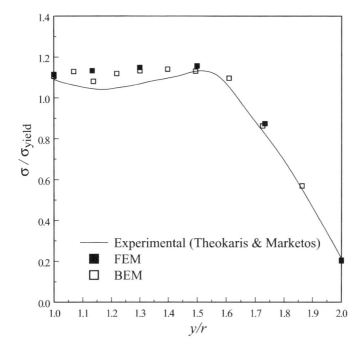

Figure 6.23: Perforated strip: stress variation on net section of plate.

to the overall dimension of the problem. For these problems, the boundary element method is an efficient alternative to the finite element method.

Recently Ochiai and Kobayashi [32, 33] proposed a method referred to as triple-reciprocity to transform the domain integrals in elastoplastic analysis into boundary ones. This method is at early stage of development, however results presented by the authors are encouraging.

References

[1] Banerjee, P.K. and Raveendra, S.T., Advanced boundary element of two- and three-dimensional problems of elastoplasticity, *International Journal for Numerical Methods in Engineering*, **23**, 985-1002, 1986

[2] Banerjee, P.K. and Raveendra, S.T., New boundary element formulation for 2-D elastoplastic analysis, *Journal of Engineering Mechanics, ASCE*, **113**, 252-265, 1987.

[3] Bui, H.D., Some remarks about the formulation of three-dimensional thermoelastoplastic problems by integral equations. *International Journal of Solids and Structures*, **14**, 935-939, 1978.

[4] Cathie, D.N., On the implementation of elasto-plastic boundary element analysis, in Proceedings of second international seminar on advances in boundary elements, Southampton, 318-334, 1980.

[5] Chandra, S. and Mukherjee, S., A boundary element analysis of metal extrusion processes, *ASME, Journal of Applied Mechanics*, **54**, 336-340, 1987.

[6] Chaudonneret, M., Méthode des équations intégrales appliquées a la résolution de problémes de viscoplasticité, *Journal of Mecanique Appl.*, **1**, 113-132, 1977.

[7] Cisilino, A.P., Aliabadi, M.H. and Otegui, J.L., A direct formulation for three-dimensional elastoplastic analysis, *Boundary Element XIX*, 169-179, 1997.

[8] Cisilino, A.P., Aliabadi, M.H. and Otegui, J.L., A three-dimensional boundary element formulation for the elastoplastic analysis of cracked bodies, *International Journal of Numerical Methods for Engineering*, **42**, 237-256, 1998.

[9] Cisilino, A.P. and Aliabadi, M.H., A boundary element method for three-dimensional elastoplastic problems, *Engineering Computations*, **15**, 41011-1030, 1988.

[10] Cisilino, A.P., Boundary element formulations for 3D analysis of fatigue crack growth in linear and nonlinear fracture mechanics., PhD thesis, Weessex Institute of Technology, University of Wales, 1997.

[11] Cisilino, A.P., *Linear and Nonlinear Crack Growth using Boundary Elements*, Topics in Engineering, Vol 36, Computational Mechanics Publications, 2000.

[12] Corradi, S., Aliabadi, M.H. and Marchetti, M., A variable stiffness dual boundary element method for mixed-mode elastoplastic crack problems, *Theoretical and Applied Fracture Mechanics*, **25**, 43-49, 1996.

[13] Drucker, D.C. *Introduction to Mechanics of Deformable Solids*, McGraw-Hill, New York, 1967.

[14] Guiggiani, M. and Gigante, A., A general algorithm for multidimensional cauchy principal value integrals in the boundary element method, *Journal of Applied Mechanics*, **57**, 906-915, 1990.

[15] Gupta,A., Delgado, H.E. and Sullivan, J., A three-dimensional BEM solution for plasticity using regression interpolation within the plastic field. *International Journal for Numerical Methods in Engineering*, **33**, 1997-2014, 1992.

[16] Henry, D.P. and Banerjee, P.K., A variable stiffness type boundary element formulation for axisymmetric elastoplastic media, *International Journal for Numerical Methods in Engineering*, **26**, 1005-1027, 1988.

[17] Leitão, V., Aliabadi, M.H., Cook, R. and Rooke, D.P., Boundary element analysis of fatigue crack growth in the presence of residual stresses, in *Localized Damage II, VolI*, Computational Mechanics Publications, Southampton, 489-510, 1992.

[18] Leitão, V., Aliabadi, M.H., Rooke, D.P. and Cook, R., Residual stress fields effect on fatigue crack growth, in *Boundary Elements XIV, Vol2*, Computational Mechanics Publications, Southampton, Elsevier Applied Science, London, 331-350, 1992.

[19] Leitão, V., Aliabadi, M.H. and Rooke, D.P., The dual boundary element formulation for elastoplastic fracture mechanics, *International Journal for Numerical Methods in Engineering* **38**, 315-333, 1995.

[20] Leitão, V.M.A., An improved boundary element formulation for nonlinear fracture mechanics., PhD thesis, Wessex Institute of Technology, University of Portsmouth, 1993.

[21] Letiao, V.M.A., *Boundary Elements in Nonlinear Fracture Mechanics*, Topics in Engineering Vol21, Computational Mechanics, Southampton, 1994.

[22] Leitão, V., Aliabadi, M.H. Rooke, D.P and Cook, R., Boundary element methods for the analysis of crack growth in the presence of residual stress fields, *J. Materials and Engineering Performance*, **7(3)**,352-360, 1998.

[23] Leitão, V.M.A and Aliabadi, M.H.. Boundary element methods for analysis of crack growth in nonlinear fracture, International Journal of Materials and Product *Technology*, **15**, 104-116, 2000

[24] LUSAS. *Fnite Element System*, version 11, Finite Element Analysis Ltd., Kingston Upon Thames.

[25] Maier, G. and Polizotto, C., A Galerkin approach to boundary element elastoplastic analysis, *Computer Methods in Applied Mechanics and Engineering*, **60**, 175-194, 1987.

[26] Martin,D.F., Elastoplastic contact problems: boundary element method, WIT, University of Wales, 1997.

[27] Matsumoto, T. and Yuuki, R., Accurate boundary element analysis of two-dimensional elastoplastic problems, *boundary element methods in applied mechanics*, Pergamon Press, Oxford, 205-214, 1988.

[28] Mendelson, A., *Plasticity: Theory and Applications*, Macmillan, New York, 1968.

[29] Mendelson, A., *Boundary integral methods in elasticity and plasticity*, Report No. NASA TN D-7418, NASA, 1973.

[30] Mendelson, A. and Albers, L.U., Application of boundary integral equations to elastoplastic problems, *Boundary integral equation methods: computational applications in applied mechanics*, ASME, New York, 47-84, 1975.

[31] Mukherjee, S., Corrected boundary integral equations in planar thermoelastoplasticity problems by integral equations, *International Journal of Solids and Structures*, **13**, 331-335, 1977.

[32] Ochiai, Y. and Kobayashi, T., Initial strain formulation without internal cells for elastoplastic analysis by triple reciprocity BEM, *International Journal for Numerical Methods in Engineering*, **50**, 1877-1892, 2001

[33] Ochiai, Y., Formulation for three-dimensional elastoplastic analysis without internal cells by triple-reciprocity BEM, *Advances in Boundary Element Techniques*, Hoggar, Geneva, 217-224, 2001.

[34] Okada, H, Rajiyah, H. and Atluri, S.N., A full tangent stiffness field-boundary-element formulation for geometric and material non-linear problems in solid mechanics, *International Journal for Numerical Methods in Engineering*, **29**, 15-35, 1990.

[35] Okada, H., Rajiyah, H. and Atluri, S.N., Some recent developments in finite-strain elastoplasticity using field-boundary element method, *Computers and Structures*, **30**, 275-288, 1988.

[36] Prager, R.M. and Hodge, P.G., *Theory of Perfectly Plastic Solids*, Dover Publications, New York, 1951.

[37] Ricardella, P.C. *An implementation of the boundary integral technique for planar problems in elasticity and elastoplasticity*, Report No. SM-73-10, Dept. Mech. Engng, Carnegie Mellon University, Pittsburg, 1973.

[38] Swedlow, J.L. and Cruse, T.A., Formulation of boundary integral equations for three-dimensional elastoplastic flow, *International Journal of Solids and Structures*, **7**, 1673-1683, 1971.

[39] Telles, J.C.F. and Brebbia, C., On the application of the boundary element method to plasticity, *Applied Mathematical Modelling*, **3**, 466-470, 1979.

[40] Telles, J.C.F., *The Boundary Element Method Applied to Inelastic Problems*, Lecture Note Series, Springer-Verlag, Berlin, 1983.

[41] Telles,J.C.F. and Brebbia, C.A., Boundary elements: new developments in elasto-plastic analysis, *Applied Mathematical Modelling*, **5**, 376-382, 1981.

[42] Telles, J.C.F. and Brebbia, C.A., Elastic/visoplastic problems using boundary elements, *International Journal of Mechanical Sciences*, **24,** 605-618, 1982.

[43] Telles, J.C.F. and Carrer, J.A.M., Implicit procedures for the solution of elasto-plastic problems by the boundary element method, *Mathematical and Computer Modelling*, **15,** 303-311, 1991.

[44] Theocaris, P.S. and Marketos, E., Elastic-plastic analysis of perforated thin strips of a strain-hardening material. *Journal of Mechanics and Physics of Solids*, **12**, 377-390, 1964.

[45] Zienkiewicz, O.C and Cormeau, I.C., Viscoplasticity, plasticity and creep in elastic solids, a unified numerical solution approach, *International Journal for Numerical Methods in Engineering*, **8**, 821-845, 1974.

Chapter 7

Contact Mechanics

7.1 Introduction

Contact mechanics is an important topic within the general field of solid mechanics, and has received much attention due to its inherent complexity and common occurrence in engineering. The progress in contact mechanics relies increasingly on the use of robust numerical methods and improved computing power to solve complex multi-body problems efficiently [4]. It may be argued from the numerical point of view that as the contact is on a boundary, a boundary-type rather than a domain-type, solution procedure such as the Boundary Element Method (BEM) would be more suitable to the analyses of these types of problems.

The first application of the BEM to contact problems can be traced back to the early 1980s [7]. Later works [15, 30] used both continuous and discontinuous quadratic elements to solve contact problems. A fully incremental formulation was developed in [19], where it was shown that linear elements should be used in place of higher order elements in the contact area [20]. Application of the method to axisymmetric and three-dimensional problems can be found in [2, 8, 10]. Extension to elastoplastic problems is reported in [24].

All the above formulations are based on the so-called 'direct constraint method'. Other prominent formulations of the BEM to contact problems include the penalty function approach [31] and mathematical programming [12, 16]. All the above works make use of conforming discretization, whereby the contact areas are modelled in such a way that for every node on one body, there must be a matching node on the other. The use of conforming discretization is increasingly being regarded as an unnecessary limitation, and therefore attention has focused on the development of non-conforming discretization method [5, 29, 25, 28].

In this chapter, the boundary element method is used for multi-body contact modelling. A direct constraint approach is presented where the solution is obtained directly and explicitly from contact considerations and the global equilibrium of the problem alone. This technique is based on iterative and incremental loading procedures, and can deal with both conforming or non-conforming types of contact. An automatic updating procedure is also employed in order to model the continuously changing boundary conditions occurring inside the contact region.

Initially, the basic concepts of the contact mechanics and numerical modelling of contact problems are briefly described. Next, an iterative and incremental loading procedure for both non-frictional and frictional problems, as developed by Man, Aliabadi and Rooke [19, 20, 22, 23] is described in detail. Application of the BEM to elastoplastic contact problems [24] is described next, followed by the development of new BEM formulations using non-conforming discretization to model the contacting bodies. Three-dimensional contact problems are also described.

Throughout this chapter μ is used to represent the coefficient of friction.

7.2 Basic Contact Mechanics

7.2.1 Friction

The resistance of two surfaces to slide associated with the physical mechanism is referred to as *friction*. Experimental evidence has shown that *friction* characteristics can be affected by several factors, such as topology, changes in material properties, environmental effects, and temperature, among others. For simplicity, Columb's friction law is generally used in most engineering analyses, in which the tangential traction is related to the normal traction through the coefficient of friction μ. It is assumed that relative sliding of two bodies in contact will occur when the tangential traction at any point in the interface exceeds the product of the normal traction at that point and the constant coefficient of friction μ.

7.2.2 Classification of Contact

Contact problems may be categorized as:

- Frictionless Contact: an idealized contact mode which can be associated with well lubricated smooth surfaces. It is assumed that contacting surfaces are able to slide over each other without any resistance along the tangential direction (i.e. parallel to the contact interface). Only normal stress is developed along the contact interface under the assumption of frictionless contact, and equilibrium is maintained by transferring the normal compressive stress. In the absence of friction, stresses in the tangential direction are zero and the evolution of contact is independent of loading history.

- Frictional Contact: a realistic contact mode in which the sliding movement in the tangential direction of a point in contact is restricted by the frictional forces (tangential shear stresses), as the point of contact, which in turn depends on the normal component of forces (normal stresses) exerted at the same point. The relationship between the tangential and normal components of forces results in a nonlinear behaviour between the sliding movement of the contact surfaces and the external load. In frictional contact, the contact conditions are either *Stick* (i.e. no relative tangential displacements) or *Slip* (i.e. sliding against resistance in the tangential direction).

- Conforming Contact: refers to the type of problem in which the contacting surfaces 'match' each other in the unloaded state. For example, a flat punch resting on a flat foundation, or a perfect-fit pin in a lug, are conforming contact problems. A key feature of this type of contact problem is that the extent of the contact area is independent of the load. For this reason, the load history is not important in this case.

- Non-conforming Contact: refers to problems with different profiles for the contacting surfaces. The main feature is that the size of the initial contact area will change once the bodies are subjected to external loading/unloading. The changes in the contact area are problem-dependant, and can relate to several factors, including initial profiles, material properties, rate of loading, direction of loading, etc. This type of problem can rarely be solved analytically, and is the most challenging for computational methods.

- Progressive or Advancing Contact: is relevant to most contact problems, as in most non-conforming contact problems, the bodies initially touch at a point or along a line, and the area of contact grows with increasing load.

- Receding Contact: refers to a type of problems in which the contact is initially over an appreciable area, and when loaded it will deform in a way such that the contact area is decreased. For example, a perfectly fit pin in a hole will initially touch around the whole circumference, but when loaded perpendicular to its axis, the contact area is reduced as a gap develop between the pin and the hole.

In practice, other types of contact in addition to the above will exist. For example, rolling contact can be related to a wide range of contact problems concerned with rolling motion, such as a railway wheel and rail track.

7.2.3   Modes of Contact

For a given contact state, the contact conditions of a contact node-pair (a and b say) may be represented by any one of the three modes shown in Table 7.1, where t_t and t_n are the tangential and normal tractions, and u_t and u_n are the tangential and normal displacements, respectively, expressed in local coordinates. Definitions in Table 7.1

Table 7.1: Modes of contact

Separation	Slip	Stick
$t_t^a - t_t^b = 0$	$t_t^a - t_t^b = 0$	$t_t^a - t_t^b = 0$
$t_n^a - t_n^b = 0$	$t_n^a - t_n^b = 0$	$t_n^a - t_n^b = 0$
$t_t^a = 0$	$t_t^a \pm \mu t_n^a = 0$	$u_t^a + u_t^b = 0$
$t_n^a = 0$	$u_n^a + u_n^b = gap^{ab}$	$u_n^a + u_n^b = gap^{ab}$

can be summarized as follows:

- Separation Mode: defined as node pairs remaining apart.

- Slip Mode: defined as node pairs not restrained in the tangential direction, but free to slide over each other.

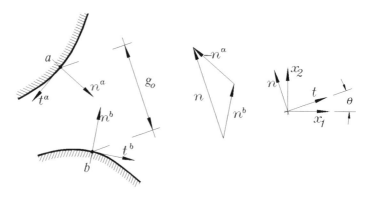

Figure 7.1: Definition of average normal.

- Stick Mode: defined as node pairs restrained in the normal and tangential directions and not having experienced any slip during loading.

- Partial Slip Mode: this was introduced to describe an intermediate stage between the two extreme cases in which node pairs are either in phase (adhesion) or out of phase (sliding) under the influence of the external load. However, in some advancing contact problems, the nodes which become restrained in the tangential traction direction for the current load increment may have undergone some slip during previous load increments. In such a situation, the concept of *Partial Slip* was introduced [19].

7.2.4   Local Coordinate System

Variables along the boundary outside the contact zone are represented in the global coordinate system defined by the Cartesian axis, whereas variables inside the contact region are analyzed in the local coordinate system.

Consider a node pair, a and b, with positions at (x^a, y^a) and (x^b, y^b) on two boundaries which are potentially coming into contact. The unit normal vectors $\mathbf{n}$ and the unit tangent vectors $\mathbf{s}$ at a and b are denoted by $\mathbf{n}^a$, $\mathbf{n}^b$ and $\mathbf{t}^a$, $\mathbf{t}^b$, respectively, as shown in Figure 7.1. An average normal $\tilde{\mathbf{n}}$ can be defined by

$$\tilde{\mathbf{n}}^a = \frac{\mathbf{n}^a - \mathbf{n}^b}{|\,\mathbf{n}^a - \mathbf{n}^b\,|} = -\tilde{\mathbf{n}}^b \tag{7.1}$$

7.2.5   Definition of Normal Gap

The normal distance between a node pair is defined by an initial normal gap g_o, as shown in Figure 7.2, from which the normal gap g_o is calculated as follows:

$$\begin{aligned} g_o &= |\,dy\,|\cos\theta + |\,dx\,|\sin\theta \\ &= |\,dy\,|\tilde{n}_y^a + |\,dx\,|\tilde{n}_x^a \end{aligned} \tag{7.2}$$

where $\tilde{n}_y^a$ and $\tilde{n}_x^a$ are the normal and tangential components of the average normal vector $\tilde{\mathbf{n}}$.

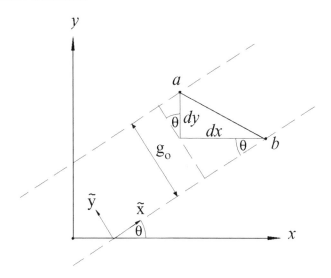

Figure 7.2: Definition of initial normal gap.

7.2.6   Numerical Modelling Concept

The numerical process of development of the contact area can be simulated discretely in the following way. As the load is applied gradually to a node pair a, b initially in separation with a gap between them, the nodes will separately undergo deformation and eventually come into contact for some particular load (i.e. the gap reduces to zero). If the external load continues to increase after nodes a, b have made contact, then a, b will be enveloped by the contact area. However, if the external load ceases to increase after nodes a, b have just made contact then node ab will become the edge of the contact area. Numerically this is the most desirable state to be reached, since contact boundaries are discretized into nodal points. An important criterion for an edge node-pair is to be just barely in contact with an external load which is just large enough to hold them in place. The second criterion which an edge node-pair must fulfil is that traction continuity in the normal and tangential directions must be maintained without violating any of the contact modes. It was suggested [21] that such criteria can be fulfilled only if the slip contact condition is prescribed for the edge node-pair, since in this mode both normal and tangential tractions are zero. Many contact problems are of the non-conforming type, that is the bodies have different initial profiles in the potential contact region. Non-conforming type problems are load dependent (since the actual extent of the contact area is not known a priori), and a method has to be employed that can evaluate the load increment corresponding to each increase or decrease in the area of contact. However, within each newly developed contact area, the contact status may not now automatically satisfy the equilibrium state. In order to satisfy both the load step and the contact state simultaneously, an iterative procedure is always required to overcome this nonlinear aspect of the problem.

In the absence of friction, the contact problem can be solved iteratively without the need for small incremental load steps. The iterative process consists of iterations for finding a contact area which must be consistent with the applied load. If the contact stresses for all the contact pairs are in equilibrium and no geometrical incompatibilities

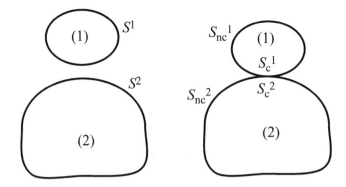

Figure 7.3: Contact region S_c and non-contact region S_{nc}.

occur, either inside or outside the contact zone, the correct contact solution has been found. When friction is taken into consideration, iterative techniques are required to calculate the size of the contact area and the extent of both the sliding and stick regions for a given external load. Load incremental procedures are also required to solve the problem in several load increments so that an accurate final solution can be obtained by following the loading history.

The size of each load increment may be obtained using a load scaling technique, in which a load increment must be scaled in such a way that a new element contact pair is either just established or just released by the total load. Since contact area is a function of external load, the size of a load increment, in this case, is therefore governed by the element size. For this reason, if a load increment technique is to be used the element size must not be too large, otherwise an accurate load history will not be obtained. Alternatively, a constant-load-increment technique may be employed so that a small pre-determined, constant load is added step by step until the final load is reached. Whilst both load incremental techniques can be used to solve a nonlinear contact problem, the load scaling technique is considered to be more efficient. The constant-load-increment technique is slow, and convergence cannot generally be guaranteed if the chosen load increment is too large.

7.3   Elastostatic Contact

Let the boundary of two elastic bodies 1 and 2 be represented by S^1 and S^2. Generally, when the bodies come into contact only parts of each boundary are in contact with each other, therefore their total boundaries may be divided into regions of contact boundary S_c and non-contact boundary S_{nc}, as shown in Figure 7.3, i.e.

$$
\begin{aligned}
S^1 &= S_{nc}^1 + S_c^1 \\
S^2 &= S_{nc}^2 + S_c^2.
\end{aligned}
\tag{7.3}
$$

A contact boundary S_c may contains regions of STICK (st), regions of SLIP (sl) and regions of SEPARATION (sp). Hence S_c for bodies $1, 2$ may be expressed as

$$
S_c^{1,2} = S_{st}^{1,2} + S_{sl}^{1,2} + S_{sp}^{1,2}.
\tag{7.4}
$$

If two bodies subjected to an external load are in contact over an area S_c, then the deformation can be described by two coupled integral equations, one for each body, to give

$$C_{ij}^1(\mathbf{x}')u_j^1(\mathbf{x}') + \int_{S_{nc}^1} T_{ij}^1(\mathbf{x}',\mathbf{x})u_j^1(\mathbf{x})dS^1 + \int_{S_c^1} T_{ij}^1(\mathbf{x}',\mathbf{x})u_j^1(\mathbf{x})dS^1$$

$$= \int_{S_{nc}^1} U_{ij}^1(\mathbf{x}',\mathbf{x})t_j^1(\mathbf{x})dS^1 + \int_{S_c^1} U_{ij}^1(\mathbf{x}',\mathbf{x})t_j^1(\mathbf{x})dS^1 \qquad (7.5)$$

$$C_{ij}^2(\mathbf{x}')u_j^2(\mathbf{x}') + \int_{S_{nc}^2} T_{ij}^2(\mathbf{x}',\mathbf{x})u_j^2(\mathbf{x})dS^2 + \int_{S_c^2} T_{ij}^2(\mathbf{x}',\mathbf{x})u_j^2(\mathbf{x})dS^2$$

$$= \int_{S_{nc}^2} U_{ij}^2(\mathbf{x}',\mathbf{x})t_j^2(\mathbf{x})dS^2 + \int_{S_c^2} U_{ij}^2(\mathbf{x}',\mathbf{x})t_j^2(\mathbf{x})dS^2 \qquad (7.6)$$

where S_{nc} is the area not in contact.

For a numerical solution to the problem, the boundary integral equations (7.5) and (7.6) of the boundaries of body 1 and body 2 are discretized separately. The two resulting sets of matrices can be written as

$$\mathbf{H}^1\mathbf{u}^1 = \mathbf{G}^1\mathbf{t}^1 \qquad \mathbf{H}^2\mathbf{u}^2 = \mathbf{G}^2\mathbf{t}^2 \qquad (7.7)$$

where vectors $\mathbf{u}^{1,2}$ and $\mathbf{t}^{1,2}$ represent boundary values of displacements and tractions. At the contact region, the two systems of equations share the boundary variables of the problem; that is the equations are coupled and must be solved simultaneously for any given combination of external load and contact conditions. Once the boundary conditions inside and outside any region of contact are implemented, equation (7.7) is reduced to the single unified system of equations given by

$$\mathbf{A}\mathbf{x} = \mathbf{f} \qquad (7.8)$$

For linear problems, once the above system of equations has been solved, the final solution for the displacements and tractions everywhere on the boundaries can be obtained. However, contact problems may be nonlinear, and the extent of the contact region may not be known *a priori* and must be determined as part of the solution. This means some contact problems require an iterative solution procedure. During the iterative process, coefficients in $\mathbf{A}$ and $\mathbf{f}$, derived from inside the contact zone only, may change from one iteration to the next; the number of changes in matrix $\mathbf{A}$ is small, because the number of elements whose contact conditions change is usually a small fraction of the total. Under normal procedures the entire system of equations would have to be reordered for the next iteration, in order to accommodate the changes in the contact zone. Repetition of this procedure, until the final solution is found, would be inefficient and costly.

In order to solve the updated system of equations efficiently and without recourse to a reformulation of the entire system matrix, it is necessary to keep the unknowns in the potential contact zone apart from the unknowns in the non-contact region. For the potential contact zones, the equations obtained from the contact conditions have to be expressed explicitly, so that they can be separated from those outside the zone. In this way, a coefficient sub-matrix $\mathbf{D}_c$ can be set up for the variables of the contact (potential) region. This separation of unknowns enables equation (7.8) to be arranged in such a way that the **A** matrix form of the merged system is as shown below:

$$
\begin{bmatrix}
H_{nc}^1 - G_{nc}^1 & H_c^1 & -G_c^1 & & H_{nc}^2 - G_{nc}^2 \\
0 & & & H_c^2 & -G_c^2 & 0 \\
0 & & D_c & & & 0
\end{bmatrix}
\{x\} = \{f\}
\qquad (7.9)
$$

The subscripts nc and c denote non-contact and contact (potential or actual) zones, respectively.

If the total number of nodes outside the potential contact zone for body 1 and body 2 is M_{nc}^1 and M_{nc}^2, respectively, this results in $2(M_{nc}^1 + M_{nc}^2)$ linear equations, since there are two unknowns per node. Inside the contact zone at each boundary point both traction components and both displacement components are unknown. Hence, for a pair of contact nodes there are eight unknowns. To account for these eight unknowns, eight equations are provided by considering displacement compatibility and traction equilibrium at the contact interface. These compatibility and equilibrium equations can be derived explicitly for each potential contact node pair by considering the contact state of the node pair itself and its immediate neighbouring node pairs, to give systems of contact equations.

A system of contact equations can then be grouped in a matrix form, so that they can be readily incorporated into the assembly of the final matrix. A contact region may contain a combination of SEPARATION, STICK and SLIP element pairs. The order in which they occur depends on the problem involved. Systems of contact equations thus have to be formulated to cope with any possible situation. These contact equations can be described as 'boundary conditions' or 'boundary constraints' inside the contact region for all the potential contact element pairs. For example, if a SEPARATION changed to a STICK mode for node-pair a and b, then the following contact conditions, expressed in matrix form, must be obeyed:

$$
\begin{bmatrix}
1 & 0 & 0 & 0 & 1 & 0 & 0 & 0 \\
0 & 1 & 0 & 0 & 0 & 1 & 0 & 0 \\
0 & 0 & 1 & 0 & 0 & 0 & -1 & 0 \\
0 & 0 & 0 & 1 & 0 & 0 & 0 & -1
\end{bmatrix}
\begin{Bmatrix}
u_t^a \\
u_n^a \\
t_t^a \\
t_n^a \\
u_t^b \\
u_n^b \\
t_t^a \\
t_n^b
\end{Bmatrix}
=
\begin{bmatrix}
0 \\
gap^{ab} \\
0 \\
0
\end{bmatrix}
\qquad (7.10)
$$

where gap^{ab} is the normal gap between node a and b. If a and b were initially in contact, then gap^{ab} would simply be zero.

Once these contact equations are defined they are then grouped to form the co-efficient sub-matrix called **D**$_c$, as shown in equation (7.9). The final matrix is now square, and equation (7.10) is in a standard form which is solvable; the vector **x** contains all the unknowns, both inside and outside the contact region, and the vector **f** contains all the boundary conditions from outside the contact region. Solution of the matrix equation can be obtained by standard procedures and all boundary values determined. The advantage for such a structured formulation is that time is saved by avoiding reassembly of the entire matrix when only a few changes are made to the contact region (i.e. $\mathbf{D}_c$). More importantly, a high speed matrix solver exploiting this feature can be employed, such that the solution process of the system of equations is speeded up considerably (see Appendix G). This technique can substantially speed up the iterative solution process, and has been shown to be highly efficient for solving contact problems [21, 23].

7.3.1 Numerical Modelling

Discretization Strategy

In the numerical modelling of contact problems, the two bodies are brought into contact at discrete points (boundary nodes) along the common boundary (contact interface). In the initial unloaded state, node pairs may either be in contact or in separation. Under the influence of an external applied load, the bodies will respond by undergoing a deformation which may lead to different possible combinations of stick, slip and separation taking place in the potential contact region.

It is essential when solving a contact problem that relatively small elements are prescribed inside the potential contact area. This requirement is needed to ensure that both the displacements and tractions are adequately modelled when contact surfaces come into, or out of, contact. Since the gradient of these quantities can be very large towards the edge of contact, it may be difficult to find a suitable function to represent such acute variations within the contact region.

Element Type

Linear and quadratic elements were chosen to solve contact problems [19]. Isoparametric linear elements are used in the contact regions and isoparametric quadratic elements are used outside the contact region. It was reported [20] that linear variations were more suitable to represent the contact surfaces, and that converged and stable contact solutions have been obtained for both load dependent or load independent frictional contact problems. On the other hand, when quadratic elements were used, contact solutions were shown to oscillate, and often failed to converge. The cause of the oscillations, is due to inherent difficulties that generally prevent the correct contact conditions being achieved at both mid-side and end nodes simultaneously. These difficulties are eliminated by using linear elements locally, and by permitting only one change in the contact conditions per load step [20].

Load Application

The final solution of a contact problem is usually governed by either the final maximum external load or the ultimate allowable contact area. However, since in most contact problems the contact area is a function of the external loads and the friction coefficient inside the contact region, the numerical solution of the problem has to be obtained iteratively; and if it is a non-conforming contact problem with friction, then the solution of the problem may have to be approximated by way of many small load steps. The way in which load steps are applied is highly critical in view of the nonlinear nature of the problem. The minimum requirement for the contact solution to be accepted is that the contact status along the contact area must satisfy equilibrium and compatibility at all times. However, this requirement, as will be discussed next by comparing different methods of load application, is not unique since different methods of load application can lead to different contact solutions. The contact solutions obtained are dependent upon both the load and the contact history. Methods of load application may be categorized into [21]:

- Total Loading – that is solving the problem with only one total load step and performing iterations until stick and slip areas are found and satisfied. This amounts to linearizing the problem and, as such, this technique is valid only for those contact problems which are associated with the so-called node-on-node contact in the stick region.

- Pseudo Incremental Loading – in this technique, a series of increasing total loads is applied successively to the initial problem, with the contact conditions updated before each loading. For instance, consider that a total load P^m has been applied and the contact conditions obtained by using the Total Loading technique described above. Keeping these contact conditions as the initial conditions, a new total load

$$P^{m+1} = P^m + \Delta P \tag{7.11}$$

is applied and the contact conditions re-determined. The additional load ΔP may be obtained either by dividing the maximum load P_{max} into a pre-defined number of equal load steps, M say,

$$P^{max} = \sum_{m=1}^{M} \Delta P^m \tag{7.12}$$

where ΔP^m represents the equal size increment load step and m is the mth step; alternatively, ΔP^m is obtained by scaling to bring a node or an element into contact. The load increment ΔP^m is then adjusted iteratively, until the stick or slip contact conditions are satisfied.

In this approach, some history of the contact condition is retained, however it always leads to node-on-node contact in the stick region and, as shown [20], sharp peaks occur in the tangential tractions at the transition from stick to slip regions.

- Fully Incremental Loading – The 'Fully Incremental Loading' differs from the "Pseudo Incremental Loading" because the deformed geometry and the contact status inside the contact region are updated through the use of previous boundary displacement and contact traction conditions, after each load increment; so only ΔP^m rather than P^m is required at each stage of loading [19]. The advantage in this approach is that both the contact history and load history are naturally retained such that the problem is always solved for the incremental perturbations of the contact bodies due to the load increment (ΔP^m). This approach can follow the history well, and can deal with both node-on-node or non-node-on-node contact, without introducing sharp transitions in the contact solutions.

Prediction of Direction of Relative Slip

The determination of slip direction is an essential requirement for a reliable and automatic procedure to be implemented in a contact algorithm.

The first assumption to be made when a node-pair (a and b say) are brought into contact is its mode of contact. It is either in Stick mode or Slip mode, depending on the magnitude of the external load that brings them into contact and the nature of their contact. If Stick mode is assumed, then both traction equilibrium and displacement compatibility must be enforced as prescribed by the contact constraints below:

$$t_t^a - t_t^b = 0, \qquad t_n^a - t_n^b = 0, \qquad u_t^a + u_t^b = 0, \qquad u_n^a + u_n^b = g_0 \qquad (7.13)$$

However, if Slip mode is assumed then slip mode constraints must be applied, namely

$$t_t^a - t_t^b = 0, \qquad t_n^a - t_n^b = 0, \qquad t_t^a \pm \mu t_n^a = 0, \qquad u_n^a + u_n^b = g_0 \qquad (7.14)$$

Here, the Slip contact mode is controlled by the Coulomb's friction model, which is represented by

$$t_t \pm \mu t_n = 0 \qquad (7.15)$$

In the above relationship, the sign of the frictional traction t_t is unknown, *a priori*. It has to be assigned such that the resultant frictional traction always opposes the relative tangential displacement in the contact region. However, without prior knowledge the direction of the relative displacement, the sign of t_t cannot be decided and the problem cannot be solved directly.

In order to solve a frictional contact problem without violating the principles of mechanics, it must be ensured that the frictional forces between the contacting bodies always oppose the relative tangential slip. In mathematical terms, the direction of the frictional force must be anticipated prior to execution of the problem. A simple and effective technique developed in [21] to deal with this requirement is described next.

From equation (7.14), if μ is set to zero, the relationships of frictionless contact are obtained as

$$t_t^a - t_t^b = 0, \qquad t_n^a - t_n^b = 0, \qquad t_t^a = 0, \qquad u_n^a + u_n^b = g_0$$

In this state, a frictionless problem can be solved easily without any decision being required regarding the tangential tractions. It was thus proposed [19] to solve the problem in the frictionless state first, and use this contact solution to predict the behaviour of the same problem but with a non-zero coefficient of friction. That is, the sign of the tangential traction is decided by observing the relative tangential slip direction of the contact surfaces in the frictionless state, and setting the frictional tractions ($\pm \mu t_n$) to oppose that observed slip. The same problem, based on this assumption, can then be solved and checked for violation in Slip mode alone.

The advantage in this approach is that the sign of the tangential traction is automatically computed without any complicated trial and error procedures.

7.3.2   Fully Incremental Loading Technique

Consider a discrete incremental load step ΔP_j^m applied to a system which is initially in equilibrium. The system will respond to the applied load by undergoing small perturbations in displacements and tractions everywhere on the boundary, to give a new equilibrium state. The new total external discrete load is defined as

$$P_j^m = P_j^{m-1} + \Delta P_j^m \tag{7.16}$$

and the changes in displacements u_j and tractions t_j can be defined as

$$u_j^m = u_j^{m-1} + \Delta u_j^m \quad \text{and} \quad t_j^m = t_j^{m-1} + \Delta t_j^m \tag{7.17}$$

where Δt_j^m and Δu_j^m are the incremental changes in tractions and displacements, respectively, due to the increment of load ΔP_j^m. Substituting equation (7.17) into displacement boundary integral equation (2.112) gives

$$C_{ij}(u_j^{m-1} + \Delta u_j^m) + \int_S T_{ij}(u_j^{m-1} + \Delta u_j^m)dS = \int_S U_{ij}(t_j^{m-1} + \Delta t_j^m)dS \tag{7.18}$$

From the principle of superposition, an integral equation in the Δ(incremental) variables can be obtained:

$$C_{ij}\Delta u_j^m + \int_S T_{ij}\Delta u_j^m dS = \int_S U_{ij}\Delta t_j^m dS \tag{7.19}$$

The nodes on the discretized boundaries have to be organized at the contact region in such a way that they form contact node pairs, representing the contact interface. Each contact node-pair has to be coupled and treated as an independent contact system where both the incremental traction Δt and the incremental displacement Δu must satisfy both equilibrium and compatibility conditions. Discretization of two bodies (1 and 2) produces two individual systems of equations, see equations (7.7); they are given, in incremental terms, by

$$\mathbf{H}^1\Delta\mathbf{u}^1 = \mathbf{G}^1\Delta\mathbf{t}^1 \quad \text{and} \quad \mathbf{H}^2\Delta\mathbf{u}^2 = \mathbf{G}^2\Delta\mathbf{t}^2 \tag{7.20}$$

The vectors $\Delta\mathbf{u}^{1,2}$ and $\Delta\mathbf{t}^{1,2}$ contain the boundary values of incremental displacements and tractions. The solution of this incremental problem can be obtained in the usual way, once the boundary conditions outside the contact region and the contact constraints inside the contact region are applied.

In the incremental load (ΔP) approach presented in equation (7.19) contact conditions of a node pair a and b have to be expressed in a form that is in accordance with the incremental load; so that, as each ΔP^m is applied, the incremental quantities ($\Delta u, \Delta t$) inside the contact region are obtained subject to the total traction equilibrium and the total displacement compatibility at the node-pair. The constraints for the different contact conditions are as follows:

Separation mode

$$
\begin{aligned}
\Delta(t_t^a)^m - \Delta(t_t^b)^m &= -\left[(t_t^a)^{m-1} - (t_t^b)^{m-1}\right] \\
\Delta(t_n^a)^m - \Delta(t_n^b)^m &= -\left[(t_n^a)^{m-1} - (t_n^b)^{m-1}\right] \\
\Delta(t_t^a)^m &= -\left[(t_t^a)^{m-1}\right] \\
\Delta(t_n^a)^m &= -\left[(t_n^a)^{m-1}\right]
\end{aligned}
\tag{7.21}
$$

Slip mode

$$\Delta(t_t^a)^m - \Delta(t_t^b)^m = -\left[(t_t^a)^{m-1} - (t_t^b)^{m-1}\right]$$

$$\Delta(t_n^a)^m - \Delta(t_n^b)^m = -\left[(t_n^a)^{m-1} - (t_n^b)^{m-1}\right]$$

$$\Delta(t_t^a)^m \pm \mu\Delta(t_n^a)^m = -\left[(t_t^a)^{m-1} \pm \mu(t_n^a)^{m-1}\right]$$

$$\Delta(u_n^a)^m + \Delta(u_n^b)^m = g_o - \left[(u_n^a)^{m-1} + (u_n^b)^{m-1}\right] \equiv g_0^m \qquad (7.22)$$

Stick or Partial Slip mode

$$\Delta(t_t^a)^m - \Delta(t_t^b)^m = -\left[(t_t^a)^{m-1} - (t_t^b)^{m-1}\right]$$

$$\Delta(t_n^a)^m - \Delta(t_n^b)^m = -\left[(t_n^a)^{m-1} - (t_n^b)^{m-1}\right]$$

$$\Delta(u_t^a)^m + \Delta(u_t^b)^m = 0$$

$$\Delta(u_n^a)^m + \Delta(u_n^b)^m = g_o - \left[(u_n^a)^{m-1} + (u_n^b)^{m-1}\right] \equiv g_0^m \qquad (7.23)$$

The previous total tractions and total displacements (t^{m-1}, u^{m-1}) which are known make up the right-hand-sides, they are updated from load-step to load-step. The unknown quantities are the (Δ) incremental values on the left-hand-sides, as shown in equation (7.21) to (7.23); they are expressed in the average local coordinate system.

Decision on Contact State

The verification of the contact status, at any step, is based on the contact decisions shown in Tables 7.2 and 7.3 for the contact node-pair a and b. Table 7.2 is used to check whether SEPARATE pairs remain in separation and whether CONTACT pairs remain in contact. Notice that violation in Table 7.2 represents geometrical incompatibility, and must not occur at any stage of the calculations. Table 7.3 is used

Table 7.2: Examine modes of contact

Assumption	Decision	
	separate	contact
SEPARATE	$(\Delta u_n{}^a + \Delta u_n{}^b)^m < g_0^{m-1}$	$(\Delta u_n{}^a + \Delta u_n{}^b)^m \geq g_0^{m-1}$
CONTACT	$t_n^{m-1} + \Delta t_n^m \geq 0$	$t_n^{m-1} + \Delta t_n^m < 0$

to differentiate STICK and SLIP for those node pairs that are in contact. Detection is made on these node pairs for violations of tangential and normal traction equilibrium. Violation in Table 7.3 generally implies that the traction continuity over the contact region has been violated, therefore redistribution of the traction is required. Any redistribution of traction inside the contact region is achieved by means of varying the sizes of stick and slip regions. In fact, the determination of an optimum STICK/SLIP partition constitutes the second iterative process in the present numerical analysis. At this stage, if STICK or SLIP conditions on node pairs have been violated, then this must be rectified by re-setting to the new calculated contact mode. In this new contact state, ΔP^m may have to be readjusted so that the edge node pairs are maintained just in contact; then the re-calculated solutions have to be re-examined once more

Table 7.3: Examine contact status

Assumption	Decision	
	stick	slip
STICK	$\|t_t^{m-1} + \Delta t_t^m\| \leq \|\mu(t_n^{m-1} + \Delta t_n^m)\|$	$\|t_t^{m-1} + \Delta t_t^m\| \geq \|\mu(t_n^{m-1} + \Delta t_n^m)\|$
SLIP	$(t_t^{m-1} + \Delta t_t^m)^b(\Delta u_t{}^a + \Delta u_t{}^b)^m > 0$	$(t_t^{m-1} + \Delta t_t^m)^b(\Delta u_t{}^a + \Delta u_t{}^b)^m \leq 0$

over the entire contact region. Further load increment is allowed only if both of these iterative processes are completely satisfied.

Load Scaling Technique

In order to satisfy the contact conditions during a load increment, loads have to be adjusted such that the physical contact of the node pairs does not violate any of the contact conditions. The applied load takes into account the response of the tractions and displacements at the edges of the contact region.

Since the exact ΔP^m required to bring the next node pair into contact is not known *a priori*, it has to be predicted, based on the loading history and the linear elastic response of the system. A linear extrapolation technique involving two load steps has been devised [19] to enable ΔP^m to be scaled in such a way that it will satisfy the Hertz's condition that normal tractions at the edges of contact should approach zero (i.e. traction continuity at the transition from contact to free surface has been attained). This condition is important for the prediction of the scaling factor, α_i. Exploitation of the linear elastic response of the system enables the present contact state to be extrapolated into the future contact state of the problem. Linear extrapolation provides a means of calculating the load scaling factor α_i systematically, with a fast rate of convergence. From Figure 7.4, it may be deduced that

$$\frac{P' - P_i}{t' - t_i} = \frac{P_{i+1} - P_i}{t_{i+1} - t_i} \tag{7.24}$$

where P' is an arbitrary load and the subscript i denotes the ith step. P_i and t_i denote the external load and the local normal tractions at the contact interface respectively.

At the edge of the contact zone, the traction should be zero (i.e. $t_{i+1} = 0$). Equation (7.24) can now be re-defined in terms of a small incremental value of load δP and the incremental traction response δt, such that

$$\frac{\delta P}{\delta t} = \frac{\alpha_i * \delta P}{t_i} \tag{7.25}$$

where α_i is the load scaling factor defined by

$$\alpha_i = \frac{t_i}{\delta t} \tag{7.26}$$

The trial incremental load δP must be sufficiently small for the contact conditions are unaltered, and the next load increment ΔP is then given by

$$\Delta P = \alpha_i * \delta P \tag{7.27}$$

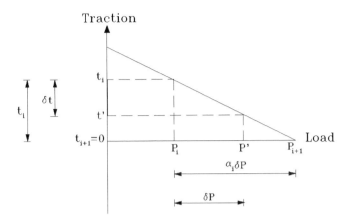

Figure 7.4: Load scaling diagram.

This load increment is then used to bring a target-node-pair (node-pair nearest to the edge of contact) into contact.

An alternative approach based on the displacement response of the system has also been used [15, 30]. The load P_{i+1} is extrapolated by means of closing the normal gap of the targeted node pair, using a similar procedure to the traction extrapolation technique. Whilst either traction or displacement extrapolation will provide a load scaling factor for the purpose of accurately closing a gap, the displacement extrapolation technique, at first glance, may seems to be more appealing because the size of the normal gap ahead of the contact zone is already known. Therefore the additional load required to close this gap can be estimated directly, based on the previous loading history. However, using this technique, traction continuity at the transition from contact to free surface cannot generally be obtained in one scaling, whereas the traction extrapolation technique can be precise. Similarly, in the case of receding contact, traction extrapolation will again be more appropriate as the normal gaps are zero inside the contact zone.

Iterative and Fully-incremental Loading Procedures

In most contact problems the contact area is a function of the external loads and the friction coefficient inside the contact region. A numerical solution of the problem has to be obtained iteratively; and if it is a non-conforming contact problem with friction, then the loading in the problem may have to be approximated by many small load steps. The way in which the load steps are applied is critical in view of the nonlinear nature of the problem. The basic requirement for the contact solution to be acceptable is that the contact status along the contact area must satisfy equilibrium and compatibility at all times. Therefore, in order to satisfy the contact conditions during a load increment, loads have to be scaled such that the physical contact of the node pairs does not violate any of the contact conditions.

The iterative and incremental process starts with a small initial load P_0 and a small contact area. The system of contact in this initial state is then solved for the tractions and displacements on the boundaries. The initial load P_0 may have to be adjusted

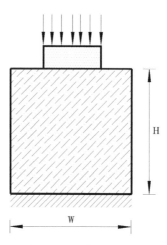

Figure 7.5: An elastic punch resting on a flat foundation.

to ensure that the surface tractions are continuous everywhere on the contact bodies. This requirement can be accomplished using the traction extrapolation load scaling technique described earlier, such that an exact load scaling factor can be predicted. Once an acceptable combination of load and contact area is found, the next step is to verify the assumed contact status for the entire contact region. The verification of the contact status of a node-pair a, b (say) is based on the following contact conditions:

- STICK node-pair remains STICK if the tangential traction is less than the traction required to cause slip, that is $| t_t | < \mu | t_n |$.

- SLIP node-pair remains SLIP if the relative tangential displacement is opposed by the tangential traction, that is $Sign(t_t) \neq Sign(\Delta u_t^a + \Delta u_t^b)$.

An intermediate contact solution can be accepted once all the conditions are satisfied. The same process is repeated for each load increment until the maximum total load or the maximum allowable contact area is reached.

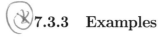7.3.3   Examples

In this section, the boundary element contact algorithm described in the previous sections is applied to certain benchmark test problems, in order to demonstrate the reliability and accuracy of the method.

Flat Punch on an Elastic Foundation

A flat punch resting on an elastic foundation is shown in Figure 7.5. The punch has a width $2w$ and height h and the foundation has width $2W$ and height H. This is a conforming type of problem, where the contact area is known in advance. However, in the presence of friction the partition between sticking and sliding zones must be found iteratively. The problem was solved in [19] using the Total Loading technique. The dimension ratios considered were $h/w = 8/5$, $H/W = 5/4$ and $w/W = 5/8$.

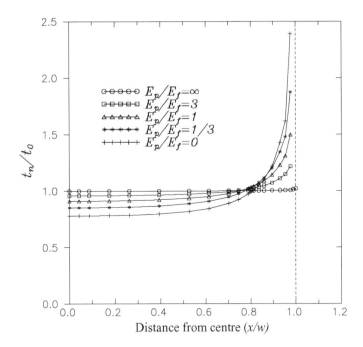

Figure 7.6: Normalized normal traction distribution of an elastic punch on a flat foundation with different values of E_p/E_f.

Constant uniform compressive load $t_o = 1.0$ Nmm^{-2} was applied on the upper face of the elastic punch and the Poisson's ratio was assumed to be $\nu = 0.35$. The results for the frictionless problem $\mu = 0$ are shown in Figure 7.6.

The Young's modulus of the punch and foundation are denoted as E_p and E_f, respectively. It can be observed that as the punch becomes highly stiff compared to the foundation $E_p/E_f = \infty$, the normal traction distribution tends to a constant equal to the applied stress. At the sharp edge of the punch, tractions become very high as a singularity exists in the elastic solution.

The results for $\mu = 0.2$ are shown in Figure 7.7. With friction the problem becomes nonlinear, as both Stick and Slip conditions will occur simultaneously on different parts of the contact surface. The tangential traction components are normalized by μt_o in the Figure. Inside the stick zone $|x/w| \leq 0.8$ the tangential components are less than the normal tractions t_n/t_o. At the point of transition from stick→slip, the normal and tangential components are equal (i.e. $t_n = t_t/\mu$). Man et al. [19] reported that when friction is involved the tangential tractions at the stick/slip partition (common node) are highly sensitive to the discretization of the stick/slip region. If the partition of the stick/slip was not correctly prescribed for a given load, discontinuous tangential tractions will occur at this common node.

Perfect-fit, Loaded Pin in a Hole in a Plate

Consider the problem of a perfect fit pin in a hole in a large plate loaded by a localized force P as shown in Figure 7.8. The radius of pin $R = 6$mm and the same material

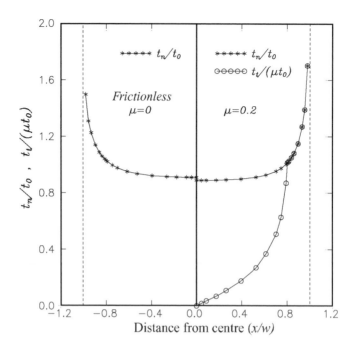

Figure 7.7: Normalized contact tractions for a punch on an elastic foundation.

is assumed for the pin and the plate with Young's modulus $E = E_{pin} = E_{plate} = 1000 \text{Nmm}^{-2}$ and Poisson's ratio $\nu = 0.3$. Contact traction solutions normalized with respect to E are presented in the form of normal and tangential distributions for $\mu = 0$ and 0.4 and are shown in Figure 7.9. In the case $\mu = 0.4$ the problem is solved iteratively to obtain both the contact area and the partition of the stick and slip region. Since both the size of the contact area and the partition are, in fact, independent of the magnitude of the point load for the perfect-fit pin, this problem can be solved in one load step. It can be seen from Figure 7.9 that the normal traction distribution is lower and the contact area is larger in the presence of friction ($\theta > 0^o$). This is attributed to the fact that friction (tangential forces) is also carrying a part of the load.

Loose-fit, Loaded Pin in a Hole in a Plate

The configuration chosen in [19] for BEM analysis is the same as that in [16] and [3], that is a pin with a radius $r = 5.999 \text{cm}$ is inserted in a hole, of radius $R = 6 \text{cm}$, in an infinite plate. Both the pin and the plate have Young's modulus $E = 2.1 \times 10^5 \text{Nmm}^{-2}$ and Poisson's ratio $\nu = 0.3$. The pin is loaded through a small cavity at its centre by the traction field derived from that of a localized force P in an infinite plate [19] as shown in Figure 7.10. For the frictionless case, the problem is linear, therefore it can be solved by applying the total load in one load step, with iterations, to find the correct contact area. The frictionless solutions obtained were found to be in good agreement with the exact solutions of [16].

When the friction is taken into account, the problem is nonlinear since it is load

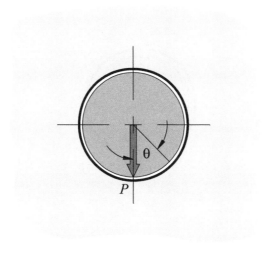

Figure 7.8: Perfect-fit, loaded pin in a hole.

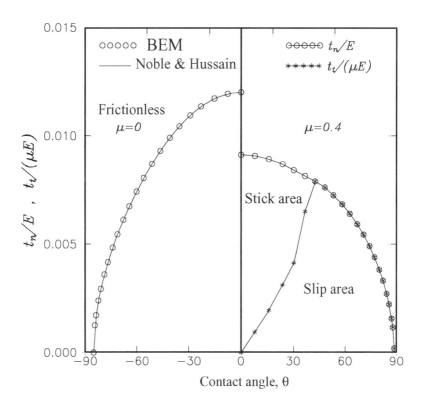

Figure 7.9: Contact tractions for a perfect fit pin in a hole.

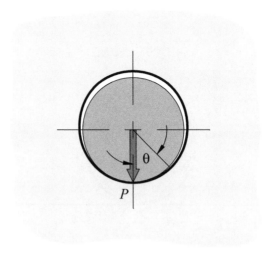

Figure 7.10: Loose-fit, loaded pin in a hole.

dependent; it must therefore be solved using the incremental loading technique, with many small load steps. In Figure 7.11, the normalized tangential tractions, $t_t/(\mu E)$, and the normalized tractions, t_n/E, have been plotted against the contact angle $(\theta < 0^o)$ for all the load steps. In the absence of load, contact is made initially at the node-pair on the line of symmetry, which remains in stick contact throughout all the subsequent loadings. In this problem, a high coefficient of friction, $\mu = 0.4$, is considered in the contact region, and consequently a large stick zone results. The incremental solutions (see Figure 7.11, $-60^o < \theta < 0^o$) show that the increasing contact area is initially in stick mode, except for the node at the edge of the contact region, which must be in Slip mode. The stick region continues to grow in size under the influence of the increasing external load, that is Partial Slip always occurs, until further slip node-pair appears at the 16th step. After the 16th step, the remaining load increments produce further slip node-pairs towards the edge of contact, that is no further Partial Slip.

Cylinder on an Elastic Foundation

Boundary element solutions of a cylinder on an elastic foundation subjected to a constant point load $P = 1500$ Nmm^{-1}, as shown in Figure 7.19, have been obtained by Man *et al.* [19]. The dimensions of the problem are $H/W = 1$, $W/R = 3.8$ and $R = 50$ mm, where R is the radius of the cylinder, and H and W are the height and width of the elastic foundation, respectively. The results for different ratios of elastic constant, $E_{cylinder}/E_{foundation} = E_c/E_f$, are shown in Figure 7.13 for frictionless condition and $\nu = 0.35$. Plane strain condition was assumed. The problem in this case is a non-conforming type with the extent of the contact region being unknown. The BEM results are in good agreement with the well known Hertz solution.

The same problem was also solved when friction is present. In this case, the problem is load dependent and highly nonlinear. A Full Incremental Loading technique was used [21] to solve this problem. Figure 7.14 shows the incremental solutions at dif-

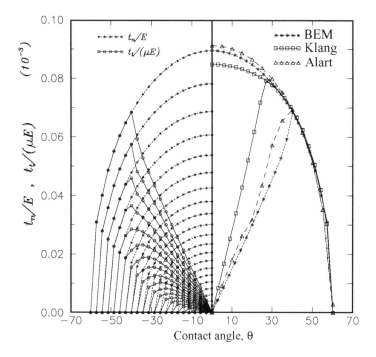

Figure 7.11: Contact tractions for a loose-fit, loaded pin with $\mu = 0.4$.

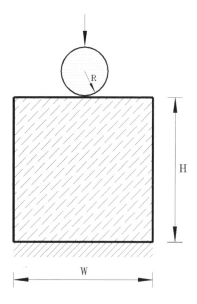

Figure 7.12: Cylinder on a flat foundation.

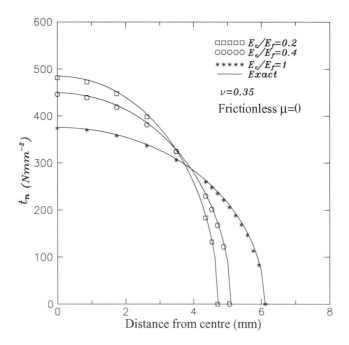

Figure 7.13: Contact tractions for a cylinder on a flat elastic foundation with $\mu = 0$.

ferent loads steps in order to observe the development of the slip zones. The solutions correspond to the case $E_c = E_f = 4000$ Nmm^{-2} with $\mu = 0.01$ and $\mu = 0.005$.

7.4   Elastoplastic Contact

In order to solve elastoplastic contact problems, it is necessary to use the elastoplastic equation (see Chapter 5) for all bodies involved in contact and couple them by enforcing the compatibility and equilibrium conditions. For example, for two bodies in contact, and using a superscript for each of them:

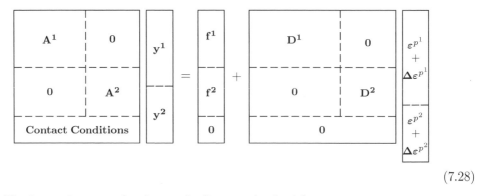

$$\tag{7.28}$$

The internal stresses for the two bodies are obtained from:

$$\sigma^1 = -\mathbf{A}'y^1 + \mathbf{f}'^1 + \mathbf{D}^{-1}(\varepsilon^{p1} + \Delta\varepsilon^{p1})$$

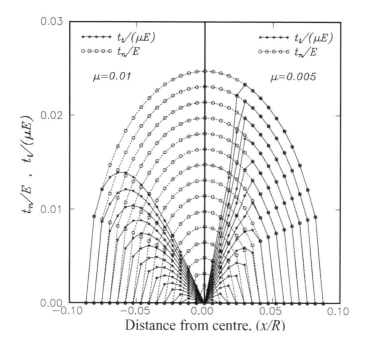

Figure 7.14: Normalized tractions for a cylinder on the flat foundation.

$$\sigma^2 \;=\; -\mathbf{A}'y^2 + \mathbf{f}'^2 + \mathbf{D}^{-1}\left(\varepsilon^{p2} + \Delta\varepsilon^{p2}\right) \tag{7.29}$$

The contact modes at each node-pair are not known in advance, and as a consequence they are first set arbitrarily to slip. Equations (7.28) and (7.29) are then solved for the current load step, where the iterative procedure to determine $\Delta\varepsilon^p$ must be carried out. Subsequently, the assumed contact modes are checked, and if they are found to be incorrect, they are updated and the system of equations is solved again, until no such violations are detected. Incremental techniques have already been introduced for the solution of plasticity and contact problems, although in the former case the load was applied through steps only after yielding had begun. In [24], an incremental strategy was developed, in order to cope from the beginning of the loading history with the nonlinear behaviour due to either frictional contact conditions and/or plasticity. This formulation is described next.

7.4.1   Incremental Formulation

Recall the boundary integral equation for elastoplastic problems, where the notation has been simplified, for the sake of clarity:

$$C_{ij}u_j + \oint_S T_{ij}u_j dS = \int_S U_{ij}t_j dS + \int_V \sigma_{ijk}\varepsilon^p_{jk} dV \tag{7.30}$$

The above equation describes a body of domain V and boundary S, which due to external loads has undergone displacements u_j, tractions t_j and plastic strains ε^p_{jk}. These fields are equilibrated, and compatible with the boundary conditions of the

problem. If they are supposed known, as a result of previous analysis, and an increment of external loads, ΔP, is considered, all tractions, displacements and plastic
strains will be perturbed, such that the following expression holds:

$$C_{ij}(u_j + \Delta u_j) + \int_S T_{ij}(u_j + \Delta u_j)dS = \int_S U_{ij}(t_j + \Delta t_j)dS + \int_V \sigma_{ijk}(\varepsilon_{jk}^p + \Delta\varepsilon_{jk}^p)dV$$

(7.31)

If the incremental quantities are now separated from the initial ones, and taking into
account the fact that the initial fields satisfy equation (7.30), an incremental boundary
integral equation is obtained

$$C_{ij}\Delta u_j + \int_S T_{ij}\Delta u_j dS = \int_S U_{ij}\Delta t_j dS + \int_V \sigma_{ijk}\Delta\varepsilon_{jk}^p dV \qquad (7.32)$$

The standard boundary element method procedure is applied next. After discretizing
the boundary S, and the parts of the domain where ε^p are expected to be non-zero,
equation (7.32) can be written in a matrix form as

$$\mathbf{H}\Delta\mathbf{u} = \mathbf{G}\Delta\mathbf{t} + \mathbf{D}\Delta\varepsilon^p \qquad (7.33)$$

and in a similar fashion, the stress equation becomes

$$\sigma = \mathbf{G}'\Delta t - \mathbf{H}'\Delta u + \mathbf{D}^*\Delta\varepsilon^p \qquad (7.34)$$

Rearranging equations (7.33) and (7.34) according to the boundary conditions, we
get

$$\mathbf{A}\Delta\mathbf{y} = \Delta\mathbf{f} + \mathbf{D}\Delta\varepsilon^p \qquad (7.35)$$

and

$$\Delta\sigma = -\mathbf{A}'\Delta\mathbf{y} + \Delta\mathbf{f}' + \mathbf{D}^*\Delta\varepsilon^p \qquad (7.36)$$

The increment of plastic strain $\Delta\varepsilon^p$ is the same vector appearing in elastoplastic
equations, and therefore, as stated in Chapter 6, it is determined iteratively. Once
convergence is achieved, all incremental quantities are added to the accumulated
values, and a new load step is allowed.

In Figure 7.15, the iterative and incremental solution procedure for elastoplastic
contact analysis is summarized in a flow chart.

The load steps are calculated using the method described in section 7.5.2. However, in the plastic range the assumption in equation (7.24) is no longer valid, and the
application of ΔP in general will give a non-zero value of the edge traction. Nevertheless, the load ΔP can still be used as a starting point for an iterative determination
of correct load increment [24].

7.4.2   Examples

In this section two examples are presented to demonstrate the application of the
elastoplastic BEM to conforming and non-conforming contact problems.

Perfect-fit Pin in a Plate

Consider a perfect-fit pin in a large plate. Figure 7.16 shows the geometry of the
problem using a half symmetry and the domain discretization of the area it is expected

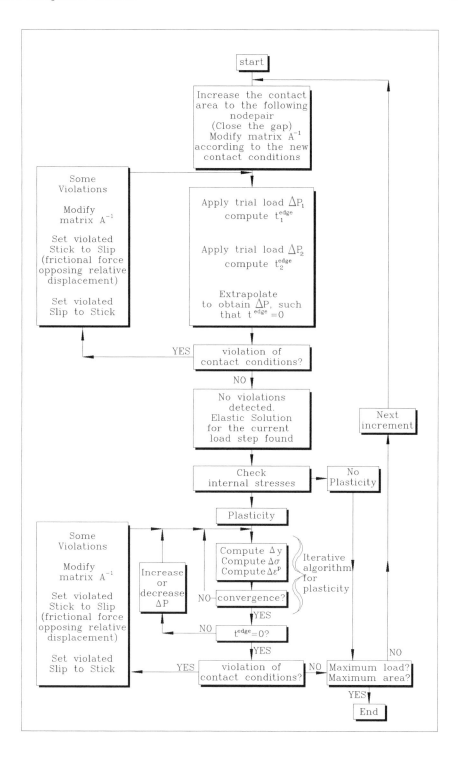

Figure 7.15: Flow diagram for fully incremental loading technique.

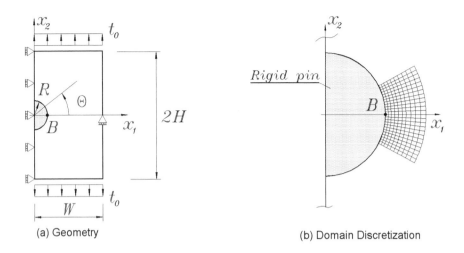

(a) Geometry   (b) Domain Discretization

Figure 7.16: Perfect-fit pin in an infinite plate.

to yield. The pin radius is $R = 25$ mm and geometry ratios are $H/W = 1$ and $W/R = 20$. The pin is assumed to be rigid, and the plate has an elastic modulus $E = 73$ GPa, Poisson's ratio $\nu = 0.33$, yield stress $\sigma_Y = 380$ MPa and the plastic modulus $H' = 600$ MPa.

The problem was analyzed by Martin and Aliabadi [24] using 106 boundary elements, of which 72 were linear elements. The expected yield region was discretized into 240 internal cells. Plain strain condition was assumed. The load at first yield was found to be $t_{oY} = 130.7$ MN/m, and the node B is where the plastic zone developed. The distribution of hoop stress $\sigma_{\theta\theta}$ normalized with respect to t_{oY} presented in Figure 7.17. The extent of plastic zone for $t_o = 1.3t_{oY}$ is shown in Figure 7.18.

Cylinder on a Flat Foundation

This is a non-conforming type problem. The solution of this example is obtained by employing an incremental loading technique, in which the contact conditions at each node pair, as well as the plastic deformations at the internal node, are compared iteratively, for each load step [24]. Consider the geometry shown in Figure 7.19 the foundation has a semi-width W and a height H. The radius of the roller is R. The ratios of $H/W = 2/1$ and $R/W = 5/9.5$ are assumed. Due to the symmetry of the problem, only half of the structure is modelled. The boundary mesh consists of 47 elements for the foundation (of which 14 are situated in the potential contact area, i.e. linear elements are used within them), and 59 elements for the cylinder, 14 of which are linear with the reminder being quadratic elements. Also shown in Figure 7.19 is the domain discretization, in the neighbourhood of the point A, the only point where the two bodies touch each other in the undeformed state. For each body, 168 linear internal cells are used, covering an area of $w \times h$ which in the case of the foundation $w = 0.085W$ and $h = 0.038H$. In order to simulate a concentrated compressive load P, a distributed load is applied vertically along a very small element on the top of the cylinder.

Both rollers and foundation are assumed to have the same material properties:

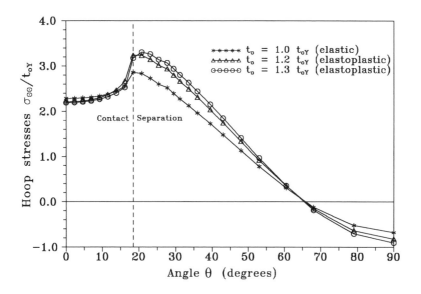

Figure 7.17: Hoop stresses vs. angle for perfect-fit pin in an infinite plate.

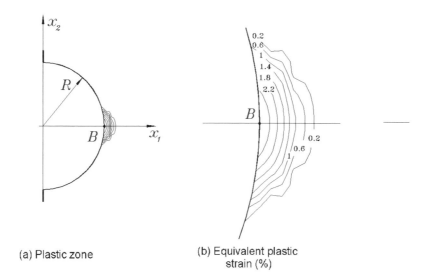

Figure 7.18: Plastic zone for $t_o = 1.3t_{oY}$.

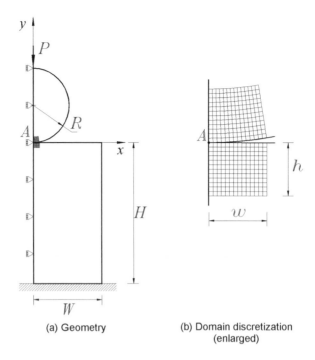

(a) Geometry   (b) Domain discretization
                                    (enlarged)

Figure 7.19: A cylinder on a flat foundation.

elastic modulus $E = 4000$ MPa; Poisson's ratio $\nu = 0.3$; yield stress $\sigma_y = 100$ MPa; plastic modulus $H' = 595$ MPa (work hardening material); coefficient of fraction $\mu = 0.05$. The problem is solved under plane strain conditions. Figure 7.20 shows the normal and tangential tractions that develop in the contact area, as the load is applied incrementally. Up to the ninth load step, the behaviour of the structure is elastic, and while the load is being applied, a small slip area develops towards the edge of the contact zone. In the tenth load step, a plastic zone begins to appear, as the yield stress is reached at point A.

Further load steps show small increases of the normal tractions in the centre of the contact area, because the yielded material is unable to significantly increase its bearing capacity. In order to equilibrate the external load, the normal tractions become larger at points close to the edge of the contact area, and then rapidly fall to zero at the edge itself. The slip area continues to extend as the load becomes higher, such that only a small stick area remains near the symmetry axis.

Figure 7.21 shows the distribution of the von Mises stresses and equivalent plastic strains within the discretized domain for the final value of the load. The results for higher values of coefficient of friction are shown in Figures 7.22 and 7.23.

Martin and Aliabadi [24] also analyzed the case of rigid foundation, i.e. $E_c = 400o$ MPa, $E_f = 12000$ MP$_a$, $\sigma_Y^c = 155$ MPa, $\sigma_Y^f = 500$ MPa, $H'_c = 635$ MPa and $H'_f = 900$ MPa. Figures 7.24 and 7.25 show the results for $\mu = 0.1$. The problem of a rigid cylinder was analysed by assuming $E_f = 4000$ MPa, $E_c = 12000$ MP$_a$, $\sigma_Y^f = 155$ MPa, $\sigma_Y^c = 500$ MPa, $H'_f = 635$ MPa and $H'_c = 900$ MPa. Figures 7.26 and 7.27 show the results for $\mu = 0.1$.

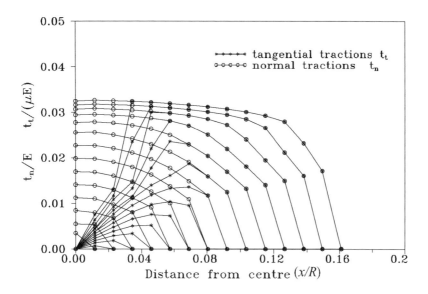

Figure 7.20: Contact tractions (elastoplastic analysis), $\mu = 0.01$.

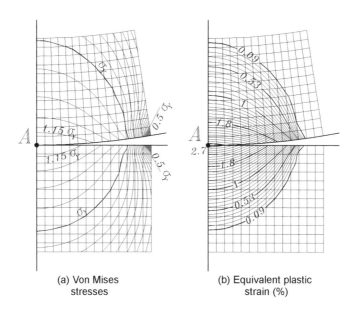

(a) Von Mises
stresses

(b) Equivalent plastic
strain (%)

Figure 7.21: Cylinder and foundation with the same material properties and $\mu = 0.01$.

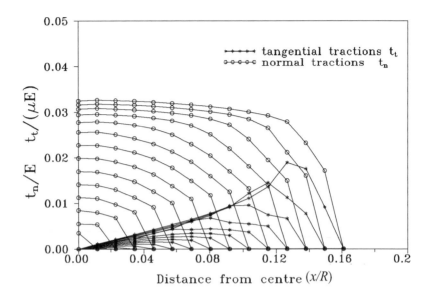

Figure 7.22: Contact tractions (elastoplastic analysis), $\mu = 0.05$.

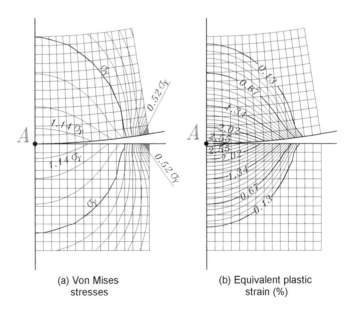

(a) Von Mises
stresses

(b) Equivalent plastic
strain (%)

Figure 7.23: Cylinder and foundation with same material, $\mu = 0.05$.

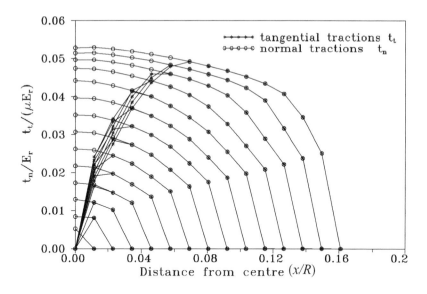

Figure 7.24: Contact tractions (elastoplastic analysis), $\mu = 0.1$.

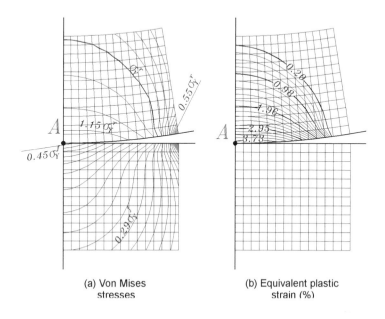

(a) Von Mises
stresses

(b) Equivalent plastic
strain (%)

Figure 7.25: Soft cylinder, $\mu = 0.1$.

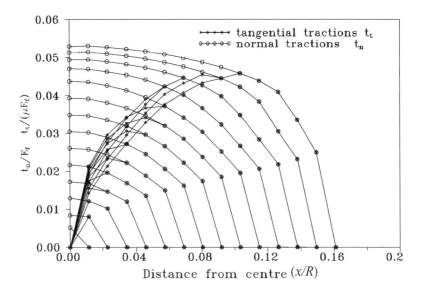

Figure 7.26: Contact tractions (elastoplastic analysis), $\mu = 0.1$.

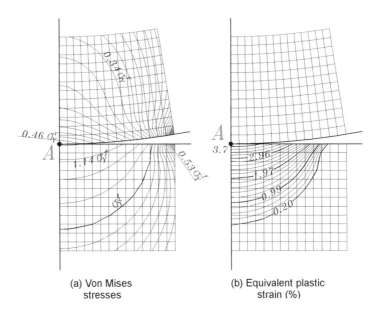

(a) Von Mises
stresses

(b) Equivalent plastic
strain (%)

Figure 7.27: Soft foundation $\mu = 0.1$.

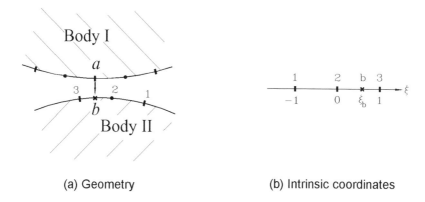

(a) Geometry	(b) Intrinsic coordinates

Figure 7.28: Non-conforming discretization, node a and point b.

7.5 Non-conforming Discretization Methods

The use of conforming discretization is increasingly being regarded as an unnecessary limitation, and therefore attention has been focused on the development of advanced formulations which do not require them. The use of non-conforming discretizations allows simpler modelling for a wider range of problems, like those involving large relative displacements. Furthermore, it reduces the modelling and data preparation, particularly in the case of non-conforming problems, in which the surfaces in contact have different shapes before the application of the loads.

The formulations presented in this section are based on [25] and [27], and describe the solution of contact problems without the use of conforming discretization. This means that in general a node within the contact zone of one body will come into contact with a point on the surface of the other body, which is not defined in the original discretization (i.e., Figure 7.28). No equations describing the displacements and tractions at that point are contained in the system of equations, and therefore the enforcement of equilibrium or compatibility constraints requires an additional set of independent equations. The extra point can be selected as an additional collocation point, and a boundary integral equation is written to express the values of displacements and tractions in terms of all nodal values of the body to which it belongs. They can then be related to the unknowns of the first node through the different contact modes, which effectively solves the contact problem. The best results are obtained when a balanced combination of compatibility and equilibrium conditions is used. More specifically, equilibrium conditions are enforced at every node of body I, and compatibility conditions are enforced at every node of body II.

7.5.1 Non-Conforming Discretization Formulations

Here two different approaches are described for non-conforming discretization. The first utilizes the local interpolation functions to approximate the $\dot{u}_t^b$, $\dot{u}_n^b$, $\dot{t}_t^b$ and $\dot{t}_n^b$; the second evaluate these parameters directly using integral equations.

Interpolation Approach

The parameters $\dot{u}_t^b$, $\dot{u}_n^b$, $\dot{t}_t^b$ and $\dot{t}_n^b$ are evaluated in terms of nodal values by employing the shape functions for $\xi = \xi_b$, as follows:

$$\dot{u}_t^b = \sum_{k=1}^{3} N_k(\xi_b)\dot{u}_t^k \qquad \dot{u}_n^b = \sum_{k=1}^{3} N_k(\xi_b)\dot{u}_n^k$$

$$\dot{t}_t^b = \sum_{k=1}^{3} N_k(\xi_b)\dot{t}_t^k \qquad \dot{t}_n^b = \sum_{k=1}^{3} N_k(\xi_b)\dot{t}_n^k \qquad (7.37)$$

In this way, the contact constraints can be enforced between node a and nodes 1, 2 and 3 of the element to which point b belongs. They become: a) Separation

$$\dot{t}_t^a = \sum_{k=1}^{3} N_k(\xi_b)\dot{t}_t^k$$

$$\dot{t}_n^a = \sum_{k=1}^{3} N_k(\xi_b)\dot{t}_n^k$$

$$\dot{t}_t^a = 0$$

$$\dot{t}_n^a = 0 \qquad (7.38)$$

b) Slip

$$\dot{t}_t^a = \sum_{k=1}^{3} N_k(\xi_b)\dot{t}_t^k$$

$$\dot{t}_n^a = \sum_{k=1}^{3} N_k(\xi_b)\dot{t}_n^k$$

$$\dot{t}_t^a = \pm\mu\dot{t}_n^a$$

$$\dot{u}_n^a = g_o - \sum_{k=1}^{3} N_k(\xi_b)\dot{u}_n^k \qquad (7.39)$$

c) Stick

$$\dot{t}_t^a = \sum_{k=1}^{3} N_k(\xi_b)\dot{t}_t^k$$

$$\dot{t}_n^a = \sum_{k=1}^{3} N_k(\xi_b)\dot{t}_n^k$$

$$\dot{u}_t^b = -\sum_{k=1}^{3} N_k(\xi_b)\dot{u}_t^k$$

$$\dot{u}_n^a = g_o - \sum_{k=1}^{3} N_k(\xi_b)\dot{u}_n^k$$

As stated earlier, apart from the two displacement equations for $\dot{u}_t^a, \dot{u}_n^a$, two more conditions are needed. Therefore additional conditions (E.3), (E.4) and (5.13) for both displacements and tractions would result in an overdetermined system of equations. Only two of them must be chosen, and it is inevitable that the final solution will contain some violations (albeit by a small margin) of those conditions not explicitly enforced (see [28]).

After some numerical experiments, in which different ways of enforcing contact conditions were tried, the best results seem to be obtained when a balanced combination of compatibility and equilibrium is used [26]. More specifically, equilibrium conditions are enforced at every node of body I, and compatibility conditions are enforced at every node of body II. It was also reported [29, 25] that the body for which the equilibrium conditions are enforced should not have a coarser mesh than that for which compatibility conditions contained in the final system of equations should be equal to or higher than the number of compatibility equations.

According to the contact modes, and making use of equation (E.3), the equilibrium conditions for node a of body I are:

a) Separation

$$
\begin{aligned}
t_t^{ia} &= 0 \\
t_n^{ia} &= 0
\end{aligned}
\tag{7.40}
$$

b) Slip

$$
\begin{aligned}
t_t^{ia} &= \pm\mu t_n^{ia} \\
t_n^{ia} &= \sum_{k=1}^{3} N^k(\xi_b) t_n^{ik}
\end{aligned}
\tag{7.41}
$$

c) Stick

$$
\begin{aligned}
t_t^{ia} &= \sum_{k=1}^{3} N_k(\xi_b) t_t^{ik} \\
t_n^{ia} &= \sum_{k=1}^{3} N_k(\xi_b) t_n^{ik}
\end{aligned}
\tag{7.42}
$$

For nodes of body II (notation a, b is maintained for convenience) the compatibility conditions are:

a) Separation

$$
\begin{aligned}
t_t^{ia} &= 0 \\
t_n^{ia} &= 0
\end{aligned}
\tag{7.43}
$$

b) Slip

$$
\begin{aligned}
t_t^{ia} &= \pm\mu t_n^{ia} \\
u_n^{ia} &= gap^{ab} - \sum_{k=1}^{3} N_k(\xi_b) u_n^{ik}
\end{aligned}
\tag{7.44}
$$

c) Stick

$$
u_t^{ia} = -\sum_{k=1}^{3} N_k(\xi_b) u_t^{ik}
$$

$$\dot{u}_n^a \; = \; g_o - \sum_{k=1}^{3} N_k(\xi_b)\dot{u}_n^k \tag{7.45}$$

Equations (7.40) to (7.45) are expressed in matrix form, and assembled in equation (7.46) as Contact Conditions.

In equation (7.29) the computation of matrices $\mathbf{A}^1, \mathbf{A}^2, \mathbf{D}^1$ and $\mathbf{D}^2$ and vectors $\mathbf{f}^1$ and $\mathbf{f}^2$ precedes the enforcement of contact conditions, and for a given node, they depend only on the boundary and domain discretization of the body to which the node belongs. For the elastoplastic analysis, quadratic 9-noded cells were used to discretize the domain [27].

Integral Equation Approach

It is a recognized fact that in order to solve a contact problem, tractions and displacements have to be computed to a high degree of accuracy. In the light of this, a formulation was proposed [24] which directly evaluates $\dot{u}_t^b$ and $\dot{u}_n^b$ by using the boundary integral displacement equation and $\dot{t}_t^b$ and $\dot{t}_n^b$ by using the traction boundary integral equation. In this manner, instead of evaluating the unknowns in terms of only nodes 1, 2 and 3, they will be expressed in terms of all nodes of body, using the original boundary discretization. Choosing point b as the collocation point, as shown in Figure 7.29, the displacement boundary integral equation for point on a smooth boundary can be written (see Chapter 2) as

$$\frac{1}{2}\dot{u}_i^b + \sum_{n=1}^{Ne} \left\{ \oint_{S_n} T_{ij}(\mathbf{x}'=b,\mathbf{x})\,\dot{u}_j(\mathbf{x})\,dS \right\}$$

$$= \sum_{n=1}^{Ne} \left\{ \int_{S_n} U_{ij}(\mathbf{x}'=b,\mathbf{x})\,\dot{t}_j(\mathbf{x})\,dS \right\} + \sum_{l=1}^{Ncel} \left\{ \int_{V_l} \Sigma_{ijkl}(\mathbf{x}'=b,\mathbf{x})\dot{\varepsilon}_{jk}^P dV \right\} \tag{7.46}$$

and similarly, for the tractions,

$$\frac{1}{2}\dot{t}_j^b = n_i(b) \sum_{n=1}^{Ne} \left\{ \oint_{S_{em}} D_{ijk}(\mathbf{x}'=b,\mathbf{x})\,\dot{t}_k(\mathbf{x})\,dS \right\}$$

$$- n_i(b) \sum_{n=1}^{Ne} \left\{ \oint S_{ijk}(\mathbf{x}'=b,\mathbf{x})\,\dot{u}_k(\mathbf{x})\,dS \right\}$$

$$n_i(b) \left[\sum_{l=1}^{Ncel} \left\{ \oint_{V_l} \Sigma_{ijkl}(\mathbf{X}'=b,\mathbf{X})\dot{u}_k(\mathbf{X})dV \right\} + \frac{1}{2}f_{ij}(\dot{\varepsilon}_{kl}^P(b)) \right] \tag{7.47}$$

where the coefficients C_{ij} take the value $\frac{1}{2}$ because point b is situated on a smooth sector of the boundary, and nel is the number of elements of body.

Equations (7.46) and (7.47) can be manipulated in the same way as in the elastoplastic formulation of the BEM. They can be written in matrix form as

$$\frac{1}{2}\dot{u}_i^b = \mathbf{g}_i^b\,\dot{\mathbf{t}}_{II} - \mathbf{h}_i^b\,\dot{\mathbf{u}}_{II} + \mathbf{d}_i^b(\varepsilon_{II}^p + \Delta\varepsilon_{II}^p) \tag{7.48}$$

and

$$\frac{1}{2}\dot{t}_i^b = \mathbf{g}_{i_t}^b\,\dot{\mathbf{t}}_{II} - \mathbf{h}_{i_t}^b\,\dot{\mathbf{u}}_{II} + \mathbf{d}_{i_t}^b(\varepsilon_{II}^p + \Delta\varepsilon_{II}^p) \tag{7.49}$$

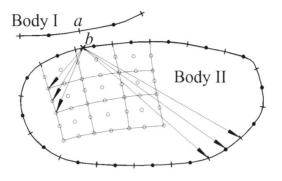

Figure 7.29: Collocation at point b.

where the superscript b denotes that the collocation point is b, and in equation (7.49) the subscript t has been used to indicate the use of the derivative kernels. Vectors $\dot{u}_{II}$, $\dot{t}_{II}$ contain the displacements and tractions of body II, respectively. Equations (7.48) and (7.49) are rearranged according to the boundary conditions, to give:

$$\frac{1}{2}\dot{u}_i^b = -\mathbf{a}_i^b\,\mathbf{y}_{II} + f_i^b + \mathbf{d}_i^b(\varepsilon_{II}^p + \Delta\varepsilon_{II}^p) \tag{7.50}$$

and

$$\frac{1}{2}\dot{t}_i^b = -\mathbf{a}_{i_t}^b\,\mathbf{y}_{II} + f_{i_t}^b + \mathbf{d}_{i_t}^b\,(\varepsilon_{II}^p + \Delta\varepsilon_{II}^p) \tag{7.51}$$

Equations (7.50) and (7.51) can be used to write the contact conditions shown in Table 7.1. For a node a of body I only equilibrium conditions are enforced, thus, in addition to the standard displacement equation, the two remaining equations are, according to the contact modes:

a) Separation

$$\dot{t}_t^a = 0$$
$$\dot{t}_n^a = 0 \tag{7.52}$$

b) Slip

$$\dot{t}_t^a = \pm\mu\dot{t}_n^a$$
$$\frac{1}{2}\dot{t}_n^a = -\mathbf{a}_{n_t}^b\mathbf{y}_{II} + f_{n_i}^b + \mathbf{d}_{i_t}^b\,(\varepsilon_{II}^p + \Delta\varepsilon_{II}^p) \tag{7.53}$$

c) Stick

$$\frac{1}{2}\dot{t}_t^a = -\mathbf{a}_{i_t}^b\mathbf{y}_{II} + f_{t_i}^b + \mathbf{d}_{i_t}^b\,(\varepsilon_{II}^p + \Delta\varepsilon_{II}^p)$$
$$\frac{1}{2}\dot{t}_n^a = -\mathbf{a}_{n_t}^b\mathbf{y}_{II} + f_{n_i}^b + \mathbf{d}_{i_t}^b\,(\varepsilon_{II}^p + \Delta\varepsilon_{II}^p) \tag{7.54}$$

while for nodes of body II only compatibility conditions are enforced, so that the additional equations are (notation a, b is maintained for convenience): a) Separation

$$\dot{t}_t^a = 0$$

$$\dot{t}_n^a = 0 \tag{7.55}$$

b) Slip

$$\dot{t}_t^a = \pm\mu \dot{t}_n^a$$

$$\frac{1}{2}\dot{u}_n^a = \mathbf{a}_n^b \mathbf{y}_I - (f_n^b + gap^{ab}) + \mathbf{d}_i^b(\varepsilon_{II}^p + \Delta\varepsilon_{II}^p)$$ (7.56)

c) Stick

$$\frac{1}{2}\dot{u}_t^a = \mathbf{a}_t^b \mathbf{y}_I - (f_t^b) + \mathbf{d}_i^b(\varepsilon_{II}^p + \Delta\varepsilon_{II}^p)$$

$$\frac{1}{2}\dot{u}_n^a = \mathbf{a}_n^b \mathbf{y}_I - (f_n^b + gap^{ab}) + \mathbf{d}_i^b(\varepsilon_{II}^p + \Delta\varepsilon_{II}^p)$$ (7.57)

The above formulation was extensively tested in [26].

7.5.2 Examples

In this section the problem of a flat punch on a foundation is used to demonstrate application of the non-conforming discretization to elastic and elastoplastic contact problems. The shape function approach and integral equation approach described above are used.

Flat Punch on a Foundation

The geometry of the problem is shown in Figure 7.30. The geometry ratios are $h/w = 2$, $H/W = 1$ and $w/W = 1/4$, with $W = 160$ mm. Both punch and foundation are assumed to be made of steel with Young's modulus $E_p = E_f = 210$ GPa and $\nu = 0.3$.

Elastic Analysis

Martin and Aliabadi [25] analysed the problem using non-conforming discretization with the shape function and integral equation approaches. The boundary element mesh used consisted of 42 elements for the foundation, of which 8 were placed within the contact area, and 33 elements for the punch, of which 11 were in the contact area.

Figure 7.31 shows the results for $\mu = 0.15$, along with the conforming discretization solutions. The elastic values of the normal and tangential tractions are normalized with respect to the applied load t_o. The partition between *stick* and *slip* zones is situated at approximately $x/w = 0.33$. The non-conforming results correspond to the shape unction approach described in Section 7.5.1. The integral equation approach also described in Section 7.5.1 gives almost identical results for most parts. A comparison between the two non-conforming approaches is shown in Figure 7.32 within a limited zone near point A. It is clear that the shape function approach presents oscillations near $x/w \geq 0.9$.

The results for $0.98 \leq x/w \leq 1$ presents oscillations due to the presence of singularity at point A, and therefore are not shown. The singularity does not, however, impair the accuracy of the solution elsewhere on the boundary.

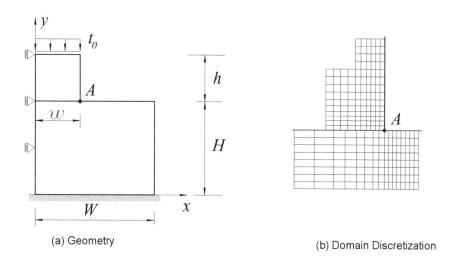

(a) Geometry   (b) Domain Discretization

Figure 7.30: Flat punch on a Foundation.

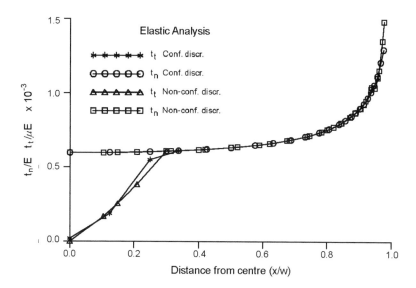

Figure 7.31: Normal and tangential tractions for a flat punch on a foundation with $\mu = 0.15$, $E_p/E_f = 1$.

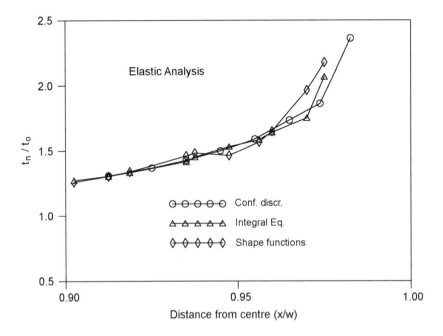

Figure 7.32: Comparisons between different non-conforming discretization and conforming discritization techniques.

Elastoplastic Analysis

Aliabadi and Martin [4] analyzed the same problem using non-conforming discretization elastoplastic BEM formulations. The elastoplastic constants for the steel are: yield stress $\sigma_Y = 490$ MPa and $H' = 3.85$ MPa. Normal and tangential tractions for the final load step $t_o = 150$ MN/m are shown in Figure 7.33. Once again, the agreement between the conforming and non-conforming discretization approaches is very good. A comparison between the shape function and integral equation techniques is shown in Figure 7.34.

Tangential forces near the edge of the punch appear to have a confining effect, which prevents the plastic strains from spreading. As a result, the shape of the plastic zone is approximately rectangular, as shown in Figure 7.35.

7.6 Three-dimensional Contact problems

Consider two elastic bodies 1 and 2 with surfaces S^A and S^B respectively (see Figure 7.36). In terms of displacements within the contact zone, in this particular case we have $u_3^1 = u_3^2$. Where frictional forces are such that an adhesive regime prevails, we have $u_i^1 = u_i^2$ with $i = 1, 2$. If conditions are such that sliding takes place, then in general, $u_i^1 \neq u_i^2$ with $i = 1, 2$. As a consequence of the two bodies being in contact, the normal tractions for each of the two bodies must be equal and opposite, that is $t_3^1 = t_3^2$. Also, these normal tractions must be compressive.

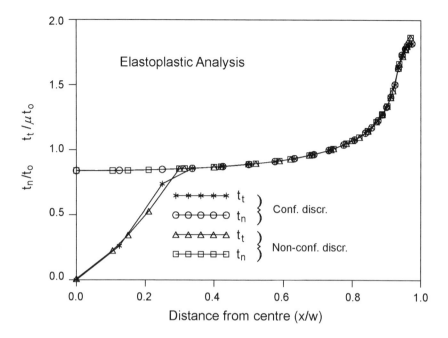

Figure 7.33: Elastoplastic contact tractions for a punch on a foundation with $\mu = 0.15$, $E_p/E_f = 1$.

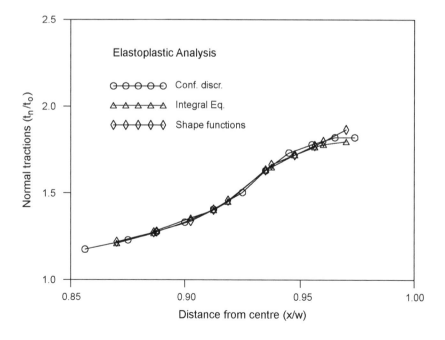

Figure 7.34: Comparsion of elastoplastic tractions.

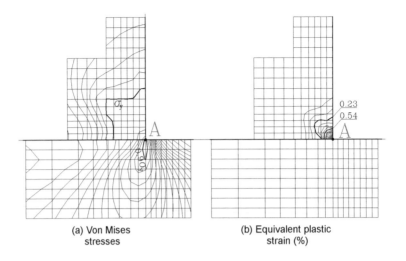

(a) Von Mises
stresses

(b) Equivalent plastic
strain (%)

Figure 7.35: Plastic stresses and strains, $\mu = 0.15$, $E_p/E_f = 1$.

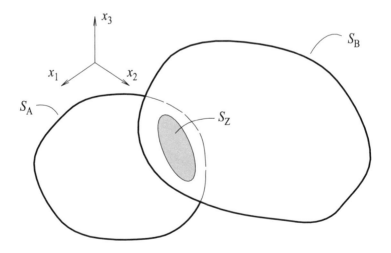

Figure 7.36: Three-dimensional contact problem.

7.6.1 Contact Conditions

The average normal $\tilde{\mathbf{n}}^k$ is defined as

$$\tilde{\mathbf{n}}^k = \frac{\mathbf{n}^{ka} - \mathbf{n}^{kb}}{\left|\mathbf{n}^{ka} - \mathbf{n}^{kb}\right|} \tag{7.58}$$

which may then be used as the x_3 axis of the local system. Using the above definition, the different contact modes can be written as

- **Slip conditions**

$$
\begin{aligned}
u_3^{k1} - u_3^{k2} &= d^{k12} \\
t_3^{k1} + t_3^{k2} &= 0 \\
t_1^{k1} &= t_1^{k2} = t_2^{k1} = t_2^{k2} = 0
\end{aligned} \tag{7.59}
$$

Here d^{kab} represents the initial normal gap at a node-pair with co-ordinates $\mathbf{x}^k$ between node a and node b, which is defined by

$$d^{k12} = \left|\left(x_2^{k1} - x_2^{k2}\right)\right| \tilde{n}_2^k + \left|\left(x_1^{k1} - x_1^{k2}\right)\right| \tilde{n}_1^k \tag{7.60}$$

where $\tilde{n}_i^k$ is the ith component of the average normal $\tilde{\mathbf{n}}$ for the kth node pair. The nodal co-ordinates x_i^{k1} are for the undeformed state. Since u_1^{k1}, u_1^{k2}, u_2^{k1} and u_2^{k2} do not appear in the slip conditions, these variables are unconstrained, which allows slip in the x_1 and x_2 directions.

- **Separation conditions**

$$t_1^{k1} = t_1^{k2} = t_2^{k1} = t_2^{k2} = t_3^{k1} = t_3^{k2} = 0 \tag{7.61}$$

The following criteria must always be observed within the contact zone:

$$t_3^{k1} \leqslant 0 \text{ (Stresses are compressive)} \tag{7.62}$$

and

$$u_3^{k1} - u_3^{k2} \leqslant 0 \text{ (No interpenetration of the bodies)} \tag{7.63}$$

In practice, the criteria need to be satisfied within a certain specified tolerance level, since due to discretization limitations for certain problems, it may prove impossible to completely satisfy both conditions at all nodes. A small tolerance prevents 'oscillation' of the numerical solution between two alternate partitions of the potential contact zone which only slightly violate one or other of the two sets of criteria.

7.6.2 Example

To demonstrate the use of the BEM for three-dimensional contact problems the two problems, of a perfect fit pin and interference fit pin are presented.

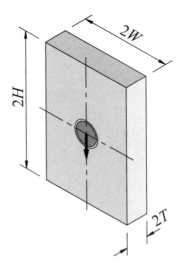

Figure 7.37: Loaded pin in a plate.

Loaded Pin in a Plate

The geometry for this problem is shown in Figure 7.37, with the following values taken for the geometric parameters: $H/W = 2$, $W/R = 16$, $T/R = 2$, and $R_{\text{hole}} = 1$. The pin is loaded by means of a uniform shear stress σ applied to its ends, and directed along the negative x_3-axis.

This model was analyzed [9] for $\sigma/E_{\text{lug}} = 1/1000$. This ratio was chosen in order to allow the elastic strains within the material to be below the levels at which plastic yield would occur. The pin is taken as having an initially perfect fit for the unloaded configuration, with $R_{\text{pin}}/R_{\text{hole}} = 1.0$, and is an example of a receding contact problem, since the contact area, expressed in terms of the angle θ defined in Figure 7.38, reduces in size upon loading. Two values for $E_{\text{pin}}/E_{\text{lug}}$ are chosen, $E_{\text{pin}}/E_{\text{lug}} = 1.0$ and $E_{\text{pin}}/E_{\text{lug}} = 3.0$. For this and all subsequent problems, the initial contact area was taken to be the intersection of the area of geometrical coincidence for the unloaded case with the symmetry plane parallel to the direction of the applied load.

Two different fit pins are presented here; a perfect fit, and a tolerance fit defined by $R_{\text{pin}}/R_{\text{hole}} = 5999/6000$. The contact area is load independent for the perfect fit case.

Perfect Fit Pin: Variation of Relative Young's Modulus

Consider a perfect fit pin with a load of $\sigma/E_{\text{lug}} = 1/1000$, with $E_{\text{pin}}/E_{\text{lug}} = 1.0$ and $E_{\text{pin}}/E_{\text{lug}} = 3.0$. This is an example of a receding contact problem, since the pin is initially in complete contact with the hole, prior to loading. The normal tractions on the surface of the pin are shown in Figures 7.39 and 7.40 for $E_{\text{pin}}/E_{\text{lug}} = 1.0$, and in Figures 7.41 and 7.42 for $E_{\text{pin}}/E_{\text{lug}} = 3.0$.

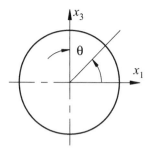

Figure 7.38: Angle around the pin.

Tolerance Fit Pin

Consider a pin with a tolerance fit defined by $R_{\text{pin}}/R_{\text{hole}} = 5999/6000$. This is an example of advancing contact, since the pin is initially in contact only for $\theta = 0°$. The normal tractions are give in Figures 7.43 and 7.44 for $E_{\text{pin}}/E_{\text{lug}} = 1.0$. It can be observed that in this case, the normal tractions vary along the length of the pin, with their minimum (since the normal tractions are compressive) value occurring at the ends of the pin (where the loading is applied) increasing to a maximum for $x_2 = 0$. This is due to a slight degree of bending which occurs in the pin. Separation occurs in all cases for the upper surface of the pin, with $\theta \gtrsim 90°$ or $\theta \lesssim -90°$. This bending is large enough to allow separation to occur at $x_2 = 0$ at the bottom of the pin ($\theta = 0°$).

7.7 Summary

In this chapter application of the boundary element method to contact problems has been described for elastic and elastoplastic problems. The applications presented includes both conforming and non-conforming contact problems with frictionless and frictional boundary conditions. A fully incremental loading technique to follow the contact history accurately was also described in detail. A recently developed non-conforming discretization techniques were also described. Although, BEM formulation is well suited to contact problems, there is still more work needed for applications to complex multi-body contact problems.

References

[1] Abascal, R., 2D transient dynamic friction contact problems I: numerical analysis. *Engineering Analysis with Boundary Elements*, **16**, 227-233, 1995.

[2] Abdul-Mihsein, Bakr, M.J. and Parker, A.P., A boundary integral equation method for axisymmetric elastic contact problems. *Computers and Structures*, **23**, 787-792, 1986.

[3] Alart, P. and Curnier, A., A mixed formulation for frictional contact problems prone to Newton like solution methods, *Computer Methods in Applied Mechanics and Engineering*, **92**, 353-375, 1991.

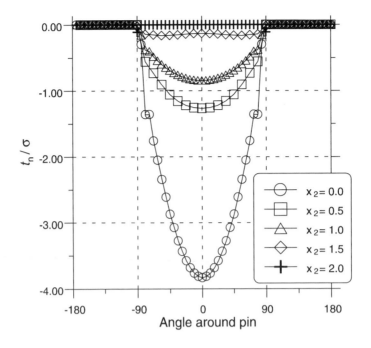

Figure 7.39: Perfect fit loaded pin, $E_{pin}/E_{lug} = 1.0$.

[4] Aliabadi, M.H., *Computational Methods in Contact Mechanics*, Elsevier Applied Science, Oxford and Computational Mechanics Publications, Southampton 1993.

[5] Aliabadi, M.H. and Martin, D., Boundary element hypersingular formulation for elastoplastic contact problems, *International Journal for Numerical Methods in Engineering*, **48**, *995-1014*, 2000.

[6] Alart, P. and Curnier, A., A mixed formulation for frictional contact problems prone to Newton like solution methods, *Computer Methods in Applied Mechanics and Engineering*, **92**, 353-375, 1991.

[7] Anderson, T., Boundary elements in two-dimensional Contact and Friction, Diss No 85, Linkoping Institute of Technology 1982.

[8] Banbridge, C., Aliabadi, M.H. and Rooke, D.P., Three-dimensional boundary element contact analysis of pin-loaded lugs. (submitted for publication).

[9] Banbridge, C., Boundary Element Analysis of Three-dimensional Contact Problems in Cracked Structures, PhD thesis, Weesex Institute of Technology, University of Wales,1997.

[10] de Lacerda, L.A. and Wrobel, L.C., Frictional contact analysis of coated axisymmetric bodies using the boundary element method, *Journal of Strain Analysis*, **35**, 423-440, 2000.

[11] Ezawa, Y. and Okamoto, N., High speed boundary element contact stress analysis using a super computer. *Proc. of the 4th International Conference* on Boundary

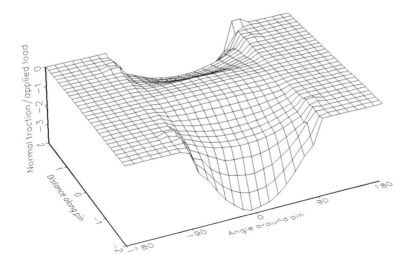

Figure 7.40: Perfect fit loaded pin, $E_{pin}/E_{lug} = 1.0$.

Element Technology, Computational Mechanics Publications, Southampton, 405-416, 1989.

[12] Gonzalez, J.A. and Abascal, R., An algorithm to solve coupled 2D rolling contact problems. *International Journal for Numerical Methods in Engineering,* **49,** 1143-1167, 2000.

[13] Gonzalez, J.A. and Abascal, R., Transient 2D rolling contact problems using BEM, *Advances in Boundary Element Techniques II,* Hoggar Press, Geneva, 311-318, 2001.

[14] Huesmann, A. and Kuhn, G., Non-conform discretization of the contact region applied to two-dimensional boundary element method. *Boundary elements XVI,* Computational Mechanics Publications, Southampton, 353-360, 1994.

[15] Karami, G, and Fenner, R.T., A two-dimensional BEM method for thermo-elastic body forces contact problems, *Proce. Boundary elements IX,* Computational Mechanics Publications, Southampton, pp417-437, 1987.

[16] Klang, M., On interior contact under friction between cylindrical elastic bodies, Thesis, Linkoping University, Linkoping, 1979.

[17] Kong, X.A., Gakawaya, A., Cardou,A. and Cloutier,L., A numerical solution of general frictional contact problems by the direct boundary element and mathematical programming approach. *Computers and Structures,* **45,** 95-112, 1991.

[18] Lee, DB.C. and Kwak, B.M., A computational method for elasto-plastic contact problems. *Computers and Structures,* **18,** 757-765, 1984.

[19] Man, K., Aliabadi, M.H. and Rooke, D.P., BEM Frictional contact analysis: An incremental loading technique, *Computers and Structures,* **47,** 893-905, 1993.

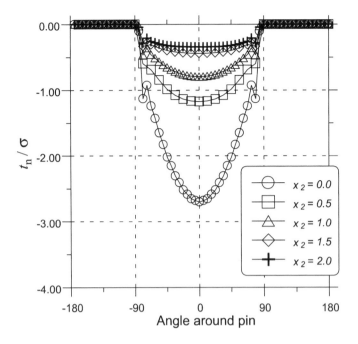

Figure 7.41: Perfect fit loaded pin, $E_{pin}/E_{lug} = 3.0$.

[20] Man, K., Aliabadi, M.H. and Rooke, D.P., BEM frictional contact analysis: modelling considerations. *Engineering Analysis with Boundary Elements*, **11**, 77-85, 1993.

[21] Man, K., Aliabadi, M.H. and Rooke, D.P., Analysis of contact friction using the boundary element method. Chapter 1: *Computational Methods in Contact Mechanics*, Elsevier Publishers, Oxford, 1-60, 1993.

[22] Man, K., Boundary element analysis of contact in fracture mechanics, PhD thesis, Wessex Institute of Technology, Portsmouth University, 1993.

[23] Man, K., *Contact Mechanics using Boundary Elements*, Topics in Engineering Vol. 22, Computational Mechanics Publications, Southampton, 1994.

[24] Martin, D and Aliabadi, M.H., Boundary element analysis of two-dimensional elastoplastic contact problems, *Engineering Analysis with Boundary Elements*, **21**, 349-360, 1998.

[25] Martin, D and Aliabadi, M.H,. A BE hyper-singular formulation for contact problems using non-conforming discretization, *Computers and Structures*, **69**, 557-565, 1998.

[26] Martin, D. and Aliabadi, M.H., Non-conforming BEM discretization in frictional elastoplastic contact problems. *Boundary Elements XIX*, Computational Mechanics Publications, Southampton, 67-75, 1997.

[27] Martin D., Elastoplastic contact problems: boundary element analysis, PhD thesis, Wessex Institute of Technology, University of Wales, 1998.

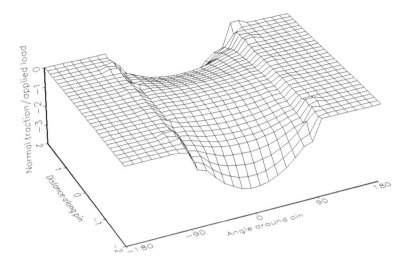

Figure 7.42: Perfect fit loaded pin, $E_{pin}/E_{lug} = 3.0$.

[28] Olukoko, O.A. and Fenner, R.T., A new boundary element approach for contact problems with friction. *International Journal for Numerical Methods in Engineering*, **36**, 2625-2642, 1993.

[29] Paris, F., Blazquez.A. and Canas, J., Contact problems with non-conforming discretizations using boundary element method.. *Computers and Structures*, **57**, 829-835, 1995.

[30] Paris, F. and Garrido, J.A., On the use of discontinuous elements in 2D contact problems., *Boundary Elements XVII*, 13-27 to 13-39, Computational Mechanics Publications, Southampton 1985.

[31] Yamazaki, K., Penalty function method using BEM. Chapter 5 in *Computational Methods in Contact Mechanics*, Elsevier Applied Sciences, Barking, 155-190, 1993.

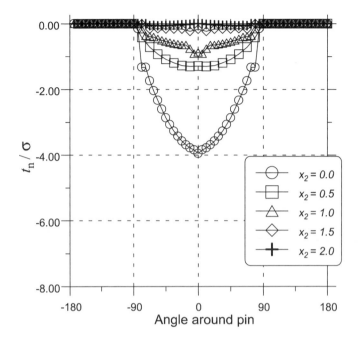

Figure 7.43: $R_{pin}/R_{hole} = 5999/6000$, $\sigma/E_{lug} = 1/1000$, $E_{pin}/E_{lug} = 1.0$.

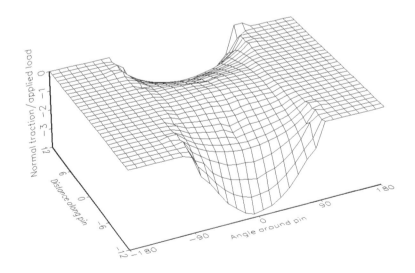

Figure 7.44: $R_{pin}/R_{hole} = 5999/6000$, $\sigma/E_{lug} = 1/1000$, $E_{pin}/E_{lug} = 1.0$.

Chapter 8

Fracture Mechanics

8.1 Introduction

In order to ensure a reasonable cost for the design and maintenance of complex engineering structures in the aerospace and automotive industries, among others, it is generally accepted that computational analysis and simulation must partially replace full scale and laboratory testing. This is important not only for designing new parts, but also in maintaining aging structures. In any method for lifetime prediction of flawed structures, the effect of the geometry of components, or structures, and its interaction with the growing crack must be considered.

An early application of the boundary element method to a crack problem was due to Cruse [63], who reported only a modest accuracy for evaluation of the stress intensity factors. Since that publication the method has improved, and has established itself as probably the most efficient and accurate general numerical technique for the evaluation of stress intensity factors and crack growth analyses.

Early applications of the BEM to general crack problems were limited, as a straightforward application of the method to crack problems leads to a mathematical degeneration if the two crack surfaces are considered co-planar [64]. Snyder and Cruse [199] later introduced a special form of fundamental solution for crack problems in anisotropic media. The fundamental solution (Green's function) contained the exact form of the traction free crack in an infinite medium, so that no modelling of the crack surfaces was required. The crack Green's function, although accurate, is limited to two-dimensional problems.

An alternative method of dealing with two co-planar crack surfaces was devised by Blandford *et al.* [36]. This approach, which is based on the multi-region formulation, is general and can be applied to both symmetric and antisymmetric problems in both 2D and 3D configurations. The multi-region method introduces artificial boundaries into the body, which connect the cracks to the boundary in such a way that each region contains a crack surface. The two regions are then joined together so that equilibrium of tractions and compatibility of displacements are enforced. The main drawback of this method is that the introduction of the artificial boundaries is not unique, and thus cannot be implemented in automatic procedures. Furthermore, the method generates a larger system of equations than is strictly necessary. Despite these drawbacks, the multi-region method was used extensively up to the early 1990s.

More recently, the Dual Boundary Element Method (DBEM), as developed by Portela, Aliabadi and Rooke [162, 164] for 2D problems and Mi and Aliabadi [134] for 3D problems, has been shown to be a general and computationally efficient way of modelling crack problems in the BEM. General mixed-mode crack problems can be solved with the DBEM, in a single region formulation, when the displacement boundary integral equation is applied on one crack surface and the traction boundary integral equation is applied on the other. The importance of this formulation for the BEM is not so much the reduction in the number of equations when compared to the multi-region method, but the striking robustness of the method for modelling crack growth [166, 135]. Crack growth processes simulated through an incremental crack extension analysis can be achieved in a simple and automatic manner with little or no remeshing of the original model. The method has since developed into a powerful tool for solving a wide range of fracture problems [20].

8.2 Basic Fracture Mechanics

8.2.1 Stress Intensity Factor

The fundamental postulate of Linear Elastic Fracture Mechanics (LEFM) is that the behaviour of cracks (i.e. whether they grow or not, and how fast they grow) is determined solely by the value of the stress intensity factors.

Irwin [111] solved several two-dimensional crack problems in linear elasticity, and showed that the stress field in the vicinity of the crack-tip was always of the same form. He showed that the stress-field component σ_{ij} at the point (r, θ) near the crack-tip is given by

$$\sigma_{ij}(r, \theta) = \frac{K}{\sqrt{2\pi r}} f_{ij}(\theta) + \text{ other terms} \qquad (8.1)$$

where the origin of the polar coordinates (r, θ) is at the crack tip and $f_{ij}(\theta)$ contains trigonometric functions. As the coordinate r approaches zero, the leading term in equation (8.1) dominates, the other terms are constant or tend to zero. The constant K in the first term is known as the 'stress intensity factor'. It therefore follows that the stress field in the vicinity of the crack tip is characterized by the 'stress intensity factor'. In general, K will be a function of the crack size and shape, the type of loading and the geometrical configuration of the structure. The stress intensity factor is often written in the following form:

$$K = Y\sigma\sqrt{\pi a} \qquad (8.2)$$

where σ is a stress, a is a measure of the crack-length and Y is a non-dimensional function of the geometry.

8.2.2 Modes of Fracture

There are three basic modes of deformation for a cracked body. These modes are characterized by the movements of the upper and lower crack surfaces with respect to each other, and are shown in figure 8.1; they are referred to as: *opening mode* (mode I), where the two crack surfaces are pulled apart; *sliding mode* or *shearing mode* (mode II), where the two crack surfaces slide over each other along the crack line; and

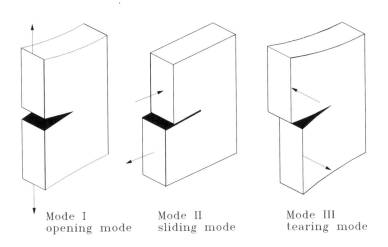

Mode I   Mode II   Mode III
opening mode   sliding mode   tearing mode

Figure 8.1: Crack deformation modes.

tearing mode (mode III), where crack surfaces slide over each other perpendicular to the crack line. From these three basic modes any crack deformation can be presented by an appropriate superposition.

The above fundamental postulate of Linear Elastic Fracture Mechanics (LEFM) implies that it is necessary to evaluate the stress intensity factor in order to be able to predict the behaviour of cracked solids. This factor, which is a function of applied loading and the geometry of the cracked component, has been evaluated for many hundreds of structural configurations and collected together in handbooks [171, 140] and databases [19].

The stress intensity factor once determined may be used in three different area:

1. the determination of the static strength of a crack in a structure (residual strength);

2. the determination of the rate of crack growth in a loaded structures subjected to variable loading (fatigue);

3. the determination of the rate of growth of a crack in a loaded structure in corrosive environment (stress corrosion).

8.2.3 Energy Balance

The first systematic study of fracture phenomena was carried out by Griffith [99], who suggested that the criterion for failure due to crack growth was determined by the balance of strain energy and surface energy. Griffith postulated that if the strain energy released by the strain field when the crack advanced a small distance was greater than the energy required to form the new surface, then unstable crack growth would occur, that is if

$$G \geq 2\Gamma \tag{8.3}$$

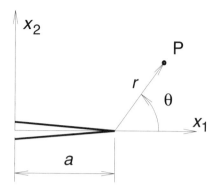

Figure 8.2: Straight crack in an infinite medium.

where G is the strain energy release rate (per unit area of crack growth) and Γ is the work required to form a unit area of the new crack surface (two surfaces). The strain energy release rate G is a function of the loading and the crack size.

By considering the elastic work to close up the tip of a crack, Irwin [111] derived a relationship between the strain energy release rate and the stress intensity factor; it was

$$G \propto K^2 \tag{8.4}$$

The constant of proportionality in the above equation is a function of the elastic constants of the material. This relationship provides a link between the crack-tip stress field and the *energy-balance* criterion for crack growth, which can now be interpreted in terms of a *critical K-values* required for crack growth.

8.2.4 Stress and Displacements Fields

Irwin [111] has shown that in general, the plane elastic state of stress in the vicinity of the crack tip can be expressed in terms of a local coordinate system (r, θ), as shown in Figure 8.2 as

$$
\begin{aligned}
\sigma_{11} &= \frac{K_I}{\sqrt{2\pi r}} \cos\left(\frac{\theta}{2}\right)\left[1 - \sin\left(\frac{\theta}{2}\right)\sin\left(\frac{3\theta}{2}\right)\right] \\
&\quad - \frac{K_{II}}{\sqrt{2\pi r}} \sin\left(\frac{\theta}{2}\right)\left[2 + \cos\left(\frac{\theta}{2}\right)\cos\left(\frac{3\theta}{2}\right)\right] \\[8pt]
\sigma_{22} &= \frac{K_I}{\sqrt{2\pi r}} \cos\left(\frac{\theta}{2}\right)\left[1 + \sin\left(\frac{\theta}{2}\right)\sin\left(\frac{3\theta}{2}\right)\right] \\
&\quad + \frac{K_{II}}{\sqrt{2\pi r}} \sin\left(\frac{\theta}{2}\right)\cos\left(\frac{\theta}{2}\right)\cos\left(\frac{3\theta}{2}\right) \\[8pt]
\sigma_{12} &= \frac{K_I}{\sqrt{2\pi r}} \sin\left(\frac{\theta}{2}\right)\cos\left(\frac{\theta}{2}\right)\cos\left(\frac{3\theta}{2}\right) \\
&\quad + \frac{K_{II}}{\sqrt{2\pi r}} \cos\left(\frac{\theta}{2}\right)\left[1 - \sin\left(\frac{\theta}{2}\right)\sin\left(\frac{3\theta}{2}\right)\right]
\end{aligned}
\tag{8.5}
$$

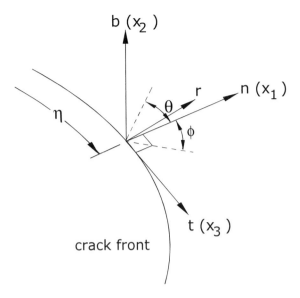

Figure 8.3: Local crack front coordinates.

for stresses, and

$$
\begin{aligned}
u_1 &= \frac{1}{4\mu}\sqrt{\frac{2r}{\pi}}\left\{K_I\left[(2\kappa-1)\cos\frac{\theta}{2}-\cos\frac{3\theta}{2}\right]\right.\\
&\quad\left.+K_{II}\left[(2\kappa+3)\sin\frac{\theta}{2}+\sin\frac{3\theta}{2}\right]\right\}
\end{aligned}
$$

$$
\begin{aligned}
u_2 &= \frac{1}{4\mu}\sqrt{\frac{2r}{\pi}}\left\{K_I\left[(2\kappa+1)\sin\frac{\theta}{2}-\sin\frac{3\theta}{2}\right]\right.\\
&\quad\left.+K_{II}\left[(2\kappa-3)\cos\frac{\theta}{2}+\cos\frac{3\theta}{2}\right]\right\}
\end{aligned} \tag{8.6}
$$

for displacements, where K_I and K_{II} are the stress intensity factors of the opening and sliding modes, respectively. The constant μ is the shear modulus and the constant κ is given by $3 - 4\nu$ for plain strain, and $(3 - \nu)/(1 + \nu)$ for plain stress, where ν is the Poisson's ratio.

For three-dimensional problems, the stress intensity factors are functions of position η on the crack front. In terms of the local spherical coordinate system shown in Figure 8.3, the stress fields in the vicinity of the crack front are given by [102]:

$$
\begin{aligned}
\sigma_n &= \frac{K_I(\eta)}{\sqrt{2\pi r}}\cos\left(\frac{\theta}{2}\right)\left[1-\sin\left(\frac{\theta}{2}\right)\sin\left(\frac{3\theta}{2}\right)\right]\\
&\quad-\frac{K_{II}(\eta)}{\sqrt{2\pi r}}\sin\left(\frac{\theta}{2}\right)\left[2+\cos\left(\frac{\theta}{2}\right)\cos\left(\frac{3\theta}{2}\right)\right]
\end{aligned}
$$

$$
\sigma_{nt} = -\frac{K_{III}(\eta)}{\sqrt{2\pi r}}\sin\left(\frac{\theta}{2}\right)
$$

$$\sigma_{bt} = \frac{K_{III}(\eta)}{\sqrt{2\pi r}} \cos\left(\frac{\theta}{2}\right)$$

$$
\begin{aligned}
\sigma_b =\;& \frac{K_I(\eta)}{\sqrt{2\pi r}} \cos\left(\frac{\theta}{2}\right)\left[1 + \sin\left(\frac{\theta}{2}\right)\sin\left(\frac{3\theta}{2}\right)\right] \\
& - \frac{K_{II}(\eta)}{\sqrt{2\pi r}} \sin\left(\frac{\theta}{2}\right)\cos\left(\frac{\theta}{2}\right)\cos\left(\frac{3\theta}{2}\right)
\end{aligned}
$$

$$\sigma_t = 2\nu\left[\frac{K_I(\eta)}{\sqrt{2\pi r}}\cos\left(\frac{\theta}{2}\right) - \frac{K_{II}(\eta)}{\sqrt{2\pi r}}\sin\left(\frac{\theta}{2}\right)\right]$$

$$
\begin{aligned}
\sigma_{nb} =\;& \frac{K_I(\eta)}{\sqrt{2\pi r}}\sin\left(\frac{\theta}{2}\right)\cos\left(\frac{\theta}{2}\right)\cos\left(\frac{3\theta}{2}\right) \\
& - \frac{K_{II}(\eta)}{\sqrt{2\pi r}}\cos\left(\frac{\theta}{2}\right)\left[1 - \sin\left(\frac{\theta}{2}\right)\sin\left(\frac{3\theta}{2}\right)\right]
\end{aligned}
\tag{8.7}
$$

where the non-singular terms have been dropped and r and θ are the polar components in the nb-plane.

The displacement fields are given by

$$
\begin{aligned}
u_n =\;& \frac{1+\nu}{E}\sqrt{\frac{2r}{\pi}}\left\{K_I(\eta)\cos\left(\frac{\theta}{2}\right)\left[(1-2\nu) + \sin^2\left(\frac{\theta}{2}\right)\right]\right. \\
& \left. + K_{II}(\eta)\sin\left(\frac{\theta}{2}\right)\left[2(1-\nu) + \cos^2\left(\frac{\theta}{2}\right)\right]\right\}
\end{aligned}
$$

$$
\begin{aligned}
u_b =\;& \frac{1+\nu}{E}\sqrt{\frac{2r}{\pi}}\left\{K_I(\eta)\sin\left(\frac{\theta}{2}\right)\left[2(1-\nu) - \cos^2\left(\frac{\theta}{2}\right)\right]\right. \\
& \left. - K_{II}(\eta)\cos\frac{\theta}{2}\left[(1-2\nu) - \sin^2\left(\frac{\theta}{2}\right)\right]\right\}
\end{aligned}
$$

$$u_t = 2\frac{1+\nu}{E}\sqrt{\frac{2r}{\pi}}K_{III}(\eta)\sin\left(\frac{\theta}{2}\right) \tag{8.8}$$

By using Taylor's series expansion, Sih and Hartranft [102] obtained the stress field around a small spherical surface of radius R centred at a point on the crack front so that $r = R\cos\phi$.

8.2.5 Residual Strength

The criterion for failure due to the unstable growth of a crack can be expressed in the following way: failure occurs if

$$K \geq K_{Ic} \text{ (plane strain condition)} \tag{8.9}$$

or

$$K \geq K_c \text{ (plane stress condition)} \tag{8.10}$$

where K_{Ic} and K_c are considered to be constants, sometimes called the *fracture toughness* of the material. From (8.2) and (8.10), the failure criterion can be written as

$$Y\sigma_c\sqrt{\pi a_c} = K_c \tag{8.11}$$

where σ_c and a_c denote the failure stress and critical crack length, respectively.

Equation (8.11) can be utilized to provide answers for two practical situations:

i) What is the maximum safe crack size at a given stress level?

ii) What is the maximum safe operating stress in the presence of a given crack-size?

If a structure is known to be cracked, the maximum safe operating stress (i.e. stress below which failure by unstable crack growth will not occur) must be determined in order to compare it with the proposed working stress. It follows from equation (8.11) that for a given crack length the critical stress depends on both the toughness and the crack length:

$$\sigma_c \propto \frac{K_c}{\sqrt{a_c}} \tag{8.12}$$

Thus, the critical stress increases as the toughness increases, and decreases as the crack length increases. It follows that a cracked structure with critical stress σ_c must be considered unsafe if the proposed operating stresses exceed σ_c. The possibility that σ_c may decrease with time, because the crack has grown longer, must be taken into account in practical assessments of long-term working stresses.

From (8.11) it follows that, for a given stress level σ_c, the critical crack-length for failure depends on the toughness K_c and the stress σ_c; in particular

$$a_c \propto \left[\frac{K_c}{\sigma_c}\right]^2 \tag{8.13}$$

Thus, the critical crack-length increases as the toughness increases, and decreases as the stress level increases. In practice, cracks must be detectable before they reach the length a_c, thus the minimum detectable size a_{min} sets a lower limit on the allowable value of the ratio (K_c/σ_c).

The above method for calculating residual strength does not take any account of stable crack-growth which occurs in some materials in thin sections. As the load is increased the crack grows in a stable manner, that is, if the load is held constant the crack ceases to grow, until the load is increased again. Assuming that the resistance to crack extension varies as the crack grows, stable crack growth can be explained in the following way. As the crack grows under raising load, the increase in the crack driving force, G, initially balances the increase in the crack growth resistance, R, and growth of the crack is stable. The point of instability is reached when the rate of increase of G equals the rate of increase of R; that is the curve of G and R as a function of crack length are tangential to each other. The instability condition is therefore $G = R$ and $dG/da = dR/da$. For engineering purposes, the R-curve is usually measured in stress intensity factor units, $K_R = \sqrt{ER}$; the instability condition becomes $K = K_R$ and $dK/da = dK_R/da$.

8.2.6 Fatigue Crack Growth

The growth of cracks which results when the applied stress is varied (fatigue) may also be described by the stress intensity factor, even though the maximum stresses

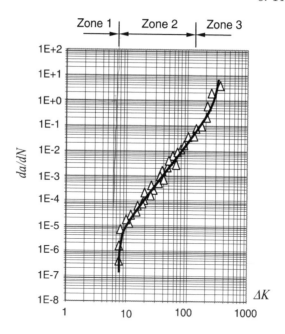

Figure 8.4: Fatigue crack growth diagram.

may be much less than the critical stress. Paris [152] postulated that the rate of growth per cycle of stress (da/dN) was a function of the stress intensity range $\Delta K (= K_{max} - K_{min})$, i.e.

$$\frac{da}{dN} = C(\Delta K)^m \tag{8.14}$$

where C and m are constants to be determined experimentally.

Figure 8.4 is a schematic log-log plot of the crack growth per load cycle, da/dN, as a function of the applied stress intensity factor range. In the first zone, crack growth goes asymptotically to zero as ΔK approaches the threshold value (ΔK_{th}). This means that for stress intensity factors below ΔK_{th} there is no crack growth (i.e. there is a fatigue limit). In zone II, $\log da/dN$ tends to vary linearly with respect to the log of ΔK given by (8.14). Finally, in zone III, a crack grows rapidly as K_{max} approaches K_c.

8.2.7   Criteria for Crack Growth Direction

Two criteria for mixed-mode loading that allow non-coplanar crack growth have been proposed; one based on the maximum principal stress by Erdogan and Sih [83] and the other on the strain energy density factor proposed by Sih [187].

Maximum Stress Criterion

There are several criteria for the prediction of crack growth direction. The maximum principal stress criterion postulates that a crack will grow in a direction perpendicular to the maximum principal stress. Considering two-dimensional combined mode I and mode II loading, the stresses $\sigma_{\theta\theta}$ and $\sigma_{r\theta}$ at the crack tip can be obtained from expressions given in Section 8.2.4 using the usual transformation; they are given by

$$
\sigma_{\theta\theta} = \frac{1}{\sqrt{2\pi r}} \cos\frac{\theta}{2} \left[K_I \cos^2\frac{\theta}{2} - \frac{3}{2} K_{II} \sin\theta \right]
$$

$$
\sigma_{r\theta} = \frac{1}{2\sqrt{2\pi r}} \cos\frac{\theta}{2} \left[K_I \sin\theta + K_{II}(3\cos\theta - 1) \right] \tag{8.15}
$$

The stress $\sigma_{\theta\theta}$ will be the maximum principal stress at $\theta_o = \theta$, where θ_o is defined by

$$
\sigma_{r\theta} = K_I \sin\theta_o + K_{II}(3\cos\theta_o - 1) = 0
$$

Thus, the crack growth direction angle θ_o is given by

$$
\theta_o = 2\tan^{-1}\left(\frac{K_I}{4K_{II}} \pm \frac{1}{4}\sqrt{\left(\frac{K_I}{K_{II}}\right)^2 + 8} \right) \tag{8.16}
$$

Strain Energy Density Criterion

The strain energy density criterion [187] states that crack growth takes place in the direction of minimum strain energy density factor $S(\theta)$. The factor $S(\theta)$, which is defined only if $r \neq 0$, is given by

$$
S(\theta) = a_{11}K_I^2 + 2a_{12}K_IK_{II} + a_{22}K_{II}^2 + a_{33}K_{III}^2 \tag{8.17}
$$

where

$$
a_{11} = \frac{1}{16\pi\mu}(3 - 4\nu - \cos\theta)(1 + \cos\theta)
$$

$$
a_{22} = \frac{1}{8\pi\mu}\sin\theta(\cos\theta - 1 + 2\nu)
$$

$$
a_{22} = \frac{1}{16\pi\mu}[4(1-\nu)(1-\cos\theta) + (3\cos\theta - 1)(1 + \cos\theta)]
$$

$$
a_{33} = \frac{1}{4\pi\mu}
$$

The theory is based on three hypotheses:

- The direction of crack growth at any point along the crack front is towards the region with the minimum value of strain energy density factor $S(\theta)$, as compared with other regions on the same spherical surface surrounding the point.

- Crack extension occurs when the strain energy density factor in the region determined by the above hypothesis, $S(\theta) = S(\theta)_{min}$ reaches a critical value, say $S_c(\theta)$.

- The length r_o of the initial crack extension is assumed to be proportional to $S(\theta)_{min}$, such that $S(\theta)_{min}/r_o$ remains constant along the new crack front.

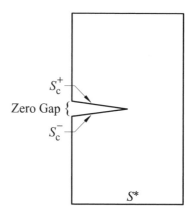

Figure 8.5: Co-planar crack surfaces.

8.3 Modelling Co-planar Surfaces

As stated earlier, the straightforward application of the boundary element method to crack problems leads to mathematical degeneration. In this section the mathematical difficulties associated with the standard application of the displacement boundary integral equation to modelling crack problems are initially outlined.

8.3.1 Displacement Integral Equation

The mathematical difficulties associated with the application of the displacement boundary integral equation to modelling crack problems have been described by Cruse [64]. Recall that Somigliana's identity for displacements at an interior point $\mathbf{X}'$ is given, in the absence of a body force, by

$$u_i(\mathbf{X}') = \int_{S^*+S_c^++S_c^-} U_{ij}(\mathbf{X}',\mathbf{x})t_j(\mathbf{x})dS - \int_{S^*+S_c^++S_c^-} T_{ij}(\mathbf{X}',\mathbf{x})u_j(\mathbf{x})dS \qquad (8.18)$$

where S_c^+ and S_c^- represent the upper and lower crack surfaces and S^* represents the remaining boundary, as shown in Figure 8.5. As can be seen from fundamental solutions given in Chapter 2, the displacement and traction on the lower and upper crack surfaces have the properties that

$$
\begin{aligned}
U_{ij}(\mathbf{X}',\mathbf{x}^+) &= +U_{ij}(\mathbf{X}',\mathbf{x}^-) \\
T_{ij}(\mathbf{X}',\mathbf{x}^+) &= -T_{ij}(\mathbf{X}',\mathbf{x}^-)
\end{aligned}
\qquad (8.19)
$$

The change in sign in tractions in (8.19) is because the direction of the normal is opposite on the two crack surfaces. By collapsing the two crack surfaces such that $S_c = S_c^+ = S_c^-$, equation (8.18) becomes

$$
\begin{aligned}
u_i(\mathbf{X}') &= \int_{S^*} U_{ij}(\mathbf{X}',\mathbf{x})t_j(\mathbf{x})dS - \int_{S^*} T_{ij}(\mathbf{X}',\mathbf{x})u_j(\mathbf{x})dS \\
&\quad + \int_{S_c} U_{ij}(\mathbf{X}',\mathbf{x})\Sigma t_j dS - \int_{S_c} T_{ij}(\mathbf{X}',\mathbf{x})\Delta u_j dS
\end{aligned}
\qquad (8.20)
$$

where

$$\begin{aligned} \Delta u_j &= u_j(\mathbf{x}^+) - u_j(\mathbf{x}^-) \\ \Sigma t_j &= t_j(\mathbf{x}^+) + t_j(\mathbf{x}^-) \end{aligned} \tag{8.21}$$

For traction free cracks, or when the crack is loaded by equal and opposite tractions $\Sigma t_j = 0$.

Following the usual BEM derivation (see Chapter 2), let $\mathbf{X}'$ tend to $\mathbf{x}^+$ on the crack surface S_c^+, therefore equation (8.21) for a traction free crack becomes

$$C_{ij}(\mathbf{x}^+)u_j(\mathbf{x}^+) + C_{ij}(\mathbf{x}^-)u_j(\mathbf{x}^-) = \int_{S^*} U_{ij}(\mathbf{x}^+,\mathbf{x})t_j(\mathbf{x})dS$$

$$- \int_{S^*} T_{ij}(\mathbf{x}^+,\mathbf{x})u_j(\mathbf{x})dS + \int_{S_c} T_{ij}(\mathbf{x}^+,\mathbf{x})\Delta u_j(\mathbf{x})dS \tag{8.22}$$

where the extra free term $C_{ij}(\mathbf{x}^-)u_j(\mathbf{x}^-)$ is due to the coincidence of the collocation point $\mathbf{x}^+$ with $\mathbf{x}^-$ on the opposite crack surface S_c^-. Collocating at $\mathbf{x}^-$ will result in the same equation as in (8.22). Hence, the final system of equations has two sets of identical rows, which corresponds to the collocation point on the top and bottom crack surfaces. Eliminating a set of rows by collocating on only one of the crack surfaces results in a system of equations with fewer independent equations than unknowns.

8.3.2 Traction Integral Equation

In a similar way to the displacement integral equation, the stress integral equation can be written for collocation points on the crack surface $\mathbf{x}^-$ as

$$\frac{1}{2}\sigma_{ij}(\mathbf{x}^-) + \frac{1}{2}\sigma_{ij}(\mathbf{x}^+) = \int_S D_{kij}(\mathbf{x}^-,\mathbf{x})t_k(\mathbf{x})dS$$

$$- \oint_S S_{kij}(\mathbf{x}^-,\mathbf{x})u_k(\mathbf{x})dS \tag{8.23}$$

Multiplying the above equation by the outward unit normal $n_i(x^-)$ and noticing that $n_i(\mathbf{x}^-) = -n_i(\mathbf{x}^+)$, the traction boundary equation on the crack surface can be written as

$$\frac{1}{2}t_j(\mathbf{x}^-) - \frac{1}{2}t_j(\mathbf{x}^+) = n_i(\mathbf{x}^-)\int_S D_{kij}(\mathbf{x}^-,\mathbf{x})t_k(\mathbf{x})dS$$

$$- n_i(\mathbf{x}^-)\oint_S S_{kij}(\mathbf{x}^-,\mathbf{x})u_k(\mathbf{x})dS \tag{8.24}$$

The fundamental solutions have the following properties:

$$\begin{aligned} D_{kij}(\mathbf{x},\mathbf{x}^+) &= +D_{kij}(\mathbf{x},\mathbf{x}^-) \\ S_{kij}(\mathbf{x},\mathbf{x}^+) &= -S_{kij}(\mathbf{x},\mathbf{x}^-) \end{aligned} \tag{8.25}$$

The two integrals in (8.24) can be written as

$$\oint_S S_{kij}(\mathbf{x}^-,\mathbf{x})u_k(\mathbf{x})dS$$

$$= \oint_{S_c^-} S_{kij}(\mathbf{x}^-,\mathbf{x})u_k^-(\mathbf{x})dS + \oint_{S_c^+} S_{kij}(\mathbf{x}^-,\mathbf{x})u_k^+(\mathbf{x})dS + \oint_{S^*} S_{kij}(\mathbf{x}^-,\mathbf{x})u_k(\mathbf{x})dS$$

$$= \oint_{S_c^+} S_{kij}(\mathbf{x}^-,\mathbf{x})\Delta u_k(\mathbf{x})dS + \int_{S^*} S_{kij}(\mathbf{x}^-,\mathbf{x})u_k(\mathbf{x})dS$$

where

$$\oint_S D_{kij}(\mathbf{x}^-,\mathbf{x})t_k(\mathbf{x})dS$$

$$= \oint_{S_c^-} D_{kij}(\mathbf{x}^-,\mathbf{x})t_k^-(\mathbf{x})dS + \oint_{S_c^+} D_{kij}(\mathbf{x}^-,\mathbf{x})t_k^+(\mathbf{x})dS + \oint_{S^*} D_{kij}(\mathbf{x}^-,\mathbf{x})t_k(\mathbf{x})dS$$

$$= \oint_{S_c^-} D_{kij}(\mathbf{x}^-,\mathbf{x})[t_k^-(\mathbf{x})+t_k^+(\mathbf{x})]dS + \int_{S^*} D_{kij}(\mathbf{x}^-,\mathbf{x})t_k(\mathbf{x})dS$$

Using the above relationships, equation (8.24) can be rewritten as

$$\frac{1}{2}t_j(\mathbf{x}^-) - \frac{1}{2}t_j(\mathbf{x}^+) \tag{8.26}$$

$$= -n_i(\mathbf{x}^-)\oint_{S_c^-} S_{kij}(\mathbf{x}^-,\mathbf{x})\Delta u_k(\mathbf{x})dS - n_i(\mathbf{x}^-)\int_{S^*} S_{kij}(\mathbf{x}^-,\mathbf{x})u_k(\mathbf{x})dS$$

$$+n_i(\mathbf{x}^-)\int_{S^*} D_{kij}(\mathbf{x}^-,\mathbf{x})t_k(\mathbf{x})dS + n_i^-(\mathbf{x})n_i(\mathbf{x}^-)\oint_{S_c^-} D_{kij}(\mathbf{x}^-,\mathbf{x})[t_k^-(\mathbf{x})+t_k^+(\mathbf{x})]dS$$

If traction equilibrium, $t_k(\mathbf{x}^-) = -t_k(\mathbf{x}^+)$, is assumed on the crack surfaces, the above equation can be written as

$$t_j(\mathbf{x}^-) + n_i(\mathbf{x}^-)\oint_{S_c^-} S_{kij}(\mathbf{x}^-,\mathbf{x})\Delta u_k(\mathbf{x})dS + n_i(\mathbf{x}^-)\oint_{S^*} S_{kij}(\mathbf{x}^-,\mathbf{x})u_k(\mathbf{x})dS$$

$$= n_i(\mathbf{x}^-)\int_{S^*} D_{kij}(\mathbf{x}^-,\mathbf{x})t_k(\mathbf{x})dS \tag{8.27}$$

The above equation, unlike the displacement integral equation, can be used for the solution of the crack problems without any difficulty. The solution will be obtained in terms of the displacement discontinuity on the crack surface.

8.4   Dual Boundary Element Method

The Dual Boundary Element Method (DBEM), as developed by Portela, Aliabadi and Rooke [164] for 2D problems and Mi and Aliabadi [134] for 3D problems, has been shown to be a general, and computationally efficient, way of modelling crack problems in the BEM. The DBEM incorporates two independent boundary integral equations, with the displacement equation applied for collocation on one of the crack surfaces, and the traction equation on the other. As a consequence, general mixed-mode crack problems can be solved in a single-region formulation. The integration path is still the same for coincident points on the crack surfaces, however, the respective boundary integral equations are now distinct.

Consider a cracked body as shown in Figure 8.5, with $S_c^\pm$ referring to upper and lower crack surfaces, respectively, and S^* being the rest of the boundary.

If the displacement integral equation is collocated on $\mathbf{x}^+$ on the upper crack surface S_c^+, it can be written as

$$C_{ij}(\mathbf{x}^+)u_j(\mathbf{x}^+) + C_{ij}(\mathbf{x}^-)u_j(\mathbf{x}^-) + \oint_S T_{ij}(\mathbf{x}^+,\mathbf{x})u_j(\mathbf{x})dS = \int_S U_{ij}(\mathbf{x}^+,\mathbf{x})t_j(\mathbf{x})dS$$

$$\tag{8.28}$$

Now, when the traction equation is used for collocation on $\mathbf{x}^-$ on the lower crack surface, it can be written on a smooth boundary as

$$\frac{1}{2}t_j(\mathbf{x}^-) - \frac{1}{2}t_j(\mathbf{x}^+)$$

$$= n_i(\mathbf{x}^-) \fint_S D_{kij}(\mathbf{x}^-, \mathbf{x})t_k(\mathbf{x})dS - n_i(\mathbf{x}^-) \fint_S S_{kij}(\mathbf{x}^-, \mathbf{x})u_k(\mathbf{x})dS \quad (8.29)$$

The main difficulty in the DBEM formulation is the development of a general, accurate modelling procedure for the integration of Cauchy and Hadamard principal value integrals which appear in the traction equation. The necessary conditions for the existence of these singular integrals, assumed in the derivation of the dual boundary integral equations, imposes certain restrictions on the choice of shape functions for the crack surfaces. In the point collocation method of solution, the displacement integral equation requires the continuity of the displacement components at the nodes (i.e. collocation points), and the traction integral equation requires the continuity of the displacement derivatives at the nodes. These requirement were satisfied in [214] by adopting the Hermitian elements, however, the solutions reported were not very accurate. Later, Watson [215] improved the accuracy of this formulation. Rudolphi et al.[175] reported unexplained oscillations in their results, while Gray et al. [97] devised a scheme based on a special integration path around the singular point for linear triangular elements. The formulations in [175] and [97] were applied to embedded cracks only. In [164] and [134], both crack surfaces were discretized with discontinuous quadratic elements; this strategy not only automatically satisfies the necessary conditions for the existence of the Hadamard integrals, but also circumvents the problem of collocating at crack kinks and crack-edge corners. Several examples including embedded, edge, kinked and curved cracks were solved accurately in [134, 164]. For other contributions in DBEM see, for example [35, 48, 54, 82, 96, 109, 130]. DBEM has been extended to analyses for anisotropic composite laminates [4, 150, 151, 204], elastoplastic problems [55, 120, 121], elastodynamic problems [85, 87, 88, 90, 228, 229, 230, 231], stiffened panels [180, 181, 182], plates and shells [2, 77], concrete [176, 177], thermoelastic problems [71, 156] and bone fracture [132]. A review of the method can be found in [20].

8.4.1   Crack Modelling Strategy

For the sake of efficiency, and to keep the simplicity of the standard boundary element, discontinuous quadratic elements are used [135, 166] for the crack modelling, as shown in Figures 8.6 and 8.7, respectively.

The general modelling strategy, developed in [166] and [135], can be summarized as follows:

- the crack boundaries are modelled with discontinuous quadratic elements, as shown in Figures 8.6 and 8.7;

- continuous quadratic elements are used along the remaining boundary of the structure, except at the intersection between a crack and an edge, where semi-discontinuous or edge-discontinuous elements are required to avoid a common node at the intersection;

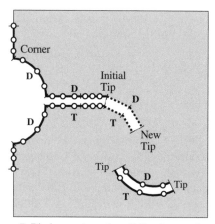

D Displacement equation   T Traction equation

Figure 8.6: Modelling strategy for two-dimensional crack problems.

- the displacement equation (8.28) is applied for collocation on one of the crack surfaces;

- the traction equation (8.29) is applied for collocation on the other crack surface;

- the usual boundary displacement equation is applied for collocation on all non-crack boundaries.

An analogous modelling strategy to the above was adopted by Cisilino and Aliabadi [54] as follows:

- Only one of the crack surfaces is discretized, and the traction equation (8.27) is applied for collocation.

- The crack surface displacements are retrieved using the expression

$$C_{ij}(\mathbf{x}^+)u_j(\mathbf{x}^+) + \int_{S_c^+} T_{ij}(\mathbf{x}^+, \mathbf{x})\Delta u_j(\mathbf{x})dS = \int_{S^*} U_{ij}(\mathbf{x}^+, \mathbf{x})t_j(\mathbf{x})dS \qquad (8.30)$$

to collocate on one of the crack surfaces (say S_c^+) and obtain u_j^+. Note that, although this will require an extra integration, there is no system of equations to be solved, since all the traction and displacement fields are already known. Finally, actual displacements u_j^- of the opposite crack surface can be easily obtained from $u_j^- = \Delta u_j - u_j^+$.

The changes required in a computer code for moving from using both displacement and traction integral equations on the crack surface to that involving only the traction equation is trivial, as the routines required for evaluation of the traction equation remain unchanged. The modelling strategy introduced in [54] clearly results in a smaller system matrix, with the reduction in size dependent on the number of nodes on the crack surface. For large three-dimensional problems such as those studied in [54] for elastostatic crack growth, and [55, 57] for elastoplastic crack growth involving more than one crack, the savings can be substantial.

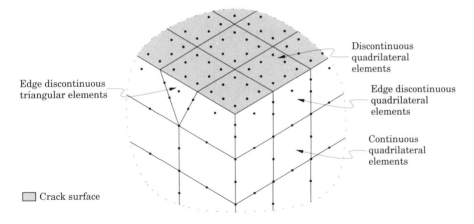

Figure 8.7: Crack modelling strategy for three-dimensional crack problems.

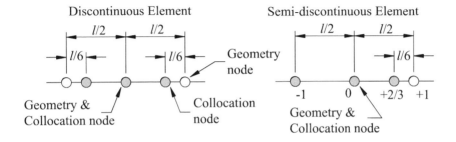

Figure 8.8: Quadratic discontinuous and semi-discontinuous elements ($\lambda = 2/3$).

Discontinuous Elements

For effective implementation of the dual boundary element method, discontinuous and semi-discontinuous elements are used for interpolating the displacement and traction fields over the crack surfaces. For these elements, the geometry is interpolated using the continuous interpolation functions described in Chapter 2.

Two-dimensional Problems The quadratic shape functions for the discontinuous element shown in Figure 8.8 are given as

$$N_1 = \eta\left(-\frac{1}{2\lambda} + \eta\frac{1}{2\lambda^2}\right)$$

$$N_2 = 1 + \eta^2\frac{1}{\lambda^2}$$

$$N_3 = \eta\left(\frac{1}{2\lambda} + \eta\frac{1}{2\lambda^2}\right) \tag{8.31}$$

For the semi-discontinuous element shown in Figure 8.8, with the discontinuous

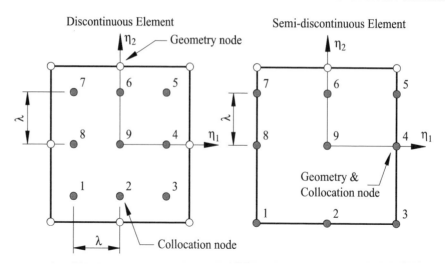

Figure 8.9: Nine node discontinuous and semi-discontinuous elements.

node at node 1, the shape functions are given as

$$N_1 = \eta \frac{1}{\lambda + \lambda^2}(-1 + \eta)$$

$$N_2 = 1 + \frac{\eta}{\lambda + \lambda^2} \left[(1 - \lambda^2) + \eta(-1 - \lambda) \right]$$

$$N_3 = \eta \frac{1}{1 + \lambda}(\lambda + \eta) \tag{8.32}$$

where λ denotes the parametric position of the nodes. The recommended value for $\lambda = 2/3$. Other values $0.5 < \lambda < 0.75$ were found not to have a significant effect on the solution.

Three-dimensional Problems For the nine noded element shown in Figure 8.9, we have

$$N^1 = \tfrac{1}{4\lambda^4}\eta_1(\eta_1 - \lambda)\eta_2(\eta_2 - \lambda) \qquad N^2 = \tfrac{1}{2\lambda^4}(\lambda^2 - \eta_1^2)\eta_2(\eta_2 - \lambda)$$

$$N^3 = \tfrac{1}{4\lambda^4}\eta_1(\eta_1 + \lambda)\eta_2(\eta_2 - \lambda) \qquad N^4 = \tfrac{1}{2\lambda^4}\eta_1(\eta_1 + \lambda)(\lambda^2 - \eta_2^2)$$

$$N^5 = \tfrac{1}{4\lambda^4}\eta_1(\eta_1 + \lambda)\eta_2(\eta_2 + \lambda) \qquad N^6 = \tfrac{1}{4\lambda^4}(\lambda^2 - \eta_1^2)\eta_2(\eta_2 + \lambda) \tag{8.33}$$

$$N^7 = \tfrac{1}{4\lambda^4}\eta_1(\eta_1 - \lambda)\eta_2(\eta_2 + \lambda) \qquad N^8 = \tfrac{1}{2\lambda^4}\eta_1(\eta_1 - \lambda)(\lambda^2 - \eta_2^2)$$

$$N^9 = \tfrac{1}{\lambda^4}(\lambda^2 - \eta_1^2)(\lambda^2 - \eta_2^2)$$

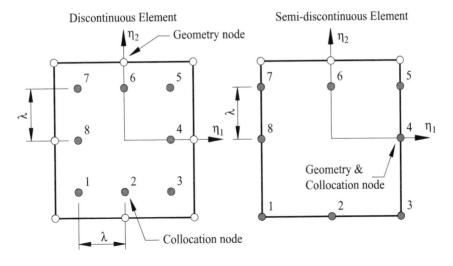

Figure 8.10: Eight node discontinuous and semi-discontinuous elements.

and for edge-discontinuous quadrilateral element,

$$N^1 = \frac{1}{2(\lambda+1)}\eta_1(\eta_1 - 1)\eta_2(\eta_2 - \lambda) \qquad N^2 = \frac{1}{(\lambda+1)}(1 - \eta_1^2)\eta_2(\eta_2 - \lambda)$$

$$N^3 = \frac{1}{2(\lambda+1)}\eta_1(\eta_1 + 1)\eta_2(\eta_2 - \lambda)(\eta_2 - \lambda) \qquad N^4 = -\frac{1}{2\lambda}\eta_1(\eta_1 + 1)(\eta_2 + 1)(\eta_2 - \lambda)$$

$$N^5 = \frac{1}{2\lambda(\lambda+1)}\eta_1(\eta_1 + 1)\eta_2(\eta_2 + 1)\eta_2(\eta_2 + 1) \qquad N^6 = \frac{1}{\lambda(\lambda+1)}(1 - \eta_1^2)\eta_2(\eta_2 + 1)$$

$$N^7 = \frac{1}{2\lambda(\lambda+1)}\eta_1(\eta_1 - 1)\eta_2(\eta_2 + 1) \qquad N^8 = -\frac{1}{2\lambda}\eta_1(\eta_1 - 1)(\eta_2 + 1)(\eta_2 - \lambda)$$

$$N^9 = -\frac{1}{\lambda}(1 - \eta_1^2)(\eta_2 + 1)(\eta_2 - \lambda)$$

For the 8-noded discontinuous elements shown in Figure 8.10, we have

$$N^1 = \frac{1}{4\lambda^3}(\lambda - \eta_1)(\lambda - \eta_2)(-\eta_1 - \eta_2 - \lambda) \qquad N^2 = \frac{1}{2\lambda^3}(\lambda^2 - \eta_1^2)(\lambda - \eta_2)$$

$$N^3 = \frac{1}{4\lambda^3}(\lambda + \eta_1)(\lambda - \eta_2)(\eta_1 - \eta_2 - \lambda) \qquad N^4 = \frac{1}{2\lambda^3}(\lambda + \eta_1)(\lambda^2 - \eta_2^2)$$

$$N^5 = \frac{1}{4\lambda^3}(\lambda + \eta_1)(\lambda + \eta_2)(\eta_1 + \eta_2 - \lambda) \qquad N^6 = \frac{1}{2\lambda^3}(\lambda^2 - \eta_1^2)(\lambda + \eta_2)$$

$$N^7 = \frac{1}{4\lambda^3}(\lambda - \eta_1)(\lambda + \eta_2)(-\eta_1 + \eta_2 - \lambda) \qquad N^8 = \frac{1}{2\lambda^3}(\lambda - \eta_1)(\lambda^2 - \eta_2^2)$$

The transition between discontinuous and continuous elements in those parts of the surface intersecting the crack face is produced with semi-discontinuous elements, as shown in Figure 8.7. The shape functions for a quadrilateral edge discontinuous element, depicted in Figure 8.10, where the side $\eta_2 = 1$ is the intersecting line, are given as

$$N^1 = \frac{1}{2(\lambda+1)}(1-\eta_1)(\lambda-\eta_2)(-\eta_1-\eta_2-1) \qquad N^2 = \frac{1}{(\lambda+1)}(1-\eta_1^2)(\lambda-\eta_2)$$

$$N^3 = \frac{1}{2(\lambda+1)}(1+\eta_1)(\lambda-\eta_2)(\eta_1-\eta_2-1) \qquad N^4 = \frac{1}{2\lambda}(1+\eta_2)(\lambda-\eta_2)(1+\eta_1)$$

$$N^5 = \frac{1}{2\lambda(\lambda+1)}(1+\eta_1)(1+\eta_2)(\lambda\eta_1+\eta_2-\lambda) \qquad N^6 = \frac{1}{(\lambda+1)}(1-\eta_1^2)(\lambda+\eta_2)$$

$$N^7 = \frac{1}{2\lambda(\lambda+1)}(1-\eta_1)(1+\eta_2)(-\lambda\eta_1+\eta_2-\lambda) \qquad N^8 = \frac{1}{2\lambda}(1+\eta_2)(\lambda-\eta_2)(1-\eta_1)$$

The shape functions for a semi-discontinuous triangular element, shown in Figure 8.11, where $\eta_2 + \eta_1 = 1$ is assumed to correspond to the intersection line, are

$$N^1 = \frac{\eta_1}{\lambda(1-2\lambda)}\left[-2\eta_1 + \frac{2(\lambda-1)}{\lambda}\eta_2 + 1\right] \qquad N^2 = \frac{4}{(\lambda)^2}\eta_1\eta_2$$

$$N^3 = \frac{\eta_2}{\lambda(1-2\lambda)}\left[-2\eta_2 + \frac{2(\lambda-1)}{\lambda}\eta_1 + 1\right] \qquad N^4 = \frac{4}{2\lambda-1}\eta_2(\lambda-\eta_1-\eta_2)$$

$$N^5 = \frac{1}{\lambda}(\lambda-\eta_1-\eta_2)(1-2\eta_1-2\eta_2) \qquad N^6 = \frac{4}{2\lambda-1}\eta_1(\lambda-\eta_1-\eta_2)$$

The functional shape functions for discontinuous triangular elements are given by:

$$N^1 = \eta_1'(2\eta_1'-1) \qquad N^2 = 4\eta_1'\eta_2'$$

$$N^3 = \eta_3'(2\eta_3'-1) \qquad N^4 = 4\eta_2'\eta_3'$$

$$N^5 = \eta_3'(2\eta_3'-1) \qquad N^6 = 4\eta_3'\eta_1'$$

where the following notation is used

$$\eta_1' = \frac{\eta_1-\lambda}{1-3\lambda} \qquad \eta_2' = \frac{\eta_2-\lambda}{1-3\lambda} \qquad \eta_3' = \frac{1-\eta_1-\eta_2-\lambda}{1-3\lambda}$$

and λ is the parametric position of the collocation nodes (see Figure 8.11).

Treatment of Singular and Hypersingular Integrals The preferred method of dealing with Cauchy principal value integrals in the BEM is the use of rigid body conditions. However, this indirect approach of calculating the integrals is not readily possible for the DBEM, and the integrals must be evaluated directly. To demonstrate this, consider the application of rigid body motion to equation (8.28):

$$\fint_{S_c^+} T_{ij}(\mathbf{x}',\mathbf{x})dS + \fint_{S_c^+} T_{ij}(\mathbf{x}',\mathbf{x})dS = -\delta_{ij} - \int_{S^*-S_c^+-S_c^-} T_{ij}(\mathbf{x}',\mathbf{x})dS \qquad (8.34)$$

As can be seen from equation (8.34), the summation of two Cauchy principal value integrals is obtained instead of one, and the left-hand side of the equation is zero, since $T_{ij}(\mathbf{x}',\mathbf{x}^+) = -T_{ij}(\mathbf{x}',\mathbf{x}^-)$.

Furthermore, for the integral involving D_{kij} and S_{kij} in equation (8.29), the rigid body consideration no longer applies, and the integrals must be evaluated directly.

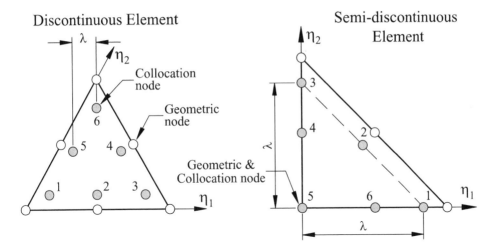

Figure 8.11: Discontinuous and semi-discontinuous elements.

Two-dimensional Kernels The Cauchy principal value integral of equation (8.28) can be expressed in the local coordinate as

$$\fint_{S_c} T_{ij}(\mathbf{x}',\mathbf{x})\, u_j(\mathbf{x})dS = \sum_{\gamma=1}^{M} u_j^\gamma \int_{-1}^{+1} \frac{f_{ij}^\gamma(\eta)}{\eta - \eta'}\, d\eta \tag{8.35}$$

where $f_{ij}^\gamma(\eta)$ is a regular function, given by the product of the fundamental solution, a shape function and the Jacobian of the coordinate transformation, multiplied by the term $(\eta - \eta')$. The integral of the right-hand side of equation (8.35) can be transformed with the aid of the first term of a Taylor's expansion of the function f_{ij}^γ, around the collocation node, to give

$$\int_{-1}^{+1} \frac{f_{ij}^\gamma(\eta)}{\eta - \eta'}\, d\eta = \int_{-1}^{+1} \frac{f_{ij}^\gamma(\eta) - f_{ij}^\gamma(\eta')}{\eta - \eta'}\, d\eta + f_{ij}^\gamma(\eta') \int_{-1}^{+1} \frac{d\eta}{\eta - \eta'} \tag{8.36}$$

The first integral of the right-hand side is now regular, and the second one can be integrated analytically to give

$$\int_{-1}^{+1} \frac{d\eta}{\eta - \eta'} = \ln\left|\frac{1 - \eta'}{1 + \eta'}\right| \tag{8.37}$$

In equation (8.36), the existence of the first-order finite-part integral requires the continuity of the first derivative of f_{ij}^γ, or at least the Hölder continuity of f_{ij}^γ, at the collocation node. For a discontinuous element, this requirement is automatically satisfied, because the nodes are internal points of the element, where f_{ij}^γ is continuously differentiable.

The Hadamard principal value integral of equation (8.29) can be expressed in the local parametric coordinate as

$$\fint_{S_c} S_{ijk}(\mathbf{x}',\mathbf{x})\, u_k(\mathbf{x})dS = \sum_{\gamma=1}^{M} u_k^\gamma \fint_{-1}^{+1} \frac{g_{ijk}^\gamma(\eta)}{(\eta - \eta')^2}\, d\eta \tag{8.38}$$

where $g_{ijk}^{\gamma}(\eta)$ is a regular function, given by the product of the fundamental solution, a shape function and the Jacobian of the coordinate transformation, multiplied by the term $(\eta - \eta')^2$. The integral of the right-hand side of equation (8.38) can be transformed with the aid of the first and second terms of a Taylor's expansion of the function g_{ijk}^{γ}, in the neighbourhood of the collocation node, to give

$$
\fint_{-1}^{+1} \frac{g_{ijk}^{\gamma}(\eta)}{(\eta - \eta')^2} d\eta = \int_{-1}^{+1} \frac{g_{ijk}^{\gamma}(\eta) - g_{ijk}^{\gamma}(\eta') - g_{ijk}^{\gamma(1)}(\eta')(\eta - \eta')}{(\eta - \eta')^2} d\eta \tag{8.39}
$$

$$
+ g_{ijk}^{\gamma}(\eta') \fint_{-1}^{+1} \frac{d\eta}{(\eta - \eta')^2} + g_{ijk}^{\gamma(1)}(\eta') \fint_{-1}^{+1} \frac{d\eta}{\eta - \eta'}
$$

where $g_{ijk}^{\gamma(1)}$ denotes the first derivative. At the collocation node, the function g_{ijk}^{γ} is required to have continuity of its second derivative, or at least a Hölder-continuous first derivative, for the finite-part integral to exist. This requirement is automatically satisfied by a discontinuous element, since the nodes are internal points of the element. Now, in equation (8.39), the first integral of the right-hand side is regular and the third integral is identical with that given in equation (8.37). The second integral of the right hand side of equation (8.39) can be integrated analytically to give

$$
\fint_{-1}^{+1} \frac{d\eta}{(\eta - \eta')^2} = -\frac{1}{1 + \eta'} - \frac{1}{1 - \eta'} \tag{8.40}
$$

In many practical crack problems the cracks are flat, but cracks do grow along curved paths, which are usually modelled as piece-wise flat. For piece-wise flat cracks, the integrals over the crack surface can be carried out analytically. Consider a discontinuous quadratic element that contains the collocation node. The local parametric coordinate system η is defined, as usual, in the range $-1 \leq \eta \leq +1$, and collocation η' is mapped onto $\mathbf{x}^{\beta}$ (source point on the crack boundary), via the continuous element shape functions N_{α}. The Cauchy principal value integral of equation (8.28) can be expressed in the local coordinate as

$$
\fint_{\Gamma_e} T_{ij}(\mathbf{x}', \mathbf{x}) \, u_j(\mathbf{x}) dS = \sum_{\gamma=1}^{M} u_j^{\gamma} \fint_{-1}^{+1} T_{ij}(\eta', \mathbf{x}(\eta)) \, N(\eta) \, J(\eta) d\eta = \mathbf{h}_i^{\gamma n} \mathbf{u}^{\gamma} \tag{8.41}
$$

where $\mathbf{u}^{\gamma}$ denotes the nodal displacement components and $J(\eta)$ is the Jacobian of the coordinate transformation. Because of the assumed flatness of the element, $J = \frac{l}{2}$, where l represents the element length, and the matrix $\mathbf{h}^{\gamma}$ is given by

$$
\mathbf{h}^{\gamma} = \frac{1 - 2\nu}{4\pi(1 - \nu)} \begin{bmatrix} 0 & -1 \\ +1 & 0 \end{bmatrix} \fint_{-1}^{+1} \frac{N_{\alpha}}{\eta - \eta'} d\eta \tag{8.42}
$$

The first order finite-part integrals are integrated analytically to give:

$$
\fint_{-1}^{+1} \frac{N_1}{\eta - \eta'} d\eta = \frac{3}{4} \left(\frac{\eta'(3\eta' - 2)}{2} \ln \left| \frac{1 - \eta'}{1 + \eta'} \right| + 3\eta' - 2 \right) \tag{8.43}
$$

$$
\fint_{-1}^{+1} \frac{N_2}{\eta - \eta'} d\eta = \frac{1}{2} \left(\frac{(3\eta' - 2)(3\eta' + 2)}{2} \ln \left| \frac{1 + \eta'}{1 - \eta'} \right| - 9\eta' \right) \tag{8.44}
$$

$$\fint_{-1}^{+1} \frac{N_3}{\eta - \eta'} d\eta = \frac{3}{4} \left(\frac{\eta'\left(3\eta' + 2\right)}{2} \ln \left| \frac{1 - \eta'}{1 + \eta'} \right| + 3\eta' + 2 \right) \tag{8.45}$$

The integral of equation (8.38) is represented by

$$\fint_{S_c} S_{ijk}(\mathbf{x}', \mathbf{x}) u_k(\mathbf{x}) dS = \sum_{\gamma=1}^{M} u_k^\gamma \fint_{-1}^{+1} S_{ijk}(\eta', \mathbf{x}(\eta)) N(\eta) J(\eta) d\eta = \overline{\mathbf{h}}_{ij}^\gamma \mathbf{u}^\gamma \tag{8.46}$$

where the matrix $\overline{\mathbf{h}}^\gamma$ is given by

$$\overline{\mathbf{h}}^\gamma = \frac{E}{4\pi \left(1 - \nu^2\right)} \frac{2}{l} \mathbf{S}' \fint_{-1}^{+1} \frac{N_\alpha}{(\eta - \eta')^2} d\eta \tag{8.47}$$

The matrix $\mathbf{S}'$ is given by

$$\mathbf{S}' = \begin{bmatrix} +n_1(2n_2^2 + 1) & -n_2(-2n_2^2 + 1) \\ +n_1(2n_1^2 - 1) & -n_2(-2n_1^2 - 1) \\ -n_2(2n_1^2 - 1) & +n_1(-2n_2^2 + 1) \end{bmatrix} \tag{8.48}$$

where n_1 and n_2 are the components of the unit outward normal to the element. The second-order finite-part integrals of equation (8.47) are integrated analytically to give

$$\fint_{-1}^{+1} \frac{N_1}{(\eta - \eta')^2} d\eta = \frac{3}{4} \left((3\eta' - 1) \ln \left| \frac{1 - \eta'}{1 + \eta'} \right| + \frac{6\eta'^2 - 2\eta' - 3}{\eta'^2 - 1} \right) \tag{8.49}$$

$$\fint_{-1}^{+1} \frac{N_2}{(\eta - \eta')^2} d\eta = \frac{1}{2} \left(9\eta' \ln \left| \frac{1 + \eta'}{1 - \eta'} \right| - \frac{18\eta'^2 - 13}{\eta'^2 - 1} \right) \tag{8.50}$$

$$\fint_{-1}^{+1} \frac{N_3}{(\eta - \eta')^2} d\eta = \frac{3}{4} \left((3\eta' + 1) \ln \left| \frac{1 - \eta'}{1 + \eta'} \right| + \frac{6\eta'^2 + 2\eta' - 3}{\eta'^2 - 1} \right) \tag{8.51}$$

Three-dimensional Problems The method for evaluation of hypersingular integrals reported here was presented by Guiggiani et al. [101], which is an extension of Guiggiani and Gigante [100], and is based on the Taylor's series expansion of the kernel, and Jacobian of transformation developed by Aliabadi, Hall and Phemister [5]. The method was successfully implemented for crack problems by Mi and Aliabadi [134]. Full details are described in Chapter 11.

Consider an element on the crack surface S_c, as shown in Figure 8.12, over which Cauchy and Hadamard principal value integrals are to be computed. As shown in the Figure, the collocation point is denoted as $\mathbf{x}^c$, and S_ε is taken to be part of S_c intersecting a sphere centred at $\mathbf{x}^c$ of radius ε. From the definition of Cauchy principal value integrals, the integral involving the T_{ij} kernel is written as

$$I = \lim_{\varepsilon \to 0} \int_{G_c - G_\varepsilon} T_{ij}[\mathbf{x}^c, \mathbf{x}(\eta_1, \eta_2)] N_\alpha(\eta_1, \eta_2) J^n(\eta_1, \eta_2) d\eta_1 d\eta_2 \tag{8.52}$$

where G_c and G_ε correspond to S_c and S_ε in the local coordinate system, respectively.

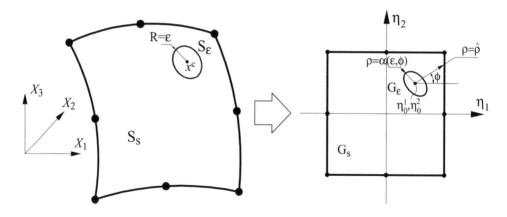

Figure 8.12: Vanishing neighbourhood around spource point $\mathbf{x}^c$ on a crack element.

The integrand can be approximated by using Taylor's series expansion, as

$$T_{ij}[\mathbf{x}^c, \mathbf{x}(\eta_1, \eta_2)]\, N_\alpha(\eta_1, \eta_2)J^n(\eta_1, \eta_2) = TNJ = \frac{T_{-1}(\theta)}{\rho^2} + O\left(\frac{1}{\rho}\right) \qquad (8.53)$$

where subscripts have been dropped for convenience, and $T_{-1}(\theta)$ denotes the first term in the Taylor expansion of T. Using (8.53), the integral I in (8.52) can be written in a polar coordinate (ρ, θ) system as

$$\begin{aligned}
I &= \lim_{\varepsilon \to 0} \int_{G_c - G_\epsilon} \left[TNJ\rho - \frac{T_{-1}(\theta)}{\rho} \right] d\rho d\theta + \lim_{\varepsilon \to 0} \int_{G_c - G_\epsilon} \left[\frac{T_{-1}(\theta)}{\rho} \right] d\rho d\theta \\
&= I_o + I_{-1}
\end{aligned} \qquad (8.54)$$

The method for evaluation of I_o and I_{-1} is described in Chapter 11.

For hypersingular integrals involving S_{kij}, a higher order series expansion approximation is applied to the integrand of equation (8.29). The integrand can be approximated using Taylor series expansion as

$$S_{kij}[\mathbf{x}^c, \mathbf{x}(\eta_1, \eta_2)]N_\alpha(\eta_1, \eta_2)J^n(\eta_1, \eta_2) = SNJ\rho = \frac{S_{-2}(\theta)}{\rho^3} + \frac{S_{-1}(\theta)}{\rho^2} + O\left(\frac{1}{\rho}\right) \quad (8.55)$$

In a similar way to the Cauchy principal value integral described above, the Hadamard principal value integral can be written (see Chapter 11) as

$$\begin{aligned}
\lim_{\varepsilon \to 0} &\left\{ \int_{G_c - G_\epsilon} S_{kij}[\mathbf{x}^c, \mathbf{x}(\eta_1, \eta_2)N_\alpha(\eta_1, \eta_2)J^n(\eta_1, \eta_2)d\eta_1 d\eta_2 + N_\alpha(\eta_o^1, \eta_o^2)\frac{b_{kij}(\mathbf{x}^c)}{\varepsilon} \right\} \\
&= \int_0^{2\pi} \int_0^{\hat\rho(\theta)} \left\{ SNJ\rho - \left[\frac{S_{-2}(\theta)}{\rho^3} + \frac{S_{-1}(\theta)}{\rho^2} \right] \right\} d\rho d\theta \\
&= \int_0^{2\pi} \left\{ S_{-1}(\theta) \ln \frac{|\hat\rho(\theta)|}{|\beta(\theta)|} - S_{-2}(\theta) \left[\frac{\gamma(\theta)}{\beta^2(\theta)} + \frac{1}{\hat\rho(\theta)} \right] \right\} d\theta \qquad (8.56)
\end{aligned}$$

The above integrals are all regular, and can be evaluated using Gaussian quadrature.

8.4.2 DBEM with Continuous Elements

Wilde, Aliabadi and Power [234] developed a technique to allow continuous elements to be employed for the discretization of the boundaries modelled with the Traction Integral Equation (TBIE). They proposed a method to increase the degree of continuity of displacement and traction fields at a source point by introducing a new interpolating scheme based upon the nodal values of all elements containing the source point. A more general technique was developed by Young [242] using an indirect strategy for evaluation of the traction equation. The method has been successfully applied to three-dimensional problems involving surface and embedded cracks [236, 242] and to crack growth analysis [237].

The derivation of the traction integral equation for non-smooth surfaces requires the use of the boundary stress integral equation presented in Chapter 2. The equation that relates the stresses $\sigma'_{i,l}$ at the source point to the derivatives of displacements $u'_{p,q}$, via Hooke's law, is

$$\sigma'_{i,l} = E_{ilpq} u'_{p,q} = E_{ilpq}^{-1} \hat{c}_{jkpq} I_{jk} \tag{8.57}$$

The traction boundary integral equation can then be written, via the Cauchy formula, as

$$t_i^m = n_l^m \sigma'_{i,l} = n_l^m E_{ilpq} \hat{c}_{jkpq}^{-1} I_{jk} \tag{8.58}$$

where the index m refers to a quantity evaluated on element m at the source point. Immediately apparent from this equation is that for source points located at elements whose orientations are different (e.g. corners or edges), more than one equation will exist. Choosing only one of these equations would lead to an incomplete representation of the problem. To produce a unique TBIE, these equations are summed in a weighted manner such that their fractional contribution is proportional to the angle subtended at the source point by the boundary element [235, 242]. Thus, finally the TBIE can be written as

$$\sum_m \phi^m t_i^m = Z_{ijk} I_{jk}$$

where

$$Z_{ijk} = \sum_m \phi^m n_l^m E_{ilpq} \hat{c}_{jkpq}^{-1} \tag{8.59}$$

In the case of crack problems, different possibilities must be considered depending on the position of the collocation point. In using continuous elements the node could be located at:

- **The crack front:** in this case the displacement integral equation can be used, and the calculation of the free terms and strongly singular integrals can be evaluated directly [242] or avoided by application of the rigid body condition [235, 237].

- **The lower and the junction of the lower crack surface and external surfaces:** the displacement integral equation is used, but all free terms and singular integrals are calculated directly as opposed to rigid body conditions.

- **The upper crack surface and the junction of the upper crack surface and external surface:** the displacement derivative integral equation is used, and it now includes two extra free terms (as in the case of discontinuous elements) due to the coincidence of the opposite crack surface to $\mathbf{x}'$, thus equation (2.175) can now be written as

$$\hat{c}^+_{iljk}u^+_{i,l} + \hat{c}^-_{iljk}u^-_{i,l} = I_{jk} + d^-_{ijk}u'_i = \hat{I}_{jk} \tag{8.60}$$

where I_{jk} includes the term d^+_{ijk} instead of d_{ijk} and the $+$ and $-$ signs denote upper and lower crack surfaces, respectively.

To allow for problems involving loading on crack surfaces, an additional term $(t^+_i - t^-_i)$ also needs to be included in the free terms. Equation (8.60) can be decomposed into symmetric and anti-symmetric components as

$$\begin{aligned}
\hat{c}^+_{iljk}u^+_{i,l} + \hat{c}^-_{iljk}u^-_{i,l} &= \frac{1}{2}(\hat{c}^+_{iljk} + \hat{c}^-_{iljk})(u^+_{i,l} + u^-_{i,l}) \\
&\quad + \frac{1}{2}(\hat{c}^+_{iljk} - \hat{c}^-_{iljk})(u^+_{i,l} - u^-_{i,l})
\end{aligned} \tag{8.61}$$

After some manipulation of the above equation (see [235, 242]), the traction integral equation can be written as

$$\sum_m \phi^m(t^{(m)+}_i - t^{(m)-}_i) + Z_{ijk}a_{pqjk}(u^+_{p,q} + u^-_{p,q}) = Z_{ijk}\hat{I}_{jk} \tag{8.62}$$

where

$$Z_{ijk} = \sum_m \phi^m n^{(m)+}_l E_{ilpq}b^{-1}_{jkpq}$$

where $a_{iljk} = \frac{1}{2}(\hat{c}^+_{iljk} + \hat{c}^-_{iljk})$ and $b_{iljk} = \frac{1}{2}(\hat{c}^+_{iljk} - \hat{c}^-_{iljk})$. Taking the summation in (8.62) only over crack elements ensures the presence of a strong crack pressure term, particularly important for problems where the crack surfaces are loaded. The displacement derivative terms in (8.62) are calculated using information from all elements containing the source point.

It is possible to evaluate the remaining displacement derivatives in equation (8.62) by differentiating displacement interpolations on one arbitrarily chosen element containing the source point. However, for the fully general case, estimates based on this approach are poor. The preferred approach consists of using the values of normal and tangential displacement derivatives and normal and shear stresses from all elements containing the source point in a weighted manner.

Investigations [235, 236, 237] have shown that the above approach with continuous elements requires a finer mesh, particularly around kinks and edges, than is required with discontinuous elements. The extension of the above approach to mixed-mode crack growth has been reported by Wilde and Aliabadi [237]. Generally, for the same number of elements, the size of the system matrix using the continuous elements approach is smaller than the equivalent approach using discontinuous elements. However, the computational effort in setting up the system matrix is much greater for the above approach of using continuous elements. Overall, in the author's opinion, the use of discontinuous elements is a more robust approach.

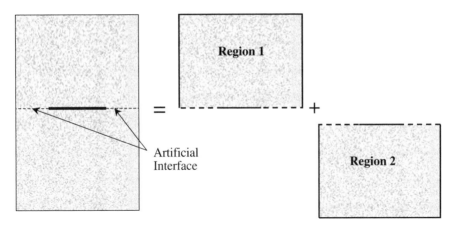

Figure 8.13: Dividing the crack region into two regions.

8.5 Multi-region Method

One way of overcoming the problem associated with co-planar crack surfaces, is to divide the cracked body into two regions, as shown in Figure 8.13, with each region containing one crack surface. The two regions are joined together such that equilibrium of tractions and compatibility of displacements are enforced along the artificial interface. The process is identical to the multi-region formulation described in Chapter 2. The main disadvantage of this approach is the introduction of the artificial boundaries, which results in a larger system of algebraic equations than is strictly necessary. There are, however, problems of cracks along the interface of dissimilar materials that can be solved effectively using this approach. Application of the BEM to interface crack problems in isotropic bimaterials can be found, for example, in [118, 141, 207, 238, 243]. Recently, the BEM has been applied to 3D interface crack problems in statics [232] and dynamics [233]. Application of BEM to interface cracks in orthotropic or anisotropic bimaterials can be found in [33, 49, 75].

8.6 Crack Green's Function Method

In Chapter 2, the use of a different fundamental solution that satisfies a traction free crack was discussed. Such an approach has the advantage that the crack surfaces do not need to be modelled, and since the correct behaviours of the stresses and displacements are embedded in the fundamental solutions, accurate values of stress intensity factors can be expected for the numerical solutions. However, this approach is restricted to two-dimensional straight-crack configurations, and fundamental solutions are algebraically complicated functions of complex variables which can be time-consuming and expensive to calculate.

The concept of a crack Green's function was introduced by Snyder and Cruse [199] into the boundary element analysis of cracked anisotropic plane elastostatic problems with traction-free cracks. The application of the crack Green's function to isotropic materials can be found in [80, 138, 241].

Once the boundary values of tractions and displacements have been obtained using the crack Green's function method, the stress intensity factors can be computed from the interior stresses or interior displacements [133]. For the stress approach, the values of stress at internal source points are computed from the Somigliana'ss stress identity (see Chapter 1) using the crack kernels D^c_{kij} and S^c_{kij}. In the vicinity of the crack tip, these kernels can be represented by

$$D^c_{kij} = \frac{1}{\sqrt{r}} D^{c*}_{kij}$$

$$S^c_{kij} = \frac{1}{\sqrt{r}} S^{c*}_{kij} \tag{8.63}$$

so that a $1/\sqrt{r}$ singularity appears as a multiplier of the modified kernels. The interior stress for the near tip fields can now be written as

$$\sigma^*_{ij}(\mathbf{X}') = \sqrt{r}\sigma_{ij}(\mathbf{X}') = \int_S \left[D^{c*}_{kij}(\mathbf{X}', \mathbf{x})t_k(\mathbf{x}) - S^{c*}_{kij}(\mathbf{X}', \mathbf{x})u_k \right] dS \tag{8.64}$$

The stress intensity factors can be obtained in terms of the Cartesian stress components by taking the limit as $\mathbf{X}' \to \mathbf{x}'_{tip}$ at the crack tip, thus

$$K_I = \lim_{r \to 0} \left(\sqrt{2\pi r}\sigma_{yy}(\mathbf{X}') \right) = \sqrt{2\pi}\sigma^*_{yy}(\mathbf{x}_{tip})$$

$$K_{II} = \lim_{r \to 0} \left(\sqrt{2\pi r}\sigma_{xy}(\mathbf{X}') \right) = \sqrt{2\pi}\sigma^*_{xy}(\mathbf{x}_{tip}) \tag{8.65}$$

Substitution of (8.64) and (8.65) into (8.63) gives

$$K_I = \sqrt{2\pi} \int_S \left[D^{c*}_{kyy}(\mathbf{x}'_{tip}, \mathbf{x})t_k - S^{c*}_{kyy}(\mathbf{x}'_{tip}, \mathbf{x})u_k \right] dS$$

$$K_{II} = \sqrt{2\pi} \int_S \left[D^{c*}_{kxy}(\mathbf{x}'_{tip}, \mathbf{x})t_k - S^{c*}_{kxy}(\mathbf{x}'_{tip}, \mathbf{x})u_k \right] dS$$

In a procedure proposed by Mews [133], the kernels U^c_{ij} and T^c_{ij} can be represented as

$$U^c_{ij} = \sqrt{r}U^{c*}_{ij} + \hat{U}^c_{ij}$$

$$T^c_{ij} = \sqrt{r}T^{c*}_{ij} + \hat{T}^c_{ij} \tag{8.66}$$

After substitution of (8.66) into the formula for the interior displacement, and after taking the limiting form at the crack tip, the stress intensity factors are given by

$$K_I = \frac{2\mu\sqrt{2\pi}}{\kappa - 1} \int_S \left[U^{c*}_{2j}(\mathbf{x}'_{tip}, \mathbf{x})t_j - T^{c*}_{2j}(\mathbf{x}'_{tip}, \mathbf{x})u_j(\mathbf{x}) \right] dS$$

$$K_{II} = \frac{2\mu\sqrt{2\pi}}{\kappa - 1} \int_S \left[U^{c*}_{1j}(\mathbf{x}'_{tip}, \mathbf{x})t_j - T^{c*}_{1j}(\mathbf{x}'_{tip}, \mathbf{x})u_j(\mathbf{x}) \right] dS$$

where $\kappa = 3 - 4\nu$ for plane strain and $(3 - 4\nu)/(1 + \nu)$ for plane stress.

   More recently, Telles, Castor and Guimaraes [212] and Castor and Telles [43] have developed techniques for numerically evaluating the crack Green's function for two- and three-dimensional problems, respectively. The crack Green's functions are evaluated from superposition of the Kelvin and a complementary solutions, as shown in Figure 8.14.

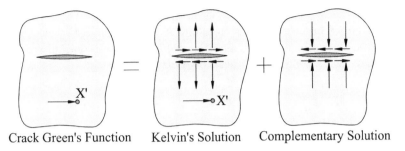

Crack Green's Function   Kelvin's Solution   Complementary Solution

Figure 8.14: Crack Green's function obtained from superpostion of Kelvin's and complementary solutions.

8.7 Displacement Discontinuity Method

Two indirect boundary integral formulations known as the displacement discontinuity and fictitious stress methods were developed by Crouch [61] and Crouch and Starfield [62]. These integral equations have been extensively used for geomechanical problems [70], and later extended to dynamic crack problems by Wen, Aliabadi and Rooke [217, 220, 220]. Application of the three-dimensional displacement discontinuity method to crack problems in infinite space can be found in [65, 191, 192, 216].

Consider a finite body with boundary S and crack boundary S_c. Let $\tilde{t}_k$ and $\tilde{t}'_k$ ($k = 1, 2$) be the traction boundary conditions on the outer boundary S and the crack surface S_c, respectively. Assuming that the fictitious loads $\phi_i(\mathbf{x})$ act at point $\mathbf{x} \in S$ and the displacement discontinuity $\Psi_i(\mathbf{x}_c)$ at the crack surface $\mathbf{x}_c \in S_c$, then at an interior point $\mathbf{X}'$ we have

$$u_k(\mathbf{X}') = \int U_{ki}(\mathbf{X}', \mathbf{x})\phi_i(x)dS + \int_{S_c} U'_{ki}(\mathbf{X}', \mathbf{x})\Psi_i(\mathbf{x}_c)dS \qquad (8.67)$$

$$t_k(\mathbf{X}') = \int T_{ki}(\mathbf{X}', \mathbf{x})\phi_i(x)dS + \int_{S_c} T'_{ki}(\mathbf{X}', \mathbf{x})\Psi_i(\mathbf{x}_c)dS$$

As the interior point approaches the boundary, i.e. $\mathbf{X}' \rightarrow \mathbf{x}'$, the boundary integral equation on a smooth boundary S is given by

$$\frac{1}{2}\phi_k(\mathbf{x}') + \int_{S_c} T_{ki}(\mathbf{X}', \mathbf{x})\phi_i(x)dS + \int_{S_c} T'_{ki}(\mathbf{X}', \mathbf{x})\Psi_i(\mathbf{x}_c)dS = \tilde{t}_k \qquad (8.68)$$

Assuming the crack lies on $x_2 = 0$ and the $\mathbf{X}'$ is on the crack surface $\mathbf{x}'_c$, the fundamental solution T'_{ki} is simplified as

$$T'_{ki} = \frac{\mu \delta_{ki}}{2\pi(1 - \nu)r^2} \qquad (8.69)$$

When the stress boundary condition on the crack surface S_c is taken into account, the integral equation can be written as

$$\int_S T_{ki}(\mathbf{x}'_c, \mathbf{x})\phi_i(\mathbf{x})dS + \fint_{S_c} T'_{ki}(\mathbf{x}', \mathbf{x}_c)\psi_i(\mathbf{x}_c)dS = \tilde{t}_k \qquad (8.70)$$

where $\tilde{t}'_k$ is the stress boundary condition on the crack surface.

The boundary S and S_c are discretized into N and M elements, with the fictitious loads acting on the outer boundary and the displacement discontinuity on the crack surface. Equations (8.68) and (8.70) can be written as

$$\sum_{n=1}^{N} A_{ki}^{\beta n} \phi_i^n + \sum_{m=1}^{M} D_{ki}^{\beta m} \psi_i^m = \tilde{t}_k^{\beta} \qquad \beta = 1, 2, ...N$$

$$\sum_{n=1}^{N} A_{ki}'^{\beta n} \phi_i^n + \sum_{m=1}^{M} D_{ki}'^{\beta m} \psi_i^m = \tilde{t}_k'^{\beta} \qquad \beta = 1, 2, ...M$$

where

$$A_{ki}^{\beta n} = \int_{S_n} T_{ki}(\mathbf{x}^{\beta}, \mathbf{x}) dS_n$$

$$D_{ki}^{\beta n} = \int_{S_c} T_{ki}(\mathbf{x}^{\beta}, \mathbf{x}_c) dS_c$$

$$A_{ki}'^{\beta n} = \int_{S_n} T_{ki}(\mathbf{x}_c^{\beta}, \mathbf{x}) dS_n$$

and from (8.69), we have

$$D_{ki}'^{\beta m} = \int_{S_c} T_{ki}'(\mathbf{x}_c^{\beta}, \mathbf{x}_c) dx_1 = \frac{\mu a_m \delta_{ki}}{\pi(1-\nu)[(x_1^{\beta} - x_1^m)^2 - a_m^2]}$$

where a_m denotes the element length.

8.8   Methods for Evaluating Stress Intensity Factors

There have been many methods devised for accurate evaluation of the Stress Intensity Factors (SIF) using the BEM. The most popular are perhaps the techniques based on the quarter-point elements, path independent integrals, energy methods, subtraction of singularity method and the weight function methods.

In this section these methods are described. A review of some of these methods can be found in Smith [195].

8.8.1   Displacement Extrapolation Method

Stress intensity factors may be computed from conventional or special crack tip elements in a number of different ways, by equating the boundary element solutions to theoretical expressions for displacements near the crack tip. The displacement fields on the crack surfaces can be rewritten, from equation (8.6), as

$$u_2(\theta = \pi) - u_2(\theta = -\pi) = \frac{\kappa+1}{\mu} K_I \sqrt{\frac{r}{2\pi}}$$

$$u_1(\theta = \pi) - u_1(\theta = -\pi) = \frac{\kappa+1}{\mu} K_{II} \sqrt{\frac{r}{2\pi}} \qquad (8.71)$$

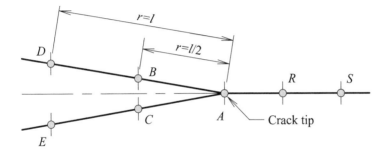

Figure 8.15: Nodes in the vicinity of crack tip (quaratic continuous elements).

For the continuous quadratic element of length l, shown in Figure 8.15, we have

$$K_I^{BC} = \frac{2\mu}{\kappa+1}\sqrt{\frac{\pi}{l}}(u_2^B - u_2^C)$$

$$K_{II}^{BC} = \frac{2\mu}{\kappa+1}\sqrt{\frac{\pi}{l}}(u_1^B - u_1^C) \qquad (8.72)$$

with $r = l/2$. Similarly, for nodes B and E,

$$K_I^{DE} = \frac{\mu}{\kappa+1}\sqrt{\frac{2\pi}{l}}(u_2^D - u_2^E)$$

$$K_{II}^{DE} = \frac{\mu}{\kappa+1}\sqrt{\frac{2\pi}{l}}(u_1^D - u_1^E) \qquad (8.73)$$

with $r = l$. Now the values of K_I and K_{II} are obtained by the linear extrapolation of K_I^{BC}, K_I^{DE} and K_{II}^{BC}, K_{II}^{DE} to the tip at A; thus

$$K_I = 2K_I^{BC} - K_I^{DE}$$
$$K_{II} = 2K_{II}^{BC} - K_{II}^{DE} \qquad (8.74)$$

An alternative way of obtaining stress intensity factors is to minimize the square root of errors between the analytical crack opening and sliding displacements, and the corresponding ones obtained numerically, i.e.

$$e_i = (\Delta u_i^{BC(Num)} - \Delta u_i^{BC(Anal)})^2 + (\Delta u_i^{DE(Num)} - \Delta u_i^{DE(Anal)})^2 \qquad i = I, II \quad (8.75)$$

where superscript Num and $Anal$ refer to numerical and analytical values, respectively. The values of stress intensity factors are obtained by minimizing the squared errors (i.e. $de_i/dK_i = 0$) to give

$$K_i = \frac{\mu}{\kappa+1}\sqrt{2\pi}\frac{\left[\sqrt{r^{BC}}\Delta u_i^{BC} + \sqrt{r^{DE}}\Delta u_i^{DE}\right]}{r^{BC} + r^{DE}} \qquad (8.76)$$

where r^{BC} and r^{DE} are distances to the crack tip associated with nodes B, C and D, E, respectively.

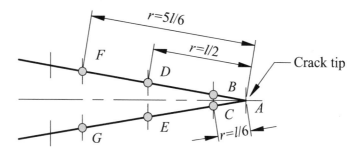

Figure 8.16: Nodes in the vicinity of the crack tip (quadratic discontinuous elements).

For the discontinuous elements shown in Figure 8.16, the stress intensity factors evaluated for nodes D, E and F, G are given by

$$K_I^{DE} = \frac{2\mu}{\kappa+1}\sqrt{\frac{\pi}{l}}(u_2^D - u_2^E)$$

$$K_{II}^{DE} = \frac{2\mu}{\kappa+1}\sqrt{\frac{\pi}{l}}(u_1^D - u_1^E) \qquad (8.77)$$

and

$$K_I^{FG} = \frac{2\mu}{\kappa+1}\sqrt{\frac{3\pi}{5l}}(u_2^F - u_2^G)$$

$$K_{II}^{FG} = \frac{2\mu}{\kappa+1}\sqrt{\frac{3\pi}{5l}}(u_1^F - u_1^G) \qquad (8.78)$$

respectively. By means of linear extrapolation from the nodes D, E and F, G to the crack tip, the stress intensity factors can be evaluated from

$$K_I = \frac{\mu}{\kappa+1}\sqrt{\frac{\pi}{l}}\left[5(u_2^D - u_2^E) - \frac{3\sqrt{15}}{5}(u_2^F - u_2^G)\right]$$

$$K_{II} = \frac{\mu}{\kappa+1}\sqrt{\frac{\pi}{l}}\left[5(u_1^D - u_1^E) - \frac{3\sqrt{15}}{5}(u_1^F - u_1^G)\right] \qquad (8.79)$$

Salgado and Aliabadi [182] adopted the stress intensity factor expressions presented in (8.76) for discontinuous elements,

$$K_I = \frac{\mu}{\kappa+1}\left(\frac{2l}{3}\right)\sqrt{2\pi}\left[\sqrt{\frac{l}{6}}\Delta u_i^{BC} + \sqrt{\frac{l}{2}}\Delta u_i^{DE}\right] \qquad (8.80)$$

Quarter-point Elements

Henshell and Shaw [104] and Barsoum [29] have shown that by moving the mid-side node of a quadratic element to a quarter-point position, as shown in Figure 8.17, the desired $\sqrt{r}$ variation for displacements can be achieved. Notice that the local coordinate η remains unchanged (i.e. $\eta = -1, 0, +1$). A similar procedure was adopted by Cruse and Wilson [66] and Smith [196] in the context of the boundary element method. Let a crack-tip element be located on the $y = 0$ plane with $x^A = 0$, $x^{B,C} = l/4$ and $x^{D,E} = l$, where l is the length of the element. From the quadratic shape function representation of the geometry (see Chapter 2), the coordinate x becomes

$$
\begin{aligned}
x &= \sum_{\alpha=1}^{3} N_\alpha(\eta)x^\alpha \\
&= \frac{1}{2}\eta(1+\eta)l + (1-\eta^2)\frac{l}{4}
\end{aligned}
$$

so that

$$
\eta = -1 + 2\sqrt{\frac{x}{l}} = -1 + 2\sqrt{\frac{r}{l}} \tag{8.81}
$$

Substitution for η in the displacement fields approximation gives the following dependence on r:

$$
u_i = u_i^A + (-3u_i^A + 4u_i^B - u_i^C)\sqrt{\frac{r}{l}} + 2(u_i^A - 2u_i^B + u_i^C)\frac{r}{l} \tag{8.82}
$$

When using the multi-region method for modelling crack problems, we also need to model the singular traction fields ahead of the crack. This is a further disadvantage of the method when compared to dual boundary elements, as attempts to model a singular field introduces additional inaccuracies into the numerical model. As shown in Chapter 2, the traction fields in the BEM are represented independently of the displacements, therefore the quarter point representation described above will also result in the traction field representation similar to (8.82), i.e.

$$
t_i = t_i^A + (-3t_i^A + 4t_i^R - t_i^S)\sqrt{\frac{r}{l}} + 2(t_i^A - 2t_i^R + t_i^S)\frac{r}{l}
$$

and, therefore, are not suitable in their present form. A suitable presentation [66] can, however, be obtained by dividing the shape functions by $\sqrt{r/l}$ to give

$$
t_i = t_i^A\sqrt{\frac{l}{r}} + (-3t_i^A + 4t_i^R - t_i^S) + 2(t_i^A - 2t_i^R + t_i^S)\sqrt{\frac{r}{l}} \tag{8.83}
$$

Further information on the use of traction singular elements can be found in Aliabadi and Rooke [16].

Similarly, the quarter-point element concept can be adopted for discontinuous elements as originally derived in [88]. The new distances of the pairs of $B - C$, $D - E$ and $F - G$ from the crack tip A shown in Figure 8.18 are: $l/36$, $9l/36$ and $25l/36$, respectively. The local coordinates of the nodes remain unchanged (i.e. $\eta = -\frac{2}{3}, 0 + \frac{2}{3}$). Fedelinski, Aliabadi and Rooke [88] used the method of minimization of the squared differences between the analytical and boundary element values of the crack opening and crack sliding displacements (see equation (8.75)) to give

$$
K_I = \frac{6\mu}{5(\kappa+1)}\sqrt{\frac{\pi}{2l}}(\Delta u_2^{BC} + 3\Delta u_2^{DE}) \tag{8.84}
$$

and

$$
K_I = \frac{6\mu}{5(\kappa+1)}\sqrt{\frac{\pi}{2l}}(\Delta u_1^{BC} + 3\Delta u_1^{DE}) \tag{8.85}
$$

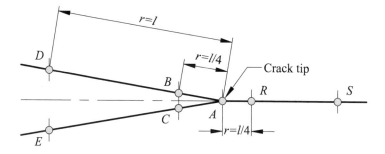

Figure 8.17: Quarter-point crack-tip elements.

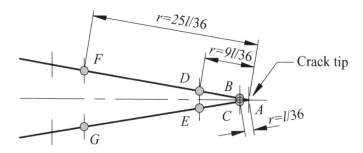

Figure 8.18: Discontinuous quarter-point elements.

For three-dimensional problems, the quarter point element is achieved as shown in Figure 8.19 by setting

$$\mathbf{x}^4 = \frac{1}{4}(3\mathbf{x}^3 + \mathbf{x}^5)$$

$$\mathbf{x}^8 = \frac{1}{4}(3\mathbf{x}^1 + \mathbf{x}^7) \tag{8.86}$$

Using the above relationships in the shape function representation of the element geometry, we obtain

$$\mathbf{x}(\eta_1, \eta_2) - \mathbf{x}(\eta_1, -1) = \mathbf{C}_2(\eta_1)(1 + \eta_2)^2 + \mathbf{C}_1(\eta_1)(1 + \eta_2) \tag{8.87}$$

where

$$\mathbf{C}_2(\eta_1) = \frac{1}{2}\left\{\frac{1}{4}\left[\mathbf{x}^3 + \mathbf{x}^1\right) - (\mathbf{x}^5 + \mathbf{x}^7) + \eta_1(\mathbf{x}^3 - \mathbf{x}^5 - \mathbf{x}^1 + \mathbf{x}^7)\right]\right\}(1 + \eta_2)^2 \tag{8.88}$$

and

$$\mathbf{C}_1(\eta_1) = \frac{1}{2}\left\{\left[\frac{\mathbf{x}^1 + \mathbf{x}^3}{2} - \frac{\mathbf{x}^5 + \mathbf{x}^7}{2} - (\mathbf{x}^2 - \mathbf{x}^6)\right](1 - \eta_1^2)\right\}(1 + \eta_2) \tag{8.89}$$

If $\mathbf{C}_1(\eta_1) = 0$, from (8.87) we have

$$r = |\mathbf{x}(\eta_1, \eta_2) - \mathbf{x}(\eta_1, -1)| = |\mathbf{C}_2(\eta_1)| (1 + \eta_2)^2$$

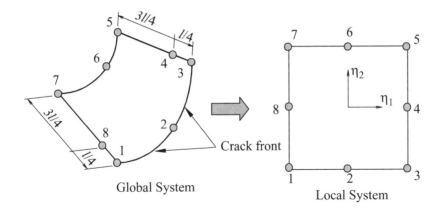

Figure 8.19: Quarter-point element for three-dimensional crack problems.

which shows that the distance of a point $\mathbf{x}$ in the element to the crack front $\eta_2 = -1$ along $\eta_1 = $ constant is proportional to $(1 + \eta_2)^2$. Providing that $\mathbf{C}_1(\eta_1) = 0$, the displacement representation over the element can be written as

$$u_i(\eta_1, \eta_2) = \sum_{\alpha=1}^{8} N_\alpha(\eta_1, \eta_2) u_i^\alpha$$
$$= L_2(\eta_1)r + L_1(\eta_1)\sqrt{r} + L_o(\eta_1) \qquad (8.90)$$

which is the desired variation. Hence, the derivation requires that $\mathbf{C}_1(\eta_1) = 0$, which implies that

$$\frac{\mathbf{x}^1 + \mathbf{x}^3}{2} - \frac{\mathbf{x}^5 + \mathbf{x}^7}{2} - (\mathbf{x}^2 - \mathbf{x}^6) = 0 \qquad (8.91)$$

The above condition puts a restriction on the geometrical position of the nodes. Geometrically, as shown in Figure 8.20, the condition requires the vector connecting the mid-side points of the line joining nodes 7 and 5 and the lines joining nodes 1 and 3 to be equal to the vector connecting nodes 6 and 2.

Unfortunately, this is not always easily achieved by meshes generated by mesh generator programs. It is possible to modify the position of the nodes to meet the geometric constraints, but this may result in a distorted crack front mesh, even for simple curved crack front problems such as the penny shaped crack. Accurate values of stress intensity factors are not generally expected from distorted crack front meshes. This restriction unfortunately makes the quarter-point element impractical for the analysis of general crack problems.

Crack-tip Interpolation Functions

An alternative approach to the quarter-point elements described in the previous sections is the use of special crack-tip elements. In this the nodal positions remain unchanged and the standard shape functions are used to approximate the geometry. However, a new set of interpolation functions which incorporate the $\sqrt{r}$ behaviour are used to represent the displacement fields on the crack surface elements adjacent to

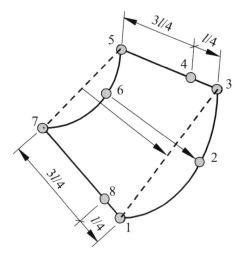

Figure 8.20: Geometrical constraint for quarter-point elements.

the crack-tip. For continuous elements there are several such functions, as described in [16]. A quadratic set which can be derived by considering the functional variation along one side of the six-node triangular finite elements [34] for the crack-tip at node 1 with $\eta = -1$, is given by

$$
\begin{aligned}
N_1 &= 1 + \frac{1+\eta}{\sqrt{2}} - \left(1 + \frac{1}{\sqrt{2}}\right)\sqrt{1+\eta} \\
N_2 &= (2+\sqrt{2})\sqrt{1+\eta} - (1+\sqrt{2})(1+\eta) \\
N &= \left(1 + \frac{1}{\sqrt{2}}\right)(1+\eta) - \left(1 + \frac{1}{\sqrt{2}}\right)\sqrt{1+\eta}
\end{aligned}
\qquad (8.92)
$$

It is worth noting that the condition $N_\alpha(\eta_i) = \delta_{\alpha i}$ for $\eta_i = -1, 0 + 1$, i.e. at nodes.

Similar types of shape functions were derived by Dirgantara and Aliabadi [78] for discontinuous quadratic elements with nodes located at $\eta = -\frac{2}{3}, 0, \frac{2}{3}$. For a crack tip located at $\eta = -1$, the modified crack-tip interpolation functions are given as

$$
\begin{aligned}
N_1 &= \frac{3}{2}\frac{(3-\sqrt{15})\eta + 2\sqrt{1+\eta} - 2}{\sqrt{15}+\sqrt{3}-6} \\
N_2 &= \frac{3(\sqrt{15}-\sqrt{3})\eta - 12\sqrt{1+\eta} + 2(\sqrt{15}+\sqrt{3})}{2(\sqrt{15}+\sqrt{3}-6)} \\
N &= \frac{3}{2}\frac{(\sqrt{3}-3)\eta + 2\sqrt{1+\eta} - 2}{\sqrt{15}+\sqrt{3}-6}
\end{aligned}
\qquad (8.93)
$$

Luchi and Poggialini [129] have developed a set of quadratic shape functions for the 8-noded quadrilateral elements, with the crack front corresponding to $\eta_2 = -1$ in the local plane (see Figure 8.21). They are given as

$$
N_\alpha = \frac{1}{4}(\eta_1\eta_1^\alpha)\left[1 - \eta_\alpha + \sqrt{2}\eta_2^\alpha\sqrt{1+\eta_2}\right]\left[\eta_1\eta_1^\alpha + \sqrt{1+\eta_2}(\sqrt{1+\eta_2^\alpha} + \eta_2^\alpha)\right]
$$

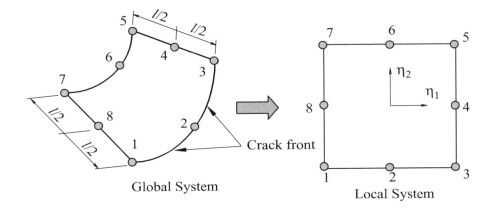

Figure 8.21: Special crack front elements.

$$\sqrt{1 + \eta_2^\alpha} + (1 + \eta_2^\alpha)\Big] \Big[\eta_1 \eta_1^\alpha + \sqrt{1 + \eta_2}\left(\frac{1}{\sqrt{2}}(1 + \eta_2^\alpha) + 1\right)$$

$$-(\frac{1}{\sqrt{2}} + 1)(1 + \eta_2^\alpha)\Big] \tag{8.94}$$

for $\alpha = 1, 3, 5, 7$ and $\eta_1^\alpha, \eta_2^\alpha = \pm 1$; and

$$N_\alpha = \frac{1}{4}(\eta_1^\alpha)^2(1 + \eta_1 \eta_1^\alpha)\Big[(2 + \sqrt{2})\sqrt{2 + \eta_2} - (1 + \sqrt{2})(1 + \eta_2)\Big]$$

$$+ \frac{1}{2}\eta_2^\alpha\Big[1 - \eta_1^\alpha + \sqrt{2}\eta_2^\alpha\sqrt{1 + \eta_2}\Big](1 - \eta_1^2) \tag{8.95}$$

for $\alpha = 2, 4, 6, 8$; $(\eta_1^\alpha, \eta_2^\alpha) = (0, \pm 1)$ for $\alpha = 2, 6$; and $(\eta_1^\alpha, \eta_2^\alpha) = (\pm 1, 0)$ for $\alpha = 4, 8$. Similar modified shape functions derived by Wilde and Aliabadi [235] for nine-noded continuous elements are given as

$$N_1 = \frac{1}{2\sqrt{2}}\eta_1(\eta_1 - 1)(\eta^* - 1)(\eta^* - \sqrt{2}) \qquad N_2 = \frac{1}{\sqrt{2}}(1 - \eta_1^2)(\eta^* - 1)(\eta^* - \sqrt{2})$$

$$N = \frac{1}{2\sqrt{2}}\eta_1(\eta_1 + 1)(\eta^* - 1)(\eta^* - \sqrt{2}) \qquad N_3 = \frac{1}{2(1 - \sqrt{2})}\eta_1(\eta_1 + 1)\eta^*(\eta^* - \sqrt{2})$$

$$N_5 = \frac{1}{2\sqrt{2}(\sqrt{2} - 1)}\eta_1(\eta_1 + 1)\eta^*(\sqrt{2} - 1) \qquad N_6 = \frac{1}{\sqrt{2}(\sqrt{2} - 1)}(1 - \eta_1^2)\eta^*(\sqrt{2} - 1)$$

$$N_7 = \frac{1}{2\sqrt{2}(\sqrt{2} - 1)}\eta_1(\eta_1 - 1)\eta^*(\sqrt{2} - 1) \qquad N_8 = \frac{1}{2(1 - \sqrt{2})}\eta_1(\eta_1 - 1)\eta^*(\eta^* - \sqrt{2})$$

$$N_9 = \frac{1}{1 - \sqrt{2}}(1 - \eta_1^2)\eta^*(\eta^* - \sqrt{2})$$

$$\tag{8.96}$$

where $\eta^* = \sqrt{1 + \eta_2}$. It is worth noting that the only restriction on the geometry for the above element is that the mid-side nodes 4 and 8 should actually bisect the lines connecting nodes 3 to 5 and 7 to 1, respectively.

Sets of special shape functions for both 8- and 9-noded discontinuous Lagrangian elements were derived by Mi and Aliabadi [136] and Cisilino and Aliabadi [54]. Provided the element side containing nodes 1, 2 and 3 (see Figure 8.22) is on the crack

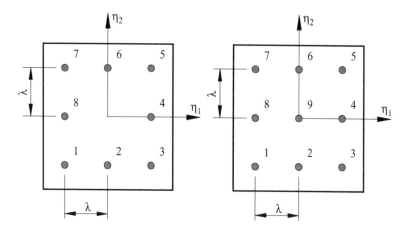

Figure 8.22: Eight- and nine-node speccial crack front elements.

front, the functional shape functions for the special crack tip elements are given as

$$N_1 = \left\{(-\lambda + \eta_1)[-\lambda p + \lambda^2(1-q) - (1-p)(1+q+2\lambda q)\right.$$
$$+(-3p + (1+\lambda)(2-q) + 2pq)\eta_1 + (\lambda^2 + \lambda(1-p)(3-q))s$$
$$\left. +(3 - 2p - 2q + pq)\eta_1 s + (\lambda(2-p) + (1-p)(1+q))s^2]\right\}/(-2\lambda D)$$

$$N_2 = \left\{(\lambda - \eta_1)(\lambda + \eta_1)[-2\lambda - 2\lambda^2 + 3\lambda p + \lambda q + \lambda^2 q - 2\lambda pq\right.$$
$$\left. +(-3\lambda + 2\lambda p + 2\lambda q - \lambda pq)s]\right\}/(\lambda^2 D)$$

$$N_3 = \left\{(-\lambda - \eta_1)[-\lambda p + \lambda^2(1-q) - (1-p)(1+q+2\lambda q)\right.$$
$$+(3p - (1+\lambda)(2-q) - 2pq)\eta_1 + (\lambda^2 + \lambda(1-p)(3-q))s$$
$$\left. +(-3 + 2p + 2q - pq)\eta_1 s + (\lambda(2-p) + (1-p)(1+q))s^2]\right\}/(-2\lambda D)$$

$$N_4 = \left\{(\lambda + \eta_1)[4\lambda + 2\lambda p(-\lambda + \lambda^2 - 2q) + 2\lambda^2(-2\lambda + q + \lambda q)\right.$$
$$\left. +4\lambda^2(-\lambda + p - q)s + (2\lambda^2(p-q) + 4\lambda(pq-1))s^2]\right\}/(4\lambda^2 D)$$

$$N_5 = \left\{-(\lambda + \eta_1)[2\lambda(1 - \lambda^2 + p - 2\lambda p + \lambda^2 p - q - \lambda q - pq + 2\lambda pq)\right.$$
$$-2\lambda(2 - 2\lambda - p + \lambda p - 3q + 2pq)\eta_1$$
$$-2\lambda^2(-3 + \lambda + p + 3q - pq)s - 2\lambda(3 - 2p - 2q + pq)\eta_1 s$$
$$\left. -2\lambda(1 - 2\lambda + p - q + \lambda q - pq)s^2\right\}/(4\lambda^2 D)$$

$$N_6 = \left\{(-\lambda - \eta_1)(\lambda - \eta_1)[-2\lambda + 2\lambda^2 + \lambda p - \lambda^2 p + 3\lambda q - 2\lambda pq\right.$$
$$\left. +(-3\lambda + 2\lambda p + 2\lambda q - \lambda pq)s]\right\}/(\lambda^2 D)$$

$$N_7 = \left\{-(\lambda - \eta_1)[2\lambda(1 - \lambda^2 + p - 2\lambda p + \lambda^2 p - q - \lambda q - pq + 2\lambda pq)\right.$$

$$+2\lambda(2 - 2\lambda - p + \lambda p - 3q + 2pq)\eta_1$$
$$-2\lambda^2(-3 + \lambda + p + 3q - pq)s + 2\lambda(3 - 2p - 2q + pq)\eta_1 s$$
$$-2\lambda(1 - 2\lambda + p - q + \lambda q - pq)s^2\} / (4\lambda^2 D)$$

$$N_8 = \{(\lambda - \eta_1)[4\lambda + 2\lambda p(-\lambda + \lambda^2 - 2q) + 2\lambda^2(-2\lambda + q + \lambda q)$$
$$+4\lambda^2(-\lambda + p - q)s + (2\lambda^2(p - q) + 4\lambda(pq - 1))s^2]\} / (4\lambda^2 D) \qquad (8.97)$$

for an eight-node element; and for nine-node element shown in Figure 8.22 as

$$N_1 = \frac{1}{2\lambda^2(q-1)(q-p)}\eta_1(\eta_1 - \lambda)(s - 1)(s - p)$$

$$N_2 = \frac{1}{\lambda^2(q-1)(q-p)}(\lambda^2 - \eta_1^2)(s - 1)(s - p)$$

$$N_3 = \frac{1}{2\lambda^2(q-1)(q-p)}\eta_1(\eta_1 + \lambda)(s - 1)(s - p)$$

$$N_4 = \frac{1}{2\lambda^2(1-q)(1-p)}\eta_1(\eta_1 + \lambda)(s - q)(s - p)$$

$$N_5 = \frac{1}{2\lambda^2(p-q)(p-1)}\eta_1(\eta_1 + \lambda)(s - q)(p - 1) \qquad (8.98)$$

$$N_6 = \frac{1}{\lambda^2(p-q)(p-1)}(\lambda^2 - \eta_1^2)(s - q)(p - 1)$$

$$N_7 = \frac{1}{2\lambda^2(p-q)(p-1)}\eta_1(\eta_1 - \lambda)(s - q)(p - 1)$$

$$N_8 = \frac{1}{2\lambda^2(1-q)(1-p)}\eta_1(\eta_1 - \lambda)(s - q)(s - p)$$

$$N_9 = \frac{1}{\lambda^2(1-q)(1-p)}(\lambda^2 - \eta_1^2)(s - q)(s - p)$$

where

$$s = \sqrt{1 + \eta_2} \qquad p = \sqrt{1 + \lambda} \qquad q = \sqrt{1 - \lambda} \qquad D = \lambda(-4\lambda + (2 + \lambda)p - s + s^3)$$

When this technique is employed, the stress intensity factors at the point P' on the crack front, as illustrated in Figure 8.23, are evaluated as

$$K_I^{P'} = \frac{E}{4(1 - \nu^2)}\sqrt{\frac{\pi}{2r}}(u_b^{P^+} - u_b^{P^-})$$

$$K_{II}^{P'} = \frac{E}{4(1 - \nu^2)}\sqrt{\frac{\pi}{2r}}(u_n^{P^+} - u_n^{P^-}) \qquad (8.99)$$

$$K_{III}^{P'} = \frac{E}{4(1 + \nu)}\sqrt{\frac{\pi}{2r}}(u_t^{P^+} - u_t^{P^-})$$

and the displacements $\mathbf{u}^{P^+}$ and $\mathbf{u}^{P^-}$ are evaluated at points P^+ and P^-, which are the first nodes away from the crack front for the upper and lower crack surfaces, respectively; u_b, u_n and u_t are the projections of $\mathbf{u}$ on the coordinate directions of the local crack coordinate system presented in Figure 8.23, and r is the distance to the crack front.

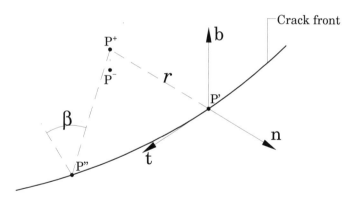

Figure 8.23: Evaluation of stress intensity factors from nodal displacements.

The stress intensity factor at another point on the crack front, e.g. point P'', can be obtained using an approximation of equation (8.99)

$$
\begin{aligned}
K_I^{P''} &= \frac{E}{4(1 - \nu^2)\sqrt{\cos \beta}} \sqrt{\frac{\pi}{2r}} (u_b^{P^+} - u_b^{P^-}) \\
K_{II}^{P''} &= \frac{E}{4(1 - \nu^2)\sqrt{\cos \beta}} \sqrt{\frac{\pi}{2r}} (u_n^{P^+} - u_n^{P^-}) \qquad (8.100) \\
K_{III}^{P''} &= \frac{E}{4(1 + \nu)\sqrt{\cos \beta}} \sqrt{\frac{\pi}{2r}} (u_t^{P^+} - u_t^{P^-})
\end{aligned}
$$

Both equations (8.99) and (8.100) are known as the one-point displacement formulae, and it can be seen from these equations that the accuracy of this technique depends strongly upon the accuracy of the displacement field calculated on the crack surfaces.

Unlike the finite element method, the implementation of the quarter-point elements into BEM codes requires some modifications. In the case of the multi-region method, the traction representations ahead of the crack tip must be modified to incorporate the singular behaviour. For the dual boundary element method, shifting of the nodes results in some minor modification to the integration routines. Therefore, the choice between the special shape functions and quarter-point elements is not so straightforward in the BEM. Both types of elements work equally well for crack problems in 2D, plates and shells. However, for 3D crack problems the correct positioning of nodes for quarter-point elements can complicate the meshing requirements.

There have been several studies into the *optimum* size of the crack tip element. However, such studies can never be conclusive as the stress intensity factors depend upon the geometrical configuration and loadings. Therefore, guidelines established based on one set of benchmark problems is not necessarily valid for another. However, experience has shown that SIF's are calculated more accurately using special shape function and quarter-point elements than conventional elements. SIF values accurate to less than 5% for 2D and less than 10% for 3D problems are achievable with relatively coarse meshes.

8.8.2 Subtraction of Singularity Method

The methods described above are based on attempts to model the singular behaviour of stresses near the crack tip. In contrast, the Subtraction of the Singularity Technique (SST) avoids the need for this task; it removes the singular fields completely. This leaves a non-singular field to be modelled numerically. This approach was first introduced in the BEM by Papamichel and Symm [149] for the analysis of a symmetrical slit in potential problems. Xanthis *et al.* [239] used this formulation to solve the same problem of a symmetrical slit using quadratic boundary elements. The extension of the method to 2D elasticity was presented by Aliabadi *et al.* [6, 7] who obtained both mode I and mode II stress intensity factors. The formulation was extended to V-notch plates in [162], and to three-dimensional problems [12]. An equivalent technique to the SST was developed by Smith and Della-Venura [198]. In their study, a two step superposition method was used to obtain the stress-intensity factors.

Here, the method is briefly described for symmetrical two-dimensional crack problems. Readers should consult the references above, or Aliabadi and Rooke [16] for a more general and detailed description of the method.

In general, the displacement and traction fields in a given crack problem can be represented as

$$
\begin{aligned}
u_j &= u_j^R + u_j^S \\
t_j &= t_j^R + t_j^S
\end{aligned}
\tag{8.101}
$$

where u_j^R, t_j^R denote regular fields and u_j^S, t_j^S denote singular fields. The stress σ_{ij}^S and displacement u_j^S fields, in the vicinity of the crack tip where singularity occurs, are known and are given in equations (8.5) and (8.6). Substituting (8.101) into the displacement integral equation, and following the usual discretization procedure, we have

$$
\mathbf{H u}^R = \mathbf{G t}^R
\tag{8.102}
$$

where, since (u_j^S, t_j^S) satisfies the governing differential equation exactly, the boundary integrals involving the singular functions cancel out. The above system of equations is now solved subject to the modified boundary conditions which do contain the singular function, that is

$$
\begin{aligned}
\bar{u}_j^R &= \bar{u}_j - \bar{u}_j^S \\
\bar{t}_j^R &= \bar{t}_j - \bar{t}_j^S
\end{aligned}
\tag{8.103}
$$

After substitution of the modified boundary conditions, the resulting system of equations for M collocation points can be written as

$$
\mathbf{AX} = \mathbf{BY} - \mathbf{BC}K_I
\tag{8.104}
$$

where $\mathbf{A}$ and $\mathbf{B}$ are $2M \times 2M$ matrices containing the appropriate coefficients from $\mathbf{H}$ and $\mathbf{G}$; $\mathbf{X}$ is a vector containing $2M$ components of the prescribed boundary conditions $\bar{u}_j$ or $\bar{t}_j$; $\mathbf{C}$ is a vector containing $2M$ components of $\bar{u}_j^S$ or $\bar{t}_j^S$; and K_I is the stress intensity factor. Since the stress intensity factor is an unknown in the solution, there are $2M$ equations with $2M + 1$ unknowns. Rearranging equation

(8.104) so that all the unknowns are collected on the left-hand side, gives

$$[\mathbf{A} \quad \mathbf{D}] \left\{ \begin{matrix} \mathbf{X} \\ K_I \end{matrix} \right\} = \{\mathbf{F}\} \tag{8.105}$$

where $\mathbf{D} = \mathbf{BC}$ and $\mathbf{F} = \mathbf{BY}$. In order to solve matrix equation (8.105), an extra equation is required, or the number of unknowns must be reduced by one; this could be achieved by setting $t_j^R = 0$ at the crack tip. This condition ensures cancellation of the singularity.

8.8.3   Energy Release Rate Method

As described in Section 8.2.3, the stress intensity factor is related to the energy release rate G. In two-dimensional crack problems, the relationship is given as

$$G_I = \frac{dU}{dS} = \begin{cases} \frac{1-\nu^2}{E} K_I^2 & \text{for plane strain} \\[2mm] \frac{1}{E} K_I^2 & \text{for plane stress} \end{cases}$$

where dU is the change in strain energy U in advancing the crack front by an incremental area of dA. The direct computation of the strain energy release rate involves two computer runs: the first run is for the original crack, where the strain energy U_1 is calculated; the second run is identical to the first except for a very small difference in crack size, and strain energy U_2 is calculated. It follows that the strain energy release rate G_I may be evaluated from

$$G_I = \frac{U_2 - U_1}{A_2 - A_1}$$

In three-dimensional problems this approach is not directly applicable, since the stress intensity factor usually varies along the crack front. This means a number of computer runs is required for each localized extension at different points on the crack front. Cruse and Meyers [67] proposed a technique based on the BEM for 3D crack problems which limited the number of computer runs to two. The first run is for the original crack front, while the second run is for the perturbed crack front, obtained by moving all the nodes on the crack front radially along the lines normal to the front. In [67] linear triangular elements were used. Tan and Fenner [205] used quadrilateral elements with quadratic variation to represent both the surface and the unknown functions. Bonnet [37, 38] derived a BEM formulation based on the material differentiation method.

8.8.4   Path Independent Integrals

The use of path independent contour integrals has also been popular in the BEM, as the stress intensity factors can generally be evaluated by a post-processing procedure [13, 25, 222, 223]. One of the most popular path independent integrals is the J−integral. Application of the J−integral to two-dimensional symmetrical crack problems can be found in Kishitani et al. [117] and Karami and Fenner [115]. Aliabadi [13] applied the J−integral and BEM to mixed-mode crack problems and decoupled the J into its symmetrical and anti-symmetrical components. It was shown [13] that accurate values of mode I and mode II stress intensity factors can be obtained from the J-integral. Man, Aliabadi and Rooke [131] utilized the mixed-mode J-integral in the study of contact forces on crack behaviour. The application of the $J-$integral to elastodynamic and thermoelastic crack problems can be found in Fedelinski $et\ al.$ [87] and Prasad $et\ al.$ [156]. Sollero and Aliabadi [202] developed an alternative method for decoupling the mixed-mode K-factors from the $J-$integral for the analysis of cracks in anisotropic composite laminates. The application of the $J-$integral to plates and shells has been reported Ahmadi and Wearing [2] and Dirgantara and Aliabadi [77], respectively.

The J-integral, being an energy approach, has the advantage that elaborate representation of the crack tip singular fields is not necessary. This is due to the relatively small contribution that the crack tip fields make to the total J (i.e. strain energy) of the body.

J-integral Method

The path independent J-integral is defined

$$J = \int_S (Wn_1 - t_j u_{j,1})\ dS \qquad (8.106)$$

where S is an arbitrary closed contour, oriented in the anti-clockwise direction, starting from the lower crack surface to the upper one and incorporating the crack tip, dS is an element of the contour S, W is the strain energy per unit volume, n_1 is the component in the x_1 direction of the outward normal to the path S, and $t_j (= \sigma_{ij} n_j)$ and $u_{j,1}$ are the components of the interior tractions and strains, respectively. The relationship between the J-integral and the stress intensity factors is given by

$$J = \frac{K_I^2 + K_{II}^2}{E'} \qquad (8.107)$$

where E' is the sheet material Young's modulus E in plane stress or $E/(1-\nu^2)$ in plane strain.

As can be seen in equation (8.107), the J-integral is related to a combination of the values of K_I and K_{II}. Thus, it is necessary to decouple it, so that the values of K_I and K_{II} can be obtained. Aliabadi [13], showed that a simple procedure, suggested by Ishikawa, Kitagawa and Okamura [112], can be used for that purpose.

Consider two points, $P(x_1, x_2)$ and $P'(x_1, -x_2)$, symmetrical with respect to the crack axis. The displacement at these points can be expressed in terms of their symmetric components u_j^I and anti-symmetric components u_j^{II} (see Figure 8.24) as

$$\left\{ \begin{array}{c} u_1^I \\ u_2^I \end{array} \right\} = \frac{1}{2} \left\{ \begin{array}{c} u_1 + u_1' \\ u_2 - u_2' \end{array} \right\}, \qquad \left\{ \begin{array}{c} u_1^{II} \\ u_2^{II} \end{array} \right\} = \frac{1}{2} \left\{ \begin{array}{c} u_1 - u_1' \\ u_2 + u_2' \end{array} \right\} \qquad (8.108)$$

and, similarly for stresses σ_{ij}^I and σ_{ij}^{II},

$$\left\{ \begin{array}{c} \sigma_{11}^I \\ \sigma_{22}^I \\ \sigma_{12}^I \end{array} \right\} = \frac{1}{2} \left\{ \begin{array}{c} \sigma_{11} + \sigma_{11}' \\ \sigma_{22} + \sigma_{22}' \\ \sigma_{12} - \sigma_{12}' \end{array} \right\}, \qquad \left\{ \begin{array}{c} \sigma_{11}^{II} \\ \sigma_{22}^{II} \\ \sigma_{12}^{II} \end{array} \right\} = \frac{1}{2} \left\{ \begin{array}{c} \sigma_{11} - \sigma_{11}' \\ \sigma_{22} - \sigma_{22}' \\ \sigma_{12} + \sigma_{12}' \end{array} \right\} \qquad (8.109)$$

Using equations (8.108) and (8.109), the J-Integral integral can be decoupled as

$$J^m = \int_S (W^m n_1 - t_j^m u_{j,1}^m)\ dS \qquad m = I, II \qquad (8.110)$$

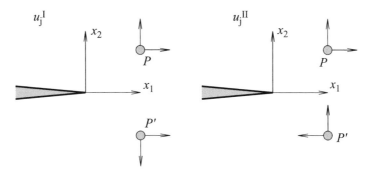

Figure 8.24: The symmetrical and anti-symmetrical components of displacement with respect to the crack axis.

and the relationship between the decoupled J-integral and the stress intensity factors is given by:

$$J^I = \frac{K_I^2}{E'} \qquad \text{and} \qquad J^{II} = \frac{K_{II}^2}{E'} \tag{8.111}$$

The implementation of the procedure described above in the DBEM is straightforward. The J-integral path can be chosen as a circular path, centred at the crack tip, starting from any crack node on one of the crack surfaces and proceeding to the corresponding node on the other surface. The possible paths are designated as S_2, S_3, S_4, etc. as illustrated in Figure 8.25. The integration path is defined by a series of internal points located at symmetrical positions with respect to the crack axis, as shown in Figure 8.25 for path S_2. The integration along the contour path can be performed with the simple Trapezoidal rule. For traction free cracks, the contribution of the integration over the part of the crack surfaces included in the contour is equal to zero, and therefore does not need to be computed.

The application of the J-integral to 3D crack problems was presented by Rigby and Aliabadi [168] and Huber and Kuhn [105]. Later, Rigby and Aliabadi [170] reported the correct decomposition of the elastic fields for 3D problems. Application of the 3D J-integral to thermoelastic crack problems can be found in dell'Ebra and Aliabadi [73].

The J-integral for 3D is defined as

$$J = \int_{\Gamma_\rho} \left(W n_1 - \sigma_{ij} \frac{\partial u_i}{\partial x_1} n_j \right) d\Gamma \tag{8.112}$$

$$= \int_{C+\omega} \left(W n_1 - \sigma_{ij} \frac{\partial u_i}{\partial x_1} n_j \right) d\Gamma - \int_{\Omega(C)} \frac{\partial}{\partial x_3} \left(\sigma_{i3} \frac{\partial u_i}{\partial x_1} \right) d\Omega$$

where Γ_ρ is a contour identical to C_ρ but proceeding in an anticlockwise direction (see Figure 8.26). The integral J is defined in the plane $x_3 = 0$ for any position on the crack front. Considering a traction free crack, the contour integral over the crack faces ω is zero.

The integral J is related to the three modes of fracture through the integrals J^I, J^{II} and J^{III} as follows:

$$J = J^I + J^{II} + J^{III} \tag{8.113}$$

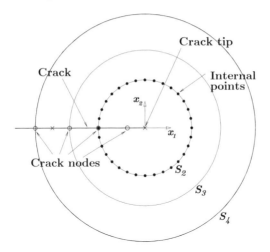

Figure 8.25: The integration paths for evaluation of the J-integral. A series of internal points located at symmetrical positions with respect to the crack axis is used to define path S_2.

Rigby and Aliabadi [170] presented the decomposition method, from which the integrals J^I, J^{II} and J^{III} in elasticity can be obtained directly from J. First, the integral J is split-up into two parts,

$$J = J^S + J^{AS} \qquad (8.114)$$

where J^S and J^{AS} are found from the symmetric and antisymmetric elastic fields about the crack plane, respectively. As the mode I elastic fields are symmetric to the crack plane, the following relationship holds:

$$J^S = J^I \quad \text{and} \quad J^{AS} = J^{II} + J^{III}$$

The integrals J^{II} and J^{III} can be obtained from J^{AS} by making an additional analysis on the antisymmetric fields. Once the integral J is obtained as separate contributions of mode I, II and III J-integrals, the stress intensity factors can be calculated as

$$
\begin{aligned}
J_1 &= J^I + J^{II} + J^{III} \\
&= \frac{1}{E'} \left(K_I^2 + K_{II}^2 \right) + \frac{1}{2\mu} K_{III}^2
\end{aligned}
\qquad (8.115)
$$

where $E' = E$ for plane stress and $E' = E/(1 - \nu^2)$ for plane strain. The stresses at points P and P' can be expressed in terms of the symmetric and antisymmetric components as

$$
\left\{
\begin{array}{c}
\sigma_{11P} \\
\sigma_{12P} \\
\sigma_{13P} \\
\sigma_{22P} \\
\sigma_{23P} \\
\sigma_{33P}
\end{array}
\right\}
=
\left\{
\begin{array}{c}
\sigma_{11P}^S \\
\sigma_{12P}^S \\
\sigma_{13P}^S \\
\sigma_{22P}^S \\
\sigma_{23P}^S \\
\sigma_{33P}^S
\end{array}
\right\}
+
\left\{
\begin{array}{c}
\sigma_{11P}^{AS} \\
\sigma_{12P}^{AS} \\
\sigma_{13P}^{AS} \\
\sigma_{22P}^{AS} \\
\sigma_{23P}^{AS} \\
\sigma_{33P}^{AS}
\end{array}
\right\}
\qquad (8.116)
$$

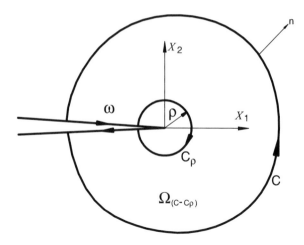

Figure 8.26: Definition of countor path for 3D $J-$integral.

and

$$
\left\{
\begin{array}{c}
\sigma_{11P'} \\
\sigma_{12P'} \\
\sigma_{13P'} \\
\sigma_{22P'} \\
\sigma_{23P'} \\
\sigma_{33P'}
\end{array}
\right\}
=
\left\{
\begin{array}{c}
\sigma_{11P'}^{S} \\
-\sigma_{12P'}^{S} \\
\sigma_{13P'}^{S} \\
\sigma_{22P'}^{S} \\
-\sigma_{23P'}^{S} \\
\sigma_{33P'}^{S}
\end{array}
\right\}
+
\left\{
\begin{array}{c}
-\sigma_{11P'}^{AS} \\
\sigma_{12P'}^{AS} \\
-\sigma_{13P'}^{AS} \\
-\sigma_{22P'}^{AS} \\
\sigma_{23P'}^{AS} \\
-\sigma_{33P'}^{AS}
\end{array}
\right\}
\tag{8.117}
$$

The symmetric and antisymmetric stress fields can be found by combining the stresses at points P and P' as follows:

$$
\left\{
\begin{array}{c}
\sigma_{11}^{S} \\
\sigma_{12}^{S} \\
\sigma_{13}^{S} \\
\sigma_{22}^{S} \\
\sigma_{23}^{S} \\
\sigma_{33}^{S}
\end{array}
\right\}
=
\frac{1}{2}
\left\{
\begin{array}{c}
\sigma_{11P} + \sigma_{11P'} \\
\sigma_{12P} - \sigma_{12P'} \\
\sigma_{13P} + \sigma_{13P'} \\
\sigma_{22P} + \sigma_{22P'} \\
\sigma_{23P} - \sigma_{23P'} \\
\sigma_{33P} + \sigma_{33P'}
\end{array}
\right\}
\tag{8.118}
$$

and

$$
\left\{
\begin{array}{c}
\sigma_{11}^{AS} \\
\sigma_{12}^{AS} \\
\sigma_{13}^{AS} \\
\sigma_{22}^{AS} \\
\sigma_{23}^{AS} \\
\sigma_{33}^{AS}
\end{array}
\right\}
=
\frac{1}{2}
\left\{
\begin{array}{c}
\sigma_{11P} - \sigma_{11P'} \\
\sigma_{12P} + \sigma_{12P'} \\
\sigma_{13P} - \sigma_{13P'} \\
\sigma_{22P} - \sigma_{22P'} \\
\sigma_{23P} + \sigma_{23P'} \\
\sigma_{33P} - \sigma_{33P'}
\end{array}
\right\}
\tag{8.119}
$$

The strains can be represented by the sum of its symmetric and anti-symmetric com-

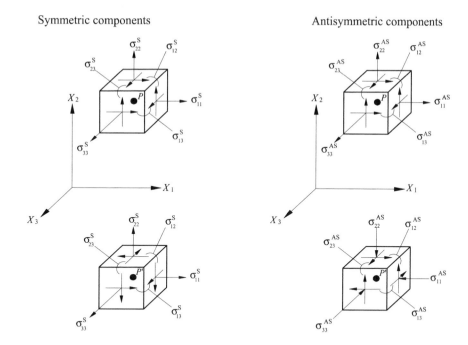

Figure 8.27: The symmetric and antisymmetric components stress with respect to the crack plane.

ponents as

$$\epsilon_{ij} = \epsilon_{ij}^S + \epsilon_{ij}^{AS} = \frac{1}{2} \left\{ \begin{array}{c} \epsilon_{11P} + \epsilon_{11P'} \\ \epsilon_{12P} - \epsilon_{12P'} \\ \epsilon_{13P} + \epsilon_{13P'} \\ \epsilon_{22P} + \epsilon_{22P'} \\ \epsilon_{23P} - \epsilon_{23P'} \\ \epsilon_{33P} + \epsilon_{33P'} \end{array} \right\} + \frac{1}{2} \left\{ \begin{array}{c} \epsilon_{11P} - \epsilon_{11P'} \\ \epsilon_{12P} + \epsilon_{12P'} \\ \epsilon_{13P} - \epsilon_{13P'} \\ \epsilon_{22P} - \epsilon_{22P'} \\ \epsilon_{23P} + \epsilon_{23P'} \\ \epsilon_{33P} - \epsilon_{33P'} \end{array} \right\} \qquad (8.120)$$

The symmetric and anti-symmetric components of the strain tensor are related to the displacements derivatives as

$$\epsilon_{ij}^\alpha = \frac{1}{2} \left(\frac{\partial u_i^\alpha}{\partial x_j} + \frac{\partial u_j^\alpha}{\partial x_i} \right) \qquad (8.121)$$

where $\alpha = S$ or AS. From now on, the sub-indices P and P' will be eliminated for simplicity, so f' will represent a field evaluated at P' and f will represent a field evaluated at P. The displacement derivatives can be obtained from the above equations as

$$\frac{\partial u_i}{\partial x_j} = \frac{\partial u_i^S}{\partial x_j} + \frac{\partial u_i^{AS}}{\partial x_j}$$

$$= \frac{1}{2} \left\{ \begin{array}{c} \frac{\partial u_1}{\partial x_j} + \frac{\partial u_1'}{\partial x_j} \\ \frac{\partial u_2}{\partial x_j} - \frac{\partial u_2'}{\partial x_j} \\ \frac{\partial u_3}{\partial x_j} + \frac{\partial u_3'}{\partial x_j} \end{array} \right\} + \frac{1}{2} \left\{ \begin{array}{c} \frac{\partial u_1}{\partial x_j} - \frac{\partial u_1'}{\partial x_j} \\ \frac{\partial u_2}{\partial x_j} + \frac{\partial u_2'}{\partial x_j} \\ \frac{\partial u_3}{\partial x_j} - \frac{\partial u_3'}{\partial x_j} \end{array} \right\} \tag{8.122}$$

Equation (8.112) can be written using the symmetric and anti-symmetric components of the fields derived above as

$$J = \int_C \left\{ \left[\int_0^{\epsilon_{ij}} (\sigma_{ij}^S + \sigma_{ij}^{AS})(\epsilon_{ij}^S + \epsilon_{ij}^{AS}) \right] n_1 - (\sigma_{ij}^S + \sigma_{ij}^{AS}) n_j \frac{\partial}{\partial x_1}(u_i^S + u_i^{AS}) \right\} d\Gamma$$
$$- \int_{\Omega(C)} \frac{\partial}{\partial x_3} \left[(\sigma_{i3}^S + \sigma_{i3}^{AS}) \frac{\partial}{\partial x_1}(u_i^S + u_i^{AS}) \right] d\Omega \tag{8.123}$$

where the definition of $W = \int_0^{\epsilon_{ij}} \sigma_{ij} d\epsilon_{ij}$ has been used.

Considering a contour C symmetric about the crack plane $x_2 = 0$, it can be shown that for any pair of symmetric points P and P', the following relationship holds for the normals $\mathbf{n}$ and $\mathbf{n}'$:

$$\mathbf{n} = (n_1, n_2, 0) \quad \text{and} \quad \mathbf{n}' = (n_1, -n_2, 0)$$

and for the integrands in equation (8.123),

$$\sigma_{ij}'^{\alpha} d\epsilon_{ij}'^{\beta} = \pm \sigma_{ij}^{\alpha} d\epsilon_{ij}^{\beta} \tag{8.124}$$

$$\sigma_{ij}'^{\alpha} n_j' \frac{\partial u_i'^{\beta}}{\partial x_1} = \pm \sigma_{ij}^{\alpha} n_j \frac{\partial u_i^{\beta}}{\partial x_1} \tag{8.125}$$

$$\sigma_{i3}'^{\alpha} \frac{\partial u_i'^{\beta}}{\partial x_1} = \pm \sigma_{i3}^{\alpha} \frac{\partial u_i^{\beta}}{\partial x_1} \tag{8.126}$$

where $\alpha, \beta = S$ or AS. The positive sign on the right-hand side of equations (8.124) to (8.126) represents the case $\alpha = \beta$, and the negative sign represents the case $\alpha \neq \beta$. As the contour C is taken symmetric about the crack plane, the integrals in equation (8.123) cancel each other at symmetric points for the case $\alpha \neq \beta$. Thus, equation (8.123) becomes

$$J = \sum_{\alpha=1}^{2} \int_C \left(W^{\alpha} n_1 - \sigma_{ij}^{\alpha} \frac{\partial u_i^{\alpha}}{\partial x_1} n_j \right) d\Gamma - \int_{\Omega(C)} \frac{\partial}{\partial x_3} \left(\sigma_{i3}^{\alpha} \frac{\partial u_i^{\alpha}}{\partial x_1} \right) d\Omega$$
$$= J^S + J^{AS} \tag{8.127}$$

where $\alpha = 1, 2$ denotes symmetric (S) and anti-symmetric (AS) components, respectively. As stated before, mode I corresponds to the integral J^S, while the integral J^{AS} is related to modes II and III. The integral J will be decoupled into mode I, II and III terms in the following section.

Decomposition of Integrands In this section the integrands of integral J will be decomposed into their mode I, II and III components. Since the symmetric components are representative of mode I, the decomposition method will involve a further decoupling of the anti-symmetric components into modes II and III.

The decomposition of stresses represented by

$$\sigma_{ij} = \sigma_{ij}^I + \sigma_{ij}^{II} + \sigma_{ij}^{III}$$

has been given by many authors [145, 190]. Recently Rigby and Aliabadi [170] gave the proper decomposition of stresses, since they showed that the expressions used in previous papers were incorrect. So, the correct decomposition of stresses is as follows:

$$
\sigma_{ij} = \sigma_{ij}^I + \sigma_{ij}^{II} + \sigma_{ij}^{III}
$$

$$
= \frac{1}{2}
\left\{
\begin{array}{c}
\sigma_{11} + \sigma'_{11} \\
\sigma_{12} - \sigma'_{12} \\
\sigma_{13} + \sigma'_{13} \\
\sigma_{22} + \sigma'_{22} \\
\sigma_{23} - \sigma'_{23} \\
\sigma_{33} + \sigma'_{33}
\end{array}
\right\}
+ \frac{1}{2}
\left\{
\begin{array}{c}
\sigma_{11} - \sigma'_{11} \\
\sigma_{12} + \sigma'_{12} \\
0 \\
\sigma_{22} - \sigma'_{22} \\
0 \\
\sigma_{33} - \sigma'_{33}
\end{array}
\right\}
+ \frac{1}{2}
\left\{
\begin{array}{c}
0 \\
0 \\
\sigma_{13} - \sigma'_{13} \\
0 \\
\sigma_{23} + \sigma'_{23} \\
0
\end{array}
\right\}
\quad (8.128)
$$

The strain decomposition is derived from the stress decomposition by the application of Hooke's Law, to give

$$
\epsilon_{ij} = \epsilon_{ij}^I + \epsilon_{ij}^{II} + \epsilon_{ij}^{III}
$$

$$
= \frac{1}{2}
\left\{
\begin{array}{c}
\epsilon_{11} + \epsilon'_{11} \\
\epsilon_{12} - \epsilon'_{12} \\
\epsilon_{13} + \epsilon'_{13} \\
\epsilon_{22} + \epsilon'_{22} \\
\epsilon_{23} - \epsilon'_{23} \\
\epsilon_{33} + \epsilon'_{33}
\end{array}
\right\}
+ \frac{1}{2}
\left\{
\begin{array}{c}
\epsilon_{11} - \epsilon'_{11} \\
\epsilon_{12} + \epsilon'_{12} \\
0 \\
\epsilon_{22} - \epsilon'_{22} \\
0 \\
\epsilon_{33} - \epsilon'_{33}
\end{array}
\right\}
+ \frac{1}{2}
\left\{
\begin{array}{c}
0 \\
0 \\
\epsilon_{13} - \epsilon'_{13} \\
0 \\
\epsilon_{23} + \epsilon'_{23} \\
0
\end{array}
\right\}
\quad (8.129)
$$

The mode I, II and III displacement derivatives can be derived from equation (8.129) by using the relationship between displacements and strains as

$$
\frac{\partial u_i}{\partial x_j} = \frac{\partial u_i^I}{\partial x_j} + \frac{\partial u_i^{II}}{\partial x_j} + \frac{\partial u_i^{III}}{\partial x_j}
\quad (8.130)
$$

where

$$
\frac{\partial u_i}{\partial x_1} = \frac{1}{2}
\left\{
\begin{array}{c}
\frac{\partial u_1}{\partial x_1} + \frac{\partial u'_1}{\partial x_1} \\
\frac{\partial u_2}{\partial x_1} - \frac{\partial u'_2}{\partial x_1} \\
\frac{\partial u_3}{\partial x_1} + \frac{\partial u'_3}{\partial x_1}
\end{array}
\right\}
+ \frac{1}{2}
\left\{
\begin{array}{c}
\frac{\partial u_1}{\partial x_1} - \frac{\partial u'_1}{\partial x_1} \\
\frac{\partial u_2}{\partial x_1} + \frac{\partial u'_2}{\partial x_1} \\
0
\end{array}
\right\}
+ \frac{1}{2}
\left\{
\begin{array}{c}
0 \\
0 \\
\frac{\partial u_3}{\partial x_1} - \frac{\partial u'_3}{\partial x_1}
\end{array}
\right\}
\quad (8.131)
$$

$$
\frac{\partial u_i}{\partial x_2} = \frac{1}{2}
\left\{
\begin{array}{c}
\frac{\partial u_1}{\partial x_2} + \frac{\partial u'_1}{\partial x_2} \\
\frac{\partial u_2}{\partial x_2} - \frac{\partial u'_2}{\partial x_2} \\
\frac{\partial u_3}{\partial x_2} + \frac{\partial u'_3}{\partial x_2}
\end{array}
\right\}
+ \frac{1}{2}
\left\{
\begin{array}{c}
\frac{\partial u_1}{\partial x_2} - \frac{\partial u'_1}{\partial x_2} \\
\frac{\partial u_2}{\partial x_2} + \frac{\partial u'_2}{\partial x_2} \\
0
\end{array}
\right\}
+ \frac{1}{2}
\left\{
\begin{array}{c}
0 \\
0 \\
\frac{\partial u_3}{\partial x_2} - \frac{\partial u'_3}{\partial x_2}
\end{array}
\right\}
\quad (8.132)
$$

$$
\frac{\partial u_i}{\partial x_3} = \frac{1}{2}
\left\{
\begin{array}{c}
\frac{\partial u_1}{\partial x_3} + \frac{\partial u'_1}{\partial x_3} \\
\frac{\partial u_2}{\partial x_3} - \frac{\partial u'_2}{\partial x_3} \\
\frac{\partial u_3}{\partial x_3} + \frac{\partial u'_3}{\partial x_3}
\end{array}
\right\}
+ \frac{1}{2}
\left\{
\begin{array}{c}
0 \\
0 \\
\frac{\partial u_3}{\partial x_3} - \frac{\partial u'_3}{\partial x_3}
\end{array}
\right\}
+ \frac{1}{2}
\left\{
\begin{array}{c}
\frac{\partial u_1}{\partial x_3} - \frac{\partial u'_1}{\partial x_3} \\
\frac{\partial u_2}{\partial x_3} + \frac{\partial u'_2}{\partial x_3} \\
0
\end{array}
\right\}
\quad (8.133)
$$

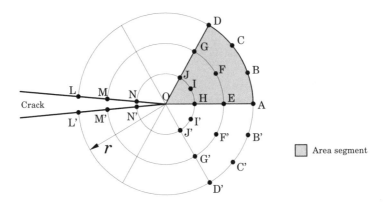

Figure 8.28: Distribution of internal points for contour and area integration.

By applying the above decoupled fields, modes I, II and III J-integrals, can be written as

$$J^m = \int_C \left(W^m n_1 - \sigma_{ij}^m \frac{\partial u_i^m}{\partial x_1} n_j \right) d\Gamma + \int_{\Omega(C)} \left(\frac{\partial \sigma_{i1}^m}{\partial x_1} + \frac{\partial \sigma_{i2}^m}{\partial x_2} \right) \frac{\partial u_i^m}{\partial x_1} d\Omega$$
$$- \int_{\Omega(C)} \sigma_{i3}^m \frac{\partial^2 u_i^m}{\partial x_1 \partial x_3} d\Omega \tag{8.134}$$

where $m = I, II, III$. The stress intensity factors can be obtained from the above integral using equation (8.115).

The strategy developed in [170] can be described, with reference to Figure 8.28, as follows:

- A plane enclosed by a circular contour of radius r, centred at a point O on the crack front, is placed perpendicular to the crack front. The area enclosed is divided into an even number of area segments.

- The points, where the internal values will be calculated are located in concentric arcs of radius $\frac{r}{3}$, $\frac{2r}{3}$ and r, as illustrated in Figure 8.28. Three internal points per segment are used in the interior arcs to avoid near singularities, as the internal points are closer to the crack elements.

- All the points are placed symmetrically with respect to the crack plane, e.g. the point B is symmetric to the point B'. By combining the results at symmetric points, the integrands for J^I, J^{II} and J^{III} can be obtained. Since all the integrands are symmetric with respect to the crack plane, the integration is carried out only for the top area and then multiplied by 2.

- The contribution to the contour integral from the segment highlighted in Figure 8.28 is found by applying the four point Newton-Cotes formula to the contour integrands at points $A - D$.

- The contribution to the area integrals from this segment is obtained via line integrals. First, line integrals are calculated for the three arcs in the segment.

A line integral L_1 is obtained by applying the Simpson's Rule to the area integrands at points $H - J$ in the inner arc. Simpson's Rule is also applied to points $E - G$ in the middle arc to produce a line integral L_2. The line integral L_3 in the outer arc is calculated from the area integrands at points $A - D$ using the four point Newton-Cotes formula. Considering that the integral of the area integrands over an area $\Omega(\varepsilon)$ will tend to zero as $\varepsilon \to 0$, a line integral $L_0 = 0$ is assumed from the integration over an arc of $r = 0$ (point O). Thus, since the arcs are equally spaced, the total area integral contribution from the area segment can be calculated from the four line integrals L_0, L_1, L_2 and L_3 by using the four point Newton-Cotes formula, yielding

$$
\begin{aligned}
J_A &= \frac{r}{8}\left(L_0 + 3L_1 + 3L_2 + L_3\right) \\
&= \frac{r}{8}\left(3L_1 + 3L_2 + L_3\right)
\end{aligned}
$$

which is equivalent to finding the integral of the area integrands I_A from

$$
J_A = \int_{\theta_0}^{\theta_1} \int_0^r I_A r \, dr \, d\theta
$$

Finally, the stress intensity factors are calculated from equation (8.115) as:

$$
\begin{aligned}
K_I &= \sqrt{E' J^I} \\
K_{II} &= \sqrt{E' J^{II}} \\
K_{III} &= \sqrt{2\mu J^{III}}
\end{aligned}
$$

8.8.5 Energy Domain Integral

Domain integral methods were developed by Shih *et al.* [188]. Similar to contour integral methods, domain integral methods are derived by applying the divergence theorem to the J-integral which produces an integral defined over a finite volume enclosing some portion of the crack front. The energy domain integral method due to Shih was chosen by Cisilino *et al.* [56] for analysis of three-dimensional elastoplastic crack problems.

Consider a notch with notch thickness h, as shown in Figure 8.29(a). It is argued (heuristically) that $h \to 0$ is the sharp crack configuration of interest. The surface of the notch consists of faces S_A and S_B, with normals m_k along x_2', and a face S_t with a normal in the $x_1' - x_3'$ plane. Now let the notch face S_t with normal m_k in the $x_1' - x_3'$ plane advance $\Delta a \cdot l_k$ in the x_k' direction, i.e.

$$
\Delta a \cdot l_k m_k = \delta l(\eta) \tag{8.135}
$$

Furthermore, if l_k is restricted to lie along S_t and to be a function of x_1' and x_3' only (i.e. $l_2 \equiv 0$), the following expression for energy release rate per unit crack advance can be obtained:

$$
\bar{J}\Delta a = \Delta a \int_{S_t} \left(\sigma_{ij}' \frac{\partial \dot{u}_j'}{\partial x_k'} - W\delta_{ki} \right) l_k m_i \, dS \tag{8.136}
$$

where W is the total strain energy density (elastic and plastic), σ_{ij}' and $\dot{u}_j'$ are the Cartesian components of stress and displacement expressed in the local system x'.

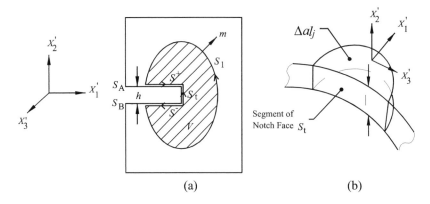

(a)   (b)

Figure 8.29: (a) Schematic of a section of body and volume V containing a notch; (b) schematic of notch face.

To develop a (volume) domain integral, the simply connected volume, V, enclosed by the surfaces S_t, S^+, S^- and S_1 (see Figure 8.29(b)) is considered. Furthermore, the functions q_k are introduced:

$$q_k = \begin{cases} l_k & \text{on} & S_t \\ 0 & \text{on} & S_1 \end{cases} \tag{8.137}$$

together with the requirement that q_k be sufficiently smooth in the volume V. Using (8.137), expression (8.136) can be rewritten in the form of an integral over the closed surface S ($S = S_1 + S^+ + S^- - S_t$) plus crack face terms:

$$\bar{J} = \int_S \left(\dot{\sigma}^*_{ij} \frac{\partial \dot{u}'_j}{\partial x'_k} - W\delta_{ki} \right) m_i q_k dS - \int_{S^++S^-} \sigma'_{2j} \frac{\partial \dot{u}'_j}{\partial x'_k} m_2 q_k dS \tag{8.138}$$

To arrive at (8.138) the results $m_1 = 0$, $m_3 = 0$ and $m_2 = \pm 1$ were used on the crack faces. It may be noted that $l_2 = q_2 = 0$ everywhere. In the absence of crack face tractions, the last term in (8.138) vanishes.

Finally, assuming that W does not depend explicitly on x'_1, the divergence theorem can be applied to the closed surface integral (8.138), obtaining in the case of traction free crack faces

$$\bar{J} = \int_V \left(\sigma'_{ij} \frac{\partial \dot{u}'_j}{\partial x'_k} - W\delta_{ki} \right) \frac{\partial q_k}{\partial x'_i} dV \tag{8.139}$$

Consistent with the path independence of J, equation (8.139) is domain-independent, so any volume V can be chosen for the purpose of evaluating $\bar{J}$.

Finally, letting $h \to 0$, the desired expression for the energy decrease when a local segment of crack front advances by $\Delta a l_k$ in its plane is obtained. The point-wise energy release ratio is given by

$$J(\eta) = \frac{\bar{J}\Delta a}{\Delta a \int_{L_c} l_k(\eta) m_k(\eta) d\eta} \tag{8.140}$$

where the term in the denominator is the increase of the crack area due to the virtual crack advance.

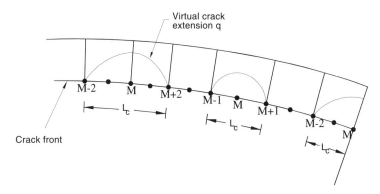

Figure 8.30: Schematic of the crack front region in the $x_1 - x_3$ plane, illustrating virtual crack extensions for a corner node, a mid-node and a node on the surface.

Boundary Element Implementation

Expression (8.140) allows the computation of J at any point η located on the crack front. In each case, this requires the evaluation of a volume integral within closed domains that enclose a segment of the crack front L_c. A natural choice here is to make η coincident with the element nodes on the crack front, while L_c is taken as the element or element sides at which points η lie (see Figure 8.30).

As depicted in Figure 8.30, three different cases should be considered, depending on whether the node of interest M is in the middle of an element side (mid-node), it is shared by two elements (corner node), or it is located on the external surface (surface node). If node M is a mid-node or surface node, L_c (the segment of the crack front over which $\bar{J}$ is computed) spans over one element, connecting nodes $M - 1$, M and $M + 1$ and nodes $M - 2$, $M - 1$ and M, respectively. On the other hand, if M is a corner node, L_c spans over two elements, connecting nodes from $M - 2$ to $M + 2$.

Volume discretization is designed to have a web-style geometry around the crack front, while the integration volumes are taken to coincide with the different rings of cells. This is illustrated in Figure 8.31, where one of the model faces has been removed to show the crack and the J-integration domain discretizations.

In [57], for the sake of simplicity, the direction of the virtual crack advance q_1 was taken to be constant along each of the crack front segments L_c. The direction of q_1 is given in each case by the normal vector x'_1 at the crack front position of node M where the J-integral is being computed. Note that this definition of q_1 is equivalent to the local system (x'_1, x'_2, x'_3) defined at node M, constant along L_c. The definition of q_1 in this way is particularly useful when dealing with curved crack fronts, because it allows the problem to be treated in Cartesian coordinates for any geometry.

If Gaussian integration is used, the discretized form of (8.139) is given by

$$\bar{J} = \sum_{\text{cells in V}} \sum_{p=1}^{m} \left\{ \left(\sigma'_{ij} \frac{\partial u'_j}{\partial x'_1} - W \delta_{1i} \right) \frac{\partial q_1}{\partial x'_i} \det \left(\frac{\partial x'_j}{\partial s_k} \right) \right\}_p w_p \tag{8.141}$$

where m is the number of Gaussian points per cell, and w_p are the weighting factors.

Finally, if it is assumed that the material exhibits a nonlinear elastic behaviour,

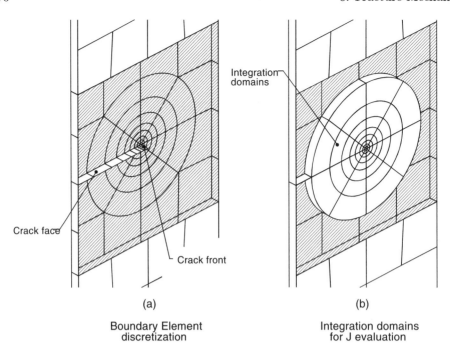

Figure 8.31: Boundary element discretization and integration domains for J-integral evaluation.

the total stress energy density W can be split into elastic and plastic components

$$W = W^e + W^p = \int_0^{\varepsilon_{kl}^e} \sigma_{ij} d\varepsilon_{ij}^e + \int_0^{\varepsilon_{kl}^p} S_{ij} d\varepsilon_{ij}^p \qquad (8.142)$$

where S_{ij} is the deviatoric stress.

8.8.6   Weight Function Method

Conventional methods for the evaluation of stress intensity factors described above rely on the repeated solution of the elasticity equations for each different applied load distribution. An alternative approach which eliminates the repeated solution is the weight function method. Weight functions can be considered as generalizations of Green's functions. Their use for the calculation of stress intensity factors for cracked structures can be efficient and economical since, once the function is determined for a given crack geometry, the stress intensity factor may be calculated for any loading on any boundary by means of simple integration.

The weight function concept was introduced by Bueckner [32] using the so-called *fundamental fields*. Rice [167] showed that the weight functions could equally well be determined by differentiating known elastic solutions for displacement fields with respect to the crack length. A detailed description of the Green's function and weight function methods can be found in Aliabadi and Rooke [16].

Rice [167] showed that if the displacement field $\mathbf{u}^{(a)}$ and the stress intensity factor $K_I^{(a)}$ are known for any symmetrical load on a given geometry with a crack to total

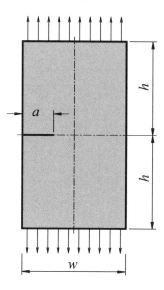

Figure 8.32: Rectangular plate with a single edge crack.

length l, then the stress intensity factor K_I for any other symmetrical loading can be obtained from

$$K_I = \int_S \mathbf{t}.\mathbf{H}_I(\mathbf{x}',l)dS \qquad (8.143)$$

where S is the boundary on which the tractions $\mathbf{t}$ are applied, and $\mathbf{x}'$ is a point on the boundary. The function $\mathbf{H}_I$ is called the weight function, and is given by

$$\mathbf{H}_I(\mathbf{x}',l) = \frac{E'}{2K_I^{(a)}} \left(\frac{\partial u_1^{(a)}}{\partial l}\hat{\imath} + \frac{\partial u_2^{(a)}}{\partial l}\hat{\jmath} \right) \qquad (8.144)$$

where $E' = E/(1 - \nu^2)$ for plane strain and $E' = E$ for plane stress.

The application of the BEM to the weight function method can be traced back to Heliot, Labbens and Pellisier-Tanon [103]. They developed a technique known as *polynomial influence functions,* which is based on an approximate polynomial distribution method. These correspond to the terms of polynomial expansion of the stress field acting on the crack faces; the influence functions were functions of the surface length and depth of the crack. In their study of a semi-elliptical surface crack, numerical crack face weight functions were obtained for a polynomial expansion, in one coordinate only, of degree four in the stress field. Therefore, five computer runs, one for each term in the polynomial, were required, and in each case the stress intensity factors were obtained by extrapolation of the local crack-tip field. Besuner [31] and Cruse [68] developed a BEM strategy for evaluating the weight functions based on Rice's derivation. In their work, a 3D BEM analysis was used to calculate average stress intensity factors for each perturbation of the crack front. The instantaneous values at a specific point and the average value along the whole crack front are not exactly equivalent for most 3D problems, since the stress intensity factors are not generally constant. Further, this technique requires many iterations to obtain a single stress intensity factor solution, and is thus computationally expensive. Cruse and

Table 8.1: $K_I/(\sigma/\sqrt{\pi a})$ for a single edge crack in a rectangular plate, $h/w = 0.5$.

a/w	Disp/extrap		SST	J-integral					Ref[59]
	D-E	D-E F-G	-	2	3	4	5	8	-
0.2	1.57	1.62	1.48	1.50	1.50	1.50	1.49	1.50	1.49
0.3	1.96	2.01	1.84	1.86	1.86	1.86	1.86	1.86	1.85
0.4	2.23	2.54	2.31	2.34	2.34	2.34	2.34	2.34	2.32
0.5	3.27	3.30	2.98	3.03	3.03	3.03	3.03	3.02	3.01
0.6	4.58	4.56	4.10	4.18	4.18	4.18	4.18	4.17	4.15

Ravendera [69] developed a two-dimensional BEM using the crack Green's function fundamental solution.

Cartwright and Rooke [42] used the procedure devised by Paris *et al.* [153] to demonstrate that the BEM can provide a more efficient representation of Buckner fundamental fields, and hence, weight function can be calculated more accurately. This formulation was extended by Aliabadi *et al.* [6, 9] to both mode I and mode II deformation, which in this formulation are independent. An improvement to this model was later reported [16] using the subtraction of singularity method. The extension to 3D was reported by Bains, Aliabadi and Rooke [26, 27] in which they derived and utilized fundamental fields for straight fronted and penny-shaped cracks.

The application of the indirect BEM to weight functions was reported [221, 224, 225, 226] for both static and dynamic weight functions. It was shown for the first time that it is possible to derive weight functions which are independent of both spatial and time variation of loading. These advantages were demonstrated for several mode, and mixed-mode, crack problems in both two- and three-dimensions.

8.9   Examples

In this section numerical examples are presented using the dual boundary element formulation. Also presented are some results using the multi-region method.

8.9.1   Edge Crack

Portela, Aliabadi and Rooke [165], analyzed a square plate with a single edge crack, as represented in Figure 8.32. The crack length is denoted by a and the width of the plate is denoted by w. The plate is subjected to the action of a uniform stress σ, applied symmetrically at the ends in the direction perpendicular to the crack.

Results have been obtained from the one, and two-point displacement extrapolation, singularity subtraction and the J-integral techniques, and compared with those published by Civelek and Erdogan [59]. Five cases were considered, $a/w = 0.2, 0.3,$ 0.4, 0.5 and 0.6. Standard quadratic discontinuous elements are used on the crack surfaces and quadratic continuous elements on the remaining boundaries. A convergence study was carried out with three different meshes of 32, 40 and 64 quadratic boundary elements, in which the crack was discretized with 4, 5 and 8 quadratic discontinuous elements on each surface. Convergence was achieved with these meshes

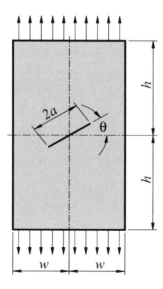

Figure 8.33: Rectangular plate with a central slant crack.

and, therefore, the mesh of 32 elements with a crack discretization graded towards the tip with the ratios 0.4, 0.3, 0.2 and 0.1, was selected for the analysis. The results obtained, presented in Table 8.2, show a high level of accuracy when compared with those of [59].

8.9.2 Centre Crack

Consider a central slant crack in a rectangular plate, as shown in Figure 8.33. The ratio between the height and width of the plate is given by $h/w = 2$. The crack has a length $2a$ and makes an angle of $\varphi = 45°$ with the direction perpendicular to the applied stress. The plate is loaded with a uniform traction $\bar{t}$, applied symmetrically at the ends.

To solve this problem, a mesh of 36 quadratic boundary elements was set up, in which six discontinuous elements were used on each face of the crack, graded from the centre towards the tips, with the ratios 0.25, 0.15 and 0.1. Accurate results for this problem were published [19]. The results obtained in [165], presented in Tables 8.3 and 8.4, show an excellent accuracy when compared with the results of [19].

8.9.3 Centre Crack Plate Loaded by Tension and Bending

A rectangular plate with a central crack loaded by edge bending and tension (as shown in Figure 8.34) was analyzed by Dirgantara and Aliabadi [77]. In this example, special crack tip elements (8.77) and the J-integral method were used. The plate is subjected to uniform tension N_o and out of plane bending load M_o at its ends. The properties of the plate are: $w/h = 2$; $M_0 = N_0 w$; $Eh/N_0 = 210\,000$; and $\nu = 0.3$.

For BEM analysis, eight quadratic elements per side of the plate and three different meshes using 8, 12 and 16 elements for each crack surface are used. Stress intensity factors are evaluated using both Crack Opening Displacements (COD) and

Table 8.2: $K_I/(\sigma/\sqrt{\pi a}\,)$ for a single edge crack in a rectangular plate, $h/w = 2$.

a/w	Displ.	SST	J-integral					Ref[19]
	D-E	-	2	3	4	5	8	-
0.2	0.53	0.52	0.52	0.52	0.52	0.52	0.52	0.52
0.3	0.55	0.54	0.54	0.54	0.54	0.54	0.54	0.54
0.4	0.59	0.57	0.57	0.57	0.57	0.58	0.58	0.57
0.5	0.63	0.61	0.62	0.61	0.61	0.62	0.62	0.61
0.6	0.69	0.66	0.67	0.66	0.67	0.67	0.67	0.66

Table 8.3: $K_{II}/(\sigma/\sqrt{\pi a})$ for a single edge crack in a rectangular plate, $h/w = 2$.

a/w	Displ.	SST	J-integral					Ref[19]
	D-E	-	2	3	4	5	8	-
0.2	0.52	0.51	0.50	0.50	0.50	0.50	0.51	0.51
0.3	0.53	0.52	0.51	0.51	0.51	0.51	0.52	0.52
0.4	0.54	0.53	0.52	0.52	0.52	0.53	0.53	0.53
0.5	0.56	0.55	0.54	0.54	0.54	0.54	0.55	0.55
0.6	0.58	0.57	0.56	0.56	0.56	0.56	0.57	0.57

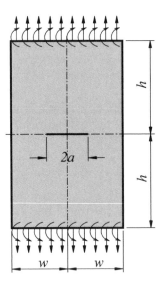

Figure 8.34: Centre crack in a rectangular plate subjected to uniform tension and bending.

Table 8.4: Normalized stress resultant intensity factors for a plate with a centre crack subjected to bending and tension.

	$K_1^{bending}/M_o\sqrt{\pi a}$		$K_1^{tension}/N_o\sqrt{\pi a}$	
	COD	J-integral	COD	J-integral
8 elements	0.870	0.873	1.157	1.195
12 elements	0.907	0.910	1.187	1.189
16 elements	0.910	0.910	1.187	1.189
[19]	0.909		1.186	

Table 8.5: Effect of crack tip element size on the stress resultant intensity factors.

$l/2d$	$K_1^{bending}/M_o\sqrt{\pi a}$	$K_1^{tension}/N_o\sqrt{\pi a}$
0.125	0.859	1.148
0.100	0.871	1.157
0.050	0.895	1.178
0.025	0.907	1.187
0.015	0.901	1.187
[19]	0.910	1.186

$J-$ integral techniques. In Table 8.4, normalized K_1 evaluated using both special shape functions and $J-$ integral techniques for $a/w = 0.5$ is presented for different meshes. The results show that both techniques can achieve accurate results ($< 1\%$ The effect of different mesh sizes to the accuracy of special crack tip element technique was also studied [77]. Five different meshes using 6, 8, 10, 12 and 16 elements for each crack surface are used, with ratios between crack tip element length l and crack length $2s$ are taken as 0.125, 0.10, 0.050, 0.025 and 0.15. All models are discretized with 8 boundary elements per side of the plate. Table 8.5 shows that 0.3% accuracy can be achieved on model with $l/2a = 0.025$ and 0.015.

8.9.4   Double Edge Crack in a Composite Laminate

One of the first applications of the BEM to cracks in anisotropic materials was due to Snyder and Cruse [199]. The multi-region method and quarter-point elements were used by Tan [210] to solve several crack problems in orthotropic materials. Sollero and Aliabadi [202, 203] presented the multi-region formulation together with a mixed-mode J-integral for crack problems in orthotropic and anisotropic materials. Doblare [79] has also used the multi-region method. Sladek and Sladek [191] presented BEM results for 3D crack problems in anistropic materials. More recently, Bush [41] used the BEM to analyse the fracture of particle reinforced composite materials. Selvadurai [186] reported a study of the behaviour of a penny shaped matrix crack which may occur at an isolated fibre which is frictionally constrained. In this study, an incremental technique was used to examine the progression of the self-similar growth of the matrix crack. Sollero and Aliabadi [204] for the first time presented the dual boundary element method for anisotropic materials. Later, Pan and Amadei [150]

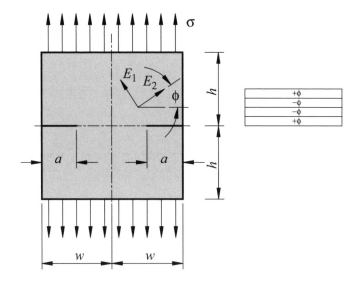

Figure 8.35: Double edge crack in a rectangular sheet.

and Pan and Yuan [151] presented dual boundary element formulations for two- and three-dimensional anisotorpic problems.

The double edge crack problem shown in Figure 8.35 was analyzed using the J-integral with multi-region method [202] and the DBEM [204]. The laminate consisted of four graphite-epoxy laminae with $E_1 = 144.8$ GPa, $E_2 = 11.7$ GPa, $G_{12} = 9.66$ GPa and $\nu_{12} = 0.21$. The laminate is subjected to uniform tensile stress σ at its ends. In Table 8.6 normalized stress intensity factors $K_I/\sigma\sqrt{\pi a}$ are presented for $h/w = 1.0$ and $a/w = 0.5$. Also included in Table 8.6 are the finite element results obtained by Chu and Hong [53].

8.9.5   Penny-Shaped Crack

Consider a penny-shaped crack of radius a in a cylindrical bar of radius R and height h, as shown in Figure 8.36, subjected to uniform stress σ at its ends. The penny-shaped crack is placed in the centre of the bar with the normal to the crack parallel to the axis of the bar. To allow comparison with the analytical solution for a penny-shaped crack in an infinite domain, the ratios of $R/a = 10$ and $h/R = 6$ were chosen [134]. Material constants were $E = 1000$ GPa and $\nu = 0.3$. The distribution of discontinuous 8-noded elements over the crack surface for the coarse mesh is shown in Figure 8.37

The problem was analyzed using the dual boundary element method with both discontinuous and continuous elements [134, 235]. The stress intensity factors were evaluated using crack opening displacements (one point formula and two point extrapolation formula), and both standard and special crack-tip elements were used. In Table 8.6, the normalized values of stress intensity factors $K_I/\sigma\sqrt{\pi a}$ are presented. The exact solution for the problem is $K_I/\sigma\sqrt{\pi a} = 2/\pi = 0.637$. As can be seen, the special crack-tip elements yield the best results.

Table 8.6: Normalized stress intensity factors for a double edge crack in a symmetric angle ply composite laminate.

$\pm\phi$	Multi-Region BEM[202]	DBEM[204]	FEM[53]
0	1.158	1.163	1.164
10	1.161	1.167	1.167
20	1.179	1.187	1.190
30	1.216	1.225	1.228
40	1.290	1.302	1.301
50	1.464	1.476	1.471
60	1.757	1.772	1.772
70	2.032	2.051	2.050
80	2.158	2.179	2.186
90	2.221	2.252	2.215

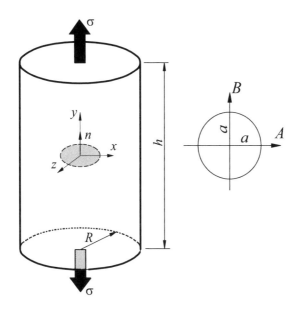

Figure 8.36: Penny-shaped crack subjected to unifrom stress.

Table 8.7: Stress intensity factors using the crack opening displacements for a penny shaped crack.

Shape Functions	Discontinuous Elem.	Continuous Elem.	
-	COD(1-point)	COD(1-point)	COD (2-point ext.)
Standard	0.643	0.544	0.532
Special	0.634	0.611	0.640

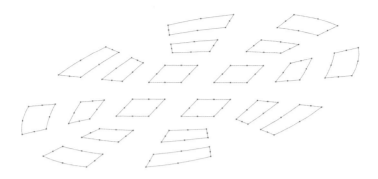

Figure 8.37: Penny-shaped crack with 40 elements.

Table 8.8: Percentage errors for penny shaped crack, $\alpha = 45^0$.

r/a	%Error (K_I)	%Error(K_{II}^A)	%Error(K_{III}^B)
0.3	1.3	0.7	0.2
0.5	0.3	0.2	2.1
0.75	0.9	0.7	9.4

8.9.6 Inclined Penny-Shaped Crack

Consider the penny-shaped crack inclined at an angle 45° in a cylindrical bar, as shown in Figure 8.38. The J-integral technique using decomposition of elastic fields was used to solve the problem [170]. The crack radius is a and the ratios are $a/R = 0.1$ and $h/R = 3$. Different J-integral contours of radius $r/a = 0.3, 0.5$ and 0.75 were tested to assess the sensitivity of the solutions to the choice of integration path.

The exact solution for the stress intensity factor are given by

$$K_I = \frac{2}{\pi}\sigma \cos^2 \alpha \sqrt{\pi a}$$

$$K_{II} = \frac{4}{\pi(2-\nu)}\sigma \sin \alpha \cos \alpha \cos \theta \sqrt{\pi a}$$

$$K_{II} = \frac{4(1-\nu)}{\pi(2-\nu)}\sigma \sin \alpha \cos \alpha \sin \theta \sqrt{\pi a}$$

Table 8.7 presents the percentage errors of the stress intensity factor for $\alpha = 45^o$, $\nu = 0.3$ for different contour paths. By inspection of the results, it is clear that the ratio $r/a = 0.75$ is too large for accurate K_{III} values.

8.9.7 Centre Crack Panel Elastoplastic Analysis

One of the first attempts in applying elastoplastic boundary elements to fracture mechanics was made by Morjaria and Mukherjee [139]. Tan and Lee [209] used the

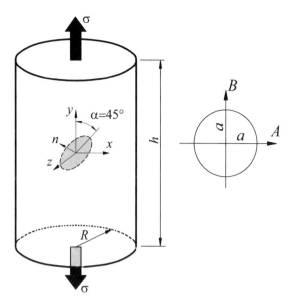

Figure 8.38: An inclined penny-shaped crack in a cylindrical bar subjected to uniform stress.

the BEM to study the behaviour of an internally pressurized thick-walled cylinder with symmetrical cracks. Leitão, Aliabadi and Rooke [119] demonstrated the efficiency of the BEM for evaluating several different nonlinear J-integrals. Application of the BEM to the analysis of cracks in the presence of residual stress fields due to pre-stress and cold expansion has been reported by Leitão and Aliabadi [123, 124] . The dual boundary element formulations for two- and three-dimensional problems were reported by Leitão, Aliabadi and Rooke [120] and Cisilino and Aliabadi [55].

In order to validate the energy integral presented in Section 8.85, Cisilino and Aliabadi [58] analysed an example with two-dimensional characteristics. It consists in a Centre-Cracked Panel (CCP) specimen in plane strain condition and under a uniaxial remote tension σ (see Figure 8.39). Normal displacements of the model faces are restricted in order to simulate the plane strain condition. The crack length is $a = 10$ mm, the specimen width $b = 2a$ and the model thickness is $t = a/10$. Material properties are Young's modulus $E = 100\,000$ MPa and Poisson's ratio $\nu = 0.3$.

The problem was analyzed in the elastoplastic regime assuming elastic-perfectly-plastic behaviour, with a yield stress $\sigma_y = 1000$ MPa. Paths very close to the crack tip were not studied. A highly strained zone develops in the vicinity of the crack tip which requires very fine modelling for accurate representation of the phenomena, and this was considered to be computationally impractical. The mesh is shown in Fig. 8.31. The boundary discretization consists of 140 elements and 661 nodes, while 36 cells and 522 nodes are used for the domain discretization. Four rings of internal cells with radii ranging from 20% to 75% of the crack length are used for evaluation of the energy integral. Remote uniaxial tension was applied in 12 increments, up to a load level that made the crack plastic zone spread up to a size of 75% of the crack length. In Figure 8.40 the average values obtained using the energy integral

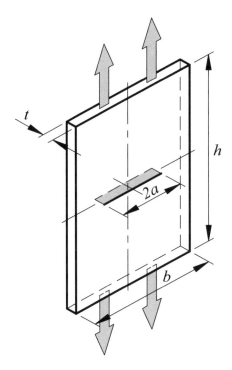

Figure 8.39: Centre-cracked specimen.

are plotted proportional to $\sqrt{J}$ in normalized form against increasing load, together with reference results obtained using the Irwin's approach; and the Dugdale's strip yield model. Very good agreement is obtained between computed values and those obtained using the Irwin approach, the maximum difference between them is around 3% at the maximum applied load level.

8.9.8   Elliptical Crack Subjected to Impact Load

Nishimura *et al.* [142, 143] used the time domain method to solve crack problems. The method was used for stationary and growing straight cracks in 2D, and plane cracks in a 3D infinite domain. Later, Zhang and Achenbach [246] improved the crack modelling used by Nishimura *et al.* by using spatial square-root functions near the crack tip. They analysed collinear cracks in an infinite domain. Hirose [106] and Hirose and Achenbach [107, 108] applied the formulation based on the traction equation with piecewise linear temporal functions to both constant and growing penny-shaped cracks. Zhang and Gross [247] used the two state conservation integral of elastodynamics, which leads to non-hypersingular traction integral equations. The unknowns in this approach are the crack opening displacements and their derivatives. Dominguez and Gallego [81] used a mixed variation of boundary values in which tractions were assumed to be constant and displacements linear in time. In this work mixed-mode crack problems were solved with multi-region BEM and quarter-point elements. Application of a Laplace transform method was presented by Sladek and Sladek [192], who analyzed a penny-shaped crack in an infinite elastic domain sub-

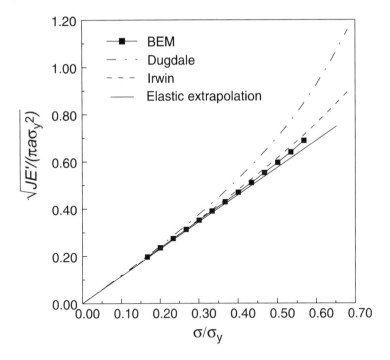

Figure 8.40: Variation of the J-integral with the applied load in the CCP specimen.

jected to harmonic and impact loads on crack surfaces. Tanaka *et al.* [211] used a similar method to claculate dynamic SIF's. The effect of varying the number of boundary elements and the number of parameters in the Durbin and Hosono methods of inverting was investigated. Polyzos *et al.* [155] analyzed a rectangular plate of viscoelastic material using the multi-region method. The dual reciprocity method was used by Balas, Sladek and Sladek [28] for symmetric crack problems. Chirino *et al.* [52] used the multi-region method and presented a comparison of the time-domain, Fourier transform and dual reciprocity methods.

The dual boundary element method for time domain, Laplace and dual reciprocity were presented by Fedelinski, Aliabadi and Rooke [85, 88, 90] for two-dimensional mixed-mode problems. In their studies they used the J-integral and quarter-point elements. The comparison between the three approaches in terms of accuracy, CPU and storage requirments was reported in [89]. The three-dimensional dual boundary element formulations for time domain, Laplace and dual reciprocity were presented by Wen, Aliabadi and Rooke [230, 231, 228]. Comparsions between the three methods in terms of accuracy, CPU and storage requirments were carried out in [232].

Consider a rectangular bar of cross-section $2a \times 2b$ and height $2h$ containing a centrally located central elliptical crack subjected to uniform load $\sigma_o H(t)$ at the ends as shown in Figure 8.41. The dimensions of the bar are $h = 15$ cm, $a = 9$ cm and $b = 2$ cm. The material constants are as follows: bulk modulus $K = 165$ GPa, shear modulus $G = 77$ GPa and density 7.9 mg/m^3. The problem was analyzed [232] using 40 quadratic 8-node elements on the boundary and 20 quadratic 8-node discontinuous elements on the crack surfaces. The dual boundary element formulations with Time Domain Method (TDM), Laplace Transform Method (LTM) and Dual Reciprocity

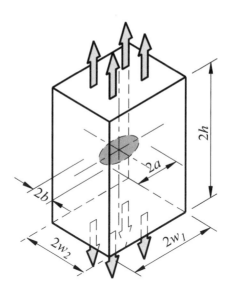

Figure 8.41: Elliptical crack in a rectangular bar.

Method (DRM) were used and the mode I stress intensity factors were evaluated using the crack opening displacements. The normalized time increment Δt was taken as 0.1 $(t = c_1 t/h)$ for both the TDM and DRM, and 50 times steps were calculated. For the LTM, the number of transform parameters is $L = 25$ and the unit time $t_o = c_1 t/h$. The dynamic stress intensity factors corresponding to a point at the end of the minor axis is plotted as $\sqrt{\pi} K_I(t)/\sigma_o$ against time in Figure 8.42. Also presented in the Figure are solutions reported by Chen and Sih [51] and Nishioka [144].

8.9.9   Interface Cracks

The use of Hetenyi's fundamental solution in the BEM, which avoids modelling the interface of two different materials, was introduced by Yuuki and Cho [243]. The use of a multi-region method for interface cracks has been reported by Lee and Choi [118], and Tan and Gao [207, 208] developed a quarter-point element to model interface cracks between materials. dePaulo and Aliabadi [75] used the BEM together with the J-integral to analyze interface cracks in isotropic and orthortopric dissimilar materials.

The problem of interface cracks for dissimilar materials presents an oscillatory behaviour for stress and displacement fields in the vicinity of the crack tip, regardless of the material (isotropic or orthotropic). The oscillation in the displacement field results in an overlapping of crack surfaces that is treated, in some works, as a contact zone. However, since this zone is confined to a very small region near the crack tip, it is often ignored in the analysis.

In the region near the crack tip, the displacements and stresses are parametrized through a complex stress intensity factor $K = K_I + iK_{II}$, where K_I and K_{II} do not in general represent opening and sliding modes in bimaterial fracture [49]. The modulus $\mid K \mid = \sqrt{K_I^2 + K_{II}^2}$, however, is uniquely related to the energy release rate of the crack, as in the homogeneous case.

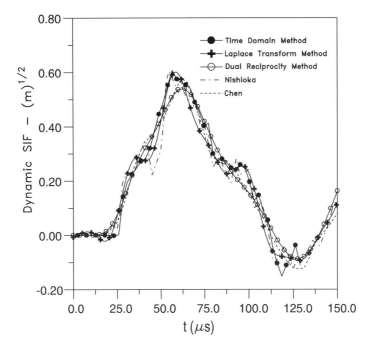

Figure 8.42: Dynamic stress intensity factors for an elliptical crack.

Table 8.9: $|K|/\sigma\sqrt{\pi a}$ for a single edge crack.

E_1/E_2	$a/w = 0.3$		$a/w = 0.4$	
	Ref.[75]	Ref.[243]	Ref.[75]	Ref.[243]
1	1.66	1.66	2.11	2.11
4	1.67	1.68	2.11	2.13
100	1.71	1.71	2.16	2.16

Consider an interface crack in an isotropic finite plate under uniform stress, as shown in Figure 8.43. The boundary element method was used by Yuuki and Cho [243] and later by dePaulo and Aliabadi [75], who used the J-integral method, to analyze this problem. The results for a non-dimensional stress intensity factor are presented in Table 8.8 for various ratios of E_1/E_2.

8.10 Crack Growth

The early attempts to model crack growth in mixed-mode conditions were by Ingraffea, Blandford and Ligget [110] and Grestle[98] for two- and three-dimensional problems, respectively. In their work the multi-region boundary element method was used. A similar method was used by Cen and Maier [44] to simulate crack growth in concrete structures. Aliabadi and co-workers have used the dual boundary

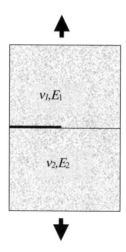

Figure 8.43: An edge crack along an interface in isotropic dissimilar material.

element method to automatically simulate crack in two- and three-dimensional elastostatic, elastoplastic, elastodynamics, thermoelstic problems and plates and shells [54, 57, 73, 91, 122, 135, 157, 166]. Application to composite materials, reinforced aircraft panels, concrete, and rock can be found in [23, 178, 182, 183, 237].

For the sake of efficiency, and to keep the simplicity of the standard boundary element, the DBEM formulation uses discontinuous quadratic elements for crack modelling. A general modelling strategy was developed by Portela, Aliabadi and Rooke[166] for two-dimensional problems, and by Mi and Aliabadi[135, 137] for three-dimensional problems.

8.10.1   Two-dimensional Modelling

A simple, yet robust, strategy that allows the DBEM to effectively model general edge or embedded crack problems; crack tips, crack-edge corners and crack kinks is described below.

In an incremental crack extension analysis, each new crack-extension increment is modelled with new discontinuous boundary elements. For two-dimensional problems, it becomes apparent that remeshing of the existing model is not required in the DBEM. The new discontinuous boundary elements of the crack-extension increment will generate new equations and update those already existing with new unknowns. In other words, new boundary elements will generate new rows and new columns in the matrix of the system of equations. Assuming that the crack extension is traction-free, the right-hand side of the system of equations is only extended for the positions corresponding to the new unknowns introduced. This procedure is illustrated schematically in Figure 8.44. If the **LU** decomposition method is adopted for solution of the system of equations, a very efficient incremental analysis can be carried out. For each increment of the analysis, only new rows and new columns need to be **LU**-decomposed. The existing rows and columns, already decomposed, are brought from the previous iteration into the current one.

The basic steps of this computation cycle, repeatedly executed for any number of crack-extensions, are summarized below:

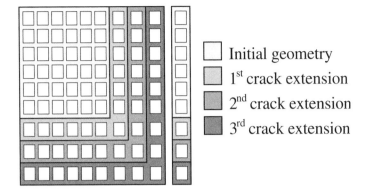

Figure 8.44: Automatic updating of the system matrix.

1. carry-out a DBEM stress analysis of the cracked structure;

2. compute the stress intensity factors;

3. compute the direction of the crack growth using either the maximum stress criterion or the minimum strain density criterion;

4. extend the crack by adding new elements;

5. repeat the processes until the maximum crack length is reached.

Crack Extension Direction   The maximum principal stress criterion (8.16) postulates that the growth of the crack will occur in a direction perpendicular to the maximum principal stress. As a continuous criterion, the criterion does not take into account the discreteness of the numerical modelling of the crack extension procedure. In other words, the crack path is defined continuously by the trajectory of the maximum principal stress, evaluated locally by (8.15). Therefore, the incremental extension of a crack in a general mixed-mode deformation field, computed by equation (8.16), is always defined locally in the same direction, whatever length of crack extension Δa is considered. As a consequence, uniqueness of the crack path can not be assured with different sizes of crack-extension increment. Hence, in an incremental analysis, the tangent direction of the crack-path, predicted by (8.16), must be corrected to give the direction of the actual crack-extension increment.

Portela, Aliabadi and Rooke[166] developed a simple procedure to correct the crack direction to ensure that, for different analyses of the same problem with different crack-extension increments, a unique final crack path is achieved. The procedure applied to define the direction of the nth crack-extension increment introduces a correction angle β to the tangent direction $\theta_{t(n)}$ predicted by the maximum principal stress criterion, as shown in Figure 8.45. Using geometric relationships, this correction angle is given by $\beta = \theta_{t(n+1)}/2$ in which $\theta_{t(n+1)}$ is the direction of the next crack-extension increment, also evaluated with the maximum principal stress criterion. For the current nth crack-extension increment, the ith iteration for $\theta_{t(n)}$ can be summarized as follows[166]:

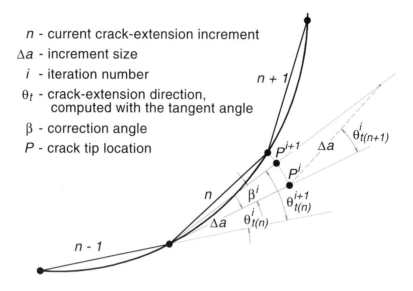

Figure 8.45: Incremental crack-extension direction.

- step 1, for the *1st* iteration only, evaluate the crack-path tangent direction $\theta^i_{t(n)}$ with the maximum principal stress criterion, equation (8.16);

- step 2, along the direction computed in the previous step, extend the crack one increment Δa to P' and evaluate the new stress intensity factors;

- step 3, with the new stress intensity factors and the maximum principal stress criterion, equation (8.16), evaluate the next crack-path direction $\theta^i_{t(n+1)}$;

- step4, define the correction angle $\beta^i = \theta^i_{t(n+1)}/2$, measured from the increment defined in the second step;

- step5, correct the direction of the crack-extension increment, defined in the second step, to its new direction given by $\theta^{i+1}_{t(n)} = \theta^i_{t(n)} + \beta^i$ so the crack tip is now at P^{i+1};

- step6, starting from the second step, repeat the above steps sequentially while $\mid \beta^{i+1} \mid < \mid \beta^i \mid$.

When the size of the crack-extension increment, Δa tends to zero, the angle $\theta_{t(n+1)}$ also tends to zero and so does the correction angle β. This means that in the limit, the direction of the increment crack-extension tends to the direction of the tangent of the continuous crack path.

Crack Growth Analysis in a Cruciform Plate   A cruciform plate with a corner crack as shown in Figure 8.46, was studied [156] for different sets of thermal and mechanical boundary conditions. The initial ratio of a/L is equal to 0.2, $L = 0.4$ m and the angle between the crack and vertical axis is 45^0. The crack was grown in 11 increments [156], where the crack length of each crack increment is equal to $0.3a$.

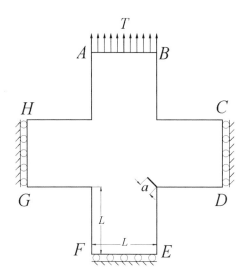

Figure 8.46: Geometry of cruciform plate with corner crack.

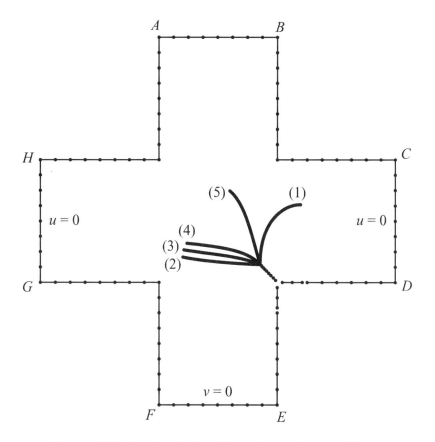

Figure 8.47: Crack path for different boundary conditions.

Table 8.10: Boundary conditions for the cruciform plate.

BC Set	Temperature in $^{\circ}C$				Traction (T Pa)
	AB	CD	EF	GH	AB
1	10	0	-10	0	0
2	0	0	0	0	10
3	10	0	-10	0	10
4	20	0	-20	0	10
5	10	-5	-10	-5	10

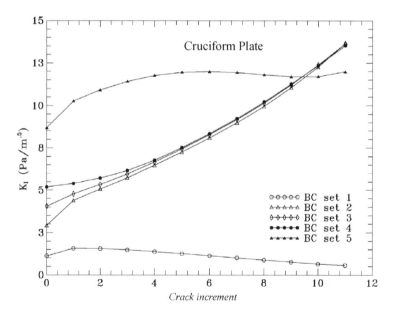

Figure 8.48: Mode I stress intensity factors vs. crack increments in cruciform plate.

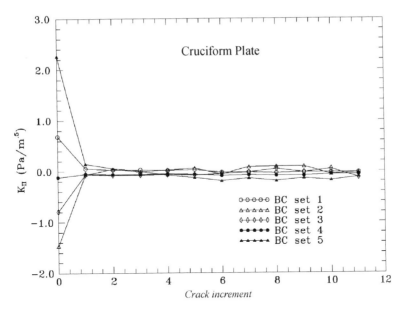

Figure 8.49: Mode II stress intensity factors vs. crack increments in cruciform plate.

This problem is solved for five different boundary condition sets, which are given in Table 8.9. The material properties used were $E = 218\,400$ Pa, $\nu = 0.3$, the coefficient of linear expansion$-1.67E - 5$ per $^\circ C$. The constants for the Paris model are $C = 4.62E - 12$ and $n = 3.3$. The crack path for the five boundary conditions are shown in Figure 8.47. Path (1) is due to thermal load only, and Path (2) is due to mechanical load only. The remaining paths are due to combined mechanical and thermal loadings. The stress intensity factors are presented in Figures 8.48 and 8.49 for mode I and mode II respectively. As can be seen, in all loading conditions the mode II stress intensity factors tend to zero after one step. The mode I stress intensity factors are more dependent on the temperature boundary conditions; for temperature condition only (set I), a real crack would virtually stop as ($K_I \simeq 0$) after a few steps.

Crack Growth Analysis in Stiffened Panels

Consider a finite panel with riveted stiffeners as presented in Figure 8.50. The dimensions in the Figure are given in millimeters (mm). The panel contains a crack emanating from a rivet hole in its centre. The corresponding rivet and the central stiffener are assumed to be broken. The panel contains a second crack emanating from the internal circular opening. The sheet is 2.3 mm thick and the stiffeners are made of Aluminium alloy A2024-T3 with the following mechanical properties: Young's Modulus= 78500 MPa; Poisson's ratio = 0.32; Shear Modulus = 29000 MPa and the crack growth properties are given by the Paris law coefficients: $C = 0.183 \times 10^{-11}$ and $m = 3.284$. The stiffeners cross sectional properties are: Area = 300 mm^2; Second moment of inertia = 1800 mm^4. The rivets are considered to be rigid and have a diameter $\phi = 6.4$ mm and a pitch $p = 25$ mm.

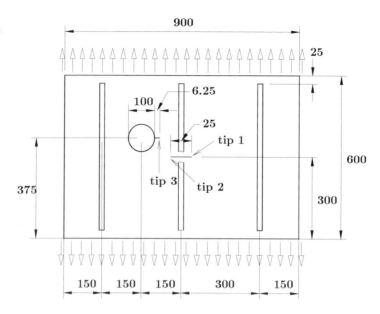

Figure 8.50: Cracked panel for incremental crack extension analysis example.

The panel is subjected to tensile stresses of value 10 MPa, applied at the top and bottom edges. The stiffeners are subjected to the same stresses applied at their extreme points. Fatigue crack growth is simulated, assuming that the load is applied in cycles of constant amplitude with a stress ratio equal to zero [180.

The increment extension size was taken as $\Delta a = 12.5$ mm, which is twice the size of the crack-tip elements used in the discretization. The crack growth paths are presented in Figure 8.51. It is noticeable how crack tips 2 and 3 interact, changing direction and bending towards each other.

In order to assess the influence of the increment extension size Δa on the direction of crack growth and on the relative crack increment sizes, the incremental analysis was repeated, considering first five crack extension increments and $\Delta a = 25$ mm and then twenty crack extension increments and $\Delta a = 6.25$ mm.

The influence of Δa on the crack growth directions can be assessed by comparing the crack paths obtained, which are presented in Figure 8.51. It can be seen that the three paths are virtually identical for the cases with $\Delta a = 6.25$ mm and $\Delta a = 12.5$ mm. For the case with $\Delta a = 25$ mm, the path for crack 1 is also identical to the others while for crack 2 it is very similar. Higher discrepancies are observed for crack tip 3, the reason being the more pronounced curvature of that path.

The influence of Δa on the relative crack increment sizes can be evaluated by comparing the crack growth diagrams presented in Figure 8.52. It can be seen that, in this case, results for cracks 1 and 3 are remarkably similar. For crack 2, results obtained with $\Delta a = 25$ mm tend to deviate slightly from the values obtained with $\Delta a = 6.25$ mm and $\Delta a = 12.5$ mm. That tendency is only observed for values of $a > 75$ mm.

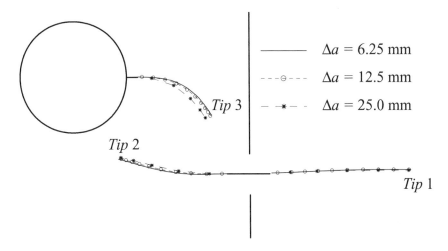

Figure 8.51: Crack paths obtained with different values of the crack extension increment Δa.

It should be noted that $\Delta a = 25$ mm is a very large value for the crack extension increment. It is equal to four times the initial size of crack 3, and therefore is not a recommended choice. Results obtained indicate that the iterative scheme introduced in the previous section consistently predicted both the trajectory and the relative growth rates, even for considerably different values of the parameter Δa.

Single Notched Shear Concrete Beam   Consider the single notched beam shown in Figure 8.53. The dual boundary element method was used by Saleh and Aliabadi [176] to study crack growth in this concrete beam. To model the cracking process in concrete, the fictitious crack model was adopted. In this non-linear model, the fracture zone is replaced by closing forces (cohesive forces) acting normal to both crack surfaces as shown in Figure 8.54. In the figure, the crack mouth opening displacement is denoted as Δu_c and its critical crack opening displacement as Δu^{cr}.

From Figure 8.53, it can be seem that the loading conditions are non-symmetric, which implies that the crack propagation from the notch will include both opening and sliding crack displacements. In the experiments [24], the load was applied at point C of the steel beam AB, and was controlled by a feedback mechanism with crack mouth sliding displacement measured along the vertical direction between the notch surfaces. The materials properties of the concrete used in the analysis [176] and experiments [24] are: $E = 24\,800$ MPa, $\nu = 0.18$, the ultimate tensile strength $f' = 2.8$ MPa and $G = 100$ N/m. The thickness of the beam is 0.156 m.

The BEM results are compared with two finite element results [3, 174]. In the BEM analysis, 37 quadratic elements were used. Figure 8.55 presents the load P against the crack mouth opening displacements. The displacement at point C are presented in Figure 8.56.

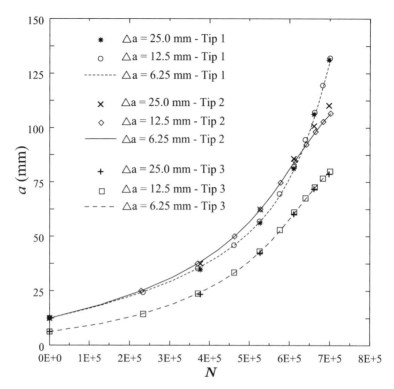

Figure 8.52: Crack growth diagrams obtained with different values of the crack extension increment Δa.

8.10.2 Three-dimensional Modelling

The minimum strain energy density criterion described in Section 8.2.7 gives the crack growth direction angle θ_o in the local coordinate plane perpendicular to the crack front as shown in Figure 8.3. The crack growth direction angle is obtained by minimizing the strain energy intensity factor $S(\theta)$ of equation (8.17) with respect to θ. The stationary points of S are calculated by solving $dS/d\theta = 0, (-\pi < \theta < \pi)$ using the numerical bisection method. Finally, $S_{\min}$ is obtained by comparing the values of S at the stationary points where $d^2 S/d\theta^2 > 0$.

For three-dimensional crack growth problems a relationship is required between the maximum incremental size and the other increment sizes along the crack front. Here two methods to solve this problem as presented by Mi and Aliabadi [135] are described. The first is based on the strain energy density criterion, and the second is based on the Paris law.

Use of Strain Energy Density Criterion The third hypothesis of the strain energy density criterion presented in Section 8.2.7 states that the length, r_o, of the initial crack extension is assumed to be proportional to $S_{\min}$ such that $S_{\min}/r_o$ remains constant along the new crack front. Therefore, the incremental size along the crack front can be decided as follows. First, the max$\{S_{\min}\}$ value is selected from among the values of $S_{\min}$ evaluated at a set of discrete points along the crack front. Then, the incremental size at the point corresponding to the max$\{S_{\min}\}$ is defined as $\Delta a_{\max}$,

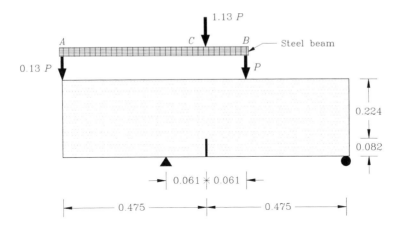

Figure 8.53: Single notched shear concrete beam.

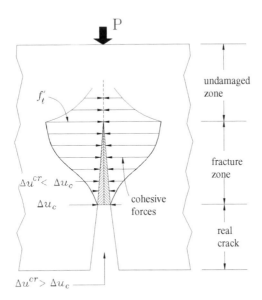

Figure 8.54: The fictitious crack model in which the fracture zone is replaced by cohesive forces.

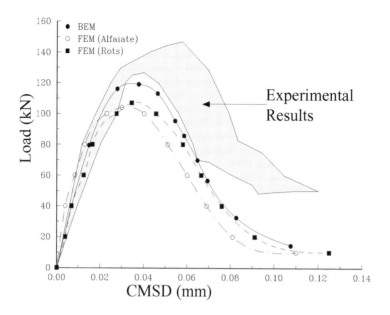

Figure 8.55: Load P against crack mouth opening displacement (CMSD) of single notched shear concrete beam.

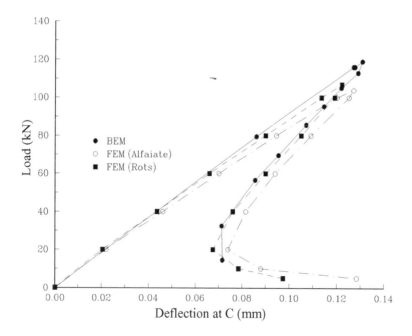

Figure 8.56: Load P against deflection at point C of single notched shear beam.

a value chosen beforehand. The general incremental size at any point on the crack front is determined by

$$\Delta a = \Delta a_{\max} \left(\frac{S_{\min}}{\max\{S_{\min}\}} \right) \tag{8.145}$$

where $S_{\min}$ is the minimum strain energy density evaluated at the front point under consideration.

Use of Paris' Law   During fatigue crack growth, the relation between the increment size and the number of load cycles may be represented by the Paris law, which states that

$$\frac{da}{dN} = C(\Delta K_{eff})^m \tag{8.146}$$

where da/dN is the rate of change of crack length with respect to the loading cycles, C and m are constants that depend upon material, load frequency, environment and mean load and

$$\Delta K_{eff} = K_{eff}^{\max} - K_{eff}^{\min} = K_{eff}^{\max}(1 - R) \tag{8.147}$$

in which $R = K_{eff}^{\min}/K_{eff}^{\max}$; K_{eff} stands for the effective stress intensity factor and is an empirical function of stress intensity factors. Since linear elasticity is considered, R can be written as

$$R = \frac{K_{eff}^{\min}}{K_{eff}^{\max}} = \frac{\sigma_{\min}}{\sigma_{\max}} \tag{8.148}$$

The maximum amount of crack extension increment $\Delta a_{\max}$ is taken to correspond to the point along the crack front, where $\Delta K_{eff}^{\max}$ occurs. From (8.146)

$$\Delta a \approx C(\Delta K_{eff})^m \Delta N$$
$$\Delta a_{\max} \approx C(\Delta K_{eff}^{\max})^m \Delta N$$

From the above equations, the following relationship is deduced

$$\frac{\Delta a}{\Delta a_{\max}} = \left(\frac{\Delta K_{eff}}{\Delta K_{eff}^{\max}} \right)^m$$

Therefore, the incremental size at any point along the crack front is evaluated from

$$\Delta a = \Delta a_{\max} \left(\frac{\Delta K_{eff}}{\Delta K_{eff}^{\max}} \right)^m \tag{8.149}$$

Boundary Remeshing Strategy   Mi and Aliabadi [137] and later Cisilino and Aliabadi [54] developed a strategy for updating the model geometry. The strategy consists of two parts. The first relates to the crack extension itself and is simply done by adding new elements along the crack front, whose dimensions and orientation are respectively given by the crack extension increment Δa and propagation vectors computed at geometrical points Q' (see Figure 8.57).

The second part is concerned with the mesh modification at the crack tip areas. Consider a region close to the crack tip, as illustrated in Figure 8.58. It can be seen that, in general, after a crack extension the tip at A has moved to a new position (point B) which may not coincide with an element node. Hence the local redefinition

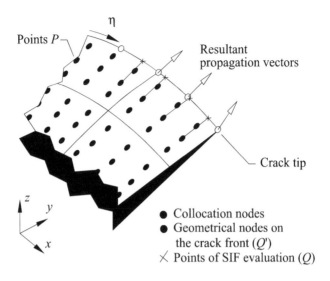

Figure 8.57: Crack front propagation vector.

of the boundary element mesh will be necessary. To tackle this problem, a local remeshing strategy has been devised [54, 137]. The remeshing strategy is illustrated in Figure 8.58, and can be summarized as follows:

1. After the new position for the crack tip is determined (point B), all elements contained in the dashed circle are removed from the mesh.

2. Nodal positions defining the undiscretized patch are identified (points C in Figure 8.58).

3. Points C, together with the previous and actual positions of the crack tip (points A and B, respectively), are employed for the discretization of the tip area into triangles by using the Delauny algorithm.

4. Vertices of the resulting triangles are subsequently used to define discontinuous triangular elements. The use of discontinuous elements makes the approach general, since no care has to be taken regarding the common nodes at crack edges.

The above strategy, which allows local remeshing into the existing model, minimizes the extra computation necessary to solve new configurations. Furthermore, the rediscretization strategy alters only a small portion of the model dicretisation, therefore only a few elements in the BEM system matrix need to be updated for the solution of the new configuration.

Crack Growh of an Embedded Elliptical Crack   An elliptical crack of semi-major axis a and semi-minor axis b in a cylindrical bar of radius R and height h subjected to a tensile stress σ, as shown in Figure 8.59, is considered. The ratios $b/a = 0.5$, $R/a = 10$ and $h/R = 6$ were chosen, and Young's modulus was taken to

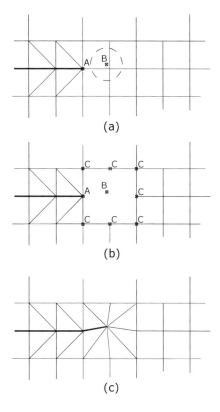

Figure 8.58: Crack tip remeshing procedure.

be 200kNmm^{-2} and Poisson's ratio as 0.3. The centre of the ellipse is coincident with the centre of the bar, and the normal to the crack surface is parallel to the axis of the bar. In a pure mode I problem (i.e. $\gamma = 90^\circ$), the crack grows in its original plane, as as shown in Figure 8.60.

It can be observed from these figures that the results are in good agreement with the generally accepted concept that the crack front always tends to grow into that shape which gives a constant mode I stress intensity along the crack front, in this case circular.

A mixed-mode crack-growth problem can be simulated for a crack which makes an angle $\gamma = 45^\circ$. In this case the crack growth is generated by a constant amplitude cyclic tensile load applied to the ends of the bar. The crack growth path obtained in [135] is shown in Figure 8.61.

Two Equal Co-planar Semi-circular Cracks Consider a prismatic bar containing two identical and symmetrical coplanar semi-circular edge cracks of radius a. The bar is subjected to a remote tensile stress σ at its ends. Its dimensions, scaled to the original crack radius a, are shown in Figure 8.62. The initial distance between the two adjacent cracks tips (B_1 and A_2 in Figure 8.62) is equal to $0.4a$. The following Paris law is employed to estimate the crack growth:

$$\frac{\Delta a}{\Delta N} = 3 \cdot 10^{-11} \Delta K^{2.92}$$

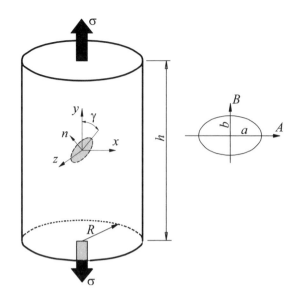

Figure 8.59: Elliptical crack in a cylindrical bar.

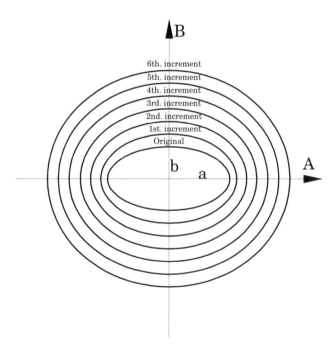

Figure 8.60: Crack growth path for mode I crack growth of an embedded elliptical crack.

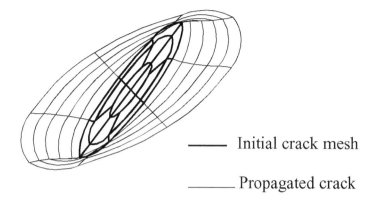

—— Initial crack mesh

—— Propagated crack

Figure 8.61: Crack growth path for the inclined elliptical crack.

with a mean load ratio $R = 0$.

The evolution of the crack shapes is shown in Figure 8.62 for seven propagation increments [54]. The crack profiles in Figure 8.62 are such that the same number of loading cycles is taken to develop from one contour to the next. The reference number of cycles ΔN_0 is fixed at 750 cycles. Crack coalescence takes place between the first and second propagation increments, and the transition from two cracks to one crack was assumed to occur when the cracks overlap. Once the cracks coalesce the resultant crack tends rapidly to a semi-elliptical shape.

Figure 8.63 shows the crack discretization for the initial configuration and the seven propagation increments.

Crack growth curves for the outer distance and maximum depth are compared with those obtained by Kishimoto et al. [116] using finite elements in Figure 8.64. The general evolution of the crack profiles are generally in good agreement, but they differ for the outer distance during the last propagation increments.

The stress intensity factors ΔK_I for the growing cracks are shown in Figure 8.65. For each crack the normalized values of the stress intensity factors are plotted as a function of the angle between the horizontal axis and a radial line from the centre of the initial crack. It can be seen that the value of the stress intensity factor at adjacent crack tips increases as the cracks approach each other. It rapidly increases at the contact zone in the early coalescence; it stays high while the single crack shape is sharply concave, and finally starts decreasing as the crack adopts a regular crack front. The evolution of the ratio $\Delta K_{max}/\Delta K_{min}$ is plotted in Figure 8.66, together with results from [116]. This ratio also reaches its maximum value during coalescence, after which it starts decreasing and tends towards unity (i.e. an iso-K configuration).

Two Offset Semi-circular Parallel Cracks   Consider a prismatic bar containing two identical offset semi-circular parallel planar edge cracks. The dimensions of the bar, as well as the relative positions of the cracks, scaled to the original crack radius a, are shown in Figure 8.67. The bar is subjected to a remote tensile stress σ at its

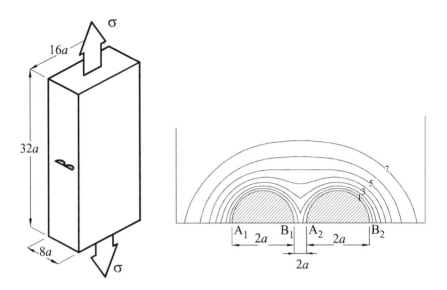

Figure 8.62: A prismatic bar with two equal semi-circular coplanar cracks under remote tension. Model geometry and predicted crack shape.

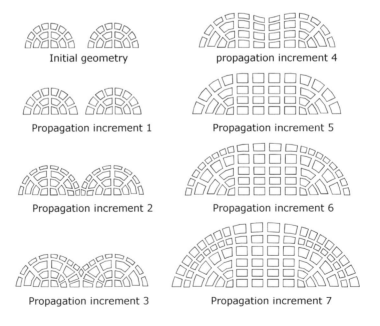

Figure 8.63: Crack discretization of the twin coplanar cracks.

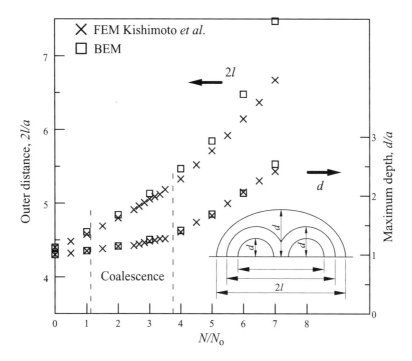

Figure 8.64: Crack growth curves predicted by the finite and boundary element methods.

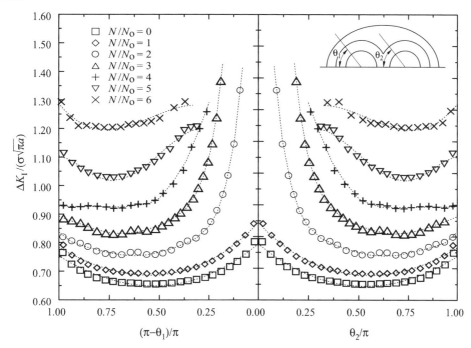

Figure 8.65: Variation of ΔK_I along the crack fronts for each propagation step.

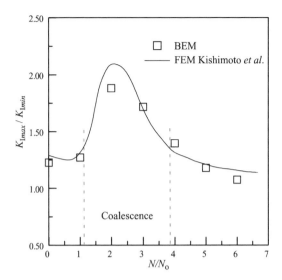

Figure 8.66: Change in the ratio $K_{\max}/K_{\min}$ along the crack front during crack development predicted by finite and boundary element methods.

ends. Crack growth is estimated using the same Paris law as in the previous examples [54].

Figure 8.68 is a rear view of the specimen where some of the boundary elements on its lateral face have been removed to show the cracks more clearly. It illustrates the evolution of the crack shapes for 5 propagation increments. In this example the reference number of cycles ΔN_0 is 3000 cycles and the propagation increments are not constant. Also shown in Figure 8.68 are the discontinuous triangular elements introduced to the model during the rediscretization process after each crack extension. The sub-Figure in the top right-hand corner shows the crack propagation on the free surface.

The evolution of the stress intensity factor components ΔK_I, ΔK_{II} and ΔK_{III} are plotted for both cracks in Figures 8.69, 8.70 and 8.71. Since the cracks now propagate out of plane, it is no longer suitable to represent the position on the crack front as a function of the angle θ, as before. In this example the position on the crack front is represented by the normalized distance given by the ratio of the distance η_i measured from the A_i crack tips (see Figure 8.67) over the total crack front length l. The behaviour of ΔK_I is almost unaffected by the presence of the second crack for the two first crack profiles when the adjacent crack tips do not pass over each other. However, this is not the case after the third increment, since a shielding effect takes place and ΔK_I values dramatically decrease for the adjacent tips. In contrast to what happens to ΔK_I, the values of ΔK_{II} are earlier influenced by the presence of the second crack. Their absolute values achieve a maximum to start with, and decrease after a second crack increment. The asymmetric evolution in the values of ΔK_{II} makes the cracks grow towards each other. On the other hand, the absolute value ΔK_{III} monotonously increases throughout the propagation process. However, these values are small compared to ΔK_I and hence are not significant.

The results obtained are from a qualitative point of view, in good agreement with experiments reported by Soboyejo et al. [200], who analysed a specimen with similar

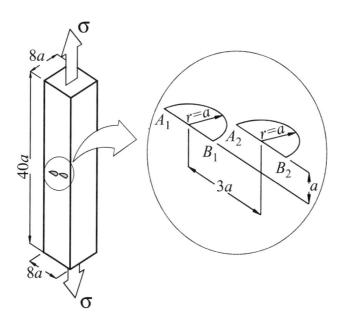

Figure 8.67: A prismatic bar with two equal semicircular out-of-plane parallel cracks under remote tension.

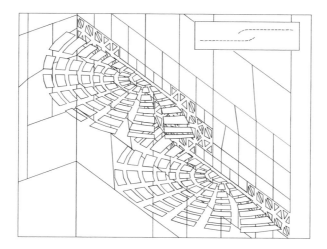

Figure 8.68: Predicted crack profiles for the third example.

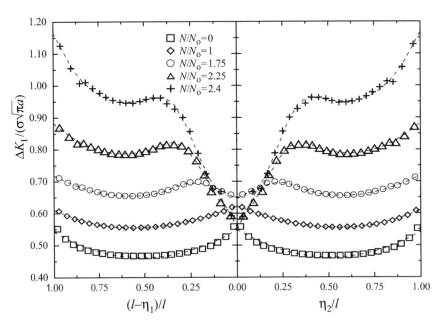

Figure 8.69: Variation of ΔK_I along the crack fronts for each propagation step.

characteristics to that reported in this example. In their work, Soboyejo et al. also observed the deviation of adjacent cracks tips as they approach each other (see Figure 8.72).

8.11 Summary

In this chapter boundary element formulations for the analysis of crack problems in fracture mechanics were presented. Also presented was a new generation of the BEM formulations referred to as the dual boundary element method. The DBEM is shown to be particulary efficient for modelling crack growth. This feature was shown in several problems, including elastostatic, thermoelastic, elastoplastic and elastodynamic problems. The crack growth is simulated along *a priori* unknown paths with little (3D) or no (2D, plates and shells) remeshing requirments. Thus, the BEM advantages over finite elements are clear, particularly for the automatic modelling of crack growth.

References

[1] Alessandri, C. and Dielo, A., On a simple 2-D model for mode II failure of plated concrete specimens. *Computational Engineering with Boundary Elements*, Computational Mechanics Publications, Southampton,111-126. 1990.

[2] Ahmadi-Brooghani, S.Y. and Wearing, J.L., The application of the dual boundary element method in linear elastic crack problems in plate bending. *Boundary*

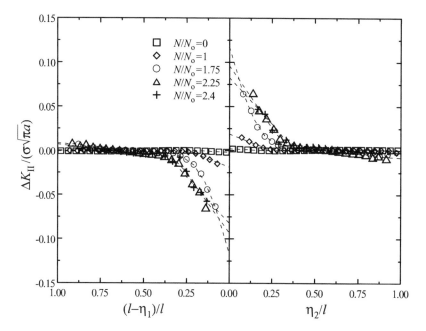

Figure 8.70: Variation of ΔK_I along the crack fronts for each propagation step.

Element Methods XVII, Computational Mechanics Publications, Southampton, 429-438, 1996.

[3] Alfaiate, J., Pires, E.B. and Martines, J.A.C,. Non-prescribed discrete crack evolution in concrete: algorithm and numerical test, *Localized Damage III,* Computational Mechanics Publications, Southampton, 185-192, 1994.

[4] Alburquerque,E.L., Sollero,P. and Aliabadi,M.H. Dual boundary element method in dynamic fracture mechanics of laminate composites, *Boundary Element Technology II,* Hoggar Press, Geneva, 379-85, 2001.

[5] Aliabadi, M.H., Hall, W.S. and Phemister, T.G., Taylor expansions for singular kernels in the boundary element method. *International Journal for Numerical Methods in Engineering,* **21**, 2221-2236, 1985.

[6] Aliabadi, M.H., Rooke, D.P. and Cartwright, D.J., An improved boundary element formulation for calculating stress intensity factors: Application to aerospace structures, *Journal of Strain analysis,* **22**, 203-207, 1987.

[7] Aliabadi, M.H., An enhanced boundary element method for determining fracture parameters. *Numerical Methods in Fracture Mechanics,* Pineridge Press, Swansea,*27-39, 1987.*

[8] Aliabadi, M.H., Cartwright, D.J. and Rooke, D.P., Fracture mechanics weight functions by the removal of singular fields using boundary element analysis, *International Journal of Fracture,* **40**, 271-284, 1989.

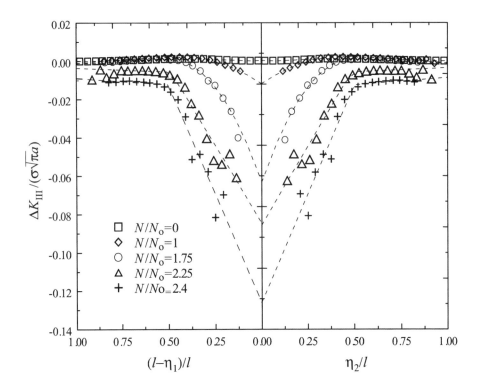

Figure 8.71: Variation of ΔK_{III} along the crack fronts for each propagation step.

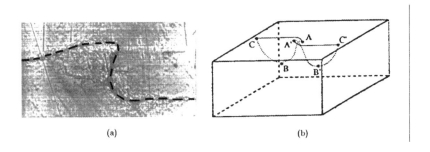

Figure 8.72: Deviations of adjacent crack tips before coalescence. (a) Coalescence region at surface of specimen, (b) Schematic illustration (After Soboyejo *et al.*).

[9] Aliabadi, M.H. and Rooke, D.P., Weight functions for crack problems using boundary element analysis, *Engineering Analysis.*, **6**,19-29, 1989.

[10] Aliabadi, M.H. and Rooke, D.P., Improved modelling of crack-tip fields in weight function analysis, *International Journal of Fracture*, **40**, R73-R75, 1989.

[11] Aliabadi, M.H., Boundary element analysis of the three-hole cracked specimen. *Boundary Element Technology*, Computational Mechanics Publications, Southampton, 33-48, 1989.

[12] Aliabadi, M.H. and Rooke, D.P., A new procedure for calculating three-dimensional stress intensity factors using boundary elements, *Adv. in BEM*, vol3, 123- 131, 1989.

[13] Aliabadi, M.H., Evaluation of mixed mode stress intensity factors using path independent integral, *Proc. 12th Int.Conf on Bound. Element Methods*, Computational Mechnaics Publications, Southampton, 281-292, 1990.

[14] Aliabadi, M.H. and Rooke, D.P., Bueckner weight functions for cracks near a half-plane, *Engineering Fracture Mechanics*, **37**, 437-446, 1990.

[15] Aliabadi, M.H. and Cartwright, D.J., Boundary element analysis of strip yield cracks, *Engineering Analysis*, **8**, 9-12, 1991.

[16] Aliabadi, M.H. and Rooke, D.P., *Numerical Fracture Mechanics*. Kluwer Academic Publishers, Dordrecht and Computational Mechanics Publications, Southampton 1991.

[17] Aliabadi, M.H. and Portela, A., Dual boundary element incremental analysis of crack growth in rotating disc. *Boundary Element Technology VII*, Computational Mechanics Publications, Southampton, 607-616, 1992.

[18] Aliabadi, M.H. and Mi, Y., Application of the three-dimensional boundary element method to quasi-static and fatigue crack propagation. *Handbook of Fatigue Crack Propagation in Metallic Structures*, Elsevier Academic Publishers, Oxford, 397-432, 1994.

[19] Aliabadi, M.H., Database of Stress Intensity Factors, Computational Mechanics Publications, Southampton *1996.*

[20] Aliabadi, M.H., A new generation of boundary element methods in fracture mechanics, *International Journal of Fracture*, **86**, 91-125, 1997.

[21] Aliabadi,M.H., Boundary element formulations in fracture mechanics, *Applied Mechanics Review*, **50,** 83-96, 1997.

[22] Aliabadi, M.H., Boundary element methods for crack dynamics, *J. Key Engineering Materials*, **145(1&2)**, 323-328, 1998.

[23] Aliabadi, M.H. and Sollero, P., Crack growth analysis in homogeneous orthotropic laminates, *Composites Science and Technology*, **58**, 1697-1703, 1998.

[24] Arrea, M. and Ingraffea, A.R., Mixed-mode crack propagation in mortar and concrete, Report 81-13, Dept. Struct. Eng., Cornell Univ., *Ithaca* 1982.

[25] Bainbridge,C.M., Aliabadi, M.H. and Rooke, D.P., A path-independent integral for stress intensity factors in three dimensions, *Journal of Strain Analysis*, **32(6)**, 411-423, 1997.

[26] Bains, R.S., Aliabadi, M.H., Fracture mechanics weight functions in three-dimensions: subtraction of fundamental fields, *International Journal for Numerical Methods in Engineering*, **35**, 179-202, 1992.

[27] Bains, R.S., Aliabadi, M.H. and Rooke, D.P., Weight functions for curved crack front using BEM, *Journal of Strain Analysis*, **28**, 67-78, 1993.

[28] Balas,S, Saldek,J and Sladek,V. *Stress Analysis by Boundary Element Methods*, Elsevier, Amsterdam, 1989.

[29] Barsoum, R.S., On the use of isoparametric finite elements in linear fracture mechanics, *International Journal for Numerical .Methods in Engineering*, **10**, 25-37, 1976.

[30] Beer, G., An efficient numerical method for modelling initiation and propagation of cracks along material interfaces. *International Journal for Numerical Methods in Engineering*, **36**, 3579-94, 1993.

[31] Besuner, P.M., The influence function method for fracture mechanics and residual fatigue life analysis of cracked components under complex stress fields. *Nuclear Engineering and Design*, **43**, 115-154, 1977.

[32] Bueckner, H.F., The propagation of cracks and the energy of elastic deformation, *Trans. ASME*, **80E**, 1225-1230, 1958.

[33] Beuth, J.L., Separation of crack extension modes in orthotropic delamination models. *International Journal of Fracture*, **77**, 305-321, 1996.

[34] Balckburn, W.S., The Calculation of stress intesity factors at crack tips using special finite elements. In the mathematical finite elements, Academic Press, 327-336, 1973.

[35] Blackburn, W.S. and Hall, W.S., The dual boundary element method for growing cracks allowing for crack curvature near the tip. *Boundary Element Method XVI*, Computational Mechanics Publications, Southampton, 413-422, 1994.

[36] Blandford, G.E., Ingraffea, A.R. and Liggett, J.A., Two-dimensional stress intensity factor computations using the boundary element method. *International Journal for Numerical Methods in Engineering*, **17**, 387-404, 1981.

[37] Bonnet, M., Shape differentiation and boundary integral equations: application to energy methods in fracture mechanics. *Boundary Element Methods XVI*, Computational Mechanics Publications, Southampton, 373-380, 1994.

[38] Bonnet, M., Computation of energy release rate using material differentiation of elastic BIE for 3-D elastic fracture. *Engineering Analysis with Boundary Elements*, **15**, 137-150, 1995.

[39] Bui, H.D., An integral equation method for solving the problem of a plane crack of arbitrary shape. *Journal of Mechanics and Physics of Solids*, **25**, 29-39, 1975.

[40] Burstow, M.C. and Wearing, J.L., An analysis of fracture problems using a combined boundary element finite element approach. *Boundary Element XIV*, Computational Mechanics Publications, 351-366, 1992.

[41] Bush, M.B., Cracking and fracture of particle reinforced composite materials. *Boundary Element Methods XVI*, Computational Mechanics Publications, Southampton, 381-388, 1994.

[42] Cartwright, D.J. and Rooke, D.P., An efficient boundary element model for calculating Green's function in fracture mechanics, *International.Journal of Fracture*, **27**, R43-R50, 1985.

[43] Castor, G.S. and Telles, J.C.F., The 3-D BEM implementation of a numerical Green's function for fracture mechanics applications, *International Journal for Numerical Methods Engineering*, **48**, 1199-1214, 2000.

[44] Cen, Z. and Maier, G., Bifurcation and instabilities in fracture of cohesive-softening structures:a boundary element analysis. *Fatigue and Fracture of Engineering Materials and Structures*, **15**, 911-928, 1992.

[45] Chan, H.C.M., Li, V. and Einstein, H.H., A hybridized displacement discontinuity and indirect boundary element method to model fracture propagation. *International Journal of Fracture*, **45**, 263-282, 1990.

[46] Chan, K.S. and Cruse, T.A., Stress intensity factors for anisotropic compact-tension specimens with inclined cracks. *Engineering Fracture Mechanics*, **23**, 863-874, 1986.

[47] Chandra, A., Haung, Y., Wei, X. and Hu, K.X., A hybrid micro-macro formulation for micro-macro clusters in elastic components. *International Journal for Numerical Methods Engineering*, **38**, 1215-1236, 1995.

[48] Chang, C. and Mear, M.E., A boundary element method for two-dimensional linear elastic fracture analysis. *International Journal of Fracture*, **74**, 219-251, 1995.

[49] Charalambides, P.G., Zhang, W., An energy method for calculation of the stress intensities in orthotropic bimaterial fracture, *International Journal of Fracture*, **76**, 97-120, 1996.

[50] Chen, W.H. and Chen, T.C., An efficient dual boundary element technique for two-dimensional fracture problem with multiple cracks. *International Journal for Numerical Methods in Enginerring*, **38**, 1739-1756, 1995.

[51] Chen, E.P. and Sih, G.C., Transient response of cracks to impact loads. *Elastodynamic Crack Problems*, Noordhoof, Leiden, 1977.

[52] Chirino, F., Gallego, R., Saez, A. and Dominguez, J., Comparative study of three boundary element approaches to transient dynamic crack problems, *Engineering Analysis with Boundary Elements*, **13**, 11-19, 1994.

[53] Chu, S.J. and Hong, C.S., Application of the J_k integral to mixed mode crack problems for anisotropic composite laminates, *Engineering Fracture Mechanics*, **8**, 49-58, 1972.

[54] Cisilino, A.P. and Aliabadi, M.H., Three-dimensional BEM analysis for fatigue crack growth in welded components, *International Journal for Pressure Vessel and Piping,* **70**, 135-144, 1997.

[55] Cisilino, A.P., Aliabadi, M.H. and Otegui, J.L., A three-dimensional boundary element formulation for the elastoplastic analysis of cracked bodies, *International.Journal for Numerical Methods in Engineering,* **42**, 237-256, 1998.

[56] Cisilino, A.P., Aliabadi, M.H. and Otegui, J.L., Energy domain integral applied to solve center and double-edge crack problems in three-dimensions, *Theoretical and Applied Fracture Mechanics,* **29**, 181-194, 1998.

[57] Cisilino, A.P. and Aliabadi, M.H., Three-dimensional boundary element analysis of fatigue crack growth in linear and non-linear fracture problems, *Engineering Fracture Mechanics,* **63**, 713-733, 1999.

[58] Cisilino, A.P. and Aliabadi, M.H., BEM implementation fo the energy domain integral for the elastoplastic analysis of 3D fracture problems, *International Journal of Fracture,* **96**, 229-245, 1999.

[59] Civelek, M.B. and Erdogan, F., Crack problems for a rectangular sheet and an infinite strip, *International Journal of Fracture,* **19**, 139-159, 1982.

[60] Corradi, S, Aliabadi, M.H. and Marchetti, M., A variable stiffness dual boundary element method for mixed-mode elastoplastic crack problems, *Theoretical . & Applied Fracture Mechanics,* **25**,43-49, 1996.

[61] Crouch, S.L., Solution of plane elasticity problems by the displacement discontinuity method, *International.Journal for Numerical Methods in Engineering,* **10**,301-343, 1976.

[62] Crouch, S.L. and Starfield, A.M., *Boundary Element Methods in Solid Mechanics.* George Allen and Unwin Publishers, London 1983.

[63] Cruse, T.A. and Van Buren, W., Three dimensional elastic stress analysis of a fracture specimen with an edge crack. *International Journal for Fracture Mechanics,* **7**, 1-15, 1971.

[64] Cruse, T.A., Numerical evaluation of elastic stress intensity factors by the boundary-integral equation method. *Surface Cracks: Physical Problems and Computational Solutions,* Ed. by J.L.Swedlow, pp153-170, American Society of Mechanical Engineers, New York 1972.

[65] Cruse, T.A., Boundary-integral equation method for three-dimensional elastic fracture mechanics. AFSOR-TR-0813 1975.

[66] Cruse, T.A. and Wilson, T.A., Boundary-integral equation method for elastic fracture mechanics analysis. AFOSR-TR-780355, Pratt and Whitney Aircraft Group 1977.

[67] Cruse, T.A. and Meyers, G.J., Three-dimensional fracture mechanics analysis, *Journal of Structural Division, American Society of Civil Engineers,* **103**, 309-320, 1977.

[68] Cruse, T.A. and Besuner, P.M., Residual life calculation for surface cracks in complex structural details. *Journal of Aircraft*, **12**, 369-375, 1979.

[69] Cruse, T.A. and Raveendra, S.T., A general procedure for fracture mechanics weight function evaluation based on the boundary element method. *Computational Mechanics*, **3**, 157-166, 1988.

[70] Das, S. A numerical method for determination of source time functions for general three-dimensional rupture propagation. *Geophysics Journal of Royal Astronautics Society*, **62**, 591–604, 1980.

[71] dell'Erba, D.N., Aliabadi, M.H. and Rooke, D.P., Dual boundary element method for three-dimensional thermoelastic crack problems, *International Journal of Fracture*, **94**, 89-101, 1998.

[72] dell'Erba, D.N. and Aliabadi, M.H., Three-dimensional thermo-mechanical fatigue crack growth using BEM, *International Journal of Fatigue*, **22**, 261-273, 2000.

[73] dell'Erba, D.N. and Aliabadi, M.H., On the solution of three-dimensional thermoelastic mixed-mode edge crack problems by the dual boundary element method, *Engineering Fracture Mechanics*, **66**, 269-285, 2000.

[74] dell'Erba, D.N. and Aliabadi, M.H.,. BEM analysis of fracture problems in three-dimensional thermoelasticity using J-integral, *International Journal of Solids and Structures*, **38**, 4609-4630, 2001.

[75] de Paula, F.A. and Aliabadi, M.H., Boundary element analysis of interface cracks in dissimilar orthotropic materials using a pth independent contour integrals, *Boundary Elements 19*, Computational Mechnaics Publications, Southampton, 359-366, 1997.

[76] Denda, M. and Dong, Y.F., A boundary element method for multiple crack problems based on micromechanical and complex variables. *Boundary Element Techniques IX*, Computational Mechanics Publications, Southampton, 227-234, 1994.

[77] Dirgantara, T. and Aliabadi, M.H., Crack growth analysis of plates loaded by bending and tension using dual boundary element method, *International Journal of Fracture*, **105**, 27-47, 2000.

[78] Dirgantara, T. and Aliabadi, M.H., Dual boundary element formulation for fracture analysis of shear deformable shells, *International Journal of Solids and Structures*, **38**, 7769-7800, 2001.

[79] Doblare, M., Espiga, F, Garcia, L. and Alcantud, M., Study of crack propagation in orthotropic materials by using the boundary element method. *Engineering Fracture Mechanics*, **37**, 953-967, 1990.

[80] Dowrick, G., Stress intensity factor in patched and orthogonally stiffened sheets. PhD Thesis, Southampton University, 1986.

[81] Dominguez, J. and Gallego, R., Time domain boundary element method for dynamic stress intensity factor computations. *International.Journal for Numerical Methods in Engineering,* **33**, 635-647, 1992.

[82] Dominguez, J., Ariza, M.P. and Gallego, R., Flux and traction boundary elements without hypersingular integrals, *International.Journal for Numerical Methods in Engineering,* **48,** 111-135, 2000.

[83] Erdogan, F and Sih, G.C., On the crack extension in plates under plane loading and transverse shear, *Journal of Basic Engineering,* **86**, 519-527, 1963.

[84] Fares, N. and Li, V.C., An indirect boundary element method for 2- finite/infinite regions with multiple displacement discontinuities. *Engineering Fracture Mechanics,* **26**, 127-141, 1987

[85] Fedelinski, P., Aliabadi, M.H and Rooke, D.P., The dual boundary element method for cracked structures subjected to inertial forces, *Bound. Elem.Abs.,* **4(4),**150-152, 1993.

[86] Fedelinski, P., Aliabadi, M.H. and Rooke, D.P., The dual boundary element method in dynamic fracture mechanics, *Engineering Analysis.,* **12**, 203-210, 1993.

[87] Fedelinski, P., Aliabadi, M.H. and Rooke, D.P., The dual boundary element method: J-integral for dynamic stress intensity factors, *International Journal of Fracture,* **65**, 369-381, 1994.

[88] Fedelinski, P., Aliabadi, M.H and Rooke, D.P., A single-region time domain BEM for dynamic crack problems, *International Journal of Solids and Structure,* **32**(24), 3555-3571, 1995.

[89] Fedelinski, P., Aliabadi, M.H. and Rooke, D.P., Boundary element formulations for the dynamic analysis of cracked structures, *Engineering Analysis* **17,**45-56, 1996.

[90] Fedelinski, P., Aliabadi, M.H and Rooke, D.P., The Laplace transform DBEM method for mixed-mode dynamic crack analysis, *Computers.& Structures,* **59,**1021-1031, 1996.

[91] Fedelinski, P., Aliabadi, M.H. and Rooke. D.P., A time-domain DBEM for rapidly growing cracks. *International.Journal for Numerical Methods in Engineering,* **40,**1555-1572, 1997.

[92] Forth, S.C. and Keat, W.D., Three-dimensional nonplanar fracture model using the surface integral method. *International Journal of Fracture,* **77,** 243-262, 1996.

[93] Gallego, R. and Dominguez, J., Dynamic crack propagation analysis by moving singular boundary elements. *Journal of Applied Mechanics,* ASME, **59**, 158-162, 1992.

[94] Gangming, L. and Yougyuan, Z., Improvement of singular element for crack problems in three-dimensional boundary element method. *Engineering Fracture Mechanics,* **31**, 993-999, 1988.

[95] Giumaraes, S. and Telles, J.C.F., On a critical appreciation of the hyper-singular boundary element formulation applied to fracture mechanics. *Boundary Element Technology VIII*, Computational Mechanics Publications, Southampton, 241-251, 1993.

[96] Gray, L.J. and Giles, G.E., Application of thin cavity method to shield calculations in electroplating. *Boundary Element X*, Vol. 2, Computational Mechanics Publications, Southampton, 441-452, 1988.

[97] Gray, L.J., Martha, L.F. and Ingraffea, A.R., Hypersingular integrals in boundary element fracture analysis. *International.Journal for Numerical Methods in Engineering*, **29**, 1135-1158, 1990.

[98] Grestle, W.H., Finite and boundary element modelling of crack propagation in two- and three-dimensions using interactive computer graphics. PhD Thesis, Cornell University, Ithaca, USA 1986.

[99] Griffith, A.A., The phenomena of rupture and flow in solids. *Phil. Trans.* **A221**, 163-198, 1921.

[100] Guaggiani, M. and Giggante, A., A general algorithm for multidimensional Cauchy principal value integrals in the boundary element method, *Journal of Applied Mechanics*, **57**, 906-915, 1990.

[101] Guiggiani, M., Krishnasamy, G., Rizzo, F.J. and Rudolphi, T.J., A general algorithm for the numerical solution of hypersingular boundary integral equations, *Journal of Applied Mechanics*, **59**, 604-614, 1992.

[102] Hartranft, R.J. and Sih, G.C., Stress singularity for a crack with an arbitrary curved crack front. *Engineering Fracture Mechanics*, **9**, 705-718, 1977.

[103] Heliot, J., Labbens, R.C. and Pellissier-Tannon., Semi-elliptical cracks in a cylinder subjected to stress gradients, *Fracture Mechanics*, STP 677 ,341-364, 1979.

[104] Henshell, R.D. and Shaw, K.G., Crack tip finite elements are unnecessary. *International.Journal for Numerical Methods in Engineering*, **9**,495-507, 1975.

[105] Huber, O., Nickel J. and Kuhn G., On the decomposition of the J-integral for 3D crack problems. *International Journal of .Fracture*, **64**, 339-348, 1993

[106] Hirose, S., Scattering from an elliptic crack by the time-domain boundary integral equation method. *Advances in Boundary Elements*, Vol. 3, , Computational Mechanics Publications, Southampton, 99-110, 1989.

[107] Hirose, S and Achenbach, J.D., Time-domain boundary element analysis of elastic wave interaction with a crack. *International.Journal for Numerical Methods in Engineering*, **28**, 629-644, 1991.

[108] Hirose, S. and Achenbach, J.D., Acoustic emission and near-tip elastodynamic fields of growing penny-shaped crack, *Engineering Fracture Mechanics*, **39**, 21-36, 1991..

[109] Hong, H. and Chen, J., Derivations of integral equations of elasticity. *Journal of Engineering Mechanics*, American Society of Civil Engineers, **114**, 1028-1044, 1988.

[110] Ingraffea, A.R., Blandford, G.E. and Ligget, J.A., Automatic modelling of mixed-mode fatigue and quasi-static crack propagation using the boundary element method. *Proc. Fracture Mechanics, 14th Symposium, ASTM STP 791*, 407-426, ASTM 1983.

[111] Irwin, G.R., Analysis of stresses and strains near the end of a crack transverseing a plate, ASME, *Journal of Applied Mechanics*, **24**, 361-364, 1957.

[112] Ishikawa, H., Kitagawa, H. and Okamura, H., J-integral of a mixed mode crack and its application. *Proc. 3rd Int. Conf on Mechanical Behaviour of Materials*, Pergamon Press, Oxford, vol3, 447-455, 1980.

[113] Jia,Z.H., Shippy, D.J. and Rizzo, F.J., On the computation of two-dimensional stress intensity factors using the boundary element method. *International.Journal for Numerical Methods in Engineering*, **26**, 2739-2753, 1988.

[114] Kamel, M. and Liaw, B.M., Boundary element formulation with special kernels for an anisotropic plate containing an elliptical hole or a crack. *Engineering Fracture Mechanics*, **39**, 695–711, 1991.

[115] Karami, G. and Fenner, R.T., Analysis of mixed mode fracture and crack closure using the boundary integral equation method. *International Journal of Fracture*, **30**, 13-29, 1986.

[116] Kishimoto, K., Soboyejo, W.O., Smith, R.A. and Knott, J.F., A numerical investigation of the interaction and coalescence of twin coplanar semi-elliptical fatigue cracks, *International Journal of Fatigue*, **11**, 91-96, 1989.

[117] Kishitani, K., Hirai, T and Murakami, K., J-integral calculations with boundary elements. *Boundary Elements*, Springer Verlag, Berlin, 481-493, 1983.

[118] Lee, K.Y. and Choi, H.J., Boundary element analysis of stress intensity factors for bimaterial cracks, *Engineering Fracture Mechanics*, **29**, 461-472, 1988.

[119] Leitão, V. and Aliabadi, M.H., Contour integrals for an elastoplastic boundary element method formulation, *International Journal of Fracture*, **64**, R97-R103, 1993.

[120] Leitão, V., Aliabadi, M.H. and Rooke, D.P., The dual boundary element formulation for elastoplastic fracture mechanics, *International.Journal for Numerical Methods in Engineering*. **38**, 315-333, 1995.

[121] Leitão, V, Aliabadi, M.H. and Rooke, D.P., Elastoplastic dual boundary elements: application to crack face contact, *Computers. & Structures*, **54**, 443-454, 1995.

[122] Leitão, V., Aliabadi, M.H. and Rooke, D.P., Elastoplastic simulation of crack growth: dual boundary element formulation, *International Journal of Fatigue*, **17**(5), 353-364, 1995.

[123] Leitão, V., Aliabadi, M.H. Rooke, D.P and Cook, R., Boundary element methods for the analysis of crack growth in the presence of residual stress fields, *Journal of Materials and Engineering Performance*, **7(3)**, 352-360, 1998.

[124] Leitão, V.M.A and Aliabadi, M.H., Boundary element methods for analysis of crack growth in nonlinear fracture, *International Journal of Materials and Product Technology*, **15**, *104-116, 2000*.

[125] Le Van, A. and Royer, J., Integral equations for three-dimensional problems. *International Journal of Fracture*, **31**,125-142, 1986.

[126] Liang, R.Y.K. and Li, Y.N., Simulation of nonlinear fracture process zone in cementatious material - a boundary element approach. *Computational Mechanics* **7**, 413-427, 1991.

[127] Liaw, B.M. and Kamel, M., A crack approaching a curvilinear hole in an anisotropic plane under general loadings. *Engineering Fracture Mechanics*, **40**, 25-35, 1991.

[128] Liu, S.B. and Tan, C.L., Two-dimensional boundary element contact mechanics analysis of angled crack problems. *Engineering Fracture Mechanics*, **42**, 273-288, 1992.

[129] Luchi, M.L. and Rizzuti, S., Boundary elements for three-dimensional elastic crack analysis. *International.Journal for Numerical Methods in Engineering*, **24**, 2253-2271, 1987.

[130] Lutz, E.D., Ingraffea, A.R. and Gray, L.J. Use of simple solutions for boundary integral methods in elasticity and fracture analysis. *International.Journal for Numerical Methods in Engineering*, **35**, 1737-1751, 1992.

[131] Man, K, Aliabadi, M.H. and Rooke, D.P., Stress intensity factors in the presence of contact stresses, *Engineering Fracture Mechanics*, **51**(4),591-601, 1995.

[132] Martinez, M. and Aliabadi, M.H., Fracture mechanics of bone using the dual boundary element method. *Simulation and Modelling in Bioengineering*, Computational Mechanics Publications, Southampton, 175-188, 1996.

[133] Mews, M., Calculation of stress intensity factors for various crack problems with the boundary element method, *Proc.9th Int. Conf. on BEM*, Vol2, 259-278, 1987.

[134] Mi, Y. and Aliabadi, M.H., Dual boundary element method for three-dimensional fracture mechanics analysis, *Engineering.Analysis*, **10**(2),161-171, 1992.

[135] Mi, Y. and Aliabadi, M.H., Three-dimensional crack growth simulation using BEM, *Computers and Structures.*, **52**, 871-878, 1994.

[136] Mi, Y and Aliabadi, M.H., Discontinuous Crack-tip elements: application to 3D boundary element method, *International Journal of Fracture*, **67**, R67-R71, 1994.

[137] Mi, Y. and Aliabadi, M.H., An Automatic procedure for mixed-mode crack growth analysis, *Communications in Numerical. Methods*, **11**,167-177, 1995.

[138] Mir-Mohammad-Sadegh, A. and Altiero, N.J., Solution of the problem of a crack in a finite plane region, using an indirect boundary-integral method. *Engineering Fracture Mechanics*, **11**, 831-837, 1979.

[139] Morjaria,M. and Mukherjee,S., Numerical analysis of planar, time dependent inelastic deformation of plates with cracks by the boundary element method, *International Journal of Solids and Structures*, **17**, 127-1, 1981.

[140] Murakami, Y., *Stress intensity factors handbook*, Pergamon 1987.

[141] Miyazaki, N., Ikeda, T., Soda, T., Munakata, T., Stress intensity factor analysis of interface crack using boundary element method: application of contour integral method, *Engineering Fracture Mechanics*, **45**, 599-610, 1993.

[142] Nishimura, N., Guo, Q.C. and Kobayashi, S., Boundary integral equation methods in elastodynamic crack problems. *Boundary Elements IX*, Vol. 2, Computational Mechanics Publications, Southampton, 279-291, 1987.

[143] Nishimura, N., Guo, Q.C. and Kobayashi, S. Elastodynamic crack analysis by BIEM. *Boundary Elements in Applied Mechanics*, Pergamon Press, Oxford, 245-254, 1988.

[144] Nishioka, T., Recent developments in computational dynamic fracture mechanics, In *Dynamic Fracture Mechanics*, Computational Mechanics Publication, Southampton 1995.

[145] Nikishkov G.P. and Atluri, S.N., An equivalent domain integral method for computing crack tip integral parameters in non-elastic, thermomechanical fracture. *Engineering Fracture Mechanics*, **26**, 851-867, 1987.

[146] Nisitani, H., Two-dimensional stress problem solved using electric digital computer. *Bulletin of Japanese Society of Mechanical Engineers*, **11**,14-23, 1968.

[147] Nisitani, H., Solutions of notch problems by body force method, *Stress analysis of Notched Problems*, Noordhoff, Leyden, 1-68, 1978.

[148] Nisitani, H., *Computational and Experimental Fracture Mechanics*. Computational Mechanics Publications, Southampton 1994.

[149] Papamichel, N. and Symm, G.T., Numerical techniques for two-dimensional Laplacian problems. *Computer Methods in Applied Mechanical Engineering*, **6**,175-194, 1975.

[150] Pan, E. and Amadei, B., Fracture mechanics of cracked 2-D anisotropic media with a new formulation of the boundary element method. *International Journal of Fracture*, **77**, 161-174, 1996.

[151] Pan, E. and Yuan, F.G., Boundary element analysis of three-dimensional cracks in anisotorpic solids, *International.Journal for Numerical Methods in Engineering*, **48**, 211-237, 2000.

[152] Paris, P.C., The fracture mechanics approach to fatigue, In fatigue, an interdisciplinary approach, 107-132, 1964.

[153] Paris, P.C., McMeeking,R.M. and Tada,H., The weight function method for determining stress intensity factors: In *Cracks and Fracture*, STP 601, American Society for Testing of Materials, 471-489, 1976.

[154] Pekau, O.A. and Batta, V., Seismic crack propagation analysis of concrete structures using boundary elements. *International.Journal for Numerical Methods in Engineering,* **35**,1547-1564, 1992.

[155] Polyzos, D., Stamos, A.A and Beskos, D.E., BEM computation of DSIF in cracked viscoelastic plates. *Communications in Numerical Methods in Engineering,* **10**,81-87, 1994.

[156] Prassad, N.N.V. and Aliabadi, M.H. and Rooke, D.P., Incremental crack growth in thermoelastic problems, *International Journal of Fracture,* **66**, R45-R50, 1994.

[157] Prasad, N.N.V., Aliabadi, M.H and Rooke, D.P., The dual boundary element method for thermoelastic crack problems, *International Journal of Fracture,* **66,** 255-272, 1994.

[158] Prasad, N.N.V., Aliabadi, M.H and Rooke, D.P., Effect of thermal singularities on stress intensity factors. *Theoretical & Applied Fracture Mechanics,* **24** 203-215, 1996.

[159] Prasad, N.N.V. and Aliabadi, M.H., Thermo-mechanical fatigue crack growth, *International Journal of Fatigue,* **18,** 349-361, 1996.

[160] Prasad, N.N.V., Aliabadi, M.H and Rooke, D.P., The dual boundary element method for transient thermoelastic crack problems, *International Journal of Solids & Structures.,* **33**, 2695-2718, 1996.

[161] Proenca, S.P.B. and Venturini, W.S., Application of damage mechanics models to reinforced concrete structures, In *Continuous Damage and Fracture,* Elsevier, Paris, 233-242, 2000.

[162] Portela, A., Aliabadi, M.H. and Rooke, D.P., Dual boundary element analysis of pin- loaded lugs, *Boundary Element Technology VI, 381-392, 1991.*

[163] Portela, A., Aliabadi, M.H. and Rooke, D.P., Efficient boundary element analysis of sharp notched plates, *International.Journal for Numerical Methods in Engineering,* **32**, 445-470, 1991

[164] Portela,A., and Aliabadi,M.H., and Rooke,D.P., The dual boundary element method : efficient implementation for cracked problems, *International.Journal for Numerical Methods in Engineering,* **33**, 1269- 1287, 1992.

[165] Portela, A., Aliabadi, M.H. and Rooke, D.P., Dual boundary analysis of cracked plates: singularity subtraction technique, *International Journal of Fracture,* **55**,17-28, 1992.

[166] Portela, A., Aliabadi, M.H. and Rooke, D.P., Dual boundary element incremental analysis of crack propagation, *Computers & Structures*, **46**, 237-247, 1993.

[167] Rice, J.R., Some remarks on elastic crack-tip stress fields, *International Journal of Solids and Structures*, **8**, 529-546, 1970.

[168] Rigby, R. and Aliabadi, M.H., Mixed mode J-integral method for analysis of 3D fracture problems using BEM, *Engineering.Analaysis*, **11**(3), 239-256, 1993.

[169] Rigby, R. and Aliabadi, M.H., Stress intensity factors for cracks at attachment lugs, *Engineering Failure Analysis*, **4**, 133-146, 1997.

[170] Rigby, R.H. and Aliabadi, M.H., Decomposition of the mixed-mode J-integral-revisited. *International Journal of Solids and Strucutres*, **35**, 2073-2099, 1998.

[171] Rooke, D.P. and Cartwright, D.J., *Compendium of stress intesity factors*, HMSO 1976.

[172] Rooke, D.P., Rayaprolu, D and Aliabadi, M.H., Crack-line and edge Green's function for stress intensity factors of inclined edge cracks, *Fatigue & Fracture of Engineering Materials & Structures*, **15**, 441-461, 1992.

[173] Rooke, D.P. and Aliabadi, M.H., Weight function for crack problems using boundary element analysis, *International Journal of Engneering Sciences*, **32**,155-166, 1994.

[174] Rots, J.G. and Blaauwendraad, J., Crack models for concrete: discrete or smeared? fixed, multi-directional or rotating?, HERON, **34**, 1-59, 1989.

[175] Rudolphi, T.J., Krishnasamy, G., Schmerr, L.W. and Rizzo, F.J., On the use of strongly singular integral equations for crack problems. *Boundary Element X*, Vol. 3, Computational Mechanics Publications, 249-264, 1988.

[176] Saleh, A.L. and Aliabadi, M.H., Crack growth analysis in concrete using boundary element method, *Engineering Fracture Mechanics*, **51**(4), 533-545, 1995.

[177] Saleh,A.L. and Aliabadi,M.H., Boundary element analysis of the pull-out behaviour of an anchor bolt embedded in concrete, *Mechanics of Cohesive-frictional Materials* **1**, 235-250, 1996.

[178] Saleh, A.L. and Aliabadi, M.H., Crack growth analysis in reinforced concrete using BEM, *Journal of Engineering Mechanics, ASCE*, **124**, 949-958, 1998.

[179] Salgado, N.K. and Aliabadi, M.H., Stress intensity factors for a crack near broken/intact stiffener. *International Journal of Fracture*,**74**,R71-74, 1996.

[180] Salgado, N.K and Aliabadi, M.H., The application of the dual boundary element method to the analysis of cracked stiffened panels, *Engineering Fracture Mechanics*, **54**, 91-105, 1996.

[181] Salgado, N.K. and Aliabadi, M.H., The analysis of mechanically fastened repairs and lap joints, *Fatigue & Fracture of Engineering Materials & Structures*, **20**, 583-593, 1997.

[182] Salgado, N.K. and Aliabadi, M.H., The boundary element analysis of cracked stiffened sheets, reinforced by adhesively bonded patches. *International.Journal for Numerical Methods in Engineering*, **42**, 195-217, 1998.

[183] Salgado, N.K. and Aliabadi, M.H**.,** Boundary element analysis of fatigue crack propagation in stiffened panels, *AIAA Journal of Aircraft*, **35**(1), 122-130, 1998.

[184] Selvadurai, A.P.S. and Au, M.C., Boundary element modelling of dilatant interface. *Boundary Element VIII*, Springer Verlag, Berlin, 735-749, 1986.

[185] Selvadurai, A.P.S., Mechanical of conoidal matrix crack development at the ends of reinforcing cylindrical fibers. *Localized Damage II*, Vol. 2, Computational Mechanics Publications, Southampton,177-190, 1992.

[186] Selvadurai, A.P.S., Matrix crack extension at fractionally constrained fiber. *Journal of Engineering Materials Technology*, **116**, 398-402, 1994.

[187] Sih, G.C., Strain energy density factor applied to mixed mode crack problems, *International Journal of Fracture,* **10**, 305-321, 1974.

[188] Shih, C.F. and Moran, B., Needleman, A., Energy release rate along a three-dimensional crack front in a thermally stressed body, *International Journal of Fracture*, **30,** 79-102, 1986.

[189] Shilko, S.V. and Shcherbakov, S.V., Boundary element method in modelling failure of compressed metal-polymeric adhesive joints, *Mechanical. Behaviour of Adhesive Joints*, 339-250, 1987.

[190] Shivakumar K.N. and Raju I.S., An equivalent domain integral for three-dimensional mixed mode fracture problems. NASA CR-182021, 1990.

[191] Sladek, V. and Sladek, J., Three dimensional crack analysis for an anisotropic body, *Applied Mathematical Modelling*, **6,** 374-382, 1982.

[192] Sladek, J. and Sladek, V., Dynamic stress intensity factors studied by boundary integro-differential equations. *International.Journal for Numerical Methods in Engineering*, **23**, 339-345, 1991.

[193] Sladek, J. and Sladek, V., Boundary element analysis for an interface crack between dissimilar elastoplastic materials. *Computational Mechanics*, **16**, 396-405, 1995.

[194] Sladek, J. and Sladek, V., Nonsingular traction BIEs for crack problems in elastodynamics, *Computational Mechanics,* **25,** 590-599, 2000.

[195] Smith, R.N.L., From rags to riches ? - development in the BEM for the solution of elastic fracture problems. *Boundary Elements X*, Vol3, Computational Mechanics Publications, Southampton, 155-176, 1988.

[196] Smith, R.N.L., The solution of mixed-mode fracture problems using the boundary element method. *Engineering Analysis with Boundary Elements*, **5,** 75-80, 1988.

[197] Smith, R.N.L. and Aliabadi, M.H., Boundary integral equation methods for the solution of cracked problems. *Mathematical and .Computer Modelling* **15**, 285-293, 1991.

[198] Smith, R.N.L. and Della-Ventura, D., A superposition method for the solution of crack problems. *Communications in Numerical Methods in Engineering*, **11**, 243-254, 1995.

[199] Snyder, M.D. and Cruse, T.A., Boundary-integral analysis of anisotropic cracked plates. *International Journal of Fracture*, **11**,315-328, 1975.

[200] Soboyejo, W.O., Kishimoto, K., Smith, R.A. and Knott, J.F., A study of the interaction and coalesence of two coplanar fatigue cracks in bending, *Fatigue and Fract. of Engng Mat. and Structures,* **12,** 167-174, 1989.

[201] Soni, M.L. and Stern, M., On the computation of stress intensity factors in fiber composite media using a contour integral method. *International Journal of Fracture,* **12,** 331-344, 1976.

[202] Sollero, P. and Aliabadi, M.H., Fracture mechanics analysis of anisotropic composite laminates by the boundary element method, *International Journal of Fracture,* **64**, 269-284, 1994.

[203] Sollero, P., Aliabadi, M.H. and Rooke, D.P., Anisotropic analysis of cracks emanating from circular holes in composite laminates using the boundary element method, *Engineering Fracture Mechanics*, **49**,213-224, 1994.

[204] Sollero, P. and Aliabadi, M.H.**,** Anisotropic analysis of cracks in composite laminates using the dual boundary element method, *Composite & Structures*, **31**(3), 229-234, 1995.

[205] Tan, C.L. and Fenner, R.T., Three-dimensional stress analysis by the boundary integral equation method. *Journal of Strain Analysis,* **13**, 213-219, 1978.

[206] Tan,C.L. and Fenner,R.T., Elastic fracture mechanics analysis by the boundary integral equation method. *Proceedings of the Royal Society of London,* **A369**, 243-260, 1979.

[207] Tan, C.L. and Gao, Y.L., Treatment of bimaterial interface problems using the boundary element method, *Engineering Fracture Mechanics*, **36,** 919-932, 1990.

[208] Tan, C.L. and Gao, Y.L., Axisymmetric boundary integral equation analysis of interface cracks between dissimilar materials. *Computational Mechanics,* **7,** 381-396, 1991.

[209] Tan, C.L. and Lee, K.H., Elastic-plastic stress analysis of a cracked thick-walled cylinder. *Journal of Strain Analysis*, 50-57, 1983.

[210] Tan, C.L. and Gao, Y.L., Boundary Integral equation fracture mechanics analysis of plane orthotropic bodies. *International Journal of Fracture*, **53**, 343-365, 1992.

[211] Tanaka, M. Nakamura, M., Aoki, K. and Matsumoto, T., Computation of dynamic stress intensity factors using the boundary element method based on Laplace transformation and regularized boundary integral equation. *Japanese Society of Mechanical Engineers, International Journal, Series A,* **36,**252-258, 1993.

[212] Telles, J.C.F., Castor, G.S. and Guimaraes, S., A numerical Green's function approach for boundary elements applied to fracture mechanics. *International.Journal for Numerical Methods in Engineering,* **38,** 3259-3274, 1995.

[213] Venturini, W.S., A new boundary element formulation for crack analysis. *Boundary Element Methods XVI,* Computational Mechanics Publications, Southampton, 405-412, 1994.

[214] Watson, J.O., Hermitian cubic and singular elements for plane strain. *Developments in Boundary Element Methods 4,* Elsevier Applied Science Publishers, Barking,1-28, 1986.

[215] Watson, J.O., Singular boundary elements for the analysis of cracks in plane strain. *International.Journal for Numerical Methods in Engineering* **38,** 2389-2412, 1995.

[216] Weaver, J., Three-dimensional crack analysis, *International Journal of Solids and Structures,* **13,** 321-330, 1977.

[217] Wen, P.H, Aliabadi, M.H and Rooke, D.P., An indirect boundary element method for three-dimensional dynamic problems, *Engineering Analysis,* **16,**351-362, 1995.

[218] Wen, P.H., Aliabadi, M.H. and Rooke, D.P., A contour integral for the evaluation of stress intensity factors. *Applied Mathematical Modelling,* **19,** 450-455 1995.

[219] Wen, P.H., Aliabadi, M.H. and Rooke, D.P., The influence of waves on dynamic stress intensity factors (Two Dimensional Problems), *Archives of Applied Mechanics,* **66,**326-355, 1996.

[220] Wen, P.H., Aliabadi, M.H. and Rooke, D.P., The influence of waves on dynamic stress intensity factors (Three Dimensional Problems), *Archives of Applied Mechanics ,***66,** 385-394, 1996.

[221] Wen, P.H. and Aliabadi, M.H., Application of the weight function method to elastodynamic fracture mechanics, *International Journal of Fracture,* **76,** 193-206, 1996.

[222] Wen, P.H, Aliabadi, M.H. and Rooke, D.P., A contour integral method for dynamic stress intensity factors, *Theoretical and Applied Fracture Mechanics,* **27,** 29-41, 1997.

[223] Wen, P.H., Aliabadi, M.H, and Rooke, D.P., A contour integral for three-dimensional crack elastostatic analysis, *Engineering Analysis* **20,** 101-111, 1997.

[224] Wen, P.H., Aliabadi, M.H. and Rooke, D.P., Mixed-mode weight functions in three-dimensional fracture mechanics: Dynamic, *Engineering Fracture Mechanics*, **59**, 577-587, 1998.

[225] Wen, P.H., Aliabadi, M.H. and Rooke, D.P., Mixed-mode weight functions in three-dimensional fracture mechnaics: Static, *Engineering Fracture Mechanics*, **59**, 563-575, 1998.

[226] Wen, P.H., Aliabadi, M.H. and Rooke, D.P., A variational technique for boundary element analysis of 3D fracture mechanics weight functions: Static, *International.Journal for Numerical Methods in Engineering*, **42**, 1409-1423, 1998.

[227] Wen, P.H., Aliabadi, M.H. and Rooke, D.P., A variational technique for boundary element analysis of 3D fracture mechanics weight functions: Dynamic, *International.Journal for Numerical Methods in Engineering*, **42**, 1425-1439, 1998.

[228] Wen, P.H., Aliabadi, M.H. and Rooke, D.P., Cracks in three dimensions: a dynamic dual boundary element analysis, *Computer Methods in Applied Mechanics and Engineering*, **167**, 139-151, 1998.

[229] Wen,P.H., Aliabadi,M.H. and Rooke,D.P., Three-dimensional dynamic fracture analysis with the dual reciprocity method in Laplace domain, *Engineering Analysis with Boundary Elements*, **23**, 51-58, 1999.

[230] Wen, P.H., Aliabadi, M.H. and Young, A., A time-dependent fromulation of dual boundary element method for 3D dynamic crack problems, *International.Journal for Numerical Methods in Engineering*, **45**, 1887-1905, 1999.

[231] Wen, P.H., Aliabadi, M.H. and Rooke, D.P., A mass-matrix formulation for three-dimensional dynamics fracture mechanics, *Computer Methods in Applied Mechanics and Engineering*, **173**, 365-374, 1999.

[232] Wen, P.H., Aliabadi, M.H. and Young, A., Dual boundary element methods for three-dimensional dynamic crack problems., *Journal of Strain Analysis*, **34**, 373-394, 1999.

[233] Wen, P.H. and Aliabadi, M.H., Evaluation of stress intensity factors for 3D interfacial cracks by boundary element method, *Journal of Strain Analysis*, **34**, 209-215, 1999.

[234] Wilde, A.J., Aliabadi, M.H. and Power, H., Application of a $C^{(0)}$ continuous element to the development of hypersingular integrals, *Communications in Boundary Elements*, **7**(3), 109-114, 1996.

[235] Wilde, A. and Aliabadi, M.H., Dual boundary element method for geomechanics problems using continuous elements. *Boundary Elements XVIII*, Computational Mechanics Publications, Southampton, 449-464, 1996.

[236] Wilde, A. and Aliabadi, M.H., Boundary element analysis of geomechanical fracture, *International Journal for Numer. & Analytical Methods in Geomechanics*, *23, 1195-1214, 1999.*

[237] Wilde, A. and Aliabadi, M.H., A 3-D Dual BEM formulation for the analysis of crack growth, *Computational Mechanics*, **23**, 250-257, 1999.

[238] Xiao, F. and Hui, C.Y., A boundary element method for calculating the K fields for cracks along a bimaterial interface, *Computational Mechanics,* **15,** 58-78, 1994.

[239] Xanthis, L.S., Bernal, M.J.M and Atkinson, C., The treatment of singularities in the calculation of stress intensity factors using the integral equation method. *Applied Mechanics for Engineers,* **26**, 285-304, 1981

[240] Yan, A.M. and Nguyen-Dang, H., Multiple cracked fatigue crack growth by BEM. *Computational Mechanics,* **16**, 273-280, 1995.

[241] Young, A., Cartwright, D.J. and Rooke, D.P., The boundary element method for analysing repair patches of cracked finite sheets. *Aeronautical Journal,* **92**, 416-421, 1988.

[242] Young, A., A single-domain boundary element method for 3-D elastostatics crack analysis using continuous elements. *International.Journal for Numerical Methods in Engineering,* **39**, 1265-1294, 1996.

[243] Yuuki, R. and Cho, S.B., Efficient boundary element analysis of stress intensity factors for interface cracks in dissimilar materials, *Engineering Fracture Mechanics,* **34,** 179-188, 1989.

[244] Zamani, N. and Sun, W., A direct method for calculating the stress intensity factor in BEM. *Engineering Analysis with Boundary Elements,* **11,** 285-292, 1993.

[245] Zeng, Z.J. and Dai, S.H., Line spring boundary element method for a surface cracked plate. *Engineering Fracture Mechanics,* **36**, 853-858, 1990.

[246] Zhang, C.H. and Achenbach, J.D., Time-domain boundary element analysis of dynamic near-tip fields for impact-loaded collinear cracks. *Engineering Fracture Mechanics,* **32**, 899-909, 1989.

[247] Zhang, C and Gross, D., A non-hypersingular time-domain BIEM for 3-D transient elastodynamic crack analysis. *International.Journal for Numerical Methods in Engineering,* **36,** 2997-3017, 1993.

Chapter 9

Sensitivity Analysis and Shape Optimization

9.1 Introduction

Boundary element methods can be used to determine the shape of components which are required to conform to certain structural criteria. Generally, shape optimization involves an iterative process, starting with the analysis of an initial shape and proceeding via several improving shapes before converging to the optimum shape, as defined by the prescribed structural criteria. During the iterative process, design sensitivities are evaluated for each change in design. Design sensitivities measure the sensitivity of the design change in term, of one of the design parameters; it is mathematically represented as the derivative of an objective function with respect to the design parameter. The computation of the design sensitivities generally forms a major part of the shape optimization procedure.

The boundary element method is well suited to shape optimization problems because of its inherent accuracy in evaluating displacements and stresses, as well as its robustness in dealing with changes in the geometry of the structure during the optimization process. There are three main approaches in computing design sensitivities: finite differences, material derivative and implicit differentiation. In the first approach, the finite difference approximation is used in the simplest method of computing the sensitivity values. However, it is a relatively crude approximation, and hence not generally suitable for an effective shape optimization method. In the second approach, material derivatives of the variational state equations are used together with the adjoint variable technique. The adjoint variable method employs an adjoint system to obtain an explicit sensitivity expression in terms of design variables. The adjoint solution generally corresponds to concentrated force solutions which are difficult to implement in BEM analysis as they lead to unbounded integrals. Often the concentrated forces are replaced with statistically equivalent distributed tractions. Applications of the material derivative method can be found, for example, in the works of Burczynski and Fedelinski [5], Choi [8], Long et al. [21] and Meric [30]. In the third approach, design sensitivities are obtained directly by differentiation of the discretized boundary integral equation. Applications of implicit differentiation can be found, for example, in Aliabadi and Mellings [1], Barone and Yang [3], Choi and

Kwak [9], Erman and Fenner [13], Kane *et al.* [16]. Other contributions can be found [5, 19, 31, 32, 33, 22, 37].

Another important application of the optimization and sensitivity analysis is in inverse analysis [7, 17, 18, 26, 36, 38]. BEM is an ideal technique to use for shape identification and flaw location, since it is purely concerned with boundary responses [28, 29, 22, 39]. Also, modification of the flaw boundary affects only the discretization of the crack boundary, allowing faster mesh updates and the reuse of large parts of the previous system matrices.

In this chapter, boundary element sensitivity formulations for potential and elasticity, based on the Implicit Differentiation Method (IDM), are described. Applications to strucutral optimization, shape identification and bone remodelling are presented. The dual boundary element formulation is also presented for flaw identification problems.

9.2   Shape Optimization Methods

In shape optimization methods, it is necessary to minimize an object function $F(\mathbf{Z})$ subjected to some constraint function $\chi_j(\mathbf{Z}) \leq h_j$, $j = 1, M$, which contains all M constraints to be used in the minimization [14, 15, 42].

In most practical problems, the objective function $F(\mathbf{Z})$ to be optimized is highly non-linear. A general algorithm for non-linear optimization starts by evaluating the objective function for an initial design variable vector Z^o. Using an optimization algorithm, a search direction d^o is computed. The unconstrained optimization method is treated iteratively and general update to the design variable is given by

$$\mathbf{Z}^{k+1} = \mathbf{Z}^k + \alpha^k \mathbf{d}^k \tag{9.1}$$

where the line step parameter α^k is a scalar multiplier determining the amount of change in $\mathbf{Z}$ to find the minimum design point along the search direction $\mathbf{d}^k$.

The above process is repeated until the error function converges, i.e.

$$F(\mathbf{Z}^{k+1}) - F(\mathbf{Z}^k) \leq \varepsilon \tag{9.2}$$

where ε is a predefined tolerance.

Quasi-Newton methods and, in particular, BFGS update are generally recommended as the most reliables technique for obtaining the search direction. It can be defined as

$$\mathbf{d}^k = -\mathbf{H}^k \nabla F(\mathbf{Z}^k)$$

where $\mathbf{H}^k$ is the kth approximation to the inverse Hessian matrix, and is given by

$$\mathbf{H}^{k+1} = \left[\mathbf{I} - \frac{\mathbf{P}^k \mathbf{q}^k}{(\mathbf{P}^k)^T \mathbf{q}^k}\right] \mathbf{H}_k \left[\mathbf{I} - \frac{\mathbf{q}^k (\mathbf{P}^k)^T}{(\mathbf{P}^k)^T \mathbf{q}^K}\right] + \frac{\mathbf{P}^k (\mathbf{q}^k)^T}{(\mathbf{P}^k)^T \mathbf{q}^k} \tag{9.3}$$

where $\mathbf{I}$ denotes an identity matrix and $()^T$ denotes a transpose with

$$\mathbf{P}^k = \mathbf{Z}^{k+1} - \mathbf{Z}^k$$

and

$$\mathbf{q}^k = \nabla F(\mathbf{Z}^{k+1}) - \nabla F(\mathbf{Z}^k)$$

Generally, the initial value $\mathbf{H}^o$ is taken to be the identity matrix, i.e. $\mathbf{H}^o = \mathbf{I}$.

It is reported [27, 37] that scaling the constraints and making them the same order improves the optimization. Scaling the objective function to order unity, in the regions of interest, has also been recommended [37].

9.3 Design Sensitivity Analysis

The method presented in this chapter involves the direct differentiation of the boundary integral equation with respect to the design variable $\mathbf{Z}_m$. Formulations presented include potential and elasticity problems.

9.3.1 Derivative Potential Formulation

In this section, a formulation based on the derivative of the potential boundary integral equation with respect to an arbitrary design parameter is first presented, followed by discretization and numerical implementation. Consider an arbitrary design variable Z_m the variation of which influences the response or shape of the structural. Differentiating the potential integral equation with respect to the design variable gives

$$C_{,m}(\mathbf{x}')\theta(\mathbf{x}') + C(\mathbf{x}')\theta_{,m}(\mathbf{x}')$$

$$+ \int_S \left[q_{,m}^*(\mathbf{x}',\mathbf{x})\theta(\mathbf{x}) + q^*(\mathbf{x}',\mathbf{x})\theta_{,m} \right] dS + \int_S q^*(\mathbf{x}',\mathbf{x})\theta(\mathbf{x})(dS)_{,m}$$

$$= \int_S \left[\theta_{,m}(\mathbf{x}',\mathbf{x})q(\mathbf{x}) + \theta(\mathbf{x}',\mathbf{x})q_{,m}^* \right] dS + \int_S \theta(\mathbf{x}',\mathbf{x})q(\mathbf{x})(dS)_{,m} \qquad (9.4)$$

where $()_{,m}$ denotes the derivative with respect to Z_m. The discretized form of the derivative potential equation (9.4) can be written as

$$C_{,m}(\mathbf{x}')\theta(\mathbf{x}') + C(\mathbf{x}')\theta_{,m}(\mathbf{x}') + \sum_{n=1}^{N_e}\sum_{\alpha=1}^{}[\theta^{n\alpha}V_{,m}^{n\alpha} + \theta_{,m}^{n\alpha}V^{n\alpha}]$$

$$= \sum_{n=1}^{N_e}\sum_{\alpha=1}^{}[q^{n\alpha}W_{,m}^{n\alpha} + q_{,m}^{n\alpha}W^{n\alpha}] \qquad (9.5)$$

where N_e denote the total number of elements, α the local node number and

$$V^{n\alpha} = \int_{-1}^{1}\int_{-1}^{1} q^*(\mathbf{x}^c,\mathbf{x}(\eta_1,\eta_2))N_\alpha(\eta_1,\eta_2)J^n(\eta_1,\eta_2)d\eta_1 d\eta_2$$

$$V_{,m}^{n\alpha} = \int_{-1}^{1}\int_{-1}^{1} q_{,m}^*(\mathbf{x}^c,\mathbf{x}(\eta_1,\eta_2))N_\alpha(\eta_1,\eta_2)J^n(\eta_1,\eta_2)d\eta_1 d\eta_2$$

$$+ \int_{-1}^{1}\int_{-1}^{1} q^*(\mathbf{x}^c,\mathbf{x}(\eta_1,\eta_2))N_\alpha(\eta_1,\eta_2)J_{,m}^n(\eta_1,\eta_2)d\eta_1 d\eta_2$$

$$W^{n\alpha} = \int_{-1}^{1}\int_{-1}^{1} \theta^*(\mathbf{x}^c,\mathbf{x}(\eta_1,\eta_2))N_\alpha(\eta_1,\eta_2)J^n(\eta_1,\eta_2)d\eta_1 d\eta_2$$

$$W_{,m}^{n\alpha} = \int_{-1}^{1}\int_{-1}^{1} \theta_{,m}^*(\mathbf{x}^c,\mathbf{x}(\eta_1,\eta_2))N_\alpha(\eta_1,\eta_2)J^n(\eta_1,\eta_2)d\eta_1 d\eta_2$$

$$+ \int_{-1}^{1} \int_{-1}^{1} \theta^*(\mathbf{x}^c, \mathbf{x}(\eta_1, \eta_2)) N_\alpha(\eta_1, \eta_2) J_{,m}^n(\eta_1, \eta_2) d\eta_1 d\eta_2$$

for three-dimensional problems. Similarly,

$$V^{n\alpha} = \int_{-1}^{1} q^*(\mathbf{x}^c, \mathbf{x}(\eta)) N_\alpha(\eta) J^n(\eta) d\eta$$

$$V_{,m}^{n\alpha} = \int_{-1}^{1} q_{,m}^*(\mathbf{x}^c, \mathbf{x}(\eta)) N_\alpha(\eta) J^n(\eta) d\eta + \int_{-1}^{1} q^*(\mathbf{x}^c, \mathbf{x}(\eta)) N_\alpha(\eta) J_{,m}^n(\eta) d\eta$$

$$W^{n\alpha} = \int_{-1}^{1} \theta^*(\mathbf{x}^c, \mathbf{x}(\eta)) N_\alpha(\eta) J^n(\eta) d\eta$$

$$W_{,m}^{n\alpha} = \int_{-1}^{1} \theta_{,m}^*(\mathbf{x}^c, \mathbf{x}(\eta)) N_\alpha(\eta) J^n(\eta) d\eta + \int_{-1}^{1} \theta^*(\mathbf{x}^c, \mathbf{x}(\eta)) N_\alpha(\eta) J_{,m}^n(\eta) d\eta$$

for two-dimensional problems. In the above equations, the fundamental solutions and their derivatives are given by

$$\theta^*(\mathbf{x}', \mathbf{x}) = \frac{1}{4\pi r}$$

$$\theta_{,m}^*(\mathbf{x}, \mathbf{x}') = -\frac{1}{4\pi r^2} r_{,m}$$

$$q^*(\mathbf{x}', \mathbf{x}) = -\frac{1}{4\pi r^2} \frac{\partial r}{\partial n}$$

$$q_{,m}^*(\mathbf{x}, \mathbf{x}') = \frac{1}{4\pi r^3} \left[3 \frac{\partial r}{\partial n} r_{,m} - (n_k r_k)_{,m} \right] \tag{9.6}$$

for three-dimensional problems, and

$$\theta^*(\mathbf{x}', \mathbf{x}) = \frac{1}{2\pi} \ln(\frac{1}{r})$$

$$\theta_{,m}^*(\mathbf{x}, \mathbf{x}') = -\frac{1}{2\pi r} r_{,m}$$

$$q^*(\mathbf{x}', \mathbf{x}) = -\frac{1}{2\pi r} \frac{\partial r}{\partial n}$$

$$q_{,m}^*(\mathbf{x}, \mathbf{x}') = \frac{1}{2\pi r^2} \left[2 \frac{\partial r}{\partial n} r_{,m} - (n_k r_k)_{,m} \right] \tag{9.7}$$

for two-dimensional problems.

The above discretized equations form a system of equations, written as

$$\mathbf{H}_{,m}\boldsymbol{\theta} + \mathbf{H}\boldsymbol{\theta}_{,m} = \mathbf{G}_{,m}\mathbf{q} + \mathbf{G}\mathbf{q}_{,m}$$

The potential $\boldsymbol{\theta}$ and flux $\mathbf{q}$ values can be computed using the BEM as described earlier. Derivative boundary conditions can be found from the differentiation of the usual boundary conditions, prescribed earlier, since it is assumed that they are independent of the position. Hence the derivative boundary conditions are given by

$$\begin{aligned} \theta_{,m}(\mathbf{x}) &= 0 \quad \mathbf{x} \in S_\theta \\ q_{,m}(\mathbf{x}) &= 0 \quad \mathbf{x} \in S_q \end{aligned} \tag{9.8}$$

The derivative equations are used to provide a system of equations for each design variable. Use of the boundary conditions and the computed boundary potential and flux values allow the system to be written in matrix form as

$$\mathbf{A}\,\mathbf{X}_{,m} = \mathbf{F}_{,m} - \mathbf{A}_{,m}\mathbf{X} \tag{9.9}$$

where $\mathbf{X}_{,m}$ is the vector of unknown flux and potential derivatives, $(\mathbf{F}_{,m} - \mathbf{A}_{,m}\mathbf{X})$ is the vector formed by application of boundary conditions, and $\mathbf{A}$ is the matrix of coefficients identical to that used in the standard BEM analysis.

9.3.2 Derivative Displacement Formulation

In this section, a formulation based on the derivatives of the displacement boundary integral equation with respect to an arbitrary design parameter is first presented, followed by discretisation and numerical implementation. Consider an arbitrary design variable Z_m, the variation of which influences the response or shape of the structure. Differentiating the displacement integral equation with respect to the design variable gives

$$C_{ij,m}(\mathbf{x}')u_j(\mathbf{x}') + C_{ij}(\mathbf{x}')u_{j,m}(\mathbf{x}')$$

$$+ \int_S [T_{ij,m}(\mathbf{x}',\mathbf{x})u_j(\mathbf{x}) + T_{ij}(\mathbf{x}',\mathbf{x})u_{j,m}]\,dS + \int_S T_{ij}(\mathbf{x}',\mathbf{x})u_j(\mathbf{x})(dS)_{,m}$$

$$= \int_S [U_{ij,m}(\mathbf{x}',\mathbf{x})t_j(\mathbf{x}) + U_{ij}(\mathbf{x}',\mathbf{x})t_{j,m}]\,dS + \int_S U_{ij}(\mathbf{x}',\mathbf{x})t_j(\mathbf{x})(dS)_{,m} \tag{9.10}$$

The discretized form of the derivative displacement equation (9.10) gives

$$C_{ij,m}(\mathbf{x}')\,u_j(\mathbf{x}') + C_{ij}(\mathbf{x}')\,u_{j,m}(\mathbf{x}') + \sum_{\gamma=1}^{N_e}\sum_{\alpha=1} u_{j,m}^{n\alpha} P_{ij}^{n\alpha} + \sum_{\gamma=1}^{N_e}\sum_{\alpha=1} u_j^{n\alpha} P_{ij,m}^{n\alpha}$$

$$= \sum_{\gamma=1}^{N_e}\sum_{\alpha=1} t_{j,m}^{n\alpha} Q_{ij}^{n\alpha} + \sum_{\gamma=1}^{N_e}\sum_{\alpha=1} t_j^{n\alpha} Q_{ij,m}^{n\alpha} \tag{9.11}$$

where

$$P_{ij}^{n\alpha} = \int_{-1}^{+1}\int_{-1}^{+1} T_{ij}(\mathbf{x}',\mathbf{x}(\eta_1,\eta_2)) N_\alpha(\eta_1,\eta_2)\, J^n(\eta_1,\eta_2))\, d\eta_1 d\eta_2$$

$$P_{ij,m}^{n\alpha} = \int_{-1}^{+1}\int_{-1}^{+1} T_{ij,m}(\mathbf{x}',\mathbf{x}(\eta_1,\eta_2))\, N_\alpha(\eta_1,\eta_2)\, J^n(\eta_1,\eta_2)\, d\eta_1 d\eta_2$$

$$+ \int_{-1}^{+1}\int_{-1}^{+1} T_{ij}(\mathbf{x}',\mathbf{x}(\eta_1,\eta_2))\, N_\alpha(\eta_1,\eta_2) J^n_{,m}(\eta_1,\eta_2)\, d\eta_1 d\eta_2$$

$$Q_{ij}^{n\alpha} = \int_{-1}^{+1}\int_{-1}^{+1} U_{ij}(\mathbf{x}',\mathbf{x}(\eta_1,\eta_2)) N_\alpha(\eta_1,\eta_2)\, J^n(\eta_1,\eta_2))\, d\eta_1 d\eta_2$$

$$Q_{ij,m}^{n\alpha} = \int_{-1}^{+1}\int_{-1}^{+1} U_{ij,m}(\mathbf{x}', \mathbf{x}(\eta_1, \eta_2))\, N_\alpha(\eta_1, \eta_2)\, J^n(\eta_1, \eta_2)\, d\eta_1 d\eta_2$$

$$+ \int_{-1}^{+1}\int_{-1}^{+1} U_{ij}(\mathbf{x}', \mathbf{x}(\eta_1, \eta_2))\, N_\alpha(\eta_1, \eta_2) J_{,m}^n(\eta_1, \eta_2)\, d\eta_1 d\eta_2$$

for three-dimensional problems, and

$$P_{ij}^{n\alpha} = \int_{-1}^{+1} T_{ij}(\mathbf{x}', \mathbf{x}(\eta)) N_\alpha(\eta)\, J^n(\eta)\, d\eta$$

$$P_{ij,m}^{n\alpha} = \int_{-1}^{+1} T_{ij,m}(\mathbf{x}', \mathbf{x}(\eta))\, N_\alpha(\eta)\, J^n(\eta)\, d\eta + \int_{-1}^{+1} T_{ij}(\mathbf{x}', \mathbf{x}(\eta))\, N_\alpha(\eta) J_{,m}^n(\eta)\, d\eta$$

$$Q_{ij}^{n\alpha} = \int_{-1}^{+1} U_{ij}(\mathbf{x}', \mathbf{x}(\eta)) N_\alpha(\eta)\, J^n(\eta)\, d\eta$$

$$Q_{ij,m}^{n\alpha} = \int_{-1}^{+1} U_{ij,m}(\mathbf{x}', \mathbf{x}(\eta)\, N_\alpha(\eta)\, J^n(\eta)\, d\eta + \int_{-1}^{+1} U_{ij}(\mathbf{x}', \mathbf{x}(\eta))\, N_\alpha(\eta) J_{,m}^n(\eta)\, d\eta$$

Derivatives of the fundamental solutions are given for tow-dimensional problems as

$$U_{ij,m}(\mathbf{x}', \mathbf{x}) = \frac{1+\nu}{4\pi E(1-\nu)r}\left[(r_{i,m}r_{,j} + r_{,i}r_{j,m}) - ((3-4\nu)\delta_{ij} + 2r_{,i}r_{,j})r_{,m}\right]$$

$$T_{ij,m}(\mathbf{x}', \mathbf{x}) = \frac{-1}{4\pi(1-\nu)r^2}\left[2\frac{\partial r}{\partial n}\{r_{i,m}r_{,j} + r_{,i}r_{j,m} - ((1-2\nu)\delta_{ij} + 4r_{,i}r_{,j})r_{,m}\}\right.$$
$$+ (n_k r_k)_{,m}((1-2\nu)\delta_{ij} + 2r_{,i}r_{,j}) + (1-2\nu)(n_{i,m}r_j - n_{j,m}r_i$$
$$\left. + n_i r_{j,m} - n_j r_{i,m}) - 2(1-2\nu)(n_i r_{,j} - n_j r_{,i})r_{,m}\right]$$

where $r_{i,m} == \partial r_i/\partial z_m = x_{i,m}(x) - x_{i,m}(x')$ and $r_{,m} = r_{,i}r_{i,m}$. The terms $x_{i,m}$ represent the effects of a change in coordinate i at $\mathbf{x}$ due to a change in design variable z_m,

The above discretized equations form a system of equations, written as

$$\mathbf{H}_{,m}\mathbf{u} + \mathbf{H}\mathbf{u}_{,m} = \mathbf{G}_{,m}\mathbf{t} + \mathbf{G}\mathbf{t}_{,m}$$

The displacement $\mathbf{u}$ and traction $\mathbf{t}$ values can be computed using the usual BEM, as described earlier. Derivative boundary conditions can be found from the differentiation of the boundary conditions prescribed earlier, since it is assumed that they are independent of the position. Hence, the derivative boundary conditions are given by

$$\begin{aligned} u_{j,m}(\mathbf{x}) &= 0 \quad \mathbf{x} \in S_u \\ t_{j,m}(\mathbf{x}) &= 0 \quad \mathbf{x} \in S_t \end{aligned} \qquad (9.12)$$

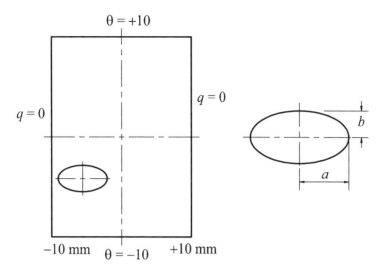

Figure 9.1: Plate with an internal cavity.

The derivative equations are used to provide a system of equations for each design variable. Use of the boundary conditions and the computed boundary displacement and traction values allow the system to be written in matrix form as

$$\mathbf{A}\,\mathbf{X}_{,m} = \mathbf{F}_{,m} - \mathbf{A}_{,m}\mathbf{X} \tag{9.13}$$

where $\mathbf{X}_{,m}$ is the vector of unknown traction and displacement derivatives, $(\mathbf{F}_{,m}\text{-}\mathbf{A}_{,m}\mathbf{X})$ is the vector formed by the application of boundary conditions, and $\mathbf{A}$ is the matrix of coefficients identical to that used in the standard BEM analysis.

9.4 Numerical Examples

In this section two examples are presented to demonstrate the use of both derivative potential and derivative displacement integral equations.

9.4.1 Sensitivities of a Square Plate with an Elliptical Hole

Consider a square plate with an elliptical cavity, as shown in Figure 9.1, subjected to potential and flux boundary conditions. The sensitivities for this problem were computed [27] for changes made to the x-coordinate of the centre of the hole, x_o. The mesh used consisted of eight quadratic elements on the external boundary and four on the elliptical hole. In Figure 9.2 potential derivatives, taken with respect to the $x-$coordinates, are presented for each node along the base of the plate. The results presented include those obtained using the implicit differentiation method described above, together with finite difference solutions for different steps. As can be seen from the figure, the forward difference results with a step size of 0.001, agree well with the implicit differentiation approach, however, as the step size is reduced to 0.0001, the forward difference results become unstable with rather high errors.

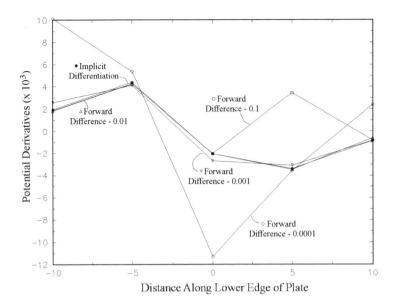

Figure 9.2: Derivative potentials taken with respect to a change in the central x-coordinate of the elliptical cavity.

9.4.2   Sensitivities of Elastic Ring

To further assess the accuracy of the direct differentiation method for evaluating design sensitivities, an elastic ring of internal radius a, and external radius b, subjected to unit external pressure (see Figure 9.3) was studied [28, 27]. The analytical values of the design sensitivities for radial displacement are given as

$$u_{r,a} = -\frac{2b^2(a^2 + b^2)}{E(b^2 - a^2)^2} \qquad \text{for } r = a$$

The values of these derivatives, normalized with respect to the Young's modulus E, are presented in Table 9.1. Presented in the table are the values obtained using both the Implicit Differential Method (IDM) and the Finite Difference Method (FDM). The results correspond to the case of the external boundary fixed at $b = 10$ and the internal boundary at $a = 1, 2$ and 5. The ring was discretized with four quadratic boundary elements on each surface [26, 27].

9.4.3   A Plate Subjected to Uniaxial Tension

Sensitivity analysis for a rectangular plate shown in Figure 10.4 was carried out in [25]. The discretized model is shown in Figure 9.5. The design variable used was the value of the $x-coordinate$ at the point Q. The material properties are: Young's modulus $E = 1 \times 10^3$MPa and Poisson's ratio, $\nu = 0.3$. The results obtained using the Implicit Differentiation Method (IDM) are shown in Table 9.2 for points P, Q and R. Also presented are the finite difference method results for different size steps. The step size h given is in millimeters.

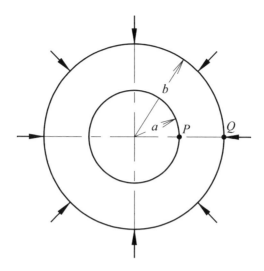

Figure 9.3: An elastic ring.

Table 9.1: Displacement derivatives $u_{r,a}E/p$.

a	Point	Exact	IDM.	FDM(0.01)	FDM(0.001)	FDM(0.0001)
1	P	-0.4081	-0.4075	-0.4094	-0.4065	-0.4098
1	Q	-2.061	-2.059	-2.059	-2.057	-2.051
2	P	-0.8681	-0.8667	-0.8695	-0.8643	-0.8568
2	Q	-2.257	-2.254	-2.255	-2.253	-2.198
5	P	-3.556	-3.547	-3.556	-3.545	-3.362
5	Q	-4.444	-4.434	-4.442	-4.435	-4.536

Table 9.2: Comparison of displacement derivatives (mm/mm) and stress derivatives (MPa/mm) using implicit differentiation method and finite difference method.

	Point	IDM	FDM($h = 0.01$)	FDM($h = 0.001$	FDM($h = 0.0001$)
	P	-0.941	-0.89	-0.90	-1.0
$u_{x,m}$	Q	-1.60	-1.53	-1.60	-2.0
	R	-0.941	-0.89	-0.90	-1.0
	P	0.978	0.95	1.0	1.0
$u_{y,m}$	Q	0.0	0.0	0.0	0.0
	R	-0.98	-0.95	-1.0	-1.0
	P	-0.584	-060	-0.60	-1.0
$\sigma_{y,m}$	Q	-4.63	-4.9	-4.7	-5.0
	R	-0.584	-0.60	-0.60	-1.0

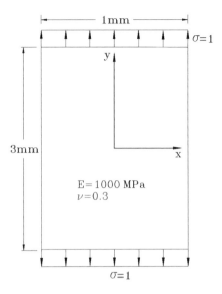

Figure 9.4: Plate subjected to uniaxial stress.

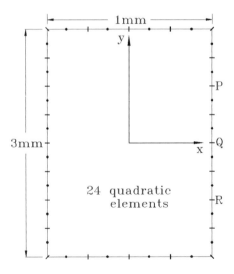

Figure 9.5: BEM model for a plate subjected to uniaxial tension.

As can be seen, the implicit differentiation method is in good agreement with the finite difference method when step size is 0.001.

9.5 Structural Optimization

In designing engineering components, different factors such as localized high stress, weight, etc. have to be taken into consideration. For example, one way of smoothing out local stress concentrations is to minimize the square of the deviation of the stresses from the desired uniform average stress, which would be related to the maximum permissible stress in the material. In this situation, the objective function can be written as

$$F(\mathbf{Z}) = \int (\sigma_v - \sigma_a)^2 ds \qquad (9.14)$$

where σ_v denotes the local von Mises stress, σ_a is the desired average stress, and s is the distance measured along the boundary.

9.5.1 Optimum Shape of a Tank with Internal Pressure

Tafreshi and Fenner [37] analyzed several structural optimization problems using the implicit differentiation approach. One of the problems they analyzed concerned the optimal shape of a thick-walled tank under internal pressure. Because of the symmetry of the tank, shown in Figure 9.6, only one quarter of the problem was modelled using the BEM. The aim was to find the optimum shape of the inner boundary EFA. The design variables are radii of the nodal points along the boundary EF; the edge FA was allowed to move with the same amount of movement as point F in the horizontal direction. The outer boundary was fixed. The initial, constant thickness, and optimum shape for the problem are shown in Figure 9.6.

9.6 Shape Identification

Non-Destructive Testing (NDT) is nowadays used to detect cracks and faults in structural components. In these methods, a comparison is made between the responses of the actual structure and the results expected from a defect free test example. Any significant difference between the two measurements indicated the presence of a flaw in the component.

Flaw identification methods identify unknown internal boundaries (i.e. cracks or cavities) in bodies, by the use of measured values on the boundary or inside a body. These responses are then compared with computed values from a body, with the same external boundary, that contains a crack of a known shape, size and location. Iterative modification of this internal boundary would reveal an approximation to the true location, shape and size of the flaw.

In shape identification methods, it is assumed that measurements of either potential or flux in potential analysis and displacement or stress in elasticity analysis are available at a number of locations in the specimen. These points are referred to as 'sensor points' and may be either boundary or internal points. It is necessary, therefore, to create a model with nodal points at sensor points. Numerical results are obtained using the BEM at the sensor points for an assumed shape or position of a flaw or cavity in the (initial) configuration. These results will differ from those

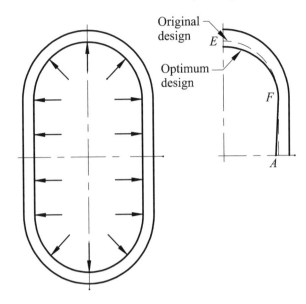

Figure 9.6: Tank with internal pressure, initial and optimum designs.

obtained in the experimental setup in response to the actual flaw. The difference in the two results originates from two categories; the error, or noise, in the experimental measurements; and the error in the assumed flaw identity. The first of these two errors is considered to be minimal; problems caused by the presence of these errors cannot be eliminated easily. The error due to the assumed shape or location of the crack or cavity is, however, a critical one, and this must be minimized. The error function $F(\mathbf{Z})$ definition is based on the sum of the square errors, i.e.

$$F(\mathbf{Z}) = \frac{\sqrt{\sum_{n=1}^{N}(\epsilon^{\tau}(\mathbf{X}^{n}))^{2}}}{N} \tag{9.15}$$

where

$$\epsilon^{\tau}(\mathbf{X}^{n}) = \left\{ \begin{array}{ll} \theta^{a} - \theta^{\tau} & \text{for potential values measured at sensor points } \mathbf{X}^{n} \\ q^{a} - q^{\tau} & \text{for flux values measured at sensor points } \mathbf{X}^{n} \end{array} \right.$$

or

$$\epsilon^{\tau}(\mathbf{X}^{n}) = \left\{ \begin{array}{ll} u_{i}^{a} - u_{i}^{\tau} & \text{for displacement values measured at sensor points } \mathbf{X}^{n} \\ t_{i}^{a} - t_{i}^{\tau} & \text{for traction values measured at sensor points } \mathbf{X}^{n} \end{array} \right.$$

In the above expressions, superscript a denotes values corresponding to the configuration with the actual crack or cavity, superscript τ denotes the iteration step and N is the total number of sensor points.

First order optimization methods such as BFGS quasi-Newton methods are often used in the identification method. In this first order optimization method, it is necessary to evaluate the gradient of the error function $\nabla \mathbf{F}$ with respect to each design variable, defined as

$$\nabla F(\mathbf{Z}) = \frac{\partial F(\mathbf{Z})}{\partial Z_{m}} \tag{9.16}$$

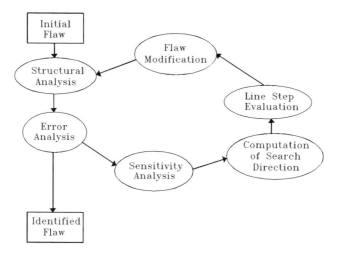

Figure 9.7: Flaw identification procedure.

where Z_m is the mth component of the design vector. In order to evaluate this function, derivatives of the potential and flux values or displacement and traction values at the sensor points are required. These values, known as design sensitivities, can be evaluated using the implicit differentiation method described earlier in this chapter.

The iterative processes for shape identification is presented in a schematic form in Figure 9.7. At each step of the optimization a new crack, cavity or boundary is defined and the geometry is updated. The process is repeated until the error is below a required tolerance level, or no changes are made to the design variables.

9.6.1 Identification of an Elliptical Cavity

A square plate with a hole of unknown position, size and shape, as shown in Figure 9.8, was studied in [28]. In this example 16 internal sensors, as shown in Figure 9.8 were used at which all the components of the stress tensor were known. The model requires the identification of four design variables, a, b, x_o and y_o (see Figure 9.9) The actual and the initial hole are shown in Figure 9.10. The convergence of design variables are presented in Figure 9.11. Also analysed in [28] were the effects of noisy date. Random errors of 1%, 2%, 5% and 10% were introduced. As shown in Figure 9.12 the results are not significantly affected by the introduction of the noisy data.

In this problem, the only boundary that has changed in the iterative procedure is the internal cavity. The external boundary is independent of the design variables, and hence the derivative of the jump term on the external boundary will vanish, i.e.

$$C_{ij,m}(\mathbf{x}') = 0 \tag{9.17}$$

The derivatives of the jump term on the boundary of the cavity to be identified can

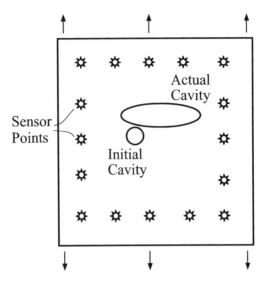

Figure 9.8: Initial and final elliptical holes.

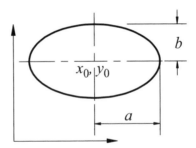

Figure 9.9: Elliptical cavity with four design variables, a, b, x_o, y_o.

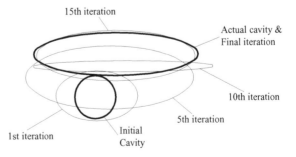

Figure 9.10: Identification of an elliptical cavity.

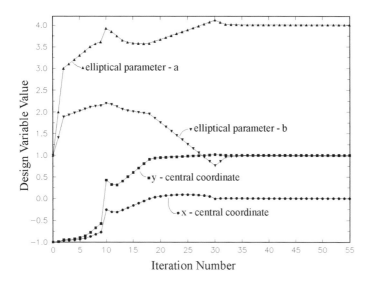

Figure 9.11: Convergence of design variables.

be evaluated from the derivative of the expressions for rigid body condition, i.e.

$$
\begin{aligned}
C_{ij,m}(\mathbf{x}') &= -\sum_{n=1}^{N}\sum_{\alpha=1}\int_{-1}^{1} T_{ij,m}(\mathbf{x}',\mathbf{x})N_{\alpha}(\eta)J^{n}(\eta)d\eta \\
&\quad -\sum_{n=1}^{N}\sum_{\alpha=1}\int_{-1}^{1} T_{ij}(\mathbf{x}',\mathbf{x})N_{\alpha}(\eta)J_{,m}^{n}(\eta)d\eta
\end{aligned}
\tag{9.18}
$$

In the above equation, all terms can be computed easily, except for the singular integral and the derivative jump term $C_{ij,m}(\mathbf{x}')$. Using the same row sum technique as used with the rigid body condition, both of these terms may be evaluated.

9.6.2 Acoustic Scattering in Fluid-Solid

In this section a boundary element formulation based on the implicit differentiation method is described for the sensitivity analysis of structures immersed in an inviscid fluid and illuminated by harmonic incident plane waves [23]. The formulation is then coupled to an optimization technique to solve an inverse problem of flaw identification. An error function is introduced as a measure of the difference between the computed and observed acoustic pressures at sensor points on the boundary, and minimized to give the flaw position and shape. The numerical response is obtained by enforcing normal equilibrium and compatibility between the external Helmholtz equation and the internal Cauchy-Navier equation. In this case, the BEM has the further advantage of satisfying the Sommerfield radiation condition at infinity automatically.

The minimization is performed using a first order, non-linear, unconstrained optimization technique, in which the error function gradient is computed by the implicit differentiation method.

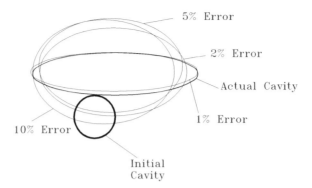

Figure 9.12: Identification of internal cavity with noisy date.

Coupled Derivative System of Equations

To evaluate the gradient it is necessary to differentiate boundary integral equations for the Helmholtz equation (see Volume I) and the Cauchy-Navier equation presented in Section 5.3.2. The differentiation of the boundary conditions on the external S_e and internal boundaries S_i gives

$$
\begin{aligned}
q_{,m}(\mathbf{x}) &= \rho_e \omega^2 \mathbf{u}_{,m}(\mathbf{x}) \cdot \mathbf{n}(\mathbf{x}) & \mathbf{x} \in S_e \\
\mathbf{t}_{,m}(\mathbf{x}) &= -p_{,m}\mathbf{n}(\mathbf{x}) & \mathbf{x} \in S_e \\
\mathbf{t}_{,m}(\mathbf{x}) &= \mathbf{0} & \mathbf{x} \in S_i
\end{aligned}
\tag{9.19}
$$

The differentiation of the discretized Helmholtz boundary integral equation for the external boundary gives

$$
c(\mathbf{x}')\theta_{,m}(\mathbf{x}') + \sum_{n=1}^{N_e}\sum_{\alpha=1} \theta_{,m}^n \int_{-1}^{+1} q^*(\mathbf{x}',\mathbf{x})N_\alpha(\eta)J^n(\eta)d\eta
$$

$$
= \sum_{n=1}^{N_e}\sum_{\alpha=1} q_{,m}^n \int_{-1}^{+1} \theta^*(\mathbf{x}',\mathbf{x})N_\alpha(\eta)J^n(\eta)d\eta
\tag{9.20}
$$

It is worth noting that $C_{,m} = 0$, $\theta_{,m}^* = 0$, $q_{,m}^* = 0$, $J_{,m}^l = 0$ and $p_{inc,m} = 0$ for every, $\mathbf{x} \in S_e$ and denoting with $(\ )_{,m}$ the partial derivative with respect to Z_m. Differentiation of the discretized boundary integral equation for Cauchy-Navier (5.56) gives

$$
C_{ij,m}(\mathbf{x}')u_j(\mathbf{x}') + C_{ij}(\mathbf{x}')u_{j,m}(\mathbf{x}')
$$

$$
+ \sum_{n=1}^{N_e}\sum_{\alpha=1} u_{j,m}^\alpha \int_{-1}^{+1} T_{ij}(\mathbf{x}',\mathbf{x},s)N_\alpha(\eta)J^n(\eta)d\eta
$$

$$
- \sum_{n=1}^{N_e}\sum_{\alpha=1} t_{j,m}^\alpha \int_{-1}^{+1} U_{ij}(\mathbf{x}',\mathbf{x},s)N_\alpha(\eta)J^n(\eta)d\eta
$$

$$
= \sum_{n=1}^{N_e}\sum_{\alpha=1} t_j^\alpha \int_{-1}^{+1} U_{ij,m}(\mathbf{x}',\mathbf{x},s)N_\alpha(\eta)J^n(\eta)d\eta
$$

$$+ \sum_{n=1}^{N_e} \sum_{\alpha=1}^{} t_j^\alpha \int_{-1}^{+1} U_{ij}(\mathbf{x}', \mathbf{x}, s) N_\alpha(\eta) J_{,m}^n(\eta) d\eta$$

$$- \sum_{n=1}^{N_e} \sum_{\alpha=1}^{} u_j^\alpha \int_{-1}^{+1} T_{ij,m}(\mathbf{x}', \mathbf{x}, s) N_\alpha(\eta) J^n(\eta) d\eta$$

$$- \sum_{n=1}^{N_e} \sum_{\alpha=1}^{} u_j^\alpha \int_{-1}^{+1} T_{ij}(\mathbf{x}', \mathbf{x}, s) N_\alpha(\eta) J_{,m}^n(\eta) d\eta \tag{9.21}$$

where s denotes the Laplace parameter and $C_{ij,m}(\mathbf{x}') = 0$ if the design variables are chosen in order to keep the geometry smooth at any node of the design boundary. The derivatives of the fundamental solutions in elastodynamics are required. The new kernels obtained in the boundary element sensitivity analysis using implicit differentiation are also singular. However, it can be shown that the order of singularity does not increase through differentiation with respect to the design variable Z_m.

The derivatives of the fundamental solution for displacements and tractions are given as follows:

$$U_{ij,m}(\mathbf{x}', \mathbf{x}, s) = \frac{1}{2\pi\rho_i c_t^2} [r_{,m}(\psi_{,r}\delta_{ij} - \chi_{,r}r_{,i}r_{,j}) - \chi(r_{,i}r_{,j})_{,m}] \tag{9.22}$$

$$T_{ij,m}(\mathbf{x}', \mathbf{x}, s) = \frac{1}{2\pi} \left\{ \left(\psi_{,rr} + \frac{\chi}{r^2} - \frac{\chi_{,r}}{r} \right)(\delta_{ij}r_{,n} + r_{,j}n_i)r_{,m} + \left(\psi_{,r} - \frac{\chi}{r} \right)(\delta_{ij}r_{,nm}) \right.$$

$$+ r_{,jm}n_i + r_{,j}n_{i,m}) + r_{,m}\left(\frac{2}{r^2}\chi - \frac{2\chi_{,r}}{r} \right)(n_j r_{,i} - 2r_{,i}r_{,j}r_{,n})$$

$$- \frac{2}{r}\chi(n_{j,m}r_{,i} + n_j r_{,im} - 2r_{,im}r_{,j}r_{,n} - 2r_{,i}r_{,jm}r_{,n} - 2r_{,i}r_{,j}r_{,nm})$$

$$- 2\chi_{,rr}r_{,i}r_{,j}r_{,n}r_{,m} - 2\chi_{,r}r_{,im}r_{,j}r_{,n} - 2\chi_{,r}r_{,i}r_{,jm}r_{,n}$$

$$- 2\chi_{,r}r_{,i}r_{,j}r_{,nm} + \left(\frac{c_l^2}{c_t^2} - 2 \right) \left[\left(\psi_{,rr} - \chi_{,rr} + \frac{\chi}{r^2} - \frac{\chi_{,r}}{r} \right) r_{,i}n_j r_{,m} \right.$$

$$\left. + \left(\psi_{,r} - \chi_{,r} - \frac{\chi}{r} \right)(r_{,im}n_j + r_{,i}n_{j,m}) \right] \right\} \tag{9.23}$$

The terms $\chi \ \chi_{,r} \ \chi_{,rr} \ \psi_{,r}$ and $\psi_{,rr}$ are defined Section 5.3.2.

The identification method developed in [23] is based on measurements (i.e. acoustic pressures or fluxes as well as displacements or tractions) taken at various sensor points, for different known incident plane waves. It was shown [20] that, in these hypotheses, the scattering obstacle is uniquely determined.

Identification of Cavity Location

The incident plane wave considered has the following analytic expression:

$$p_{inc}(\mathbf{x}) = p_0 exp[-ikrcos(\phi - \alpha)]$$

where $i = \sqrt{-1}$ and (r, ϕ) are the polar coordinates of $\mathbf{x}$.

An internal elliptical hole (see Figure 9.13) is located, for different dimensionless wave numbers kR and different incident angles α, with pressure at the sensor points computed numerically. The dimensions and the properties of fluid and scatterer are shown in Table 9.3.

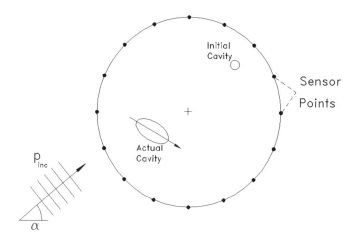

Figure 9.13: Initial and final elliptical hole.

Table 9.3: Material Properties.

Brass Cylinder	
Mass density	$\rho = 8500 \text{Kg/m}^3$
Young's modulus	$E = 10.5 \times 10^{11} \text{Pa}$
Poisson's	$\nu = 0.33$
Radius	$R = 1.0\text{m}$
Acoustic medium: water	
Mass density	$\rho = 988 \text{Kg/m}^3$
Sound speed	$c = 1486 \text{m/s}$

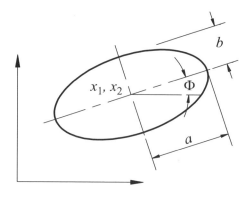

Figure 9.14: Design variables.

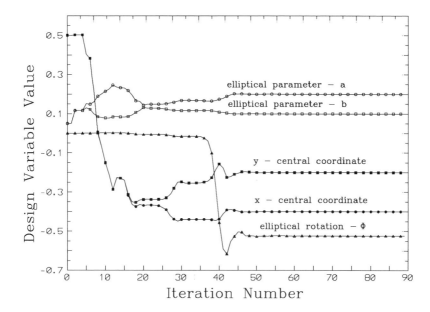

Figure 9.15: Convergence of the design variable $kR = 1$ and four incident waves.

An internal elliptical hole with parameters: $x_1 = -0.4$, $x_2 = -0.2$, $a = 0.2$, $b = 0.1$ and $\Phi = -\pi/6$ (see Figure 9.14) is identified using 16 sensor points, for different dimensionless wave numbers and different incident angles. The initial guess was a circular cavity with radius 0.05 and centred at $(0.5, 0.5)$, as shown in Figure 9.13.

Four incident waves ($\alpha = 0$, $\alpha = \pi/2$, $\alpha = \pi$, $\alpha = 3/2\pi$) are used to compare the results for $kR = 0.5$, 1, 2. Figure 9.15 shows the convergence of design variables, and Figure 9.16 presents normalized errors in the case $kR = 0.5$, 1, 2 and four incident waves. It can be seen that after about 50 iterations the process has converged. The higher errors, shown in Figure 9.16 for $kR = 0.5$ and $kR = 2$ show a lower capacity of these frequencies to find the final cavity. For $kR = 2$, the optimization procedure gives an accurate solution $(-0.399, -0.200, 0.199, 0.0999, -0.520)$, but needs more constrain to avoid a higher number of local minima. For $kR = 0.5$ instead, the solution $(-0.432, -0.227, 0.169, 0.114, -0.010)$ is less accurate.

For $kR = 1$ and four incident waves, the change in the cavity shape and location after every 10 iterations is shown in Figure 9.17.

Figure 9.18 presents the normalized error versus the iterations for $kR = 1$ and different incident angles. For the given shape and location of the final cavity, the results are quite good for all the incident angles considered with an exception $\alpha = \pi/2$, which stops the optimization at $(-0.452, -0.222, 0.186, 0.106, -0.278)$. .

9.7 Flaw Identification

In this section the dual boundary element sensitivity formulation for flaw identification is described.

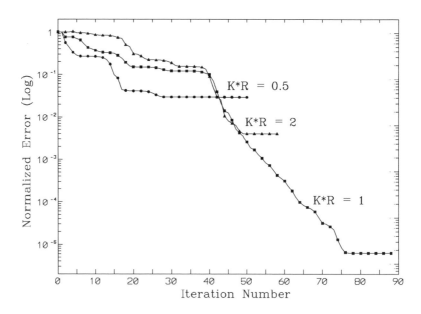

Figure 9.16: Convergence of the normalized error for different dimensionless wave numbers.

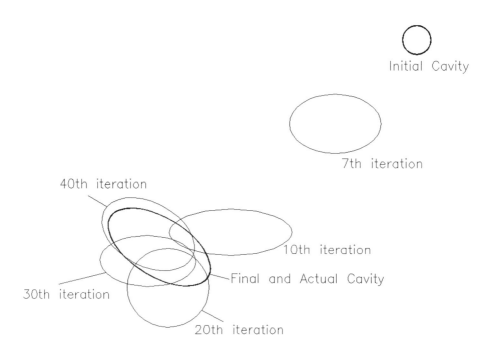

Figure 9.17: Identification of an elliptical cavity.

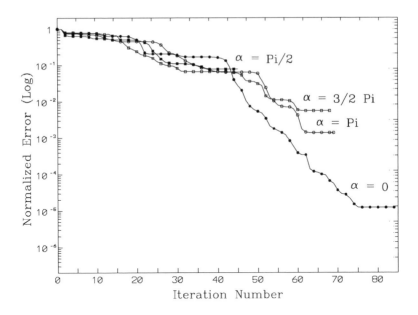

Figure 9.18: Convergence of the normalized error for different incident angles -$kR = 1$.

9.7.1   Potential Formulation

Direct differentiation of the discretized DBEM equations gives new equations in terms of the potential, flux, potential derivative and flux derivative nodal values. Differentiation of the potential equation on a smooth crack surface gives [26]

$$\frac{1}{2}\theta(\mathbf{x}^-) + \frac{1}{2}\theta(\mathbf{x}^+) + \sum_{n=1}^{N_e}\sum_{\alpha=1}^{N_e}[\theta^{n\alpha}V_{,m}^{n\alpha} + \theta_{,\alpha}^{n\alpha}V^{n\alpha}]$$

$$= \sum_{n=1}^{N_e}\sum_{\alpha=1}^{N_e}[q^{n\alpha}W_{,m}^{n\alpha} + q_{,m}^{n\alpha}W^{n\alpha}] \qquad (9.24)$$

The differentiation of the flux equation gives

$$\frac{1}{2}q_{,m}(\mathbf{x}^-) - \frac{1}{2}q_{,m}(\mathbf{x}^+) + n_{i,m}(\mathbf{x}')\sum_{n=1}^{N_e}\sum_{\alpha=1}^{N_e}\theta^{n\alpha}H^{n\alpha} \qquad (9.25)$$

$$+n_i(\mathbf{x}')\sum_{n=1}^{N_e}\sum_{\alpha=1}^{N_e}q_{,m}^{n\alpha}K^{n\alpha} + n_i(\mathbf{x}')\sum_{n=1}^{N_e}\sum_{\alpha=1}^{N_e}q^{n\alpha}K_{,m}^{n\alpha}$$

$$= n_{i,m}(\mathbf{x}')\sum_{n=1}^{N_e}\sum_{\alpha=1}^{N_e}q^{n\alpha}K^{n\alpha} + n_i(\mathbf{x}')\sum_{n=1}^{N_e}\sum_{\alpha=1}^{N_e}\theta_{,m}^{n\alpha}H^{n\alpha} + n_i(\mathbf{x}')\sum_{n=1}^{N_e}\sum_{\alpha=1}^{N_e}\theta^{n\alpha}H_{,m}^{n\alpha}$$

where

$$H^{n\alpha} = \fint_{-1}^{+1}\fint_{-1}^{+1} q_{,i}^* N_\alpha J^n \, d\eta_1, d\eta_2$$

$$K^{n\alpha} = \int_{-1}^{+1}\int_{-1}^{+1} \theta^*_{,i} \, N_\alpha \, J^n \, d\eta_1 d\eta_2$$

$$K^{n\alpha}_{,m} = \int_{-1}^{+1}\int_{-1}^{+1} \left(\theta^*_{,im} \, N_\alpha J^n + \theta^*_{,i} \, N_\alpha J^n_{,m} \right) d\eta_1 d\eta_2$$

$$H^{n\alpha} = \oint_{-1}^{+1}\oint_{-1}^{+1} q^*_{,i} \, N_\alpha \, J^n \, d\eta_1 d\eta_2$$

$$H^{n\alpha}_{,m} = \oint_{-1}^{+1}\oint_{-1}^{+1} \left(q^*_{,im} \, N_\alpha \, J^n + q^*_{,i} \, N_\alpha \, J^n_{,m} \right) d\eta_1 d\eta_2$$

for three-dimensional problems. Similar expressions can be written for two-dimensional problems.

In these equations the derivatives of the fundamental solutions u^*, q^*, $u^*_{,i}$ and $q^*_{,i}$ are given by

$$\theta^*_{,im}(\mathbf{x}', \mathbf{x}) = -\frac{1}{4\pi \, r^3} \left(3\, r_{,i} r_{,m} - r_{i,m}\right)$$

$$\theta^*_{,im}(\mathbf{x}', \mathbf{x}) = \frac{1}{4\pi \, r^4} \left(3\frac{\partial r}{\partial n} \left[5\, r_{,i} \, r_{,m} - r_{i,m} \right] - 3(r_l \, n_l)_{,m} \, r_{,i} - 3\, n_i \, r_{,m} + n_{i,m} \, r \right)$$

for three-dimensional problems, and

$$u^*_{,im}(\mathbf{x}', \mathbf{x}) = -\frac{1}{2\pi \, r^2} \left(2\, r_{,i} r_{,m} - r_{i,m}\right)$$

$$q^*_{,im}(\mathbf{x}', \mathbf{x}) = -\frac{1}{2\pi \, r^3} \left(2\frac{\partial r}{\partial n} \left[-4 r_{,i} \, r_{,m} + r_{i,m} \right] + 2(r_l \, n_l)_{,m} \, r_{,i} + 2\, n_i \, r_{,m} - n_{i,m} \, r \right)$$

for two-dimensional problems.

9.7.2 Elasticity Formulation

The derivative displacement integral equation for collocation points on the crack surface can be written as

$$\frac{1}{2} u_{i,m}(\mathbf{x}^+) + \frac{1}{2} u_{i,m}(\mathbf{x}^-) + \sum_{n=1}^{N_e}\sum_{\alpha=1}^{N_e} u^{n\alpha}_{j,m} P^{n\alpha}_{ij} + \sum_{n=1}^{N_e}\sum_{\alpha=1}^{N_e} u^{n\alpha}_j P^{n\alpha}_{ij,m}$$

$$= \sum_{n=1}^{N_e}\sum_{\alpha=1}^{N_e} t^{n\alpha}_{j,m} Q^{n\alpha}_{ij} + \sum_{n=1}^{N_e}\sum_{\alpha=1}^{N_e} t^{n\alpha}_j Q^{n\alpha}_{ij,m} \tag{9.26}$$

and the derivative traction equation as

$$\frac{1}{2} t_{i,m}(\mathbf{x}^+) - \frac{1}{2} t_{i,m}(\mathbf{x}^-) + n_{j,m}(\mathbf{x}') \sum_{n=1}^{N_e}\sum_{\alpha=1}^{N_e} u^{n\alpha}_k S^{n\alpha}_{kij}$$

$$+n_j(\mathbf{x}') \sum_{n=1}^{N_e} \sum_{\alpha=1} u_{k,m}^{n\alpha} S_{kij}^{n\alpha} + n_j(\mathbf{x}') \sum_{n=1}^{N_e} \sum_{\alpha=1} u_j^{n\alpha} S_{kij,m}^{n\alpha}$$

$$= n_{j,m}(\mathbf{x}') \sum_{n=1}^{N_e} \sum_{\alpha=1} t_k^{n\alpha} R_{kij}^{n\alpha} + n_j(\mathbf{x}') \sum_{n=1}^{N_e} \sum_{\alpha=1} t_{k,m}^{n\alpha} R_{kij}^{n\alpha} + n_j(\mathbf{x}') \sum_{n=1}^{N} \sum_{\alpha=1} t_k^{n\alpha} R_{kij,m}^{n\alpha}$$

(9.27)

where

$$S_{kij}^{n\alpha} = \int_{-1}^{+1} S_{kij} N_\alpha J^n \, d\eta$$

$$S_{kij,m}^{n\alpha} = \int_{-1}^{+1} S_{kij,m} N_\alpha J^n \, d\eta + \int_{-1}^{+1} S_{kij} N_\alpha J_{,m}^n \, d\eta$$

$$R_{kij}^{n\alpha} = \int_{-1}^{+1} D_{kij} N_\alpha J^n \, d\eta$$

$$R_{kij,m}^{n\alpha} = \int_{-1}^{+1} D_{kij,m} N_\alpha J^n \, d\eta + \int_{-1}^{+1} D_{kij} N_\alpha J_{,m}^n \, d\eta$$

and

$$D_{kij,m}(\mathbf{x}',\mathbf{x}) = -\frac{1}{4\pi(1-\nu)r^2}[(2r_{,i}r_{,k} + (1-2\nu)\delta_{ki})r_{j,m}$$
$$+ (2r_{,j}r_{,k} + (1-2\nu)\delta_{jk})r_{i,m} + (2r_{,i}r_{,j} + (1-2\nu)\delta_{ji})r_{k,m}$$
$$-2((1-2\nu)(\delta_{ki}r_{,j} + \delta_{kj}r_{,i} - \delta_{ij}r_{,k}) + r_{,i}r_{,j}r_{,k})r_{,m}$$

(9.28)

$$S_{kij,m}(\mathbf{x}',\mathbf{x}) = \frac{E}{4\pi(1-\nu)r^3}\left[2\frac{\partial r}{\partial n}\{(1-2\nu)\delta_{ij} - 4r_{,i}r_{,j})r_{k,m}\right.$$
$$+ (\nu\delta_{ik} - 4r_{,i}r_{,k})r_{j,m} + (\nu\delta_{jk} - 4r_{,j}r_{,k})r_{i,m}$$
$$-4((1-2\nu)\delta_{ij}r_{,k} + \nu(\delta_{ik}r_{,j} + \delta_{jk}r_{,i}) - 6r_{,i}r_{,j}r_{,k})r_{,m}\}$$
$$+ (n_l r_l)_{,m}((1-2\nu)\delta_{ij}r_{,k} + \nu(\delta_{ik}r_{,j} + \delta_{jk}r_{,i}) - 4r_{,i}r_{,j}r_{,k})$$
$$-2\{\nu(4r_{,j}r_{,k}r_{,m} - (r_{j,m}r_{,k} + r_{,j}r_{k,m})) + (1-2\nu)\delta_{jk}r_{,m}\}n_i$$
$$-2\{\nu(4r_{,i}r_{,k}r_{,m} - (r_{i,m}r_{,k} + r_{,i}r_{k,m})) + (1-2\nu)\delta_{ik}r_{,m}\}n_j$$
$$-2\{(1-2\nu)(4r_{,i}r_{,j}r_{,m} - (r_{i,m}r_{,j} + r_{,i}r_{j,m})) - (1-4\nu)\delta_{ij}r_{,m}\}n_k$$
$$+ (2\nu r_{,j}r_{,k} + (1-2\nu)\delta_{jk})n_{i,m}r + (2\nu r_{,i}r_{,k} + (1-2\nu)\delta_{ik})n_{j,m}r$$
$$\left. + (2(1-2\nu)r_{,i}r_{,j} - (1-4\nu)\delta_{ij})n_{k,m}r\right]$$

(9.29)

for two-dimensional problems.

In the flaw identification method, the system of equations is solved with the LU decomposition method, in which the $\mathbf{A}$ matrix is decomposed into an upper and a lower triangular matrix. The system is then solved using forward and backward substitution. Since the $\mathbf{A}$ matrix in (9.9) is identical to that in the matrix generated by the direct boundary integral equations (see Chapter 2) the decomposition can be preserved, rather than the matrix itself, and this can then be reused for each of the derivative analyses [26, 28].

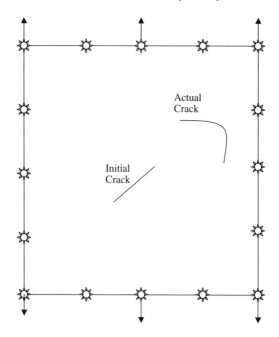

Figure 9.19: Identification of a curved crack.

9.8 Examples

In this section two examples are presented to demonstrate the application of the dual boundary sensitivity formulation for crack identification.

9.8.1 Curved Crack Identification

A square plate contains a curved crack as shown in Figure 9.19. The plate is subjected to tension at its ends. The crack was initially modelled by a single element on each face, and the BFGS method was used to minimize the error [28]. The traction sensor data is known at the 16 sensor points shown in the figure. The model is refined with more elements on each crack surface as an optimal crack is located. The process is repeated until the error function is small enough. The crack positions after iterations 1, 3, and 5 elements, as well as the final crack position, are shown in Figure 9.19.

In the flaw identification problem, the only boundary changed in the iterative procedure will be the internal crack. The external boundary shape is independent of the design variable, on the external boundary and hence $C_{,m}(\mathbf{x}') = 0$. For nodes on the crack boundary, jump terms at both $C_{ij}(\mathbf{x}^-)$ and $C_{ij}(\mathbf{x}^+)$ will be 0.5, since discontinuous elements are used. Hence, the derivatives of the jump terms on the crack boundaries are also zero.

9.8.2 Identification of an Elliptical Crack

A cylinder of radius R and height $2R$ is assumed to contain an elliptical crack (see Figure 9.21). The centre of the elliptical crack (x_0, y_0, z_0) is to be identified together

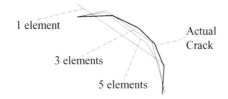

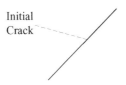

Figure 9.20: Identification of curved crack.

with the elliptical parameters a and b defined by

$$\frac{(x - x_0)^2}{a^2} + \frac{(y - y_0)^2}{b^2} = 1 \quad \text{and} \quad z = z_0 \tag{9.30}$$

The initial design is chosen with the design variables initially being

$$a = 0.4,\ b = 0.1,\ x_0 = -0.25,\ y_0 = 0.25 \text{ and } z_0 = -0.5$$

and flux values are measured at the sensor points. In this example the sensor points were located at the two ends of the cylinder where the temperature was prescribed.

The crack was again identified successfully [29], as shown in Figure 9.22.

9.9   Bone Remodelling

Martinez, Aliabadi and Power [24] showed that sensitivity analysis and design optimization procedures can provide an efficient means to simulate bone remodelling. If bone remodelling can be represented as structural analysis, then it is reasonable to assume that a boundary value problem can be formulated in the same manner as a boundary problem in linear elastostatics, but it is necessary to prescribe the boundary conditions for a specific time period [10]. As the body evolves to new shapes, the stress and strain must be varied quasi-statically. At any instant the bone behaves as an elastic body, but the moving boundaries associated with bone growth cause local stress and strain to redistribute themselves slowly with time.

Martinez, Aliabadi and Power [24] adopted the surface bone remodelling theory proposed by Cowin and Van Buskirk [11] in their boundary element work. This theory assumes that the bone tissue can be modelled as a linear elastic body whose free surface moves according to an additional constitutive equation relating to the changes from a mechanical state to the remodelling equilibrium. Cowin and Van

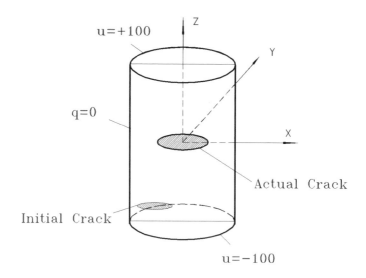

Figure 9.21: Identification of an elliptical crack.

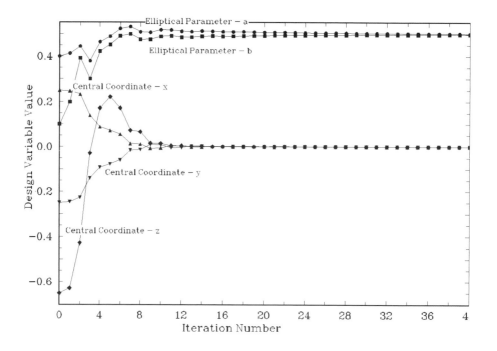

Figure 9.22: Convergence of design parameters in the identification of an elliptical crack.

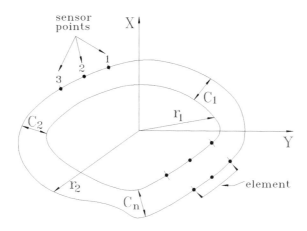

Figure 9.23: General description of a bone model.

Buskirk [11] proposed that the velocity of the external surface can be related to strains via

$$U(\mathbf{x}) = F_{ij}(\mathbf{x})[\varepsilon_{ij}(\mathbf{x}) - \varepsilon_{ij}^R(\mathbf{x})] \qquad (9.31)$$

where F_{ij} are the surface strain remodelling rate constants, $\varepsilon_{ij}(\mathbf{x})$ is the actual strain tensor at point $\mathbf{x}$, $\varepsilon_{ij}^R(\mathbf{x})$ is the remodelling equilibrium strain tensor at point $\mathbf{x}$ and U is the normal velocity of the remodelling surface point at $\mathbf{x}$. If the strain state at $\mathbf{x}_s$ is such that $\varepsilon_{ij} = \varepsilon_{ij}^R$, then the velocity of the surface is zero, $U(\mathbf{x}_s) = 0$ and no remodelling occurs. If the right-hand side of (9.31) is positive, the surface will be growing by deposition of bone material. If, on the other hand, the right-hand side of equation (9.31) is negative, the surface will be resorbing. The right-hand side of (9.31) can also be expressed in terms of stress, as

$$U(\mathbf{x}) = \chi_{ij}(\mathbf{x})[\sigma_{ij}(\mathbf{x}) - \sigma_{ij}^R(\mathbf{x})] \qquad (9.32)$$

where χ_{ij} are surface stress remodelling rate coefficients at point $\mathbf{x}$, σ_{ij} is the actual stress tensor and σ_{ij}^R is the remodelling equilibrium stress at $\mathbf{x}$.

The initial shape of the bone is defined using a set of m design variables, denoted by $\mathbf{Z}_m = (\mathbf{z}_1, \mathbf{z}_2, , , , \mathbf{z}_m)$. Next, a set of boundary nodal points is selected as the so-called sensor points $(\mathbf{x}^n, n = 1, ...N)$. It is assumed that the stress values of the remodelling equilibrium state are known, from experiment say, at the sensor points. In Figure 9.31, a general transverse cross section of a bone describing the sensor points, elements and design variables is shown. Numerical stress values are obtained at the sensor points from the boundary element solutions for the initial configuration of loaded bone. Generally, stress values from the BEM and stress values from the remodelling equilibrium are different, and lead to the definition of an error function $F(\mathbf{Z})$. This allows the difference at sensor node $\mathbf{x}^n$ to be defined as

$$S_{ij}(\mathbf{x}^n) = \sigma_{ij}^{BEM}(\mathbf{x}^n) - \sigma_{ij}^R(\mathbf{x}^n) \qquad (9.33)$$

where σ_{ij}^R denotes the stress at remodelling equilibrium state. From (9.33), the error

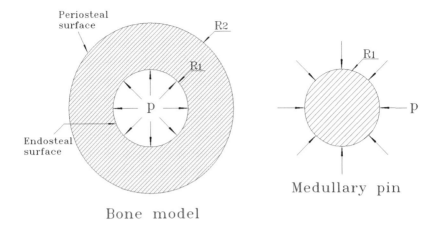

Figure 9.24: Medullary Pin Problem.

function is defined as

$$F(Z_m) = \frac{\sqrt{\sum_{n=1}^{N} \sum_{i,j=1}^{2} (S_{ij}(\mathbf{x}^n))^2}}{N} \tag{9.34}$$

9.9.1 Medullary Pin

The movement of the internal (endosteal) and external (periosteal) surfaces of a transverse section of a bone model can be seen as a consequence of the problem caused by a force-fitted medullary pin [12]. Figure 9.24 shows the geometry of the bone and medullary pin problem. The diaphysical region is modelled as a hollow circular or elliptical cylinder [24]. Both internal and external boundaries are discretized into eight quadratic elements, as shown in Figure 9.25. The material properties of the bone used are $E = 17.12$ GPa, $\nu = 0.28$. As the main aim was to find out the bone shape from initial to final equilibrium, a unit pressure was applied ($p = 1$), that is to say, unit pressure on the endosteal with the periosteal free of loading. Figure 9.26 shows a summary of results presented in [24, 25]. The second and third columns, show the initial and final bone shapes, respectively. The fourth and fifth columns present the radii of the initial and final bone shapes, respectively. The last column in the figure presents the number of iterations required in the sensitivity analysis to reach the final bone shapes.

Bone Ingrowth into a Thread of a Screw-implant

In this example the final shape and bony ingrowth into the thread of a screw is presented [25]. The placement of a screw into a cylindrical hole of a smaller size in the bone causes the sharp edge of the screw thread to penetrate into the bone tissue. A two-dimensional model of a typical region of the bone around a screw thread is shown in Figure 9.27. The boundary element model consisted of 16 quadratic elements. The bone boundary represented by AB will be moving into the threads of the screw and its position and shape represent the objective of the problem, as growth of bone tissue

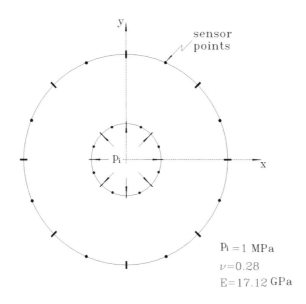

Figure 9.25: Circular transverse section of a bone model discretized.

allows the engagement of the bone tissue to the screw. The sensor points are taken at nodal points along AB (see Figure 9.28), with their $y-coordinates$ being the design variables. In order for the growth not to overlap the screw, the profile of the thread was defined to be the constraint. The material properties of the bone are taken as $E = 17.12$ GPa, $\nu = 0.28$. The dimensions for the screw are $H = 0.866$ mm and $\beta = 60^0$.

Figures 9.29 and 9.30 show the iteration steps and final shapes reached for the bone tissue into the thread of a screw for two different load levels, $\sigma = 17.5$ MPa and 40 MPa. The total number of iterations carried out to reach the final shape are indicated on the graphs as $n = f$.

In Table 9.3, comparisons are made between the results obtained using the optimization approach [24] and Sadegh et $al.$ [35] who did not used optimization procedures. The boundary element procedure of Sadegh et $al.$ [35] can be regarded a zeroth order optimization. As can be seen from the results, the increase in loading results in an increase in maximum depth values. An interesting point to note is that while the sensitivity approach takes only a few iterations to converge, the solutions in [35] require iterations in order of hundred to reach the final solution.

9.10   Summary

In this chapter the application of the boundary element method to shape optimization and identification was presented. Formulations were presented for both potential and elastostatic problems. Unlike the finite element method, the BEM does not require extensive remeshing to update the model after each change to the boundary shape. The dual boundary element formulation was also extended to deal with flaw identifi-

Examples	Initial Shape	Final Shape	Surface Movements		Iterations
			Start	End	
1	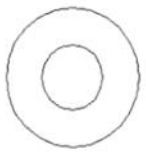	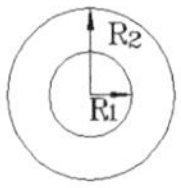	$R_1 = 4$ $R_2 = 8$	$R_1 = 5$ $R_2 = 10$	2
2	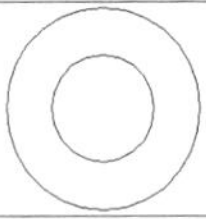	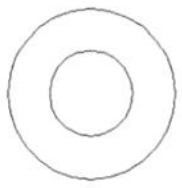	$R_1 = 6$ $R_2 = 12$	$R_1 = 5$ $R_2 = 10$	71
3	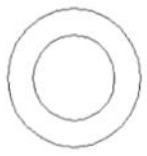	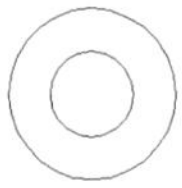	$R_1 = 5$ $R_2 = 8$	$R_1 = 5$ $R_2 = 10$	3
4	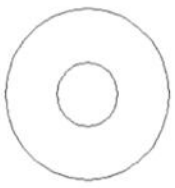	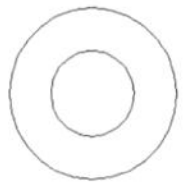	$R_1 = 4$ $R_2 = 10$	$R_1 = 5$ $R_2 = 10$	32
5	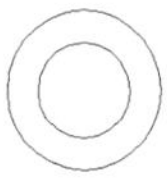	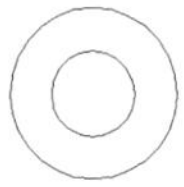	$R_1 = 5.5$ $R_2 = 9.5$	$R_1 = 5$ $R_2 = 10$	89
6	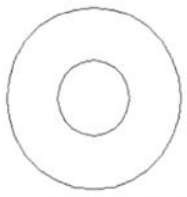	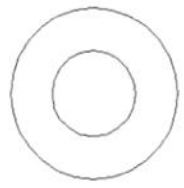	$R_1 = 4.5$ $R_2 = 10.5$	$R_1 = 5$ $R_2 = 10$	61
7	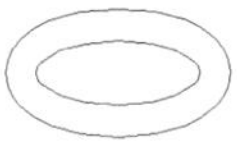	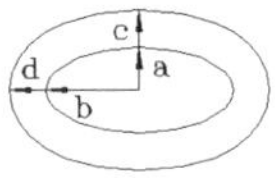	$a = 1.0$ $b = 0.5$ $c = 2.0$ $d = 1.5$	$a = 2.0$ $b = 1.0$ $c = 3.5$ $d = 3.0$	16

Figure 9.26: Different possible movements of the endosteal and periosteal surface of a bone model.

Table 9.4: Comparison of bony ingrowth d into the thread of screw-type implant.

Load (MPa)	Sadegh *et al.*[35]	Optimization [25]
10	0.216	0.205
12.5	0.311	0.311
15	0.519	0.517
17.5	0.701	0.722
19	0.779	0.781
25	0.805	0.826
40	0.840	0.861

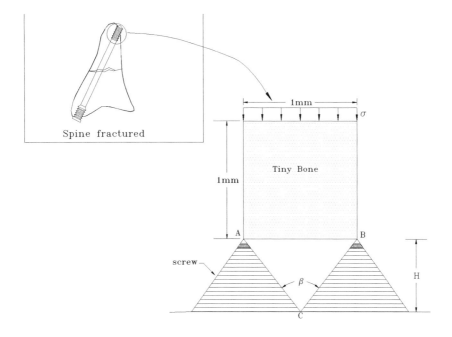

Figure 9.27: Bony ingrowth into a thread of a screw-implant.

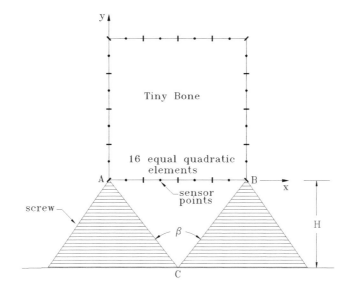

Figure 9.28: Discretized boundary element model.

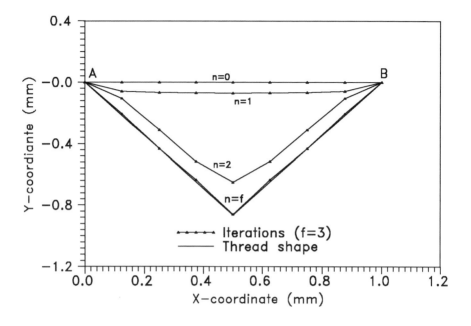

Figure 9.29: Bony ingrowth into the thread of a screw-implant with 17.5 MPa loading.

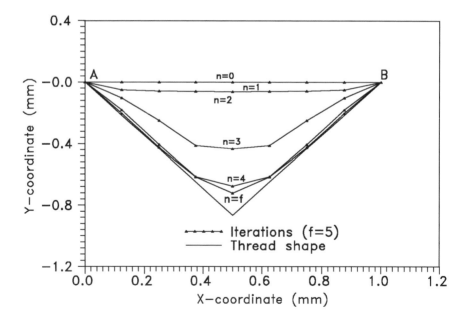

Figure 9.30: Bony ingrowth into the thread of a screw-implant with 40 MPa loading.

cation problems. The formulations presented here included the implicit differentiation method coupled with first order optimization algorithms.

More recently, the BEM has been combined with genetic algorithms to analyse shape optimization problems [2]. Application of evolutionary algorithms and the boundary element method to identification problems can be found in [6].

References

[1] Aliabadi, M.H. and Mellings, S.C., Boundary element formulations for sensitivity analysis and crack identification, *Boundary Integral Formulations for Inverse Analysis*, Computational Mechanics Publications, Southampton, 1997.

[2] Annicchiarico, W. and Cerrolaza, M., An evolutionary approach for the shape optimization of general boundary element models, *Advances in Boundary Element Techniques II*, Hoggar, Geneva, 289-296, 2001.

[3] Barone, M.R. and Yang, R.J., A boundary element approach for recovery of shape sensitivities in three dimensional elastic solids, *Computer Methods in Applied Mechanics and Engineering*, **74**, 69-82, 1989.

[4] Bonnet, M., Regularized BIE formulations for first- and second order shape sensitivity of elastic fields. *Computers and Structures*, **56**, 799-811, 1995.

[5] Burczynski, T. and Fedelinski, P., Boundary elements in shape design sensitivity analysis and optimal design of vibrating structures, *Engineering Analysis with Boundary Elements*, **9**, 195-201, 1992.

[6] Burczynski, T. and Dlugosz, A., Application of boundary element method in identification problems of thermoelasticity, *Advanves in Boundaru Element Techniques II*, Hoggar, Geneva, 563-570, 2001.

[7] Bezerra, L.M. and Saigal, S A., boundary element formulation for the inverse elastostatic problems (IESP) of flaw detection, *International Journal of Numerical Methods in Engineering*, **28**, 2795-2911, 1989.

[8] Choi, K.K. and Haug, E.J., Shape design sensitivity analysis of elastic structures, *Journal of Structural Mechanics*, **11**, 231-269, 1983.

[9] Choi, J.O. and Kwak, B.M., Boundary integral equation method for shape optimization of elastic structures, *International Journal for Numerical Methods in Engineering*, **26**, 1579-1595, 1988.

[10] Cowin, S.C. and Van Buskirk, W.C., Internal bone remodelling induced by a medullary pin, *Journal of Biomechanics*, **11**, 269-275, 1978.

[11] Cowin, S.C. and Van Buskirk, W.C. Surface remodelling induced by a medullary pin, *Journal of Biomechanics*, **12**, 269-276, 1979.

[12] Cowin, S.C., *Mechanical Properties of Bone*, CRC Press, Inc. Boca Raton, Florida, 1989.

[13] Erman, Z. and Fenner, R.T., Three-dimensional design optimization using the boundary integral equation method, *Journal of Strain Analysis*, **31**, 289-298, 1996.

[14] Fletcher, R., A Review of Methods for Unconstrained Optimization, *Optimization*, Academic Press, London, 1969.

[15] Greig, D.M., *Optimization*, Longman, London, 1980.

[16] Kane, J.H., Zhao, G., Wang, H, and Prasad,G., Boundary formulations for three-dimensional continuum structural shape sensitivity analysis, *ASME, Journal of Applied Mechanics*, **59**, 827-834, 1992.

[17] Kassab, A.J. Mosley, F.J. and Daryapurkar, A.B., Non-destructive detection of cavities by an inverse elastostatic boundary element method, *Engineering Analysis with Boundary Elements*, **13**, 44-55, 1994.

[18] Kubo, S. Ohji, K and Sakagami, T., A simple scheme for identification of a semi-elliptical surface crack from D.C. potential readings, *International Journal of Applied Electromagnetic in Materials*, **2**, 81-90, 1991.

[19] Kwak, B.M., A review on shape optimal design and sensitivity analysis, *Journal of Structural Mechanics and Earthquake Engineering*, JSCE No 483/I-26, 159s-174s, 1994.

[20] Lax, P.D. and Philipps, R.S., *Scattering Theory*, Academic Press, New York, 1967.

[21] Long, A. Unzueta, J. Scaeidt, E, Alvarez, A and Anza, J.J., A general related variational approach to shape optimum design, *Advances in Engineering Software*, **16**,135-142, 1993.

[22] Nishuimura, N. and Kobayashi, S., A boundary integral equation method for an inverse problem related to crack detection, *International Journal of Numerical Methods in Engineering*, **7**, 59-65, 1990.

[23] Mallardo, V. and Aliabadi, M.H., A BEM sensitivity and shape identification analysis for acoustic scattering in fluid-solid problems. *International Journal for Numerical Methods in Engineering*, **41**, 1527-1541, 1998.

[24] Martinez, M, Aliabadi, M.H. and Power, H., Bone remodelling using sensitvity analysis, *Journal of Biomechanics*, **31**, 1059-1062, 1998.

[25] Martinez, M.J., A boundary element sensitivity formulation for bone remodelling, PhD thesis, Wessex Institute of Technology, University of Wales, 1998.

[26] Mellings,S.C. and Aliabadi, M.H., Dual boundary element formulation for inverse potential problems in crack identification, *Engineering Analysis with Boundary Elements*, **12**, 275-281, 1993.

[27] Mellings, S.C., Flaw identification using the inverse dual boundary element method. Phd thesis, Wessex Institute of Technology, University of Portsmouth, 1994.

[28] Mellings, S.C. and Aliabadi, M.H., Flaw identification using the boundary element method, *International Journal of Numerical Methods in Engineering*, **38**, 399-419, 1995.

[29] Mellings, S.C. and Aliabadi, M.H., Three-dimensional flaw identification using inverse analysis, *International Journal of Engineering Sciences*, **34**, 453-469, 1996.

[30] Meric, R.A., Shape optimization of thermoelastic solids, *Journal of Therm. Sciences*, **11**, 187-206, 1988.

[31] Mitra, A.K. and Das, S., Solution of inverse problems by using the boundary element method, *Boundary Element Technology VII*, 721-731, 1992.

[32] Mota Soares, C.A., Leal, R.P. and Choi, K.K., Boundary elements in shape optimal design of structural components. *Proceedings of NATO ASI on Computer Aided Optimal Design: Structural and Mechanical Systems*, Springer-Verlag, Berlin, 605-631, 1987.

[33] Mukherjee, S. and Chandra, A., A boundary element formulation for design sensitivities in problems involving both geometric and material nonlinearities. *Mathematical and Computer Modelling*, **15**, 245-255, 1991.

[34] Pollard, J.E. and Kassab, A.J., Automated solution of an inverse heat conduction problems for the non-destructive detection of sub-surface cavities using boundary elements, *Boundary Element Technology VII*, 441-456, 1992.

[35] Sadegh, A.M., Luo, G.M., Cowin, S.C., Bone ingrowth: an application of the boundary element method to bone remodelling at the implant interface, *Journal of Biomechanics*, **26**, 167-182, 1993.

[36] Sakagami, T. Kubo, S. Hashimoto, T. Yamawaki, H. and Ohji, K., Quantitative measurement for two-dimensional inclined cracks by the electric-potential CT method with multiple current applications, *JSME International Journal Series*, *1*, **31**, 76-86, 1988.

[37] Tafreshi, A. and Fenner, R.T., Design sensitivity analysis using the boundary element method, *Journal of Strain Analysis*, **28**, 283-291, 1993.

[38] Tanaka, M. and Masuda, Y., Boundary element method applied to some inverse problems, *Engineering Analysis*, **3**, 138-143, 1986.

[39] Tanaka, M. Nakamura, N. and Nakano, N., Identification of cracks or defects by means of the elastodynamic BEM, *Advances in Boundary Elements*, *Vol3*, 183-194, 1989.

[40] Tosaka, N., Utani, A., New filter theory - boundary element method and its application to inverse problem, *Inverse Problems in Engineering Mechanics*, Balkema, Rotterdam, 453-460, 1994.

[41] Ulrich, T.W., Moslehy, F.A. and Kassab, A.J., A BEM based pattern search solution for a class of inverse elastostatic problems, *International Journal of Solids and Structures*, **33**, 2123-2131, 1996.

[42] Vanderplaats, G.N., Structural optimization - past, present and future, *AIAA Journal*, **20**, No 2, 992-1000, 1981.

[43] Zabaras,N., Morellas,V and Schnur,D., Spatially regularised solution of inverse elasticity problems using BEM, *Communications in Applied Numerical Methods*, **5**, 547-553, 1989.

[44] Zeng, X. and Saigal, S., An inverse formulation with boundary elements, *ASME, Journal of Applied Mechanics*, **59**, 835-840, 1992.

Chapter 10

Assembled Structures

10.1 Introduction

In this chapter applications of the boundary element method to built-up structures such as those of panels reinforced with stringers and ribs, lap joints and patches are described. There are many types of built-up structures requiring analysis of plates with beam attachments. For example, stiffened panels are the basic construction cell used in aircraft. They consist of a thin metal sheet, referred to as the panel skin, reinforced by longitudinal and transversal stiffeners. The stiffeners can be either machined to form an integral stiffened panel, attached to it by means of fasteners, or alternatively, bonded to the sheet. An early application of the two-dimensional boundary element method to plates reinforced with stringers can be found in Dowrick, Cartwright and Rooke [2] using constant elements. Later, Salgado and Aliabadi [7] presented a formulation using higher order elements. Latif and Aliabadi [5] used a similar formulation to analyse limit loads in reinforced concrete beams. Wen, Aliabadi and Young [14, 13] presented boundary element formulations for shear deformable plates and shells reinforced with longitudinal and transverse beams.

In some other engineering applications it may be necessary to analyse plates fastened with rivets or adhesively bonded. It is generally difficult to analyse mechanically fastened panels with the finite element method due to the necessity for modelling the rivet and the interaction between the panel and rivet. Young, Cartwright and Rooke [21] presented a boundary element formulation for adhesively bonded repair patches. Salgado and Aliabadi [8, 9] presented boundary element formulations for mechanically fastened and adhesively bonded repair patches. For the adhesively bonded repair patches [9], the dual reciprocity method was implemented to transfer the domain integrals due to the adhesive to the boundary. Widagdo and Aliabadi [19, 20] presented boundary element formulations for analyses of mechanically attached and adhesively bonded anisotropic composite patches. Wen, Aliabadi and Young [15, 16, 17] have presented boundary element formulations for mechanically attached and adhesively bonded patches to shear deformable plates and shells.

In this chapter, the formulations presented in [7, 8, 9] for two-dimensional problems and [14, 13, 16] for shear deformable plates are described. Several examples are presented to illustrate application of the BEM to reinforced panels.

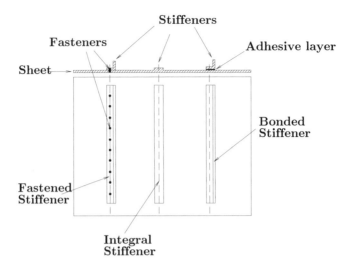

Figure 10.1: A stiffened panel.

10.2   Structures Reinforced with Beams

An illustration of a stiffened panel is presented in Figure 10.1, with a mechanically fastened stiffener on the left, an integral stiffener at the centre and a bonded stiffener on the right. Bonded and integral stiffeners are continuously attached to the skin, while mechanically fastened stiffeners are attached only at discrete points.

10.2.1   Two-dimensional Formulations

The sheet is assumed to be thin, so that the interaction forces with the stiffeners can be treated as action-reaction body forces. As the distributed body forces are confined to straight lines inside the domain, integrals in the displacement integral equation reduce to line integrals over the body forces loci. The boundary element equation for a thin sheet with M stiffeners continuously bonded to it can thus be written as

$$C_{ij}(\mathbf{x}')u_j(\mathbf{x}') = \int_S U_{ij}(\mathbf{x}',\mathbf{x})t_j(\mathbf{x})dS - \fint_S T_{ij}(\mathbf{x}',\mathbf{x})u_j(\mathbf{x})dS$$

$$+\frac{1}{h}\sum_{m=1}^{M}\int_{S_{S_m}} U_{ij}(\mathbf{x}',\mathbf{X})b_j^{S_m}(\mathbf{X})dS_{S_m} \tag{10.1}$$

where S_{S_m} stands for the stiffeners loci, $b_j^{S_m}$ represents the unknown stiffener attachment forces and h is the sheet thickness. Similarly, the displacements of an interior point $\mathbf{X}'$ inside the sheet domain are given by

$$u_j(\mathbf{X}') = \int_S U_{ij}(\mathbf{X}',\mathbf{x})t_j(\mathbf{x})\ dS - \int_S T_{ij}(\mathbf{X}',\mathbf{x})u_j(\mathbf{x})\ dS$$

$$+\frac{1}{h}\sum_{m=1}^{M}\int_{S_{S_m}} U_{ij}(\mathbf{X}',\mathbf{X})b_j^{S_m}(\mathbf{X})\ dS_{S_m} \tag{10.2}$$

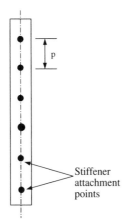

Figure 10.2: Discretly attached stringer.

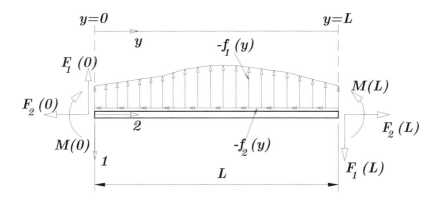

Figure 10.3: An isolated stiffener subjected to in-plane loads.

Mechanically fastened stiffeners, unlike the bonded stiffeners, are not continuously attached to the sheet. They are attached at discrete points, where the fasteners are located, as illustrated in Figure 10.2.

For BEM analysis, instead of being continuously distributed along the attachment lines, the interaction forces are considered as point forces, applied at the fastener loci. The integrals over the stiffener line reduce to a summation, i.e.

$$\int_{S_{S_m}} (\text{kernel}) b_j^{S_m} dS_{S_m} \text{ reduces to } \sum_{k=1}^{N^{S_{S_m}}} (\text{kernel}) b_j^{S_m}(\mathbf{x}^k) \tag{10.3}$$

where $N^{S_{S_m}}$ denotes the number of nodes in the mth stiffener locus, and $b_j^{S_m}(\mathbf{x}^k)$ are the nodal values of concentrated attachment forces at node k.

Stiffeners Modelling

Consider now the problem of an isolated stiffener of length L subjected to the following set of in-plane loads, as illustrated in Figure 10.3:

- Distributed loads $-f_j(y)$ acting throughout the length in both longitudinal and transverse directions;

- Concentrated loads $F_j(0)$ and $F_j(L)$ acting at the end points in both directions;

- Moments $M(0)$ and $M(L)$ acting at both end points.

If the stiffener is in equilibrium under the action of the above forces, its relative displacements $\Delta v_j(y)$ with respect to a rigid body motion of the stiffener are given by

$$\Delta v_1(y) = v_1(y) - v_1(0) = y\Upsilon(0) + \frac{1}{A_S\mu_S}\left\{yF_1(0) + \int_0^y (y-\eta)f_1(\eta)d\eta\right\}$$

$$-\frac{1}{I_S E_S}\left\{\frac{1}{2}y^2 M(0) + \frac{1}{6}y^3 F_1(0) + \int_0^y \frac{1}{6}(y-\eta)^3 f_1(\eta)d\eta\right\} \qquad (10.4)$$

and

$$\Delta v_2(y) = v_2(y) - v_2(0) = \frac{1}{A_S E_S}\left\{yF_2(0) + \int_0^y (y-\eta)f_2(\eta)d\eta\right\} \qquad (10.5)$$

where the indices 1 and 2 denote the directions of the stiffeners' local coordinate system as shown in Figure 10.3, y is an arc length parameter ($0 \le y \le L$), $v_i(0)$ and $\Upsilon(0)$ are rigid body translations and rotation of the reference point ($y = 0$), A_S is the stiffeners', cross-sectional area, I_S is the stiffeners' cross-sectional second moment of inertia, E_S and μ_S are the stiffeners material Young's modulus and shear modulus, respectively.

For the stiffener to be in equilibrium, the following equations have to be satisfied:

$$F_2(L) - F_2(0) = \int_0^L f_2(y)dy \qquad (10.6)$$

$$F_1(L) - F_1(0) = \int_0^L f_1(y)dy \qquad (10.7)$$

$$M(L) - M(0) - LF_1(0) = \int_0^L (L-y)f_1(y)dy \qquad (10.8)$$

Displacement Compatibility

The displacement compatibility conditions between the stiffners and the sheet are expressed, with respect to a reference point $\mathbf{X}^0$ at the same stiffener locus ($\mathbf{X}^0 \in S_{S_n}$), as

$$\Delta u_j(\mathbf{X}') - \Delta u_j^{S_m}(\mathbf{X}') = \frac{h_{A_m}}{\mu_{A_n}}\Delta \tau_j^{A_m}(\mathbf{X}') \qquad (10.9)$$

where h_{A_n} is the thickness of the adhesive layer, μ_{A_n} is the coefficient of shear deformation of the adhesive material, $\tau_j^{A_m}$ is the shear stress at the adhesive, $\Delta u_j(\mathbf{X}') =$

$u_j(\mathbf{X}') - u_j(\mathbf{X}^0)$, $\Delta u_j^{S_m}(\mathbf{X}') = u_j^{S_m}(\mathbf{X}') - u_j^{S_m}(\mathbf{X}^0)$, $\Delta \tau_j^{A_m}(\mathbf{X}') = \tau_j^{A_m}(\mathbf{X}') - \tau_j^{A_m}(\mathbf{X}^0)$.

The displacement compatibility conditions (10.9) for points at the stiffeners' attachment region are based on the assumption that the displacement u_j of a point $\mathbf{X}'$ ($\mathbf{X}' \in S_{S_m}$) at the sheet and $u_j^{S_n}$ of a corresponding point at the nth stiffener, has to be compatible with the shear deformation of the adhesive layer connecting the sheet to the stiffener. For the line stiffeners, the adhesive shear stresses $\tau_j^{A_m}$ are equal in value to the attachment forces $b_j^{S_m}$ divided by the width of the adhesive line w_{A_n}. The displacement compatibility equation can be written in terms of the body forces as

$$\Delta u_j(\mathbf{X}') - \Delta u_j^{S_m}(\mathbf{X}') = \Phi_{A_n} \Delta b_j^{S_m}(\mathbf{X}') \tag{10.10}$$

where $\Delta b_j^{S_m}(\mathbf{X}') = b_j^{S_m}(\mathbf{X}') - b_j^{S_m}(\mathbf{X}^0)$ and

$$\Phi_{A_m} = \frac{h_{A_m}}{w_{A_m} \mu_{A_m}} \tag{10.11}$$

is the coefficient of shear deformation of the adhesive.

Taking the reference point $\mathbf{X}^0$ to coincide with the stiffener starting point ($y = 0$), the relative displacements $\Delta u_j^{S_m}$ in equation (10.10) can be expressed as a function of the unknown interaction forces $b_j^{S_m}$, by using expressions (10.4) and (10.5). The relationship between the relative displacements and forces expressed in terms of the sheet and the stiffeners' coordinate systems is given by

$$\Delta u_i^{S_m} = \Theta_{ij}^{S_m} \Delta v_j^{S_m} \tag{10.12}$$

and

$$b_i^{S_m} = \Theta_{ij}^{S_m} f_j^{S_m} \tag{10.13}$$

where transformation matrix is given by

$$\Theta^{S_m} = \begin{bmatrix} +\cos\varphi^{S_m} & -\sin\varphi^{S_m} \\ +\sin\varphi^{S_m} & +\cos\varphi^{S_m} \end{bmatrix} \tag{10.14}$$

with φ^{S_m} denoting the angle between the sheet direction x_2 and the nth stiffener axis (as shown in Figure 10.3).

The relative displacements $\Delta u_j(\mathbf{X}') = u_j(\mathbf{X}') - u_j(\mathbf{X}^0)$ in the sheet can be written with the help of equation (10.2) as

$$\Delta u_j(\mathbf{X}') = \int_S \left[U_{ij}(\mathbf{X}',\mathbf{x}) - U_{ij}(\mathbf{X}^0,\mathbf{x}) \right] t_j(\mathbf{x}) \, dS$$

$$- \int_S \left[T_{ij}(\mathbf{X}',\mathbf{x}) - T_{ij}(\mathbf{X}^0,\mathbf{x}) \right] u_j(\mathbf{x}) \, dS$$

$$+ \frac{1}{h} \sum_{m=1}^M \int_{S_{S_m}} \left[U_{ij}(\mathbf{X}',\mathbf{X}) - U_{ij}(\mathbf{X}^0,\mathbf{X}) \right] b_j^{S_m}(\mathbf{X}) \, dS_{S_m} \tag{10.15}$$

The stiffeners' loci S_S can be discretized with straight continuous elements.

Consider a discretization involving N^S nodes at the sheet boundary and N^{S_S} nodes at the stiffeners' attachment loci. Before introduction of the boundary conditions, the total number of boundary unknowns is given by the usual $2N^S$ displacements and

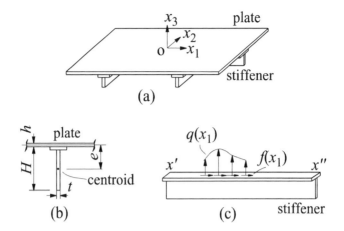

Figure 10.4: Stiffened plate and interaction forces in the attachment region.

tractions. There are, however, $2N^{Ss} + N$ additional unknowns, which comprise two attachment forces $b_j^{S_n}$ per node in S_{S_n} plus one stiffener rigid body rotation coefficient $\Upsilon^{S_n}(0)$ per stiffener (see equation (10.4)).

A system of linear equations is constructed by applying the equations presented above in the following manner [7]:

- The displacement equation (10.1) to all sheet boundary nodes.

- The displacement compatibility equation (10.10) to all nodes at the stiffeners' loci, with the exception of the starting node in each stiffener. This node is used as the reference point in relation to which equations (10.4), (10.5) and (10.2) are written.

- The equilibrium equations (10.6), (10.7) and (10.8) for each stiffener.

10.2.2 Plate Bending Formulations

Consider a stiffened plate as shown in Figure 10.4(a)(b) with stiffener m ($m = 1, 2, ...M$) at $x_2^m =$ constant; the axial direction of the beam is along x_1- axis, and the coordinates of ends of the beam are denoted as x_1' and x_1''. It is assumed that the effect of torsional deflection around axis x is much too small and can be ignored compared with the bending around axis x_2; in this case there are two interaction forces $q_m(x_1)$ and $f_m(x_1)$, as shown in Figure 10.4(c), acting along the beam [14]. From the configuration shown in Figure 10.5, the displacement along the beam axis can be written as follows:

$$u_m^b(x_1) = u_1(x_1, x_2^m) + e_m \frac{dw_3}{dx_1} \tag{10.16}$$

where e_m denotes the distance from the mid–plane of the plate to the centroid of the beam, and u_1 and w_3 denote the longitudinal displacement and deflection of plate along x_1 at $x_2 = x_2^m$. According to the elastic beam theory, the deflection of beam

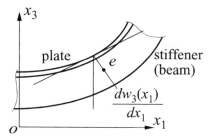

Figure 10.5: Deformation of stiffened plate.

should satisfy the equation

$$[EI]_m \frac{d^4 w_3}{dx_1^4} = q_m(x_1) + e_m \frac{df_m(x_1)}{dx_1} \tag{10.17}$$

where $[EI]_m$ is the beam bending stiffness. From simple beam theory, the following relationship holds along the axial direction:

$$\frac{d^2 u_m^b}{dx_1^2} = -\frac{f_m(x_1)}{[EA]_m} \tag{10.18}$$

where A_m is the area of beam section. Substituting equation (10.18) into equations (10.16) and (10.17) results in

$$([EI]_m + [EA]_m e_m^2) \frac{d^4 w_3}{dx_1^4} = q_m(x_1) - [EA]_m e_m \frac{d^3 u_1}{dx_1^3} \tag{10.19}$$

The interaction forces and moment can be approximated by interpolation functions as

$$q_m(x_1) = \sum_{n=1}^{N} L_n(x_1) q_m^n, \quad f_m(x_1) = \sum_{n=1}^{N} L_n(x_1) f_m^n \tag{10.20}$$

where N denotes the number of nodes on the beam, and q_m^n and f_m^n are the nodal values of interaction forces. The Lagrangian shape functions $L_n(x_1)$ are given by

$$L_n(x_1) = \prod_{p=1,p\neq n}^{N} (x_1 - x_{1p}) / \prod_{p=1,p\neq n}^{N} (x_{1n} - x_{1p}) \tag{10.21}$$

The in-plane displacements $u_1(x_1)$ are unknown at the two ends of the beam, so

$$u_1(x_1, x_2^m) = \sum_{n=1}^{N+2} L_n(x_1) u_1^n \tag{10.22}$$

where $x_{1(N+1)} = x'$ and $x_{1(N+2)} = x''$ as shown in Figure 10.4(c), and u_1^n are the nodal values of displacement. Substituting equations (10.20) and (10.22) into (10.19) and solving the differential equation (10.19), the deflection of the beam can be obtained as

$$w_3(x_1) = \frac{1}{[EI_s]_m} \sum_{n=1}^{N} Q_n(x_1) q_m^n - \frac{A_m e_m}{I_s} \sum_{n=1}^{N+2} P_n(x_1) u_1^n \tag{10.23}$$

where $I_s = I + A_m e_m^2$. The coefficients in the deflection equation (2.13) are given as

$$Q_n(x_1) = [\bar{L}_n(x_1) + c_0 + c_1 x_1 + c_2 x_1^2 + c_3 x_1^3] \tag{10.24}$$

where

$$\bar{L}_n(x_1) = \int\int\int\int L_n(x_1)dx_1 dx_1 dx_1 dx_1$$

and constants c_0, c_1, c_2 and c_3 can be determined by applying the boundary conditions of the beam at the two ends x' and x''. These coefficients can be obtained from

$$Q_n(x') = Q_n(x'') = 0, \quad \left.\frac{d^2 Q_n}{dx_1^2}\right|_{x_1 = x'} = \left.\frac{d^2 Q_n}{dx_1^2}\right|_{x_1 = x''} = 0 \tag{10.25}$$

for a simply supported beam, and

$$Q_n(x') = Q_n(x'') = 0, \quad \left.\frac{dQ_n}{dx_1}\right|_{x_1 = x'} = \left.\frac{dQ_n}{dx_1}\right|_{x_1 = x''} = 0 \tag{10.26}$$

for a clamped beam. In the same way, solution $P_n(x_1)$ can be written as

$$P_n(x_1) = [\bar{L}_n(x_1) + c_0' + c_1' x_1 + c_2' x_1^2 + c_3' x_1^3] \tag{10.27}$$

where

$$\bar{L}_n(x_1) = \int L_n(x_1)dx_1$$

where constants c_0', c_1', c_2' and c_3' can be determined by applying boundary conditions of the beam at two ends x' and x''. A for different support conditions at the two ends. Similarly, axial beam displacements can be determined from equation (10.18), as

$$u_m^b(x_1) = \frac{1}{[EA]_m} \sum_{n=1}^N F_n(x_1) f_m^n + c_0 + c_1 x_1$$

$$= u_1(x_1) + e_m \frac{dw_3}{dx_1} \tag{10.28}$$

where

$$F_n(x_1) = -\bar{L}_n(x_1) - \left(\bar{L}_n(x') - \bar{L}_n(x'')\right) x_1 + \frac{1}{2}\left(\bar{L}_n(x') + \bar{L}_n(x'')\right)$$

$$\bar{L}_n(x_1) = \int\int L_n(x_1)dx_1 dx_1$$

and c_0, c_1 can be determined by displacement conditions at the two ends

$$c_0 = \frac{1}{x'' - x'}[x'' u_1(x') - x' u_1(x'') + x'' e w_3'(x') - x' e w_3'(x'')]$$

$$c_1 = \frac{1}{x'' - x'}[u_1(x'') - u_1(x') + e w_3'(x'') - e w_3'(x')]$$

where $w_3' = dw_3/dx_1$. Equation (10.28) can be arranged to give

$$\frac{1}{[EA]_m} \sum_{n=1}^N F_n(x_1) f_m^n = u_1(x_1) - c_0 - c_1 x_1$$

$$+e_m \left[\frac{1}{[EI_s]_m} \sum_{n=1}^{N} \frac{dQ_n(x_1)}{dx_1} q_m^n - \frac{Ae_m}{I_s} \sum_{n=1}^{N+2} \frac{dP_n(x_1)}{dx_1} u_1^n \right] \qquad (10.29)$$

Taking the collocation point at each of the nodes on the beam $x_1 = x_1^k, k = 1, 2, ..., N$, equations (10.23) and (10.29) can be rearranged and written in matrix form as

$$\mathbf{Q}\mathbf{q}_m = \mathbf{w} + \mathbf{P}\mathbf{u} \qquad (10.30)$$

and

$$\mathbf{F}\mathbf{f}_m = \mathbf{G}\mathbf{q} + \mathbf{H}\mathbf{u} \qquad (10.31)$$

Finally, the nodal values of interaction forces $\mathbf{q}_m^n$ and $\mathbf{f}_m^n$ can be obtained

$$\mathbf{q}_m = \mathbf{Q}^{-1}\mathbf{w} + \mathbf{Q}^{-1}\mathbf{P}\mathbf{u} \qquad (10.32)$$

and

$$\mathbf{f}_m = \mathbf{F}^{-1}\mathbf{H}\mathbf{Q}^{-1}\mathbf{w} + \mathbf{F}^{-1}(\mathbf{G}\mathbf{Q}^{-1}\mathbf{P} + \mathbf{H})\mathbf{u} \qquad (10.33)$$

where $m = 1, 2, ..., M$.

Modelling of Stiffened Plates

From equations (10.32) and (10.33), it follows that the interaction forces and moments can be represented in terms of the displacements along the beam. Assuming there is only uniform pressure load q_0 acting on the plate,

$$f_{1m}^s(x_1, x_2) = -\sum_{m=1}^{M} f_m(x_1)\delta(x_2 - x_2^m), \quad f_{1m}^s(x_1, x_2) = 0 \qquad (10.34)$$

and

$$f_{1m}^p(x_1, x_2) = f_{2m}^p(x_1, x_2) = 0, \quad f_{3m}^p(x_1, x_2) = q_0 - q_m(x_1)\delta(x_2 - x_2^m) \qquad (10.35)$$

where superscripts s and p refer to stiffeners and plate, respectively, x_2^m and $\delta(x_2 - x_2^m)$ denote the position of stiffener $m = 1, 2, ...M$ and a Dirac delta function, respectively. Using quadratic elements, the boundary integral equation for plate bending (see Chapter 3) can be written in a discretized form as

$$C_{\alpha\beta}^b(\mathbf{x}')u_\beta(\mathbf{x}') + \sum_{n=1}^{L_0} \sum_{l=1}^{3} u_\beta^l \int_{-1}^{1} T_{\alpha\beta}^b[\mathbf{x}', \mathbf{x}(\eta)]N_l(\eta)J^n(\eta)d\eta$$

$$- \sum_{n=1}^{L_0} \sum_{l=1}^{3} t_\beta^l \int_{-1}^{1} U_{\alpha\beta}^b[\mathbf{x}', \mathbf{x}(\eta)]N_l(\eta)J^n(\eta)d\eta$$

$$= -\sum_{m=1}^{M} \int_{x'}^{x''} U_{\alpha1}^b(\mathbf{x}', x_1)f_m(x_1)dl \qquad (10.36)$$

where $\alpha, \beta = 1, 2, 3$, L_0 is the total number of elements, $N_l(\eta)$ $(l = 1, 2, 3)$ are the shape function and $J^n(\eta) = dS/d\eta$ is the Jacobian of transformation. Similarly, the displacement integral equation for in-plane displacements becomes

$$C_{ik}^m(\mathbf{x}')w_k(\mathbf{x}') + \sum_{n=1}^{L_0} \sum_{l=1}^{3} w_k^l \int_{-1}^{1} T_{ik}^m[\mathbf{x}', \mathbf{x}(\eta)]N_l(\eta)J^n(\eta)d\eta$$

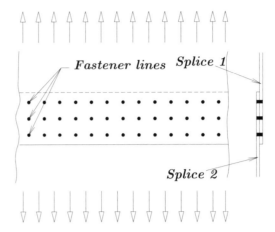

Figure 10.6: A lap joint with three lines of fasteners attaching the overlapping splices.

$$-\sum_{n=1}^{L_0}\sum_{l=1}^{3}p_k^l\int_{-1}^{1}U_{ik}^m[\mathbf{x}',\mathbf{x}(\eta)]N_l(\eta)J^n(\eta)d\eta$$

$$=q_0\int_{V}U_{i3}^m(\mathbf{x}',\mathbf{X})dV(\mathbf{X})-\sum_{m=1}^{M}\int_{x'}^{x''}U_{i3}^m(\mathbf{x}',x_1)q_m(x_1)dl \qquad (10.37)$$

where $q_m(x_1), f_m(x_1)$ are represented in equation (10.20) and can be expressed in terms of deflection and displacements of a stiffener from equations (10.32) and (10.33). The domain integrals in (10.37) can be evaluated as shown in [4]. After the collocation point passes through all the collocation nodes on the boundary S and on the line $x_2 = x_2^m$ (in this case $c_{\alpha\beta}^b = \delta_{\alpha\beta}, c_{ik}^m = \delta_{ik}$ in the above equations), equations (10.36) and (10.37) give the following linear equations in matrix form:

$$\left\{\begin{array}{cc}\mathbf{H}^s & \mathbf{H}^w \\ \mathbf{H}^u & \mathbf{H}^p\end{array}\right\}\left\{\begin{array}{c}\mathbf{u} \\ \mathbf{w}\end{array}\right\}=\left\{\begin{array}{cc}\mathbf{G}^s & \mathbf{0} \\ \mathbf{0} & \mathbf{G}^p\end{array}\right\}\left\{\begin{array}{c}\mathbf{t} \\ \mathbf{p}\end{array}\right\}+\left\{\begin{array}{c}\mathbf{0} \\ \mathbf{b}\end{array}\right\} \qquad (10.38)$$

where $\mathbf{u} = \{\mathbf{u}_1, \mathbf{u}_2\}^\top, \mathbf{w} = \{\mathbf{w}_1, \mathbf{w}_2, \mathbf{w}_3\}^\top; \mathbf{H}^s, \mathbf{G}^s, \mathbf{H}^p$ and $\mathbf{G}^p$ are boundary element influence matrices for plane stress elasticity and plate bending problems, respectively, $\mathbf{H}^w, \mathbf{H}^u$ are coupling matrices caused by the interaction between the beam and the plate, and $\mathbf{b}$ denotes domain integrals for distributed load in the domain. Considering the boundary conditions on S, the unknown of boundary values and rivet forces along $x_2 = x_2^m$ can be obtained by solving the above linear system.

10.3 Patch Attachments

Bolted or riveted joints and adhesively bonded joints are common in many engineering structures. For example, aircraft fuselage lap joints are constructed by overlapping two portions of the skin and fastening them together, as illustrated in Figure 10.6. Accurate determination of the fastener forces plays an important role in the design of mechanically fastened repairs.

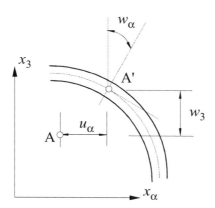

Figure 10.7: Coordinate system of shear deformable plate and definitions of displacement, deflection and rotation.

10.3.1 Two-dimensional Riveted Joints Model

Consider a sheet to which M patches are attached by means of fasteners. If the sheet and patches are thin, two-dimensional elastostatics theory can be used to study the problem. The point force model adopted for the fasteners and their shear flexibility are the same as those described in the earlier section [8]. Displacement compatibility between the sheet, patches and fasteners is imposed.

Let $b_j(\mathbf{X}^d)$ represent the unknown point forces exchanged between the sheet and the patches at the fastener attachment points $\mathbf{X}^d$. If those forces are treated as discrete body forces, the boundary integral displacement equation for a source point $\mathbf{x}'$ at the boundary of the sheet can be written as

$$C_{ij}(\mathbf{x}')u_j(\mathbf{x}') = \int_S U_{ij}(\mathbf{x}',\mathbf{x})t_j(\mathbf{x})dS - \int_S T_{ij}(\mathbf{x}',\mathbf{x})u_j(\mathbf{x})dS$$

$$+ \frac{1}{h}\sum_{m=1}^{M}\sum_{d=1}^{D_m}U_{ij}(\mathbf{x}',\mathbf{X}^d)b_j^m(\mathbf{X}^d) \qquad (10.39)$$

where D_m is the number of fasteners used to attach the mth patch to the sheet.

The displacements of a point $\mathbf{x}'$ at the boundary of the mth patch are given by

$$C_{ij}^{P_m}(\mathbf{x}')u_j^{P_m}(\mathbf{x}') = \int_{S_{P_m}} U_{ij}(\mathbf{x}',\mathbf{x})t_j^{P_m}(\mathbf{x})dS_{P_m} - \int_{S_{P_m}} T_{ij}(\mathbf{x}',\mathbf{x})u_j^{P_m}(\mathbf{x})dS_{P_m}$$

$$- \frac{1}{h_{P_m}}\sum_{d=1}^{D_m}U_{ij}(\mathbf{x}',\mathbf{X}^d)b_j^m(\mathbf{X}^d) \qquad (10.40)$$

where $u_j^{P_m}$ and $t_j^{P_m}$ are displacements and tractions at the patch boundary, h_{P_m} is the thickness of the patch, $C_{ij}^{P_m}$ is a coefficient whose value depends on the position of the source point $\mathbf{x}'$ with respect to the patch boundary S_{P_m} and can be determined by rigid body movement considerations.

The displacements u_j and $u_j^{P_m}$ of a point $\mathbf{X}^d$ ($\mathbf{X}^d \in V_{P_m}$) at the sheet and a corresponding point at the mth patch have to be compatible with the shear deformation of the fastener connecting the sheet to the m-th patch at the point $\mathbf{X}^d$. This condition gives rise to M sets of equations expressed by the following relationship:

$$u_j(\mathbf{X}^d) - u_j^{P_m}(\mathbf{X}^d) = \Phi_{A_m} b_j^m(\mathbf{X}^d) \tag{10.41}$$

where Φ_{A_m} is the coefficient of shear deformation of the fasteners used to attach the sheet to the mth patch. The displacements $u_j(\mathbf{X}')$ of an internal sheet point are given by

$$u_j(\mathbf{X}') = \int_S U_{ij}(\mathbf{X}',\mathbf{x})t_j(\mathbf{x})dS - \int_S T_{ij}(\mathbf{X}',\mathbf{x})u_j(\mathbf{x})dS$$

$$+\frac{1}{h}\sum_{m=1}^{M}\sum_{d=1}^{D_m}U_{ij}(\mathbf{X}',\mathbf{X}^d)b_j^m(\mathbf{X}^d) \tag{10.42}$$

and the displacements $u_j^{P_m}(\mathbf{X}')$ of an internal patch point by

$$u_j^{P_m}(\mathbf{X}') = \int_{S_{P_m}} U_{ij}(\mathbf{X}',\mathbf{x})t_j^{P_m}(\mathbf{x})dS_{P_m} - \int_{S_{P_m}} T_{ij}(\mathbf{X}',\mathbf{x})u_j^{P_m}(\mathbf{x})dS_{P_m}$$

$$-\frac{1}{h_{P_m}}\sum_{d=1}^{D_m}U_{ij}(\mathbf{X}',\mathbf{X}^d)b_j^m(\mathbf{X}^d) \tag{10.43}$$

After solution of the system of equations, stresses at internal points $\mathbf{X}'$ in the sheet are calculated from the following integral equation:

$$\sigma_{ij}(\mathbf{X}') = \int_S D_{kij}(\mathbf{X}',\mathbf{x})t_k(\mathbf{x})dS - \int_S S_{kij}(\mathbf{X}',\mathbf{x})u_k(\mathbf{x})dS$$

$$+\frac{1}{h}\sum_{m=1}^{M}\sum_{d=1}^{D_m}D_{kij}(\mathbf{X}',\mathbf{X}^d)b_k^m(\mathbf{X}^d) \tag{10.44}$$

Similarly, stresses at internal points $\mathbf{X}'$ in the patch are calculated from:

$$\sigma_{ij}^{P_m}(\mathbf{X}') = \int_{S_{P_m}} D_{kij}(\mathbf{X}',\mathbf{x})t_k^{P_m}(\mathbf{x})dS_{P_m} - \int_{S_{P_m}} S_{kij}(\mathbf{X}',\mathbf{x})u_k^{P_m}(\mathbf{x})dS_{P_m}$$

$$-\frac{1}{h_{P_m}}\sum_{d=1}^{D_m}D_{kij}(\mathbf{X}',\mathbf{X}^d)b_k^m(\mathbf{X}^d) \tag{10.45}$$

When the source point $\mathbf{X}'$ and the field point $\mathbf{X}^d$ are the same ($\mathbf{X}' = \mathbf{X}^d$) in equations (10.42), (10.43), (10.44) and (10.45), the fundamental solutions $U_{ij}(\mathbf{X}',\mathbf{X}^d)$ and $D_{kij}(\mathbf{X}',\mathbf{X}^d)$ become singular. This difficulty is dealt with in [7], by using a distributed force model.

10.3.2   Plate Bending Formulations

Let us define w_α as rotations in the x_α direction, w_3 as deflection of the sheet along the x_3-axis (see Figure 10.7), and $\bar{z}_k$ ($k = 1,2,3$) as the distribution of body forces per unit area; $\bar{z}_\alpha(\alpha = 1,2)$ represents body forces of moment and $\bar{z}_3$ the out-of-plane body force along the axis x_3.

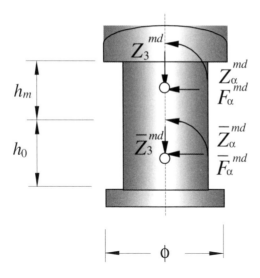

Figure 10.8: Geometry of rivet and applied forces.

Let $\bar{F}_\alpha^{md}(\mathbf{X}^d)$ and $\bar{Z}_k^{md}(\mathbf{X}^d)$ represent the five concentrated forces (unknown) on the middle plane of sheet in the area of attachment m at rivet geometry central point $\mathbf{X}^d$ (see Figure 10.8) and $A_m^d (= \pi\phi^2/4)$ denotes the section area of rivet and ϕ denotes the diameter of rivet, thus

$$\bar{f}_\alpha^m(\mathbf{X}) = \bar{f}_\alpha^0 + \frac{\bar{F}_k^{md}(\mathbf{X}^d)}{A_m^d}, \quad \bar{z}_k^{md}(\mathbf{X}) = \bar{z}_k^0 + \frac{Z_k^{md}(\mathbf{X}^d)}{A_m^d} \qquad \mathbf{X} \in A_m^d$$

where $\bar{f}_\alpha^0, \bar{z}_k^0$ are applied body forces. Using the above relationships, the boundary integral equation for plate bending (see Chapter 3) can be rewritten as

$$C_{\alpha\beta}^b(\mathbf{x}')u_\beta(\mathbf{x}') = \int_S U_{\alpha\beta}^b(\mathbf{x}',\mathbf{x})t_\beta(\mathbf{x})dS - \int_S T_{\alpha\beta}^b(\mathbf{x}',\mathbf{x})u_\beta(\mathbf{x})dS$$

$$+\frac{1}{h}\int_V U_{\alpha\beta}^b(\mathbf{x}',\mathbf{X})\bar{f}_\beta^0(\mathbf{X})dV + \frac{1}{h}\sum_{m=1}^M\sum_{d=1}^{D_m} U_{\alpha\beta}^b(\mathbf{x}',\mathbf{X}^d)\bar{F}_\beta^{md}(\mathbf{X}^d) \qquad (10.46)$$

and similarly, the integral equation for in-plane displacements can be rewritten as

$$C_{ik}^m(\mathbf{x}')w_k(\mathbf{x}') = \int_S U_{ik}^m(\mathbf{x}',\mathbf{x})p_k(\mathbf{x})dS - \int_S T_{ik}^m(\mathbf{x}',\mathbf{x})w_k(\mathbf{x})dS$$

$$+\int_V U_{ik}^m(\mathbf{x}',\mathbf{X})\bar{z}_k^0(\mathbf{X})dV + \sum_{m=1}^M\sum_{d=1}^{D_m} U_{ik}^m(\mathbf{x}',\mathbf{X}^d)\bar{Z}_k^{md}(\mathbf{X}^d) \qquad (10.47a)$$

where M, D_m represent the number of patches on the sheet and the number of rivet on attachment m.

When the collection point $\mathbf{X}'$ is taken to lie on the centre of rivet area A_m^d, the free terms $C_{\alpha\beta}^s$ and C_{ik}^s are equal to $\delta_{\alpha\beta}$ and δ_{ik}, respectively, and the displacement equation in the domain V can be written as

$$u_\alpha(\mathbf{X}') = \int_S U^m_{\alpha\beta}(\mathbf{X}', \mathbf{x}) t_\beta(\mathbf{x}) dS - \int_S T^m_{\alpha\beta}(\mathbf{X}', \mathbf{x}) u_\beta(\mathbf{x}) dS$$

$$+ \frac{1}{h} \int_V U^m_{\alpha\beta}(\mathbf{X}', \mathbf{X}) \bar{f}^0_\beta(\mathbf{X}) dV + \frac{\bar{F}^{md}_\beta}{A^d_m h} \sum_{m=1}^M \int_{A^d_m} U^m_{\alpha\beta}(\mathbf{X}', \mathbf{X}^d) dA \qquad (10.48)$$

and

$$w_i(\mathbf{X}') = \int_S U^m_{ik}(\mathbf{X}', \mathbf{x}) p_k(\mathbf{x}) dS - \int_S T^m_{ik}(\mathbf{X}', \mathbf{x}) w_k(\mathbf{x}) dS$$

$$+ \int_V U^m_{ik}(\mathbf{X}', \mathbf{X}) \bar{z}^0_k(\mathbf{X}) dV + \frac{\bar{Z}^{md}_k}{A^d_m} \sum_{m=1}^M \int_{A^d_m} U^m_{ik}(\mathbf{X}', \mathbf{X}^d) dA \qquad (10.49)$$

The displacements of a point $\mathbf{x}'$ at the boundary of the mth patch of thickness h_m can be represented using boundary integral equations similar to (10.46) and (10.47a), as shown by Wen, Aliabadi and Young [15].

The displacements, rotations and deflections on the rivets $\mathbf{X}^d$ at the sheet and a corresponding point at patch m are compatible with the shear deformation of the rivet. Therefore M sets of connection conditions can be written as

$$u^d_\alpha - u^{md}_\alpha = \Phi^K \bar{F}^{md}_\alpha \pm \frac{h + h_m}{2} w^d_\alpha \qquad \alpha = 1, 2 \qquad (10.50)$$

$$w^d_k - w^{md}_k = \Phi^V \bar{Z}^{md}_k \qquad k = 1, 2, 3 \qquad (10.51)$$

$$d = 1, , 2...D_m; \quad m = 1, 2, ..., M$$

where u^d_α, w^d_k are the displacement of the sheet on rivet $\mathbf{X}^d$, and $\bar{F}^{md}_\alpha$ and $\bar{Z}^{md}_k$ represent the body forces in the section area A^d_m of rivet m. For the patches on the top surface of the sheet, the sign "+" is selected. Φ^K is defined as the coefficients of shear deformation of the fasteners which are used to connect the sheet and patch, and can be determined from the empirical equation derived by Swift [12]

$$\Phi^K = \frac{1}{E_R \phi} \left\{ a_0 + a_1 \phi \left[\frac{1}{h} + \frac{1}{h_m} \right] \right\} \qquad (10.52)$$

where E_R is the Young's modulus of rivet material, h and h_m are the thicknesses of sheet and patch, as shown in Figure 10.8. Two constants a_0 and a_1 are chosen as 5 and 0.8, respectively, to correspond to Aluminium rivets [12]. Φ^V are coefficients of bending and the axial deformation of rivet, and in the following analysis they are ignored, i.e. $\Phi^V = 0$.

Consider the equilibrium of fasteners, there are $5 \times \sum_{m=1}^M D_m$ equations related to the concentrate forces on rivet $\mathbf{X}^d$:

$$\bar{F}^{md}_\alpha + F^{md}_\alpha = 0$$

$$\bar{Z}^{md}_3 + Z^{md}_3 = 0$$

$$\bar{Z}^{md}_\alpha + Z^{md}_\alpha \pm \frac{h + h_m}{2} F^{md}_\alpha = 0 \qquad (10.53)$$

for $\alpha = 1, 2$.

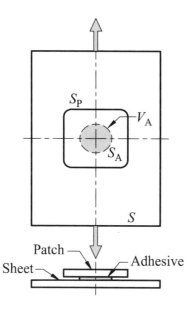

Figure 10.9: A patch adhesively attached to a sheet.

10.3.3 Adhesively Bonded Patches

Figure 10.9 presents a finite sheet to which a patch is attached by means of a layer of adhesive. The sheet boundary is denoted by S and the patch boundary by S_P. The adhesive is placed over a patch sub-region called the attachment region and denoted by V_A with boundary S_A. When the sheet deforms under the action of tractions and displacement constraints applied at its boundary, interaction forces develop over the attachment region. If the structures remain plane after deformation (i.e. the sheet, patch and adhesive layer are thin), two-dimensional elastostatics theory can be used to study the problem. In that case, the attachment interaction forces can be treated as unknown (action-reaction) body forces exchanged by the sheet and patch over the attachment sub-region. The displacements at the sheet and patch have to be compatible with the shear deformation of the adhesive layer connecting them.

Consider a configuration similar to that presented in Figure 10.9, but with an arbitrary number (M) of patches adhesively bonded to the sheet. The boundary integral displacement equation for a source point $\mathbf{x}'$ at the sheet is given by:

$$C_{ij}(\mathbf{x}')u_j(\mathbf{x}') = \int_S U_{ij}(\mathbf{x}',\mathbf{x})t_j(\mathbf{x})dS - \int_S T_{ij}(\mathbf{x}',\mathbf{x})u_j(\mathbf{x})dS$$

$$+\frac{1}{h}\sum_{m=1}^M \int_{V_{A_m}} U_{ij}(\mathbf{x}',\mathbf{x})b_j^{A_m}(\mathbf{x})dV_{A_m}(\mathbf{x}) \qquad (10.54)$$

where $b_j^{A_m}$ represents the attachment forces exchanged over the m-th attachment region V_{A_m}.

Similarly, the displacement of a point $\mathbf{x}'$ at the mth patch is given by

$$C_{ij}^{P_m}(\mathbf{x}')u_j^{P_m}(\mathbf{x}') = \int_{S_{P_m}} U_{ij}(\mathbf{x}',\mathbf{x})t_j^{P_m}(\mathbf{x})dS_{P_m} - \int_{S_{P_m}} T_{ij}(\mathbf{x}',\mathbf{x})u_j^{P_m}(\mathbf{x})dS_{P_m}$$

$$-\frac{1}{h_{P_m}}\int_{V_{A_m}}U_{ij}(\mathbf{x}',\mathbf{x})b_j^{A_m}(\mathbf{x})dV_{A_m} \tag{10.55}$$

where $u_j^{P_m}$ and $t_j^{P_m}$ are displacements and tractions at the patch, h_{P_m} is the thickness of the patch, $C_{ij}^{P_m}$ is a coefficient whose value depends on the position of the source point $\mathbf{x}'$ with respect to the patch boundary S_{P_m} and can be determined by rigid body movement considerations.

The difference Δu_j in the displacement u_j of a point $\mathbf{x}'$ ($\mathbf{x}' \in V_{A_m}$) at the sheet and $u_j^{P_m}$ of a corresponding point at the mth patch has to be compatible with the shear deformation of the adhesive layer connecting the sheet to the mth patch. This condition gives rise to M sets of equations expressed by the following relationship:

$$\Delta u_j^{P_m}(\mathbf{x}') = \left[u_j(\mathbf{x}') - u_j^{P_m}(\mathbf{x}')\right] = \frac{h_{A_m}}{\mu_{A_m}}\tau_j^{A_m}(\mathbf{x}') \tag{10.56}$$

where h_{A_m} is the thickness of the adhesive layer, μ_{A_m} is the coefficient of shear deformation of the adhesive material, and $\tau_j^{A_m}$ is the shear stress at the adhesive, which is equal in value to the attachment forces $b_j^{A_m}$. The displacement compatibility equation can therefore be written in terms of the body forces as

$$u_j(\mathbf{x}') - u_j^{P_m}(\mathbf{x}') = \frac{h_{A_m}}{\mu_{A_m}}b_j^{A_m}(\mathbf{x}') \tag{10.57}$$

Most equations presented above require integrals to be carried out over the patch attachment domains V_{A_m}. Transformation of these domain integrals into boundary integrals can be accomplished through the use of the Dual Reciprocity Method [9]. The attachment forces are unknown body forces acting within the attachment regions domain V_{A_m}. They can be approximated using the DRM as a sum of unknown coefficients $\alpha^{A_m d}$ multiplied by coordinate functions $f^{A_m d}(\mathbf{x}^{A_m d},\mathbf{x})$:

$$b_j^{A_m}(\mathbf{x}) = \sum_{d=1}^{D_m}\alpha_j^{A_m d}f^{A_m d}(\mathbf{x}^{A_m d},\mathbf{x}) \tag{10.58}$$

The points $\mathbf{x}^{A_m d}$ are DRM collocation points (see Chapter 2). The coordinate function $f^{A_m d}(\mathbf{x}^{A_m d},\mathbf{x}) = c + r(\mathbf{x}^{A_m d},\mathbf{x})$ was chosen, where c is a constant and $r(\mathbf{x}^{A_m d},\mathbf{x})$ is the distance between the DRM collocation point $\mathbf{x}^{A_m d}$ and the field point $\mathbf{x}$.

Substituting the body forces in equation (10.58) with the right-hand side of equation (10.58) and using the reciprocity theorem, the displacement equation for a source point $\mathbf{x}'$ at the sheet can be re-written as

$$C_{ij}(\mathbf{x}')u_j(\mathbf{x}') = \int_S U_{ij}(\mathbf{x}',\mathbf{x})t_j(\mathbf{x})dS - \int_S T_{ij}(\mathbf{x}',\mathbf{x})u_j(\mathbf{x})dS$$

$$+\frac{1}{h}\sum_{m=1}^{M}\sum_{d=1}^{D_m}\alpha_k^{A_m d}\left[C_{ij}^{A_m}(\mathbf{x}')\widehat{u}_{kj}^{A_m d}(\mathbf{x}') + \int_{S_{A_m}}T_{ij}(\mathbf{x}',\mathbf{x})\widehat{u}_{kj}^{A_m d}(\mathbf{x})dS_{A_m}\right.$$

$$\left. - \int_{S_{A_m}}U_{ij}(\mathbf{x}',\mathbf{x})\widehat{t}_{kj}^{A_m d}(\mathbf{x})dS_{A_m}\right]$$

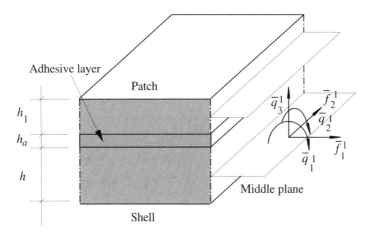

Figure 10.10: Force distributions for adhesive patch.

$$+\frac{1}{h}\sum_{n=1}^{N}\int_{S_{S_n}}U_{ij}(\mathbf{x}',\mathbf{x})b_j^{S_n}(\mathbf{x})dS_{S_n} \tag{10.59}$$

where $\widehat{u}_{kj}^{A_m d}$ and $\widehat{t}_{kj}^{A_m d}$ are the particular displacements and tractions which correspond to the function $f^{A_m d} = c + r$, as given in Chapter 2, and the coefficient $C_{ij}^{A_m}$ is a constant whose value depends on the position of the collocation point with respect to the boundary of the attachment region S_{A_m}.

The displacement of a point at the mth patch as:

$$C_{ij}^{P_m}(\mathbf{x}')u_j^{P_m}(\mathbf{x}') + \int_{S_{P_m}}T_{ij}(\mathbf{x}',\mathbf{x})u_j^{P_m}(\mathbf{x})dS_{P_m} - \int_{S_{P_m}}U_{ij}(\mathbf{x}',\mathbf{x})t_j^{P_m}(\mathbf{x})dS_{P_m}$$

$$= -\frac{1}{h_{P_m}}\sum_{d=1}^{D_m}\alpha_k^{A_m d}\left[C_{ij}^{A_m}(\mathbf{x}')\widehat{u}_{kj}^{A_m d}(\mathbf{x}') + \int_{S_{A_m}}T_{ij}(\mathbf{x}',\mathbf{x})\widehat{u}_{kj}^{A_m d}(\mathbf{x})dS_{A_m}\right.$$

$$\left.+\int_{S_{A_m}}U_{ij}(\mathbf{x}',\mathbf{x})\widehat{t}_{kj}^{A_m d}(\mathbf{x})dS_{A_m}\right] \tag{10.60}$$

Equations (10.59) and (10.60) are written in terms of the DRM coefficients $\alpha_j^{A_m d}$ instead of the patch attachment forces $b_j^{A_m d}[9]$. The coefficients $\alpha_j^{A_m d}$ have no physical meaning and are related to the attachment forces through equation (10.58). The displacement compatibility equation (10.57) can be also written, in terms of the DRM coefficients $\alpha_j^{A_m d}$, as

$$u_j(\mathbf{x}') - u_j^{P_m}(\mathbf{x}') = \frac{h_{A_m}}{\mu_{A_m}}\sum_{d=1}^{D_m}\alpha_j^{A_m d}f^{A_m d}(\mathbf{x}^{A_m d},\mathbf{x}') \tag{10.61}$$

The extension of the above approach to shear deformable plates and shells has been developed by Wen, Aliabadi and Young [16, 17]. In this case, there are five force distributions, as shown in Figure 10.10, to consider.

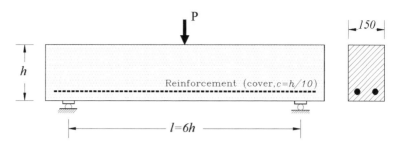

Figure 10.11: Three-point bending reinforced concrete beam.

10.4 Examples

10.4.1 Three-point Bending Concrete Beam

Aliabadi and Saleh [1] used the boundary element method to analyse two types of concrete (grades 2 and 4). The geometry and loading of the beam are shown in Figure 10.11. The reinforcement cover, which is the distance of the bar from the bottom beam surface, is equal to $1/10$ of the beam depth ($c = 0.1h$) for each case. The material properties of the grade 2 concrete are: compressive strength $f_c = 29.38$ MPa, Young's modulus -34300 MPa, fracture energy $\mu_F = 134$ N/m and $f'_t = 2.28$ MPa.

The results obtained for stress along reinforcements just before crack initiation are presented in Figures 10.12 and 10.13 for different steel areas A_s.

10.4.2 Single or Double Circular Patch Bonded to a Sheet

A square sheet of side $L = 200$ mm is subjected to biaxial stresses $\sigma = 1$ GPa as shown in Figure 10.14. The sheet has thickness 1.5 mm, Young's modulus $E = 70$ GPa and $\nu = 0.3$. A single circular patch of radius $R = 30$ mm, 1.5 mm thick, of the same material as the sheet and with an unloaded boundary, is bonded to the centre of the sheet by means of an adhesive layer of thickness $h_A = 0.15$ mm and shear modulus $\mu_A = 0.6$ GPa (detail on the right-hand side of Figure 10.14). The adhesive layer locus encloses the whole patch domain (i.e. the attachment region boundary coincides with the patch boundary). The single patch problem was analysed [8] using different discretizations, each corresponding to a different number of internal points (45, 69, 109 and 145), regularly distributed throughout the attachment region. Figure 10.15 shows the discretization containing 145 internal points. Displacement boundary conditions were applied to the sheet in order to constrain rigid body movements. Patch rigid body movements were naturally constrained by the imposition of displacement compatibility with the sheet.

Results for the attachment shear forces (τ) obtained in [8] are presented in Figure 10.16 normalized with respect to the sheet far field stresses ($\sigma_x = \sigma_y = 1$ GPa). On the right-hand side of Figure 10.16, results obtained for different discretizations are compared with the analytic solution for an infinite sheet subjected to hydrostatic traction. It can be seen that good agreement was obtained even for relatively coarse internal points grids. Refining the boundary mesh was found not to affect the results

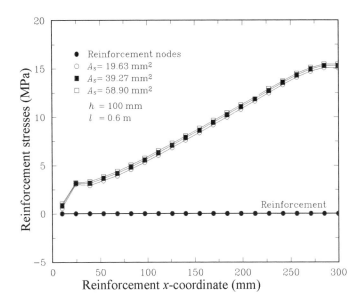

Figure 10.12: Stresses along the reinforcement just before crack initiation.

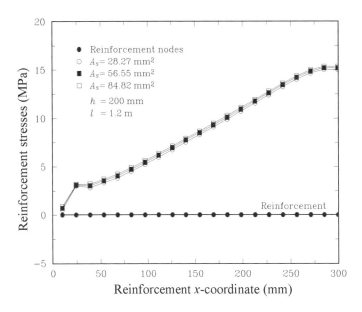

Figure 10.13: Stresses along the reinforcement just before the crack initiation.

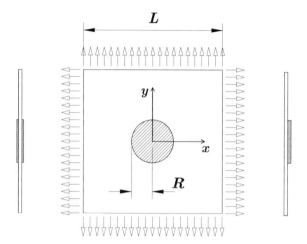

Figure 10.14: A large square sheet bonded with a single circular patch (on the right), or double circular patches (on the left).

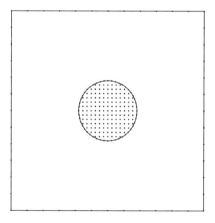

Figure 10.15: Boundary element mesh.

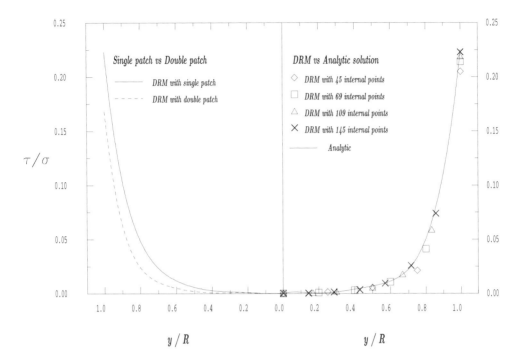

Figure 10.16: Normalized attachment shear stresses for a circular patch bonded to a large square sheet. On the right, results are compared to analytical solution for infinite sheet. On the left, comparison is made between attachment shear stresses resulting from the use of a double patch instead of a single patch.

significantly (i.e. doubling the number of elements resulted in the maximum shear stress changing less than 1%).

If, instead of one patch 1.5 mm thick, two patches 0.75 mm thick are bonded either side of the sheet, the attachment forces reduce slightly, as shown in the left-hand side of Figure 10.16. For the analysis of this case, two patches with identical external boundaries were modelled. The boundary and internal points discretization is identical to that presented in Figure 10.15.

10.4.3 Simply Supported Plate with a Stiffener

Consider a stiffened plate (see Figure 10.17) subjected to uniform pressure load q_0. The geometric dimensions are chosen as $h = 0.01$ m and $a = 1.0$ m. Material constants are $E/q_0 = 1.7 \times 10^7$, $\nu = 0.3$. The boundary conditions are defined as

$$u_2 = 0, \quad w_3 = 0, \quad t_1 = p_1 = p_2 = 0 \quad \text{at} \quad x = \pm a/2$$

and

$$u_1 = 0, \quad w_3 = 0, \quad t_2 = p_1 = p_2 = 0 \quad \text{at} \quad y = \pm a/2.$$

A BEM mesh with 32 quadratic boundary elements is used [17] and the number of nodes on the beam is $N = 3$. The numerical results of deflection and in–plane

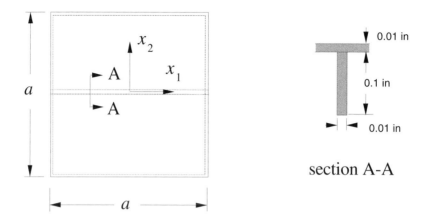

Figure 10.17: McBean plate and section of a stiffener.

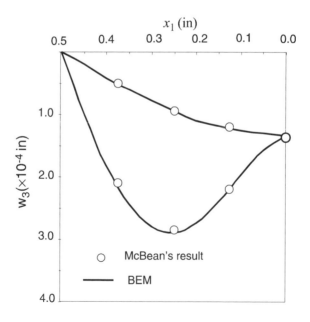

Figure 10.18: Variation of deflection along x_1 for McBean's plate.

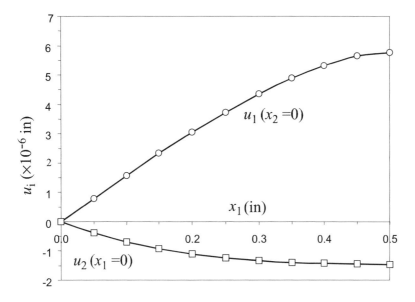

Figure 10.19: Membrane displacements (in).

displacement by BEM, together with the finite element results, are shown in Figures 10.18 and 10.19.

10.4.4 Reinforced Cylindrical Shell

A cylindrical shell with two frames (see Figure 10.20) subjected to uniform pressure load q_0 is analysed. The geometric parameters are chosen as: thickness of shell $h = 0.05$ m, width and length of shell $b_1 = b_2 = 1.0$ m and curvature $k = k_{11} = 0.1/$m, $k_{22} = 0$ (see Figure 10.20). Material constants are chosen as $E/q_0 = 1.7 \times 10^7$, $\nu = 0.33$. The cross-sectional area of frames is $A = 0.01 \times 0.2$ m^2 and the centroidal moment of inertia $I = 3.33 \times 10^{-4}$ m^4. The displacements and deflection are defined on the boundary as

$$w_3 = u_2 = 0 \quad \text{at} \quad x_2 = \pm b_2, \quad \text{or} \quad x_1 = b_1.$$

A BEM mesh with 40 quadratic boundary elements and 67 DRM domain points is used to model the shell. The beam is modelled using $N = 3, 5$ and 9. The numerical result of deflection along the axis x_1 are shown in Figure 10.21. Due to the symmetry of the structure, only one-quarter of the shell is modelled using finite element analysis. The total number of rectangular elements by the finite element method is 441. The results for deflection by the BEM and FEM are shown (see Figure 10.21) to be in good agreement.

10.4.5 Fastened Single Patch Repair

A square plate to which a square patch is attached on one side is shown in Figure 10.22. Both plate and patch have a thickness of $h = h_1 = 1.5$ mm and are made of

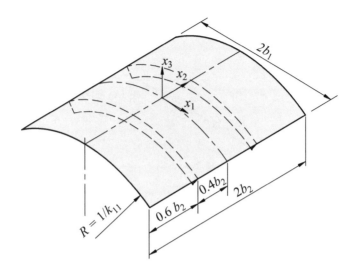

Figure 10.20: Geometry of stiffened shell.

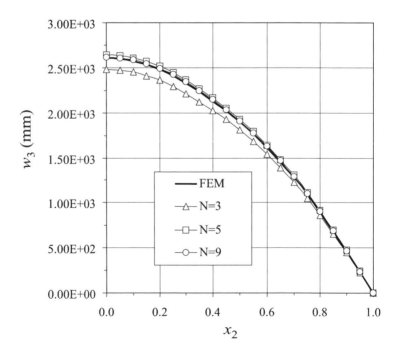

Figure 10.21: Deflection of shell on the symmetric line $x_1 = 0$.

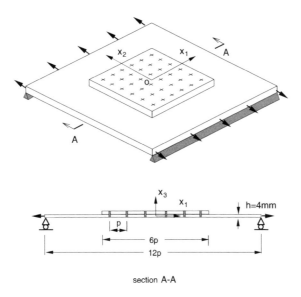

Figure 10.22: Riveted repair sheet.

aluminium with modulus of elasticity $E = 70$ GPa and Poisson's ratio $\nu = 0.3$. The patch is attached with 36 rivets over the region -152.4 mm $\leq x_1 \leq 152.4$ mm, -152.4 mm $\leq x_2 \leq 152.4$ mm. The distribution of rivet and geometry of the patched plate as modelled with the BEM are shown in Figure 10.23. The fastener diameter is $\phi = 4$ mm and the fastener pitch is $p = 25.4$ mm. Boundary conditions are given by

$$p_1 = p_2 = t_1 = t_2 = w_3 = 0 \qquad x_2 = \pm152.4 \text{ mm}$$

$$p_1 = p_2 = t_1 = t_2 = t_3 = 0 \qquad x_1 = \pm152.4 \text{ mm}$$

for plate and free boundary conditions on the patch.

The BEM mesh with 24 quadratic continuous elements on the sheet outer boundary, 16 continuous elements on the patch boundary, and 36 domain points (rivets point) are used. The deflections along the x_2 and x_1 axes, the moment and membrane forces along the x_2 axis in the domain of the sheet, subjected to uniform pressure load $q_0(= 1$ GPa) along the direction x_3 in the domain of plate and membrane load $N_0(= 1.5$ GPa) along the direction x_2 at two end of plate, are shown in Figures 10.24 against diameter of rivet.

Maximum values of deflection occur on the boundary ($x_1 = 152.4$mm, $x_2 = 0$) for both loadings see Figure 10.25.

10.4.6   Cylindrical Shell Panels with a Patch.

A rectangular cylindrical clamped shell panel subjected to uniform pressure q_0 is shown in Figure 10.26. Material constants for the shell and the patch are the same with Young's modulus $E = 210$ GPa and Poisson's ratio $\nu = 0.3$. The shell is of dimensions $-6p \leq x_1 \leq 6p, -9p \leq x_2 \leq 9p$ and thickness $h_0 = 1.6$ mm, p is the pitch of rivet and chosen as 25.4 mm. Diameter of rivet $\phi = 4$ mm and thickness $h_1 = h_0$. The panel curvature coefficients are $k_{11} = 1/60p, k_{22} = 0$ and the uniform

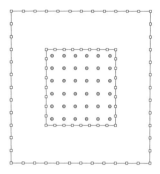

Figure 10.23: Boundary element mesh and distribution of rivets.

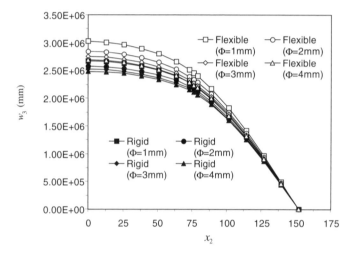

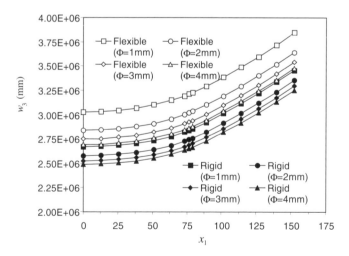

Figure 10.24: Deflection of sheet under uniform pressure q_0.

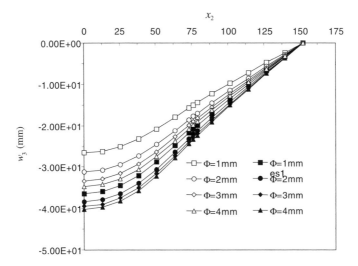

Figure 10.25: Deflection under membrane force along the x_2 direction.

load $q_0 = 1\text{MPa}$. Four sides of the panel are clamped, that is $u_\alpha = w_i = 0$ on all boundaries.

The boundaries of the panel and attachment are divided into 24 and 20 continuous quadrature elements, respectively, as shown in Figure 7.19. Here 98 domain points including 54 rivet points are used, and distribution of the domain point is shown in this Figure. The numerical results for deflections of the shell along x_2 are plotted in Figure 10.27. The results for the repaired panel and for the unrepaired panel are shown in the same Figure.

10.5   Summary

In this chapter the application of the boundary element method to assembled structures was presented. It was shown that the BEM is an efficient method for analysing mechanically fastened and adhesively bonded attachments. The fastener attachments can be considered flexible or rigid with elastic or elasto-plastic behaviour [8]. The BEM formulation requires only the boundary of the problem to be discretized. A set of internal points is used to define the position of the fasteners.

For the adhesive bonded structures, the DRM was used to avoid discretization of the patch attachment domain into internal cells. The formulation presented does not require the storage and inversion of the DRM coefficient matrix. The internal DRM collocation points are also used for displacement compatibility conditions.

References

[1] Aliabadi, M.H. and Saleh, A.L., A boundary element method for analysis of cracking in plain and reinforced concrete, in *Computational Fracture Mechanics in Concrete Technology*, WIT Press, Southampton, 44-100, 1999.

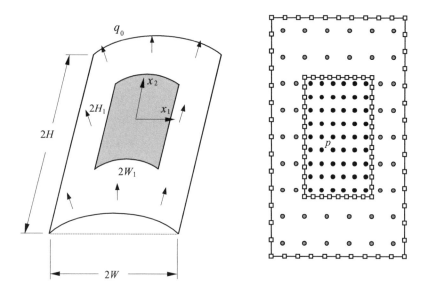

Figure 10.26: Fastener patched cylindrical shell panel under uniform pressure q_0 and boundary element mesh.

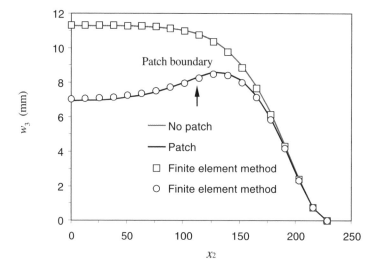

Figure 10.27: Deflection w_3 of a shell along the longest side of the shell.

[2] Dowrick, G., Cartwright, D.J. and Rooke, D.P. Boundary effects for a reinforced cracked sheet using the boundary element method, *Theoretical and Applied Mechanics*, **12**, 251-260, 1990.

[3] McBean, M.S., Analysis of stiffened plates by finite element method, Ph.D. Thesis, Stanford University, 1968.

[4] Rashed, Y.F., Aliabadi, M.H., and Brebbia, C.A., The boundary element method for Reissner plates resting on elastic foundations, *Plate Bending Analysis with Boundary Elements*, Computational Mechanics Publications, Southampton, 1998.

[5] Saleh, A.L and Aliabadi, M.H., Boundary element analysis of the pull-out behaviour of an anchor bolt embedded in concrete, *Mechanics of cohesive-frictional material.* **1**, 235-250, 1996.

[6] Saleh, A.L. and Aliabadi, M.H., Crack growth analysis in reinforced concrete using BEM, *Journal of Engineering Mechanics, ASCE*, **124**, 949-958, 1998.

[7] Salgado, N.K. and Aliabadi, M.H., The application of the dual boundary element method to the analysis of cracked stiffened panels, *Engineering Fracture Mechanics*, **54, 91-105,** 1996.

[8] Salgado, N.K and Aliabadi, M.H., The analysis of mechanically fastener repairs and lap joints, *Fatigue & Fracture of Engineering Materials & Structures*, **20,** 583-593, 1997.

[9] Salgado, N.K. and Aliabadi, M.H., A dual reciprocity method for the analysis of adhesively patched sheets, *Communications in Numerical Methods in Engineering*, **13**, 397-405, 1997.

[10] Salgado, N.K., Boundary Element Methods for Damage Tolerance Design of Aircraft Structures., PhD thesis, Wessex Institute of Technology, University of Wales, 1997.

[11] Salgado, N., *Boundary Element Method for Damage Tolerance Design of Aircraft Structures*, Topics in Engineering Vol. 33, 1998.

[12] Swift, T., The effects of fastener flexibility and stiffener geometry on the stress intensity in stiffened cracked sheets, *Proc. Conf. on Prospects of Fracture Mechanics*, Delft, 419-457, 1974.

[13] Wen, P.H., Aliabadi, M.H. and Young, A., Boundary element analysis of shear deformable stiffened plates. to appear in *J.Strain Analysis, 2002.*

[14] Wen, P.H., Aliabadi, M.H. and Young, A., Stiffened cracked plates analysis by dual boundary element method, *International Journal of Fracture*, **106,** 245-258, 2000.

[15] Wen, P.H., Aliabadi, M.H. and Young, A., Boundary element analysis of cracked panels with mechanically fastened repair patches (to appear in *Journal of Strain Analysis, 2002*).

[16] Wen, P.H., Aliabadi, M.H. and Young, A,. Boundary element analysis of curved panels with mechanically fastened repair patches (to appear in *Computer Methods in Engineering Sciences, 2002*).

[17] Wen, P.H., Aliabadi, M.H. and Young, A., Boundary element analysis of flat cracked panels with adhesively bonded patches (to appear).

[18] Wen,P.H., Aliabadi,M.H. and Young,A., Boundary element analysis of curved cracked panels with adhesively bonded patches (to appear).

[19] Widagdo, D and Aliabadi, M.H., Boundary element analysis of cracked panels repaired by mechanically fastened composite patches, *Engineering Analysis with Boundary Elements*, **25**, 339-345, 2001.

[20] Widagdo, D and Aliabadi, M.H., Boundary element analysis of composite repair patches, *Advances in Boundary Element Techniques II*, Hoggar Press, Geneva, 327-5, 2001.

[21] Young, A., Cartwright, D.J. and Rooke, D.P., Numerical study of balanced patch repairs to cracked sheets, *Aeronautical Journal*, **39,** 327-334, 1988.

Chapter 11

Numerical Integration

The boundary integral equation itself is a statement of exact solution to the problem posed. Therefore, the errors are due primarily to discretization and numerical integration. No matter how refined the discretization, accurate and stable solutions can only be obtained if the integrations are sufficiently accurate. The special difficulty which arises is that the kernel of the integration equation is often singular, that is, it becomes infinite when the integration variable and collocation point coincide. Generally, the singularity presents no difficulty if the surface is formed of plane facets (constant elements), since the integrals can then be expressed analytically in closed forms; aerodynamicists have used this approach successfully for many years. However, in most practical structural problems the numerical procedure needs to be made sufficiently sophisticated by using curved boundary elements.

Depending on the nature of the kernel, and the relative position of the collocation point with respect to the element on which integration is being carried out, five types of integrals can be identified. They include: regular, near singular, weakly singular, strongly singular and hypersingular.

11.1   Integration of Regular Integrals

Regular integrals are referred to integrals which are not singular and their integrand does not vary sharply in the region of integration. This type of integral applies to all kernels provided $\mathbf{x}'$ and $\mathbf{x}$ do not belong to the same element. The common method for the integration of regular integrals is the Gaussian quadrature.

The one-dimensional BEM integrals can be evaluated numerically using the Gauss-Legendre formula,

$$\int_{-1}^{1} f(\eta)d\eta = \sum_{s=1}^{S} A_s f(\eta_s) \tag{11.1}$$

where A_s and η_s are respectively the weights and abscissas (see Table 11.1), and S is the number of Gauss points.

The two-dimensional integrals can be evaluated by a repeated application of (11.1)

491

Table 11.1: Abscissas and weights for Gauss-Legendre quadrature.

S	η_s	A_s
2	$\pm0.57735\ 02691\ 89626$	$1.0000\ 00000\ 00000$
4	$\pm0.33998\ 10435\ 84856$	$0.65214\ 51548\ 62546$
	$\pm0.86113\ 63115\ 94053$	$0.34785\ 48451\ 37454$
6	$\pm0.23861\ 91860\ 83197$	$0.46791\ 39345\ 72691$
	$\pm0.66120\ 93864\ 66265$	$0.36076\ 15730\ 48139$
	$\pm0.93246\ 95142\ 03152$	$0.17132\ 44923\ 79170$
8	$\pm0.18343\ 46424\ 95650$	$0.36268\ 37833\ 78362$
	$\pm0.52553\ 24099\ 16329$	$0.31370\ 66458\ 77887$
	$\pm0.79666\ 64774\ 13627$	$0.22238\ 103445\ 3374$
	$\pm0.96028\ 98564\ 97536$	$0.10122\ 853629\ 0376$
10	$\pm0.14887\ 43389\ 81631$	$0.29552\ 42247\ 14753$
	$\pm0.43339\ 53941\ 29247$	$0.26926\ 67193\ 09996$
	$\pm0.67940\ 95682\ 99024$	$0.21908\ 63625\ 15982$
	$\pm0.86506\ 33666\ 88985$	$0.14945\ 13491\ 50581$
	$\pm0.97390\ 65285\ 17172$	$0.06667\ 13443\ 08688$
12	$\pm0.12523\ 34084\ 11469$	$0.24914\ 70458\ 13403$
	$\pm0.36783\ 14989\ 98180$	$0.23349\ 25365\ 38355$
	$\pm0.58731\ 79542\ 86617$	$0.20316\ 74267\ 23066$
	$\pm0.76990\ 26741\ 94305$	$0.16007\ 83285\ 43346$
	$\pm0.90411\ 72563\ 70475$	$0.10693\ 93259\ 95318$
	$\pm0.98156\ 06342\ 46719$	$0.04717\ 53363\ 86512$

Table 11.2: Rules for one-dimensional integration of regular integrals.

Condition	Gauss Quadrature
$d > 5.5l$	2
$5.5l \geq d > 1.5l$	4
$1.5l \geq d$	6

Table 11.3: Rules for two-dimensional integration of regular integrals.

Condition	Gauss Quadrature
$d > 1.32l$	4×4
$1.32l \geq d \geq 0.8l$	8×8
$1.32l \geq d \geq 0.8l$	12×12
$0.8l \geq d \geq 0.3l$	16×16

along, η_1 and η_2 directions, that is

$$\int_{-1}^{1} \int_{-1}^{1} f(\eta_1, \eta_2) d\eta_1 d\eta_2 = \sum_{s_1=1}^{S_1} \sum_{s_2=1}^{S_2} A_{s_1} A_{s_2} f(\eta_{s_1}, \eta_{s_2}) \tag{11.2}$$

where S_1 and S_2 are the number of Gauss points in the S_1 and S_2 directions, respectively.

For two-dimensional problems, a simple set of rules based on the distance between the collocation point and the element centre, d, and the element length, was reported in Brebbia *et al.* [9] and reproduced in Table 11.2.

The evaluation of regular integrals takes most of the computational time for setting up the coefficient matrices. Therefore, it is necessary to optimize the integration processes as much as possible. Lachat and Watson [28] developed a method which related the order of the quadrature to the maximum upper bound error in the integration. They varied the number of integration points depending on the ratio of the distance between the element and the order of the kernel (i.e. $O(1/r)$, $O(1/r^2)$, etc.). Rigby [35] carried out extensive tests and recommended a set of rules for integration to an accuracy of $\log(\max(\text{error})) < -5$. Table 11.3 lists the rules based on the shortest distance between the collocation point and the element d and the largest side of the element l.

11.2 Near Singular Integration

Integrals are said to be near singular when the collocation point is close to the element under consideration. Special treatment is required as the integrand varies sharply as the source point approaches the element. Several techniques have been developed to deal with this type of integration. They include element subdivision [28, 22], adaptive Gaussian integration [10], the variable transformation technique [47, 20, 13] and semi-analytical integration based on series expansion [12, 41]. Kane *et. al* [24] proposed an algorithm for minimizing the number of integration points for three-dimensional problems

11.2.1 Transformation of Variable Technique

In this section, the element subdivision, and regularization and transformation techniques are described. The former is a simple approach which is widely used for treatment of near-weak-singular and near-strong-singular integrals, while the more elaborate latter approach is also suitable for near-hypersingular integrals.

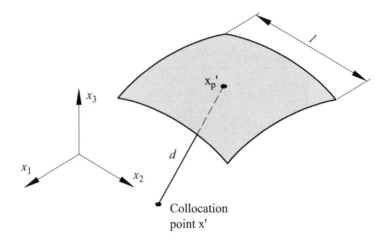

Figure 11.1: Shortest distance from the collocation point to the integration element.

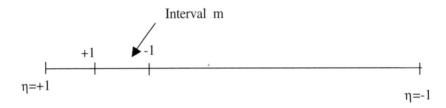

Figure 11.2: Subdivision on the integration interval into M subintervals with $-1 \le \bar{\eta}_s \le +1$.

Element Subdivision Technique

In this approach the element of the integration is divided into M subintervals of equal length. The order of quadrature can be varied within each subinterval. Lachat and Watson [28] related the number of subdivisions to the upper bound of error in the Gauss-Legendre formula.

Dividing the interval of integration into M intervals (see Figure 11.2) leads to

$$\int_{-1}^{1} f(\eta) d\eta = \frac{1}{M} \sum_{m=1}^{M} \sum_{s=1}^{S} A_s f(\bar{\eta}_s) \tag{11.3}$$

with

$$\bar{\eta}_s = \frac{1}{M} [M - 2m + 1 + \eta_s]$$

where the factor $1/M$ is the Jacobian of $d\bar{\eta}/d\eta$. In (11.3) the same order of quadrature is used within each interval. However, one may wish to vary the order of the quadrature in each interval, so that

$$\int_{-1}^{1} f(\eta) d\eta = \frac{1}{M} \sum_{m=1}^{M} \sum_{s=1}^{S(m)} A_{s(m)} f(\bar{\eta}_{s(m)}) \tag{11.4}$$

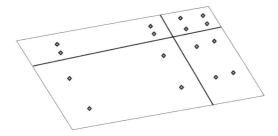

Figure 11.3: Subdivison of the integration interval.

where

$$\bar{\eta}_{s(m)} = \frac{1}{M}\left[M - 2m + 1 + \eta_{s(m)}\right] \tag{11.5}$$

A similar approach can be used for three-dimensional problems (see Figure 11.3); the area integrals can be written as

$$\int_{-1}^{1}\int_{-1}^{1} f(\eta_1,\eta_2)d\eta_1 d\eta_2 = \frac{1}{ML}\sum_{m=1}^{M}\sum_{l=1}^{L}\sum_{s_1=1}^{S_1}\sum_{s_2=1}^{S_2} A_{s_1} A_{s_2} f(\bar{\eta}_{s_1},\bar{\eta}_{s_2}) \tag{11.6}$$

where

$$\bar{\eta}_{s_1} = \frac{1}{M}\left[M - 2m + 1 + \eta_{s_1}\right]$$

$$\bar{\eta}_{s_2} = \frac{1}{L}\left[L - 2l + 1 + \eta_{s_2}\right]$$

where M and L denote the number of subdivisions in the η_1 and η_2 directions, respectively.

Regularization and Transformation Technique

Another relatively simple way of dealing with near singularities is to divide the element under consideration into several triangles with the point $\mathbf{x}'_p$ (projection of $\mathbf{x}'$ with minimal distance) as the vertex, as shown Figure 11.4. Each triangle is than mapped into a square over which Gauss integration is applied. The transformation results in a concentration of Gauss points near the projection point $\mathbf{x}'_p$. The condensation of Gauss points ensures that the rapid variation of the integrands around $\mathbf{x}'_p$ can be accurately modelled. This variable of transformation technique will be described in detail in a later section on weakly singular integration.

The above approach is adequate for the potential integral equation as it contains near-weak singular integrals at most. However, the displacement integral equation has near-strong integrals due to the traction fundamental solution. Fiedler [13] proposed a regularization procedure, to be used together with the variable transformation approach described above, to further weaken the near-strong singular integrals. This method was successfully applied to near-strong singular and near-hypersingular integrals by Wilde and Aliabadi [48].

Consider a traction free rigid boundary condition, i.e.

$$C_{ij}(\mathbf{x}') + \int_S T_{ij}(\mathbf{x}',\mathbf{x})dS(\mathbf{x}) = 0 \tag{11.7}$$

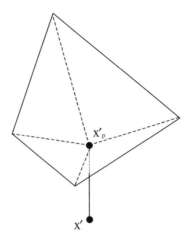

Figure 11.4: Splitting the element into triangular elements with $\mathbf{x}'_p$ as the vertex.

Multiplying (11.7) by $u_j(\mathbf{x}'_p)$ and adding to the displacement integral equation gives

$$
\begin{aligned}
C_{ij}(\mathbf{x}')u_j(\mathbf{x}') \;=\;& C_{ij}(\mathbf{x}')u_j(\mathbf{x}'_p) + \int_S U_{ij}(\mathbf{x}',\mathbf{x})t_j(\mathbf{x})dS \\
& - \int_S T_{ij}(\mathbf{x}',\mathbf{x})[u_j(\mathbf{x})-u_j(\mathbf{x}'_p)]dS
\end{aligned}
\tag{11.8}
$$

In (11.8) the near-strong singularity has reduced to near-weak singularity.

Similar regularization using rigid body motion can be applied to the flux and stress integral equations to weaken the near-hypersingular integrand. The triangular to square transformation (see Figure 11.5) can then follow to further weaken the near-strong singularity. However, there still remains a near weak-singularity. To overcome this, an additional transformation is used to further concentrate the Gauss points near $\mathbf{x}'_p$, as shown in Figure 11.5. The transformation is given as

$$
\eta''_1 = \frac{a}{\sqrt{(\eta'_1 + 1)^2 + \gamma^2}} + b
$$

The constants a and b are determined from the condition that the boundary maps onto itself, leading to

$$
a = \frac{-2\alpha\sqrt{4+\alpha^2}}{\sqrt{4+\alpha^2} - \alpha}
$$

$$
b = \frac{\sqrt{4+\alpha^2} + \alpha}{\sqrt{4+\alpha^2} - \alpha}
$$

The parameter α controls how far the Gauss points are concentrated near $\mathbf{x}'_p$, such as

$$
\alpha = \left| \frac{\mathbf{x}' - \mathbf{x}'_p}{\left(\dfrac{\partial \mathbf{x}}{\partial \eta''_1}\right)_{\eta''=-1}} \right|
$$

which includes a normalizing factor based on the size of the element.

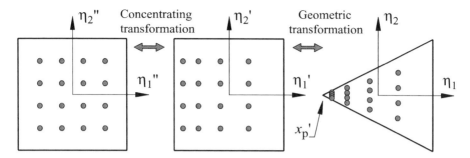

Figure 11.5: Geometric and concentrating transformations for near singular integration.

11.2.2 Singularity Subtraction Technique

Cruse and Aithal [12] proposed a semi-analytical approach for treating the near-singular integrals in the BEM. Their approach, which is analogous to that proposed in [1], was shown to produce accurate results for low-orders of Gaussian quadrature. In tests carried out [12], the singularity subtraction approach was shown to be more accurate than the element subdivision approach, with computational savings of around 40%. The extension of this approach to near-hypersingular integration was reported by Mi and Aliabadi [41]. This approach will be described in detail for weak, strong and hypersingular integrals in later sections.

11.3 Weakly Singular Integration

This type of integral occur when the collocation node lies within the integration element and the singularity of the kernel is of order $O(\ln(1/r))$ and $O(1/r)$ for two- and three-dimensional problems, respectively. There are three basic ways of dealing with this type of integral: Weighted Gaussian Integration, Transformation of Variable, Subtraction of Singularity. Analytical integrations for potential and elasticity kernels using contact element can be found in [19].

11.3.1 Two-dimensional Problems

Weighted Gaussian Integration

Gaussian formulae with a weight function $\ln(1/\eta)$ have been used extensively for the treatment of $\ln(1/r)$ kernels in the BEM. Thus

$$\int_0^1 f(\eta) \ln\left(\frac{1}{\eta}\right) d\eta = \sum_{s=1}^S w_s f(\eta_s) \tag{11.9}$$

where w_s and η_s are weights and abscissas, respectively, for the Gaussian quadrature formula $\ln(1/\eta)$ as the weighting function; they are given in Stroud and Secrest [45], and also listed in Table 11.4.

Table 11.4: Abscissas and weights for Gauss quadrature with $\ln(1/\eta)$ weighting factor.

S	η_s	w_s
2	0.11200 88061 66976	0.71853 93190 30384
	0.60227 69081 18738	0.28146 06809 69615
4	0.04144 84801 99383	0.38346 40681 45135
	0.24547 49143 20602	0.38687 53177 74762
	0.55616 54535 60278	0.19043 51269 50142
	0.84898 23945 32985	0.03922 54871 29960
6	0.02163 40058 44117	0.23876 36625 78547
	0.12958 33911 54950	0.30828 65732 73946
	0.31402 04499 14765	0.24531 74265 63210
	0.53865 72173 51802	0.14200 87565 66476
	0.75691 53373 77402	0.05545 46223 24886
	0.92266 88513 72120	0.01016 89586 92932
8	0.01332 02441 60892	0.16441 66047 28002
	0.07975 04290 13895	0.23752 56100 23306
	0.19787 10293 26188	0.22684 19844 31919
	0.35415 39943 51909	0.17575 40790 06070
	0.52945 85752 34917	0.11292 40302 46759
	0.70181 45299 39099	0.05787 22107 17782
	0.84937 93204 41106	0.02097 90737 42133
	0.95332 64500 56359	0.00368 64071 04028

For curved boundaries using quadratic or higher order elements, the distance r can be written as

$$r(\mathbf{x}', \mathbf{x}(\eta)) = (\eta - \eta')r_d \qquad (11.10)$$

where η' is the position of collocation point in the local η coordinates and r_d denotes the non-singular part of the distance r. For flat elements $r_d = l/2$, where l is the length of the element. The integral involving the $\ln(1/r)$ term can be written as

$$\int_{-1}^{1} \ln\left(\frac{1}{r}\right) N_\alpha(\eta) J^n(\eta) d\eta = \int_{-1}^{1} \ln\frac{1}{(\eta - \eta')} N_\alpha(\eta) J^n(\eta) d\eta - \int_{-1}^{1} \ln(r_d) N_\alpha(\eta) J^n(\eta) d\eta$$

$$(11.11)$$

the first integral on the right-hand side of (11.11) can be integrated using the weighted Gaussian formula (11.9), and the second integral which is non-singular can be integrated using the Gauss-Legendre formula (11.1).

Since the quadrature data for the logarithmic weighted formula is generated in the interval $\eta \in [0, 1]$, the first integral on the right-hand side of (11.11) needs to be transformed. For singularity at $\eta' = +1$, the transformation is given as

$$\eta' = 1 - 2\eta$$

and for singularity at $\eta' = -1$ as

$$\eta' = 2\eta - 1$$

For quadratic and higher order elements, the singularity can be within the interval of integration. In this case, the distance r can be redefined such that

$$r = (\eta - \eta')r_R \qquad -1 \le \eta \le \eta'$$
$$r = (\eta - \eta')r_L \qquad \eta' \le \eta \le +1$$

The first integral on the right-hand side of (11.11) can be written as a sum of two integrals, i.e.

$$\int_{-1}^{1} \ln \frac{1}{(\eta - \eta')} N_\alpha(\eta) J^n(\eta) d\eta = \int_{\eta'}^{1} \ln \frac{1}{(\eta - \eta')r_L} N_\alpha(\eta) J^n(\eta) d\eta$$

$$+ \int_{-1}^{\eta'} \ln \frac{1}{(\eta - \eta')r_R} N_\alpha(\eta) J^n(\eta) d\eta \quad (11.12)$$

Again, as before, the above integrals need to be transformed to an interval $[0, 1]$ so that formula (11.9) can be used. The transformations are given by

$$\eta' = (\eta' + 1)(1 - \eta) - 1 \qquad -1 \le \eta \le \eta' \qquad (11.13)$$
$$\eta' = (1 - \eta')\eta - \eta' \qquad \eta' \le \eta \le 1 \qquad (11.14)$$

The main difficulty with the weighted Gaussian integration is that the components of the integral have to be separated into singular and non-singular parts. This may be a simple procedure for two-dimensional potential problems, but it is more complicated if the logarithmic singularity is not explicit, as in the case of axisymmetric BEM formulations. Smith [37] presents a direct Gauss quadrature formula for logarithmic singularities in two-dimensional, isoparametric elements, which implicity considers the integral to include a sum of singular and non-singular terms. Several tests performed in [37] confirmed the accuracy and efficiency of the direct Gauss quadrature formula for curved boundary elements of up to third order (cubic).

Transformation of Variable Technique

A transformation of variable may be found in such a way that the resulting Jacobian of transformation either cancels the singularity exactly or weakens the effect of the singularity sufficiently so that a relatively low order quadrature could be used to carry out the integration.

A transformation of variable technique has been proposed by Telles [47] for the integration of weakly singular integrals in two dimensions. The simplest transformation is given by

$$\eta = \bar{\eta} + \frac{1}{2}(1 - (\bar{\eta})^2) \qquad (11.15)$$

for singularity at $\eta = 1$. The Jacobian of transformation is given by

$$J = \frac{d\eta}{d\bar{\eta}} = (1 - \bar{\eta}) \qquad (11.16)$$

In order to illustrate the effect of such transformation, consider

$$\int_{-1}^{1} \ln(1 - \eta) d\eta = 2 \int_{-1}^{1} \ln \left(\frac{1 - \eta}{2} \right) (1 - \bar{\eta}) d\bar{\eta} \qquad (11.17)$$

where the integrand of the integral on the right-hand side of (11.17) is now regular.

A transformation based on a third degree polynomial was also developed by Telles [47] to deal with singularities with an interval of integration of $[-1, 1]$. Consider, a third degree polynomial:

$$\eta(\xi) = a\xi^3 + b\xi^2 + c\xi + d \tag{11.18}$$

By imposing the following conditions:

$$\left(\frac{d^2\eta}{d\xi^2}\right)_{\eta'} = 0$$

$$\left(\frac{d\eta}{d\xi}\right)_{\eta'} = 0$$

$$\eta(1) = 1$$

$$\eta(-1) = -1$$

where η' is a point in which an integral is singular.

The constants in (11.18) are evaluated as

$$a = \frac{1}{(1 + \bar{\xi}^2)}$$

$$b = -\frac{3\bar{\xi}}{(1 + \bar{\xi}^2)}$$

$$c = \frac{3\bar{\xi}^2}{(1 + \bar{\xi}^2)}$$

$$d = -b$$

where $\bar{\xi}$ is the value of ξ which satisfies $\eta(\bar{\xi}) = \bar{\eta}$, and is given by

$$\bar{\xi} = \sqrt[3]{[\bar{\eta}(\bar{\eta}^2 - 1) + |\bar{\eta}^2 - 1|]} + \sqrt[3]{[\bar{\eta}(\bar{\eta}^2 - 1) - |\bar{\eta}^2 - 1|]} + \bar{\eta} \tag{11.19}$$

Using the above transformation, we have

$$\int_{-1}^{1} f(\eta)d\eta = \int_{-1}^{1} f\left\{\frac{\left[(\xi - \bar{\xi})^3 + \bar{\xi}(\bar{\xi}^2 + 3)\right]}{(1 + 3\bar{\xi}^2)}\right\} \frac{3(\xi - \bar{\xi}^2)}{(1 + 3\bar{\xi}^2)}d\xi$$

A feature of the above transformation is that it automatically concentrates the integration points near to the singularity.

Singularity Subtraction Technique

A classical way of dealing with kernel singularities is to subtract them out in such a way that leaves the remaining integrand regular, and the subtracted singular part can be integrated analytically. For example,

$$\int_{-1}^{1} F(\eta)d\eta = \int_{-1}^{1} [F(\eta) - F^*(\eta)]\,d\eta + \int_{-1}^{1} F^*(\eta)d\eta \tag{11.20}$$

where $F^*(\eta)$ is a function which has the same singularity as $F(\eta)$ but in a simpler form which can be integrated exactly, and $[F(\eta) - F^*(\eta)]$ is not singular and therefore can be integrated accurately by numerical quadrature. Aliabadi, Hall and Phemister [1] developed a technique based on series expansion of the kernel for the treatment of singularities in three-dimensional problems. Later, Aliabadi [3] and Aliabadi and Hall [8] used the method for two-dimensional problems. It is possible to express vector **r** from a variable **x** to a fixed point **x**′ as a series expansion in terms of η arising from the shape function representation of the boundary. The expansion is about η' which corresponds to **x**′, thus

$$\mathbf{r} = \mathbf{x}(\eta) - \mathbf{x}(\eta') = \mathbf{x}_\eta \delta\eta + \frac{1}{2}\mathbf{x}_{\eta\eta}\delta\eta^2 + \ldots \tag{11.21}$$

where $\delta\eta = \eta - \eta'$. The derivatives

$$\mathbf{x}_\eta = \left(\frac{\partial\mathbf{x}}{\partial\eta}\right)_{\eta=\eta'} = \sum_{\alpha=1}\left(\frac{\partial N_\alpha(\eta')}{\partial\eta}\right)\mathbf{x}^\alpha \tag{11.22}$$

$$\mathbf{x}_{\eta\eta} = \left(\frac{\partial^2\mathbf{x}}{\partial\eta^2}\right)_{\eta=\eta'} = \sum_{\alpha=1}\left(\frac{\partial^2 N_\alpha(\eta')}{\partial\eta^2}\right)\mathbf{x}^\alpha \tag{11.23}$$

are evaluated at η'.

As **x**′ is taken to correspond to nodal coordinates, similarly η' corresponds to nodal values in $\eta-$ coordinates. The expansion is limited, depending on the shape function representation. For example, if we use quadratic shape functions

$$\mathbf{r} = \mathbf{x}_\eta \delta\eta + \frac{1}{2}\mathbf{x}_{\eta\eta}\delta\eta^2 \tag{11.24}$$

since $\mathbf{x}_{\eta\eta\eta} = 0$ and $\mathbf{x}_\eta$ and $\mathbf{x}_{\eta\eta}$ are evalauted at colocation points $\eta' = -1, 0, +1$. Thus from (11.24), the distance r may be derived from

$$r^2 = \mathbf{r}.\mathbf{r} = \delta\eta^2(d_o + d_1\delta\eta + d_2\delta\eta^2)$$

so that

$$r = \delta\eta(d_o + d_1\delta\eta + d_2\delta\eta^2)^{1/2}$$

where $d_o = \mathbf{x}_\eta.\mathbf{x}_\eta$, $d_o = \frac{1}{2}\mathbf{x}_\eta.\mathbf{x}_{\eta\eta}$ and $d_2 = \frac{1}{2}\mathbf{x}_{\eta\eta}.\mathbf{x}_{\eta\eta}$. It is worth noting that for flat isoparametric elements, $d_1 = d_2 = 0$ and $\mathbf{x}_\eta = l/2$. Similarly, the shape function N_α may be expanded in a limit form as

$$N_\alpha^* = N_o + N_\eta\delta\eta + \frac{1}{2}N_{\eta\eta}\delta\eta^2$$

where subscript α has been dropped for convenience. The Jacobian term $J(\eta)$ cannot, however, be expanded in a limit as

$$J^* = J_o + J_\eta\delta\eta + \frac{1}{2}J_{\eta\eta}\delta\eta^2 + \ldots$$

where N_o and J_o denote the values of the shape function and Jacobian evaluated at η', and subscript η denotes differentiation with respect to η. All the derivatives are evaluated at $\eta = \eta'$.

The integral involving $\ln(1/r)$ can now be expressed as

$$\int_{-1}^{1} N_\alpha(\eta) J^n(\eta) \ln\left(\frac{1}{r}\right) d\eta = \int_{-1}^{1} \ln\left(\frac{1}{r}\right) [N_\alpha J^n - N_\alpha^* J^{n*}] d\eta$$

$$+ \int_{-1}^{1} \ln(\frac{1}{r}) N_\alpha^* J^{n*} d\eta \qquad (11.25)$$

where the first integral is regular and can be integrated using the Gauss-Legendre formula in (11.1). The second integral on the right-hand side of (11.25) can be integrated analytically

$$\int_{-1}^{1} \ln\left(\frac{1}{r}\right) N_\alpha^*(\eta) J^{*n}(\eta) d\eta = \sum_{n=0}^{4} I_n \qquad (11.26)$$

where

$$I_o = -N_o J_o \left\{ \int_{-1}^{1} \ln \delta\eta \, d\eta + \frac{1}{2} \int_{-1}^{1} \ln \Delta \, d\eta \right\}$$

$$I_1 = -(N_o J_\eta + N_\eta J_o) \left\{ \int_{-1}^{1} \delta\eta \ln \delta\eta \, d\eta + \frac{1}{2} \int_{-1}^{1} \delta\eta \ln \Delta \, d\eta \right\}$$

$$I_2 = -\left(N_\eta J_\eta + \frac{1}{2} N_o J_{\eta\eta} + \frac{1}{2} N_{\eta\eta} J_o \right) \left\{ \int_{-1}^{1} \delta\eta^2 \ln \delta\eta \, d\eta + \frac{1}{2} \int_{-1}^{1} \delta\eta^2 \ln \Delta \, d\eta \right\}$$

$$I_3 = -\left(\frac{1}{2} N_\eta J_{\eta\eta} + \frac{1}{2} N_{\eta\eta} J_\eta \right) \int_{-1}^{1} \delta\eta^3 \ln \delta\eta \, d\eta + \frac{1}{2} \int_{-1}^{1} \delta\eta^3 \ln \Delta \, d\eta$$

$$I_4 = -\frac{1}{4} N_{\eta\eta} J_{\eta\eta} \left\{ \int_{-1}^{1} \delta\eta^4 \ln \delta\eta \, d\eta + \frac{1}{2} \int_{-1}^{1} \delta\eta^4 \ln \Delta \, d\eta \right\}$$

where $I_o, I_1, I_2, ...$ correspond to first, second, third order expansions and $\Delta = d_o + d_1 \delta\eta + d_2 \delta\eta^2$. Integrals in (11.26) can be evaluated in closed forms, i.e.

$$L_1^n = \int \delta\eta^n \ln \delta\eta \, d\eta = \frac{\delta\eta^{n+1}}{n+1} \ln \delta\eta - \frac{\delta\eta^{n+1}}{(n+1)^2} \qquad (11.27)$$

$$L_2^n = \int \delta\eta^n \ln \Delta \, d\eta = \frac{\delta\eta^{n+1}}{n+1} \ln \Delta - \frac{2d_2}{n+1} A_{n+2} - \frac{d_1}{n+1} A_{n+1}$$

where

$$A_o = \frac{2}{\sqrt{4d_o d_1 - d_1^2}} \tan^{-1} \frac{2d_2 \delta\eta + d_1}{\sqrt{4d_o d_2 - d_1^2}}$$

$$A_1 = \frac{1}{2d_2} \ln \Delta - \frac{d_1}{2d_2} A_o$$

$$A_n = \int \frac{\delta\eta^n}{\Delta} d\eta = \frac{\delta\eta^{n-1}}{(n-1)d_2} - \frac{d_1}{d_2} A_{n-1} - \frac{d_o}{d_2} A_{n-2}$$

11.3.2 Three-dimensional Problems

Weighted Gaussian Integration

Special weight function formulae for the three-dimensional kernel $1/r$ were developed by Cristescu and Lobignac [11] for square and triangular elements. Their formula was shown by Aliabadi and Hall [6] to be accurate only for plane squares. An extension to more general plane elements was developed [6], in which the limitations of the weighted Gaussian formulae for 3D problems were also highlighted.

Transformation of Variable Technique

Lachat and Watson [28] developed a triangle to square transformation such that the Jacobian of the transformation exactly cancels out the singularity, as shown by Aliabadi, Hall and Hibbs [5]. A similar method was also used by Lean and Wexler [29]. For example, consider a triangle, shown in Figure 11.33, which has a singularity at $(\eta_1, \eta_2) = (-1, -1)$ in the local (η_1, η_2) plane; the singular integral is transformed in the following manner

$$\int_{-1}^{1} \int_{-1}^{\eta_1} K[\mathbf{x}', \mathbf{x}(\eta_1, \eta_2)] N_\alpha((\eta_1, \eta_2) J^n(\eta_1, \eta_2) d\eta_2 d\eta_1$$

$$= \int_{-1}^{1} \int_{-1}^{1} K[\mathbf{x}', \mathbf{x}(\eta_1, \eta_2)] N_\alpha((\eta_1, \eta_2) J^n(\eta_1, \eta_2) J^C(\eta_1, \eta_2) du dv \quad (11.28)$$

where

$$\eta_1 = u$$
$$\eta_2 = \frac{1}{2}[(1+u)v - (1-u)] \quad (11.29)$$

and the $K[\mathbf{x}', \mathbf{x}(\eta_1, \eta_2)]$ is the kernel of $O(1/r)$, such as Kelvin's displacement fundamental solution $U_{ij}(\mathbf{x}', \mathbf{x})$ presented in Chapter 2. The Jacobian of this transformation $J^C(\eta_1, \eta_2) = (1 + \eta_1)/2$ exactly cancels out the singularity. Quadrilateral elements are divided into a number of triangular elements, as shown in Figure 11.7; the number depends on the position of the singularity in the element. For example, if the singularity corresponds on the corner node $(-1, -1)$ in the local plane (see Figure 11.8a), then

$$\int_{-1}^{1} \int_{-1}^{1} K N_\alpha J^n d\eta_1 d\eta_2$$

$$= \int_{-1}^{1} \int_{-1}^{\eta_1} K N_\alpha J^n d\eta_2 d\eta_1 + \int_{-1}^{1} \int_{-1}^{\eta_1} K N_\alpha J^n d\eta_1 d\eta_2 \quad (11.30)$$

By applying the transformation

$$\eta_1 = u_1$$
$$\eta_2 = \frac{1}{2}[(1+u_1)v_1 - (1-u_1)] \quad (11.31)$$

to the first integral on the right-hand side of (11.30), and

$$\eta_1 = \frac{1}{2}[(1+u_2)v_2 - (1-u_2)]$$
$$\eta_2 = u_2 \quad (11.32)$$

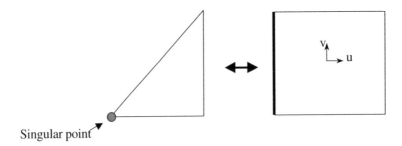

Figure 11.6: Transformation of triangle to square.

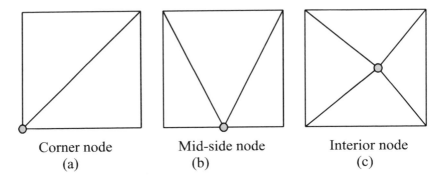

| Corner node | Mid-side node | Interior node |
| (a) | (b) | (c) |

Figure 11.7: Subdivision of quadrilateral element into triangular sub-elements.

to the second integral, each triangular element is transformed into a square (see Figure 11.8). Thus,

$$I \quad = \quad \int_{-1}^{1} \int_{-1}^{1} K N_\alpha J^n d\eta_1 d\eta_2 \tag{11.33}$$

$$= \quad \frac{1}{2} \int_{-1}^{1} \int_{-1}^{1} K N_\alpha J^n J_1(u_1) du_1 dv_1 + \frac{1}{2} \int_{-1}^{1} \int_{-1}^{1} K N_\alpha J^n J_2(v_2) du_1 dv_2$$

where the Jacobian factors $J_1(u_1) = (1+u_1)$ and $J_2(v_2) = (1+v_2)$ cancel out the $1/r$ singularity. As a consequence of these transformations, the integration points within the triangles are more concentrated near the singular points, as shown in Figure 11.9, for a 4×4 Gaussian integration.

In general, for each triangle sub-element (say k) the transformation may be obtained from

$$\eta_1 \quad = \quad \sum_{\alpha=1}^{4} N_\alpha(u_k, v_k) \eta_1^{\alpha k}$$

$$\eta_2 \quad = \quad \sum_{\alpha=1}^{4} N_\alpha(u_k, v_k) \eta_2^{\alpha k} \tag{11.34}$$

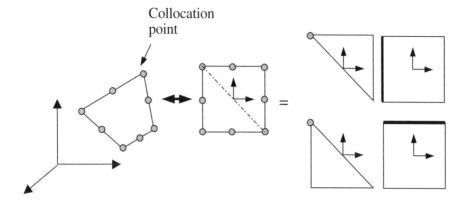

Figure 11.8: Systematic use of the transformation of variable technique.

where $N_\alpha(u_k, v_k)$ are the shape functions for a four noded square, given as

$$N_1(u_k, v_k) = \frac{1}{4}(1 + u_k)(1 + v_k) \qquad N_3 = \frac{1}{4}(1 - u_k)(1 - v_k)$$

$$N_2(u_k, v_k) = \frac{1}{4}(1 - u_k)(1 + v_k) \qquad N_4 = \frac{1}{4}(1 + u_k)(1 - v_k)$$

and $(\eta_1^{\alpha k}, \eta_2^{\alpha k})$ are intrinsic coordinates of the triangular sub-element at the vertex where the collocation point is located. Therefore, the integrals in (11.33) are

$$I = \sum_{k=1}^{K} \sum_{s_2=1}^{S_1} \sum_{s_2=1}^{S_1} A_{s_1} A_{s_2} K[\mathbf{x}', \mathbf{x}(\eta_{s_1}, \eta_{s_2})] N_\alpha(\eta_{s_1}, \eta_{s_2}) J^n(\eta_{s_1}, \eta_{s_2}) J^C(\eta_{s_1}, \eta_{s_2})$$

(11.35)

where K denotes the total number of triangular subelements.

The triangular to square transformation described above will exactly cancel the singularity in fundamental solutions which are of $O(1/r)$. However, the transformation will only reduce the singularity in strong singular integrals from $O(1/r^2)$ to $O(1/r)$. But as shown in Chapter 2, the condition of rigid body translation can be used to circumvent further treatment of this integral. Nevertheless, the use of the transformation technique is recommended for the strong singular fundamental solution when $\mathbf{x}'$ belongs to the element under consideration, as it will serve to improve the accuracy of the so-called off-diagonal entries, that is $\mathbf{x}' \neq \mathbf{x}$, but both $\mathbf{x}'$ and $\mathbf{x}$ belong to the same element.

Subtraction of Singularity Method

This method, which is based on series expansion of the fundamental solution, shape function and the Jacobian, was introduced by Aliabadi *et al.*[1, 2] for the treatment of weakly singular kernels in three-dimensional problems. They wrote the integrals as follows:

$$\int_{-1}^{1} \int_{-1}^{1} K N_\alpha J d\eta_1 d\eta_2$$

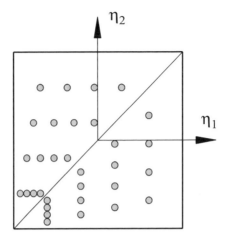

Figure 11.9: Integration points for a quadrilateral element divided into two triangular subelements.

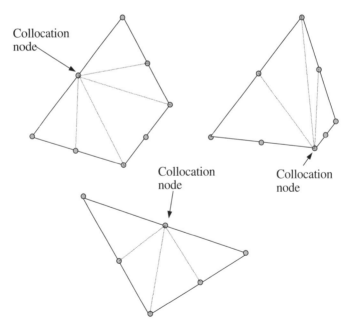

Figure 11.10: Subdivision into triangular elements.

$$= \int_{-1}^{1} \int_{-1}^{1} [K N_\alpha J - K^* N_\alpha^* J^*] \, d\eta_1 d\eta_2 + \int_{-1}^{1} \int_{-1}^{1} K N_\alpha J \, d\eta_1 d\eta_2 \quad (11.36)$$

where the approximate integrand denoted by $*$ has the same singularity as the original integrand, but is of simpler form which can be integrated analytically or semi-analytically. In (11.36) the first integral on the right-hand side is non-singular, and can be integrated accurately using standard numerical quadrature.

The weakly singular integrals have a singularity of $O(1/r)$, and the first order series expansion of $K^* N^* J^*$ is sufficient to remove the singularity. However, as shown in [1], higher order approximations can result in a reduction in computational time as a lower order quadrature can be used to integrate the first integral on the right-hand side of (11.36). They also demonstrated that the singularity subtraction technique can be employed together with the variable transformation technique to allow for the reduction of weak singular integrals to one-dimensional non-singular integrals [?, 18]. Comparisons among weighted Gaussian formula, transformation of variable technique and the singularity subtraction technique can be found in [7].

First order approximation

The vector $\mathbf{r}$ from a variable $\mathbf{x}$ to a fixed point $\mathbf{x}'$ can be written as a series expansion in terms of η as

$$\mathbf{r} = \mathbf{x}(\eta_1, \eta_2) - \mathbf{x}(\eta_1', \eta_2') = \mathbf{x}_{\eta_1} \delta \eta_1 + \mathbf{x}_{\eta_2} \delta \eta_2 + O(\rho^2) \quad (11.37)$$

where $\delta \eta_i = \eta_i - \eta_i'$ $(i = 1, 2)$, $\rho^2 = \delta \eta_1^2 + \delta \eta_2^2$ and

$$\mathbf{x}_{\eta_i} = \frac{\partial \mathbf{x}(\eta_1', \eta_2')}{\partial \eta_i} = \sum_{\alpha=1} \frac{\partial N_\alpha(\eta_1', \eta_2')}{\partial \eta_i} \mathbf{x}^\alpha \qquad i = 1, 2 \quad (11.38)$$

Expression (11.36) can be written in a polar coordinate system centred at (η_1, η_2) as

$$\eta_1 - \eta_1' = \rho \cos \theta \quad (11.39)$$
$$\eta_2 - \eta_2' = \rho \sin \theta \quad (11.40)$$

so that

$$d\eta_1 d\eta_2 = \rho \, d\rho \, d\theta \quad (11.41)$$
$$\mathbf{r} = \rho(\mathbf{x}_{\eta_1} \cos \theta + \mathbf{x}_{\eta_2} \sin \theta) + O(\rho^2) = \rho \mathbf{a}(\theta) + O(\rho^2) \quad (11.42)$$

Hence,

$$r^2 = \mathbf{r}.\mathbf{r} = A^2(\theta)\rho^2 + O(\rho^3) \quad (11.43)$$

where

$$A^2(\theta) = \mathbf{a}(\theta).\mathbf{a}(\theta) \quad (11.44)$$

The distance r and $1/r$ can now be written as

$$r = \rho A(\theta) + O(\rho^2) \quad (11.45)$$
$$\frac{1}{r} = \frac{1}{\rho A(\theta)} + O(1) \quad (11.46)$$

From (11.42) and (11.46), we have

$$
\begin{aligned}
r_{,i} &= \frac{r_i}{r} = \frac{r_i}{A(\theta)\rho}[1 + O(\rho)] \\
&= \frac{1}{A(\theta)} \left[(\mathbf{x}_{\eta_1})_i \cos \theta + (\mathbf{x}_{\eta_2})_i \sin \theta \right] + O(\rho) \\
&= \frac{a_i}{A(\theta)} + O(\rho) \quad (11.47)
\end{aligned}
$$

where r_i denotes the ith component of vector r. Similarly, the shape function and Jacobian can be written as

$$N_\alpha^*(\eta_1, \eta_2) = N_o + O(\rho) \tag{11.48}$$
$$J^*(\eta_1, \eta_2) = J_o + O(\rho) \tag{11.49}$$

Using the above transformation, the product of the kernel, shape function and Jacobian can be written as

$$K[\mathbf{x}', \mathbf{x}(\eta_1, \eta_2)] N_\alpha^*(\eta_1, \eta_2) J^*(\eta_1, \eta_2) = \frac{K_{-1}(\theta)}{\rho} + O(1) \tag{11.50}$$

Therefore, the weakly singular integral can be written as

$$\int_S K[\mathbf{x}', \mathbf{x}(\eta_1, \eta_2)] N_\alpha(\eta_1, \eta_2) J^n(\eta_1, \eta_2) d\eta_1 d\eta_2 = \int_S [KNJ\rho - K_{-1}(\theta)] d\rho d\theta$$
$$+ \int_S K_{-1}(\theta) d\rho d\theta \tag{11.51}$$

where the first integral on the right-hand side of (11.51) is non-singular and can be integrated using the Gauss quadrature, and the second integral is also non-singular and can be integrated analytically with respect to ρ and numerically with respect to θ.

11.4   Strongly Singular

In the flux, displacement and stress boundary element integral equations, strongly singular integrals exist, in which the integrands are unbounded to an extent that the usual Riemann integration is not possible. In such situations, the result of the *integration* means simply the finite-part of the divergent integral. The necessary continuity conditions for the existence of principal-value integrals are also the necessary conditions for the existence of the finite-part integrals.

Readers are recommended to consult books by Mikhlin [42] and Kupradze [25] for detailed discussions on strongly singular integrals. The methods presented here represent an extension of their work to the boundary element method.

In the next section some basic definitions [34] are presented.

11.4.1   Basic Definitions

Consider an improper integral, defined in the interval $a \le \eta \le b$, in which the integrand, unbounded at an interior point $a < \eta' < b$, has the singular part given by $1/(\eta - \eta')$. Let us assume that the regular part of the integrand, function $f(\eta)$, satisfies a Hölder or Lipschitz continuity condition which is expressed as $f(\eta) \in C^{0,\alpha}$; this means that there are constants $|B| < \infty$ and $0 < \alpha \le 1$, such that the following inequality holds:

$$|f(\eta) - f(\eta')| \le B|\eta - \eta'|^\alpha \tag{11.52}$$

A Cauchy principal-value integral of the function $f(t) \in C^{0,\alpha}$ is defined, for an interior singular point, as

$$\fint_a^b \frac{f(\eta)}{\eta - \eta'} d\eta = \lim_{\varepsilon \to 0} \left\{ \int_a^{\eta - \varepsilon} \frac{f(\eta)}{\eta - \eta'} d\eta + \int_{\eta + \varepsilon}^b \frac{f(\eta)}{\eta - \eta'} d\eta \right\} \tag{11.53}$$

in which the neighbourhood ε is symmetric about the singular point. Although a Cauchy principal-value integral is defined only for an interior singular point, it can be evaluated separately on both sides of the singularity by the following pair of limit expressions:

$$\int_a^x \frac{f(\eta)}{\eta - \eta'} d\eta = \lim_{\varepsilon \to 0} \left\{ \int_a^{x-\varepsilon} \frac{f(\eta)}{\eta - \eta'} d\eta - f(\eta') \ln \varepsilon \right\} \qquad (11.54)$$

$$\int_x^b \frac{f(\eta)}{\eta - \eta'} d\eta = \lim_{\varepsilon \to 0} \left\{ \int_{x+\varepsilon}^b \frac{f(\eta)}{\eta - \eta'} d\eta + f(\eta') \ln \varepsilon \right\} \qquad (11.55)$$

These limit expressions, which can replace definition (11.53), must be taken together, as any single contribution does not exist alone.

In a BEM mesh, the type of finite-part integrals that is dealt with is determined by the local position of the node where collocation is carried out. The local position of a node is defined by its local intrinsic coordinates η, in the range $-1 \leq \eta \leq +1$. Internal nodes have their local coordinate $\eta \neq \pm 1$, while end nodes have $\eta = \pm 1$. Hence, discontinuous elements have only internal nodes, while continuous elements have two end nodes and some possible internal nodes. Collocation at internal nodes gives rise to two-sided finite part integrals, while collocation at end nodes gives rise to one-sided finite-part integrals.

Consider a finite-part integral obtained by collocation of any BEM strongly singular integral at an internal node. The piece-wise polynomial character of the shape functions implies continuous differentiability of the regular part of the integrals [34], that is $f(\eta) \in C^\infty$. Therefore, when collocation is done at internal nodes of an element, all the continuity requirements for the existence of finite-part integrals, of any order, are implicitly satisfied.

For the case of collocation at an end node, common to two continuous boundary elements, one-sided finite part integrals are defined. Bearing in mind the equivalence between finite-part and principal-value integrals, these element contributions must be assembled into a two-sided finite-part integral, with the range defined on both boundary elements, since any single contribution of a one-sided principal-value integral does not exist alone. Note that assembly is the process by means of which continuity conditions of the field variable are imposed in the discrete system. Thus, when collocation is carried out at a node common to two elements, the continuity requirements for the existence of finite-part integrals can be analysed by considering that the collocation node constitutes an internal singular point, in the interval defined by both elements that share the collocation node. Therefore, if $C^{0,\alpha}$ continuous elements are used, only first order finite-part integrals can be defined by collocation at the end nodes; $C^{1,\alpha}$ continuous boundary elements are required. Basic definitions of finite-part integrals are presented in the following sections, and can also be found in Mangler [38], Martin and Rizzo[39] and Portela *et al.* [34].

Two-sided finite-part integrals

The finite-integral of the first order of a function, $f(\eta)$, with a continuous first derivative, $f(\eta) \in C^1$, is given by

$$\int_a^b \frac{f(\eta)}{\eta - \eta'} d\eta = \int_a^b \frac{f(\eta) - f(\eta')}{\eta - \eta'} d\eta + f(\eta') \int_a^b \frac{1}{\eta - \eta'} d\eta \qquad (11.56)$$

in which

$$\fint_a^b \frac{d\eta}{\eta - \eta'} = [\ln | \eta - \eta' |\,]_a^b = \ln | \frac{b - \eta'}{\eta' - a} | \qquad (11.57)$$

The continuity requirement imposed on the first derivative of the function $f(\eta)$ is a necessary condition for the Taylor's expansion to be possible. If the function $f(\eta)$ has a degree of continuity not high enough to allow the required Taylor's expansion, then a sufficient condition for the existence of the finite-part integral in equation (11.56) is that $f(\eta)$ satisfies a Hölder or Lipschitz continuity condition, as defined in (11.52). The Hölder condition (11.52) also guarantees that the first integral of the right-hand side of equation (11.56) is, at most, weakly singular.

One-sided finite-part integral

The finite-part integral of the first order of a function $f(\eta) \in C^1$ is defined as

$$\fint_a^{\eta'} \frac{f(\eta)}{\eta - \eta'} d\eta = \int_a^{\eta'} \frac{f(\eta) - f(\eta')}{\eta - \eta'} d\eta + f(\eta') \fint_a^{\eta'} \frac{1}{\eta - \eta'} d\eta \qquad (11.58)$$

$$\fint_{\eta'}^b \frac{f(\eta)}{\eta - \eta'} d\eta = \int_{\eta'}^b \frac{f(\eta) - f(\eta')}{\eta - \eta'} d\eta + f(\eta') \fint_{\eta'}^b \frac{1}{\eta - \eta'} d\eta \qquad (11.59)$$

in which

$$\fint_a^{\eta'} \frac{d\eta}{\eta - \eta'} = \text{finite part of } [\ln | \eta - \eta' |\,]_a^{\eta'} = -\ln | a - \eta' | \qquad (11.60)$$

$$\fint_{\eta'}^b \frac{d\eta}{\eta - \eta'} = \text{finite part of } [\ln | \eta - \eta' |\,]_{\eta'}^b = +\ln | b - \eta' | \qquad (11.61)$$

If $f(\eta) \notin C^1$, a sufficient condition for the existence of the finite-part integrals in equations (11.58) and (11.59) is a $f(\eta) \in C^{0,\alpha}$. When equation (11.56) is derived with the aid of equations (11.58) and (11.59), it becomes evident that the existence of the finite-part integral requires the continuity of the function $f(\eta)$ at the singular point η'.

11.4.2   Two-dimensional Problems

Weighted Gaussian Integration

Kutt [26] presented weighted Gaussian formulae for the computation of Cauchy and Hadamard finite part integrals. Kutt's quadrature can be written as

$$\fint_{\eta'}^r \frac{f(\eta)}{(\eta - \eta')^\lambda} d\eta \cong (r - \eta')^{1-\lambda} \left\{ \sum_{s=1}^S w_s f[(r - \eta')\eta_s + \eta'_{\_}] + \frac{f(\eta') \ln | r - \eta' |}{(\lambda - 1)!} \right\} \qquad (11.62)$$

for $r > \eta'$ and

$$(-1)^\lambda \fint_{-\eta'}^{-r} \frac{f(\eta)}{(\eta - \eta')^\lambda} d\eta \cong -(r - \eta')^{1-\lambda} \left\{ \sum_{s=1}^S w_s f[(\eta' - r)\eta_s + \eta'_{\_}] + \frac{f(\eta') \ln | r - \eta' |}{(\lambda - 1)!} \right\} \qquad (11.63)$$

for $r > \eta'$. The Gaussian weights w_s and integration points η_s are listed in Table 11.5, for $\lambda = 1$.

Table 11.5: Gaussian weights and points for $\lambda = 1$..

S	η_s	w_s
2	-0.13188 13079 12987	-1.30930 73414 59543
	0.63188 13079 12986	1.30930 73414 59543
4	-0.01835 88088 36002	-2.59362 86894 20635
	0.19973 81810 49987	1.67202 98013 53350
	0.58395 30474 65078	0.67543 34270 27654
	0.90998 72556 45613	0.24616 54710 39631
8	-0.00324 25015 97297	-3.93063 68100 68551
	0.05349 07660 72146	1.73740 63109 00222
	0.17782 73263 92695	0.85763 45422 73069
	0.35071 78785 52550	0.53836 07734 82498
	0.54581 95204 68488	0.35975 75915 92037
	0.73342 70806 57186	0.23726 76990 94551
	0.88498 30533 70117	0.14146 22854 96869
	0.97745 43783 79800	0.05874 76072 29312

Subtraction of Singularity Method

Consider a discontinuous quadratic element that contains the collocation node. The local parametric coordinate system η is defined, as usual, in the range $-1 \le \eta \le +1$, and collocation η' is mapped onto $\mathbf{x}'$ via the continuous element shape functions. A Cauchy principal value integral of equation can be expressed in the local coordinate as

$$\fint_S K(\mathbf{x}', \mathbf{x}) u(\mathbf{x}) dS = u^n \fint_{-1}^{1} \frac{h(\eta)}{\eta - \eta'} d\eta \tag{11.64}$$

where u^n denotes the value of the function at node n, $h(\eta)$ is a regular function, given by the product of the fundamental solution, a shape function and the Jacobian of the coordinate transformation, multiplied by the term $\eta - \eta'$. The kernel $K(\mathbf{x}', \mathbf{x})$ is a kernel with $O(1/r)$ singularity. It can be considered as a representative of fundamental solutions $T_{ij}(\mathbf{x}', \mathbf{x})$ and $S_{kij}(\mathbf{x}', \mathbf{x})$. The integral of the right-hand side of equation (11.64) can be transformed with the aid of the first term of a Taylor's expansion of the function h, around the collocation node, to give

$$\fint_{-1}^{1} \frac{h(\eta)}{\eta - \eta'} d\eta = \int_{-1}^{1} \frac{h(\eta) - h(\eta')}{\eta - \eta'} d\eta + h(\eta') \fint_{-1}^{1} \frac{d\eta}{\eta - \eta'} \tag{11.65}$$

Now, the first integral of the right-hand side is regular, and the second can be integrated analytically to give

$$\fint_{-1}^{1} \frac{d\eta}{\eta - \eta'} = \ln \left| \frac{1 - \eta'}{1 + \eta'} \right| \tag{11.66}$$

In equation (11.66), the requirements of continuity at the collocation point are automatically satisfied with the discontinuous elements, because the nodes are internal points of the elements, where $h(\eta)$ is continuously differentiable.

For singularity at the end nodes (i.e. $\eta' = \pm 1$), the procedure developed by Guiggiani and Casalini [14] will be described. They denoted the product of the

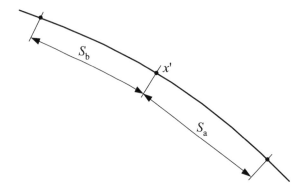

Figure 11.11: Boundary segments sharing the collocation point.

kernel and shape function KN_α as $g^a(\eta)$ and $g^b(\eta)$ on boundary segments S_a and S_b, respectively (see Figure 11.11). These functions are written in terms of regular functions $d_a(\eta)$ and $d_b(\eta)$ as follows:

$$g^a(\eta) \;=\; \frac{d^a(\eta)}{\eta - 1} \tag{11.67}$$

$$g^b(\eta) \;=\; \frac{d^b(\eta)}{\eta - 1} \tag{11.68}$$

Equation (11.64) can now be written as

$$I = \int_S K(\mathbf{x}',\mathbf{x})u(\mathbf{x})dS = \lim_{\varepsilon \to 0^+} \left\{ \int_{-1}^{1-\Delta\xi_b} \frac{d^b(\eta)}{\eta - 1} J_b(\eta)d\eta + \int_{-1+\Delta\xi_a}^{1} \frac{d^a(\eta)}{\eta + 1} J_a(\eta)d\eta \right\} \tag{11.69}$$

where, in general, $J_b(\eta) \neq J_a(\eta)$. The terms $\Delta\xi_a$ and $\Delta\xi_b$ are related to ε via

$$\varepsilon = J_b(1)\Delta\xi_b + O(\Delta\xi_b^2) = J_a(-1)\Delta\xi_a + O(\Delta\xi_a^2)$$

Using the above relationships, equation (11.64) can be rewritten as

$$I = \lim_{\varepsilon \to 0^+} \left\{ \int_{-1}^{1-(\varepsilon/J_b(1))} \frac{d^b(\eta)}{\eta - 1} J_b(\eta)d\eta + \int_{-1+(\varepsilon/J_a(-1))}^{1} \frac{d^a(\eta)}{\eta + 1} J_a(\eta)d\eta \right\} \tag{11.70}$$

Denoting, $h(\eta) = d(\eta)J(\eta)$, and noting that $h^b(1) = h^a(-1)$ from the existence of the Cauchy principal value integral, I can be rewritten as

$$
\begin{aligned}
I \;=\; & \lim_{\varepsilon \to 0^+} \left\{ \int_{-1}^{1-(\varepsilon/J_b(1))} \frac{h^b(\eta) - h^b(1)}{\eta - 1}d\eta + \int_{-1}^{1-(\varepsilon/J_b(1))} \frac{h^b(1)}{\eta - 1}d\eta \right. \\
& \left. \int_{-1+(\varepsilon/J_a(-1))}^{1} \frac{h^a(\eta) - h^a(-1)}{\eta + 1}d\eta + \int_{-1+(\varepsilon/J_a(-1))}^{1} \frac{h^a(-1)}{\eta + 1}d\eta \right\} \\
\;=\; & \lim_{\varepsilon \to 0^+} \left\{ \int_{-1}^{1-(\varepsilon/J_b(1))} \frac{h^b(\eta) - h^b(1)}{\eta - 1}d\eta + \int_{-1+(\varepsilon/J_a(-1))}^{1} \frac{h^a(\eta) - h^a(-1)}{\eta + 1}d\eta \right. \\
& \left. \left[h^b(1)\ln|\eta - 1|\right]_{-1}^{1-(\varepsilon/J_b(1))} + \left[h^a(-1)\ln|\eta + 1|\right]_{-1+(\varepsilon/J_a(-1))}^{1} \right\} \tag{11.71}
\end{aligned}
$$

which simplifies to

$$\oint_S K(\mathbf{x}', \mathbf{x}) u(\mathbf{x}) dS(\mathbf{x}) = \int_{-1}^{1} \frac{h^b(\eta) - h^b(1)}{\eta - 1} d\eta + \int_{-1}^{1} \frac{h^a(\eta) - h^a(-1)}{\eta + 1} d\eta$$
$$- h^b \ln|J_b(1)| + h^a(-1) \ln|J_a(-1)| \qquad (11.72)$$

Stokes' Theorem

An alternative way of evaluating the singular integrals is by utilizing Stokes' theorem. This approach has been used by several researchers, as reviewed in [46]. Stokes's theorem states that

$$\int_{S_m} \Im_{ij} \{f\} dS = e_{ijk} \oint_C f dx_k \qquad (11.73)$$

where e_{ijk} is the permutation tensor, C is the line which bounds the open surface S_m, and the differential operator is given by $\Im_{ij}\{f\} = n_i f_{,i} - n_j f_{,i}$. Stoke's theorem for line integrals may be derived as a special case of (11.73), where $f(\mathbf{x})$ does not vary with x_3 and S_m projection of the 2D (x_1, x_2) boundary are denoted by PQ through x_3, i.e.

$$\int_{P_{ij}}^{Q} \{f\} dS = e_{ij3} [f]_P^Q \equiv e_{ij3}(f(Q) - f(P)) \qquad (11.74)$$

The stokes theorem was used in [49] to evaluate the strongly singular integrals for elastostatics.. The traction fundamental solution for two-dimensional elastostatic problems (plane strain condition) presented in Chapter 2, can be rewritten as

$$T_{ij}(\mathbf{x}', \mathbf{x}) = \frac{1}{4\pi(1-\nu)} \{(1-2\nu)\Im_{ij}(\log r) + \Im_{mj}(r_{,i} r_{,m})\} - \frac{\delta_{ij}}{2\pi} \frac{r_{,m} n_m}{r} \qquad (11.75)$$

Using Stokes' theorem (11.74) we have

$$\int_{S_m} T_{ij} ds = \frac{1}{4\pi(1-\nu)} \{(1-2\nu)e_{ij3} \log r + e_{mj3} r_{,i} r_{,m}\}_Q^P + \frac{1}{2\pi} \delta_{ij} \Theta(\mathbf{x}') \qquad (11.76)$$

Using the identity $n_m dS_m = e_{mk3} dr_k$, the solid angle term Υ reduces to [49]

$$\Theta(\mathbf{x}') = -\int_{S_m} \frac{r_{,m} n_m}{r} dS_m = \int_{S_m} \frac{r_2 dr_1 - r_1 dr_2}{r^2} = \left[\tan^{-1}\left(\frac{r_1}{r_2}\right)\right]_P^Q$$

which may be evaluated directly in terms of radial vectors directed at the ends P and Q of the arc S_m. For example, $r^P r^Q \cos \Theta = r_1^P r_1^Q + r_2^P r_2^Q$ and $r^P r^Q \sin \Theta = r_1^P r_2^Q + r_2^Q r_1^P$.

11.4.3 Three-dimensional Problems

The method for the evaluation of hypersingular integrals, reported in this section was presented by Guiggiani *et al.* [16], which is an extension of Guiggiani and Gigante[15] and is based on the Taylor's series expansion of the kernel, and Jacobian transformation developed by Aliabadi, Hall and Phemister [1].

Higher order approximation

The vector **r** can be written as

$$\mathbf{r} = \mathbf{x}_{\eta_1}\delta\eta_1 + \mathbf{x}_{\eta_2}\delta\eta_2 + \mathbf{x}_{\eta_1\eta_2}\delta\eta_1\delta\eta_2 + \frac{1}{2}\mathbf{x}_{\eta_1\eta_1}\delta\eta_1^2 + \frac{1}{2}\mathbf{x}_{\eta_2\eta_2}\delta\eta_2^2 + O(\rho^3) \qquad (11.77)$$

or

$$\mathbf{r} = \rho\mathbf{a}(\theta) + \rho^2\mathbf{b}(\theta) + O(\rho^3) \qquad (11.78)$$

where

$$\mathbf{b}(\theta) = \mathbf{x}_{\eta_1\eta_2}\sin\theta\cos\theta + \frac{1}{2}\mathbf{x}_{\eta_1\eta_1}\cos^2\theta + \frac{1}{2}\mathbf{x}_{\eta_2\eta_2}\sin^2\theta$$

so that

$$\begin{aligned}
\frac{1}{r} &= \left[A^2(\theta)\rho^2 + 2\mathbf{a}(\theta)\mathbf{b}(\theta)\rho^3 + O(\rho^4)\right]^{-1/2} \\
&= \frac{1}{A(\theta)\rho}\left[1 + \frac{2\mathbf{a}(\theta).\mathbf{b}(\theta)}{A^2(\theta)}\rho + O(\rho^2)\right]^{-1/2} \\
&= \frac{1}{A(\theta)\rho}\left[1 - \frac{\mathbf{a}(\theta).\mathbf{b}(\theta)}{A^2(\theta)}\rho + O(\rho^2)\right] \qquad (11.79)
\end{aligned}$$

and

$$\begin{aligned}
r_{,i} &= \frac{1}{A(\theta)\rho}\left[1 - \frac{\mathbf{a}(\theta).\mathbf{b}(\theta)}{A^2(\theta)}\rho + O(\rho^2)\right]\left[\rho a_i + \rho^2 b_i + O(\rho^3)\right] \\
&= \frac{a_i}{A(\theta)}\left[\frac{b_i}{A(\theta)} - \frac{(\mathbf{a}.\mathbf{b})}{A^3(\theta)}a_i\right]\rho + O(\rho^2) \qquad (11.80)
\end{aligned}$$

Similarly, the shape function and Jacobian can be expanded as

$$\begin{aligned}
N_\alpha(\eta_1,\eta_2) &= N_o + N_{\eta_1}\delta\eta_1 + N_{\eta_2}\delta\eta_2 + O(\rho^2) \\
&= N_o + \rho N_{\eta_1}\cos\theta + \rho N_{\eta_2}\sin\theta + O(\rho^2) \qquad (11.81)
\end{aligned}$$

and

$$\begin{aligned}
J(\eta_1,\eta_2) &= J_o + J_{\eta_1}\delta\eta_1 + J_{\eta_2}\delta\eta_2 + O(\rho^2) \\
&= J_o + \rho J_{\eta_1}\cos\theta + \rho J_{\eta_2}\sin\theta + O(\rho^2) \qquad (11.82)
\end{aligned}$$

Subtraction of Singularity Method

Consider an element on the surface S_s, as shown in Figure 11.12, over which Cauchy and Hadamard principal value integrals are to be computed. As shown in the Figure, the collocation point is denoted as $\mathbf{x}^c$, and S_ε is taken to be part of S_s intersecting a sphere centred at $\mathbf{x}^c$ of radius ϵ. In the local intrinsic coordinate system (η_1,η_2), $S_s, \mathbf{x}^c$ and S_ϵ correspond to $G_s, (\eta_1',\eta_2')$ and G_ε, respectively. From the definition of Cauchy principal value integrals, the integral involving the strongly singular kernel is written as

$$I = \lim_{\epsilon\to 0}\int_{G_s-G_\varepsilon} K[\mathbf{x}^c, \mathbf{x}(\eta_1,\eta_2)]N_\alpha(\eta_1,\eta_2)J^n(\eta_1,\eta_2)d\eta_1 d\eta_2 \qquad (11.83)$$

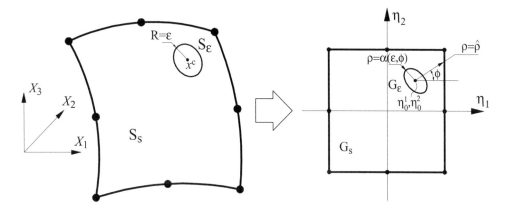

Figure 11.12: Vanishing neighbourhood around source point $\mathbf{x}^\omega$.

The integrand can be approximate using Taylor's series expansion, as

$$K[\mathbf{x}^c, \mathbf{x}(\eta_1, \eta_2)] N_\alpha(\eta_1, \eta_2) J^n(\eta_1, \eta_2) = KNJ = \frac{K_{-1}(\theta)}{\rho^2} + O(\frac{1}{\rho}) \qquad (11.84)$$

where subscripts have been dropped for convenience, and $T_{-1}(\theta)$ denotes the first term in the Taylor expansion. Using (11.84), the integral I in (11.83) can be written in a polar coordinate (ρ, θ) system as

$$I = \lim_{\epsilon \to 0} \int_{G_s - G_\epsilon} \left[KNJ\rho - \frac{K_{-1}(\theta)}{\rho} \right] d\rho d\theta + \lim_{\epsilon \to 0} \int_{G_s - G_\epsilon} \left[\frac{K_{-1}(\theta)}{\rho} \right] d\rho d\theta$$

$$I_o + I_{-1} \qquad (11.85)$$

The integrand of the first integral on the right-hand side of (11.85) is regular (since $KNJ\rho - K_{-1}/\rho = O(1)$). In the limit, as $\varepsilon \to 0$, the integral I_0 is obtained as

$$I_o = \int_{G_S} \left[KNJ\rho - \frac{K_{-1}}{\rho} \right] d\rho d\theta \qquad (11.86)$$

To evaluate the above integral, G_s is divided into four triangles as shown in Figure 11.13, so that (11.85) is then rewritten as

$$I_o = \sum_{i=1}^4 \int_{\theta_i}^{\theta_{i+1}} \int_0^{\hat{\rho}(\theta)} \left[TNJ\rho - \frac{T_{-1}}{\rho} \right] d\rho d\theta \qquad (11.87)$$

where $\rho = \hat{\rho}(\theta)$ is the polar coordinate equation of the boundary G_s. The integrals for each triangle in (11.87) are evaluated using the Gaussian quadrature.

For calculation of the second integral in (11.85), let $\rho = \alpha(\varepsilon, \theta)$ denote the equation for the mapping of the cut surface of the sphere centred at $\mathbf{x}^c$ with radius ε, common to G_s and G_ε. From (11.79), we have

$$\epsilon = r = A\rho \left[1 + \frac{\mathbf{a} \cdot \mathbf{b}}{A^2} \rho + O(\rho^2) \right]^{1/2} = A\rho + \frac{\mathbf{a} \cdot \mathbf{b}}{A} \rho^2 + O(\rho^3) \qquad (11.88)$$

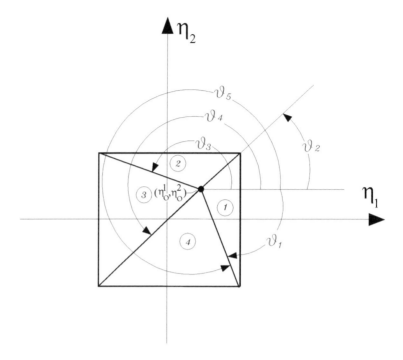

Figure 11.13: Element subdivision.

The expansion of powers of ϵ is obtained using the reversion of (11.88), to give

$$\rho = \alpha(\varepsilon, \theta) = \beta(\theta)\varepsilon + \gamma(\theta)\varepsilon^2 + O(\varepsilon^3) \tag{11.89}$$

where $\beta(\theta) = 1/A(\theta)$ and $\gamma(\theta) = -\mathbf{a}.\mathbf{b}./A^4(\theta)$. Using the expression in (11.89), the second integral on the right-hand side of (11.85) becomes

$$
\begin{aligned}
I_{-1} &= \lim_{\epsilon \to 0} \int_{G_s - G_\varepsilon} \left[\frac{K_{-1}(\theta)}{\rho} \right] d\rho d\theta = \lim_{\varepsilon \to 0} \int_0^{2\pi} \int_{\alpha(\varepsilon, \theta)}^{\hat{\rho}(\theta)} \left[\frac{K_{-1}(\theta)}{\rho} \right] d\rho d\theta \\
&= \int_0^{2\pi} K_{-1}(\theta) \ln \frac{| \hat{\rho}(\theta) |}{| \beta(\theta) |} d\theta
\end{aligned}
\tag{11.90}
$$

where the condition $\int_0^{2\pi} K_{-1}(\theta) d\theta = 0$ is used. The integral in the above equation is regular, and can be evaluated using Gaussian integration.

Stokes' Theorem

Stokes' theorem (11.73) can be utilized for integration of strongly singular integrals in three dimensions. For example, the Kelvin's traction fundamental solution presented in Chapter 2 can be rewritten as

$$T_{ij}(\mathbf{x}', \mathbf{x}) = \frac{1}{8\pi(1 - \nu)} \left\{ (2\nu - 1)\Im_{ij} \left(\frac{1}{r} \right) + \Im_{ij} \left(\frac{r_{,m} n_m}{r^2} \right) \right\} \tag{11.91}$$

Using stokes' theorem (11.73) gives [49]

$$\int_{S_m} T_{ij}(\mathbf{x}',\mathbf{x})ds = \frac{1}{8\pi(1-\nu)}\oint_C \frac{1}{r}[(1-2\nu)e_{lij}+e_{lmj}r_{,i}r_{,m}]dr_l + \frac{1}{4\pi}\delta_{ij}\Theta(\mathbf{x}') \quad (11.92)$$

Using the identity $n_m dS_m = e_{mk3}dr_k$, the solid angle term Υ reduces to [49]

$$\Theta(\mathbf{x}') = -\int_{S_m}\frac{r_{,m}n_m}{r}dS_m$$

11.5 Hypersingular Integrals

11.5.1 Basic Definitions

Consider an improper integral, defined in the interval $a \leq \eta \leq b$, in which the integrand, unbounded at an interior point $a \leq \eta \leq b$, has the singular part given by $1/(\eta-\eta')^2$. Consider also that the regular part of the integrand, function $f(\eta)$, has a Hölder-continuous first derivative. This condition, which is expressed as $f(\eta) \in C^{1,\alpha}$, means that there are constants $\mid B \mid < \infty$ and $0 < \alpha \leq 1$, such that the following inequalities holds:

$$\mid f(\eta) - f(\eta') - f'(\eta-\eta') \mid \leq B\mid \eta-\eta'\mid^{\alpha+1} \quad (11.93)$$

The Hadamard [17] principal-value integral of the function $f(t) \in C^{1,\alpha}$ is defined, for an interior singular point, as

$$\fint_a^b \frac{f(\eta)}{(\eta-\eta)^2}d\eta = \lim_{\varepsilon\to\infty}\left\{\fint_a^{\eta'-\varepsilon}\frac{f(\eta)}{(\eta-\eta)^2}d\eta + \int_{\eta'+\varepsilon}^b \frac{f(\eta)}{(\eta-\eta)^2}d\eta - \frac{2f(\eta)}{\varepsilon}\right\} \quad (11.94)$$

in which the neighbourhood ε is symmetric about the singular point. Although the Hadamard principal-value integral is defined only for an interior singular point, it can be evaluated separately on both sides of the singularity by the following pair of limit expressions:

$$\fint_a^{\eta'}\frac{f(\eta)}{(\eta-\eta)^2}d\eta = \lim_{\varepsilon\to\infty}\left\{\fint_a^{\eta'-\varepsilon}\frac{f(\eta)}{(\eta-\eta)^2}d\eta - \frac{2f(\eta)}{\varepsilon} - f'(\eta)\ln\varepsilon\right\} \quad (11.95)$$

$$\fint_{\eta'}^b\frac{f(\eta)}{(\eta-\eta)^2}d\eta = \lim_{\varepsilon\to\infty}\left\{\fint_{\eta'+\varepsilon}^0\frac{f(\eta)}{(\eta-\eta)^2}d\eta - \frac{2f(\eta)}{\varepsilon} + f'(\eta)\ln\varepsilon\right\} \quad (11.96)$$

These limit expressions must be taken together, as any single contribution does not exist alone.

If the function $f(\eta) \in C^2$, then the Hadamard principal-value integral (11.94) is equivalent to a second-order two-sided integral, as reported by Ioakimidis [21]. In addition, the integrals (11.95) and (11.96) are equivalent to second-order, one-sided, finite-part integrals.

The differentiation of the Cauchy principal-value integral must be carried out by means of Leibnitz's rule. Thus, the application of Leibnitz's rule to differentiate equation (11.53) leads to

$$
\begin{aligned}
\frac{d}{d\eta} \int_a^b \frac{f(\eta)}{\eta - \eta'} d\eta &= \lim_{\varepsilon \to 0} \frac{d}{d\eta} \left\{ \int_a^{\eta - \varepsilon} \frac{f(\eta)}{\eta - \eta'} d\eta + \int_{\eta + \varepsilon}^b \frac{f(\eta)}{\eta - \eta'} d\eta \right\} \\
&= \lim_{\varepsilon \to 0} \left\{ \int_a^{\eta - \varepsilon} \frac{f(\eta)}{(\eta - \eta')^2} d\eta + \int_{\eta + \varepsilon}^b \frac{f(\eta)}{(\eta - \eta')^2} d\eta \right. \\
&\qquad \left. - \frac{f(\eta' - \varepsilon) + f(\eta' + \varepsilon)}{\varepsilon} \right\}
\end{aligned}
\tag{11.97}
$$

The continuity conditions of $f(\eta)$ imply that $f(\eta' - \varepsilon) = f(\eta' + \varepsilon) = f(\eta')$, when $\varepsilon \to 0$. Therefore, equation (11.97) becomes

$$
\frac{d}{d\eta} \int_a^b \frac{f(\eta)}{\eta - \eta'} d\eta = \lim_{\varepsilon \to 0} \frac{d}{d\eta} \left\{ \int_a^{\eta - \varepsilon} \frac{f(\eta)}{(\eta - \eta')^2} d\eta + \int_{\eta + \varepsilon}^b \frac{f(\eta)}{(\eta - \eta')^2} d\eta - \frac{2f(\eta')}{\varepsilon} \right\} \tag{11.98}
$$

which is identical to equation (11.94). Consequently, the following relationship is established:

$$
\frac{d}{d\eta} \int_a^b \frac{f(\eta)}{\eta - \eta'} d\eta = \fint_a^b \frac{f(\eta)}{(\eta - \eta')^2} d\eta \tag{11.99}
$$

which means that the differentiation of the Cauchy principal-value integral results in a Hadamard principal-value integral.

Two-sided finite-part integrals

The finite part integral of a function, $f(\eta) \in C^2$ is defined by

$$
\begin{aligned}
\fint_a^b \frac{f(\eta)}{(\eta - \eta')^2} d\eta &= \int_a^b \frac{f(t) - f(\eta') - f^{(1)}(\eta')(\eta - \eta')}{(\eta - \eta')^2} d\eta \\
&\quad + f(\eta') \fint_a^b \frac{d\eta}{(\eta - \eta')^2} + f^{(1)}(\eta') \fint_a^b \frac{d\eta}{\eta - \eta'}
\end{aligned}
\tag{11.100}
$$

in which the last integral is evaluated with the result of equation (11.57), and the second integral on the right-hand side of (11.100) is given by

$$
\fint_a^b \frac{1}{(\eta - \eta')^2} d\eta = \left[\frac{-1}{\eta - \eta'} \right]_a^b = \frac{1}{a - \eta'} - \frac{1}{b - \eta'} \tag{11.101}
$$

11.5.2   Two-dimensional Problems

The Hadamard principal value integral can be expressed in the local parametric coordinate as

$$
\fint_{S_c} K(\mathbf{x}', \mathbf{x}) \, u(\mathbf{x}) dS = \sum_{\gamma=1}^M u^n \fint_{-1}^{+1} \frac{g^\gamma(\eta)}{(\eta - \eta')^2} d\eta \tag{11.102}
$$

where $K(\mathbf{x}'\mathbf{x})$ is of $O(1/r^2)$, and is representative of fundamental solutions such as $D_{kij}(\mathbf{x}', \mathbf{x})$; $g(\eta)$ is a regular function, given by the product of the fundamental solution, a shape function and the Jacobian of the coordinate transformation, multiplied by the term $(\eta - \eta')^2$. The integral of the right-hand side of equation (11.102) can be transformed with the aid of the first and second terms of a Taylor's expansion of the function g, in the neighbourhood of the collocation node, to give

$$\fint_{-1}^{+1} \frac{g^\gamma(\eta)}{(\eta - \eta')^2} d\eta = \int_{-1}^{+1} \frac{g^\gamma(\eta) - g^\gamma(\eta') - g^{\gamma(1)}(\eta')(\eta - \eta')}{(\eta - \eta')^2} d\eta$$

$$+ \quad g^\gamma(\eta') \fint_{-1}^{+1} \frac{d\eta}{(\eta - \eta')^2}$$

$$+ \quad g^{\gamma(1)}(\eta') \int_{-1}^{+1} \frac{d\eta}{\eta - \eta'} \qquad (11.103)$$

where $g^{(1)}$ denotes the first derivative. At the collocation node, the function g is required to have continuity of its second derivative, or at least a Hölder-continuous first derivative, for the finite-part integral to exist. This requirement is automatically satisfied by a discontinuous element, since the nodes are internal points of the element. Now, in equation (11.103), the first integral on the right-hand side is regular and the third integral is identical to that given in equation (11.101). The second integral on the right-hand side of equation (11.103) can be integrated analytically to give

$$\fint_{-1}^{+1} \frac{d\eta}{(\eta - \eta')^2} = -\frac{1}{1 + \eta'} - \frac{1}{1 - \eta'} \qquad (11.104)$$

11.5.3   Three-dimensional Problems

For hypersingular integrals involving singularities of $O(1/r^3)$, higher order series expansion is applied to the integrand of equation (8.29). Following the expansions presented in Section 11.4.3, we have

$$\frac{1}{r^3} = \frac{1}{\rho^3 A^3} - 3\frac{1}{\rho^2} \frac{\mathbf{a}(\theta).\mathbf{b}(\theta)}{A^5} + O\left(\frac{1}{\rho}\right)$$

The integrand can be approximated using Taylor series expansion as

$$K[\mathbf{x}^c, \mathbf{x}(\eta_1, \eta_2)] N_\alpha(\eta_1, \eta_2) J^n(\eta_1, \eta_2) = K N_\alpha J\rho = \frac{K_{-2}(\theta)}{\rho^3} + \frac{K_{-1}(\theta)}{\rho^2} + O\left(\frac{1}{\rho}\right) \qquad (11.105)$$

In a similar way to the Cauchy principal value integral described above, the Hadamard principal value integral can be written as [16, 40]:

$$\lim_{\epsilon \to 0} \left\{ \int_{G_s - G_\epsilon} K[\mathbf{x}^c, \mathbf{x}(\eta_1, \eta_2) N_\alpha(\eta_1, \eta_2) J^n(\eta_1, \eta_2) d\eta_1 d\eta_2 + N_\alpha(\eta_o^1, \eta_o^2) \frac{b_{kij}(\mathbf{x}^c)}{\epsilon} \right\}$$

$$= \int_0^{2\pi} \int_0^{\hat\rho(\theta)} \left\{ K N_\alpha J\rho - \left[\frac{K_{-2}(\theta)}{\rho^3} + \frac{K_{-1}(\theta)}{\rho^2} \right] \right\} d\rho d\theta$$

$$= \int_0^{2\pi} \left\{ K_{-1}(\theta) \ln \frac{|\hat\rho(\theta)|}{|\beta(\theta)|} - K_{-2}(\theta) \left[\frac{\gamma(\theta)}{\beta^2(\theta)} + \frac{1}{\hat\rho(\theta)} \right] \right\} d\theta \qquad (11.106)$$

The above integrals are all regular and can be evaluate a using Gaussian quadrature.

If the function $f(\eta)$ has a degree of continuity not high enough to allow the required Taylor's expansion, then a sufficient condition for the existence of the finite-part integral in equation (11.100) is that $f(\eta)$ has a Hölder-continuous first derivative. The Hölder condition also guarantees that the first integral on the right-hand side of (11.100) is, at most, weakly singular.

11.6   Summary

In this chapter several prominent methods for the evalution of integrals in the boundary element method were described. Other important contributions can be found in [31, 32, 36, 44].

References

[1] Aliabadi, M.H., Hall, W.S. and Phemister, T.G., Taylor expansions for singular kernels in the boundary element method. *International Journal for Numerical Methods in Engineering*, **21**, 2221-2236, 1985.

[2] Aliabadi, M.H., Hall, W.S. and Phemister, T.G., Taylor expansions in the boundary element methods for Neumann problems, *Boundary Elements VII*, 12-31 to 12-39, 1985.

[3] Aliabadi, M.H., Exact evaluation of the integrals in two-dimensional boundary element method, Report No. EMR/10/1, Engineering Materials, Southampton University, April, 1985.

[4] Aliabadi, M.H. and Hall, W.S., Analytical removal of singularities and one-dimensional integration of three-dimensional boundary element method kernels. *Engineering Analysis*, **4**, 21-24, 1987.

[5] Aliabadi, M.H., Hall, W.S. and Hibbs., T.T., Exact singularity cancelling for the potential kernel in the boundary element method, *Communications in Applied Numerical Methods in Engineering.*, **3**, 123-128. 1987.

[6] Aliabadi, M.H. and Hall, W.S., Weighted Gaussian methods for three-dimensional boundary element kernel integration, *Communications in Applied Numerical Methods in Engineering.* **4**, 89-96, 1987.

[7] Aliabadi, M.H. and Hall, W.S., The regularising transformation integration method for boundary element kernels. Comparison with series expansion and weighted Gaussian integration, *Engineering Analysis.*, **6**, 66-71, 1989.

[8] Aliabadi, M.H. and Hall, W.S., Two-dimensional boundary element kernel integration using series expansions, *Engineering Analysis.*, **6**, 140-143, 1989.

[9] Brebbia, C.A., Telles, J.C.F. and Wrobel, L.C., *Boundary Element Techniques*, Springer-Verlag, Berlin, 1984.

[10] Chaudouet, A and Afzali, M., CASTOR 3D: three dimensional boundary element analysis computer code, *Boundary Elements*, Springer-Verlag, Berlin, 461-470, 1983.

[11] Cristescu, M and Loubignac, G., Gaussian quadrature formula for functions with singularities in 1/R over traingles and quadrangles, in *Recent Advances in the Boundary Element Method*, Pentech Press, London, 375-390, 1978.

[12] Cruse, T.A. and Aitha, R., Non-singular boundary integral equation implementation, *International Journal for Numerical Methods in Engineering*, **36**, 237-254, 1993.

[13] Fiedler, C., A new regularization method for computing interior displacements and stresses in 3D-BEM, *Boundary Element Technology IX*, Computational Mechanics Publications, Southampton, 287-297, 1995.

[14] Guiggiani, M. and Casalini, P., Direct computation of Cauchy principal value integrals in advanced boundary elements, *International Journal for Numerical Methods in Engineering*, **24**, 1711-1720, 1987.

[15] Giggiani, M and Gigante, A., A general algorithm for multidimensional Cauchy principal value integral in the boundary element method. *Journal of Applied Mechanics*, **57**, 906-915, 1990.

[16] Guiggiani, M., Krishnasamy, G., Rudolphi, T.J. and Rizzo, F.J., A general algorithm for numerical solution of hypersingular equations. *Journal of Applied Mechanics*, **59**, 604-614, 1992.

[17] Hadamard, J., *Lectures on Cauchy's problems in linear partial differential equations*, Yale University Press, New Haven, 1923.

[18] Hall, W.S. and Hibbs, T.T., Subtraction, expansion and regularizing transformation method for singular kernel integrations in elastostatics, *Mathematical and Computer Modelling*, **15**, 313-323, 1991.

[19] Hall, W.S., *The Boundary Element Method*, Kluwer Academic Publishers, Dordrecht, 1994.

[20] Hayami, K., A projection transformation method for nearly singular surface boundary element integrals, *in Lecture Notes in Engineering*, Vol 73, Springer Verlag, Berlin, 1992.

[21] Ioakimidis, N.I., Application of finite-part integrals to the singular integral equations of crack problems in plane and three-dimensional elasticity, *Acta Mechanics*, **45**, 31-47, 1982.

[22] Jun, L., Beer, G. and Meek, J.L., The application of double exponential formulas in the boundary element method, in *Boundary Elements VII*, Springer Verlag, Berlin, 1029-1042, 1985.

[23] Jun, L., Beer, G. and Meek, J.L., Efficient evaluation of integrals or order $1/r, 1/r^2, 1/r^3$ using Gauss quadrature. *Engineering Analysis*, **2**, 118-123, 1985.

[24] Kane, J.H., Gupta, A. and Saigal, S., Reusable intrinsic sample point (RISP) algorithm for the efficient numerical integration of three-dimensional curved boundary elements, *International Journal for Numerical Methods in Engineering*, **28**, 1661-1676, 1980.

[25] Kupradze, V.D., *Potential Methods in the Theory of Elasticity*. Israel Program for Scientific Translations, Jerusalem, 1965.

[26] Kutt, H.R., Quadrature formulae for finite-part integrals, CSIR Special report, National Research Institute for Mathematical Sciences, WISK 178, September 1975.

[27] Lacaht, J.C., A further development of the boundary integral technique for elastostatics, PhD thesis, University of Southampton, England, 1975.

[28] Lachat, J.C. and Watson, J.O., Effective numerical treatment of boundary integral equations: a formulation for elastostatics, *International Journal for Numerical Methods in Engineering*, **10**, 991-1005, 1976.

[29] Lean, M.H. and Wexler, A., Accurate numerical integration of singular boundary element kernels over boundaries with curvature, *International Journal for Numerical Methods in Engineering*, **21**, 211-228, 1985.

[30] Liu, Y.J. and Rudolphi, T.J., Some identities for fundamental solutions and their applications to weakly-singular boundary element formulations. *Engineering Analysis with Boundary Elements*, **8**, 301-311, 1991.

[31] Lutz, E., Exact Gaussian quadrature methods for near singular integrals in the boundary element method.. *Engineering Analysis with Boundary Elements*, **9**, 233-245, 1992.

[32] Lutz, E. and Gray, L.J., Analytical evaluation of singular boundary integrals without CPV, *Communications in Applied Numerical Methods in Engineering*, **9**, 909-915 1993.

[33] Portela, A, and Aliabadi, M.H., The dual boundary element method : efficient implementation for cracked problems, *International Journal for Numerical Methods in Engineering*, **33**, 1269- 1287, 1992.

[34] Portela, A, Aliabadi, M.H. and Rooke, D.P., Dual boundary element method. Chapter 1, in Advanced formulation in boundary element methods,*Elsevier Applied Science Pub., Oxford* 1993.

[35] Rigby. R.H., Boundary element analysis of cracks in aerospace structures, PhD thesis, Wessex Institute of Technology, University of Portsmouth, 1995.

[36] Rudolphi, T.J., The use of simple solutions in the reqularization of hypersingular integral equations, *Mathematical and Computer Modelling*, **15**, 269-278, 1991

[37] Smith, R.N.L., Direct Gauss quadrature formulae for logarithmic singularities on isoparametric elements, *Engineering Analysis with Boundary Elements*, **24**, 161-167, 2000.

[38] Mangler, K.W., Improper integrals in theoretical aerodynamics, Report No. AERO2424, Royal Aircraft Establishment, Farnborough, June 1991.

[39] Martin, P.A. and Rizzo, F.J., On boundary integral equations for crack problems, *Proc. Royal Society*, **A 421**,341-355, 1989.

[40] Mi, Y and Aliabadi,M.H., Dual boundary element method for three-dimensional fracture mechanics analysis, *Engineering Analysis with Boundary Elements*, **10**, 161-171, 1992.

[41] Mi, Y and Aliabadi, M.H., A Taylor expansion algorithm for integration of 3D near-singular integrals, *Communications. in Applied Numerical Methods in Engineering.* **12**,51-62, 1996.

[42] Mikhlin, S.G., *Integral Equations and Their Applications to Certain Problems in Mechanics*, Pergamon Press, London, 1957.

[43] Mikhlin, S.G., *Multidimensional Singular Integrals and Integral Equations*, Pergamon Press, New York, 1957.

[44] Nagarajan, A. and Mukherjee, S., A mapping method for numerical evaluation of two-dimensional integrals with 1/r singularity, *Computational Mechanics*, **12**, 19-27, 1993.

[45] Stroud, A.H and Secrest, D., *Gaussian Quadrature Formulas*, Prentice-Hall, New York, 1966.

[46] Tanaka, M, Sladek, V. and Sladek, J., Regularization techniques applied to boundary element methods, **47**, 457-499, 1994.

[47] Telles, J.C.F., A self-adaptive coordinate transformation for efficient numerical evaluation of general boundary element integrals, *International Journal for Numerical Methods in Engineering*, **24**, 959-973, 1987.

[48] Wilde, A. and Aliabadi, M.H., Boundary element analysis of geomechanical fracture, *International Journal for Numerical & Analytical Methods in Geomechanics*, **23**, 1195-1214, 1999.

[49] Young, A., Improved numerical method for traction boundary integral equation by application of Stokes' theorem, *International Journal for Numerical Methods in Engineering*, **40**, 3141-3161, 1997.

Appendix A

Complex Stress Functions

In this appendix, derivation of fundamental solutions using complex functions is presented. Initially, some basic definitions of stresses and displacements in terms of Airy and Muskhelishvili's stress functions are presented.

A.1 Airy Stress Functions

The equations of equilibrium are automatically satisfied if

$$\sigma_{xx} = \frac{\partial^2 \psi}{\partial y^2}, \quad \sigma_{yy} = \frac{\partial^2 \psi}{\partial x^2}, \quad \sigma_{xy} = -\frac{\partial^2 \psi}{\partial x \partial y} \tag{A.1}$$

where ψ is known as an Airy Stress function. Substituting the above equation and the strain-displacement equation (2.46) into the stress-strain relationship (2.45), and differentiating twice, leads to the compatibility equation for plane stress:

$$\frac{\partial^4 \psi}{\partial y^2} + 2\frac{\partial^4 \psi}{\partial x^2 \partial y^2} + \frac{\partial^4 \psi}{\partial y^4} = 0 \tag{A.2}$$

The same equation can be obtained for plane strain; it may be written as

$$\nabla^4 \psi = \nabla^2 (\nabla^2 \psi) = 0 \tag{A.3}$$

and is called the 'biharmonic equation' as $\nabla^2 = (\partial^2/\partial^2 x + \partial^2/\partial^2 y)$ is the harmonic operator.

A.2 Muskhelishvili Stress Functions

The Airy stress functions defined in the previous section are limited to bodies with smooth boundaries, and represent a special case of the more general complex stress functions developed by Muskhelishvili [3]. Following his work on complex functions, stresses for the two-dimensional equations of elasticity may be defined in terms of two functions, the Airy stress function ψ and some other function φ, such that

$$\sigma_{11} = \frac{\partial^2 \psi}{\partial x^2} - 2\frac{\partial^2 \varphi}{\partial x \partial y}$$

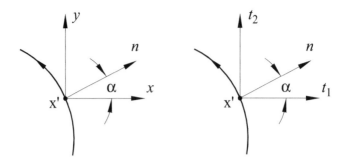

Figure A.1: Outward normal to the boundary.

$$\sigma_{22} = \frac{\partial^2 \psi}{\partial y^2} + 2\frac{\partial^2 \varphi}{\partial x \partial y}$$

$$\sigma_{12} = \frac{\partial^2 \psi}{\partial x \partial y} - \frac{\partial^2 \varphi}{\partial y^2} + \frac{\partial^2 \varphi}{\partial x^2}$$

The functions ψ and φ are given in terms of two analytical functions $\phi(z)$ and $\chi(z)$ of the complex variable $z(= x + iy)$, such that

$$\psi = \mathrm{Re}[\bar{z}\phi(z) + \chi(z)]$$
$$\varphi = \mathrm{Im}[\bar{z}\phi(z) + \chi(z)]$$

where Re and Im denote the real and imaginary parts, respectively ($\bar{z}$ is the complex conjugate of z). The stresses in terms of ϕ and χ are given by

$$\sigma_{11} + \sigma_{22} = 2[\phi'(z) + \overline{\phi'(z)}] = 4\,\mathrm{Re}[\phi'(z)]$$
$$\sigma_{11} - \sigma_{22} + 2i\sigma_{12} = 2[z\overline{\phi''(z)} + \overline{\chi'(z)}] \qquad \text{(A.4)}$$

The displacements are given by

$$2\mu(u_1 + iu_2) = \kappa\phi(z) - z\overline{\phi'(z)} - \overline{\chi(z)} \qquad \text{(A.5)}$$

where $'$ denotes differentiation with respect to z and — denotes complex conjugate, and κ is equal to $(3 - 4\nu)$ for plane strain and $(3 - \nu)/(1 + \nu)$ for plane stress. The surface tractions in the x and y directions (see Figure A.1) are given by

$$t_1 = \sigma_{11}n_1 + \sigma_{12}n_2 = \sigma_{11}\cos\alpha + \sigma_{12}\sin\alpha$$
$$t_2 = \sigma_{12}n_1 + \sigma_{22}n_2 = \sigma_{12}\cos\alpha + \sigma_{22}\sin\alpha$$

where n_1 and n_2 are the Cartesian components of the outward normal **n** and α is the angle between **n**, and the x-axis. The above equation can be rewritten in the form

$$t_1 + it_2 = (\sigma_{11} + i\sigma_{12})\cos\alpha + i(\sigma_{22} - i\sigma_{12})\sin\alpha \qquad \text{(A.6)}$$

The traction fields t_x and t_y can be expressed in terms of the complex stress function $\phi(z)$ and $\chi(z)$ by substituting (A.4) into (A.6) to give

$$t_1 + it_2 = e^{i\alpha}[\phi'(z) + \overline{\phi'(z)}] - e^{-i\alpha}[z\overline{\phi''} + \overline{\chi'(z)}] \qquad \text{(A.7)}$$

A.3 Complex Stress Function Method

An alternative way of deriving the fundamental solutions for two-dimensional problems is through the use of complex functions described in Section 2.2.10. Fundamental solutions for a semi-infinite sheet, for circular and elliptical cut-outs and for a straight crack in two-dimensional sheets are also available. The fundamental solutions U_{ij} and T_{ij} are derived using the complex function representation (A.6) and (A.7), as given by

$$(u_1 + iu_2) = \frac{1}{2\mu}\left[\kappa\phi(z) - z\overline{\phi}'(z) - \overline{\chi}(z)\right] \tag{A.8}$$

$$t_1 + it_2 = e^{i\alpha}[\phi'(z) + \overline{\phi}'(z)] - e^{-i\alpha}[z\overline{\phi}'' + \overline{\chi}'(z)] \tag{A.9}$$

The stress function $\phi(z)$ and $\chi(z)$ are given by

$$\begin{aligned}
\phi &= \phi^{Kelvin} + \phi^{Complementary} \\
\chi &= \chi^{Kelvin} + \chi^{Complementary}
\end{aligned} \tag{A.10}$$

For a point force at z_0, with components e_1 and e_2 in the x and y directions, the stress functions are given by

$$\begin{aligned}
\phi^{Kelvin} &= A\ln(z - z_0) \\
\chi^{Kelvin} &= -\kappa\bar{A}\ln(z - z_0) - A\bar{z}_0/(z - z_o)
\end{aligned} \tag{A.11}$$

where $A = (e_1 + ie_2)/(2\pi(1 + \kappa))$, $z - z_o = re^{i\theta}$.

Substituting (A.11) into (A.8) and (A.9) gives

$$2\mu(u_1 + iu_2) = 2\kappa A\ln|r| - \bar{A}e^{2i\theta} \tag{A.12}$$

$$t_1 + it_2 = \frac{1}{r}\left\{e^{i\alpha}[Ae^{-i\theta} + \bar{A}e^{i\theta}] + e^{-i\alpha}[\bar{A}e^{3i\theta} + \kappa Ae^{i\theta}]\right\} \tag{A.13}$$

The fudamental solutions U_{ij} and T_{ij} can be obtained by separating displacement and traction fields in (A.12) and (A.13) respectively into real and imaginary parts. The resulting U_{ij} and T_{ij} are the same as those given in (2.90) and (2.91).

The fundamental solutions for a point force in a semi-infinite body (i.e $x > 0$) which satisfies the traction-free condition over the boundary of a half-plane $x = 0$, are due to Melan [2] (corrected by Kurshin [1]) and are given as

$$\phi^{Complementary} = \phi^{semi-inf} = \kappa A\ln(z + \bar{z}_o) - \frac{A(z_o + \bar{z}_o)}{(z + \bar{z}_o)^2}$$

$$\begin{aligned}
\chi^{Complementary} &= \chi^{semi-inf} = -\bar{A}\left[\ln(z + \bar{z}_o) - \frac{(z_o + \bar{z}_o)}{(z + \bar{z}_o)} + \frac{(z_o + \bar{z}_o)}{(z + \bar{z}_o)^2}\right] \\
&\quad - A\kappa\frac{\bar{z}_o}{(z + \bar{z}_o)}
\end{aligned} \tag{A.14}$$

The complementary stress functions for a stress-free circular hole of radius ρ centred at the origin are given by

$$\phi^{Complementary} = \phi^{crcular} = \kappa A\ln(1 - \frac{\rho}{z\bar{z}_o}) - \frac{\bar{A}(z_o - \frac{\rho}{\bar{z}_o})\rho}{\bar{z}_o^2(z - \frac{\rho}{\bar{z}_o})}$$

$$\chi^{Complementary} = \chi^{crcular} = -A\ln(1 - \frac{\rho}{z\bar{z}_o}) + \frac{A\rho}{zz_o} - \frac{\rho}{z}\frac{\partial\phi^{Circular}}{\partial z} \tag{A.15}$$

For the stress-free elliptical hole with semi-minor axis b and semi-major axis c, the complementary stress functions are given by

$$\phi^{Complementary} = \phi^{elliptical} = -A\left\{\ln\left(\frac{s_o - \xi}{s_o}\right) - \kappa\ln\left(\frac{t_o - \xi}{t_o}\right)\right\}$$
$$-\bar{A}\left\{\frac{\rho M\xi^2 - z_o\xi + \rho}{\rho\xi^2 - \bar{z}_o\xi + \rho M} + \frac{(\rho M t_1^2 - z_o t_1 + \rho)\xi}{\rho t_1(\xi_1 - t_1)(t_o - t_1)} - \frac{1}{M}\right\}$$

$$\chi^{Complementary} = \chi^{elliptical} = -\bar{A}\left\{\ln\left(\frac{t_o - \xi}{t_o}\right) - \kappa\ln\left(\frac{s_o - \xi}{s_o}\right)\right\} -$$
$$A\left\{\frac{\rho\xi^2 - \bar{z}_o\xi + \rho M}{\rho M\xi^2 - z_o\xi + \rho} + \frac{(\rho s_1^2 - \bar{z}_o s_1 + \rho M)\xi}{\rho M s_1(\xi - s_1)(s_o - s_1)} - M\right\}$$
$$+\frac{\xi(\xi^2 + M)}{(1 - M\xi^2)}\frac{\partial\phi^{elliptical}}{\partial\xi} \tag{A.16}$$

where ξ is related to z via the mapping function $z = \rho\left[\frac{1}{\xi} + M\xi\right]$ with $M = (c - b)/(c + b)$ and $\rho = (c + b)/2$. The parameters s_o and s_1 are obtained from

$$s_1, s_o = \frac{1}{2M\rho}\left(z_o \pm \sqrt{z_o^2 - 4\rho^2 M}\right) \tag{A.17}$$

and the parameters t_o and t_1 from

$$t_1, t_o = \frac{1}{2\rho}\left(\bar{z}_o \pm \sqrt{\bar{z}_o^2 - 4\rho^2 M}\right) \tag{A.18}$$

where the sign in (A.17) and (A.18) is chosen such that $|s_1| < |s_o|$ and $|t_1| < |t_o|$ respectively.

The complementary stress function for a straight crack lying along x−axis with crack tips are $z = \pm a$ can be obtained from the elliptical hole solution (A.16) by setting $M = 1$, to give

$$\phi^{Complementary} = \phi^{crack} = -A\left\{\ln\left(\frac{s_o - \xi}{s_o}\right) - \kappa\ln\left(\frac{\bar{s}_o - \xi}{\bar{s}_o}\right)\right\}$$
$$-\bar{A}\left\{\frac{\xi(\rho\bar{s}_o^2 - z_o\bar{s} + \rho)}{\rho\bar{s}_o(\bar{s}_o - \bar{s}_1)(\xi - \bar{s}_o)}\right\}$$
$$\chi^{Complementary} = \chi^{crack} = \bar{\phi}^{crack} + \frac{\xi(\xi^2 + 1)}{(1 - \xi^2)}\frac{\partial\phi^{crack}}{\partial\xi}$$

The advantage of using the fundamental solution for a half-plane, a circular hole, an elliptical hole or a straight crack in an appropriate problem, is that the first integral in (2.102) vanishes identically on these boundaries (S_o say), because these solutions automatically satisfy traction-free boundary conditions.

References

[1] Kurshin, L.M., Mixed-mode boundary value problem of the theory of elasticity for a quadran. *Applied Mathematics & Mechanics (English Edition)*, **23**, 1403-1408, 1959.

[2] Melan, E., Der spannungszustand der durch eine einzelkraft im Innern beanspruchten Halbscheibc; Z.Angew. Math. Mech., **12**, 343-346, 1932.

[3] Muskhelishvili, N.I., *Some Basic Problems of the Mathematical Theory of Elasticity*, Noordhoff, Leyden 1953.

Appendix B

Limiting Process for Boundary Stress Equation

Similar limiting processes considered in Section 2.3.7 can be employed to obtain the boundary form of the derivative displacement integral equation. In this appendix, the derivations are presented for three-dimensional problems. Similar procedures can also be used for the simpler two-dimensional problem. In this analysis the augmented spherical surface (radius ε) is considered to be composed of two constituent surfaces S'_ε and S^*_ε, the surface S^*_ε being the component of the spherical surface confined by the hemispherical component (S^*_ε is part of $r = \varepsilon$ which is between the surface S and the tangent planes to S at the collocation point $\mathbf{X}'$. S^*_ε only exists if S is curved and has area of order ε^3). For a general point, these surface are shown in Figure B.1. As can be seen, S^*_ε exist only by virtue of the surface curvature at the source point. In the limit as $\varepsilon \to 0$, the augmented boundary composed of $S - S_\varepsilon + S'_\varepsilon + S^*_\varepsilon$ tends towards the actual boundary S. Defining

$$
\begin{aligned}
I_S &= \lim_{\varepsilon \to 0} \left[\int_S \{T_{ij,k}(\mathbf{x}',x)u_k(\mathbf{x}) - U_{ij,k}(\mathbf{x}',\mathbf{x})t_k(\mathbf{x})dS\} \right] \\
&= I_{S-S_\varepsilon} + I_{S'_\varepsilon} + I_{S^*_\varepsilon}
\end{aligned}
\tag{B.1}
$$

The integration over S'_ε can be written as

$$
I_{S'_\varepsilon} = \lim_{\varepsilon \to 0} \left[\int_{S'_\varepsilon} \{T_{ij,k}(\mathbf{x}',x)u_k(\mathbf{x}) - U_{ij,k}(\mathbf{x}',\mathbf{x})t_k(\mathbf{x})dS\} \right]
\tag{B.2}
$$

where

$$
\begin{aligned}
dS &= \varepsilon^2 \sin\theta d\theta d\phi \\
\mathbf{r} &= \mathbf{x}' - \mathbf{x} = \varepsilon\mathbf{n}
\end{aligned}
\tag{B.3} \tag{B.4}
$$

and $\mathbf{n}$ is the outward unit normal to S'_ε. On the continuous part of the surface near to the source point, the displacement should be continuous and differentiable, and take the form

$$
u_i = u'_i + u'_{i,j}r_j + O(r^2)
\tag{B.5}
$$

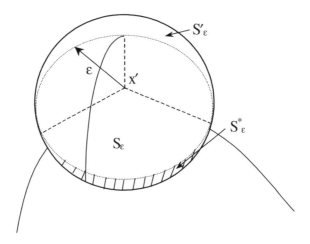

Figure B.1: Augmented surface for limiting process.

where superscript $'$ denotes evaluation at the source point. Combining (B.4) with (B.5) gives

$$u_i = u_i' + \varepsilon n_k u_{i,j}' + O(\varepsilon^2) \tag{B.6}$$

Substituting (B.4) into the kernel expressions of the fundamental solutions, we have

$$
\begin{aligned}
U_{ij,k}(\mathbf{x}', \mathbf{x}) &= \varepsilon^{-2} \hat{U}_{ij,k}(\mathbf{n}) \\
T_{ij,k}(\mathbf{x}', \mathbf{x}) &= \varepsilon^{-3} \hat{T}_{ij,k}(\mathbf{n})
\end{aligned} \tag{B.7}
$$

where

$$\hat{U}_{ij,k}(\mathbf{n}) = \frac{(1+\nu)}{8\pi(1-\nu)E} \left[3n_i n_j n_k + (3-4\nu)\delta_{ij}n_k - \delta_{jk}n_i - \delta_{ki}n_{jj}\right]$$

$$\hat{T}_{ij,k}(\mathbf{n}) = \frac{-1}{4\pi(1-\nu)} \left[6n_i n_j n_k + (1-2\nu)\delta_{ij}n_k - (1+\nu)\delta_{ik}n_j - (2-\nu)\delta_{kj}n_i\right]$$

Substituting (B.3), (B.6) and (B.7) into (B.2), and applying Hooke's law ($t_p' = n_q' E_{pqil} u_{i,l}'$), gives

$$I_{S_\varepsilon'} = \lim_{\varepsilon \to 0} \left[\varepsilon^{-1} \bar{B}_{ijk} u_i'\right] + c_{iljk} u_{i,l}' + O(\varepsilon) \tag{B.8}$$

with

$$\bar{B}_{ijk} = \int_{S_\varepsilon'} \hat{T}_{ijk}(\mathbf{n}) \sin\theta d\theta d\phi \tag{B.9}$$

and

$$c_{iljk} = \int_{S_\varepsilon'} \left[\hat{T}_{ij,k}(\mathbf{n})n_l' - \hat{U}_{pj,k}(\mathbf{n})n_q' E_{pqil}\right] \sin\theta d\theta d\phi \tag{B.10}$$

where $E_{ijkl} = (\delta_{ik}\delta_{jl} + \delta_{ik}\delta_{jl} + \frac{2\nu}{1-2\nu}\delta_{ik}\delta_{jl})$.

Now, in order to evaluate the integral I_{S-S_ε}, it is necessary to expand the kernels in terms of the polar coordinate (ρ, ψ) defined on the tangent plane centred at $\mathbf{x}'$, to give

$$
\begin{aligned}
T_{ij,k} &= \rho^{-3} T_{ijk}^{(3)}(\psi) + \rho^{-2} T_{ijk}^{(2)}(\psi) + O(\rho^{-1}) \\
U_{ij,k} &= \rho^{-2} U_{ijk}^{(2)}(\psi) + O(\rho^{-1})
\end{aligned}
\tag{B.11}
$$

where

$$
\begin{aligned}
U_{ijk}^{(2)}(\psi) &= \frac{(1+\nu)}{8\pi(1-\nu)E} \left\{ 3\rho_{,i}\rho_{,j}\rho_{,k} - \delta_{jk}\rho_{,i} - \delta_{ki}\rho_{,j} + (3-4\nu)\delta_{ij}\rho_{,k} \right\} \\
T_{ijk}^{(2)}(\psi) &= \frac{\kappa}{8\pi(1-\nu)} \left\{ \frac{9}{2}\rho_{,i}\rho_{,j}\rho_{,k} + \frac{1}{2}(1-2\nu)\delta_{ij}\rho_{,k} - \frac{1}{2}(1+4\nu)\rho_i\delta_{jk} \right. \\
&\quad \left. - \frac{1}{2}(5-4\nu)\delta_{ki}\rho_{,j} + \frac{3}{2}(1-2\nu)(n_i'\rho_{,j}n_k' + \rho_{,j}n_j'n_k') \right\} \\
T_{ijk}^{(3)}(\psi) &= \frac{1}{8\pi(1-\nu)} \left\{ \left[(1-2\nu)\delta_{ij} + 3\rho_{,i}\rho_{,j} \right] n_k' - (1-2\nu) \left[\delta_{jk} + 3\rho_{,j}\rho_{,k} \right] n_i' \right. \\
&\quad \left. + (1-2\nu) \left[\delta_{ki} - 3\rho_{,k}\rho_{,i} \right] n_j' \right\}
\end{aligned}
$$

On the surface S, $r_j = \rho_j + O(\rho^2)$, and since $\rho_j = \rho_{,j}\rho$, (B.5) takes the form

$$
u_i = u_i' + u_{i,j}'\rho_{,j}\rho + O(\rho^2)
\tag{B.12}
$$

The differential area in this case is given by

$$
dS = \rho\, d\rho\, d\psi
\tag{B.13}
$$

Substituting (B.13), (B.12) and (B.11) into I_{S-S_ε} gives

$$
\begin{aligned}
I_{S-S_\varepsilon} &= \lim_{\varepsilon \to 0} \left[\varepsilon^{-1} B_{ijk} u_i' + \log \varepsilon \left\{ \alpha_{ijk} u_i' + \beta_{iljk} u_{i,l}' + \gamma_{jik} t_i' \right\} \right] \\
&\quad + \oint_{S-S_\varepsilon} T_{ij,k}(\mathbf{x}', \mathbf{x}) u_i(\mathbf{x}) dS + \int_{S-S_\varepsilon} U_{ij,k}(\mathbf{x}', \mathbf{x}) t_i(\mathbf{x}) dS
\end{aligned}
\tag{B.14}
$$

and

$$
\begin{aligned}
B_{ijk} &= \int_{S_\varepsilon} T_{ijk}^{(3)} d\psi \\
\alpha_{ijk} &= \int_{S_\varepsilon} T_{ijk}^{(2)} d\psi \\
\beta_{iljk} &= \int_{S_\varepsilon} T_{ijk}^{(3)} \rho_j d\psi \\
\gamma_{jik} &= -\int_{S_\varepsilon} U_{ijk}^{(2)} d\psi
\end{aligned}
$$

where $T_{ijk}^{(3)}$, $T_{ijk}^{(2)}$ and $U_{ijk}^{(2)}$ are given in [6] and [5].

The integral expression for $I_{S^*_\varepsilon}$ can be evaluated by introducing an additional off-plane coordinate ω, as shown in Figure B.2.

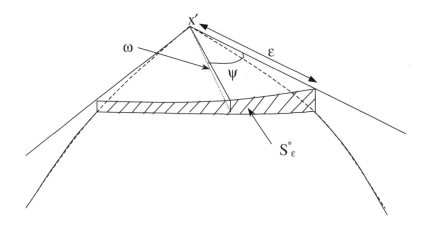

Figure B.2: Coordinate system for surface S_ε^*.

The differential surface area S_ε^* is given by

$$dS = \varepsilon^2 \cos\omega\, d\omega\, d\psi \tag{B.15}$$

and the limits of the integration for a particular element, say m, area $\psi_1^m \le \psi \le \psi_2^m$; $0 \le \omega \le \omega_2^m$. It can be shown that

$$\tan\omega_2^m = \frac{\frac{1}{2}\varepsilon^2 \kappa^m(\psi)}{\varepsilon} = \frac{\varepsilon\kappa^m(\psi)}{2}$$

where $\kappa^m(\psi)$ denotes the curvature on element m at the collocation point $\mathbf{x}'$. As the angle ω is small,

$$\omega_2^e = \tan^{-1}\left(\frac{\varepsilon\kappa^m(\psi)}{2}\right) \simeq \frac{\varepsilon\kappa^m(\psi)}{2}$$

Substituting the kernels in (B.7), together with the expression for the differential area (B.15), into $I_{S_\varepsilon^*}$ gives

$$I_{S_\varepsilon^*} = \sum_{m=1}^{M} \int_{\psi_1^m}^{\psi_2^m} \left[\lim_{\varepsilon\to 0} \frac{1}{\varepsilon} \int_0^{\frac{1}{2}\varepsilon\kappa^m(\psi)} \left(\hat{T}_{ij,k}(\psi,\omega)u_i' + O(\varepsilon)\right) \cos\omega\, d\omega \right] d\psi$$

where M is the total number of elements surrounding the source point. The limit within the square brackets is of the form

$$\lim_{\varepsilon\to 0} \frac{1}{\varepsilon} \int_0^{\frac{1}{2}\varepsilon\kappa^m(\psi)} f(\psi,\omega)\, d\omega$$

which, according to the mean limit theorem, can be shown to be equal to $\frac{1}{2}\kappa^m(\psi)f(\psi,0)$. Finally, $I_{S_\varepsilon^*}$ can be expressed as

$$
\begin{aligned}
I_{S_\varepsilon^*} &= \sum_{m=1}^{M} \int_{\psi_1^m}^{\psi_2^m} \frac{1}{2}\kappa^m \hat{T}_{ij,k}(\psi,0)u_i'\, d\psi + O(\varepsilon) \\
&= d_{ijk}u_i' + O(\varepsilon)
\end{aligned}
$$

B.1 Enriched Element Approach

The resulting integral equation can be written [6, 3], as

$$\left(c_{iljk}(\mathbf{x}') + \delta_{ij}\delta_{lk}\right) u'_{i,l} = \lim_{\varepsilon \to 0} \left[\frac{1}{\varepsilon}\left\{\bar{B}_{ijk}(\mathbf{x}') + B_{ijk}(\mathbf{x}')\right\} u'_i\right] \qquad (\text{B.16})$$

$$- \lim_{\varepsilon \to 0}\left[\log\varepsilon\left\{\alpha_{ijk}(\mathbf{x}')u'_i + \beta_{iljk}(\mathbf{x}')u'_{i,l} + \gamma_{ijk}(\mathbf{x}')t'_i\right\}\right]$$

$$- \fint_{S-S_\varepsilon} T_{ij,k}(\mathbf{x}',\mathbf{x})u_i dS(\mathbf{x}) + \int_{S-S_\varepsilon} U_{ij,k}(\mathbf{x}',\mathbf{x})t_i(\mathbf{x})dS - d_{ijk}(\mathbf{x}')u'_i$$

where superscript $'$ denotes values at the source point, $c_{iljk}(\mathbf{x}')$ and $d_{ijk}(\mathbf{x}')$ are $O(1)$. The presence of the second free term d_{ijk} is due to the boundary curvature at the source point.

Generally, the use of Hadamard finite part integrals requires the displacement and traction fields to be Hölder continuous. This is a stringent condition and there has been much discussion about the circumstances under which this condition may be weakened [1, 2]. By the definition of finite part integrals, all the singular terms appearing in (B.16) must cancel if the assumed forms for displacement derivatives and tractions are Hölder continuous. Inspection of (B.16) shows that the coefficients of ε^{-1} terms only involve numerical displacements at the source point. For this term to cancel, the only physical condition satisfied by displacements alone is continuity, i.e.

$$\bar{B}_{ijk}(\mathbf{x}') + B_{ijk}(\mathbf{x}') = 0$$

The coefficient of the $\log\varepsilon$ term involves displacements, tractions and displacement derivatives at the source point, and since there is no physical property linking all three, the displacement component may be considered independently and, as before, assumed to be continuous, and therefore $\alpha_{ijk}(\mathbf{x}') = 0$. The relationship between the remaining components may be established by considering Hooke's law, at the source point. The stress-strain relationship in can be rewritten as

$$\sigma'_{ij} = E_{ijkl}u'_{k,l} \qquad (\text{B.17})$$

where E_{ijkl} is the elastic tensor with symmetry properties $E_{ijkl} = E_{jikl} = E_{ijlk} = E_{klji}$ and is defined by $E_{ijkl} = \frac{E}{2(1+\nu)}(\delta_{ik}\delta_{jl} + \delta_{ik}\delta_{jl} + \frac{2\nu}{1-2\nu}\delta_{ik}\delta_{jl})$. Notice that the Kronecker delta has been used repeatedly to change indices (e.g. $u_{i,j} = \delta_{im}u_{m,j} = \delta_{im}\delta_{jl}u_{m,l}$) in the derivation of (B.17). Since $t'_i = \sigma'_{ij}n'_j$, then

$$t'_i = n'_j E_{ijkl}u'_{k,l} \qquad (\text{B.18})$$

Generally, the field variables are approximated with piecewise continuous interpolation functions, and hence there will be a different set of displacement derivatives for each element adjoining a node. Young [6] replaced these separate displacement derivatives with a unique displacement derivative by adopting a special interpolation function.

Inspection of the standard shape functions for continuous elements reveals that a unique value of displacement is assigned to nodes shared between elements and thus u'_i is implicitly continuous. However, the continuity condition on the displacement derivatives are not, in general, satisfied by the standard shape function, so a new element representation should be prescribed in which these requirements are embedded [6, 3].

Denoting the local coordinates of the source point as (η'_1, η'_2), the shape function N_α can be expressed in terms of the polar radius ρ about the local source point,

$$N_\alpha(\eta_1, \eta_2) = N_\alpha(\eta'_1, \eta'_2) + \frac{\partial N_\alpha(\eta'_1, \eta'_2)}{\partial \rho_j} \rho_j + O(\rho^2) \tag{B.19}$$

The interpolated displacement field is therefore of the form

$$
\begin{aligned}
u_i(\mathbf{x}) &= \sum_\alpha u_i^\alpha N_\alpha(\eta'_1, \eta'_2) + \sum_\alpha u_i^\alpha \frac{\partial N_\alpha(\eta'_1, \eta'_2)}{\partial \rho_j} \rho_j + O(\rho^2) \\
&= u'_i + \bar{u}'_{i,j} \rho_j + O(\rho^2)
\end{aligned}
\tag{B.20}
$$

where $\bar{u}'_{i,j}$ are displacement derivatives at the source point contained in the first order term of the standard interpolation,

$$\bar{u}'_{i,j} \rho_j = \sum_\alpha u_i^\alpha \frac{\partial N_\alpha(\eta'_1, \eta'_2)}{\partial \rho_j} \rho_j \tag{B.21}$$

The above displacement derivatives involve derivatives of shape functions, and as such are element dependent, i.e. not single-valued. The continuity conditions require that they are unique, thus the special interpolation scheme replaces $\bar{u}'_{i,j}$ with unique values of the displacement derivatives $\tilde{u}'_{i,j}$ which are consistent with (B.18). Removing $\bar{u}'_{i,j}$ and adding $\tilde{u}'_{i,j}$ in the shape function approximation of the displacement field over an element gives

$$\check{u}_i(\eta_1, \eta_2) = \bar{u}_i(\eta_1, \eta_2) - \bar{u}'_{i,j} \rho_j + \tilde{u}'_{i,j} \rho_j \tag{B.22}$$

Now, it only remains to ensure that the traction at the source point t'_i is consistent with $\check{u}'_{i,j}$. This was done in [6] by subtracting the traction term due to the standard interpolation at the source point, $\bar{t}'_i$, and adding in a new term, $\tilde{t}'_i(\eta'_1, \eta'_2)$, based on equation (B.18),

$$
\begin{aligned}
\check{t}'_i(\eta_1, \eta_2) &= \bar{t}'_i(\eta_1, \eta_2) - \bar{t}'_i(\eta'_1, \eta'_2) + \tilde{t}'_i(\eta'_1, \eta'_2) \\
&= \bar{t}'_i(\eta_1, \eta_2) - \bar{t}'_i(\eta'_1, \eta'_2) + n'_j E_{ijkl} \tilde{u}'_{k,l}
\end{aligned}
\tag{B.23}
$$

In summary, provided the displacement derivative is $C^{(0)}$ continuous and the traction is related to the displacement derivatives at the collocation point by (B.18), then all singular terms will cancel, no linear elastic properties are violated and the displacement derivative BIE. It is sufficient to assume simple continuity as opposed to Hölder continuity, since the interpolations are based on well-behaved polynomial functions. The *enriched* continuous elements, comprising of the special interpolation functions (B.22) and (B.23), replace the standard continuous elements containing the source point. In all other cases, the standard continuous elements are satisfactory. A special localizing function $L(\eta_1, \eta_2)$ was recommended in [89] to restrict the effect to the region surrounding the source point such that,

$$
\begin{aligned}
\check{u}_i(\eta_1, \eta_2) &= \bar{u}_i(\eta_1, \eta_2) + \left[-\bar{u}'_{i,j} \rho_j + \tilde{u}'_{i,j} \rho_j \right] L(\eta_1, \eta_2) \\
\check{t}'_i(\eta_1, \eta_2) &= \bar{t}'_i(\eta_1, \eta_2) + \left[-\bar{t}'_i(\eta'_1, \eta'_2) + n'_j E_{ijkl} u'_{k,l} \right] L(\eta_1, \eta_2)
\end{aligned}
$$

where $L = 1$ at $\mathbf{x}'$ and $L = 0$ on the edges of the element surrounding $\mathbf{x}'$. The above method was used in [3] to evaluate the stresses on the boundary using different localizing functions.

References

[1] Cruse, T.A. and Richardson,J., Non-singular Somiglinana stress identities in elasticity, *International Journal for Numerical Methods in Engineering*, **39**, 3273-3304, 1996.

[2] Martin, P. and Rizzo, F., Hypersingular integrals: how smooth must the density be? *International Journal for Numerical Methods in Engineering*, **39**, 687-704, 1996.

[3] Wilde, A.J. and Aliabadi, M.H., Direct evalaution of boundary stresses in the 3D BEM of elastostatics, *Communications in Numerical Methods in Engineering*, **14**, 505-517, 1998.

[4] Wilde, A.J. A hypersingular dual boundary element formulation for three-dimensional fracture analysis, PhD thesis, WIT, University of Wales, 1998.

[5] Wilde, A.J., *A Dual Boundary Element Formulation for Three-dimensional Fracture Mechanics*, Topics in Engineering Series Vol. 37, WIT Press, Southampton, 2000.

[6] Young, A., A single-domain boundary element method for 3D elastostatic crack analysis using continuous elements, *International Journal for Numerical Methods in Engineering*, **39**, 1265-1293, 1996.

Appendix C

Limit Integrals For Shallow Shells

In this appendix, the limits and jump terms are evaluated for the integral equations of shallow shells. Further details can be found in [1]. From Figure 3.12, the following relationships can be obtained:

$$r = \varepsilon; \qquad r_{,n} = 1; \qquad dS = \varepsilon d\varphi; \tag{C.1}$$

$$r_{,1} = n_1 = \cos\varphi; \qquad r_{,2} = n_2 = \sin\varphi; \tag{C.2}$$

$$\int_{S^+} \cdots dS = \int_0^\pi \cdots \varepsilon d\varphi \tag{C.3}$$

and

$$\lim_{\varepsilon \to 0} A(\lambda\varepsilon) = \frac{-1}{2};$$

$$\lim_{\varepsilon \to 0} \lambda^2 \varepsilon^2 K_0(\lambda\varepsilon) = 0;$$

$$\lim_{\varepsilon \to 0} \lambda\varepsilon K_1(\lambda\varepsilon) = 1;$$

$$\lim_{\varepsilon \to 0} \lambda^2 \varepsilon^2 B(\lambda\varepsilon) = 0 \tag{C.4}$$

C.1 The Displacement Integral Equations

Recalling displacement integral equation (3.111), we have

$$w_i(\mathbf{x}') + \lim_{\varepsilon \to 0} \int_{S - S_\varepsilon + S^+} T_{ik}^b(\mathbf{x}', \mathbf{x}) w_k(\mathbf{x}) \, dS = \lim_{\varepsilon \to 0} \int_{S - S_\varepsilon + S^+} U_{ik}^b(\mathbf{x}', \mathbf{x}) p_k(\mathbf{x}) \, dS$$

$$- \lim_{\varepsilon \to 0} \int_V U_{i3}^b(\mathbf{x}', \mathbf{X}) \left\{ B[(1 - \nu) k_{\alpha\beta} + \nu k_{\phi\phi} \delta_{\alpha\beta}] u_{\alpha,\beta} - B(k_{11}^2 + k_{22}^2 + 2\nu k_{11} k_{22}) w_3 \right\} dV$$

$$+ \lim_{\varepsilon \to 0} \int_V U_{i3}^b(\mathbf{x}', \mathbf{X}) q_3(\mathbf{X}) dV \tag{C.5}$$

Equation (C.5) can be written in the following form:

$$w_i(\mathbf{x'}) + I_1^b = I_2^b - (I_3^b + I_4^b) \tag{C.6}$$

The boundary values of w_i are assumed to satisfy Hölder continuity. The boundary integrals I_{23}^b contain weakly singular kernels, and the integrals I_3^b and I_4^b are domain integrals which also contain weakly singular kernels, and will lead to no jump terms.

The integral I_1^b contains T_{ij}^b which has a strong singularity of $O(r^{-1})$. The integral I_1^b can be written as follows:

$$
\begin{aligned}
I_1^b &= \lim_{\varepsilon \to 0} \int_{S-S_\varepsilon+S_\varepsilon^+} T_{ij}^b(\mathbf{x'},\mathbf{x})w_j(\mathbf{x})\,dS \\
&= \lim_{\varepsilon \to 0} \int_{S-S_\varepsilon} T_{ij}^b(\mathbf{x'},\mathbf{x})w_j(\mathbf{x})dS + \lim_{\varepsilon \to 0} \int_{S_\varepsilon^+} T_{ij}^b(\mathbf{x'},\mathbf{x})[w_j(\mathbf{x}) - w_j(\mathbf{x'})]\,dS \\
&\quad + w_j(\mathbf{x'})\lim_{\varepsilon \to 0} \int_{S+} T_{ij}^b(\mathbf{x'},\mathbf{x})dS
\end{aligned}
\tag{C.7}
$$

where S^+ is the boundary a semi-circular domain around the point $\mathbf{x'}$, S_ε is the original boundary which is replaced by S^+, and S is the rest of the boundary, as shown in Figure 3.12.

The second term on the right-hand side (RHS) of (C.7) is zero in the limit as $\varepsilon \to 0$. The first term on the RHS of (C.7) forms a Cauchy principal value integral. By considering the relationships in equations (C.1)-(C.3) and the limits in equation (C.4), the last term on the RHS of (C.7) leads to the following jump term:

$$w_j(\mathbf{x'})\lim_{\varepsilon \to 0} \int_{S+} T_{ij}^b(\mathbf{x'},\mathbf{x})dS = -\frac{1}{2}\delta_{ij}w_j(\mathbf{x'}) \tag{C.8}$$

The integral equation for membrane contributions can be written as

$$u_\alpha(\mathbf{x'}) + \lim_{\varepsilon \to 0} \int_{S-S_\varepsilon+S_\varepsilon^+} T_{\alpha\beta}^m(\mathbf{x'},\mathbf{x})u_\beta(\mathbf{x})dS$$

$$+ \lim_{\varepsilon \to 0} \int_V U_{\alpha\beta}^m(\mathbf{x'},\mathbf{X})B\left[k_{\alpha\beta}(1-\nu) + \nu\delta_{\alpha\beta}k_{SS}\right]w_3(\mathbf{X})dV$$

$$= \lim_{\varepsilon \to 0} \int_{S-S_\varepsilon+S_\varepsilon^+} U_{\alpha\beta}^m(\mathbf{x'},\mathbf{x})t_\beta(\mathbf{x})dS + \lim_{\varepsilon \to 0} \int_V U_{\alpha\beta}^m(\mathbf{x'},\mathbf{X})f_\beta^b(\mathbf{X})dV \tag{C.9}$$

Equation (3.113) can be written in the following form:

$$u_\theta(\mathbf{x'}) + I_7^m + I_8^m = I_9^m + I_{10}^m \tag{C.10}$$

The boundary values of u_β are assumed to satisfy Hölder continuity. The integral I_9^m contains weakly singular kernels and the integrals I_8^m and I_{10}^m are domain integrals which also contain weakly singular kernels. These integrals will lead to no jump terms, and they exist in the limit in normal sense.

The integral I_7^m contains $T_{\alpha\beta}^m$ which has a strong singularity of $O(r^{-1})$. This integral was discussed in Chapter 2 with one term of the Taylor series expansion for the integrand to give:

$$
\begin{aligned}
I_7^m &= \lim_{\varepsilon \to 0} \int_{S-S_\varepsilon+S^+} T_{\alpha\beta}^m\left(\mathbf{x}',\mathbf{x}\right) u_\beta\left(\mathbf{x}\right) dS \\
&= \lim_{\varepsilon \to 0} \int_{S-S_\varepsilon} T_{\alpha\beta}^m\left(\mathbf{x}',\mathbf{x}\right) u_\beta\left(\mathbf{x}\right) dS \\
&\quad + \lim_{\varepsilon \to 0} \int_{S^+} T_{\alpha\beta}^m\left(\mathbf{x}',\mathbf{x}\right) [u_\beta(\mathbf{x}) - u_\beta(\mathbf{x}')] dS \\
&\quad + u_\beta(\mathbf{x}') \lim_{\varepsilon \to 0} \int_{S^+} T_{\alpha\beta}^m\left(\mathbf{x}',\mathbf{x}\right) dS
\end{aligned}
\tag{C.11}
$$

The second term on the RHS in equation (C.11) was shown to vanish as $\varepsilon \to 0$ and the first term on the RHS forms a Cauchy principal value integral. Again, by considering the relationships in equations (C.1)-(C.3), the jump term appear from the last term on the RHS as follows:

$$
u_\beta(\mathbf{x}') \lim_{\varepsilon \to 0} \int_{S^+} T_{\alpha\beta}^m(\mathbf{x}',\mathbf{x}) dS = -\frac{1}{2}\delta_{\alpha\beta} u_\beta(\mathbf{x}')
\tag{C.12}
$$

C.2 The Bending Stress Resultant Integral Equations

The bending stress resultant integral equation is given as

$$
M_{\alpha\beta}\left(\mathbf{x}'\right) + \lim_{\varepsilon \to 0} \int_{S-S_\varepsilon+S^+} S_{\alpha\beta\gamma}^b(\mathbf{x}',\mathbf{x}) w_\gamma(\mathbf{x}) dS
$$

$$
+ \lim_{\varepsilon \to 0} \int_{S-S_\varepsilon+S^+} S_{\alpha\beta 3}^b(\mathbf{x}',\mathbf{x}) w_3(\mathbf{x}) dS
$$

$$
= \lim_{\varepsilon \to 0} \int_{S-S_\varepsilon+S^+} D_{\alpha\beta\gamma}^b(\mathbf{x}',\mathbf{x}) p_\gamma(\mathbf{x}) dS + \lim_{\varepsilon \to 0} \int_{S-S_\varepsilon+S^+} D_{\alpha\beta 3}^b(\mathbf{x}',\mathbf{x}) p_3(\mathbf{x}) dS
$$

$$
\lim_{\varepsilon \to 0} \int_V S_{\alpha\beta k}^b(\mathbf{X}',\mathbf{X}) f_k^b dV
\tag{C.13}
$$

Equation (C.13) can be written in the following form:

$$
M_{\alpha\beta}\left(\mathbf{x}'\right) + I_{11}^b + I_{12}^b = I_{13}^b + I_{14}^b + I_{15}^b
\tag{C.14}
$$

The boundary values of w_i are assumed to be $C^{1,\alpha}$, $(0 < \alpha < 1)$, to allow for a Taylor series expansion for the integrands up to two terms. In the following, each of the integrals in equation (C.14) will be expanded and considered individually.

The integral I_{11}^b

The integral I_{11}^b can be written as

$$
\begin{aligned}
I_{11}^b &= \lim_{\varepsilon \to 0} \int_{S-S_\varepsilon+S^+} S_{\alpha\beta\gamma}^b(\mathbf{x}',\mathbf{x}) w_\gamma(\mathbf{x}) dS = \lim_{\varepsilon \to 0} \int_{S-S_\varepsilon} S_{\alpha\beta\gamma}^b(\mathbf{x}',\mathbf{x}) w_\gamma(\mathbf{x}) dS \\
&\quad + \lim_{\varepsilon \to 0} \int_{S^+} S_{\alpha\beta\gamma}^b(\mathbf{x}',\mathbf{x}) [w_\gamma(\mathbf{x}) - w_\gamma(\mathbf{x}') - w_{\gamma,\theta}(\mathbf{x}')(x_\theta(\mathbf{x}) - x_\theta(\mathbf{x}'))] dS
\end{aligned}
$$

$$+ \quad w_\gamma(\mathbf{x}') \lim_{\varepsilon \to 0} \int_{S+} S^b_{\alpha\beta\gamma}(\mathbf{x}', \mathbf{x}) dS$$

$$+ \quad w_{\gamma,\theta}(\mathbf{x}') \lim_{\varepsilon \to 0} \int_{S+} S^b_{\alpha\beta\gamma}(\mathbf{x}', \mathbf{x})(x_\theta(\mathbf{x}) - x_\theta(\mathbf{x}')) dS \tag{C.15}$$

It has to be noted that the integral I^b_{11} contains the kernel $S^b_{\alpha\beta\gamma}$, which is hypersingular of $O(\frac{1}{r^2} + \ln r)$, therefore two terms of the Taylor series expansion for the integrand are appropriate.

In (C.15), the second term on the RHS of (C.15) tends to zero as $\varepsilon \to 0$. The first and third terms together form a Hadamard finite part integral. By considering the relationships in equations (C.1)-(C.3) and the limits in equation (C.4), the last term on the RHS of (C.15) leads to the following jump terms:

$$w_{\gamma,\theta}(\mathbf{x}') \lim_{\varepsilon \to 0} \int_{S+} S^b_{\alpha\beta\gamma}(\mathbf{x}', \mathbf{x})(x_\theta(\mathbf{x}) - x_\theta(\mathbf{x}')) dS$$

$$= \quad -\frac{D(1+\nu)(1-\nu)}{16}(w_{\beta,\alpha}(\mathbf{x}') + w_{\alpha,\beta}(\mathbf{x}') + w_{\gamma,\gamma}(\mathbf{x}')\delta_{\alpha\beta}) \tag{C.16}$$

Now, the integral I^b_{11} can be written as follows:

$$I^b_{11} = \fint_S S^b_{\alpha\beta\gamma}(\mathbf{x}', \mathbf{x}) w_\gamma(\mathbf{x}) dS$$

$$- \frac{D(1+\nu)(1-\nu)}{16}(w_{\beta,\alpha}(\mathbf{x}') + w_{\alpha,\beta}(\mathbf{x}') + w_{\gamma,\gamma}(\mathbf{x}')\delta_{\alpha\beta}) \tag{C.17}$$

where $\fint$ denotes the Hadamard finite part integral.

The integral I^b_{12}

The kernel $S^b_{\alpha\beta3}$ contains a strong singularity of $O(r^{-1})$, and only one term of the Taylor series expansion is needed, that is

$$I^b_{11} = \lim_{\varepsilon \to 0} \int_{S-S_\varepsilon+S+} S^b_{\alpha\beta3}(\mathbf{x}', \mathbf{x}) w_3(\mathbf{x}) dS = \lim_{\varepsilon \to 0} \int_{S-S_\varepsilon} S^b_{\alpha\beta3}(\mathbf{x}', \mathbf{x}) w_3(\mathbf{x}) dS$$

$$+ \quad \lim_{\varepsilon \to 0} \int_{S+} S^b_{\alpha\beta3}(\mathbf{x}', \mathbf{x})[w_3(\mathbf{x}) - w_3(\mathbf{x}')] dS$$

$$+ \quad w_3(\mathbf{x}') \lim_{\varepsilon \to 0} \int_{S+} S^b_{\alpha\beta3}(\mathbf{x}', \mathbf{x}) dS \tag{C.18}$$

In (C.18), the second term on the RHS tends to zero as $\varepsilon \to 0$. The first term on the RHS forms a Cauchy principal value integral. By considering the relationships in equations (C.1)-(C.3) and the limits in (C.4), the jump terms that appear from the last term on the RHS vanish. So that the integral I^b_{12} can be written as follows:

$$I^b_{12} = \fint_S S^b_{\alpha\beta3}(\mathbf{x}', \mathbf{x}) w_3(\mathbf{x}) dS \tag{C.19}$$

The integral I^b_{13}

The integrand $D^b_{\alpha\beta\gamma}$ is strongly singular. Using the first term of the Taylor series expansion of $M_{\gamma\theta}$, the following form can be written:

$$I^b_{13} = \lim_{\varepsilon \to 0} \int_{S-S_\varepsilon+S+} D^b_{\alpha\beta\gamma}(\mathbf{x}', \mathbf{x}) p_\gamma(\mathbf{x}) dS = \lim_{\varepsilon \to 0} \int_{S-S_\varepsilon} D^b_{\alpha\beta\gamma}(\mathbf{x}', \mathbf{x}) p_\gamma(\mathbf{x}) dS$$

$$+ \quad \lim_{\varepsilon \to 0} \int_{S+} D^b_{\alpha\beta\gamma}(\mathbf{x}',\mathbf{x})n_\theta(\mathbf{x})[M_{\gamma\theta}(\mathbf{x}) - M_{\gamma\theta}(\mathbf{x}')]dS$$

$$+ \quad M_{\gamma\theta}(\mathbf{x}') \lim_{\varepsilon \to 0} \int_{S+} D^b_{\alpha\beta\gamma}(\mathbf{x}',\mathbf{x})n_\theta(\mathbf{x})dS \tag{C.20}$$

In the above integrals, the second term on the RHS tends to zero as $\varepsilon \to 0$. The first term on the RHS forms a Cauchy principal value integral. Considering the relationships in equations (C.1)-(C.3) and the limits in (C.4), the last term on the RHS leads to the following jump terms:

$$M_{\gamma\theta}(\mathbf{x}') \lim_{\varepsilon \to 0} \int_{S+} D^b_{\alpha\beta\gamma}(\mathbf{x}',\mathbf{x})n_\theta(\mathbf{x})dS$$

$$= \left(\frac{(3\nu - 1)}{16} M_{\gamma\gamma}(\mathbf{x}')\delta_{\alpha\beta} - \frac{2(\nu - 3)}{16} M_{\alpha\beta}(\mathbf{x}') \right) \tag{C.21}$$

Now, the integral I^b_{13} can be written as follows:

$$I^b_{13} = \int_S D^b_{\alpha\beta\gamma}(\mathbf{x}',\mathbf{x})p_\gamma(\mathbf{x})dS$$

$$+ \left(\frac{(3\nu - 1)}{16} M_{\gamma\gamma}(\mathbf{x}')\delta_{\alpha\beta} - \frac{2(\nu - 3)}{16} M_{\alpha\beta}(\mathbf{x}') \right) \tag{C.22}$$

The integral I^b_{14}
The integrand $D^b_{\alpha\beta 3}$ is weakly singular and leads to no jump term.
The integral I^b_{15}
This is a domain integral which contains a weakly singular kernel, and therefore leads to no jump term.

C.3   The Shear Stress Resultant Integral Equation

The shear stress resultant integral equation (3.143) can be written as

$$Q_\beta(\mathbf{x}') + \lim_{\varepsilon \to 0} \int_{S-S_\varepsilon+S+} S^b_{3\beta\gamma}(\mathbf{x}',\mathbf{x})w_\gamma(\mathbf{x})dS + \lim_{\varepsilon \to 0} \int_{S-S_\varepsilon+S+} S^b_{3\beta 3}(\mathbf{x}',\mathbf{x})w_3(\mathbf{x})dS$$

$$= \lim_{\varepsilon \to 0} \int_{S-S_\varepsilon+S+} D^b_{3\beta\gamma}(\mathbf{x}',\mathbf{x})p_\gamma(\mathbf{x})dS + \lim_{\varepsilon \to 0} \int_{S-S_\varepsilon+S+} D^b_{3\beta 3}(\mathbf{x}',\mathbf{x})p_3(\mathbf{x})dS$$

$$+ \lim_{\varepsilon \to 0} \int_V D^b_{3\beta k}(\mathbf{x}',\mathbf{X})f^b_k dV \tag{C.23}$$

Equation (3.140) can be written in the following form:

$$Q_\beta(\mathbf{x}') + I^b_{16} + I^b_{17} = I^b_{18} + I^b_{19} + I^b_{20} \tag{C.24}$$

In the following, each of the integrals I^b_{16} to I^b_{20} will be considered individually.

The integral I_{16}^b

This integrand $S_{3\beta\gamma}^b$ is strongly singular. Using the first term of the Taylor series expansion of w_γ, the following form can be written:

$$
\begin{aligned}
I_{16}^b \;=\;& \lim_{\varepsilon\to 0}\int_{S-S_\varepsilon+S+} S_{3\beta\gamma}^b(\mathbf{x}',\mathbf{x})w_\gamma(\mathbf{x})dS = \lim_{\varepsilon\to 0}\int_{S-S_\varepsilon} S_{3\beta\gamma}^b(\mathbf{x}',\mathbf{x})w_\gamma(\mathbf{x})dS \\
&+ \lim_{\varepsilon\to 0}\int_{S+} S_{3\beta\gamma}^b(\mathbf{x}',\mathbf{x})[w_\gamma(\mathbf{x}) - w_\gamma(\mathbf{x}')]dS \\
&+ w_\gamma(\mathbf{x}')\lim_{\varepsilon\to 0}\int_{S+} S_{3\beta\gamma}^b(\mathbf{x}',\mathbf{x})dS
\end{aligned}
\tag{C.25}
$$

In the above integrals, the second term on the RHS tends to zero as $\varepsilon \to 0$. The first term forms a Cauchy principal value integral. Considering the relationships in equations (C.1)-(C.3) and the limits in (C.4), the jump terms that appear from the last term on the RHS can be written in the following form:

$$
w_\gamma(\mathbf{x}')\lim_{\varepsilon\to 0}\int_{S+} S_{3\beta\gamma}^b(\mathbf{x}',\mathbf{x})dS = -\frac{D(1-\nu)\lambda^2}{8}w_\beta(\mathbf{x}')
\tag{C.26}
$$

Then the integral I_{16}^b can be written in the following form:

$$
I_{16}^b = \fint_S S_{3\beta\gamma}^b(\mathbf{x}',\mathbf{x})w_\gamma(\mathbf{x})dS - \frac{D(1-\nu)\lambda^2}{8}w_\beta(\mathbf{x}')
\tag{C.27}
$$

The integral I_{17}^b

The integrand $S_{3\beta3}^b$ is a hypersingular kernel of order $O(\frac{1}{r^2}+\ln r)$. Using the first two terms of the Taylor series expansion of w_3, the following form can be written:

$$
\begin{aligned}
I_{17}^b \;=\;& \lim_{\varepsilon\to 0}\int_{S-S_\varepsilon+S+} S_{3\beta3}^b(\mathbf{x}',\mathbf{x})w_3(\mathbf{x})dS = \lim_{\varepsilon\to 0}\int_{S-S_\varepsilon} S_{3\beta3}^b(\mathbf{x}',\mathbf{x})w_3(\mathbf{x})dS \\
&+ \lim_{\varepsilon\to 0}\int_{S+} S_{3\beta3}^b(\mathbf{x}',\mathbf{x})[w_3(\mathbf{x}) - w_3(\mathbf{x}') - w_{3,\alpha}(\mathbf{x}')(x_\alpha(\mathbf{x}) - x_\alpha(\mathbf{x}'))]dS \\
&+ w_3(\mathbf{x}')\lim_{\varepsilon\to 0}\int_{S+} S_{3\beta3}^b(\mathbf{x}',\mathbf{x})dS \\
&+ w_{3,\alpha}(\mathbf{x}')\lim_{\varepsilon\to 0}\int_{S+} S_{3\beta3}^b(\mathbf{x}',\mathbf{x})(x_\alpha(\mathbf{x}) - x_\alpha(\mathbf{x}'))dS
\end{aligned}
\tag{C.28}
$$

In the above integrals, the second term on the RHS tends to zero as $\varepsilon \to 0$. The first and third terms together form a Hadamard finite part integral. Considering the relationships in equations (C.1)-(C.3) and the limits in (C.4), the last term on the RHS leads to the following jump terms:

$$
\begin{aligned}
& w_{3,\alpha}(\mathbf{x}')\lim_{\varepsilon\to 0}\int_{S+} S_{3\beta3}^b(\mathbf{x}',\mathbf{x})(x_\alpha(\mathbf{x}) - x_\alpha(\mathbf{x}'))dS \\
&= -\frac{D(1-\nu)\lambda^2}{8}u_{3,\beta}(\mathbf{x}')
\end{aligned}
\tag{C.29}
$$

Now, the integral I_{17}^b can be written as follows:

$$
I_{17}^b = \fint_S S_{3\beta3}^b(\mathbf{x}',\mathbf{x})w_3(\mathbf{x})dS - \frac{D(1-\nu)\lambda^2}{8}w_{3,\beta}(\mathbf{x}')
\tag{C.30}
$$

The integral I_{18}^b

The integralnd $D_{3\beta S}^b$ is weakly singular and leads to no jump term.

The integral I_{19}^b

This integrand $D_{3\beta 3}^b$ is strongly singular. Using the first term of the Taylor series expansion of Q_θ, the following form can be written:

$$
\begin{aligned}
I_{19}^b &= \lim_{\varepsilon \to 0} \int_{S-S_\varepsilon+S^+} D_{3\beta 3}^b(\mathbf{x}',\mathbf{x})p_3(\mathbf{x})dS = \lim_{\varepsilon \to 0} \int_{S-S_\varepsilon} D_{3\beta 3}^b(\mathbf{x}',\mathbf{x})p_3(\mathbf{x})dS \\
&+ \lim_{\varepsilon \to 0} \int_{S^+} D_{3\beta 3}^b(\mathbf{x}',\mathbf{x})n_\theta(\mathbf{x})[Q_\theta(\mathbf{x}) - Q_\theta(\mathbf{x}')]dS \\
&+ Q_\theta(\mathbf{x}') \lim_{\varepsilon \to 0} \int_{S^+} D_{3\beta 3}^b(\mathbf{x}',\mathbf{x})n_\theta(\mathbf{x})dS
\end{aligned}
\tag{C.31}
$$

In the above integrals, the second term on the RHS tends to zero as $\varepsilon \to 0$. The first term on the RHS forms a Cauchy principal value integral. By considering the relationships in equations (C.1)-(C.3) and the limits in (C.4), the jump terms that appear from the last term on the RHS can be written in the following form:

$$
Q_\theta(\mathbf{x}') \lim_{\varepsilon \to 0} \int_{S^+} D_{3\beta 3}^b(\mathbf{x}',\mathbf{x})n_\theta(\mathbf{x})dS = \frac{Q_\beta(\mathbf{x}')}{4}
\tag{C.32}
$$

Then the integral I_{19}^b can be written in the following form:

$$
I_{19}^b = \fint_S D_{3\beta 3}^b(\mathbf{x}',\mathbf{x})p_3(\mathbf{x})dS(\mathbf{x}) + \frac{Q_\beta(\mathbf{x}')}{4}
\tag{C.33}
$$

The integral I_{20}^b

This domain integral contains a weakly singular kernel and therefore will lead to no jump term.

References

[1] Dirgantara, T., Boundary element analysis of cracks in shear deformable plates and shells, PhD Thesis, Department of Engineering, Queen Mary, University of London, 2000.

Appendix D

Particular Solutions For Shear Deformable Plates

Particular solutions needed for the dual reciprocity technique as applied to plates and shells were derived by Wen, Aliabadi and Young [1], and are given in the following sections.

D.1   Plate Bending

The governing equation for the shear deformable plate bending problem can be written as

$$\hat{\mathbf{w}} = \mathbf{H}\mathbf{e}\varphi \tag{D.1}$$

where particular solutions of displacement $\hat{\mathbf{w}} = \{\hat{w}_1, \hat{w}_2, \hat{w}_3\}^\top$, $\mathbf{e} = \{e_1, e_2, e_3\}^\top$ is an arbitrary constant vector, and components of matrix $\mathbf{H}$ are

$$H_{\alpha\beta} = 2\delta_{\alpha\beta}\bigtriangledown^4 - [(1+\nu)\bigtriangledown^2 + (1-\nu)\lambda^2)]\frac{\partial^2}{\partial x_\alpha \partial x_\beta}$$

$$H_{3\alpha} = -H_{\alpha 3} = -(1-\nu)(\bigtriangledown^2 - \lambda^2)\frac{\partial}{\partial x_\alpha}$$

$$H_{33} = (\bigtriangledown^2 - \lambda^2)[2\bigtriangledown^2 - (1-\nu)\lambda^2]/\lambda^2 \tag{D.2}$$

The function φ can be defined from equation (D.1) such that

$$D(1-\nu)(\bigtriangledown^2 - \lambda^2)\bigtriangledown^4 \varphi + F(r) = 0 \tag{D.3}$$

If $e_1 = 0, e_2 = 0$ and $e_3 = 1$, the particular solution used in equation (3.127) can be written as

$$\hat{w}_{m\alpha} = -\frac{1}{D}\frac{\partial\psi}{\partial x_\alpha}$$

$$\hat{w}_{m3} = \frac{1}{(1-\nu)D\lambda^2}[2\bigtriangledown^2 \psi - (1-\nu)\lambda^2\psi] \tag{D.4}$$

where

$$\bigtriangledown^4\psi(r) + F(r) = 0 \tag{D.5}$$

The tractions on the boundary can be obtained by

$$\hat{p}_{m\alpha} = \hat{M}_{\alpha\beta} n_\beta, \quad \hat{p}_{m3} = \hat{Q}_\alpha n_\alpha \tag{D.6}$$

If radial basis function $F(r) = 1 + r$, The function $\psi(r)$ can be solved from equation (D.5)

$$\psi(r) = -\left(\frac{r^4}{64} + \frac{r^5}{225}\right) \tag{D.7}$$

and the rotations and deflection can be deduced,

$$\hat{w}_1^l = -\left(\frac{1}{16} + \frac{r}{45}\right)\frac{x_1 r^2}{D}$$

$$\hat{w}_2^l = -\left(\frac{1}{16} + \frac{r}{45}\right)\frac{x_2 r^2}{D}$$

$$\hat{w}_3^l = -\left(\frac{1}{2} + \frac{2r}{9}\right)\frac{r^2}{(1-\nu)\lambda^2 D} + \left(\frac{1}{64} + \frac{r}{225}\right)\frac{r^4}{D} \tag{D.8}$$

The particular solutions of moments $\hat{M}_{\alpha\beta}$ and shear forces $\hat{Q}_\beta$ can be determined by equations (3.89)-(3.90) to give

$$\hat{M}_{11}^l = -\left[\left(\frac{1}{8} + \frac{r}{15}\right)(x_1^2 + \nu x_2^2) + (1+\nu)\left(\frac{r^2}{16} + \frac{r^3}{45}\right)\right]$$

$$\hat{M}_{12}^l = -(1+\nu)\left(\frac{1}{8} + \frac{r}{15}\right)(x_1 x_2)$$

$$\hat{M}_{22}^l = -\left[\left(\frac{1}{8} + \frac{r}{15}\right)(\nu x_1^2 + x_2^2) + (1+\nu)\left(\frac{r^2}{16} + \frac{r^3}{45}\right)\right]$$

$$\hat{Q}_1^l = -\frac{x_1}{2}\left(1 + \frac{2r}{3}\right)$$

$$\hat{Q}_2^l = -\frac{x_2}{2}\left(1 + \frac{2r}{3}\right) \tag{D.9}$$

and the tractions on the boundary can be obtained from the relationships in equation (D.6).

For the derivative of function $F_{,\alpha} = x_\alpha/r$, the solution $\psi^\alpha(r)$ can be found

$$\psi^\alpha(r) = -\frac{r^3 x_\alpha}{45} \tag{D.10}$$

and particular solutions $\overline{w}_{mk}$ for $\alpha = 1$ are

$$\overline{w}_1^l = -(3x_1^2 + r^2)\frac{r}{45D}$$

$$\overline{w}_2^l = -\frac{x_1 x_2 r}{15D}$$

$$\overline{w}_3^l = -[30 - (1-\nu)\lambda^2 r^2]\frac{r x_1}{45(1-\nu)\lambda^2 D} \tag{D.11}$$

and the particular solutions of moments $\overline{\hat{M}}_{\alpha\beta}$ and shear forces $\overline{\hat{Q}}_\beta$ are

$$\overline{\hat{M}}_{11}^l = -\frac{x_1}{15}\left[\nu\left(\frac{x_1^2}{r} + 3r\right) + \left(\frac{x_2^2}{r} + r\right)\right]$$

$$\overline{\hat{M}}^l_{12} = -(1-\nu)\frac{x_2}{15}\left(\frac{x_1^2}{r}+r\right)$$

$$\overline{\hat{M}}^l_{22} = -\frac{x_1}{15}\left[\nu\left(\frac{x_1^2}{r}+3r\right)+\left(\frac{x_2^2}{r}+r\right)\right]$$

$$\overline{\hat{Q}}^l_{1} = -\frac{1}{3}\left(\frac{x_1^2}{r}+r\right)$$

$$\overline{\hat{Q}}^l_{2} = -\frac{1}{3}\frac{x_1 x_2}{r} \tag{D.12}$$

and , for $\alpha = 2$, we have

$$\overline{\hat{w}}^l_{1} = -\frac{x_1 x_2 r}{15D}$$

$$\overline{\hat{w}}^l_{1} = -(3x_2^2 + r^2)\frac{r}{45D}$$

$$\overline{\hat{w}}^l_{3} = -[30 - (1-\nu)\lambda^2 r^{2\cdot}\frac{r x_2}{45(1-\nu)\lambda^2 D} \tag{D.13}$$

and the particular solutions of moments $\overline{\hat{M}}_{\alpha\beta}$ and shear forces $\overline{\hat{Q}}_{\beta}$ are

$$\overline{\hat{M}}^l_{11} = -\frac{x_2}{15}\left[\nu\left(\frac{x_1^2}{r}+r\right)+\left(\frac{x_2^2}{r}+3r\right)\right]$$

$$\overline{\hat{M}}^l_{12} = -(1-\nu)\frac{x_1}{15}\left(\frac{x_2^2}{r}+r\right)$$

$$\overline{\hat{M}}^l_{22} = -\frac{x_2}{15}\left[\nu\left(\frac{x_1^2}{r}+r\right)+\left(\frac{x_2^2}{r}+3r\right)\right]$$

$$\overline{\hat{Q}}^l_{1} = -\frac{1}{3}\frac{x_1 x_2}{r}$$

$$\overline{\hat{Q}}^l_{2} = -\frac{1}{3}\left(\frac{x_2^2}{r}+r\right) \tag{D.14}$$

D.2   Two-dimensional Plane Stress

An expression for a particular solution for displacement $\hat{u}^l_\alpha$ can be found in polar coordinates with the use of the Galerkin vector $G_{\alpha\beta}$ as

$$\hat{u}^l_\alpha(r) = G_{\beta\alpha,\gamma\gamma}(r) - \frac{1+\nu}{2}G_{\gamma\alpha,\beta\gamma}(r) \tag{D.15}$$

where $G_{\alpha\beta}$ satisfies

$$\nabla^4 G_{\beta\alpha} + \frac{2}{(1-\nu)B}\frac{x_1}{r}\delta_{\gamma\beta} = 0 \tag{D.16}$$

and a solution is determined by

$$G_{\beta\alpha} = -\frac{r^3 x_1}{45(1-v)B}\delta_{\alpha\beta} \tag{D.17}$$

Substituting equation (D.17) into equation (D.15), the particular solution for displacements can then be arranged as

$$\hat{u}_1^l = -\frac{2}{(1-\nu)B}\left[\frac{rx_1}{3} - \frac{1+\nu}{30}\left(\frac{x_1^3}{r} + 3x_1r\right)\right]$$

$$\hat{u}_2^l = \frac{(1+\nu)}{15(1-\nu)B}\left(\frac{x_1^2x_2}{r} + x_2r\right) \tag{D.18}$$

and using strain displacement relationships in equation (3.10), the strain are obtained as

$$\hat{\varepsilon}_{11}^l = -\frac{2}{(1-\nu)}\left[\left(\frac{x_1^2}{r} + \frac{r}{3}\right) - \frac{1+\nu}{30}\left(-\frac{x_1^4}{r^3} + \frac{6x_1^2}{r} + 3r\right)\right]$$

$$\hat{\varepsilon}_{12}^l = -\frac{2}{(1-\nu)}\left[\frac{x_1x_2}{6r} - \frac{1+\nu}{30}\left(-\frac{x_1^3x_2}{r^3} + \frac{3x_1x_2}{r}\right)\right]$$

$$\hat{\varepsilon}_{22}^l = \frac{2}{(1-\nu)}\frac{1+\nu}{30}\left(-\frac{x_1^2x_2^2}{r^3} + 2r\right) \tag{D.19}$$

The particular solution for membrane stress resultant can be derived by substituting equation (D.19) into the stress resultant-strain relationships in equation (3.16) to give

$$\hat{N}_{11}^l = B\left[(1-\nu)\hat{\varepsilon}_{11}^l + \nu\hat{\varepsilon}_{\alpha\alpha}^l\right]$$

$$\hat{N}_{12}^l = B(1-\nu)\hat{\varepsilon}_{12}^l$$

$$\hat{N}_{22}^l = B\left[(1-\nu)\hat{\varepsilon}_{22}^l + \nu\hat{\varepsilon}_{\alpha\alpha}^l\right] \tag{D.20}$$

and the particular solutions of traction are obtained from

$$\hat{t}_\alpha^l = \hat{N}_{\alpha\beta}^l n_\beta \tag{D.21}$$

In the same way, particular solutions for displacement $\overline{\tilde{u}}_k^l$ can be obtained as follows:

$$\overline{\tilde{u}}_1^l = \frac{(1+\nu)}{15(1-\nu)B}\left(\frac{x_2^2x_1}{r} + x_1r\right) \tag{D.22}$$

$$\overline{\tilde{u}}_2^l = -\frac{2}{(1-\nu)B}\left[\frac{rx_2}{3} - \frac{1+\nu}{30}\left(\frac{x_2^3}{r} + 3x_2r\right)\right]$$

and the strains are

$$\overline{\tilde{\varepsilon}}_{11}^l = \frac{2}{(1-\nu)}\frac{1+\nu}{30}\left(-\frac{x_1^2x_2^2}{r^3} + 2r\right)$$

$$\overline{\tilde{\varepsilon}}_{12}^l = -\frac{2}{(1-\nu)}\left[\frac{x_1x_2}{6r} - \frac{1+\nu}{30}\left(-\frac{x_2^3x_1}{r^3} + \frac{3x_1x_2}{r}\right)\right]$$

$$\overline{\tilde{\varepsilon}}_{22}^l = -\frac{2}{(1-\nu)}\left[\left(\frac{x_2^2}{r} + \frac{r}{3}\right) - \frac{1+\nu}{30}\left(-\frac{x_2^4}{r^3} + \frac{6x_2^2}{r} + 3r\right)\right] \tag{D.23}$$

The particular solution for membrane stress resultant are

$$\overline{\hat{N}}_{11}^l = B\left[(1-\nu)\hat{\varepsilon}_{11}^l + \nu\hat{\varepsilon}_{\alpha\alpha}^l\right]$$

$$\overline{\hat{N}}_{12}^l = B(1-\nu)\hat{\varepsilon}_{12}^l$$

$$\overline{\hat{N}}_{22}^l = B\left[(1-\nu)\hat{\varepsilon}_{22}^l + \nu\hat{\varepsilon}_{\alpha\alpha}^l\right] \tag{D.24}$$

and finally, the particular solutions of traction are obtained from

$$\bar{t}_\alpha^l = \overline{\hat{N}}_{\alpha\beta}^l n_\beta \tag{D.25}$$

References

[1] Wen, P.H., Aliabadi, M. H. and Young, A., Plane stress and plate bending coupling in BEM analysis of Shallow shells, *International Journal for Numerical Methods in Engineering*, **48**, 1107-1125, 2000.

Appendix E

Fundamental Solutions for Elastodynamics

In this appendix the fundamental solutions of elastodynamics for displacements and stresses are presented.

E.1 Fundamental Solutions for Time Domain

For three-dimensional problems the displacement fundamental solution is given [5] as

$$U_{ij}(\mathbf{X}', \mathbf{X}, t') = \frac{1}{4\pi\mu} (\psi \delta_{ij} - \chi r_{,i} r_{,j}) \tag{E.1}$$

The traction fundamental solution in 3D is given [5] as

$$T_{ij}(\mathbf{X}', \mathbf{X}, t') = \frac{1}{4\pi} \left[\left(\psi_{,r} - \frac{\chi}{r} \right) \left(\frac{\partial r}{\partial n} \delta_{ij} + r_{,j} n_i \right) \right.$$

$$-2\frac{\chi}{r} \left(n_j r_{,i} - 2r_{,i} r_{,j} \frac{\partial r}{\partial n} \right) - 2\chi_{,r} r_{,i} r_{,j} \frac{\partial r}{\partial n}$$

$$\left. + \left(\frac{c_1^2}{c_2^2} - 2 \right) \left(\psi_{,r} - \chi_{,r} - 2\frac{\chi}{r} \right) r_{,i} n_j \right] \tag{E.2}$$

where

$$\psi = \frac{c_2^2}{r^3} t' [H(t' - r/c_2) - H(t' - r/c_1)] + \frac{1}{r} \delta(t' - r/c_2)$$

$$\chi = 3\psi - \frac{2}{r} \delta(t' - r/c_2) - \frac{c_2^2}{c_1^2} \frac{1}{r} \delta(t' - r/c_1)$$

$$\psi_{,r} = \frac{\partial \psi}{\partial r} = -\frac{\chi}{r} - \frac{1}{r^2} \left[\delta(t' - r/c_2) + \frac{r}{c_2} \dot\delta(t' - r/c_2) \right]$$

$$\chi_{,r} = \frac{\partial \chi}{\partial r} = -\frac{3\chi}{r} - \frac{1}{r^2} \left[\delta(t' - r/c_2) + \frac{r}{c_2} \dot\delta(t' - r/c_2) \right]$$

$$+\frac{c_2^2}{c_1^2}\frac{1}{r^2}\left[\delta(t'-r/c_1)+\frac{r}{c_1}\dot{\delta}(t'-r/c_1)\right]$$

$$\psi_{,rr}=\frac{\partial^2\psi}{\partial r^2}=\frac{4\chi}{r^2}+\frac{1}{r^3}\left(3\delta(t'-r/c_2)+\frac{3r}{c_2}\dot{\delta}(t'-r/c_2)+\frac{r^2}{c_2^2}\ddot{\delta}(t'-r/c_2)\right)$$

$$-\frac{c_2^2}{c_1^2}\frac{1}{r^3}\left(\delta(t'-r/c_1)+\frac{r}{c_2}\dot{\delta}(t'-r/c_1)\right)$$

$$\chi_{,rr}=\frac{\partial^2\chi}{\partial r^2}=\frac{12\chi}{r^2}+\frac{1}{r^3}\left[5\delta(t'-r/c_2)+\frac{5r}{c_2}\dot{\delta}(t'-r/c_2)+\frac{r^2}{c_2^2}\ddot{\delta}(t'-r/c_2)\right]$$

$$-\frac{c_2^2}{c_1^2}\frac{1}{r^3}\left[5\delta(t'-r/c_1)+\frac{5r}{c_2}\dot{\delta}(t'-r/c_1)+\frac{r^2}{c_2^2}\ddot{\delta}(t'-r/c_1)\right]$$

For two-dimensional problems, the displacement fundamental solution can be obtained from (5.12) by integration along the third spatial dimension, that is

$$U_{ij}^{2D}=\int_{-\infty}^{\infty}U_{ij}^{3D}dx_3$$

The final expression can be written [4] as

$$U_{ij}(\mathbf{X},\mathbf{X'},t')=\frac{1}{2\pi\rho}\left\{\frac{1}{c_1}H\left(\frac{c_1t'}{r}-1\right)\left[\left(2\left(\frac{c_1t'}{r}\right)^2-1\right)S_1^{-\frac{1}{2}}\left(\frac{r_{,i}r_{,j}}{r}\right)-\left(\frac{\delta_{ij}}{r}\right)S_1^{\frac{1}{2}}\right]\right.$$

$$\left.+\frac{1}{c_2}H\left(\frac{c_2t'}{r}-1\right)\left[-\left(2\left(\frac{c_2t'}{r}\right)^2-1\right)S_2^{-\frac{1}{2}}\left(\frac{r_{,i}r_{,j}}{r}\right)+\left(\frac{\delta_{ij}}{r}\right)\left(\frac{c_2t'}{r}\right)^2S_2^{-\frac{1}{2}}\right]\right\}$$

(E.3)

In (E.3) the Heaviside function ensures the causality condition (i.e. the term multiplying it will vanish if the wave has not reached the field point).

The traction fundamental solution can be derived from (E.3) as

$$T_{ij}(\mathbf{X},\mathbf{X'},t')=\frac{\mu}{2\pi\rho r}\left\{\frac{1}{c_1}H\left(\frac{c_1t'}{r}-1\right)\left[S_1^{-\frac{3}{2}}\left(\frac{A_1}{r}\right)+\left(2\left(\frac{c_1t'}{r}\right)^2-1\right)S_1^{-\frac{1}{2}}\left(\frac{2A_2}{r}\right)\right]\right.$$

$$\left.-\frac{1}{c_2}H\left(\frac{c_2t'}{r}-1\right)\left[S_2^{-\frac{3}{2}}\left(\frac{A_3}{r}\right)+\left(2\left(\frac{c_2t'}{r}\right)^2-1\right)S_2^{-\frac{1}{2}}\left(\frac{2A_2}{r}\right)\right]\right\}$$

(E.4)

where

$$S_1=\left(\frac{c_1t'}{r}\right)^2-1$$

$$S_2=\left(\frac{c_2t'}{r}\right)^2-1$$

$$A_1=(\lambda/\mu)n_ir_{,j}+2r_{,i}r_{,j}\frac{\partial r}{\partial n}$$

$$A_2=n_ir_{,j}+n_jr_{,i}+\frac{\partial r}{\partial n}(\delta_{ij}-4r_{,i}r_{,j})$$

$$A_3=\frac{\partial r}{\partial n}(2r_{,i}r_{,j}-\delta_{ij})-n_jr_{,i}$$

The fundamental solutions D_{kij} and S_{kij} for stress integral equation are given [6]
as

$$D_{ijk}(\mathbf{X}, \mathbf{X}', t') = \frac{1}{4\pi}\left[2\left(\chi_{,r} - 2\frac{\chi}{r}\right)r_{,i}r_{,j}r_{,k} + 2\frac{\chi}{r}\delta_{ij}r_{,k}\right.$$

$$\left. - \left(\psi_{,r} - \frac{\chi}{r}\right)(\delta_{ik}r_{,j} + \delta_{jk}r_{,i}) - \frac{\lambda}{\mu}\left(\psi_{,r} - \chi_{,r} - 2\frac{\chi}{r}\right)\delta_{ij}r_{,k}\right] \quad \text{(E.5)}$$

and

$$S_{ijk}(\mathbf{X}, \mathbf{X}', t') = \frac{\mu}{4\pi}\left\{\frac{\partial r}{\partial n}\left[4\left(\chi_{,rr} - 5\frac{\chi_{,r}}{r} + 8\frac{\chi}{r^2}\right)r_{,i}r_{,j}r_{,k}\right.\right. \quad \text{(E.6)}$$

$$\left(\psi_{,rr} - \frac{\psi_{,r}}{r} - 3\frac{\chi_{,r}}{r} + 6\frac{\chi}{r^2}\right)(\delta_{ik}r_{,j} + \delta_{jk}r_{,i}) + 2\Upsilon\delta_{ij}r_{,k}\right] + 2\Upsilon r_{,i}r_{,j}n_k$$

$$- \left(\psi_{,rr} - \frac{\psi_{,r}}{r} - 3\frac{\chi_{,r}}{r} + 6\frac{\chi}{r^2}\right)(r_{,j}n_i + r_{,i}n_j)r_{,k} - 2\left(\frac{\psi_{,r}}{r} - \frac{\chi}{r^2}\right)(\delta_{kj}n_i + \delta_{ki}n_j)$$

$$+ \left(4\frac{\chi}{r^2} + \frac{4\lambda}{\mu}\left(\frac{\chi_{,r}}{r} + 2\frac{\chi}{r^2} - \frac{\psi_{,r}}{r}\right) + \frac{\lambda^2}{\mu^2}\left(\chi_{,rr} + 4\frac{\chi_{,r}}{r} + 2\frac{\chi}{r^2} - \psi_{,rr} - 2\frac{\psi_{,r}}{r}\right)\right)\delta_{ij}n_k\right\}$$

with $\Upsilon = \left(2\frac{\chi_{,r}}{r} - 4\frac{\chi}{r^2} + \frac{\lambda}{\mu}\left(\chi_{,rr} + \frac{\chi_{,r}}{r} - 4\frac{\chi}{r^2} - \psi_{,rr} + \frac{\psi_{,r}}{r}\right)\right)$, for three-dimensional
problems, and

$$D_{kij}(\mathbf{X}, \mathbf{X}', t') = \frac{-\mu}{2\pi\rho r}\left\{\frac{1}{c_1}H\left(\frac{c_1t'}{r} - 1\right)\left[S_1^{-\frac{3}{2}}\left(\frac{B_1}{r}\right) + \left(2\left(\frac{c_1t'}{r}\right)^2 - 1\right)\right.\right.$$

$$\left.\times S_1^{-\frac{1}{2}}\left(\frac{2B_2}{r}\right)\right] - \frac{1}{c_2}H\left(\frac{c_2t'}{r} - 1\right)\left[S_2^{-\frac{3}{2}}\left(\frac{B_3}{r}\right) + \left(2\left(\frac{c_2t'}{r}\right)^2 - 1\right)S_2^{-\frac{1}{2}}\left(\frac{2B_2}{r}\right)\right]\right\} \quad \text{(E.7)}$$

$$S_{kij}(\mathbf{X}, \mathbf{X}', t') = \frac{\mu^2}{2\pi\rho r^2}\left\{-\frac{1}{c_1}H(\frac{c_1t'}{r} - 1)\left[S_1^{-\frac{5}{2}}\left(\frac{3D_1}{r}\right) - S_1^{-\frac{3}{2}}\left(\frac{2D_2}{r}\right)\right.\right.$$

$$\left. - \left(2\left(\frac{c_1t'}{r}\right)^2 - 1\right)S_1^{-\frac{1}{2}}\left(\frac{4D_3}{r}\right)\right]$$

$$+ \frac{1}{c_2}H(\frac{c_2t'}{r} - 1)\left[S_2^{-\frac{5}{2}}\left(\frac{3D_4}{r}\right) - S_2^{-\frac{3}{2}}\left(\frac{2E_5}{r}\right)\right.$$

$$\left.\left. - \left(2\left(\frac{c_2t'}{r}\right)^2 - 1\right)S_2^{-\frac{1}{2}}\left(\frac{4E_3}{r}\right)\right]\right\}$$

for two-dimensional problems, where

$$S_1 = \left(\frac{c_1t'}{r}\right)^2 - 1$$

$$S_2 = \left(\frac{c_2t'}{r}\right)^2 - 1$$

$$B_1 = \frac{\lambda}{\mu}\delta_{ij}r_{,k} + 2r_{,i}r_{,j}r_{,k}$$

$$B_2 = r_{,i}\delta_{jk} + r_{,j}\delta_{ik} + r_{,k}\delta_{ij} - 4r_{,i}r_{,j}r_{,k}$$

$$B_3 = 2r_{,i}r_{,j}r_{,k} - r_{,i}\delta_{jk} - r_{,j}\delta_{ik}$$

and

$$D_1 = \left(\frac{\lambda}{\mu} \delta_{ij} + 2r_{,i}r_{,j} \right) \left(\frac{\lambda}{\mu} n_k + 2r_{,i}n_i r_{,k} \right)$$

$$D_2 = -n_k \left[\frac{\lambda}{\mu} \delta_{ij} (2 + \frac{\lambda}{\mu}) + 2r_{,i}r_{,j} \right] - 2n_i r_{,j}r_{,k} - 2n_j r_{,j}r_{,k}$$

$$\quad\quad -2r_{,i}n_i (r_{,i}\delta_{jk} + r_{,j}\delta_{ik} + r_{,k}\delta_{ij} - 6r_{,i}r_{,j}r_{,k})$$

$$D_3 = n_i (-\delta_{jk} + 4r_{,j}r_{,k}) + n_j (-\delta_{ik} + 4r_{,i}r_{,k}) + n_k (-\delta_{ij} + 4r_{,i}r_{,j})$$

$$\quad\quad +4r_{,i}n_i (r_{,i}\delta_{jk} + r_{,j}\delta_{ik} + r_{,k}\delta_{ij} - 6r_{,i}r_{,j}r_{,k})$$

$$D_4 = r_{,i}n_i (4r_{,i}r_{,j}r_{,k} - r_{,i}\delta_{jk} - r_{,j}\delta_{ik}) - n_i r_{,j}r_{,k} - n_j r_{,i}r_{,k}$$

$$D_5 = n_i (\delta_{jk} - 2r_{,j}r_{,k}) + n_j (\delta_{ik} - 2r_{,i}r_{,k}) - 2n_k r_{,i}r_{,j}$$

$$\quad\quad -2r_{,i}n_i (r_{,i}\delta_{jk} + r_{,j}\delta_{ik} + r_{,k}\delta_{ij} - 6r_{,i}r_{,j}r_{,k})$$

E.2   Convoluted Fundamental Solutions

In this section the convoluted fundamental solutions of elastodynamics for displacements and stresses are presented.

E.2.1   Two-dimensional Problems

The following simplifying notation is used:

$$\varphi_\gamma = \frac{r}{c_\gamma \Delta\tau (M - m + 1)} \tag{E.8}$$

$$\chi_\gamma = \frac{r}{c_\gamma \Delta\tau (M - m)} \tag{E.9}$$

$$\psi_\gamma = \frac{r}{c_\gamma \Delta\tau (M - m - 1)} \tag{E.10}$$

where c_γ is the velocity of the wave; the subscript γ denotes the number of the wave, that is $\gamma = 1$ corresponds to the longitudinal and $\gamma = 2$ to the shear wave; and r is the distance from the source point to a field point.

For the piecewise constant interpolation of tractions, the convoluted fundamental solution $\tilde{U}_{ij}^{Mm}(\mathbf{x}', \mathbf{x})$ has the form [1, 2]

$$\tilde{U}_{ij}^{Mm}(\mathbf{x}', \mathbf{x}) = \int_{\tau^{n-1}}^{\tau^n} U_{ij}^M (\mathbf{x}', \mathbf{x}) d\tau$$

$$= \sum_{\gamma=1}^{2} \frac{1}{4\pi \rho c_\gamma^2} \left[\delta_{ij} \left(\ln \frac{1 + \sqrt{1 - \varphi_\gamma^2}}{\varphi_\gamma} - \ln \frac{1 + \sqrt{1 - \chi_\gamma^2}}{\chi_\gamma} \right) \right.$$

$$\left. + (-1)^\gamma (\delta_{ij} - 2r_{,i}r_{,j}) \left(\frac{\sqrt{1 - \varphi_\gamma^2}}{\varphi_\gamma^2} - \frac{\sqrt{1 - \chi_\gamma^2}}{\chi_\gamma^2} \right) \right] \tag{E.11}$$

where δ_{ij} is the Kronecker delta.

For the piecewise linear interpolation of displacements, the convoluted fundamental solution $\tilde{T}_{ij}^{Mm}(\mathbf{x}', \mathbf{x})$ has the form [2, 4]

$$\tilde{T}_{ij}^{Mm}(\mathbf{x}', \mathbf{x}) = \int_{\tau^{m-1}}^{\tau^m} T_{ij}^M(\mathbf{x}', \mathbf{x}) M^1 d\tau + \int_{\tau^m}^{\tau^{m+1}} T_{ij}^M(\mathbf{x}', \mathbf{x}) M^2 d\tau$$

$$= \sum_{\gamma=1}^{2} \frac{\mu}{2\pi\rho c_\gamma^2} \frac{1}{c_\gamma \Delta\tau} \left\{ \frac{2A_\gamma}{3} \left[\frac{(1-\varphi_\gamma^2)^{\frac{3}{2}}}{\varphi_\gamma^3} - 2\frac{(1-\chi_\gamma^2)^{\frac{3}{2}}}{\chi_\gamma^3} + \frac{(1-\psi_\gamma^2)^{\frac{3}{2}}}{\psi_\gamma^3} \right] \right.$$

$$\left. -B_\gamma \left[\frac{\sqrt{1-\varphi_\gamma^2}}{\varphi_\gamma} - 2\frac{\sqrt{1-\chi_\gamma^2}}{\chi_\gamma} + \frac{\sqrt{1-\psi_\gamma^2}}{\psi_\gamma} \right] \right\} \qquad \text{(E.12)}$$

The coefficients of the convoluted fundamental solution are

$$A_\gamma = -(-1)^\gamma \left[n_j r_{,i} + n_i r_{,j} + \frac{\partial r}{\partial n} (\delta_{ij} - 4r_{,i} r_{,j}) \right]$$

$$B_1 = \frac{\lambda}{\mu} n_j r_{,i}, + 2\frac{\partial r}{\partial n} r_{,i} r_{,j}$$

$$B_2 = \frac{\partial r}{\partial n} (\delta_{ij} - 2r_{,i} r_{,j}) + n_i r_{,j}$$

where λ and μ are the Lamé constants. For the linear interpolation functions, the contribution of the integration over the time interval before and after the time node is taken into account.

For the piecewise constant interpolation of tractions, the convoluted fundamental solution $\tilde{S}_{kij}^{Mm}(\mathbf{x}', \mathbf{x})$ has the form

$$\tilde{S}_{kij}^{Mm}(\mathbf{x}', \mathbf{x}) = \int_{\tau^{m-1}}^{\tau^m} S_{kij}^M(\mathbf{x}', \mathbf{x}) d\tau$$

$$= \sum_{\gamma=1}^{2} \frac{(-1)^\gamma \mu}{2\pi\rho c_\gamma^2} \frac{1}{r} \left\{ -C_\gamma \left[\frac{1}{\sqrt{1-\varphi_\gamma^2}} - \frac{1}{\sqrt{1-\chi_\gamma^2}} \right] + 2D \left[\frac{\sqrt{1-\varphi_\gamma^2}}{\varphi_\gamma^2} - \frac{\sqrt{1-\chi_\gamma^2}}{\chi_\gamma^2} \right] \right\} \quad \text{(E.13)}$$

The coefficients of the convoluted fundamental solution are:

$$C_1 = \frac{\lambda}{\mu} \delta_{ij} r_{,k} + 2r_{,i} r_{,j} r_{,k}$$

$$C_2 = 2r_{,i} r_{,j} r_{,k} - r_{,i} \delta_{jk} - r_{,j} \delta_{ik}$$

$$D = r_{,i} \delta_{jk} + r_{,j} \delta_{ik} + r_{,k} \delta_{ij} - 4r_{,i} r_{,j} r_{,k}$$

For the piecewise linear interpolation of displacements the convoluted fundamental solution $\tilde{T}_{kij}^{Mm}(\mathbf{x}', \mathbf{x})$ has the form:

$$\tilde{D}_{kij}^{Mm}(\mathbf{x}', \mathbf{x}) = \int_{\tau^{m-1}}^{\tau^m} D_{kij}^M(\mathbf{x}', \mathbf{x}) M^1 d\tau + \int_{\tau^m}^{\tau^{m+1}} D_{kij}^M(\mathbf{x}', \mathbf{x}) M^2 d\tau$$

$$= \sum_{\gamma=1}^{2} \frac{(-1)^{\gamma}\mu^2}{2\pi\rho c_{\gamma}^2} \frac{1}{rc_{\gamma}\Delta\tau} \left\{ 2(E_{\gamma}+F_{\gamma})\left[\frac{\sqrt{1-\varphi_{\gamma}^2}}{\varphi_{\gamma}} - 2\frac{\sqrt{1-\chi_{\gamma}^2}}{\chi_{\gamma}} + \frac{\sqrt{1-\psi_{\gamma}^2}}{\psi_{\gamma}} \right] \right.$$

$$-\frac{4}{3}G\left[\frac{(1-\varphi_{\gamma}^2)^{\frac{3}{2}}}{\varphi_{\gamma}^3} - 2\frac{(1-\chi_{\gamma}^2)^{\frac{3}{2}}}{\chi_{\gamma}^3} + \frac{(1-\psi_{\gamma}^2)^{\frac{3}{2}}}{\psi_{\gamma}^3} \right] \qquad (A.13)$$

$$\left. +E_{\gamma}\left[\frac{\varphi_{\gamma}}{\sqrt{1-\varphi_{\gamma}^2}} - 2\frac{\chi_{\gamma}}{\sqrt{1-\chi_{\gamma}^2}} + \frac{\psi_{\gamma}}{\sqrt{1-\psi_{\gamma}^2}} \right] \right\} \qquad (E.14)$$

The coefficients of the convoluted fundamental solution are:

$$E_1 = \left(\frac{\lambda}{\mu}\delta_{ij} + 2r_{,i}r_{,j} \right)\left(\frac{\lambda}{\mu}n_k + 2\frac{\partial r}{\partial n}r_{,k} \right)$$

$$E_2 = \frac{\partial r}{\partial n}(4r_{,i}r_{,j}r_{,k} - r_{,i}\delta_{jk} - r_{,j}\delta_{ik}) - n_i r_{,j}r_{,k} - n_j r_{,i}r_{,k}$$

$$F_1 = -n_k\left[\frac{\lambda}{\mu}\delta_{ij}(2+\frac{\lambda}{\mu}) + 2r_{,i}r_{,j} \right] - 2n_i r_{,j}r_{,k} - 2n_j r_{,i}r_{,k}$$

$$-2\frac{\partial r}{\partial n}(r_{,i}\delta_{jk} + r_{,j}\delta_{ik} + r_{,k}\delta_{ij} - 6r_{,i}r_{,j}r_{,k})$$

$$F_2 = n_i(\delta_{jk} - 2r_{,j}r_{,k}) + n_j(\delta_{ik} - 2r_{,i}r_{,k}) - 2n_k r_{,i}r_{,j}$$

$$-2\frac{\partial r}{\partial n}(r_{,i}\delta_{jk} + r_{,j}\delta_{ik} + r_{,k}\delta_{ij} - 6r_{,i}r_{,j}r_{,k})$$

$$G = n_i(-\delta_{jk} + 4r_{,j}r_{,k}) + n_j(-\delta_{ik} + 4r_{,i}r_{,k}) + n_k(-\delta_{ij} + 4r_{,i}r_{,j})$$

$$+4\frac{\partial r}{\partial n}(r_{,i}\delta_{jk} + r_{,j}\delta_{ik} + r_{,k}\delta_{ij} - 6r_{,i}r_{,j}r_{,k})$$

In evaluating the above variables φ_{γ}, χ_{γ} and ψ_{γ}, the causality condition must be satisfied. That is, if r is greater than the distance travelled by the wave at the given time then the value of the variable φ_{γ}, χ_{γ} and ψ_{γ} is greater than 1. In this case, the terms in equations (E.11)-(E.14) which contain that variable must be put equal to zero.

E.2.2   Three-dimensional Problems

The functions $U_{ij\gamma}^{M-m+1}$ and $D_{kij\gamma}^{M-m+1}$ can be deduced by substitution of solutions $\psi_{\gamma}(U), \chi_{\gamma}(U)$, which are listed below, into equations (E.1) and (E.5), and $T_{ij\gamma}^{M-m+1}$ and $S_{kij\gamma}^{M-m+1}$ can be obtained in terms of $\psi_{\gamma}(T), \chi_{\gamma}(T)$ from equations (E.2) and (E.6). There are five cases after integration of the kernels with respect to time τ, where $A = (M-m+1)$ and $B = (M-m) = A-1$ [5]

   Case 1. $r < Bc_2\Delta t$ and $r > Ac_1\Delta t$

$$\psi_{\gamma}(U) = \psi_{\gamma}(T) = \chi_{\gamma}(U) = \chi_{\gamma}(T) = 0 \qquad (E.15)$$

Case 2. $Bc_1\Delta t < r < Ac_1\Delta t$ and $r > Ac_2\Delta t$

$$\psi_1(U) = \frac{c_2^2}{2r^3}\left[\frac{r^2}{c_1^2} - A^2(\Delta t)^2 \right]; \quad \chi_1(U) = 3\psi_1(U) - \frac{c_2^2}{c_1^2}\frac{1}{r}; \quad \psi_{1,r}(U) = -\frac{1}{r}\chi_1(U)$$

$$(E.16)$$

$$\chi_{1,r}(U) = -\frac{3}{r}\chi^{(1)}(U) + \frac{c_2^2}{c_1^2}\frac{1}{r^2} + \frac{c_2^2}{c_1^3 r \Delta t} \tag{E.17}$$

$$\psi_2(U) = 0; \quad \chi_2(U) = 0; \quad \psi_{2,r}(U) = 0 \tag{E.18}$$

$$\chi_{2,r}(U) = -\frac{c_2^2}{c_1^3 r \Delta t}; \quad \psi_1(T) = \frac{c_2^2}{2r^3}\left([A - \frac{2r}{3c_1\Delta t}\cdot\frac{r^2}{c_1^2} - \frac{1}{3}A^3(\Delta t)^2\right) \tag{E.19}$$

$$\chi_1(T) = 3\psi_1(T) - \frac{c_2^2}{rc_1^2}\left[A - \frac{r}{c_1\Delta t}\right]; \quad \psi_{1,r}(T) = -\frac{\chi_1(T)}{r} \tag{E.20}$$

$$\chi_{1,r}(T) = -3\frac{\chi^{(1)}(T)}{r} + \frac{c_2^2}{c_1^2}\frac{1}{r^2}A; \quad \psi_{1,rr}(T) = \frac{4}{r^2}\psi_1 - \frac{c_2^2}{c_1^2 r^3}A \tag{E.21}$$

$$\chi_{1,rr}(T) = \frac{12}{r^2}\psi_1 - 5\frac{c_2^2}{c_1^2 r^3}A - \frac{c_2^2}{c_1^4 r(\Delta t)^2} \tag{E.22}$$

$$\psi_2(T) = \frac{c_2^2}{2r^3}\left(\left[\frac{2r}{3c_1\Delta t} - B\right]\left(\frac{r}{c_1}\right)^2 - \frac{1}{3}A^3(\Delta t)^2 - A^2(\Delta t)^2\right) \tag{E.23}$$

$$\chi_2(T) = 3\psi_2(T) + \frac{c_2^2}{rc_1^2}\left[B - \frac{r}{c_1\Delta t}\right]; \quad \psi_{2,r}(T) = -\frac{1}{r}\chi_2(T) \tag{E.24}$$

$$\chi_{2,r}(T) = -\frac{3}{r}\chi_2(T) - \frac{c_2^2}{c_1^2}\frac{1}{r^2}B; \quad \psi_{2,rr}(T) = \frac{4}{r^2}\chi_2 + \frac{c_2^2}{c_1^2 r^3}B \tag{E.25}$$

$$\chi_{2,rr}(T) = \frac{12}{r^2}\chi_1 + 5\frac{c_2^2}{c_1^2 r^3}B + \frac{2c_2^2}{c_1^4 r(\Delta t)^2} \tag{E.26}$$

$$\psi_3(T) = 0; \quad \chi_3(T) = 0 \tag{E.27}$$

$$\psi_{3,r}(T) = 0; \quad \chi_{3,r}(T) = 0; \quad \psi_{3,rr}(T) = 0 \tag{E.28}$$

$$\chi_{3,rr}(T) = -\frac{c_2^2}{c_1^4 r(\Delta t)^2} \tag{E.29}$$

Case 3. $Bc_1\Delta t < r < Ac_2\Delta t$

$$\psi_1(U) = \frac{1}{2r}\left(1 + \frac{c_2^2}{c_1^2}\right); \quad \chi_1(U) = 3\psi_1(U) - \left(2 + \frac{c_2^2}{c_1^2}\right)\frac{1}{r} \tag{E.30}$$

$$\psi_{1,r}(U) = -\frac{1}{r}\chi_1(U) - \frac{1}{r^2} - \frac{1}{c_2 r \Delta t} \tag{E.31}$$

$$\chi_{1,r}(U) = -\frac{3}{r}\chi^{(1)}(U) + \frac{c_2^2}{c_1^2}\frac{1}{r^2} - \frac{1}{r^2} - \frac{1}{c_2 r \Delta t} + \frac{c_2^2}{c_1^3 r \Delta t} \tag{E.32}$$

$$\psi_2(U) = 0; \quad \chi_2(U) = 0; \quad \psi_{2,r}(U) = \frac{1}{c_2 r \Delta t} \tag{E.33}$$

$$\chi_{2,r}(U) = \frac{1}{c_2 r \Delta t} - \frac{c_2^2}{c_1^3 r \Delta t} \tag{E.34}$$

$$\psi_1(T) = \frac{c_2^2}{2r^3}\left(\left[A - \frac{2r}{3c_1\Delta t}\right]\frac{r^2}{c_1^2} - \left[A - \frac{2r}{3c_2\Delta t}\right]\frac{r^2}{c_2^2}\right) \tag{E.35}$$

$$+\frac{1}{r}[A-\frac{r}{c_2\Delta t}\dot{}]$$

$$\chi_1(T)=3\psi_1(T)-\frac{2}{r}[A-\frac{r}{c_2\Delta t}\dot{}]-\frac{c_2^2}{c_1^2 r}[A-\frac{r}{c_2\Delta t}\dot{}] \tag{E.36}$$

$$\psi_{1,r}(T)=-\frac{1}{r}\chi_1(T)-\frac{1}{r^2}A \tag{E.37}$$

$$\chi_{1,r}(T)=-3\frac{\chi_1(T)}{r}-\left(1-\frac{c_2^2}{c_1^2}\right)\frac{1}{r^2}A \tag{E.38}$$

$$\psi_{1,rr}(T)=\frac{4}{r^2}\chi_1+\frac{3}{r^3}A-\frac{c_2^2}{c_1^2 r^3}; \tag{E.39}$$

$$\chi_{1,rr}(T)=\frac{12}{r^2}\chi_1+\frac{5}{r^3}\left(1-\frac{c_2^2}{c_1^2}\right)A+(\frac{1}{c_2^2}-\frac{c_2^2}{c_1^4})\frac{1}{r(\Delta t)^2} \tag{E.40}$$

$$\psi_2(T)=\frac{c_2^2}{2r^3}\left([\frac{r}{c_1\Delta t}-B\dot{}\frac{r^2}{c_1^2}-[\frac{r}{c_2\Delta t}-B\dot{}\frac{r^2}{c_2^2}\right. \tag{E.41}$$

$$\left.-\frac{1}{3}\left(\frac{r}{c_1}\right)^3+\frac{1}{3}\left(\frac{r}{c_2}\right)^3\right)+\frac{1}{r}[\frac{r}{c_2\Delta t}-B\dot{} \tag{E.42}$$

$$\chi_2(T)=3\psi_2(T)+\frac{2}{r}[B-\frac{r}{c_2\Delta t}\dot{}]+\frac{c_2^2}{c_1^2 r}[B-\frac{r}{c_1\Delta t}\dot{}] \tag{E.43}$$

$$\psi_{2,r}(T)=-\frac{1}{r}\chi_2(T)+\frac{1}{r^2}B \tag{E.44}$$

$$\chi_{2,r}(T)=-\frac{3}{r}\chi_2(T)+\left(1-\frac{c_2^2}{c_1^2}\right)\frac{1}{r^2}B \tag{E.45}$$

$$\psi_{2,rr}(T)=\frac{4}{r^2}\chi_1-\frac{3}{r^3}B+\frac{c_2^2}{c_1^2 r^3}B-\frac{2}{c_2^2 r(\Delta t)^2} \tag{E.46}$$

$$\chi_{2,rr}(T)=\frac{12}{r^2}\chi_2-\frac{5}{r^3}B\left(1-\frac{c_2^2}{c_1^2}\right)-\frac{2}{r(\Delta t)^2}\left(\frac{1}{c_2^2}-\frac{c_2^2}{c_1^4}\right) \tag{E.47}$$

$$\psi_3(T)=0;\quad \chi_3(T)=0 \tag{E.48}$$

$$\psi_{3,r}(T)=0;\quad \chi_{3,r}(T)=0 \tag{E.49}$$

$$\psi_{3,rr}(T)=\frac{1}{c_2^2 r(\Delta t)^2} \tag{E.50}$$

$$\chi_{3,rr}(T)=\left(\frac{1}{c_2^2}-\frac{c_2^2}{c_1^4}\right)\frac{1}{r(\Delta t)^2} \tag{E.51}$$

Case 4. $Ac_2\Delta t<r<Bc_1\Delta t$

$$\psi_1(U)=\frac{c_2^2}{2r^3}[B^2-A^2\dot{}(\Delta t)^2;\quad \chi_1(U)=3\psi_1(U) \tag{E.52}$$

$$\psi_{1,r}(U)=-\frac{1}{r}\chi_1(U);\quad \chi_{1,r}(U)=-\frac{3}{r}\chi_1(U) \tag{E.53}$$

$$\psi_2(U)=0;\quad \chi_2(U)=0;\quad \psi_{2,r}(U)=0;\quad \chi_{2,r}(U)=0 \tag{E.54}$$

$$\psi_1(T) = \frac{c_2^2}{6r^3}[4B^2 - A^3](\Delta t)^2; \quad \chi_1(T) = 3\psi_1(T) \tag{E.55}$$

$$\psi_{1,r}(T) = -\frac{1}{r}\chi_1(T); \quad \chi_{1,r}(T) = -\frac{3}{r}\chi_1(T) \tag{E.56}$$

$$\psi_{1,rr}(T) = \frac{4}{r^2}\chi_1(T); \quad \chi_{1,rr}(T) = \frac{12}{r^2}\chi_1(T) \tag{E.57}$$

$$\psi_2(T) = -\frac{c_2^2}{6r^3}[B^2 + 2A^2](\Delta t)^2; \quad \chi_2(T) = 3\psi_2(T) \tag{E.58}$$

$$\psi_{2,r}(T) = -\frac{1}{r}\chi_2(T); \quad \chi_{2,r}(T) = -\frac{3}{r}\chi_2(T) \tag{E.59}$$

$$\psi_{2,rr}(T) = \frac{4}{r^2}\psi_2(T); \quad \chi_{2,rr}(T) = \frac{12}{r^2}\psi_2(T) \tag{E.60}$$

$$\psi_3(T) = 0; \quad \chi_3(T) = 0 \tag{E.61}$$

$$\psi_{3,r}(T) = 0; \quad \chi_{3,r}(T) = 0 \tag{E.62}$$

$$\psi_{3,rr}(T) = 0; \quad \chi_{3,rr}(T) = 0 \tag{E.63}$$

Case 5. $Bc_2\Delta t < r < Ac_2\Delta t$ and $r < Ac_2\Delta t$

$$\psi_1(U) = \frac{c_2^2}{2r^3}\left(B^2(\Delta t)^2 - \frac{r^2}{c_2^2}\right) + \frac{1}{r} \tag{E.64}$$

$$\chi_1(U) = 3\psi_1(U) - \frac{2}{r} \tag{E.65}$$

$$\psi_{1,r}(U) = -\frac{1}{r}\chi_1(U) - \frac{1}{r^2} - \frac{1}{c_2 r \Delta t} \tag{E.66}$$

$$\chi_{1,r}(U) = -\frac{3}{r}\chi_1(U) - \frac{1}{r^2} - \frac{1}{c_2 r \Delta t} \tag{E.67}$$

$$\psi_2(U) = 0; \quad \chi_2(U) = 0 \tag{E.68}$$

$$\psi_{2,r}(U) = 0; \quad \chi_{2,r}(U) = 0; \tag{E.69}$$

$$\psi_1(T) = \frac{c_2^2}{2r^3}\left[B^2(\Delta t)^2 + \frac{1}{3}B^3(\Delta t)^2\right. \tag{E.70}$$

$$\left. -\left[A - \frac{r}{c_2\Delta t}\right]\left(\frac{r}{c_2}\right)^2 - \frac{1}{3}\left(\frac{r}{c_2}\right)^3 + \frac{1}{r}\left[A - \frac{r}{c_2\Delta t}\right]\right]$$

$$\chi_1(T) = 3\psi_1(T) - \frac{2}{r}\left[A - \frac{r}{c_2\Delta t}\right] \tag{E.71}$$

$$\psi_{1,r}(T) = -\frac{1}{r}\chi_1(T) - \frac{1}{r^2}A \tag{E.72}$$

$$\chi_{1,r}(T) = -\frac{3}{r}\chi_1(T) - \frac{1}{r^2}A \tag{E.73}$$

$$\psi_{1,rr}(T) = \frac{4}{r^2}\chi_1(T) + \frac{3}{r^3}A + \frac{1}{c_2^2 r(\Delta t)^2} \tag{E.74}$$

$$\chi_{1,rr}(T) = \frac{12}{r^2}\chi_1(T) + \frac{5}{r^3}A + \frac{1}{c_2^2 r(\Delta t)^2} \tag{E.75}$$

$$\psi_2(T) = \frac{c_2^2}{2r^3}\left(\left[B - \frac{2r}{3c_2\Delta t}\right]\left(\frac{r}{c_2}\right)^2 - \frac{1}{3}B^3(\Delta t)^2\right)$$ (E.76)

$$+\frac{1}{r}\left[\frac{r}{c_2\Delta t} - B\right]$$

$$\chi_2(T) = 3\psi_2(T) + \frac{2}{r}\left[B - \frac{r}{c_2\Delta t}\right]$$ (E.77)

$$\psi_{2,r}(T) = -\frac{1}{r}\chi_2(T) + \frac{1}{r^2}B$$ (E.78)

$$\chi_{2,r}(T) = -\frac{3}{r}\chi_2(T) + \frac{1}{r^2}B$$ (E.79)

$$\psi_{2,rr}(T) = \frac{4}{r^2}\chi_2(T) - \frac{3}{r^3}B - \frac{2}{c_2^2r(\Delta t)^2}$$ (E.80)

$$\chi_{2,rr}(T) = \frac{12}{r^2}\chi_2(T) - \frac{5}{r^3}B - \frac{2}{c_2^2r(\Delta t)^2}$$ (E.81)

$$\psi_3(U) = 0; \quad \chi_3(U) = 0$$ (E.82)

$$\psi_{3,r}(T) = 0; \quad \chi_{3,r}(T) = 0$$ (E.83)

$$\psi_{3,rr}(T) = \frac{1}{c_2^2r(\Delta t)^2}; \quad \chi_{3,rr}(T) = \frac{1}{c_2^2r(\Delta t)^2}$$ (E.84)

E.3 Fundamental Solutions for Laplace Transform Domain

The Laplace transforms of the fundamental solutions $\bar{U}_{ij}(\mathbf{X}, \mathbf{X}', s)$ and $\bar{T}_{ij}(\mathbf{X}, \mathbf{X}', s)$ in equation (5.36) can be obtained from (E.1) and (E.2):

$$\bar{U}_{ij}(\mathbf{X}', \mathbf{X}, s) = \frac{1}{4\pi\mu}(\psi\delta_{ij} - \chi r_{,i}r_{,j})$$ (E.85)

and

$$\bar{T}_{ij}(\mathbf{X}', \mathbf{X}, s) = \frac{1}{4\pi}\left[\left(\psi_{,r} - \frac{\chi}{r}\right)\left(\frac{\partial r}{\partial n}\delta_{ij} + r_{,j}n_i\right)\right.$$

$$-2\frac{\chi}{r}\left(n_j r_{,i} - 2r_{,i}r_{,j}\frac{\partial r}{\partial n}\right) - 2\chi_{,r}r_{,i}r_{,j}\frac{\partial r}{\partial n}$$

$$\left.+\left(\frac{c_1^2}{c_2^2} - 2\right)\left(\psi_{,r} - \chi_{,r} - 2\frac{\chi}{r}\right)r_{,i}n_j\right]$$ (E.86)

and solutions $\bar{S}_{kij}(\mathbf{X}, \mathbf{X}', s)$ and $\bar{D}_{kij}(\mathbf{X}, \mathbf{X}', s)$ can be obtained from (E.5) and (E.6):

$$\bar{D}_{kij}(\mathbf{X}, \mathbf{X}', s) = \frac{1}{4\pi}\left[2\left(\chi_{,r} - 2\frac{\chi}{r}\right)r_{,i}r_{,j}r_{,k} + 2\frac{\chi}{r}\delta_{ij}r_{,k}\right.$$

$$\left.-\left(\psi_{,r} - \frac{\chi}{r}\right)(\delta_{ik}r_{,j} + \delta_{jk}r_{,i}) - \frac{\lambda}{\mu}\left(\psi_{,r} - \chi_{,r} - 2\frac{\chi}{r}\right)\delta_{ij}r_{,k}\right]$$ (E.87)

and

$$\bar{S}_{kij}(\mathbf{X}, \mathbf{X}', s) = \frac{\mu}{4\pi} \left\{ \frac{\partial r}{\partial n} \left[4 \left(\chi_{,rr} - 5\frac{\chi_{,r}}{r} + 8\frac{\chi}{r^2} \right) r_{,i} r_{,j} r_{,k} \right.\right.$$

$$- \left(\psi_{,rr} - \frac{\psi_{,r}}{r} - 3\frac{\chi_{,r}}{r} + 6\frac{\chi}{r^2} \right) (\delta_{ik} r_{,j} + \delta_{jk} r_{,i})$$

$$+ 2 \left(2\frac{\chi_{,r}}{r} - 4\frac{\chi}{r^2} + \frac{\lambda}{\mu} \left(\chi_{,rr} + \frac{\chi_{,r}}{r} - 4\frac{\chi}{r^2} - \psi_{,rr} + \frac{\psi_{,r}}{r} \right) \right) \delta_{ij} r_{,k} \right]$$

$$+ 2 \left(2\frac{\chi_{,r}}{r} - 4\frac{\chi}{r^2} + \frac{\lambda}{\mu} \left(\chi_{,rr} + \frac{\chi_{,r}}{r} - 4\frac{\chi}{r^2} - \psi_{,rr} + \frac{\psi_{,r}}{r} \right) \right) r_{,i} r_{,j} n_k$$

$$- \left(\psi_{,rr} - \frac{\psi_{,r}}{r} - 3\frac{\chi_{,r}}{r} + 6\frac{\chi}{r^2} \right) (r_{,j} n_i + r_{,i} n_j) r_{,k}$$

$$+ \left(4\frac{\chi}{r^2} + \frac{\lambda}{\mu} \left(4\frac{\chi_{,r}}{r} + 8\frac{\chi}{r^2} - 4\frac{\psi_{,r}}{r} \right) \right.$$

$$+ \frac{\lambda^2}{\mu^2} \left(\chi_{,rr} + 4\frac{\chi_{,r}}{r} + 2\frac{\chi}{r^2} - \psi_{,rr} - 2\frac{\psi_{,r}}{r} \right) \right) \delta_{ij} n_k$$

$$\left. - 2 \left(\frac{\psi_{,r}}{r} - \frac{\chi}{r^2} \right) (\delta_{kj} n_i + \delta_{ki} n_j) \right\} \tag{E.88}$$

For three-dimensional problems, functions ψ and its derivatives are given by

$$\psi = \frac{e^{-z_2}}{r} + \frac{1+z_2}{(z_2)^2}\frac{e^{-z_2}}{r} - \frac{c_2^2}{c_1^2}\frac{1+z_1}{(z_1)^2}\frac{e^{-z_1}}{r} \tag{E.89}$$

$$\psi_{,r} = \frac{\partial \psi}{\partial r} = -\frac{\chi}{r} - (1+z_2)\frac{e^{-z_2}}{r^2} \tag{E.90}$$

$$\psi_{,rr} = 4\frac{\chi}{r^2} + \left(3 + 3z_2 + (z_2)^2\right)\frac{e^{-z_2}}{r^3} - \frac{c_2^2}{c_1^2}(1+z_1)\frac{e^{-z_1}}{r^3} \tag{E.91}$$

for function χ and its derivatives:

$$\chi = 3\psi - 2\frac{e^{-z_2}}{r} - \frac{c_2^2}{c_1^2}\frac{e^{-z_1}}{r} \tag{E.92}$$

$$\chi_{,r} = \frac{\partial \chi}{\partial r} = -3\frac{\chi}{r} - (1+z_2)\frac{e^{-z_2}}{r^2} + \frac{c_2^2}{c_1^2}(1+z_1)\frac{e^{-z_1}}{r^2}$$

$$- \frac{c_2^2}{c_1^2}\left(5 + 5z_1 + (z_2)^2\right)\frac{e^{-z_2}}{r^3} \tag{E.93}$$

$$\chi_{,rr} = 12\frac{\chi}{r^2} + \left(5 + 5z_2 + (z_2)^2\right)\frac{e^{-z_2}}{r^3} \tag{E.94}$$

where $z_1 = sr/c_1$ and $z_2 = sr/c_2$.

For two-dimensional problems ψ, and its derivatives are given by

$$\psi = \left[-\frac{c_2}{c_1}K_1(z_1) + K_1(z_2) \right]\frac{1}{z_2} + K_0(z_2) \tag{E.95}$$

$$\psi_{,r} = \left\{ \frac{c_2}{c_1} \Big[K_0(z_1)z_1 + 2K_1(z_1) \Big] \right.$$
$$\left. - \Big[K_0(z_2)z_2 + 2K_1(z_2) \Big] - K_1(z_2)z_2^2 \right\} \frac{1}{rz_2} \tag{E.96}$$

$$\psi_{,rr} = \left\{ -\frac{c_2}{c_1} \Big[3K_0(z_1)z_1 + K_1(z_1)(z_1^2 + 6) \Big] \right.$$
$$\left. + \Big[3K_0(z_2)z_2 + K_1(z_2)(z_2^2 + 6) \Big] + \Big[K_0(z_2)z_2 + K_1(z_2) \Big] z_2^2 \right\} \frac{1}{z_2 r^2} \tag{E.97}$$

for function χ and its derivatives:

$$\chi = -(\frac{c_2}{c_1})^2 \Big[K_0(z_1) + \frac{2}{z_1} K_1(z_1) \Big] + K_0(z_2) + \frac{2}{z_2} K_1(z_2) \tag{E.98}$$

$$\chi_{,r} = \left\{ (\frac{c_2}{c_1})^2 \Big[K_1(z_1)z_1 + 2[K_0(z_1) + \frac{2}{z_1} K_1(z_1)] \Big] \right.$$
$$\left. - \Big[K_1(z_2)z_2 + 2[K_0(z_2) + \frac{2}{z_2} K_1(z_2)] \Big] \right\} \frac{1}{r} \tag{E.99}$$

$$\chi_{,rr} = \left\{ -(\frac{c_2}{c_1})^2 \Big[K_0(z_1)z_1^2 + 3K_1(z_1)z_1 + 6[K_0(z_1) + \frac{2}{z_1} K_1(z_1)] \Big] \right.$$
$$\left. + \Big[K_0(z_2)z_2^2 + 3K_1(z_2)z_2 + 6[K_0(z_2) + \frac{2}{z_2} K_1(z_2)] \Big] \right\} \frac{1}{r^2} \tag{E.100}$$

The terms K_0 and K_1 are the modified Bessel functions of the second kind of orders zero and one, respectively.

E.3.1 Behaviour of Transformed Fundamental Solutions

The integration of the transformed fundamental solutions requires an understanding of their behaviour in the neighbourhood of the collocation point. The fundamental solutions can be expressed by functions ψ and χ and their derivatives with respect to r, as shown in the previous section.

The leading terms for the three-dimensional fundamental solutions can be obtained from (E.89)-(E.93) as

$$\lim_{r \to 0}(r\psi) = \frac{1}{2}\left(1 + \frac{c_2^2}{c_1^2}\right) \tag{E.101}$$

$$\lim_{r \to 0}(r\chi) = -\frac{1}{2}\left(1 - \frac{c_2^2}{c_1^2}\right) \tag{E.102}$$

$$\lim_{r \to 0}(r^2\psi_{,r}) = -\frac{1}{2}\left(1 + \frac{c_2^2}{c_1^2}\right) \tag{E.103}$$

$$\lim_{r \to 0}(r^2\chi_{,r}) = \frac{1}{2}\left(1 - \frac{c_2^2}{c_1^2}\right) \tag{E.104}$$

$$\lim_{r \to 0}(r^3\psi_{,rr}) = \frac{c_2^2}{c_1^2} - 1 \tag{E.105}$$

Substituting the dominant terms (E.89)-(E.93) and (E.94)-(E.100) into their corresponding fundamental solutions gives the fundamental solutions for elastostatics (see Chapter 2). Furthermore, from these limits, it is clear that the singularities in the fundamental solutions are exactly the same as in the static problem. Thus, the transformed fundamental solutions and kernel functions can be written in two parts:

$$\bar{U}_{ij}(\mathbf{x}',\mathbf{x},s) = U_{ij}^{static}(\mathbf{x}',\mathbf{x})+\bar{U}_{ij}^{m}(\mathbf{x}',\mathbf{x},s) \tag{E.106}$$

$$\bar{T}_{ij}(\mathbf{x}',\mathbf{x},s) = T_{ij}^{static}(\mathbf{x}',\mathbf{x})+\bar{T}_{ij}^{m}(\mathbf{x}',\mathbf{x},s) \tag{E.107}$$

and

$$\bar{D}_{kij}(\mathbf{x}',\mathbf{x},s) = D_{kij}^{static}(\mathbf{x}',\mathbf{x})+\bar{D}_{kij}^{m}(\mathbf{x}',\mathbf{x},s) \tag{E.108}$$

$$\bar{S}_{ij}(\mathbf{x}',\mathbf{x},s) = S_{kij}^{static}(\mathbf{x}',\mathbf{x})+\bar{S}_{kij}^{m}(\mathbf{x}',\mathbf{x},s) \tag{E.109}$$

where $\bar{U}_{ij}^{m}$, $\bar{T}_{ij}^{m}$, $\bar{D}_{kij}^{m}$ and $\bar{S}_{kij}^{m}$ denote the modified terms, which are functions of the Laplace transformed parameter and contain only weak singularities.

These functions for two dimensional problems depend upon modified zero- and first-order Bessel functions of the second kind. When the parameter of the Bessel function $z = sr/c_\alpha$ tends to zero (s or r tends to zero, or c_α tends to infinity), the Bessel functions have the following forms:

$$K_0(z) = -\ln\left(\frac{z}{2}\right) - \gamma + O[z^2 \ln(z)] \tag{E.110}$$

$$K_1(z) = \frac{1}{z} + \frac{z}{2}\left[\ln(\frac{z}{2}) + \gamma - \frac{1}{2}\right] + O[z^3 \ln(z)] \tag{E.111}$$

where $\gamma = 0.5772156649$.

The functions ψ and χ and their derivatives, expressed by Bessel functions for small parameters, defined by equations (E.110) and (E.111), have the following form:

$$\psi = -\frac{1}{2}\left[1 + \left(\frac{c_2}{c_1}\right)^2\right]\ln r - \frac{1}{2}\left[\ln\left(\frac{s}{2c_2}\right) + \left(\frac{c_2}{c_1}\right)^2 \ln\left(\frac{s}{2c_1}\right)\right]$$
$$-\frac{1}{2}\gamma\left[1 + \left(\frac{c_2}{c_1}\right)^2\right] - \frac{1}{4}\left[1 - \left(\frac{c_2}{c_1}\right)^2\right] + O[z^2 \ln(z)] \tag{E.112}$$

$$\chi = -\frac{1}{2}\left[1 - \left(\frac{c_2}{c_1}\right)^2\right] + O[z^2 \ln(z)] \tag{E.113}$$

$$\psi_{,r} = -\frac{1}{2r}\left[1 + \left(\frac{c_2}{c_1}\right)^2\right] - \frac{z_2^2}{2r}\left[\ln\left(\frac{z_2}{2}\right) + \gamma - \frac{1}{2}\right] + O\left[\frac{z^2}{r}\ln(z)\right] \tag{E.114}$$

$$\chi_{,r} = \left(\frac{c_2}{c_1}\right)^2 \frac{z_2^2}{2r}\left[\ln\left(\frac{z_2}{2}\right) + \gamma - \frac{1}{2}\right] - \frac{z_2^2}{2r}\left[\ln\left(\frac{z_2}{2}\right) + \gamma - \frac{1}{2}\right] + O\left[\frac{z^2}{r}\ln(z)\right] \tag{E.115}$$

$$\psi_{,rr} = \frac{1}{2r^2}\left[1 + \left(\frac{c_2}{c_1}\right)^2\right] - \frac{1}{r^2}\left\{\left(\frac{c_2}{c_1}\right)^2 \frac{z_1^2}{2}\left[\ln\left(\frac{z_2}{2}\right) + \gamma - \frac{1}{2}\right] + \frac{z_2^2}{2}\right\} + O\left[\frac{z^2}{r}\ln(z)\right] \tag{E.116}$$

$$\chi_{,rr} = \left\{ -\left(\frac{c_2}{c_1}\right)^2 \frac{z_2^2}{2}\left[\ln\left(\frac{z_2}{2}\right) + \gamma - \frac{3}{2}\right] + \frac{z_2^2}{2}\left[\ln\left(\frac{z_2}{2}\right) + \gamma - \frac{3}{2}\right]\right\} + O\left[\frac{z^2}{r}\ln(z)\right]$$

(E.117)

The dominant terms of the functions have the following forms:

$$\psi = -\frac{1}{2}\left[1 + \left(\frac{c_2}{c_1}\right)^2\right]\ln r + E_s + O(1)$$

$$= -\frac{3-4\nu}{4(1-\nu)}\ln(r) + E_s + O(1) \tag{E.118}$$

$$\chi = -\frac{1}{2}\left[1 - \left(\frac{c_2}{c_1}\right)^2\right] + O[z^2\ln(z)]$$

$$= -\frac{1}{4(1-\nu)} + O[z^2\ln(z)] \tag{E.119}$$

$$\psi_{,r} = -\frac{1}{2r}\left[1 + \left(\frac{c_2}{c_1}\right)^2\right] + O\left[\frac{z^2}{r}\ln(z)\right]$$

$$= -\frac{3-4\nu}{4(1-\nu)}\frac{1}{r} + O\left[\frac{z^2}{r}\ln(z)\right] \tag{E.120}$$

$$\chi_{,r} = O\left[\frac{z^2}{r}\ln(z)\right] \tag{E.121}$$

$$\psi_{,rr} = \frac{1}{2r^2}\left[1 + \left(\frac{c_2}{c_1}\right)^2\right] + O\left[\frac{z^2}{r}\ln(z)\right]$$

$$= \frac{3-4\nu}{4(1-\nu)}\frac{1}{r} + O\left[\frac{z^2}{r}\ln(z)\right] \tag{E.122}$$

$$\chi_{,rr} = O\left[\frac{z^2}{r}\ln(z)\right] \tag{E.123}$$

where

$$E_s = -\frac{1}{2}\left[\ln\left(\frac{s}{2c_2}\right) + \left(\frac{c_2}{c_1}\right)^2\ln\left(\frac{s}{2c_2}\right)\right]$$

The terms containing $\ln(r)$, $1/r$ and $1/r^2$ are singular as $r \to 0$, and the term E_s is singular as $s \to 0$; the remaining terms which are of $O[z^2\ln(z)]$ are not singular as $z \to 0$. It is worth noting that for the two-dimensional fundamental solution of elastostatics, U_{ij} differs from the transformed solution $\bar{U}_{ij}$ by a constant because of the term $E_s + O(1)$. This additional term has no influence on the limiting solution for finite domains. For infinite domains the static solution is obtained if the applied traction along the internal boundaries are in equilibrium. Therefore, the static solution can be obtained from the present method by assuming, for example, a vanishingly small mass density of the material, thereby decreasing the parameter z of the Bessel function towards zero. It is also worth noting that, because the transformed and static fundamental solutions are singular, the calculation of their values and subsequent subtraction can lead to large round-off errors, particularly for integration points close to collocation points. In order to obtain accurate integrals, the dominant terms are subtracted from the Bessel functions. The exact differences between transformed and static fundamental solutions are obtained after modifications have been applied to functions $\psi, \psi_{,r}, \psi_{,rr}, \chi, \chi_{,r}$ and $\chi_{,rr}$.

References

[1] Dominguez, J., *Boundary Elements in Dynamics*, Computational Mechanics Publications, Southampton 1993.

[2] Fedelinski, P., Aliabadi, M.H. and Rooke, D.P., A single region time domain BEM for dynamic crack problems, *International Journal Solids and Structures*, **32**, 3555-3571, 1995.

[3] Fedelinski, P., Aliabadi, M.H. and Rooke, D.P., The Laplace transform DBEM method for mixed-mode dynamic crack analysis, *Computers and Structures*, **59**, 1021-1031, 1996.

[4] Israil, A.S.M. & Banerjee, P.K., Interior stress calculations in 2-D time-domain transient BEM analysis, *International Journal of Solids and Structures*, **27**, 915-927, 1991.

[5] Wen,P.H., Aliabadi,M.H. and Rooke,D.P., The influence of waves on dynamic stress intensity factors (Three Dimensional Problems), *Archives of .Applied. Mechanics.* ,**66**,385-394, 1996.

[6] Wen, P.H., Aliabadi, M.H. and Rooke, D.P., Cracks in three dimensions: a dynamic dual boundary element analysis, *Computer Methods in Applied Mechanics and Engineering*, **167**, 139-151, 1998.

[7] Wen, P.H., Aliabadi, M.H. and Young, A., A time-dependent formulation of dual boundary element method for 3D dynamic crack problems, *International Journal of Numerical Methods in Engineering*, **45**, 1887-1905, 1999.

Appendix F

Shape Functions of Brick Cell Elements

F.1   Continuous cells

The 27-noded isoparametric cell is shown in Figure F.1(a). In the case of continuous cells, functionality as well as geometry are interpolated using the same set of shape functions. These shape functions are given by

$$\Phi^1 = \frac{1}{8}\zeta_1(\zeta_1 - 1)\zeta_2(\zeta_2 - 1)\zeta_3(\zeta_3 - 1)$$

$$\Phi^2 = \frac{1}{4}(1 - \zeta_1^2)\zeta_2(\zeta_2 - 1)\zeta_3(\zeta_3 - 1)$$

$$\Phi^3 = \frac{1}{8}\zeta_1(\zeta_1 + 1)\zeta_2(\zeta_2 - 1)\zeta_3(\zeta_3 - 1)$$

$$\Phi^4 = \frac{1}{4}\zeta_1(\zeta_1 + 1)(1 - \zeta_2^2)\zeta_3(\zeta_3 - 1)$$

$$\Phi^5 = \frac{1}{8}\zeta_1(\zeta_1 + 1)\zeta_2(\zeta_2 + 1)\zeta_3(\zeta_3 - 1)$$

$$\Phi^6 = \frac{1}{4}(1 - \zeta_1^2)\zeta_2(\zeta_2 + 1)\zeta_3(\zeta_3 - 1)$$

$$\Phi^7 = \frac{1}{8}\zeta_1(\zeta_1 - 1)\zeta_2(\zeta_2 + 1)\zeta_3(\zeta_3 - 1)$$

$$\Phi^8 = \frac{1}{4}\zeta_1(\zeta_1 - 1)(1 - \zeta_2^2)\zeta_3(\zeta_3 - 1)$$

$$\Phi^9 = \frac{1}{2}(1 - \zeta_1^2)(1 - \zeta_2^2)\zeta_3(\zeta_3 - 1)$$

$$\Phi^{10} = \frac{1}{4}\zeta_1(\zeta_1 - 1)\zeta_2(\zeta_2 - 1)(1 - \zeta_3^2)$$

$$\Phi^{11} = \frac{1}{2}(1 - \zeta_1^2)\zeta_2(\zeta_2 - 1)(1 - \zeta_3^2)$$

$$\Phi^{12} = \frac{1}{4}\zeta_1(\zeta_1 + 1)\zeta_2(\zeta_2 - 1)(1 - \zeta_3^2)$$

$$\Phi^{13} = \frac{1}{2}\zeta_1(\zeta_1 + 1)(1 - \zeta_2^2)(1 - \zeta_3^2)$$

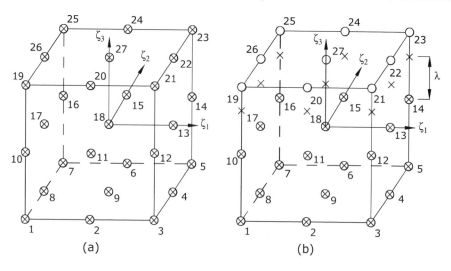

Figure F.1: Continuous (a) and face-discontinuos (b) cells.

$$\Phi^{14} = \frac{1}{4}\zeta_1(\zeta_1 + 1)\zeta_2(\zeta_2 + 1)(1 - \zeta_3^2)$$

$$\Phi^{15} = \frac{1}{2}(1 - \zeta_1^2)\zeta_2(\zeta_2 + 1)(1 - \zeta_3^2)$$

$$\Phi^{16} = \frac{1}{4}\zeta_1(\zeta_1 - 1)\zeta_2(\zeta_2 + 1)v(1 - \zeta_3^2)$$

$$\Phi^{17} = \frac{1}{2}\zeta_1(\zeta_1 - 1)(1 - \zeta_2^2)(1 - \zeta_3^2)$$

$$\Phi^{18} = (1 - \zeta_1^2)(1 - \zeta_2^2)(1 - \zeta_3^2)$$

$$\Phi^{19} = \frac{1}{8}\zeta_1(\zeta_1 - 1)\zeta_2(\zeta_2 - 1)\zeta_3(\zeta_3 + 1)$$

$$\Phi^{20} = \frac{1}{4}(1 - \zeta_1^2)\zeta_2(\zeta_2 - 1)\zeta_3(\zeta_3 + 1)$$

$$\Phi^{21} = \frac{1}{8}\zeta_1(\zeta_1 + 1)\zeta_2(\zeta_2 - 1)\zeta_3(\zeta_3 + 1)$$

$$\Phi^{22} = \frac{1}{4}\zeta_1(\zeta_1 + 1)(1 - \zeta_2^2)\zeta_3(\zeta_3 + 1)$$

$$\Phi^{23} = \frac{1}{8}\zeta_1(\zeta_1 + 1)\zeta_2(\zeta_2 + 1)\zeta_3(\zeta_3 + 1)$$

$$\Phi^{24} = \frac{1}{4}(1 - \zeta_1^2)\zeta_2(\zeta_2 + 1)\zeta_3(\zeta_3 + 1)$$

$$\Phi^{25} = \frac{1}{8}\zeta_1(\zeta_1 - 1)\zeta_2(\zeta_2 + 1)\zeta_3(\zeta_3 + 1)$$

$$\Phi^{26} = \frac{1}{4}\zeta_1(\zeta_1 - 1)(1 - \zeta_2^2)\zeta_3(\zeta_3 + 1)$$

$$\Phi^{27} = \frac{1}{2}(1 - \zeta_1^2)(1 - \zeta_2^2)\zeta_3(\zeta_3 + 1) \tag{F.1}$$

F.2   Face-discontinuous cells

The face-discontinuous cells are used in crack problems as discussed in Chapter 8. Geometrical shape functions, are the same to those used for continuous cells and given in expressions (F.1). The face $\zeta_3 = 1$ is assumed to be the discontinuous face, functional shape functions Φ^α are given by [1]

$$\Phi^1 = \frac{1}{4}\zeta_1(\zeta_1 - 1)\zeta_2(\zeta_2 - 1)\zeta_3\frac{(\zeta_3 - 1)}{(1 + \lambda)}$$

$$\Phi^2 = \frac{1}{2}(1 - \zeta_1^2)\zeta_2(\zeta_2 - 1)\zeta_3\frac{(\zeta_3 - 1)}{(1 + \lambda)}$$

$$\Phi^3 = \frac{1}{4}\zeta_1(\zeta_1 + 1)\zeta_2(\zeta_2 - 1)\zeta_3\frac{(\zeta_3 - 1)}{(1 + \lambda)}$$

$$\Phi^4 = \frac{1}{2}\zeta_1(\zeta_1 + 1)(1 - \zeta_2^2)\zeta_3\frac{(\zeta_3 - 1)}{(1 + \lambda)}$$

$$\Phi^5 = \frac{1}{4}\zeta_1(\zeta_1 + 1)\zeta_2(\zeta_2 + 1)\zeta_3\frac{(\zeta_3 - 1)}{(1 + \lambda)}$$

$$\Phi^6 = \frac{1}{2}(1 - \zeta_1^2)\zeta_2(\zeta_2 + 1)\zeta_3\frac{(\zeta_3 - 1)}{(1 + \lambda)}$$

$$\Phi^7 = \frac{1}{4}\zeta_1(\zeta_1 - 1)\zeta_2(\zeta_2 + 1)\zeta_3\frac{(\zeta_3 - 1)}{(1 + \lambda)}$$

$$\Phi^8 = \frac{1}{2}\zeta_1(\zeta_1 - 1)(1 - \zeta_2^2)\zeta_3\frac{(\zeta_3 - 1)}{(1 + \lambda)}$$

$$\Phi^9 = (1 - \zeta_1^2)(1 - \zeta_2^2)\zeta_3\frac{(\zeta_3 - 1)}{(1 + \lambda)}$$

$$\Phi^{10} = -\frac{1}{4}\zeta_1(\zeta_1 - 1)\zeta_2(\zeta_2 - 1)\frac{(\zeta_3 + 1)(\zeta_3 - \lambda)}{\lambda}$$

$$\Phi^{11} = -\frac{1}{2}(1 - \zeta_1^2)\zeta_2(\zeta_2 - 1)\frac{(\zeta_3 + 1)(\zeta_3 - \lambda)}{\lambda}$$

$$\Phi^{12} = -\frac{1}{4}\zeta_1(\zeta_1 + 1)\zeta_2(\zeta_2 - 1)\frac{(\zeta_3 + 1)(\zeta_3 - \lambda)}{\lambda}$$

$$\Phi^{13} = -\frac{1}{2}\zeta_1(\zeta_1 + 1)(1 - \zeta_2^2)\frac{(\zeta_3 + 1)(\zeta_3 - \lambda)}{\lambda}$$

$$\Phi^{14} = -\frac{1}{4}\zeta_1(\zeta_1 + 1)\zeta_2(\zeta_2 + 1)\frac{(\zeta_3 + 1)(\zeta_3 - \lambda)}{\lambda}$$

$$\Phi^{15} = -\frac{1}{2}(1 - \zeta_1^2)\zeta_2(\zeta_2 + 1)\frac{(\zeta_3 + 1)(\zeta_3 - \lambda)}{\lambda}$$

$$\Phi^{16} = -\frac{1}{4}\zeta_1(\zeta_1 - 1)\zeta_2(\zeta_2 + 1)\frac{(\zeta_3 + 1)(\zeta_3 - \lambda)}{\lambda}$$

$$\Phi^{17} = -\frac{1}{2}\zeta_1(\zeta_1 - 1)(1 - \zeta_2^2)\frac{(\zeta_3 + 1)(\zeta_3 - \lambda)}{\lambda}$$

$$\Phi^{18} = -(1 - \zeta_1^2)(1 - \zeta_2^2)\frac{(\zeta_3 + 1)(\zeta_3 - \lambda)}{\lambda}$$

$$\Phi^{19} = \frac{1}{4}\zeta_1(\zeta_1 - 1)\zeta_2(\zeta_2 - 1)\frac{(\zeta_3 + 1)\zeta_3}{(\lambda + 1)\lambda}$$

$$\Phi^{20} = \frac{1}{2}(1 - \zeta_1^2)\zeta_2(\zeta_2 - 1)\frac{(\zeta_3 + 1)\zeta_3}{(\lambda + 1)\lambda}$$

$$\Phi^{21} = \frac{1}{4}\zeta_1(\zeta_1 + 1)\zeta_2(\zeta_2 - 1)\frac{(\zeta_3 + 1)\zeta_3}{(\lambda + 1)\lambda}$$

$$\Phi^{22} = \frac{1}{2}\zeta_1(\zeta_1 + 1)(1 - \zeta_2^2)\frac{(\zeta_3 + 1)\zeta_3}{(\lambda + 1)\lambda}$$

$$\Phi^{23} = \frac{1}{4}\zeta_1(\zeta_1 + 1)\zeta_2(\zeta_2 + 1)\frac{(\zeta_3 + 1)\zeta_3}{(\lambda + 1)\lambda}$$

$$\Phi^{24} = \frac{1}{2}(1 - \zeta_1^2)\zeta_2(\zeta_2 + 1)\frac{(\zeta_3 + 1)\zeta_3}{(\lambda + 1)\lambda}$$

$$\Phi^{25} = \frac{1}{4}\zeta_1(\zeta_1 - 1)\zeta_2(\zeta_2 + 1)\frac{(\zeta_3 + 1)\zeta_3}{(\lambda + 1)\lambda}$$

$$\Phi^{26} = \frac{1}{2}\zeta_1(\zeta_1 - 1)(1 - \zeta_2^2)\frac{(\zeta_3 + 1)\zeta_3}{(\lambda + 1)\lambda}$$

$$\Phi^{27} = (1 - \zeta_1^2)(1 - \zeta_2^2)\frac{(\zeta_3 + 1)\zeta_3}{(\lambda + 1)\lambda} \tag{F.2}$$

References

[1] Cisilino, A.P., Aliabadi, M.H. and Otegui, J.L., A three-dimensional boundary element formulation for the elastoplastic analysis of cracked bodies, *International Journal of Numerical Methods for Engineering*, **42**, 237-256, 1998.

Appendix G

Fast Solver For Contact Problems

G.1 LU decomposition

Suppose the matrix $[D]$ is decomposed into a product of two matrices, such that

$$D = L \cdot U \tag{G.1}$$

where L is lower triangular (i.e. has elements only on the diagonal and below) and U is upper triangular (ie. has elements only on the diagonal and above). An important advantage of the LU decomposition method is that it is particularly efficient for solutions with any number of right-hand sides, and for matrix inversion. It will be shown next that these features can be exploited in the high speed matrix solver for contact problems.

G.1.1 High Speed Solver for Contact Problems

For most contact problems, contact regions are usually small, therefore the coefficient matrix D_c in equation (7.9) is also small. D_c can be updated from stage to stage very efficiently, as described in Section 2; therefore relatively little time is used for re-establishing sub-matrix D_c. However, the process for solving the system of equations takes the most time. In order to overcome this problem, the unique feature of the structured matrix is exploited so that time is saved by avoiding reassembly of the entire matrix when only a few changes are made to the contact region. Most importantly, the solution of the matrix equations can be obtained using a new high speed matrix solver without re-solving the entire system of equations. The high speed solver technique is based on the same principle as the Sherman-Morrison and Woodbury formulas [2], where the solution of a square matrix D_0 (original matrix) is 're-used' to obtain the solution of matrix D_s (updated matrix). This method is particularly efficient if matrix D_s has only a few changes when compared with the original matrix D_0. In fact, this is exactly the situation when contact conditions inside the contact zone have to be changed from one iteration to the next. The changes in the coefficient matrix only occur in D_c, and they are usually very few when compared with the rest of the system matrix.

The above technique has been used successfully to solve contact problems [1, 3]. It has been shown that a general change in matrix D_i can be expressed in terms of the previous matrix D_{i-1} as

$$D_i = D_{i-1} + e_i v_i^T \tag{G.2}$$

where e_i is a column vector which contains only one component that is 'unity' and all the others are 'zero' and v_i is a column vector which contains the difference between the components in the original and those in the new matrix. The superscript T means transpose. The inverse of matrix D_i can be expressed as

$$D_i^{-1} = D_{i-1}^{-1} - \frac{D_{i-1}^{-1} e_i v_i^T D_{i-1}^{-1}}{1 + v_i^T D_{i-1}^{-1} e_i} \tag{G.3}$$

General solutions of the systems of equations

$$D_i x^i = b \quad \text{and} \quad D_i y_j^i = e_j \tag{G.4}$$

are given by

$$x^i = D_i^{-1} b \quad \text{and} \quad y_j^i = D_i^{-1} e_j \tag{G.5}$$

respectively. Thus, from (G.3),

$$
\begin{aligned}
x^i = D_i^{-1} b &= D_{i-1}^{-1} b - \frac{D_{i-1}^{-1} e_i v_i^T D_{i-1}^{-1} b}{1 + v_i^T D_{i-1}^{-1} e_i} \\
&= x^{i-1} - \left[\frac{y_i^{i-1} v_i^T}{1 + v_i^T y_i^{i-1}} \right] x^{i-1} \\
&= x^{i-1} - C(i-1) x^{i-1} \tag{G.6}
\end{aligned}
$$

Similarly

$$y_j^i = D_i^{-1} e_j = y_j^{i-1} - C(i-1) y_j^{i-1} \tag{G.7}$$

Generally, when there are many changes, then D_s can be represented by

$$D_s = D_0 + e_1 v_1^T + e_2 v_2^T + \dots + e_s v_s^T \tag{G.8}$$

where s is the total number of changes in going from matrix D_0 to D_s. The solution for a total of s changes has to be deduced in successive steps for all the changes occurring in matrix D_s. Initially, the solution of the original system of equations is given by the

Zero Order

$$
\begin{aligned}
D_0 x^0 &= b, \\
x^0 &= D_0^{-1} b
\end{aligned}
$$

The solution of the original system of equations with one single change is given by the

$\boxed{1^{st} \text{ Order}}$

$$(D_0 + e_1 v_1^T)x^1 = b$$
$$D_1 x^1 = b$$
$$x^1 = x^0 - \left[\frac{y_1^0 v_1^T}{1 + v_1^T y_1^0}\right]x^0$$
$$\equiv x^0 - C(0)x^0$$

where

$$y_1^0 = D_0^{-1} e_1$$

The solution of the original system of equations with two changes is given by the

$\boxed{2^{nd} \text{ Order}}$

$$(D_0 + e_1 v_1^T + e_2 v_2^T)x^2 = b$$
$$(D_1 + e_2 v_2^T)x^2 = b$$
$$D_2 x^2 = b$$
$$x^2 = x^1 - \left[\frac{y_2^1 v_2^T}{1 + v_2^T y_2^1}\right]x^1$$
$$\equiv x^1 - C(1)x^1$$

where

$$y_2^1 = D_0^{-1} e_2 - C(0)D_0^{-1} e_2$$
$$y_2^1 = y_2^0 - C(0)y_2^0$$

and

$$y_2^0 = D_0^{-1} e_2$$

In general, the solution of the original system of equations with a total of s changes is given by the

$\boxed{s^{th} \text{ Order}}$

$$D_s x^{(s)} = b,$$

ie.

$$(D_{s-1} + e_s v_s^T)x^s = b.$$
$$x^s = x^{s-1} - C(s-1)x^{s-1},$$

where

$$C(s-1) = \frac{y_s^{s-1} v_s^T}{1 + v_s^T y_s^{s-1}},$$
$$y_s^{k-1} = y_s^{k-2} - C(k-2)y_s^{k-2}, \quad and \quad k = 2, ..., s$$

and

$$y_s^{(0)} = D_0^{-1} e_s.$$

For clarity, the computational process was explained with an example by Man *et al.* [3]. Consider a case where there are three changes made from matrix D_0 to D_s. From equation (G.2), D_s can be expressed as

$$D_s = D_0 + e_1 v_1^T + e_2 v_2^T + e_3 v_3^T \tag{G.9}$$

and

$$[D_s]\{x^s\} = \{b \tag{G.10}$$
$$[D_0 + e_1 v_1^T + e_2 v_2^T + e_3 v_3^T]\{x^3\} = \{b\} \tag{G.11}$$

The necessary steps to obtain the solution vector $x^{(3)}$ for this problem are illustrated in Table G.1.

Table G.1: Steps to obtain the solution vector, $A_0 = \frac{y^0 \nu_1^T}{1+v_1^T y_1^0}$, $A_1 = \frac{y_2^0 \nu_1^T}{1+v_1^T y_2^0}$ and $A_2 = \frac{y_3^0 \nu_1^T}{1+v_1^T y_3^0}$.

Step 0	Step 1	Step 2	Step 3
$LUx^0 = b$			
$LUy_1^0 = e_1$	$x^1 = x^0 - C(0)x^0$		
$LUy_2^0 = e_2$	$y_2^1 = y_2^0 - C(0)y_2^0$	$x^2 = x^1 - C(0)x^1$	
$LUy_3^0 = e_3$	$y_3^1 = y_3^0 - C(0)y_3^0$	$y_3^2 = y_3^1 - C(0)y_3^1$	$x^3 = x^2 - C(2)x^2$
$C(0) = A_0$	$C(1) = A_1$	$C(2) = A_2$	Finished
Solution: x^0	Solution: x^1	Solution: x^2	Solution: x^3

Here, x^0 is solved in Step (0) by LU decomposition. It is the solution vector of the initial problem. This step takes the longest, since matrix D_0 has to be decomposed into an L and a U matrix. However, the decomposition is only carried out once for the entire solution operation. The same LU is then repeatedly used to produce 'contribution' vectors, y_1^0, y_2^0 and y_3^0. Finally, the constant $C(0)$ has to be computed, before proceeding to the next step. Once step (0) is completed, all the remaining steps can easily be accomplished, with each intermediate step taking the intermediate solution closer to the final solution.

References

[1] Ezawa, Y. and Okamoto, N., High speed boundary element contact stress analysis using a super computer. *Proc. of the 4th International Conference* on Boundary Element Technology, Computational Mechanics Publications, Southampton, 405-416, 1989.

[2] Press,W.H., Teukolsky,S.A., Vetterling,W.T. and Flanner,B.P., *Numerical Recipes*- The Art of Scientific Computing, Cambridge University Press, Cambridge, 66-70, 1990

[3] Man, K., Aliabadi, M.H. and Rooke, D.P., BEM Frictional contact analysis: An incremental loading technique, *Computers and Structures*, **47**, 893-905, 1993.

Index

Acoustics scattering, 219-222, 439-441, 525

Airy stress function, 525

Aluminium alloy, 254, 258, 389, 474

Angular velocity, 64

Anisotropy, 74, 319, 331, 343, 359, 375, 461

Axisymmetric problems, 33, 37

Bauschinger effect, 232

Bending stiffness, 108, 122

Bessel function, 111, 222, 564, 565

Betti's reciprocal theorem, 28-29

Biharmonic equation, 525

Body forces, 38, 61-70, 129, 167, 196, 244, 462, 471, 472

Bone
  remodelling, 449
  tibial, 4, 92-95

Bony ingrowth, 452

Boundary conditions
  clamped, 127, 138, 143, 468
  flux, 164, 174
  free, 127
  impact, 211, 380
  simply supported, 127, 134, 145, 481
  thermal shock, 187

Cantilever beam, 136

Cartesian tensor notation, 5-7

Cauchy principal value, 35, 36, 40, 57, 112, 166, 171, 230, 241, 331, 509, 510

Classical plate theory, 103, 116, 147, 148

Coefficient of linear thermal expansion, 162, 186

Cohesive forces, 391

Cold expansion, 229, 257-260

Collocation, 50, 53, 69, 82, 127, 171, 200, 202, 208, 241, 246, 301, 329, 330, 337

Compatibility, 23

Composite material, 331, 359, 375, 384, 461

Concrete beam, 478, 391

Conforming contact, 270, 293

Contact problems, 269-313

Corner force, 150

Corner problem, 71-73

Coupling BEM and FEM, 81-82

Crack
  centre, 372, 373, 379
  edge, 372, 375
  elliptical, 380, 396
  Green's function, 343-344
  growth, 383-406
  interface, 382
  mouth opening displacement, 391
  opening displacement, 350, 376, 381, 391
  penny, 377

Creep, 230

Cruciform plate, 386-388

Curvature, 108, 116, 117, 122, 134, 143, 147

Cylindrical shell, 147, 483

Density, 162, 196

Design variable, 425, 427, 429

Diagonal coefficients, 51, 150

Dirac delta function, 8

Direction cosines, 19, 20, 59, 178

Discontinuous elements, 72, 333-336

Displacement
  extrapolation, 346
  integral equation, 29, 328
  out of plane, 105, 109, 110, 113, 117, 121, 123, 151, 373, 401

Displacement discontinuity method, 348

Discretization, 42, 245, 277, 301

Divergence theorem, 8

Dual boundary elements, 330-342, 372, 373, 375, 377, 379, 380, 383, 384, 392, 443-447, 448

Dual reciprocity method, 67-70, 88, 129-130, 204, 208, 215, 470, 483, 485
Durbin method, 203
Dynamic stress intensity factor, 382

Elastic-perfectly plastic, 230, 231
Element subdivision technique, 246, 492, 493-494
Elliptic integral, 34
Equilibrium, 16, 18, 19, 71, 77, 108, 113, 120, 133, 162, 174, 237, 279, 303, 451, 464, 525
Explicit procedures, 253

Fast Fourier transform, 203
Fast solvers, 573-576
Fatigue, 257, 321, 325, 390, 395
Fictitious crack, 391
Fictitious stress method, 348
Field point, 32, 37, 76, 77, 79, 111, 240
Finite difference method, 208, 431, 432, 435
Finite element method, 6, 81, 87, 104, 141, 229, 245, 265, 391, 401
Flat punch, 284, 285
Flux
    at boundary points, 437, 449,
    derivatives, 145
    integral equation, 445, 496, 508
Foundation plates, 114
Fourier transform, 75, 195, 202, 203
Fracture toughness, 375
Fredholm integral equations, 1
Friction, 270
Fundamental solution,
    anisotropic, 75-76
    axisymmetric, 33-34
    Helmholtz equation, 440-441
    Kelvin, 30-34
    Laplace equation, 427
    Laplace transform, 562-564
    plate bending, 111, 112-113, 115, 150

Gauss-Legendre formula, 491, 493
Gauss quadrature, 491-493
Graphite- epoxy laminate, 376

Hadamard
    finite parts, 509, 517
    principal value, 510, 518
Hardening, 231
Heat capacity, 162

Heat conduction, 162
Helmholtz equation, 220
Hertzian contact, 282, 288
Hessian matrix, 426
Hölder continuity, 124, 337, 341, 535
Hooke's law, 24
Hypersingular
    formulation, 40, 131, 161, 187
    integral equation, 131, 132, 380
    integral, 41, 165, 166, 246, 247, 336, 339, 340, 517-519

Identity matrix, 427
Incident angle, 441
Incremental loading, 252, 253, 278, 279, 291
Infinite regions, 38
Initial stress approach, 244
Interpolation functions,
    constant cell, 105, 138, 143
    constant element, 43
    constant time, 161, 176, 180, 181, 184, 187
    crack tip, 351-352
    cubic, 49
    linear time, 181, 184
    eight-node, 48-49
    linear, two-dimensional, 44
    nine-node, 49
    quadratic, two-dimensional, 44
    triangular, 47
Implant, 452, 455
Implicit differentiation method, 425, 431, 433, 435
Implicit procedures, 252
Inviscid fluid, 439
Isotropic hardening, 231
    Isotropic material, 24, 116, 150, 162

Jacobian, 38, 45, 50, 51, 58
J-integral method, 358-366
Jump term, 36, 40, 110, 124, 125, 128, 132, 437, 439, 539

Kelvin solution, 30-33
Kinematic hardening, 232
Kirchhoff plate formulation, 148-153
Kronecker delta, 6

Lagrangian shape functions, 44
Lamé constants, 24
Laplace

domain, 207
equation, 168
inversion method, 203
operator, 149
transform, 161, 163, 195, 201, 202, 211, 380, 562
Leibnitz formula, 241
LU decomposition, 573

Mass density, 211
Maximum stress criterion, 326-327, 385
Mixed hardening, 232
Modes of
contact, 271-272
fracture, 320-321
Moments
bending, 106, 109, 120, 121, 135, 148, 469
twisting, 106, 119, 120
Multiaxial loading, 233
Multi-region, 70, 80, 82, 113, 382
Multiple reciprocity method, 67

Navier's equation, 26
Nearly singular integral, 246, 493-497
Non-conforming
contact, 271, 286,288
discretization, 301-308
Non-homogeneous problem, 70
Nonlinear material behaviour, 230-239
Normal
direction, 6
gap, 272
Numerical integration, 491-520

Optimization, 426
Orthotropic, 59
Orthotropic system, 59

Paris law, 395
Particular integrals, 65-66
Particular solutions, 66-68, 112, 129, 205, 207, 547-554
Pasternak foundation, 10, 105, 115
Penalty function method, 269
Plastic modulus, 232
Plate bending, 105-115, 148-154
Polar coordinates, 37, 75, 247, 320, 324, 441
Polynomial influence function, 371
Poroelasticity, 161
Possion's ratio, 25

Potential integral equation, 427
Prandtl-Reuss equation, 235, 236
Pressurized cylinder, 83
Prestressing, 257
Progressive contact, 271
Pseudo body force approach, 244

Quarter point element
two-dimensional, 348-349
three-dimensional, 350-351
Quasi-Newton method, 426

Radial basis functions, 129, 548
Receding contact, 271
Reciprocal theorem, 26
Regularity condition, 38,51
Regularization and transformation technique, 495
Reimann convolution, 197
Reissner plate theory, 103, 109, 114
Remodelling, 449
Repair patch
adhesive, 475
riveted, 471
Residual stress, 257
Rethard time, 197
Rigid body motion, 33, 37, 51, 80, 128, 198, 336, 341, 439, 464, 466, 476
Rivet, 389, 461, 470
Rotation, 64, 103, 105, 109, 114, 123

Saint-Venant, 233
Screw-implant, 452, 455
Self-weight, 88, 90
Semi-discontinuous elements, 72, 336
Semi-infinite region, 39
Sensitivity methods, 427-429
Sensor point, 435
Separation mode, 271, 280
Serendipity, 48
Shape functions, 105-114
Shape identification, 435-439
Shape optimization, 426
Shear deformable plate, 105-114
Shear deformable shell, 115-134
Shear factor, 108, 122
Shear stiffness, 108, 122
Shear stress, 16, 21, 22, 105, 106, 117, 120
Slip mode, 271, 272, 279, 281, 288
Somigliana's identity, 15, 30, 40, 56
Sommerfield condition, 200

Source point, 8, 30, 33, 34, 37, 41, 43, 76, 77

Spherical shell, 138, 145

Stiffness matrix, 81

Stiffener, 389, 390, 461, 462, 464, 481

Strain,
    energy, 27, 233, 321
    energy density criterion, 327, 392, 395

Stress
    Airy, 525
    boundary, 56, 57, 61, 133, 243,
    complex, 525-530
    concentration, 2, 87, 103
    couple, 105, 106, 116, 117, 118, 119
    integral equation, 40, 56
    interior, 56, 241
    principal, 21
    resultant, 133
    resultant integral equation, 130
    strain relationship, 24

Stress intensity factors, 320, 324

Stick mode, 271,

Stokes theorem, 513, 516-517

Strongly singular integral, 246, 247, 508-513

Subparametric, 45

Subtraction of singularity method, 357, 497, 500, 511, 514

Superparametric, 45

Symmetric Galerkin formulation, 57

Taylor series, 35, 40, 41, 79, 80, 165, 171, 237, 247, 248, 325, 337, 340, 510, 511, 514

Tension stiffness, 108, 122

Thermal body force, 167

Thermal conductivity, 164, 187

Thermal loads, 187

Thermoelasticity
    coupled, 162, 163
    steady state, 163-173
    transient, 174-185
    uncoupled, 163, 172, 174

Time domain, 196-200

Thin bodies, 31, 92, 95, 96

Tibia bone, 92

Time integration, 178, 179, 208

Traction, 19, 28

Traction integral equation, 329-330

Transformation of coordinates, 58, 59, 61

Transformation of variable technique, 493

Transient elastodynamics, 196-198

Trapezoidal rule, 253, 360

Tresca, 233

Vibration, 156, 157, 224

Von mises, 233, 234, 235, 254

Unit impulse, 197

Wave
    dilatational velocity, 196, 211
    propagation, 196
    scattering, 220
    shear velocity, 196

Weakly singular integral, 497-507

Weight function method, 369-371

Weighted Gaussian integration, 497, 502, 510

Winkler foundation, 104, 105, 114, 115

X-core structure, 4, 138

Yielding, 231

Yield stress, 232, 233, 234, 237, 253